B. I. BALINSKY, Dr. Biol. Sci.

Professor Emeritus
University of the Witwatersrand
Johannesburg, South Africa

An introduction to Embryology

FOURTH EDITION

1975   W. B. SAUNDERS COMPANY
Philadelphia, London, Toronto

W. B. Saunders Company: West Washington Square
Philadelphia, PA   19105

12 Dyott Street
London, WC1A   1DB

833 Oxford Street
Toronto, Ontario M8Z 5T9, Canada

Library of Congress Cataloging in Publication Data

Balinsky, Boris Ivan, 1905–

An introduction to embryology.

Bibliography: p.

Includes index.

1. Embryology. I. Title. [DNLM: 1. Embryology. QL955 B186i]

QL955.B184   1975   591.3′3   74-17748

ISBN 0-7216-1518-X

Listed here is the latest translated edition of this
book together with the language of the translation
and the publisher.

Italian (*2nd Edition*)—Zannichelli, Bologna, Italy

Japanese (*2nd Edition*)—Iwanami Shoten, Tokyo,
                                            Japan

Spanish (*1st Edition*)—Ediciones Omega,
                                        Barcelona, Spain

An Introduction to Embryology   ISBN 0-7216-1518-X

Last digit is the print number: 9   8   7   6   5   4   3   2   1

PREFACE TO THE FOURTH EDITION

This new edition of *An Introduction to Embryology* has given me an opportunity to restructure the text and to add new and more up-to-date material. In doing this, I have included not only work published since the third edition, but also some earlier work which has gained significance in the light of new discoveries and new ideas. Some passages of the previous text that appeared to be of less interest have been removed in an endeavor to prevent the book from becoming excessively bulky. Hopefully, in this way, the governing ideas will stand out more clearly and thus will make the book better suited for the student-reader. The section dealing with the structure of DNA and the mechanism of protein synthesis has been deleted, since the material appears to be common knowledge among biologists at this stage and is included in all elementary textbooks of biology and zoology.

In working on the text, I have had the benefit of receiving helpful criticisms and advice from many persons, to whom I would like to express my gratitude. I am particularly indebted to the following individuals: Professor J. T. Bagnara, University of Arizona; Professor Anna R. Brummett, Oberlin College; Dr. B. C. Fabian, University of the Witwatersrand; and Professor C. C. Lambert, California State University, Fullerton. I sincerely appreciate the help of my colleagues, who have been kindly sending me reprints of their publications and thus have greatly facilitated the onerous task of following the current literature. I hope my correspondents realize that only a very small portion of the results of the current work can be included in a textbook.

Mr. J. Thompson is gratefully acknowledged for helping me with the photographic work for the present edition.

It has been a great pleasure over the years to work with the staff of the W. B. Saunders Company, whose cooperation and encouragement I greatly value.

Johannesburg   B. I. BALINSKY

PREFACE TO THE
THIRD EDITION

In preparing this third edition of *An Introduction to Embryology* for publication I was guided by the intention to correct those defects and shortcomings of the second edition which I have noticed myself or to which my attention was drawn by reviewers and critics, and also to incorporate into the text the new developments that have taken place in the fields of general biology and embryology since the printing of the second edition. The most significant advances made, so far as the field covered by *An Introduction to Embryology* is concerned, are the deciphering of the genetic code and an increased understanding of the interaction of the genes and cytoplasm in development. As a result, the general outlines of a comprehensive theory of ontogenetic development are gradually emerging, though as yet not clearly enough to be put down on paper; a number of links are still missing.

One of the criticisms made against this book is that obsolete or otherwise unacceptable theories and views are included in the text. Even if the inadequacy of such theories and views is explained in subsequent pages, the student, it is claimed, is confused or has to study that which is not of value to him. I feel that this criticism is not well founded. Science develops by overcoming previous mistakes; what has been the truth yesterday may be falsehood today; but equally, a similar transformation may occur tomorrow in what today is considered as established truth. Describing some of the blind alleys into which science has stumbled in the past and criticizing the weakness of some views and theories should teach the student not to regard the information given in a textbook as absolute and immutable truth, and it should develop in the student a critical approach and an enquiring spirit that are the essence of the scientific attitude.

Embryological terminology has recently been a subject under consideration by the Subcommittee on Embryological Nomenclature of the International Anatomical Nomenclature Committee. At this time, when my work on the third edition has been completed, the proposed list of terms has not yet been finalized, as it is to be submitted for approval to the next Anatomical Congress. In the meantime I have found it desirable to implement those recommended terms which are most likely to be adopted.

In the course of reviewing the text, some parts have been rearranged so as to bring them into a more logical and streamlined form. This refers in particular to the sections on oogenesis and differentiation. The section on placentation has been expanded. Several new illustrations have been added, directly supplied, in part, by the authors, to whom I would like to express my gratitude. The recognition is given in the legends of the respective figures.

I would like to use this opportunity to thank the many biologists and embryologists who have reviewed or criticized the second edition of *An Introduction to Embryology* and by so doing have helped me to improve the book. I would like to name in particular: Professor J. Bagnara, University of Arizona; Professor R. Barnes, Gettysburg College; Professor Anna R. Brummett, Oberlin College; Dr. V. V. Brunst, Roswell Park Memorial Institute, Buffalo; Professor R. F. Ruth, University of Alberta; Professor M. S. Steinberg, Princeton University; and Professor B. Strauss, University of Chicago.

In my own Department of Zoology at the University of the Witwatersrand, I have had valuable discussions concerning changes and improvements in *An Introduction to Embryology* with Dr. Vivian Gabie, and Dr. B. Fabian, senior lecturers in the department.

As in the previous editions Mr. M. J. de Kock assisted me with the preparation of new figures, and Mrs. E. J. Pienaar has been very helpful in typing the corrections and additions for the new edition and in checking the references. To all these persons I would like to express my sincere gratitude.

B. I. BALINSKY

Johannesburg

PREFACE TO THE SECOND EDITION

In the short time since the publication of the first edition of this book, embryology has made significant progress requiring a revision of many parts of the text in respect to both factual statements and interpretations. This progress is the result mainly of investigations of a biochemical nature and the wide application of electron microscopy to embryological problems. The organization of the eggs and sperm in particular has emerged in a new light because of electron microscopic investigations of the last few years. On the biochemical side, the discovery of the way in which protein synthesis is controlled by nucleic acids has made such a profound change in biological concepts that development of the egg cannot be dealt with at present without consideration of the DNA-RNA codes and the mechanism of protein synthesis.

In preparing the second edition of *An Introduction to Embryology* I have made an attempt to reflect these advanced trends and to incorporate them into the text as well as to fill the gaps which have been pointed out by reviewers and critics.

The chapter on the organization of the egg has been greatly expanded, and much more attention has been given to the spermatozoon. The mechanism of gene action in development has been presented in the light of decoding of information encoded in the nuclear DNA. New data on the mechanism of embryonic induction and on cell growth and differentiation have been included. The list of references has been increased by over 200 titles, drawn mainly from the literature that has appeared during the last three years. The current research literature has been systematically studied and utilized.

To meet the wishes of users of the book who found that the text of the first edition contained insufficient information on the morphological aspects of development, I have considerably expanded the descriptive parts of the book and have given more emphasis to mammalian development. The illustrations have been augmented by the addition of 153 new figures, some of them composite. Some of the new figures have been grouped together to illustrate a few of the developmental stages of the frog and the chick, in the section on "Stages of Development." It is hoped that this section will help the student to grasp the development of the embryo as a whole, and that it will present a background against which the development of the various organ systems can be studied in greater detail.

Some of the views and statements contained in the first edition have been challenged by critics and reviewers, and in many cases I have accepted the criticisms and made corresponding changes in the text. Occasionally these changes concern terminology. In a few cases, however, I do not agree with the proposed corrections, and here perhaps is the right place to state my reasons, in order to avoid misunderstandings.

The controversial terms which I prefer to retain without change are "blastoderm" and "blastodisc." The term "blastoderm" is used here to denote the layer of cells surrounding the blastocoele of a blastula, in whatever form the blastula is encountered—whether as a hollow sphere formed after complete cleavage, as a disc after discoidal cleavage or as a layer of cells surrounding the yolk in centrolecithal eggs. This layer of cells should have a specific name, and the term blastoderm is a very appropriate one, as it matches the terms "blastula" and "blastocoele." Also etymologically it is constructed similarly to the terms ectoderm, mesoderm and endoderm, and thus helps the student to see the continuity of structural units in the embryo. The term "blastoderm" in the sense just defined has been used before by both older and contemporary embryologists. It was used in this sense by Haeckel (*The Evolution of Man,* Vol. 1, 1874, English translation by McCabe; Watts and Co., 1923, p. 62), by Korschelt and Heider (*Text-book of the Embryology of Invertebrates,* Part 1, 1893, English translation 1895; Swan Sonnenschein and Co., p. 4), and in our times by Nelsen (*Comparative Embryology of Vertebrates,* 1953) and by Witschi (*Development of Vertebrates,* 1956). Also in the same sense the term blastoderm is used in Figure 5–1 of Patten's *Foundations of Embryology* (1958), although not in the text of that book.

If the term blastoderm is to be applied to the cellular lining of any type of blastula, and not only to blastulae of animals having discoidal cleavage as some would like to have it, then the cap of cells on the animal pole of a discoblastula should obviously be called a "blastodisc." In the past the term "blastodisc" has been used either to denote the concentration of cytoplasm on the animal pole of the uncleaved hen's egg or to denote the mass of cells into which this cytoplasm becomes subdivided during cleavage, or for both. In recent literature, usage of the term blastodisc has been extended to include the chick embryo in the primitive streak stage or even in later stages. Thus while some looseness in the application of the term seems to be in practice, I should like to confine the term to those stages in which the developing embryo does have the shape of a disc, that is, mainly to the stages of meroblastic discoidal cleavage.

In the task of preparing the second edition I have been greatly assisted by numerous teachers and research workers who have given me the benefit of their opinions. To all those I should like to express my sincere gratitude, but most especially to the following colleagues whose advice and constructive criticisms have been most helpful: Professor Joseph T. Bagnara of the University of Arizona, Professor James T. Duncan of the San Francisco State College, Professor Royal F. Ruth of the University of Alberta, Professor Nelson T. Spratt of the University of Minnesota, Professor Malcolm S. Steinberg of the Johns Hopkins University and Professor Roland Walker of the Rensselaer Polytechnic Institute.

I should like also to record may special thanks to a number of embryologists who have kindly supplied me with original drawings and photographs, which have been invaluable in improving the illustrations. The names of these colleagues appear in the legends of the corresponding figures.

In conclusion I should like again to thank my assistants in the Zoology Department of the University of the Witwatersrand, Mr. M. J. de Kock for helping me with the preparation of new figures appearing in the second edition, and Mrs. E. J. Pienaar, who has been most helpful in searching for literature when it was not at hand and in typing the new sections of the text.

B. I. BALINSKY

PREFACE TO THE FIRST EDITION

The teaching of embryology has long been an established feature at universities throughout the world, both for students in biology and students in medical sciences. Although overshadowed during a large part of the twentieth century by the rapid development of genetics and cytology, embryology has also made rapid advances, especially as an experimental science—as experimental or physiological embryology. It is realized now that embryology is a branch of biology which has a most immediate bearing on the problem of life. Life cannot be fully accounted for without an understanding of its dynamic nature, which expresses itself in the incessant production of new organisms in the process of ontogenetic development.

In the midst of the rapid changes of outlook that the experimental method has brought with it, it has been difficult to coordinate the older data of purely descriptive embryology with the new discoveries. This has hampered the teaching of embryology and is to this day reflected in the subdivision of most textbooks of embryology into two groups. Books of the first type deal with the classic "descriptive" embryology and are written mainly for the use of medical students. Short chapters on experimental embryology are appended to them, but these chapters are extremely brief and not organically connected with the description of the morphology of the developing embryos. The second type of book deals with experimental embryology or "physiological" embryology. These books are written for advanced students and the basic facts of development are more or less taken for granted, so that a student cannot profitably proceed to the study of such a book without previously making himself familiar with "descriptive" embryology from one of the books of the first type.

In the course of many years' teaching of embryology to university students I have endeavored to present embryology as a single science in which the descriptive morphological approach and the experimental physiological approach are integrated and both contribute to the understanding of the ontogenetic development of organisms. This integrated approach to development is now incorporated in the present book. Data of a more purely physiological and biochemical nature are adduced inasmuch as it is practicable to treat them in a book that does not presuppose an advanced knowledge of biochemistry in the student.

The subject of embryology is interpreted in my book in a broad sense, as the science dealing with ontogenetic development of animals, and includes therefore such topics as postembryonic development, regeneration, metamorphosis and asexual reproduction, which are seldom handled in students' textbooks at any length. Lastly, I believe that embryology cannot be presented adequately without establishing some connection with genetics, inasmuch as processes of development are under the control of genes. The connection between inheritance and development is therefore also indicated in the text.

With such a wide scope, my book can only be "an introduction to embryology." The whole field could not be covered in the same detail as is customarily given in textbooks dealing with only one aspect of the science of embryology. The student having studied this book, however, will be prepared to understand and appreciate special information in any section of the science which would be of interest to him in his further studies.

The first draft of this book was written in 1952, and duplicated copies of the manuscript were used by my students during subsequent years. This gave me an opportunity to convince myself of the usefulness of the book and also to eliminate some defects in the original text. For the present printed edition the book has been completely revised and brought up to date. An extensive study of special literature up to the end of 1958 has been carried out for this purpose (as can be seen from the list of references). Later publications could not be included in the text.

In illustrating the book I have drawn on my own experience in embryological work wherever practicable, but of course most of the illustrations have been reproduced from other sources.

In preparing the book for print I have been assisted by a number of persons to whom I should like to express my gratitude on this occasion. In the first place I wish to thank all the authors and publishers who have kindly agreed to the reproduction of figures used to illustrate this book, as well as colleagues in many countries who by sending me reprints of their publications have facilitated the arduous task of keeping track of current embryological literature.

Of my immediate collaborators and friends I am most profoundly indebted to Dr. Margaret Kalk of the University of the Witwatersrand for reading the whole text of the book and for many valuable suggestions and helpful criticism. I am indebted to Dr. H. B. S. Cooke of the same University for his expert advice on the preparation of illustrations for the book. I am very grateful for the invaluable assistance of Mrs. E. J. Pienaar, who has typed the manuscript, has assisted me in preparing the index, and has been of great help on diverse occasions during the work on the manuscript and on the proofs. I should like to thank Miss R. J. Devis, Mrs. E. du Plessis and Mr. M. J. de Kock for their help in preparing the illustrations for the book.

Last but not least I should like to express my gratitude to the staff of the W. B. Saunders Company, whose friendly encouragement has done much to bring this book to its present form.

B. I. BALINSKY

Johannesburg

CONTENTS

Part One. THE SCIENCE OF EMBRYOLOGY

Chapter 1

THE SCOPE OF EMBRYOLOGY AND ITS
DEVELOPMENT AS A SCIENCE 3

    1–1   *Ontogenetic Development as the Subject
Matter of Embryology* 3

    1–2   *The Phases of Ontogenetic Development* 4

    1–3   *Historical Review of the Main Trends of
Thought in Embryology* 7

Part Two. GAMETOGENESIS

Chapter 2

SPERMATOGENESIS 19

    2–1   *Cells in Seminiferous Tubules* 19

    2–2   *Meiosis* 20

    2–3   *Differentiation of the Spermatozoa* 24

Chapter 3

OOGENESIS 32

    3–1   *Growth of the Oocyte* 32

    3–2   *Accumulation of Food Reserves in the
Cytoplasm of the Oocyte* 42

    3–3   *Organization of the Egg Cytoplasm* 54

    3–4   *Maturation of the Egg* 60

    3–5   *The Egg Membranes* 62

Chapter 4

THE DEVELOPING EGG AND THE ENVIRONMENT 67

**Part Three.   FERTILIZATION AND THE BEGINNING OF
                EMBRYOGENESIS**

Chapter 5

FERTILIZATION 75

5–1   *Approach of the Spermatozoon to the Egg* 75

5–2   *The Reaction of the Egg* 81

5–3   *The Essence of Activation* 88

5–4   *Parthenogenesis* 90

5–5   *The Spermatozoon in the Egg Interior* 92

5–6   *Changes in the Organization of the Egg
       Cytoplasm Caused by Fertilization* 96

Chapter 6

CLEAVAGE 101

6–1   *Peculiarities of Cell Divisions in Cleavage* 101

6–2   *Chemical Changes During Cleavage* 104

6–3   *Patterns of Cleavage* 106

6–4   *Morula and Blastula* 114

6–5   *The Nuclei of Cleavage Cells* 119

6–6   *Distribution of Cytoplasmic Substances of
       the Egg During Cleavage* 125

6–7   *Role of the Egg Cortex* 130

6–8   *The Morphogenetic Gradients in the
       Egg Cytoplasm* 134

6–9   *Manifestation of Maternal Genes During
       the Early Stages of Development* 143

**Part Four.   GASTRULATION AND THE FORMATION
               OF THE PRIMARY ORGAN RUDIMENTS**

Chapter 7

MORPHOLOGICAL ASPECTS OF GASTRULATION AND
PRIMARY ORGAN FORMATION 151

7–1   *Fate Maps* 151

7–2   *Gastrulation in* Amphioxus 153

7–3   *Formation of the Primary Organ
      Rudiments in* Amphioxus 158

7–4   *Gastrulation in Amphibians* 161

7–5   *Formation of the Primary Organ
      Rudiments in Amphibians* 170

7–6   *Gastrulation and Formation of the Primary
      Organ Rudiments in Fishes* 175

7–7   *Gastrulation and the Formation of the
      Primary Organ Rudiments in Birds* 178

Chapter 8

DIVERSIFICATION OF EMBRYONIC PARTS AND ITS CONTROL DURING GASTRULATION AND PRIMARY ORGAN FORMATION 189

8–1   *General Metabolism During Gastrulation* 189

8–2   *Gene Activity During Gastrulation* 192

8–3   *Involvement of Parental Genes in the
      Control of Development* 194

8–4   *Determination of the Primary Organ Rudiments* 197

8–5   *Spemann's Primary Organizer* 202

8–6   *Analysis of the Nature of Induction* 208

8–7   *Mechanism of Action of the Inducing Substances* 218

8–8   *Gradients in the Determination of the Primary
      Organ Rudiments in Vertebrates* 220

Chapter 9

CREATION OF FORM DURING GASTRULATION AND IN SUBSEQUENT DEVELOPMENT 229

9–1   *Morphogenetic Movements* 229

9–2   *Selective Affinities of Cells as a Determining
      Factor in Cell Rearrangements* 230

9–3   *Morphogenetic Movements in Epithelia* 234

9–4   *Mechanism of Changes in the Shape of Cells
      During Morphogenesis* 240

9–5   *Morphogenetic Movements in Mesenchyme* 246

Chapter 10

EMBRYONIC ADAPTATIONS AND THE DEVELOPMENT
OF MAMMALS 253

10–1   *The Extraembryonic Structures in Reptiles
       and Birds* 254

10–2   *Mammalian Eggs* 259

10–3   *Cleavage, Blastocyst, and Development of
       Germinal Layers in Mammals* 263

10–4   *Relations Between the Embryo and the
       Maternal Body in Mammals* 273

10–5   *Placentation* 284

10–6   *Review of Placentae in Different Groups of Mammals* 289

10–7   *Physiology of the Placenta* 290

10–8   *Hormonal Control of Ovulation and
       Pregnancy* 291

Part Five.   ORGANOGENESIS

Chapter 11

GENERAL INTRODUCTION TO ORGANOGENESIS 297

11–1   *Development of General Body Form* 297

11–2   *Normal Stages of Development* 299

11–3   *The Anatomy of Representative Stages of Development
       of the Frog and Chick Embryos* 302

Chapter 12

DEVELOPMENT OF THE ECTODERMAL ORGANS
IN VERTEBRATES 330

12–1   *Development of the Central Nervous System* 330

12–2   *Development of the Eyes* 353

12–3   *The Fate of the Neural Crest Cells* 361

12–4   *The Fate of the Epidermis and the Structures
       Derived from It* 364

Chapter 13

DEVELOPMENT OF THE MESODERMAL ORGANS
IN VERTEBRATES 375

13-1   The Fate of the Somites and the Origin of
       the Somatic Muscles 375

13-2   The Axial Skeleton: Vertebral Column and
       Skull 377

13-3   Development of the Paired Limbs 382

13-4   Development of the Urinary System 394

13-5   Development of the Heart 402

13-6   Development of the Blood Vessels 409

13-7   Development of the Reproductive Organs 423

Chapter 14

DEVELOPMENT OF THE ENDODERMAL ORGANS
IN VERTEBRATES 440

14-1   The Relation Between the Archenteron and
       the Definitive Alimentary Canal 440

14-2   Development of the Mouth 448

14-3   Development of the Branchial Region 453

14-4   Development of the Accessory Organs of the
       Alimentary Canal: Lungs, Liver, Pancreas,
       Bursa Fabricii 457

14-5   Determination of the Endodermal Organs 461

Part Six.   DIFFERENTIATION AND GROWTH

Chapter 15

GENERAL CONSIDERATIONS ON GROWTH AND
DIFFERENTIATION 467

15-1   Definitions 467

15-2   Mechanisms of Cell Reproduction 469

15-3   Relation of Cell Proliferation to
       Differentiation 476

Chapter 16

DIFFERENTIATION 478

16–1   The Chemical Basis of Differentiation 478

16–2   The Messenger RNA in Metazoa 482

16–3   Selective Action of Genes in Differentiation 485

16–4   Changing Pattern of Protein Synthesis 495

16–5   Control of Differentiation by the
       Intraorganismic Environment 502

16–6   Control of the Reactive Ability of Tissues by
       the Genotype 512

16–7   Sequence of Gene Action in Development 515

Chapter 17

GROWTH 519

17–1   Measurement of Growth and Its Graphic
       Representation 519

17–2   Growth on the Cellular and Organismic Levels 521

17–3   Interpretation of Growth Curves 523

17–4   Proportional and Disproportional
       Growth of Organs 529

Part Seven.   MORPHOGENETIC PROCESSES IN THE
              LATER PART OF ONTOGENESIS

Chapter 18

METAMORPHOSIS 535

18–1   Changes of Organization During
       Metamorphosis in Amphibians 536

18–2   Causation of Metamorphosis in Amphibians 540

18–3   Tissue Reactivity in Amphibian Metamorphosis 543

18–4   Processes of Induction During Amphibian
       Metamorphosis 544

18–5   Molting and Its Relation to Metamorphosis
       in Insects 545

18–6   Causation of Molting and Metamorphosis
       in Insects 550

18–7   *Nature of the Factors Controlling Molting
         and Metamorphosis in Insects* 555

18–8   *Mechanism of Action of Insect Hormones* 557

18–9   *Final Remarks on Metamorphosis* 561

Chapter 19

REGENERATION 562

19–1   *Typical Case of Regeneration: The Renewal of
         a Limb in a Salamander* 562

19–2   *Regenerative Ability in Various Animals* 564

19–3   *Stimulation and Suppression of Regeneration* 567

19–4   *Histological Processes Concerned in
         Regeneration* 571

19–5   *Release of Regeneration* 579

19–6   *Relation of the Regenerating Parts to the
         Remainder of the Organ and to the
         Organism as a Whole* 581

19–7   *Polarity and Gradients in Regeneration* 585

19–8   *Reconstitution from Isolated Cells* 590

Chapter 20

ASEXUAL REPRODUCTION 592

20–1   *Occurrence and Forms of Asexual Reproduction* 592

20–2   *Sources of Cellular Material in Asexual
         Reproduction* 595

20–3   *Comparison of Blastogenesis and Embryogenesis* 600

REFERENCES 605

INDEX 635

Part One

THE SCIENCE OF
EMBRYOLOGY

Chapter 1

THE SCOPE OF EMBRYOLOGY AND ITS DEVELOPMENT AS A SCIENCE

1−1 ONTOGENETIC DEVELOPMENT AS THE SUBJECT MATTER OF EMBRYOLOGY

The aim of this book is to familiarize the student with the basic facts and problems of the science of **embryology.** The name "embryology" is somewhat misleading. Literally it means the study of **embryos.** The term "embryo" denotes the juvenile stage of an animal while it is contained in the egg (within the egg membranes) or in the maternal body. A young animal, once it has hatched from the egg or has been born, ceases to be an embryo and would escape from the sphere pertaining to the science of embryology if we were to keep strictly to the exact meaning of the word. Although birth or hatching from the egg is a very important occasion in the life of the animal, it must be admitted that the processes going on in the animal's body may not be profoundly different before and after the hatching from an egg, especially in some lower animals. It would be artificial to limit studies of the juvenile forms of animal life to the period before the animal is hatched from the egg or is born. It is customary, therefore, to study the life history of an animal as a whole and accordingly to interpret the scope of the science of embryology as **the study of the development of animals.**

The word "development" must be qualified in turn. In the sphere of biology with which we are concerned, the term "development" is used with two different meanings. It is used to denote the processes that are involved in the transformation of the fertilized egg, or some other rudiment derived from a parent organism, into a new adult individual. The term development may, however, also be applied legitimately to the gradual historical transformation of the forms of life, starting with simple forms which might have been the first to appear and leading to the contemporary diversity of organic life on our planet. Development of the first type may be distinguished as individual development or **ontogenetic development.** Development of the second type is the historical development of species or **phylogenetic development.** Phylogenetic development is often referred to as evolutionary development or simply **evolution.** Accordingly, we will define embryology as **the study of the ontogenetic development of organisms.** In this book we will be dealing only with the ontogenetic development of multicellular animals, the **Metazoa.**

In multicellular animals, the typical and most widespread form of ontogenetic development is the type occurring in sexual reproduction. In sexual reproduction new individuals are produced by special **generative cells** or **gametes.** These cells differ essentially from other cells of the animal, in that they go through the process of maturation or **meiosis,** as a result of which they lose half of their chromosomes and become

haploid, whereas all other cells of the parent individual, the **somatic cells,** are, as a rule, **diploid.** Once a cell has gone through the process of meiosis, it can no longer function as an integral part of the parent body but is sooner or later extruded to serve in the formation of a new individual. In multicellular animals there exist two types of sex cells: the female cells or **ova,** and the male cells or **spermatozoa.** As a rule the two cells of the opposite sexes must unite in the process of fertilization before development can start. When the two gametes (the ovum and the spermatozoon) unite, they fuse into a single cell, the **zygote,** which again has a diploid number of chromosomes. The zygote, or fertilized ovum, then proceeds to develop into a new adult animal.

Side by side with sexual reproduction there exists in many species of animals a different mode of producing new generations—asexual reproduction. In asexual reproduction the offspring are not derived from generative cells (gametes) but rather from parts of the parent's body consisting of somatic cells. The size of the part which is set aside as the rudiment of the new individual may be large or small, but in the Metazoa it always consists of more than one cell. The development of an animal by way of asexual reproduction obviously belongs in the same category as the development from an egg and should be treated as a special form of ontogenetic development. It will be dealt with in Chapter 20. To distinguish between the two forms of ontogenetic development, the term **embryogenesis** may be used to denote development from the egg, and the term **blastogenesis** may be used for the development of new individuals by means of asexual reproduction.

1–2   THE PHASES OF ONTOGENETIC DEVELOPMENT

From what has already been said it is clear that the processes leading to the development of a new individual really start before the fertilization of the egg, because the ripening of the egg and the formation of the spermatozoon, which constitute the phase of **gametogenesis,** create the conditions from which the subsequent embryogenesis takes its start. In both oogenesis and spermatogenesis, meiosis, by discarding half of the chromosomes, singles out the set of genes which are to operate in the development of a particular individual. In both sexes the initial cells giving rise to the gametes are very similar and, as a rule, not essentially different from other cells of the body except that these cells are not involved in any of the differentiations serving to support the life of the parent individual. In both sexes the first step in the production of gametes is a more or less rapid proliferation of cells by ordinary mitosis. The proliferating cells in the testes are known as the **spermatogonia;** the proliferating cells in the ovaries are called **oogonia.** Once proliferation ceases, the cells are called **spermatocytes** in the male and **oocytes** in the female. They then enter into a stage of growth and later into a stage of maturation.

Although the stage of proliferation is not essentially different in the male and female, the processes of growth and maturation in the two sexes differ to a very great extent. The cytoplasmic differentiations of the spermatozoon enable it to reach the egg by active movement and to fertilize it. On the other hand, the egg cell accumulates in its cytoplasm substances which are used up during development—either directly, by becoming transformed into the various structures of which the embryo consists, or indirectly, as sources of energy for development. The elaboration in the egg cell of cytoplasmic substances to be used by the embryo and their placing in correct positions are essential parts of what occurs during the first phase of development.

The second phase of development is **fertilization.** Fertilization involves a number of rather independent biological and physiological processes. First, the spermatozoa must be brought into proximity with the eggs if fertilization is to occur. This involves adaptations on the part of the parents which insure that they meet during the breeding season, discharge their sex cells simultaneously in cases of external fertilization, or copulate in cases of internal fertilization. Next, the spermatozoa must find the egg and penetrate into the egg cytoplasm. This entails a very finely adjusted mechanism of morphological and physicochemical reactions. The egg is then **activated** by a spermatozoon and starts developing. A further rearrangement of the organ-forming substances in the egg is among the first changes that take place in the egg after fertilization. (See Section 5–6.)

The third phase of development is the period of **cleavage.** The fertilized egg is still a single cell, since the nucleus and cytoplasm of the spermatozoon fuse with the nucleus and cytoplasm of the egg. If a complex and multicellular organism is to develop from a single cell, the egg, the latter must give rise to a large number of cells. This is achieved by a number of mitotic cell divisions following one another in quick succession. During this period the size of the embryo does not change, the **cleavage cells** or **blastomeres** becoming smaller and smaller with each division. No far-reaching changes can be discovered in the substance of the developing embryo during the period of cleavage, as if the preoccupation with the increase of cell numbers excludes the possibility of any other activity. The whole process of cleavage is dominated by the cytoplasmic organoids of the cells, the centrosomes and achromatic figures. The nuclei multiply but do not interfere with the processes going on in the cytoplasm. The result of cleavage is sometimes a compact heap of cells, but usually the cells are arranged in a hollow spherical body, a **blastula,** with a layer of cells, the **blastoderm,** surrounding a cavity, the **blastocoele.**

The fourth phase of development, that of **gastrulation,** follows. During this phase the single layer of cells, the blastoderm, gives rise to two or more layers of cells known as the **germinal layers.** The germinal layers are complex rudiments from which are derived the various organs of the animal's body. In higher animals the body consists of several layers of tissues and organs, such as the skin, the subcutaneous connective tissue, the layer of muscles, the wall of the gut, and so on. All these tissues and organs may be traced back to three layers of cells—the aforementioned germinal layers. Of these the external one, the **ectoderm,** always gives rise to the skin epidermis and the nervous system. The layer next to the first, the **mesoderm,** is the source of the muscles, the blood vascular system, the lining of the secondary body cavity (the **coelom,** in animals in which such a cavity is present), and the sex organs. In many animals, particularly the vertebrates, the excretory system and most of the internal skeleton are also derived from the mesoderm. The third and innermost germinal layer, the **endoderm,** forms the alimentary canal and the digestive glands.

The germinal layers are produced by the disappearance of a part of the blastoderm from the surface and its enclosure by the remainder of the blastoderm. The part that remains on the surface becomes ectoderm; the part disappearing into the interior becomes endoderm and mesoderm. The disappearance of endoderm and mesoderm from the surface sometimes takes the form of a folding-in of part of the blastoderm, so that the simple spherical body becomes converted into a double-walled cup, as if one side of the wall of an elastic hollow ball had been pushed in by an external force. This infolding or pushing in of the endoderm and mesoderm is known as **invagination,** and the resulting embryo is known as a **gastrula**—whence the term **gastrulation.** The way in which the endoderm and mesoderm become separated from each other in the interior of the gastrula varies a great deal in different animals and cannot be described in this general review. (See Chapter 7.) If the gastrula is formed by invagination, the cavity of the double-walled cup is called the **archenteron,** and the opening leading from this cavity to the exterior is called the **blastopore.** In animals in which the gastrula is formed in a different way—not by invagination—the cavity (archenteron) and the opening of the cavity to the exterior (blastopore) may still appear later on.

The archenteron, or part of it, eventually gives rise to the cavity of the alimentary system. The fate of the blastopore differs in the three main groups of Metazoa. In **Coelenterata** it becomes the oral opening. In **Protostomia** (including Annelida, Mollusca, Arthropoda, and allied groups) it becomes subdivided into two openings, one of which becomes the mouth and the other the anus. In **Deuterostomia** (including Echinodermata and Chordata) only the anal opening is connected in its development with the blastopore, the mouth being formed later on as an independent perforation of the body wall. The whole of the lining of the alimentary canal does not always consist of endoderm; in all groups of animals the ectoderm may be invaginated secondarily at the oral or at both oral and anal openings to become a part of the alimentary canal. The parts of the alimentary canal lined by ectoderm are known as the stomodeum (adjoining the mouth) and proctodeum (adjoining the anus).

With the formation of the three germinal layers, the process of subdivision of the embryo into parts with specific destinies commences. This subdivision is continued in the next (fifth) phase of development, the phase of **organogenesis** (organ formation). The continuous masses of cells of the three germinal layers become split up into smaller groups of cells, each of which is destined to produce a certain organ or part of the animal. Every organ begins its development as a group of cells segregated from the other cells of the embryo. This group of cells we will call the **rudiment** of the respective organ. The rudiments into which the germinal layers become subdivided are called **primary organ rudiments.** Some of these are very complex, containing cells destined to produce a whole system of organs, such as the entire nervous system or the alimentary canal. These complex primary organ rudiments later become subdivided into **secondary organ rudiments**—the rudiments of the subordinated and simpler organs and parts. The formation of the primary organ rudiments follows so closely on the processes of gastrulation that the two processes can hardly be considered separately. Dynamically they are linked into one whole and will be described in conjunction in the following pages. The chapters on organogenesis will then be concerned with the later development of the primary organ rudiments and with the formation of the secondary organ rudiments. With the appearance of primary and secondary organ rudiments, the embryo begins to show some similarity to the adult animal, or to the larva if the development includes a larval stage.

The sixth phase of development is the period of **growth** and **histological differentiation.** After the organ rudiments are formed they begin to grow and greatly increase their volume. In this way the animal gradually achieves the size of its parents. Sooner or later the cells in each rudiment become histologically differentiated; that is, they acquire the structure and physicochemical properties which enable them to perform their physiological functions. When the cells in all the organs, or at least in the vitally important organs, have become capable of performing their physiological functions, the young animal can embark upon an independent existence—an existence in which it has to procure food from the surrounding environment.

In rather rare cases (in the nematodes, for instance) the young animal emerging from the egg is a miniature copy of the adult animal and differs from the latter only in size and the degree of differentiation of the sex organs. In this case the subsequent development consists only of growth and maturation of the gonads. It is more usual, however, for animals emerging from eggs to differ from the adult to a greater extent; not only the gonads but also other organs may not be fully differentiated, or they may even be absent altogether and have to develop later. Sometimes the animal emerging from the egg possesses special organs which are absent in the adult but which are necessary for the special mode of existence of the young animal. In this case the young animal is called a **larva.** The larva may lead a different mode of life from the adult, and therein lies one of the advantages of having a larval stage in development. The larva undergoes a process of **metamorphosis** when it is transformed into an animal similar to the adult. The metamorphosis involves more or less drastic changes in the organization of the larva, depending on the degree of difference between the larva and the adult. During metamorphosis new organs may develop, so that morphogenetic processes become active again after a more or less prolonged period of larval life.

A secondary activation of morphogenetic processes may be produced in a different way. Many animals possess considerable plasticity and may be able to repair injuries sustained from the environment or caused experimentally. Lost parts may be **regenerated,** and this means that the developmental processes may sometimes be repeated in an adult or adolescent organism.

Asexual reproduction of animals involves the development of new parts and organs in animals that have already achieved the adult stage.

All morphogenetic processes occurring in the later life of the animal, after the larval stage, or even when the adult stage has been achieved, will be dealt with as constituting a seventh and last phase of development.

1-3 HISTORICAL REVIEW OF THE MAIN TRENDS OF THOUGHT IN EMBRYOLOGY*

Descriptive and Comparative Embryology.   Although the correct understanding of ontogenetic development could be achieved only after the establishment of the cell theory, fragmentary information on the development of animals has been obtained since very ancient times. Aristotle had described the development of the chick in the egg as early as 340 B.C. Many observations on development of various animals, especially of insects and vertebrates, were made in the seventeenth and eighteenth centuries. However, the data of embryology were first presented in a coherent form by Karl Ernst von Baer (1828). In his book, Ueber Entwicklungsgeschichte der Tiere, Beobachtung und Reflexion, Baer not only summed up the existing data and supplemented them by his original investigations but also made some important generalizations. The most important of these is known as **Baer's law.** The law can be formulated thus: "More general features that are common to all the members of a group of animals are, in the embryo, developed earlier than the more special features which distinguish the various members of the group" Thus the features that characterize all vertebrate animals (brain and spinal cord, axial skeleton in the form of a notochord, segmented muscles, aortic arches) are developed earlier than the features distinguishing the various classes of vertebrates (limbs in quadrupeds, hair in mammals, feathers in birds, etc.). The characters distinguishing the families, genera, and species come last in the development of the individual. The early embryo thus has a structure common to all members of a large group of the animal kingdom and may be said to represent the basic plan of organization of that particular group. The groups having a common basic plan of organization are the **phyla** of the animal kingdom.

Baer's law was formulated at a time when the theory of evolution was not recognized by the majority of biologists. It has been found, however, that the law can be

---

*Further references: Nordenskiöld, 1929; Needham, Part II, 1931; Singer, 1931; Hall, 1951; Gabriel and Fogel, 1955; Needham, 1959; Oppenheimer, 1967.

reinterpreted in the light of the evolutionary theory. In its new form the law is known as the **biogenetic law** of Müller-Haeckel. Müller propounded the law in its new form and supported it by extensive observations on the development of crustaceans (1864). Haeckel (in 1868) gave it the name of the 'biogenetic law' and contributed most to its wide application in biology.

According to Baer's law, the common features of large groups of animals develop earliest during ontogeny. In the light of the evolutionary theory, however, these features are the ones that are inherited from the common ancestor of the animal group in question; therefore, they have an ancient origin. The features that distinguish the various animals from one another are those that the animals have acquired later in the course of their evolution. Baer's law states that these features in ontogeny develop at later stages. Briefly, the features of ancient origin develop early in ontogeny; features of newer origin develop late. Hence, the ontogenetic development presents the various features of the animal's organization in the same sequence as they evolved during the phylogenetic development. **Ontogeny** is a recapitulation of **phylogeny.*** The repetition is obviously not a complete one, and the biogenetic law states that "ontogeny is a shortened and modified recapitulation of phylogeny." The shortening of the process is evident not only from the fact that what had once taken thousands of millions of years (phylogeny) is now performed in a matter of days and weeks (ontogeny), but also from the fact that many stages which occurred in the original phylogenetic development may be omitted in ontogeny. The modifications arise mainly because the embryo at any given time is a living system which has to be in harmony with its surroundings if it is to stay alive. The embryo must be adapted to its surroundings, and these adaptations often necessitate the modification of inherited features of organization. A good example of such adaptation is the placenta in mammals. The placenta is a structure developed by the embryo to establish a connection with the uterine wall of the mother and thus to provide for the nutrition of the embryo. This structure, though developed rather early in the life of the embryo, could not have existed in the adult mammalian ancestors. It is obviously an adaptation to the special conditions in which a mammalian embryo develops.

Even if the repetition of features of their ancestors in the ontogenetic development of contemporary animals is not complete, the fragmentary repetition of certain ancestral characters may still be very useful in elucidating the relationships of animals. As an example of this we may consider the formation of gill clefts or, at least, pharyngeal pouches in the ontogenetic development of all vertebrates. In the aquatic vertebrates, such as Cyclostomata and fishes, the gill clefts serve as respiratory organs. In the adult state of terrestrial vertebrates, the pharyngeal pouches have disappeared completely or have been modified out of all recognition, and the function of respiration has been taken over by other organs—the lungs. Nevertheless, the pharyngeal pouches appear in the embryo. (See Figure 391.) In amphibians whose larvae are aquatic, the pharyngeal pouches at least temporarily serve for respiration. In reptiles, birds, and mammals, the pharyngeal pouches of the embryo do not serve for respiration at all. Their formation can be explained only as an indication that the terrestrial vertebrates have been derived from aquatic forms with functional gills. The paleontological evidence fully confirms this conclusion.

The systematic position of some animals cannot be recognized from adult structure, owing to profound modification acquired as a result of adaptation to very special conditions. Here the knowledge of the development sometimes throws unexpected

---

*For a modern explanation of the facts on which the biogenetic law is based, see page 518.

light on true relationships. The adult ascidian is a sessile animal with no organs of loco-motion and a nervous system of a very primitive nature. The adult animal had been classed as a near relative of molluscs until Kowalevsky (1866) discovered that the larvae of the ascidians possess a well-developed dorsal brain and spinal cord, a definite noto-chord, and lateral bands of muscles (in short, organs that are typical for the verte-brates). The ascidians are therefore considered as belonging to the same phylum as the vertebrates, the phylum Chordata (Fig. 1).

In the adult parasitic animal *Sacculina*, the organization of the animal is very much simplified in relation to the easy life that the parasite enjoys; it is reduced practically to a shapeless sack producing eggs and a system of branched rhizoids, by means of which the parasite is attached to its host, the crab, and absorbs the host's body fluids on which it feeds. It would be impossible to place *Sacculina* in any group of the animal kingdom if its development were not known (Fig. 2). However, the larva of *Sacculina* is a typical arthropod, bearing a close similarity to the larvae of the lower crustaceans, the Ento-mostraca (Delage, 1884).

A rather similar larva is also found in the barnacles *(Cirripedia)* which, though pos-sessing jointed legs like other arthropods, have lost the segmentation of the body in the adult state.

The attachment of the starfish larva, the brachiolaria, to the substrate while it is metamorphosing into the definitive form is an indication that the free-living echino-derms have been derived from sessile forms. This conclusion is again borne out by the evidence of paleontology.

Following the principles of Baer's law and the biogenetic law, embryologists have systematically investigated the development of animals belonging to all the major

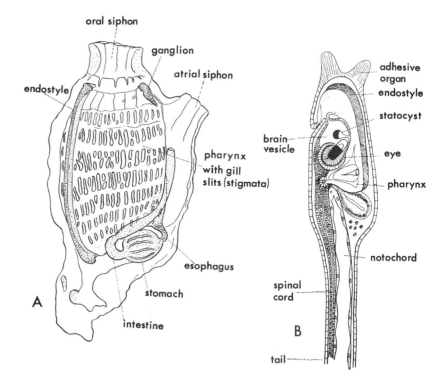

Figure 1.   *A*, Adult ascidian, *Ciona intestinalis. B,* Larva of ascidian, lateral view. (After Kowalevsky, from Korschelt, 1936.)

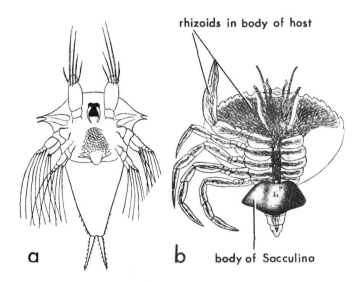

rhizoids in body of host

a

b   body of Sacculina

Figure 2. Parasitic cirripede, *Sacculina.* Nauplius larva (*a*) and adult (*b*) attached to a crab. (After Delage, from Parker and Haswell. Text-book of Zoology, Macmillan, London, and from Korschelt, 1936.)

groups of the animal kingdom. As a result of very extensive and painstaking investigations, a magnificent edifice of **comparative embryology** has been built.

Explaining Development: Theories of Preformation and Epigenesis. Neither the description of morphological transformations occurring in the embryo nor the comparison of embryos and larvae among themselves and with the adult animals exhausts all the problems presented by the ontogenetic development of animals. The fundamental problem presented by the existence of cyclical ontogenetic changes is the question: Why does ontogeny occur at all? What are the forces which produce the changes? How is it that, starting from a simple spherical cell, the process always ends in producing a highly complex and specific structure which, though varying in detail, reproduces with astonishing perseverance the same or almost the same adult form?

Attempts at solving this basic problem have been made ever since the human mind recognized the existence of development. For a long time the explanations proposed were purely speculative. Aristotle attempted to give a solution to the problem of ontogeny along the general lines of his philosophical teaching, distinguishing between the **substance** and the **form** of things. The form appears in this concept as the creative principle. Aristotle further supposed that the substance for the development of a child is provided by the mother (in the form of nutrition) but that the creative principle is supplied by the father. He thus accounted also for the necessity of fertilization. Although this treatment of the phenomena of development is completely contradictory to what we now know of the material basis of development (the parts played by the ovum and the spermatozoon), still the concept of a creative principle has turned up repeatedly in the teachings of embryologists up to the twentieth century.

In the seventeenth and eighteenth centuries, when all biological sciences developed rapidly, together with the physicochemical sciences, there existed a widespread theory explaining the ontogenetic development of animals. This was the **theory of preformation.** The theory of preformation claimed that if we see that something develops from the egg, then this something must actually have been there all the time but in an invisible form. It is common knowledge that in a bud of a tree the leaves, and sometimes also the flowers with all their parts, can be discovered long before the bud starts growing and spreading and thus exposing to view all that before was hidden inside, covered by the superficial scales of the bud. Furthermore, it was known that in a chrysalis of a butterfly the parts of the butterfly's body—the legs, the wings, etc.—can be discovered if the cuticular coat of the chrysalis is carefully removed a few days before the butterfly emerges from the chrysalis. Something of this sort was supposed to exist in the egg. All the parts of the future embryo were imagined to be already in the egg, but they were thought to be transparent, folded together, and very small, so that they could not be seen. When the embryo began to develop, these parts supposedly started to grow, unfold, and stretch themselves, become denser and therefore more readily visible. The embryo, and therefore indirectly also the future animal, was **preformed** in the egg. Hence the theory is called the theory of preformation.

When spermatozoa were discovered in the seminal fluid, the relative significance of the ova and spermatozoa had to be accounted for. It is obvious that a preformed embryo cannot be present in both the egg and the spermatozoon. The preformationists were split therefore into two rival schools, the ovists and the animalculists. (The latter name comes from the word **animalcule,** as the spermatozoa were then called.) The ovists asserted that the embryo was preformed in the egg. The spermatozoa then seemed superfluous, and in fact, they were declared parasites living in the spermatic fluid. On the other hand, the animalculists declared that the embryo was preformed in the spermatozoon and that the egg served only to supply nutrition for the developing embryo. A lively discussion arose, which ended in favor of the ovists. The victory of the ovists was due to the discovery of parthenogenetic development in some insects, e.g., the aphids (Bonnet, 1745). If the egg could develop without fertilization it was clear that the embryo could not be preformed in the spermatozoon.

The theory of preformation, although very popular in its time, did not satisfy all biologists, and opposing views, denying the existence of a preformed embryo in the egg, were proposed. The most important contribution in this field was the **theory of epigenesis,** proposed by Caspar Friedrich Wolff (1759). In favor of his theory Wolff adduced his own observations on the formation of the chick embryo. In the earliest stages of the development of the chick he could not find any parts of the future embryo. Moreover, he found that the egg was by no means devoid of any visible structure; there was a structure present, but it was different from that of the later embryo. Wolff found that the substance of which the embryo is composed is granular. Presumably the granules must have been the cells or their nuclei. These granules were later arranged into the layers which we now call the **germinal layers.** Wolff saw that by the formation of local thickenings in some parts of these layers, by thinning out in others, and by the formation of folds and pockets, the layers are transformed into the body of the embryo. He concluded, therefore, that in the early egg there does not exist a preformed embryo but only the material of which the embryo is built. This material does not represent an embryo any more than a heap of bricks represents the house that will be built of them. In both cases there had to be an architect who would use the material for a purpose that he had in mind. In the case of the developing embryo the architect was represented by a vital force, perhaps not essentially different from the "creative principle" postulated by Aristotle.

Experimental Embryology. Wolff's observations, however, could not be considered as final in deciding between the alternative theories of preformation and epigenesis. In spite of what he actually observed, it was still conceivable that organs and parts of the body of the future embryo were represented in the egg by discrete particles, qualitatively different among themselves. The granules which he saw might have been different in their properties. Even if the transformation of such qualitatively different parts into the organs of the embryo should have been more complicated than was envisaged by the crude preformistic theory, the principle of preformation might well have held true in spite of the apparent homogeneity of the material of which the embryo was supposedly made. Observation alone could not make further advances toward

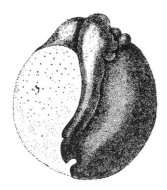

Figure 3.   Half embryo produced by W. Roux by killing one blastomere of a frog's egg with a hot needle. (After Roux, from Morgan, 1927.)

the solution of this problem, and further progress could be achieved only with the aid of experiment.

One of the experiments which is relevant to the preceding problem is the separation of the two cells into which the egg is divided at the beginning of development. If the theory of preformation is correct, we should expect that one of the two first blastomeres, containing one half of the egg material, should develop into an embryo lacking one half of its organs and parts. If, on the other hand, the substances contained in the egg are but the building material used for the construction of the embryo, then it is conceivable that half of the material might be sufficient for making a complete embryo even if it may have to be on a diminished scale, just as the bricks prepared for the construction of a big house may be used for building two houses of a smaller size.

The first embryologist to see this way of solving the problem was Wilhelm Roux (1850–1924). Accordingly, he proceeded to test one of the first two cleavage cells in the common frog for its ability to develop. To achieve his end Roux destroyed one of the two cleavage cells with a red-hot needle (1888). The embryo that was derived from the surviving cleavage cell was found to develop, at least at first, as if it were still forming a half of a complete embryo. In other words, the developing embryo was defective, as it should have been according to the theory of preformation (Fig. 3). It was found later, however, that the technique used by Roux was too crude. The damaged cleavage cell had not been removed, and it was the presence of this damaged cleavage cell, as was later found out, that caused the defects in the surviving embryo. If the two cleavage cells of the egg were separated completely, two whole and, except for their size, normal embryos could develop, one from each of the two cleavage cells. This result was first found by H. Driesch (1891), working on sea urchin eggs (Fig. 4), and later by Endres (1895) and Spemann (1901, 1903), working with eggs of newts. Eventually the experiment was repeated by Schmidt (1933) on the frog, the same animal that had served for the experiments of Roux. Schmidt found that if the two cleavage cells were completely separated, each could develop into a whole embryo (Fig. 5).

Figure 4.   Separation of the first two blastomeres in the sea urchin, resulting in the development of two whole embryos. (After Driesch, from Gabriel and Fogel, 1955.)

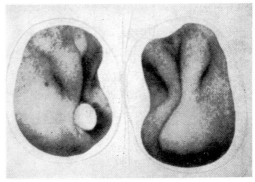

Figure 5. Separation of the first two blastomeres in the frog, resulting in the development of two whole embryos. (From Schmidt, 1933.)

The first experiments on the developing embryo were followed by many others, and soon a new science was born: **experimental embryology** (Roux, 1905).

Experimental embryology, in contrast to comparative embryology or descriptive embryology, uses experiment as a method of investigation. However, the use of the experimental method in itself does not create a science or a branch of science. New branches of science are created by novel viewpoints and novel problems set before science. It was the problem of what ontogenetic development actually is, what the driving forces behind it are, that necessitated the application of experiment after the methods of speculation and of pure observation were found to be impotent in solving the problem.

Modern Embryology — Analytical Embryology.   After the middle of the present century embryology had got caught up in the new trend that developed in biological science. Early in the century, the background for this new trend was established mainly by the work of T. H. Morgan and his school (Morgan, 1919). This work proved that the units of heredity, the genes, are arranged in linear order in the chromosomes of the cells. Analysis shows that chromosomes consist of several chemical components: deoxyribonucleic acid (DNA), ribonucleic acid (RNA), and proteins.

In an epoch-making paper published in 1953, Watson and Crick suggested that the deoxyribonucleic acid, as found in the chromosomes, consists of pairs of very elongated molecules twisted spirally around each other in a double helix. Each strand of the helix is made up of a number of units, the mononucleotides, which differ from one another only in the nitrogenous base (i.e., adenine, thymine, guanine, or cytosine) which each contains. The bases form two pairs, which structurally "fit" together, so that in the intertwined double helix adenine always links with thymine, and guanine with cytosine. Further work made it clear that the arrangement of the bases in the molecule of deoxyribonucleic acid contains a **code** for the proteins that may be synthesized by a particular species of organism. The code is essentially a series of "triplets" — groups of three bases which correspond to one amino acid in a polypeptide (protein) chain. Thus, a sequence of triplets in the DNA determines a sequence of amino acids in a protein molecule, and the section of the deoxyribonucleic acid molecule containing this sequence is the essential part of what geneticists call a "gene." The "genetic code," showing which sequences of bases correspond to which of the 20 amino acids constituting most of the proteins in the organic world, is shown in Table 1. Note that several different triplets in the DNA may code for the same amino acid.

Between the genetic code in the chromosomal DNA and the cell proteins, there are certain intermediate steps. The "message" contained in the DNA must first be copied in the form of a ribonucleic acid molecule, whose nucleotide sequence is complementary to the nucleotide sequence of the DNA (except that uridine takes the place of

TABLE 1   The Genetic Code

1st Letter \ 2nd Letter	U (Uracil*)	C (Cytosine)	A (Adenine)	G (Guanine)
U (Uracil*)	UUU UUC } Phenylalanine UUA UUG } Leucine	UCU UCC UCA UCG } Serine	UAU UAC } Tyrosine UAA } Termination of code of one gene UAG	UGU UGC } Cysteine UGA Termination of code of one gene UGG tryptophan
C (Cytosine)	CUU CUC CAU CUG } Leucine	CCU CCC CCA CCG } Proline	CAU CAC } Histidine CAA CAG } Glutamine	CGU CGC CGA CGG } Arginine
A (Adenine)	AUU AUC AUA } Isoleucine AUG Methionine**	ACU ACC ACA ACG } Threonine	AAU AAC } Asparagine AAA AAG } Lysine	AGU AGC } Serine AGA AGG } Arginine
G (Guanine)	GUU GUC GUA } Valine GUG Valine or Methionine	GCU GCC GCA GCG } Alanine	GAU GAC } Aspartic acid GAA GAG } Glutamic acid	GGU GGC GGA GGG } Glycine

*Uracil in ribonucleic acids corresponds to thymine in deoxyribonucleic acids.

**The methionine serves for initiation of a polypeptide chain. In bacteria, the amino acid is formyl-methionine. In both cases AUG serves as the beginning of a genetic message (gene).

After M. Nirenberg and collaborators and H. G. Khorana and collaborators; see Proc. Nat. Acad. Sc. U.S.A., Vol. 54, 1965, pp. 954–960 and 1378–1385.

thymine). This is the "transcription" phase. The code is then contained in an RNA molecule ("messenger" RNA). Two further kinds of ribonucleic acid are modeled on the DNA: the ribosomal RNA, which together with certain proteins forms small (± 200 Å in diameter) particles, the **ribosomes;** and the transfer RNA, which is involved in bringing the correct amino acid to the ribosome, where the amino acids become arranged and joined together in the correct sequence according to the code contained in the messenger RNA. This procedure constitutes the "translation" phase.

The importance of these discoveries for embryology derives from the following considerations. It has become evident that *all* the properties of any organism are determined in the last instance by the sequence of base triplets in the DNA molecules. Furthermore, it is accepted that the sequence of the base triplets directly determines what kinds of proteins can be produced by an organism. All other manifestations specific to any organism, whether morphological or physiological, depend more or less directly on the assortment of proteins coded for by the hereditary DNA. This new way of looking at the organic world shifts the problem of ontogenetic development directly into the realm of molecular relationships. It also makes possible, in principle, the construction of a *complete* theory of development. Such a theory would start with the triplet sequences in the DNA and would show first how these sequences are "read out" by transforming them into an array of proteins, placed and distributed in an organized way in space and time, and then would show how the proteins, acting partly on their own and partly through other chemical components, produce the complicated system that is an adult organism (animal, in the context of this book).

A whole array of new techniques has been mobilized in working toward such a theory of development. Electron microscopy has made great advances after the mid-1950's, when methods were developed for embedding tissues in plastics and for cutting ultrathin sections for the study of the fine structure of cells. Refined methods of chemical analysis, such as chromatography, electrophoresis, ultracentrifugation, and the use of radioactive tracers, have been put at the disposal of embryologists.

With the change in the theoretical background and techniques, a subtle change has permeated the work of embryologists. The aim of investigation is no longer the study of the development of any particular animal, or any group of animals, but the discovery of the basic principles and processes of development. This trend in science may perhaps be called **analytical embryology,** and this is what "modern" embryology actually is.

It must be realized that analytical embryology can proceed only on the basis of knowledge provided by descriptive embryology, because after all, it is the actual course of the transformations that have to be explained by the theory of development, of comparative embryology, because it is necessary to know of how general a significance any particular phenomenon of development is, and of experimental embryology, because it has revealed the causal relationships of many developmental processes.

In this "introduction to embryology" all the approaches to development (i.e., the descriptive and the comparative embryology and the experimental and the analytical embryology) will be dealt with together as far as possible.

Part Two

GAMETOGENESIS

Gametogenesis is the first phase in the sexual reproduction of animals. The essential process during this phase is the transformation of certain cells in the parents into specialized cells: the eggs, or ova, in the female and the spermatozoa in the male (or in the female and male organs in hermaphroditic animals).

The egg and the spermatozoon jointly constitute the material system, or jointly contain all the essential factors, which, given a suitable environment, will produce in the course of time a new individual with, in the main, the same characters as the two parent organisms. The various properties of both the paternal and maternal progenitors must thus be contained, in some form, in either the egg or the spermatozoon or both. It is obvious, however, that the characters of adult organisms are not there to be seen or directly detected in either the eggs or spermatozoa, the latter being single cells of a highly peculiar structure. It follows that the characteristics of the adult organisms are contained in the egg and sperm in a different and not immediately recognizable form. It is said, therefore, that these characteristics are encoded in the structure of the egg and sperm, just as a transmitted message may be concealed by the use of a code which makes the message unintelligible except to a person in possession of the code. It is also said that the egg and spermatozoon possess the "information" that is needed to build a new organism, meaning by this that all the specifications for the future organism are contained in the egg and sperm. Since the information is contained in the egg and spermatozoon in a form which does not bear an obvious relationship to the characters of the adult organism, we may say that it is contained in an encoded form. During the development of the egg the encoded information is decoded, that is, translated or transformed from a concealed to an overt form; characters inherent in the egg and sperm serve to determine the properties of the adult organism. The decoding or "reading" of the information contained in the gametes is then equivalent to the process of ontogenetic development, as defined on page 3.

The encoded information which is possessed by the egg and the spermatozoon may be contained in the structure (including the chemical composition) of either the nuclei or the cytoplasm of the gametes. Generally speaking, both the nucleus and the cytoplasm are indispensable for development to occur, since animal cells are fully viable only when both these parts (the nucleus and the cytoplasm) act together and in harmonious cooperation with each other.

In view of the profound differences in the structure and function of the male and female gametes, their development must be dealt with separately.

Chapter 2

SPERMATOGENESIS

2–1 CELLS IN SEMINIFEROUS TUBULES

The development of the spermatoza takes place in the male gonads, the **testes.** In vertebrates and insects, the testes are composite organs consisting of numerous **seminiferous tubules** converging toward common ducts which lead the mature sperm to the exterior.

Spermatogenesis is a continuous process, and various stages of development of the sperm may be observed in the seminiferous tubules at the same time. In the tubule there is, however, an orderly arrangement of cells undergoing different phases of development. In insects, the proximal ends of the tubules contain the spermatogonia—the cells undergoing proliferation by mitosis. Further down the tubule, the cells in the growth and maturation stages (the spermatocytes) are found. The ripe spermatozoa fill the most distal parts of the tubules. In vertebrates, the arrangement of the cells in the seminiferous tubules is somewhat different: all stages, from spermatogonia to mature spermatozoa, may be found at the same level of the tubule, but while the early stages (the spermatogonia) are located at the outer surface of the tubule in an epithelium-like arrangement next to the basement membrane, the later stages of differentiation, including ripe spermatozoa, lie nearer to the lumen of the tubule (Fig. 6).

A special feature in the testes of vertebrates is the presence in the seminiferous

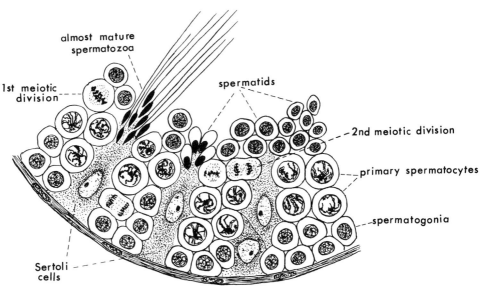

Figure 6. Diagram of part of a seminiferous tubule in a mammal. (Redrawn from Balinsky: "Vorlesungen über Histologie," München, 1947.)

19

tubules of somatic cells, which assist the developing spermatoza by anchoring the differentiating cells and possibly nourishing them during the latter part of sperm development. These cells, known as **Sertoli cells,** are tall columnar cells attached proximally to the basement membrane and reaching distally to the lumen of the tubule. The Sertoli cells have large pale nuclei with conspicuous nucleoli, thus differing from the rather dense chromatin-rich nuclei of spermatogonia and spermatocytes. The cells differentiating into spermatozoa become partially embedded in the cytoplasm of the Sertoli cells (Burgos and Fawcett, 1955), the future heads of the spermatozoa pointing toward the base of the Sertoli cells and the tails growing out toward the lumen of the seminiferous tubule. There are no Sertoli cells in the seminiferous tubules of insects, but similar cells are found in the testes of molluscs (Yasuzumi, Tanaka, and Tezuka, 1960).

The spermatogonia, as already stated, are found in vertebrates next to the basal membranes of the seminiferous tubules, and in microscopic preparations many of these may be seen to be in mitosis. While part of the spermatogonia remain in this condition and form a source of new sex cells throughout the reproductive life of the animal, some of the cells which are produced move toward the lumen of the tubule and enter the next phase of spermatogenesis: the phase of growth. The cells in this stage are called **primary spermatocytes.** The growth of the spermatocytes is actually very limited, though as a result they become perceptibly larger in volume than the spermatogonia (roughly by a factor of 2). However, the main feature of these cells is that they enter into the prophase of meiotic divisions which are of the greatest importance in the reproductive cycle of all organisms, but which can be dealt with only in their essentials here.

2-2   MEIOSIS

The somatic (body) cells in multicellular antimals have, as a rule, a **diploid** or double number of chromosomes; that is, apart from some exceptions due to the presence of sex chromosomes, every kind of chromosome in the set is represented by a pair of chromosomes of which one is derived from the male parent and one from the female parent. The total number is therefore an even one (Fig. 7). The two chromosomes of each pair are called **homologous** chromosomes. The **haploid** number of chromosomes is found in the gametes of animals and is half the number in the somatic cells. The diploid and haploid numbers of chromosomes in a few representatives of the animal kingdom are listed in the following chart.

	Diploid Number	*Haploid Number*
Man	46	23
Mouse	40	20
Frog	26	13
Fruit fly *Drosophila melanogaster*	8	4

The process of meiosis consists of two divisions of the nucleus, as a result of which the diploid number of chromosomes present immediately before is reduced to the haploid number.

The reduction of the number of chromosomes from the diploid to the haploid state can be considered as a means of preventing a continuous increase in the chromosome

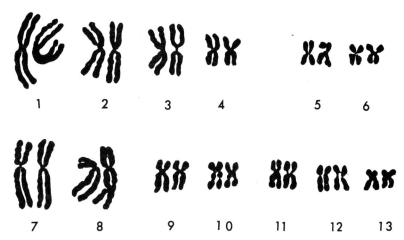

Figure 7. Set of chromosomes from a bone marrow cell of a frog. Each chromosome consists of two chromatids connected at the centromere. Homologous chromosomes are grouped together. (Redrawn after Di Berardino, King, and McKinnell, 1963.)

numbers which become doubled at each fertilization through the addition of the chromosomes of the egg and the spermatozoon, with a subsequent perpetuation of the increased number in mitotic divisions. This reduction can also be considered to be the last stage in the follow-up of the preceding sexual process which had caused a doubling of chromosome numbers in the zygote, after which the chromosome numbers again revert to the original haploid state. In many unicellular organisms the reduction of chromosome numbers immediately follows the fusion of the gametes, the very first division after the fusion being a meiotic division. The cells not involved in the sexual process are then all haploid. In multicellular animals, however, the reduction of chromosome numbers is delayed and the somatic or body cells are diploid. Reduction occurs only during the next reproductive cycle.

The mechanism by which a reduction of chromosome numbers is achieved involves a sequence of two divisions during which, however, the chromosomes divide only once. In ordinary mitosis, each chromosome splits during the prophase to form two chromatids; these separate in anaphase to become the daughter chromosomes. In this way each of the two daughter cells receives as many chromosomes as there were in the original cell. The chromosome numbers remain the same from cell generation to cell generation. In meiosis the chromosomes also split into two chromatids during the prophase, but instead of one division, producing two cells, two divisions occur in more or less quick succession, so that the available chromatids are distributed among four cells, each of which then receives only half the chromosomes originally present. The distribution of the chromosomes among the four cells is not haphazard but is the result of a peculiar feature of the meiotic prophase which must now be considered.

When the chromosomes first become visible as thin chromatin threads in the early meiotic prophase (the **leptonema** stage) (Fig. 8*a*), the number of chromosomes is a diploid one: there are as many threadlike chromosomes as could be seen during mitosis of somatic cells. It can often be noticed that the chromosomes at this stage take up a specific orientation inside the nucleus; the ends of the chromosomes converge toward one side of the nucleus, the side where the centrosome lies (the **bouquet stage**). Next, the chromosomes become sorted out in pairs: the two homologous chromosomes of each pair apply themselves to each other at first only at certain points, especially at the end where they are nearest to the centrosome, but later the two homologous chromosomes become

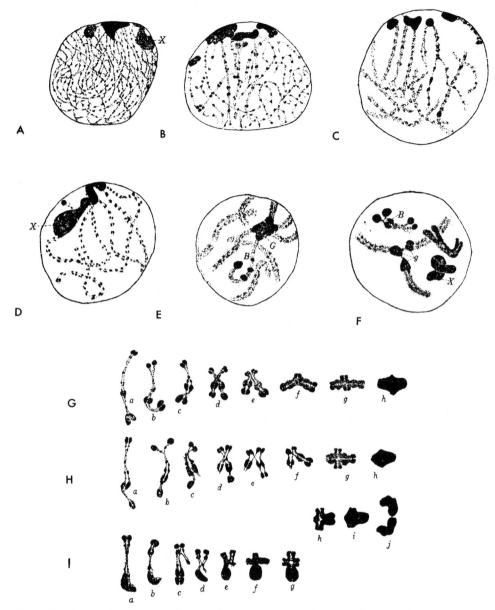

Figure 8. Stages of meiotic prophase in the spermatocytes of the grasshopper *Phrynotettix magnus*. *A*, Leptonema stage; *B* and *C*, zygonema stage; *D*, pachynema stage; *E*, diplonema stage; *F*, transition to diakinesis; *G*, *H*, *I*, chromosome bivalents condensing in diakinesis stage and showing chiasmata. (From Wenrich, 1916.)

joined along their whole length (the **zygonema stage**) (Fig. 8*b*, *c*). The joining together of the homologous chromosomes is known as the process of *synapsis* and is the most important part of the meiotic prophase. The pair of chromosomes become twisted spirally around each other and cannot be distinguished separately; they appear to form a single thick thread (**pachynema stage**) (Fig. 8*d*). The number of these thick threads corresponds to the number of pairs of chromosomes in the somatic set and is half the number of thin threads appearing in the preceding stage. Some time during the pachynema stage, each of the two conjoined chromosomes splits lengthwise to form two chromatids.

Actually, the doubling of the DNA molecule strands, which is necessary for the subsequent duplication of chromosomes, occurs earlier, before the beginning of meiotic prophase, as can be shown by measuring the amount of DNA in the nucleus. Through the earlier part of the meiotic prophase, however, the DNA molecules in each chromosome remain closely joined together, and each chromosome behaves as a single body. In the pachynema stage this is changed; the two chromatids of each chromosome, containing half of the DNA present in the chromosome at the start, become partially independent of each other, although they still continue to be linked together by their common centromere. Nevertheless, it is important to know that a pachynema chromatin thread actually consists of four chromatids closely joined together in one complex unit called a **bivalent,** because it contains a pair of chromosomes.

While the chromatids are in close proximity, breakages occur along the length of the chromatids and subsequently parts of the chromatids join up in new combinations. Sections originally belonging to different partners of the pair of chromosomes now become joined in one and the same chromatid. This is the phenomenon of **crossing over,** which enables an exchange of parts of the chromosomes to take place. The importance of crossing over stems from the fact that the hereditary material of the two parents, embodied in the two homologous chromosomes, is now reshuffled and redistributed to the four chromatids joined in synapsis. At any point, breaks and crossing over involve only one of the two chromatids of each of the chromosomes; the other chromatid in each of the chromosomes remains intact (Fig. 9). Barring some irregularities which occasionally happen, the two chromosomes exchange corresponding parts, that is, sections of the chromosomes containing the same genes. As a rule, therefore, each chromatid emerges with a full complement of genes.

After the pachynema stage, the chromatids united in each bivalent start separating from each other, as if the forces of attraction which had led to the joining of homologous chromosomes in synapsis have now been replaced by forces of repulsion. A distinct split appears between the two chromosomes in each bivalent (**diplonema stage**) (Fig. 8 *e*), and the two chromatids belonging to the same chromosome become distinguishable. The complex, however, continues to be held together, because at every location where a crossing over occurred, the chromosomes are connected by a **chiasma.** The chiasma

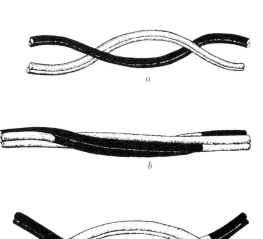

Figure 9. The crossing over of two chromosomes during synapsis (*a, b*) and the subsequent opening up of the bivalent in diakinesis (*c*). (From Morgan, 1919.)

is a link between chromatids belonging to different chromosomes of a pair, resulting from the fact that a part of a chromatid of one chromosome has been joined by crossing over to a complementary part of a chromatid of the other chromosome. As the crossing over is a reciprocal process, the chiasmata appear as X-wise links between the chromosomes (Fig. 9). Depending on how many times crossing over has occurred in any bivalent, there may be one or more chiasmata present.

Toward the end of the meiotic prophase, the chromatids joined in the bivalent contract and thicken in a very marked degree, leading to the **diakinesis** stage (Fig. 8 *f, g, h, i*). In diakinesis the chromatids of every bivalent are typically short stout rods, still united by the chiasmata. With the attainment of this stage, the chromosomes are ready to enter into the meiotic divisions.

The two meiotic divisions which follow, in the case of spermatogenesis, are very similar to ordinary mitosis insofar as the mechanism of cell division is concerned. An achromatic figure is formed, and in the first meiotic division the bivalents are placed in the equatorial plate of the achromatic figure. At the start of anaphase, the two chromosomes forming a bivalent separate from each other and are drawn to the opposite poles and thus to the two daughter cells, which are known as the **secondary spermatocytes.** In each chromosome the two chromatids are still joined by the common centromere. A short interphase follows which, however, may be very much reduced so that the chromosomes do not change substantially from their condition during the first meiotic metaphase. Even if a resting nucleus is formed, there is no duplication of chromatids and no synthesis of genic material at this stage.

In the second meiotic division, each chromosome on the equatorial plate consists of only two chromatids linked together by the centromere. The centromere now breaks, and the chromatids pass to the opposite poles. As a result of the two divisions, four cells are formed from each primary spermatocyte. These are called **spermatids.** Each spermatid receives one chromatid (now becoming an independent chromosome) from a bivalent. Because the bivalents are pairs of homologous chromosomes, it follows that the spermatids possess one chromosome for each pair of chromosomes present in somatic cells; that is, they have now a **haploid set of chromosomes** (Fig. 10).

The two homologous chromosomes joined in the bivalent are normally derived from different parents: from the father through the spermatozoon and from the mother through the egg. Thus, the usual condition is that half the chromosomes in somatic cells are maternal and half are paternal. In the bivalents, one of the component chromosomes is paternal and the other maternal. There is ample genetic evidence that during the first meiotic division the distribution of the paternal and maternal chromosomes is random, each chromosome having equal chances of being delivered into one or the other secondary spermatocyte. It follows that a secondary spermatocyte and consequently also every spermatid may have any combination of paternal and maternal chromosomes, provided that it can have only one chromosome from each pair. This, together with crossing over, provides for an almost infinite variety of combinations of paternal and maternal genes in any gamete.

2–3 DIFFERENTIATION OF THE SPERMATOZOA

The spermatids, though possessing a haploid set of chromosomes, are still not capable of functioning as male gametes. They have to undergo a process of differentiation to become the spermatozoa.

The function of the spermatozoon, for which its whole organization is adapted, is to reach the egg and, by fusing with the egg, to cause the egg to start developing and to

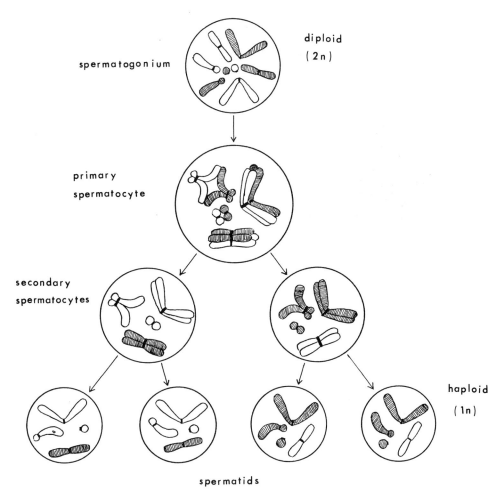

spermatogonium

diploid
(2n)

primary
spermatocyte

secondary
spermatocytes

haploid
(1n)

spermatids

Figure 10.   Distribution of chromosomes during meiotic divisions in spermatogenesis. The animal is presumed to have in its somatic cells, including the spermatogonia, four pairs of chromosomes of different shapes. The homologous chromosomes join to form four bivalents, each consisting of four chromatids, and these are subsequently distributed to the spermatids, each receiving the haploid number of chromosomes.

transmit to the developing embryo the paternal genes. In most cases this means that the spermatozoon must have a high degree of mobility, and in fact, the organization of typical spermatozoa is largely determined by the presence of a highly developed locomotory mechansim.

A typical spermatozoon consists of the following main parts: the head, the middle piece, and the tail or flagellum (Fig. 11). The anterior tip of the head is differentiated as the **acrosome,** the function of which is to enable the spermatozoon to penetrate through the egg membranes and to establish connection with the egg cytoplasm. The major part of the head is occupied by the nucleus containing the genes and is thus responsible for the transmission of hereditary characters from the male parent. The posterior part of the head also contains the centriole of the spermatozoon, which will be necessary for the initiation of cell division in the fertilized egg. The middle piece of the spermatozoon contains the base of the flagellum and—around it—the mitochondria. The latter, being carriers of oxidative enzymes and the enzymes responsible for oxidative phosphorylation, are the "power plant," supplying the flagellum with energy in a suitable form to be used for the propulsion of the spermatozoon. The tail or flagellum,

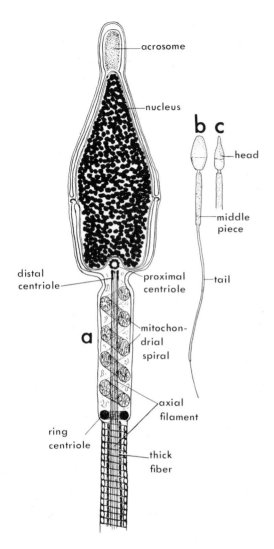

Figure 11.   A mammalian spermatozoon. *a*, Semidiagrammatic drawing (redrawn after an electron micrograph by Burgos and Fawcett: J. Biophys. & Biochem. Cytol., 1955). *b*, *c*, The same, as seen in the light microscope; the head is seen from the flattened side in *b* and from the narrow side in *c*.

usually by far the longest part of the spermatozoon, by its movements causes the spermatozoon to swim with the head (acrosome) foremost.

The changes which transform the spermatid into a spermatozoon are of a most radical nature. The nucleus of the spermatid, after the telophase of the second meiotic division, assumes the typical structure of an interphase nucleus with finely dispersed chromatin and a nuclear membrane. During transformation of the spermatid into a spermatozoon, the nucleus shrinks by losing water from the nuclear sap, and the chromosomes become closely packed into a small volume. This is necessary in order to reduce the dead weight to be carried by the locomotory apparatus of the spermatozoon and to enhance its motility. It appears that everything superfluous is removed from the nucleus, everything not directly concerned with transmission of hereditary characters, leaving only the actual material of the genes. All ribonucleic acid, which is abundant in the functioning nucleus, especially in the nucleolus, is eliminated, leaving only the deoxyribonucleoproteins present—the material of the genes. The shape of the nucleus also changes: instead of the usual spherical form the nucleus becomes elongated and narrow, an obvious adaptation for propulsion in water. In different animals the shape of the head of the spermatozoon (largely dependent on the shape of the nucleus) varies considerably. It may be ovoid and flattened from the sides (in man and bull), drawn out into a scimitar shape with a pointed tip (in rodents and amphibia), or spirally twisted like a corkscrew (in birds and some molluscs); occasionally it may be almost round (bivalve molluscs). Some of these shapes may also be interpreted as an adaptation for propulsion in water.

The acrosome of the spermatozoon is derived from the **Golgi bodies.** The Golgi body in an early spermatid consists of a series of membranes arranged concentrically around an aggregation of small vacuoles (Fig. 12). In the next stage one or more of the vacuoles start enlarging, and inside the vacuole a small dense body, the **proacrosomal granule,** appears. If more than one vacuole and granule are found, as sometimes happens, they fuse together so that eventually one big vacuole remains, containing a single large dense granule. The contents of the vacuole and the granule yield a positive staining reaction for mucopolysaccharides—the so-called periodic acid-Schiff reaction. (See Clermont and Leblond, 1955.) The vacuole with its granule now becomes closely applied to the tip of the elongating nucleus. The granule increases further and becomes the **acrosomal granule,** which forms the core of the acrosome. The vacuole loses its liquid content, and its walls become spread out over the acrosomal granule and the front half of the nucleus, covering them with a double sheath known as the **cap** of the spermatozoon (Fig. 13). The remainder of the Golgi body undergoes a gradual regression and eventually is discarded as the "Golgi rest," together with some of the spermatid's cytoplasm (Fig. 14).

The foregoing description of acrosome development is derived from a study of

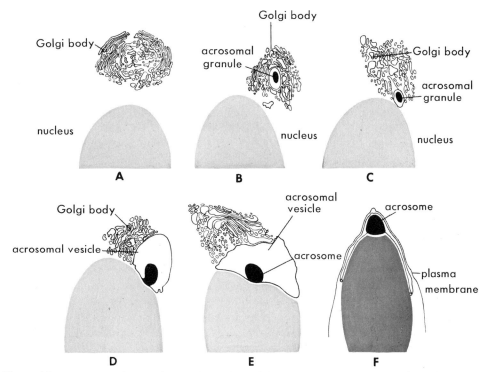

Figure 12. Sequence of stages in the formation of the acrosome and head cap from the Golgi complex during spermatogenesis in the cat. In drawings *A, B, C, D,* and *E,* only the nucleus and the Golgi body with the developing acrosome are shown; *F* represents the anterior part of the sperm head, including the plasma membrane. (Redrawn from Burgos and Fawcett, 1955.)

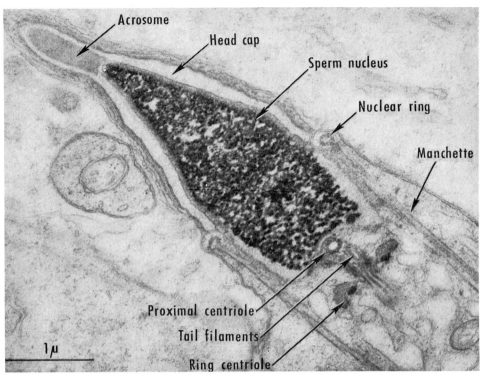

Figure 13. Electronmicrograph of late spermatid of cat. (After Burgos and Fawcett, from Bloom and Fawcett: A Textbook of Histology, Ed. 8. Saunders, 1962.)

spermatogenesis in the cat (Burgos and Fawcett, 1955). In other animals the development may be complicated by the formation of a second main component which appears between the acrosomal granule and the nucleus, protruding itself from behind into the acrosomal granule in the direction of the main axis of the spermatozoon. This **axial body** (Galtsoff and Philpott, 1960) or **acrosomal cone** (Kaye, 1962) may perhaps be the rudiment of the **acrosomal filament** developing in the spermatozoon during its approach to the egg (p. 000). As for the acrosomal granule, there is evidence that it contains a supply of enzymes which are used to dissolve the egg membranes during fertilization (L. H. Colwin and A. L. Colwin, 1961).

The centrosome of a spermatid after the second meiotic division consists of two centrioles, which can be shown with the aid of an electron microscope to have the structure of two cylindrical bodies, lying at right angles to each other. At an early stage in the differentiation of the spermatozoon, the two centrioles move to a position just behind the nucleus. A depression is formed in the posterior surface of the nucleus, and one of the two centrioles becomes placed in this depression with its axis approximately at right angles to the main axis of the spermatozoon. This is the **proximal centriole** of the spermatozoon; the other centriole, the **distal centriole,** takes up a position behind the proximal centriole with its axis coinciding with the longitudinal axis of the spermatozoon. The distal centriole then gives rise to the axial filament of the flagellum of the spermatozoon for which it serves as a starting point or basal granule (Figs. 13, 15, and 16). The axial filament of the spermatozoon has the same organization as the axial filaments of the flagella and kinocilia present in other animal (and plant) cells; that is, it has a pair of longitudinal fibers along its middle and a ring of nine pairs of longitudinal fibers surrounding it. (See Fawcett and Porter, 1954; Afzelius, 1959.) These fibers

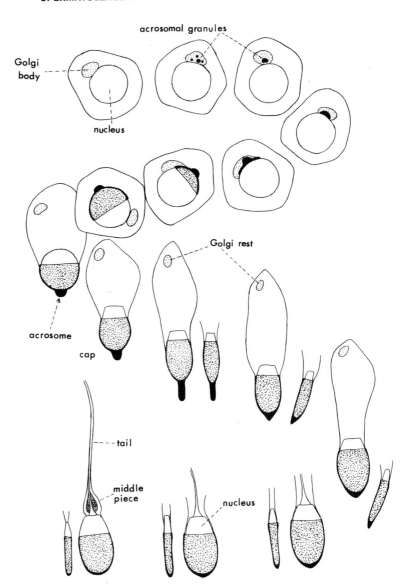

Figure 14. Transformation of the young spermatid (upper left-hand corner) into a free spermatozoon (lower left-hand corner) in the ram, as seen in preparations with periodic acid-Schiff staining. The acrosomal granules and acrosomes are shown in black; the head cap and Golgi body (including Golgi body rest) are stippled. The mass of mitochondria in middle piece of spermatozoon is indicated by very close stippling, and the nuclei, which are not stained by the periodic acid-Schiff method, are shown only in contour. The heads of ram spermatozoa are flattened; therefore in later stages the drawings show the spermatozoa as viewed from the broad side and from the narrow side. (Redrawn from Clermont and Leblond, 1955.)

are anchored in the distal centriole in the same way that the fibers of a cilium or flagellum are connected to their basal granules (Figs. 15 and 16).

The distal centriole and the proximal part of the axial filament lie in the middle piece of the spermatozoon. They are surrounded by the mitochondria, which become concentrated in this region from other parts of the cell. In the middle piece, the mitochondria lose their individuality to a certain degree by fusing together to a greater or lesser extent. In many animals (mammals in particular) the mitochondria join in one continuous body which becomes twisted spirally around the axial filament and the proximal centrosome. However, in other animals the spiral arrangement of the mitochondria is lacking, and they are joined in one or more massive clumps (mitochondrial bodies), forming the bulk of the middle piece of the spermatozoon (Figs. 55 and 57). Around the periphery of the middle piece the cytoplasm forms a condensed layer known as the **manchette** which also surrounds the posterior part of the head of the spermatozoon, where it is not covered by the cap (Figs. 13 and 16).

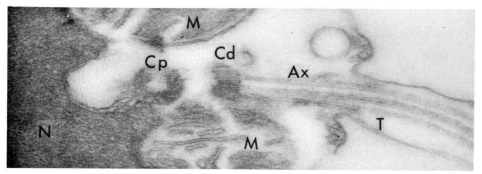

Figure 15. Electronmicrograph of posterior part of the head and middle piece of a spermatozoon of the annelid *Hydroides hexagonus*. Ax, Axial filament of tail; Cp, proximal centriole, cut transversely; Cd, distal centriole, cut longitudinally; M, mitochondrial body; N, nucleus; T, beginning of tail. (Courtesy of Dr. A. L. Colwin and Dr. L. H. Colwin, 1961.)

A dark ring is sometimes seen at the posterior end of the middle piece, forming the boundary between the middle piece and the tail (Figs. 11 and 13). The structure has been known as the "ring centriole," but it has been shown by electron microscopy that it does not in any way resemble a centriole in its structure, and that therefore it must have a completely different origin. The function of the ring is not known.

The axial filament is the main part of the tail or flagellum of the spermatozoon, and its structure is essentially similar to that found in cilia and flagella of other cells. The presence and arrangement of 10 pairs of longitudinal fibers have previously been mentioned. The axial filament is surrounded in the simpler cases by a very thin layer of cytoplasm and by the plasmalemma. Some other structures may also be present. In mammals the nine pairs of longitudinal fibers are accompanied on the outside by nine much thicker fibers which are wedge-shaped in cross section. These fibers start

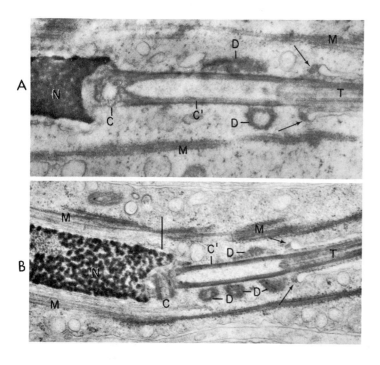

Figure 16. Electronmicrographs of the posterior head and middle piece region of late spermatids of the domestic fowl. The photographs show the two centrioles: the proximal centriole (C) in cross-section in *A* and in longitudinal section in *B* and the distal centriole (C') in longitudinal section in both photographs. D, Dense material which forms the so-called ring centriole; M, "manchette" surrounding the posterior part of the nucleus (N) and the middle piece of the spermatozoon; T, axial filament of the tail. (Courtesy of Dr. T. Nagano, 1962.)

in the middle piece of the spermatozoon but do not quite reach the tip of the flagellum (Yasuzumi, 1956; Telkka, Fawcett, and Christensen, 1961). Another set of threads—or, actually, flattened bands—surrounds the longitudinal fibers of the spermatozoon tail in mammals. For a long time the threads were thought to be coiled spirally around the central core of the tail, but careful electron microscopic studies have shown that the fiber does not form a continuous spiral but is composed of semicircular ribs articulating with each other on the opposite sides of the sperm tail (Telkka, Fawcett, and Christensen, 1961). The tip of the mammalian sperm tail (the "end piece") lacks the additional elements and consists of only the axial filament covered with cytoplasm and plasmalemma—a structure similar to the structure of the whole tail of less elaborately differentiated invertebrate spermatozoa.

A different kind of complication is present in some fishes and amphibians. In spermatozoa of these animals there is an undulating membrane which stretches along most of the length of the tail and presumably takes an active part in the locomotory activity of the sperm (Burgos and Fawcett, 1956).

Much of the cytoplasm of the spermatid becomes redundant in the spermatozoon, and it is simply discarded. As the acrosome is being formed at the anterior end of the spermatid nucleus, the cytoplasm flows away from it in the opposite direction, leaving only an extremely thin layer with the plasmalemma covering the acrosome and the nucleus. The bulk of the cytoplasm is then attached to what will be the middle piece of the spermatozoon, while the tail is growing out at the posterior end (Fig. 14). After the mitochondria have arranged themselves around the base of the axial filament of the flagellum, the remainder of the cytoplasm (containing also the "Golgi rest") is simply pinched off from the spermatozoon, leaving only a fairly narrow sleeve of cytoplasm surrounding the mitochondria in the middle piece. The detached part of the cytoplasm disintegrates.

In a few groups of animals, such as the nematodes and the decapod crustaceans, the spermatozoa do not have flagella and are therefore incapable of swimming. These spermatozoa are also in other respects profoundly different from those in other animals. When the spermatozoa of a decapod were examined with the electron microscope, it was found that they do not possess centrioles and are completely devoid of mitochondria—which makes sense, as they do not require large amounts of energy for movement (Moses, 1961). The mechansim of penetration of the nonflagellate spermatozoa into the egg must be very different from that of flagellate spermatozoa, and in Chapter 5 on fertilization only the latter will be discussed.

Chapter 3

OOGENESIS*

3-1 GROWTH OF THE OOCYTE

In the female sex the first stage of the development of gametes is similar to that found in spermatogenesis: the **oogonia**—the cells eventually giving rise to the eggs—undergo proliferation by mitotic divisions. They then become **oocytes** and enter a period of growth. Owing to the fact that the egg contributes the greater part of the substances used in development, growth plays a much greater role in oogenesis than in spermatogenesis. Also, the differentiation of the egg occurs simultaneously with growth, rather than after maturation.

The period of growth in the female gametes is a very prolonged one, and the increase in size is considerable. In frogs a young oocyte may be about 50 μ in diameter, and the fully developed egg in many species is between 1000 μ and 2000 μ in diameter. If *Rana pipiens*, in which the diameter of the mature egg is about 1500 μ, is taken as an example, the increase in size of the oocyte is by a factor of 27,000. This growth takes place over a period of three years. The young oocytes start growing after the tadpoles metamorphose into young froglets. One-year-old and two-year-old frogs do not yet have mature eggs, but by the third year the eggs are ready and the frogs may spawn for the first time. Every year a new batch of oocytes is produced as the result of oogonial divisions, but these do not mature until three years later, so that oocytes of three generations may be contained in the ovary at the same time. The growth of the oocytes is fairly slow during the first two seasons but becomes much more rapid in the summer of the frog's third year of life, so that by the autumn the eggs reach their maximum dimensions (Fig. 17) (Grant, 1953).

In other animals the growth of oocytes may proceed at a much higher rate and may take a shorter time for completion. In the hen the last rapid growth of the oocyte occurs in the 6 to 14 days preceding ovulation, and during this time the volume of the oocyte increases 200 fold. In mammals the proliferation of the oogonia is restricted to the intrauterine period of life, and all the eggs produced by a mammalian female throughout her reproductive life are derived from oocytes already present at birth. The oocytes reach their full size in 16-day-old mice. The eggs of mammals are, of course, much smaller than those of amphibians. The oocyte of a mouse grows from a size of about 20 μ in diameter into the ripe ovum of about 70 μ in diameter, an increase by a factor of only 43 as compared with an increase of 27,000 in the frog.

The fully grown egg is a relatively large cell, always larger than the average somatic cell in the animal. When large quantities of foodstuffs are stored in the egg cell, it may attain giant proportions. The yolk of a hen's egg, which represents the egg cell in this

---

*Valuable reviews of a number of aspects of oogenesis can be found in J. D. Biggers, and A. W. Schuetz (Eds.): Oogenesis. Baltimore, University Park Press, 1970.

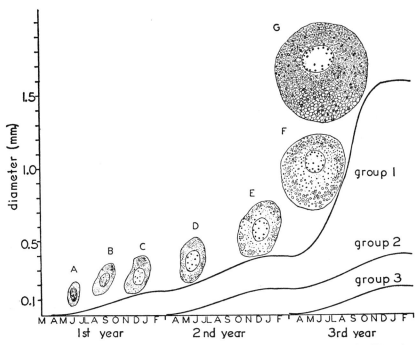

1.5

diameter (mm)

1.0

0.5

0.1

G

F

group 1

E

D

group 2

C

B

group 3

A

M AMJ JLA S O N D J F ' A M J JLA S O N D J F ' A M J JLA S O N D J F '
1st year   2nd year   3rd year

Figure 17. Growth of frog oocytes during the first three years of the female's life. The curves show the increase in diameter of three generations of oocytes; the drawings represent the changes in size and structure of oocytes of the first generation. (Modified from Grant, 1953.)

case, has an average weight of 55 gm. and is, of course, by no means the largest of its kind. Very large eggs are also found in reptiles and in some sharks. However, the quantity of active cytoplasm in such large eggs is comparatively rather small, and the nucleus, although larger than in the somatic cells, never increases in proportion to the bulk of the whole egg.

Simultaneously with the growth of the oocyte its nucleus enters into the prophase of the meiotic divisions; the homologous chromosomes pair together similarly to what occurs in the primary spermatocytes. The subsequent stages of meiosis, however, are postponed until the end of the period of growth. Instead, the nucleus of the oocyte increases in size, though not nearly to the same extent as the cytoplasm. The increase in size is due mainly to the production of large amounts of nuclear sap, so that the nuclei of advanced oocytes appear to be bloated with fluid and are often referred to as **germinal vesicles.** The chromosomes at the same time may increase in length, but the amount of deoxyribonucleic acid in the chromosomes does not increase in proportion to the enlargement of the nucleus. As a result, the nuclei in this stage are difficult to stain with the usual agents, such as methyl green or Feulgen reagent. In oocytes of animals having large eggs, the chromosomes acquire a very characteristic appearance; thin threads or loops are "thrown out" transverse to the main axis of the chromosomes, making the chromosomes look like lamp brushes, thus the name **lamp brush chromosomes** (Fig. 18). It is believed that the loops represent actual sections of the chromosome which have become completely despiralized, and that this is a favorable condition for the main activity of the genes, namely, to synthesize the messenger RNA, which is subsequently to control the synthesis of proteins in the cell. By using a radioactively labeled RNA precursor (uridine), it has actually been proved that RNA synthesis occurs on the loops of the lamp brush chromosomes. With the aid of very fine microchemical

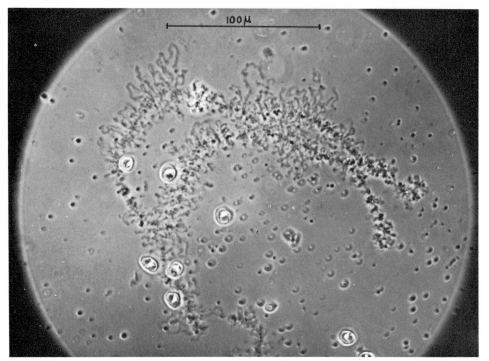

Figure 18. Lamp brush chromosomes from an oocyte of the newt, *Triturus carnifex.* (Courtesy of Professor H. G. Callan.)

methods, the chromosomal RNA in oocytes has been found to have a different base composition (that is, a different ratio of adenine and uracil to guanine and cytosine) than the base ratio of ribosomal RNA, but it fairly closely resembles the base ratio of the chromosomal DNA (with thymidine replacing uracil). (See Gall, 1963.) These observations are in keeping with the assumption that the RNA synthesized on the loops of the lamp brush chromosomes is, base for base, a copy of the chromosomal DNA, in other words, that it becomes a carrier of genetic information contained in the chromosomes. This genetic information is then presumably passed into the oocyte cytoplasm, where proteins are being synthesized in the course of the growth of the oocyte.

The nucleoli in the germinal vesicle seem to be actively involved in the metabolism of the growing oocyte, as they are concerned with the synthesis of ribosomal RNA. The nucleolus of the oocyte increases greatly in size and becomes very conspicuous against the background of the vesicular nucleus. In many animals, particularly amphibians, instead of one large nucleolus, numerous smaller nucleoli are formed in the germinal vesicle. Most of these become localized on the periphery of the nucleus, immediately underneath the nuclear membrane (Fig. 19; see also Fig. 37, p. 51).

The formation of numerous nucleoli is the result of a very peculiar phenomenon that occurs in the oocytes of at least some animals. It is the outward expression of an increase in the number of genes coding for ribosomal nucleic acids, an occurrence which is due to the need for producing large amounts of these acids during the growth of the oocyte. The genes for ribosomal nucleic acids, even in somatic cells, are present in a number of identical copies. In the frog, *Xenopus laevis,* the genes coding for the two main molecules which go into the formation of a ribosome, the 18S* and the 28S RNA's, are

---

*S, the **Svedberg unit,** is a measure of the rate of sedimentation of molecules subjected to ultracentrifugation. Indirectly, it is also an indication of the size of a molecule.

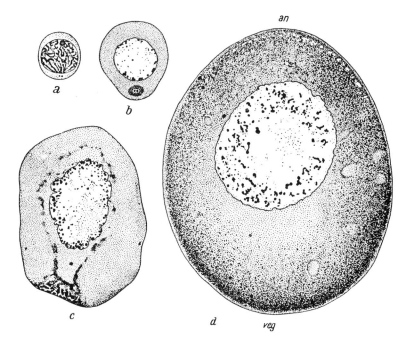

Figure 19. Four stages in the development of the amphibian oocyte. *a, Ambystoma,* young oocyte with chromosomes in the "bouquet stage." *b, c, d, Rana,* beginning of deposition of yolk. (After Witschi, from Kühn, 1955.)

represented in a haploid chromosome set about 450 times. The third gene, coding for the smaller 5S ribosomal RNA, is present in some 20,000 copies. In normal mitosis the ribosomal genes are duplicated in the same way as other genes, preserving their numbers in the daughter cells. In oocytes, however, the genes for 18S and 28S RNA's are multiplied, without mitosis taking place, by a factor of several hundred (Brown and Dawid, 1968; Brown 1973). This increase in the number of genes without mitosis is referred to as **amplification** of the genes concerned. The genes which code for the 18S and 28S rRNA's are located in a special sector of a chromosome responsible for the formation of the nucleolus, the "nucleolus organizer." The amplification of the ribosomal genes produces numerous nucleolus organizers and is responsible for the large number of nucleoli in amphibian oocytes. The genes coding for the 5S rRNA, which are scattered on different chromosomes, are not amplified, perhaps because their number is already so high in somatic chromosomes. Ribosomal gene amplification has also been recorded in a worm and a mollusc, although the degree of amplification is much lower (by a factor of 5) and does not lead to the formation of additional nucleoli (Brown and Dawid, 1968).

There are many reports, in both the older and more recent literature, that nucleic acids, in particular ribonucleic acid, pass out of the nucleus into the cytoplasm during the growth of the oocyte. In some cases large masses of nucleolar material, visible with the light microscope, have been seen passing through gaps in the nuclear membrane (Fig. 20) (Bretschneider and Raven, 1951; Logachev, 1956). In practically every animal studied with the electron microscope, electron dense material, which could be recognized as RNA, has been seen passing through the nuclear pores into the cytoplasm. (In the frog, see Balinsky and Devis, 1963.)

In the oocyte cytoplasm the RNA can be detected in the form of abundant ribosomes through either cytochemical methods or electron microscopy. Increased quantities are first seen in the vicinity of the nucleus, and later, some RNA becomes concentrated toward the periphery of the cytoplasm (Kemp, 1953). As the yolk platelets begin

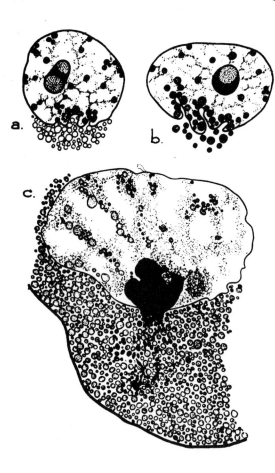

Figure 20. Extrusion of nuclear substances into the cytoplasm during oogenesis of *Limnaea stagnalis.* (From Bretschneider and Raven, 1951.)

to accumulate at the periphery of the oocyte, the area of cytoplasm rich in ribonucleic acid becomes restricted to the deeper parts of the cytoplasm.

There can be no doubt that messenger and transfer RNA are also released from the nucleus during the growth of the oocyte, although this cannot be demonstrated by morphological methods. Evidence concerning the presence of these types of RNA in the cytoplasm of the mature egg will be presented in a different connection. (See Sections 5–3 and 8–3.)

Of considerable importance is the presence of deoxyribonucleic acid—DNA—in the egg cytoplasm. Quite substantial amounts of DNA have been found in the cytoplasm of mature eggs of amphibians, insects, and sea urchins. (See Chemical Changes During Cleavage, p. 104.) Recent studies have shown that in amphibian oocytes the cytoplasmic DNA is contained in the mitochondria (Dawid, 1970). The mitochondrial DNA has a remarkable feature; the molecules are circular, like the DNA molecules of bacterial and certain viral chromosomes. There are only a few molecules of DNA per mitochondrion, perhaps even only one, and it is believed that the mitochondrial DNA does not code for the mitochondrial enzymes (there are too few molecules per mitochondrion for this), but that it may somehow be concerned with the maintenance of the mitochondrion as such. Though the amount of DNA per mitochondrion is very small, the total amount per mature oocyte may be considerable (200 times greater than the amount of DNA contained in a tetraploid set of somatic chromosomes in *Xenopus laevis*) (Brown and Dawid, 1968).

In many groups of animals, notably in the chordates, the oocytes are surrounded during their entire growth and maturation stages by special cells of the ovary, the **follicle cells.** In mammals the follicle cells are derived from the germinal epithelium of the ovaries, and initially the young oocyte is surrounded by one layer of follicle cells, which form a simple cuboidal epithelium around the oocyte. Later, the number of follicle cells increases greatly, the cells becoming arranged in several rows. As the egg approaches maturity, an eccentric cavity appears in the mass of the follicle cells. This cavity is filled with fluid secreted presumably by the cells of the follicle. The follicle at this stage is known as a **Graafian follicle** (Fig. 21). The oocyte is surrounded by follicle cells not only in mammals but in other vertebrates as well, though due to the larger size of the egg the follicle cells are not so conspicuous. It is believed that the follicle cells actively assist the growth of the oocyte by secreting substances which are taken up by the oocyte.

The structural relationships between the follicle cells and the oocytes, revealed by electron microscopic studies in recent years, are very peculiar. Originally there is a simple apposition of the follicle cells and the oocyte, similar to the one existing between ordinary epithelial cells: the cytoplasmic membranes of the adjoining cells are separated only by a narrow gap of about 80Å (Fig. 22). At certain points the two cytoplasmic membranes show a close connection in the form of **desmosomes;** here the plasma membranes of the adjoining cells are thickened, and the space in between appears to be filled by a denser substance which presumably holds the two cell membranes together (Anderson and Beams, 1960; Franchi, 1960; Wartenberg, 1962). At a later stage a wider space appears between the follicle cells and the oocyte. The follicle cells, however, retain contact with the oocytes at the points where the desmosomes could be noticed in the previous stage. At these points the cytoplasm of the follicle cell becomes drawn out into elongated processes or **microvilli,** reaching across the space separating the follicle cell and the oocyte. At the same time the surface of the oocyte also produces numerous finger-like microvilli projecting into the space between the oocyte and follicle cells (Fig. 23). The microvilli of the oocyte interdigitate with those of the follicle cells (Kemp, 1956; Sotelo and Porter, 1959) (Fig. 24). With the light microscope, individual

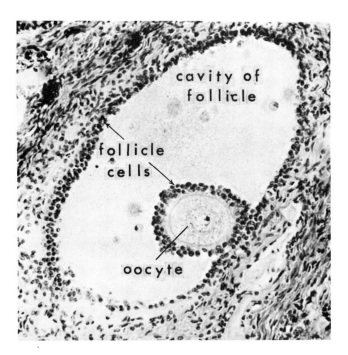

Figure 21. Graafian follicle in the ovary of a bitch.

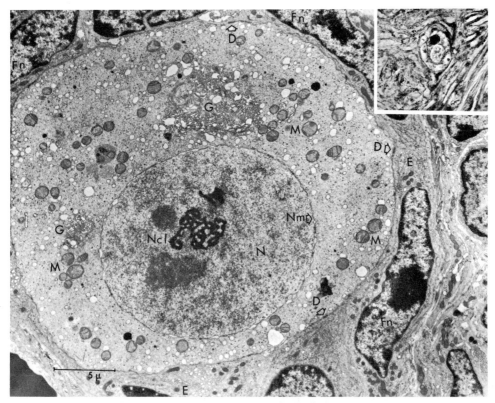

Figure 22. Young oocyte of guinea pig (electronmicrograph), showing Golgi material ("yolk nucleus"). Inset: Oocyte in a similar stage, with "yolk nucleus," as shown by ordinary microscopy. N, Nucleus; Ncl, nucleolus; Nm, nuclear membrane; G, Golgi material; M, mitochondria; D, desmosomes at junction of oocyte and follicle cells; Fn, nuclei of follicle cells; E, endoplasmic reticulum in follicle cells. (Courtesy of Professor E. Anderson.)

microvilli cannot be seen, and the zone of microvilli appears as a radially striated layer, which has long been known, in mammals, as the **zona radiata.** The presence of the microvilli greatly increases the surface area of the oocyte. In the frog oocyte the increase in surface area has been estimated to be by a factor of about 35. The increase in area appears to facilitate metabolic turnover between the oocyte and its environment.

Small inpocketings of the oocyte cytoplasm may often be observed at the base of the microvilli. These are interpreted as an indication that the oocyte is taking in fluids and dissolved substances from the space between itself and the follicle cells by means of **pinocytosis** or "cell drinking" (Press, 1959; Anderson and Beams, 1960).

Toward the time when the oocyte reaches its full size a denser material, often of fibrillar structure, appears in between the interdigitating processes of the oocyte and the follicle cells. The material becomes consolidated and eventually fills most of the space between the follicle cells and the oocyte, becoming the egg membrane. (See further on page 62.)

In the last stages of oocyte maturation the cytoplasmic processes of the oocyte and of the follicle cells may be withdrawn, and the space between the two is then taken up by the egg membrane. This is the case in all vertebrates studied so far and particularly in mammals (Odor, 1960). In bivalve molluscs, however, the microvilli of the oocyte are not withdrawn, and at the time when the egg is released from the ovary, its vitelline membrane is still perforated by long processes of the egg cytoplasm; these persist even at the beginning of the cleavage (Reverberi and Mancuso, 1961; Humphreys, 1962;

Rebhun, 1962; Pasteels and de Harven, 1962). In fishes the perforations in the egg membrane which are left after the cytoplasmic processes are withdrawn become canals ("micropiles") through which the spermatozoa can reach the egg (Sadov, 1956).

The development on the surface of the oocyte of a system of microvilli which interdigitate with cytoplasmic processes of the follicle cells is a general rule, but there are some exceptions, the most important of which are in the eggs of sea urchins. The follicle cells in these animals are closely apposed to the surface of the oocyte, and no interdigitating microvilli are present. Nevertheless, when the mature egg leaves the ovary it is surrounded by a thin layer of jelly, which must have been formed during the growth of the oocyte in the space between the oocyte and the surrounding follicle cells.

In some insects, molluscs, and annelids the relationship between the oocyte and the rest of the ovary is further complicated by the presence of special nurse cells which, together with follicle cells, take part in providing the nutrition for growing oocytes. The nurse cells are closely related in their origin to the egg cell and may be considered to be abortive oocytes. In *Drosophila* an oogonial cell gives rise, by four successive mitotic divisions, to 16 cells of which one becomes an oocyte and the other 15 become nurse cells. The nurse cells are thus siblings of the oocyte. The whole complex is surrounded by follicle cells (Fig. 25). The mode of transfer of materials from nurse cells to the oocyte is radically different from the way matter is given off to the oocyte by the follicle cells. No microvilli or cytoplasmic processes are developed at the interface between the oocyte and nurse cells. Instead gaps may appear through the cell membranes of the

Figure 23. Advanced oocyte of guinea pig (electronmicrograph), showing zona radiata and microvilli. CO, Cytoplasm of oocyte; Mv, microvilli produced by the oocyte; Fm, cytoplasmic projections from the follicle cells; F, follicle cell; ZR, zona radiata; D, desmosome at point of contact of follicle cell projection and surface of oocyte; M, mitochondria; Fn, nucleus of follicle cell; G, Golgi body. (Courtesy of Professor E. Anderson.)

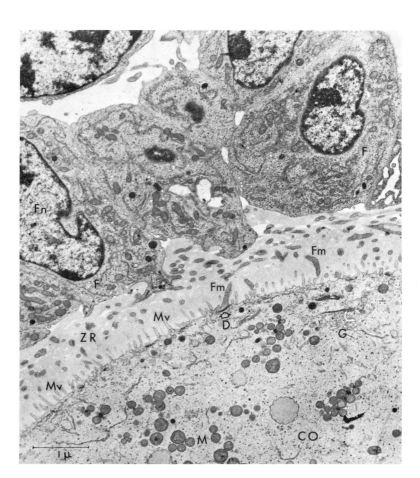

Figure 24. Young oocyte of a mammal surrounded by follicle cells. D, Desmosomes; ER, endoplasmic reticulum; FE, follicle cells; FGC, Golgi complex in follicle cells; GC, Golgi complex in oocyte; IB, inclusion body; M, mitochondria; MV, microvilli; MVB, multivesicular body; N, nucleus; NCL, nucleolus; PFE, processes of follicle cells; PO, oocyte (primary oocyte); V, vacuoles formed during pinocytosis; ZP, zona radiata. (From Anderson and Beams, 1960.)

oocyte and the nurse cell, and through such gaps the cytoplasm of the nurse cell pours into the oocyte (King and Devine, 1958). In any case, the nurse cells become used up during the growth of the oocyte (in some insects and annelids; Fig. 26), or they may be completely engulfed in the cytoplasm of the oocyte (in the snail *Helix*).

The materials used in the growth of the oocytes are to a large extent produced outside, in other parts of the body of the female, and brought to the gonads by way of the blood stream. In vertebrates the site of synthesis of egg proteins and phospholipids appears to be the liver. This may be proved by using precursors marked by a radioactive isotope atom such as P^{32} in disodium hydrogen phosphate. When such a precursor was injected into the blood of a laying hen and the amount of radioactivity was measured in different tissues, it was found that after six hours the radioactivity was highest in the liver (164.0 units / γ phosphate); the radioactivity was weaker in the blood (38.3 units / γ phosphate), and very small indeed in the growing oocytes (0.62 units / γ phosphate). These results show that the injected phosphate was mainly taken up by the liver. Twelve hours later the radioactivity in the liver was down to 101.0 units / γ phos-

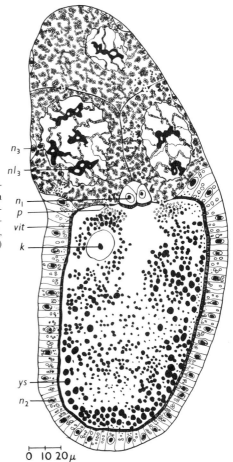

Figure 25. Diagram of an ovarian follicle of *Drosophila melanogaster*. n_1, n_2, n_3, Nuclei of various types of follicle cells; nl_3, a portion of the compound nucleolus of the nurse cell; p, pore (intercellular bridge) connecting a nurse cell to the oocyte; k, karyosome in the nucleus of the oocyte; vit, vitelline membrane; ys, yolk sphere. (Courtesy of Dr. R. C. King and Dr. E. A. Koch, 1963.)

phate but was increased slightly in the blood (to 45.4 units / γ phosphate). There was a very large increase of radioactivity in the oocytes to 4.1 units / γ phosphate. This indicates that the substances synthesized in the liver with inclusion of the injected phosphate found their way via the blood into the growing oocytes (Flickinger and Rounds, 1956). By direct chemical analysis it can be shown that a chemical constituent of the egg

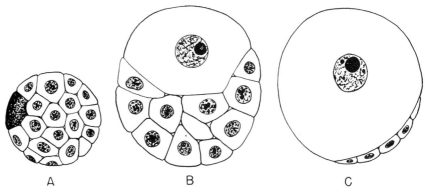

Figure 26. Three stages in the development of the oocyte of the leech *Piscicola*, which uses up the follicle (nurse) cells as it grows. (From Jörgensen, in Willier, Weiss, and Hamburger, 1955.)

yolk (phosvitin—see further p. 46) is present in the blood plasma of a laying hen (Mok, Martin, and Common, 1961). By use of serological methods (p. 500) it was found that, in the frog, substances (antigens) identical to the ones found in the growing oocytes are present in the liver and in the blood serum (Flickinger and Rounds, 1956). A similar investigation has been performed with females of the cecropia silk moth *(Platysamia cecropia)*. It was proved by using serological methods that a specific protein contained in the blood plasma of female pupae passes into the developing eggs and becomes a major part of the yolk of the egg (Telfer, 1954). As only larger molecules can be detected by serological methods, the foregoing investigations serve to indicate not only that building materials such as amino acids are brought to the oocytes from without but that macromolecules (proteins, phosphoproteins, and phospholipids) can be supplied to the oocytes in the same way.

That building materials are supplied to the oocytes through the mediation of follicle cells and nurse cells (where such exist) can also be demonstrated by using radioactively marked precursors. Females of the common fly, *Musca domestica*, were injected with tritiated (containing radioactive hydrogen, H^3) cytidine, a nucleic acid precursor, or tritiated (H^3 containing) histidine, a protein precursor. After varying intervals the animals were fixed, the gonad sectioned, and the sections covered with a photographic film. The emulsion of the film shows black dots where the film is hit by electrons shooting out from the radioactive hydrogen atoms in the tissues (Bier, 1963). Thirty minutes after the injection, the radioactively labeled amino-acid molecules are lodged predominantly in the follicle cells (Fig. 27*A*), but after three hours, radioactivity in the follicle cells disappears, and intense radioactivity is present in oocyte cytoplasm, mainly in the surface layer (Fig. 27*B*), thus showing that the labeled material has been passed from the follicle cells to the oocyte. The nucleic acid precursors, on the other hand, appear to be taken primarily into the nuclei of the nurse cells, which are intensely radioactive one hour after the injection (Fig. 27*C*). The newly synthesized radioactively labeled ribonucleic acid soon passes out of the nuclei into the cytoplasm of the nurse cells, so that in five hours the nuclei are practically free of radioactivity (Fig. 27*D*). From the cytoplasm of the nurse cells the labeled RNA then passes into the cytoplasm of the oocyte through the gaps connecting the nurse cells to the oocyte (p. 39).

A comparison of work done on the growth of avian, amphibian, and insect oocytes shows that the mechanism of food supply to the oocyte in these animals is rather similar. There is evidence, on the other hand, that in some animals the synthesis of yolk proteins occurs inside the oocytes, and consequently, the materials necessary for yolk formation enter the oocytes in a simpler form (p. 52).

3–2 ACCUMULATION OF FOOD RESERVES IN THE CYTOPLASM OF THE OOCYTE

During the growth of the oocyte not only does the amount of cytoplasm increase in quantity, but it changes in quality by the elaboration and regular distribution of various cell inclusions and specially modified parts of the cytoplasm which, as we will see, are essential for the development of the embryo.

In most animals which have been investigated with modern methods, young oocytes show a very similar and fairly simple organization. The cytoplasm in young oocytes is rather poor in complex inclusions and possesses practically none of the specialized structures found in the adult oocyte and mature egg (Fig. 28). Large amounts of ribonucleic acid are present in the cytoplasm of young oocytes. As seen with the electron microscope the cytoplasm is finely granular. The granules are almost certainly ribonucleoprotein in nature, as the cytoplasm is strongly and uniformly stained by

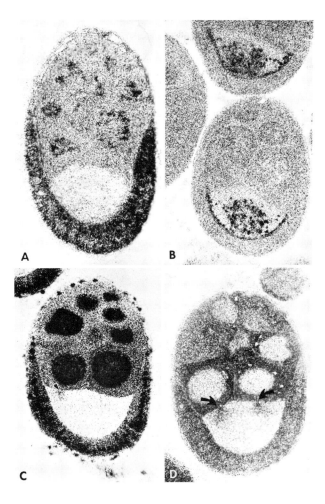

Figure 27. Passage of radioactive nucleic acid precursors and protein precursors into the growing oocyte of the common fly, *Musca domestica.* The figure shows autoradiographs of sections. *A,* Follicle 30 minutes after injection of H³-histidine. *B,* Follicles 3 hours after injection of radioactive histidine. *C,* Follicle 1 hour after injection of H³ cytidine. *D,* Follicle 5 hours after injection of radioactive cytidine. Arrows in *C* and *D* indicate the intercellular bridges through which the radioactively labeled RNA streams from the nurse cells into the oocyte. Compare with diagram of egg follicle, Figure 25. (Courtesy of Dr. K. Bier, 1963.)

pyronin or azure B (Brachet, 1941), stains used to reveal ribonucleic acid. Mitochondria are fairly scarce in young oocytes but may increase in numbers considerably during the growth of oocytes and may, in some animals (amphibia, birds), be aggregated in the form of large "mitochrondrial clouds" (Romanoff, 1960; Wartenberg, 1962; Balinsky and Devis, 1963) (Fig. 29). Because mitochondria are carriers of oxidative enzymes, the overall oxygen consumption increases during the growth of the oocyte. Young frog oocytes, before the beginning of yolk formation, have been found to absorb 0.69 cu. mm. of oxygen per cubic millimeter of oocyte. At the stage when the yolk platelets begin to form, the oxygen consumption rises to 1.5 cu. mm. per cubic millimeter. In older oocytes nearing maturity, the oxygen consumption per cubic millimeter falls to 1.2 cu. mm. This is still a high value, considering that a large part of the volume of the oocyte is filled with inert yolk; thus, the remaining active cytoplasm continues to respire at a high rate (Mestscherskaia, 1935, cited after Brachet, 1950).

Golgi bodies are found in younger oocytes around the centrosome.

In oocytes of some mammals (guinea pig, Anderson and Beams, 1960; rat, Odor, 1960), Golgi-type membranes form a large spherical mass (Fig. 22, p. 38) which later disperses. In oocytes which have advanced in growth, small Golgi bodies become scattered throughout the whole cytoplasm, or they may be located in the subcortical cytoplasm, sometimes distributed there at fairly regular intervals (frog, Balinsky and Devis,

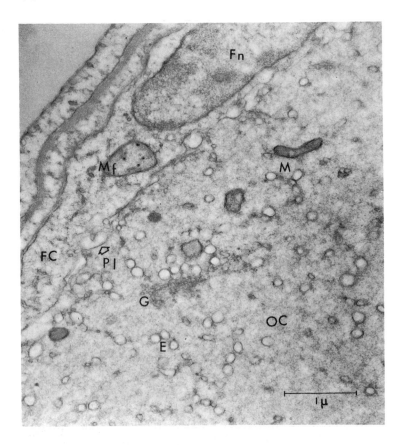

Figure 28. Surface of young frog oocyte with adjoining follicle cell (electronmicrograph). OC, Cytoplasm of oocyte; G, Golgi body; M, mitochondrion; Pl, plasmalemma of oocyte; FC, follicle cell; Fn, nucleus of follicle cell; Mf, mitochondrion in follicle cell; E. endoplasmic reticulum (vesicular form).

1963; chick, Bellairs, 1963). (Fig. 28). In mature oocytes Golgi bodies sometimes disappear completely (Odor, 1960), possibly indicating that these organoids have been used up or transformed into some other structures.

Typical endoplasmic reticulum, in the form of double lamellae with ribosomes attached to their outer surfaces, is very rarely found in young oocytes. The cytoplasm instead contains fairly numerous vesicles surrounded by a simple membrane. These vesicles are generally believed to be equivalent to or a modification of the endoplasmic reticulum. Ribosomes are often attached to the surface of the vesicles. A different kind of membranous structure is, however, often found in oocytes; it takes the form of stacks of double membranes, sometimes in parallel, and sometimes in spiral arrangement (Fig. 30). The membranes usually do not have ribosomes attached to them but instead are perforated by pores, which closely resemble the pores of the nuclear membrane (Merriam, 1959; Wischnitzer, 1960; and Rebhun, 1961). For this reason the membranes are regarded by some authors as being derived from the nuclear membrane (Kessel, 1963, 1968).

Yolk appears in the oocytes in the second period of their growth. Accordingly, many embryologists distinguish two periods of oocyte development: the previtellogenesis period and the vitellogenesis period. (See Raven, 1961.) Growth is very much accelerated during the vitellogenesis period, as indicated in Figure 17; this period is thus also a period of rapid growth.

The most usual form of food storage in the egg consists of granules of **yolk**. Yolk is not a definite chemical substance, but rather a morphological term; the chemical substance may not be the same in all cases. The main chemical components of yolk are

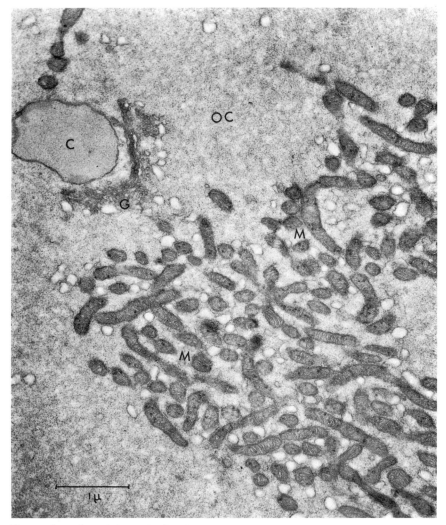

Figure 29. Edge of the "mitochondrial cloud" in the cytoplasm of a young frog oocyte (electronmicrograph). M, Mitochondria; G, Golgi body; C, cortical granule; OC, finely granular ground substance of cytoplasm.

proteins, phospholipids, and to a lesser extent, neutral fats. Depending on which components predominate, we can distinguish "protein yolk," which may contain variable amounts of lipids as well as protein, and "fatty yolk," which in addition to phospholipid and fat may contain some admixture of protein. Protein yolk and fatty yolk are present side by side in the eggs of many animals.

Protein yolk is the main form of food reserve in many invertebrates (for instance, in echinoderms) and in lower chordates (*Amphioxus*, tunicates) (Fig. 31). The amount of yolk in these animals is relatively small; in the sea urchin *Arbacia* the yolk granules take up about 27 per cent of the total volume of the egg (Harvey, 1956). The yolk granules are fine and fairly evenly distributed in the cytoplasm of the egg. Eggs with a small amount of yolk are called **oligolecithal.**

In amphibian eggs the protein yolk is found in the form of rather large granules, usually described as the **yolk platelets.** The yolk platelets have an oval shape and are flattened in one plane. The cytoplasm is densely packed with them (Fig. 32). The am-

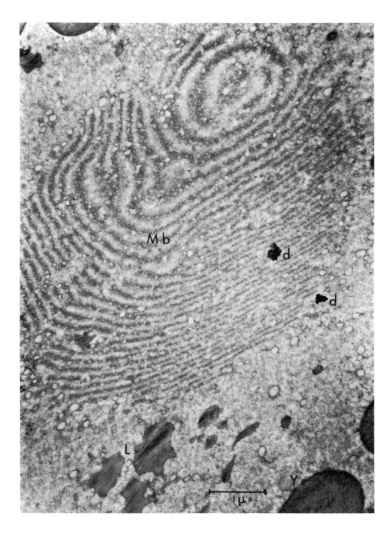

Figure 30. System of membranes in the cytoplasm of a frog oocyte. Mb, Membranes; Y, yolk; L, lipochondria. (d, Dirt on section.)

phibian egg is much larger, so that not only the relative but also the absolute amount of yolk is far in excess of that found in the eggs of *Amphioxus* or the echinoderms. The distribution of yolk in the amphibian egg is distinctly unequal: the yolk platelets are densest in the lower part of the egg, and there is relatively more cytoplasm in the upper part of the egg. Eggs of this type are known as **telolecithal.**

The yolk platelets of amphibians contain two main proteinaceous substances: phosvitin and lipovitellin. Phosvitin is a highly phosphorylated protein (phosphorus content: 8.4 per cent), with a molecular weight of 35,000. Lipovitellin is a protein with a very much larger molecule (molecular weight 400,000) and containing a very considerable amount of bound lipid (17.5 per cent). In the yolk platelets two molecules of phosvitin are associated with each molecule of lipovitellin in a structural unit, the units being arranged in the platelet in a crystalline lattice with regular hexagonal packing (Fig. 33) (Wallace, 1963). This can be clearly seen in electron micrographs of amphibian yolk (Fig. 34).

In addition to the yolk platelets the amphibian egg contains stored supplies in the form of lipid and glycogen. Lipid is distributed throughout the cytoplasm of the egg in the form of organized inclusions, the **lipochondria** (Holtfreter, 1946), which consist of

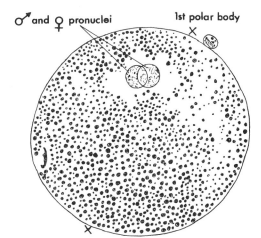

Figure 31. Oligolecithal egg of *Amphioxus* shortly after fertilization. The animal and vegetal poles are indicated by crosses. (After Kerr, 1919.)

an internal core of lipid surrounded by a thin protein coat. The lipochondria are much smaller than the yolk platelets and are roughly spherical in shape (Fig. 32). Glycogen is present in the egg cytoplasm in the form of small granules. Fat may also be stored in the egg for the nourishment of the developing embryo, but its quantity is relatively small in the lower vertebrates.

In a mature amphibian egg, protein yolk constitutes roughly 45 per cent of the dry weight, lipids 25 per cent, and glycogen 8.1 per cent. Only about 20 per cent of the dry weight of the mature egg is active cytoplasm (Barth and Barth, 1954).

Cyclostomes, elasmobranchs, ganoids, and the lungfishes have eggs with a distribu-

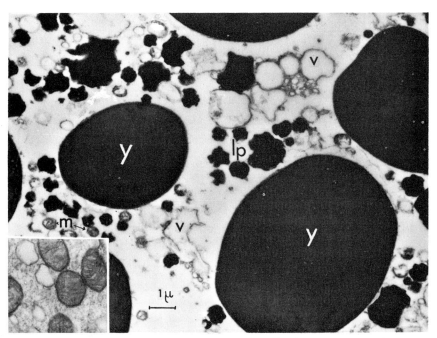

Figure 32. Electronmicrograph of cytoplasmic inclusions in an unfertilized egg of a frog, *Xenopus laevis*. y, Yolk; lp, lipochondria; m, mitochondria; v, vacuoles. Inset: A group of mitochondria under higher magnification.

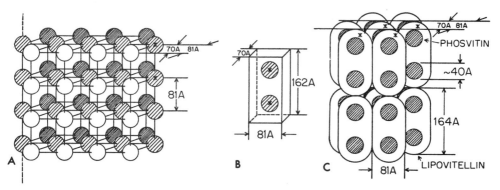

Figure 33. Diagram of crystalline structure in the main body of an amphibian yolk platelet. *A*, Arrangement of phosvitin molecules (seen as dark dots in electronmicrographs) in the lattice. *B*, A structural unit of the crystal consisting of two molecules of phosvitin and one molecule of lipovitellin. *C*, Arrangement of phosvitin and lipovitellin molecules in a simple hexagonal lattice (when viewed end on). The figures show actual dimensions in A. (Courtesy of Dr. R. Wallace and Dr. S. Karasaki.)

tion of food reserves much the same as in amphibians, though the amount of yolk may be greater, especially in some selachians. In the higher teleosts, on the other hand, the yolk becomes segregated from the cytoplasm and concentrated in the interior of the egg, while most of the cytoplasm is found in a thin surface layer covering the yolk and thickened on the uppermost side of the egg in the shape of a cytoplasmic cap. The nucleus of the egg in this case lies inside this cytoplasmic cap. The core of the yolk platelets of cyclostomes and fishes, as in amphibians, has a crystalline structure (Karasaki, 1967; Yamamoto and Oota, 1967).

In the bony fishes, fat may be present in the form of large fat droplets inside the mass of yolk. The number and size of the fat droplets is typical for different families of fishes. Sometimes only one large droplet is present, and at other times there is a large number of smaller droplets.

The yolk of a bird's or a reptile's egg, as in the case of the bony fishes, lies in a compact mass in the interior of the egg, and the cytoplasm is restricted to a thin layer on the surface, with a thickened cap of cytoplasm on the upper side. As in bony fishes, this cytoplasmic cap also contains the nucleus of the egg cell. Most of the yolk is liquid, but about 23 per cent is in the form of solid "yolk spheres." The yolk as a whole contains 48.7 per cent water, 16.6 per cent proteins, 32.6 per cent phospholipids and fats, and 1 per cent carbohydrates (Romanoff and Romanoff, 1949). The proteins of the avian yolk are chemically closely related to those of the amphibian yolk: phosvitin and lipovitellin are the main components. Part of the lipovitellin is bound to phosvitin; this part of the yolk protein is insoluble in water. The remainder of the lipovitellin (referred to by some authors as "livetin"—Flickinger, 1961) is water soluble. The fatty portion of avian yolk is predominantly neutral fat (50 per cent of the yolk dry weight), the rest being phosphatides and cholesterol.

Some of the invertebrates also have developed eggs in which the relative amount of yolk is high and more or less segregated from the cytoplasm. In cephalopods and some gastropods among the molluscs the eggs are telolecithal, much as in the lower vertebrates (Fig. 35). Arthropods, especially insects, have developed a different type of egg: the yolk is concentrated in the interior of the egg, and the cytoplasm is distributed as a thin coat on the external surface; however, there is also an island of cytoplasm in the center of the egg. This island, surrounded on all sides by yolk, contains the nucleus of the egg cell (Fig. 36). Eggs of this type are called **centrolecithal.**

For a long time it has been believed that the production of yolk in the oocyte is the result of the activity of a special body, found in the oocytes of many animals, which accordingly was called the "yolk nucleus" or "yolk nucleus of Balbiani." (See Raven, 1961.) This body was first discovered in the oocytes of spiders, but later it was shown to be present in avian oocytes and in the oocytes of some mammals and amphibians. In the early stages this body lies next to the nucleus (Fig. 37), but in later stages it breaks up and its fragments are distributed to the periphery of the oocyte.

As it is at the periphery of the oocyte that the yolk platelets appear first, the con-

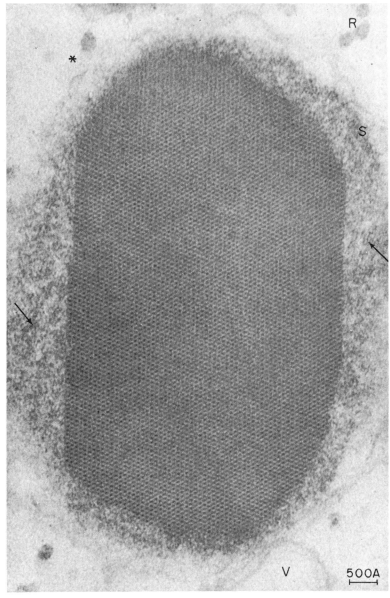

Figure 34. Electronmicrograph of a yolk platelet from an amphibian embryo, showing the crystalline structure of the main body of the platelet. R, Ribosomes in the cytoplasm; S, superficial layer of yolk platelet; V, cytoplasmic vesicle. Arrows indicate fibrillar structures in the superficial layer. (Courtesy of Dr. S. Karasaki, 1963.)

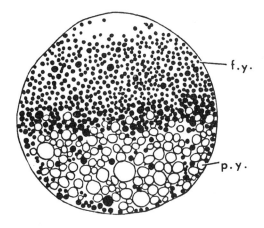

Figure 35. Telolecithal egg of the mollusc *Aplysia limacina*. f.y., Fatty yolk; p.y., protein yolk. (After Ries and Gersch, from Raven, 1958.)

clusion has been drawn that vitellogenesis (production of yolk) is initiated by the components of the "yolk nucleus." At present, it is not possible to support this conclusion, at least not in a general form. Electron microscopic investigations have shown that "yolk nuclei" in different animals are not always of a similar nature. In spiders the yolk nuclei consist of numerous stacked membranes with mitochondria and vesicular structures surrounded by these membranes or interspersed between them (Sotelo and Trujillo-Cenóz, 1957; André and Rouiller, 1957). There is no indication that these bodies have anything to do with the formation of yolk. In amphibians the name "yolk nucleus" has been applied to the mitochondrial cloud mentioned earlier (p. 43), and the same appears to be true of the avian oocyte (Romanoff, 1960), although an electron microscopic verifi-

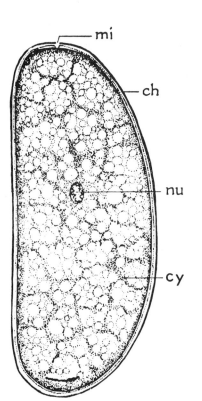

Figure 36. Insect egg, diagrammatic. nu, Nucleus; cy, cytoplasm; ch, chorion; mi, micropyle. (From Johannsen and Butt, 1941.)

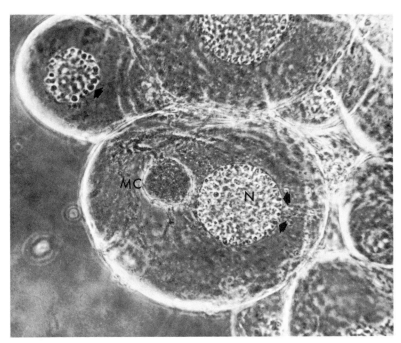

Figure 37. Phase contrast photograph of living young oocytes from the ovary of a juvenile frog. N, Nucleus; MC, mitochondrial cloud ("yolk nucleus"). Nucleoli indicated by arrows.

cation of the nature of the body in the bird is still outstanding. In amphibians *(Xenopus)*, it was found that although parts of the mitochondrial cloud reach the periphery of the oocyte, they do so too late to participate in yolk platelet formation, which starts before the breakup of the mitochondrial cloud (Balinsky and Devis, 1963). In mammals a spherical body lying in the cytoplasm next to the nucleus and having, in the living state, the same appearance as the mitochondrial cloud of the amphibian oocyte, consists of a mass of Golgi-type membranes and is now referred to as the Golgi material (Fig. 22, p. 38, Anderson and Beams, 1960). The Golgi material in mammals cannot be involved in yolk formation for the simple reason that, in the oocytes of higher mammals (p. 260), yolk, at least in the form of yolk platelets, is completely lacking.

Apart from the "yolk nucleus," several cellular organoids have been considered responsible for yolk formation in different animals by different authors. The organoids most often mentioned in this connection are the endoplasmic reticulum, the Golgi bodies, and the mitochondria. The evidence is rather contradictory. The discrepancies may perhaps be resolved if one considers the actual site of synthesis of the yolk proteins. It has been indicated previously that the yolk is not necessarily synthesized in the oocytes, but may be produced elsewhere in the body (in the liver in the case of vertebrates) and is then transported in a soluble form via the blood stream and the follicle cells and is redeposited in the oocyte. In other cases no such transport of ready-made yolk has been recorded, and it seems plausible that the yolk is actually synthesized from simple components in the oocyte by the usual process of protein synthesis, that is, on the ribosomes of the oocyte itself.

This last process very likely occurs in animals such as coelenterates, which do not have a circulatory system. In a medusa the first visible yolk particles have been found in cisternae of the Golgi bodies (Kessel, 1968b), but it is plausible that the actual synthesis occurs on the ribosomes of rough endoplasmic reticulum associated with the Golgi bod-

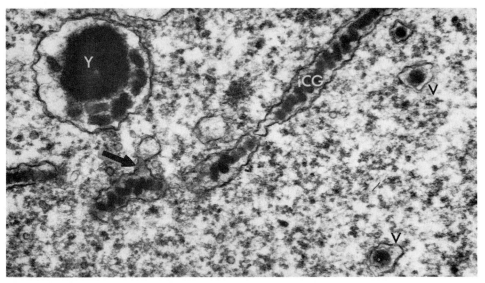

Figure 38. Formation of yolk in the oocyte of the crayfish (electronmicrograph). Yolk particles appear in the cisternae of the endoplasmic reticulum (ICG). Similar particles are found in small vacuoles (V) which perhaps are budded from the endoplasmic reticulum (arrow). In Y the yolk particles fuse to form a larger yolk granule. (Courtesy of Professor R. G. Kessel.)

ies, which serve as the "packaging" organoid, as they do in some secretory cells of adult animals. Some carbohydrate may be added in the process, the latter synthesized by the Golgi bodies themselves. In Crustaceans yolk granules actually appear in the interior of laminar or vesicular endoplasmic reticulum (Fig. 38). It has been found also that radioactively labeled leucine is taken up into the granules—a very strong indication that yolk is synthesized locally, directly from amino acids (Kessel, 1968a).

While yolk formation in connection with the endoplasmic reticulum and the Golgi bodies conforms to what is generally known about the process of protein synthesis in the cell, this cannot be claimed to be the case where yolk appears in or close to mitochondria of oocytes.

There is conclusive proof that in amphibians (Lanzavecchia, 1960; Ward, 1962; Balinsky and Devis, 1963) and fishes (Yamamoto and Oota, 1967) the yolk platelets are formed inside modified mitochondria. In this process the inner mitochondrial membranes become dislodged and eventually become arranged in concentric layers, while the interior of the mitochondrion is taken up by the main body of the yolk platelet. Occasionally a yolk crystal may be seen lying inside a swelling of a mitochondrion, which in other parts still shows a typical structure with systems of transverse cristae (Fig. 39).

A somewhat similar origin of protein yolk has been described in gastropod molluscs (Carasso and Favard, 1960). The protein yolk platelets, which in gastropods also show crystalline structure, are formed either actually inside mitochondria or in between clumps of mitochondria.

It has been indicated previously that in some animals the yolk is not synthesized in the oocytes at all but is produced elsewhere in the body (in the liver, in the case of vertebrates), and that it is then transported in a soluble form via the blood stream and the follicle cells to the oocyte, where it is finally deposited in the form of yolk platelets or yolk granules.

In this connection, some properties of the yolk proteins are of particular interest. The yolk protein phosvitin (p. 46) in insoluble in water when it is fully phosphorylated.

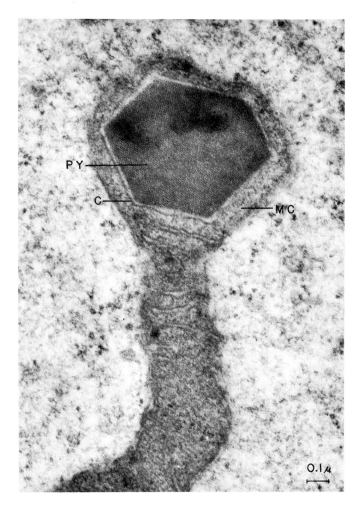

Figure 39. Formation of a yolk crystal inside a mitochondrion in a frog oocyte. PY, Yolk platelet (crystal); MC, mitochondrial cortex; C, membrane of mitochondrial crista. (Courtesy of Dr. R. T. Ward.)

If, however, part of the phosphate is removed by the action of a phosphatase, the remaining phosvitin becomes water soluble. The reverse would then also hold true; namely, if soluble partially phosphorylated phosvitin were more fully phosphorylated in the oocyte, it would become insoluble. An enzyme, protein kinase, has actually been found in the frog ovary which is capable of promoting the incorporation of phosphate into partially phosphorylated phosvitin (Wallace, 1964). The phosphate is taken from ATP. The reaction is fully reversible and takes place according to the following equation:

$$\begin{matrix} \text{partially} \\ \text{phosphorylated} \\ \text{protein} \end{matrix} + \text{ATP} \xrightarrow[\text{Mg}^{++}]{\text{protein kinase}} \begin{matrix} \text{fully} \\ \text{phosphorylated} \\ \text{protein} \end{matrix} + \text{ADP}$$

The action of protein kinase is restricted to only one component of the amphibian yolk (phosvitin). In the yolk crystal, however, the lipovitellin is bound to phosvitin, and therefore the rendering of phosvitin insoluble fixes the lipovitellin in position as well. Now it is very important that protein kinase is a mitochondrial enzyme, as this explains why the yolk platelets may be formed inside or in connection with mitochondria: it is there that the yolk proteins are converted from the soluble into the insoluble form.

Obviously, much remains to be done to explain the deposition of yolk in such animals as birds (see Bellairs, 1964), and also to clarify the origin of fatty yolk.

3-3 ORGANIZATION OF THE EGG CYTOPLASM

As the oocyte grows and as various inclusions are produced in its cytoplasm, a pattern of organization gradually emerges, which will be of importance when the eggs start developing.

The arrangement of various substances and cellular constituents in the advanced oocyte and, later, in the egg shows a polarity, that is, an unequal distribution with respect to what may be called the two opposite poles of the egg and with respect to the main axis of the egg—the line connecting the two poles. The nucleus of the egg is approximated to one pole of the egg, which is termed the **animal pole.** The opposite pole is termed the **vegetal pole,** because the accumulation of yolk at that pole serves for the nutrition of the developing embryo. When the cell undergoes meiosis, the nucleus of the egg cell approaches the animal pole of the egg, and the polar bodies are always discharged at the animal pole (p. 61). This process serves to distinguish the animal pole in oligolecithal eggs, if the concentration of the yolk at the vegetal pole is not very distinct, as is often the case, and also in centrolecithal eggs.

In amphibian oocytes the first yolk platelets appear around the periphery, in the subcortical cytoplasm. As the yolk platelets increase in size and more yolk platelets are produced, they fill the cytoplasm from the outside inward. In the last stages of the growth of the oocyte, the yolk platelets are formed around the nucleus, so that the cytoplasm becomes filled up with yolk. The yolk platelets are not all the same size. At the animal pole of the oocyte, where the polar bodies are to be given off later, the yolk platelets remain relatively small and are not packed quite so densely. At the vegetal pole they reach a larger size (about 1.5μ in length) and come to lie very close to one another, leaving little cytoplasm in between. The interior of the oocytes, the area surrounding the nucleus, is fairly closely packed with yolk platelets which are not quite as large as those at the vegetal pole (Wittek, 1952).

In bony fishes, reptiles, and birds, the thickened cytoplasmic cap containing the nucleus denotes the animal pole of the egg.

Yolk and the other substances obviously destined to serve as food for the developing embryo are not the only substances that are laid down in the oocyte during its growth. A characteristic feature of the mature eggs in many animals is the presence of granules of pigment, which are elaborated during the growth of the oocyte.

In the oocytes of the ascidian, *Styela partita,** in addition to the yolk, which accumulates at the vegetal pole, there appears in the later stages a yellow pigment in the form of granules distributed all over the surface of the oocyte in a thin cortical layer of cytoplasm. In the sea urchin *Paracentrotus lividus* a similar distribution of pigment granules is observed, but the pigment is red. We will see later that the yellow pigment of *Styela partita* and the red pigment of *Paracentrotus lividus*, although uniformly distributed over the surface of the maturing egg, will later be concentrated and come to lie in specific parts of the embryo—in the muscles and mesenchyme of the former, and in the walls of the gut of the latter. Although it does not necessarily follow that the pigment is in any way a precursor of muscular or intestinal differentiation, it may well be that the

---

*There has been some confusion among embryologists concerning the correct generic name of this animal. In previous editions of this text the animal was referred to as *Cynthia partita.* *Styela* appears to be the correct generic name.

pigment is an indicator of some specialization in the cytoplasm of the oocyte which eventually leads to specific types of differentiation of parts of the embryo.

In the eggs of most amphibians there is present a greater or lesser amount of dark brown or black pigment. Depending on the amount of pigment, the egg may appear to be light fawn, through various shades of brown, to pitch black. However, young oocytes have no pigment. Pigment granules start to be formed somewhat later than the yolk platelets, in oocytes which have grown to about one half of the final diameter. The greatest number of pigment granules becomes located in the cortical layer of cytoplasm of the mature oocyte, but quite considerable amounts of pigment are distributed in the interior.

A remarkable feature in the distribution of pigment in the amphibian egg is that it is not uniform. There is much more pigment in the animal hemisphere of the oocyte than in the vegetal hemisphere. The difference may be very marked, so that while the animal hemisphere may be dark brown or black, the vegetal hemisphere appears clear white, although in reality a small number of pigment granules is practically always present in the vegetal hemisphere as well. The transition from the dark to the light areas is fairly sharp, but there is always a zone of intermediate, gradually fading pigmentation, which can be conveniently referred to as the **marginal zone.**

In a cross section of a ripe amphibian egg (Fig. 40) it may be seen that the vegetal half of the egg is filled by a densely packed mass of yolk containing very little pigment. This mass of yolk is slightly concave on the top. The center of the egg is occupied by a roughly lens-shaped mass of cytoplasm with middle-sized yolk platelets and a moderate amount of pigment. This zone also contains the nucleus in the immature oocyte. On the outer edges of this interior mass of protoplasm lies a ring-shaped area containing large amounts of pigment. The ring is thicker on one side of the egg, where it also reaches nearer to the surface. This side of the egg corresponds, as we will see later, to the future dorsal side of the embryo and thus, in conjunction with the differences along the main axis of the egg, indicates a plane of bilateral symmetry. Lastly, on top of the interior mass of protoplasm lies, like an inverted saucer, a layer of cytoplasm of the animal hemisphere (the "animal cap"), which is rich in pigment, especially in the cortical layer, and relatively poor in yolk.

The pigment granules in themselves may perhaps not be very important for the development of the embryo; in fact, there are some species of amphibians which do not have any pigment in their eggs (the large European crested newt, *Triturus cristatus*, and some frogs making foam nests, like the African *Chiromantis xerampelina*) and yet develop in the same way as related species which have pigmented eggs. However, the uneven

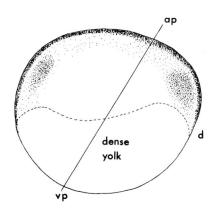

Figure 40. Distribution of yolk and pigment in the ripe egg of a frog. Median section. ap, Animal pole; vp, vegetal pole; d, dorsal side of the egg. (Modified after Lehmann, 1945, and Wittek, 1952.)

distribution of the pigment may be considered as an indicator of qualitatively different areas in the cytoplasm of the egg. This can be corroborated by some further observations. We have seen that the cytoplasm of young frog oocytes contains large amounts of ribonucleic acid. It has been observed that in oocytes of 400 to 500 μ in diameter the ribonucleic acid is especially concentrated in the surface layer of the animal hemisphere of the oocyte. With the development of yolk platelets in the subcortical layer of cytoplasm, the ribonucleic acid concentration disappears, but it seems very probable that the localization of pigment in the animal hemisphere reflects some peculiarities in cytoplasm—peculiarities which are related to the distribution of the ribonucleic acid in the preceding stage (Wittek, 1952; Kemp, 1953, p. 493).

In the elephant tusk mollusc, *Dentalium*, very distinct cytoplasmic areas can be seen in the egg even before it leaves the ovary. The fully grown oocyte contains a pigment varying in different individuals from olive green to brick red. However, the pigment does not encroach on two areas at the opposite sides of the egg, which are therefore colorless. One of the two colorless areas lies on the side of the egg which is attached to the wall of the ovary. While the oocyte is in the ovary, this part is somewhat drawn out, so that the oocyte is more or less pear-shaped. The second pigment-free area is on the free surface of the oocyte, and in the center of this area the polar bodies are given off, thus marking this side as the animal pole. The egg is released from the ovary with the germinal vesicle still intact, and it proceeds to the meiotic divisions only afterwards. After release from the ovary, the egg partially rounds up but remains slightly flattened from animal to vegetal pole. In a vertical section through the egg (Fig. 41), the light area at the vegetal pole (the pole that was attached to the wall of the ovary) is seen to be made up of clear protoplasm which contains no yolk granules. This protoplasm reaches inward to the nucleus and surrounds it, and at the outer edges it is continuous with a very thin cortical layer covering the entire egg surface. There is also a small patch of clear cytoplasm at the opposite pole which marks the place where the polar bodies will be given off. The rest of the cytoplasm is filled by rather densely packed yolk granules. The yolk-free and unpigmented cytoplasm at the vegetal pole may be termed the **vegetal polar plasm.** The pigment-free **animal polar plasm** is only partly free from yolk. The significance of the two polar plasms for the development of the embryo will be considered in a later section (Section 6–6).

Of all the developments in the cytoplasm of the oocytes, perhaps the most important are those which concern the surface layer of the oocytes. The superficial layer of cytoplasm in oocytes differs in its physical properties from the rest of the cytoplasm. Most of the oocyte cytoplasm is in the physical state of a suspension, with a continuous liquid phase (containing in solution various large and small particles, ranging in size from simple inorganic ions to protein molecules) and with larger inclusions, such as mitochondria and ribosomes, dispersed more or less at random in the continuous

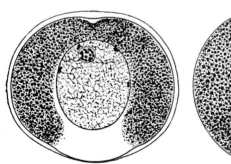
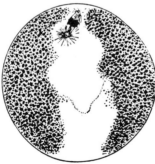

Figure 41. Oocytes of *Dentalium* just before the beginning of meiotic divisions (left) and in the stage of the first meiotic division (right). (From Wilson, 1904.)

phase. These inclusions are freely movable, under natural conditions, owing to cyto-plasmic streaming, or in experiments in which the cells are exposed to a centrifugal force. The superficial layer is, on the other hand, in a gelated state or at least in a very much more viscous state, so that its components are not readily displaced during cyto-plasmic streaming or in moderately strong centrifugation. This superficial layer of the cytoplasm is known as the **cortex,** and it plays the role of a fixture in the cell, remaining in the same position for continuous periods of growth and development. The thickness of the cortex has been measured by various means, and the results have been somewhat contradictory. It has been claimed on the grounds of centrifugation experiments that the only immovable parts of the egg are the cell membrane and the particles immedi-ately attached to the cell membrane, which would then represent the cortex of the egg (Mercer and Wolpert, 1962). It is more generally recognized, however, that in addition to the cell membrane a layer of cytoplasm, about 2 to 3 μ in thickness, is in a gelated state and thus constitutes the cortex of the egg (Hiramoto, 1957; Mitchison, 1956). The larger cytoplasmic inclusions are either embedded in or attached to the cortex and so remain in a fixed position. In addition to fixing the position of parts of the oocyte cytoplasm, the cortex is the bearer of some differentiations which are highly characteristic of the late oocyte and mature egg.

The development of the microvilli on the surface of oocytes as well as the forma-tion of the pinocytotic vesicles were mentioned earlier. Both seem to be concerned with the transport of substances from the follicle cells into the oocyte. An entirely different aspect of cortical differentiation is seen in the formation of special structures known as the **cortical granules.** These are spherical bodies, varying in diameter from 0.8 μ (in sea urchins) to 2 μ (in frogs), surrounded by a simple membrane and containing acid mu-copolysaccharides (Fig. 42). In mature oocytes the cortical granules are arranged in a layer close to the plasmalemma. Cortical granules have been found in oocytes of many animals, but not in all. They are present in sea urchins (Afzelius, 1956), frogs (Kemp, 1956), fishes (Yamamoto, 1954), bivalve molluscs (Reverberi and Mancuso, 1961; Re-bhun, 1962; Pasteels and Harven, 1962; Humphreys, 1962), some annelids (Lillie, 1911), and some mammals, such as the hamster, the rabbit (Hadek, 1963a), and man (Baca and Zamboni, 1967). They are absent, however, in other mammals, such as the rat (Sotelo and Porter, 1959; Odor, 1960) and the guinea pig (Anderson and Beams, 1960). They are also absent in gastropod molluscs (Recourt, 1961), urodele amphibians (Wartenberg and Schmidt, 1961), insects (Okada and Waddington, 1959), and birds (Bellairs, 1964).

In most cases the mucopolysaccharide content of the cortical granules is fairly homogeneous, finely granular or floccular, but in sea urchins the cortical granules show a complicated internal structure consisting of a system of concentric or possibly spirally arranged dense lamellae, separated by less dense material (Fig. 43). The corti-cal granules are known to be formed in the interior of the oocyte, and in sea urchins, frogs, and humans, their origin has been traced to the Golgi bodies (Balinsky and De-vis, 1963; Baca and Zamboni, 1967; Anderson, 1968). The cortical granules when first formed lie inside the cup-shaped space formed by the Golgi membranes (Fig. 29). La-ter, they move to the periphery to the cortical layer of cytoplasm. The cortical granules appear to play an important role at the time of fertilization, when the granules burst and their contents contribute to the accumulation of fluid around the fertilized egg (pp. 83, 86).

In the last stage of maturation, the nuclear membrane of the oocyte breaks down, and the chromosomes move to the surface at the animal pole to take part in the matu-ration divisions. At the same time the nuclear sap merges with the cytoplasm of the oocyte. It has been observed in some animals that the nuclear sap does not mix com-

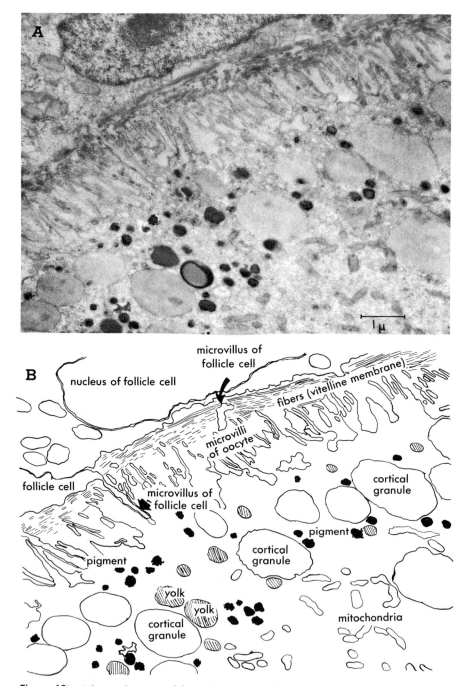

Figure 42. Advanced oocyte of frog showing interdigitating microvilli in the zona radiata and a layer of cortical granules. *A.* Electronmicrograph; *B.* contours of cell organoids corresponding to electronmicrograph.

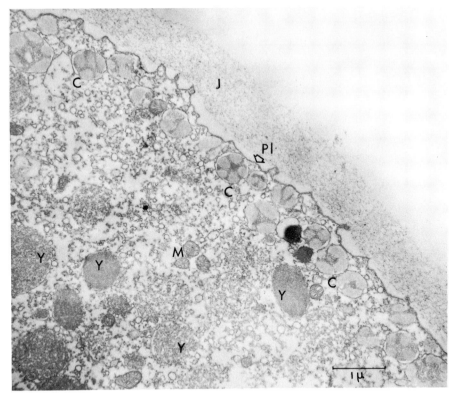

Figure 43. Surface of a ripe, unfertilized sea urchin egg showing layer of cortical granules (electronmicrograph). C, Cortical granule; Y, yolk granule; M, mitochondria; Pl, plasmalemma; J, jelly coat.

pletely with the cytoplasm, but that it forms a more or less separate mass, thus increasing the diversity of the cytoplasmic areas of the eggs.

As a general rule mature eggs are spherical in shape, though elongated eggs are not infrequently found, especially among insects. Among vertebrates, oval-shaped eggs are found in the hagfish *Myxine* and in the ganoid fishes. The elongated shape of the bird's egg, on the other hand, is not due to an elongated shape of the egg cell itself. The egg cell, which is the yellow of the egg, is in this case spherical.

Before concluding this chapter on the organization of the egg and the processes which provide for the arrangement of the various cytoplasmic substances in the egg, we must return once more to the polarity of the egg and consider its origin. The polarity of the egg is discernible from the position of the parts in the egg cell: the nucleus, yolk, and other cytoplasmic substances.

Many efforts have been made to elucidate what factors are responsible for the unequal distribution of these parts. It has been claimed that the polarity of the egg may be imposed on it by the direction of flow of the nutrient substances during the growth of the oocyte. In the molluscs and echinoderms the vegetal pole of the egg develops from that end of the oocyte which is attached to the wall of the ovary. The nutrient substances, at the expense of which the oocyte grows, presumably enter the ovary from outside, from the body cavity. Therefore, it stands to reason that greater amounts of yolk might be deposited in the part of the cell nearest to the proximal surface of the ovary, thus causing this part to become the vegetal pole. This explanation does not hold, however, for oocytes which are surrounded by follicle cells from all sides, as are the oocytes of amphibians or mammals. It has been suggested that the course of the nearest blood vessel supplying parts of the ovary with nourishment might cause the parts of the oocyte nearest to the vessel to develop into the vegetal pole. But according to the views of Child (1941), the animal pole, as the more active one, should develop from that part of the oocyte which has a better oxygen supply, and on this principle the part of the oocyte nearest to a blood vessel should become the animal pole. Actually, there does not seem to be a clear connection between the position of the animal and vegetal poles of the egg and the course of the blood vessels. In view of the differences in the structure of ovaries in different animals, it would seem rather hopeless to try to find a common factor in the environments of growing oocytes which could be held responsible for the origin of polarity of the egg.

Polarity is, however, a phenomenon which is found not only in egg cells but in other cells as well. In epithelial cells there is a distinct difference between the proximal end of the cell (the end resting on the underlying basement membrane) and the distal end (which forms the free surface of the epithelium). In nerve cells the polarity of the cell takes the form of the opposite differentiations of axon and dendrites.

In the interior of any cell, the position of the centrosome with respect to the nucleus establishes a general form of polarity, the main axis of which is the line drawn through the centrosome and the nucleus. This polarity not only affects the distribution of cytoplasmic inclusions (the Golgi bodies are often found grouped around the centrosome) but may involve the intimate structure of the nucleus itself. When the oocytes are in the early leptonema stage of the meiotic prophase, the chromosomes become arranged in a definite way, converging to that side of the nucleus which is nearest to the centrosome. This stage is known as the "bouquet stage" (Fig. 19, p. 35). Early oocytes thus already possess a polarity, and this intrinsic polarity may well serve as a basis for the distribution of cellular constituents during the growth and maturation of the oocyte.

What has been said so far refers to the polarity along the main axis of the egg (along the axis extending from the animal to the vegetal pole). A similar sort of polarity may be responsible for the difference between the side which is to develop into the dorsal side of the embryo and that which is to develop into the ventral side of the embryo. However, even less is known about the origin of this polarity than about the origin of polarity along the main axis.

3–4 MATURATION OF THE EGG

After the oocyte completes its growth, it is ready for the reduction divisions. By this time, the chromosomes of the oocyte have reached the stage of diakinesis. The meiotic division in the female gametes cannot, however, proceed in the same way as in the male gametes, that is, by division of the original cell into four cells of equal size. Such a division would largely defeat the purpose of the accumulation of food reserves during the period of growth, for each of the four cells would receive only one quarter of the stockpile of food supplies. What actually occurs is that during the meiotic divisions the cytoplasm is partitioned very unequally, the bulk of the cytoplasm with the yolk remaining intact at each division.

When the oocyte is ready to undergo reduction divisions, the nuclear membrane breaks up, and the contents of the nucleus become intermingled with the surrounding cytoplasm. The rupture of the nuclear membrane, apart from liberating the chromosomes for the subsequent division, is of great importance because substances accumulated in the nucleoplasm are released and become mixed with the rest of the cytoplasm. It is found that the ability of the egg cytoplasm to respond to the spermatozoon (to become activated – see p. 88) is in some animals dependent on the intermingling of the nuclear sap with cytoplasm after the breakdown of the nuclear membrane. (For reference see Monroy and Tyler, 1967.) On the other hand, it was found that certain substances contained in the nuclear sap may be vitally important for the normal development of the embryo in later stages (p. 146).

After the breakdown of the nuclear membrane, the chromosomes, which have become greatly contracted and concentrated toward the center of the germinal vesicle, are carried to the periphery of the oocyte. Here an achromatic figure is formed, which takes up a position perpendicular to the surface of the oocyte. The bivalents are placed on the equatorial plate and in due course separate into the two component chromosomes (Fig. 44). A bulge now appears at the surface of the oocyte, and the outer pole of the spindle with half of the chromosomes during anaphase enters into this cytoplasmic bulge. The part of the cytoplasm with the chromosomes, which is bulging out, is then pinched off from the rest of the oocyte. The resulting small cell is called the first polar body. The set of chromosomes contained in the first polar body is equivalent to the set

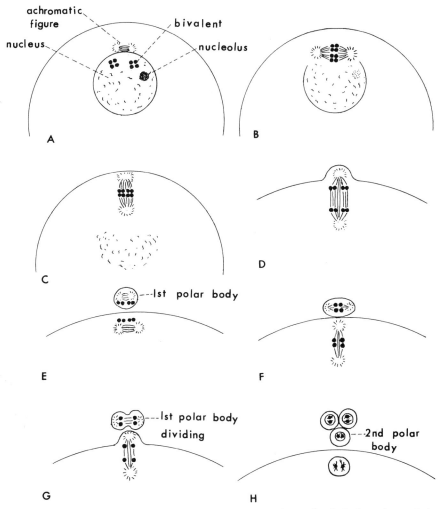

Figure 44. Reduction divisions in an oocyte. *A*, Oocyte immediately before the meiotic divisions; *B, C, D*, first meiotic divisions; *E*, first polar body separated, preparation for second meiotic division; *F, G*, second meiotic division and, simultaneously, division of first polar body; *H*, meiosis completed. (Redrawn with modifications after Wilson, 1925.)

of chromosomes remaining in the oocyte (which can now be distinguished as the **secondary oocyte**), just as the two sets of chromosomes in secondary spermatocytes are equivalent to each other. The first polar body, however, receives only a very small quantity of cytoplasm, while all the rest goes to the secondary oocyte. The second meiotic division is carried out in the same way; that is, an achromatic figure is again formed at the surface, and when division takes place, half of the chromatids are given off, together with a small quantity of cytoplasm, as the second polar body. The large cell, which inherits the bulk of the cytoplasm together with one half of the chromatids, is now the fully mature egg. The chromosome set of the egg, as the result of two meiotic divisions, is a haploid one.

In the meantime, the first polar body may sometimes divide, so that eventually four cells are produced from one oocyte: one cell is the mature egg, the second and third cells are produced by the division of the first polar body, and the fourth cell is the second polar body. These four cells can be compared to the four spermatids which are produced from a primary spermatocyte after two mitotic divisions, the difference being that each of the four spermatids later becomes a functional spermatozoon, while of the four cells produced by meiotic division in the female only one cell, the egg, is a functional gamete, and the other three cells (or two cells, if the first polar body does not divide further) later die off. Having so little cytoplasm, with no food reserves, they are not able to start developing. The similarities (and differences) between oogenesis and spermatogenesis are illustrated in Figure 45.

At this stage we should again point out that, after maturation (= reduction divisions), the spermatids must still go through the process of differentiation before they become spermatozoa, that is, functional male gametes. On the other hand, the egg, after reduction divisions, does not undergo any further changes; it is quite ready to start developing.

A further distinction between the male and the female gametes is that the male gametes, the spermatozoa, become capable of fertilizing the egg some time after they have completed the reduction divisions and subsequently have become differentiated. The eggs become capable of being fertilized sometimes *before* they have completed reduction divisions and before they have given off the polar bodies (p. 93).

3–5 THE EGG MEMBRANES

All eggs, like any other cells, are of course covered by the cell membrane or **plasmalemma.** Under the electron microscope the plasmalemma sometimes can be seen to be double, with two layers of electron-dense material about 50 Å thick separated by a less dense interval of about 60 Å. In addition to the plasmalemma, eggs of all animals, with the exception of sponges and some coelenterates, are surrounded by special egg membranes. Depending on their origin these may be subdivided in two groups: the primary membranes and the secondary membranes (Balfour, 1880). The primary egg membranes are those which develop in the ovary between the oocytes and the follicle cells in the space occupied by the interdigitating microvilli. In the initial stages of its formation the membrane shows positive reactions for mucopolysaccharides, but in later stages it is fortified by addition of further substances and may consist of fibrous proteins (Fig. 46), or in insects, it may be sclerotized and may become highly impermeable. The membrane thus formed bears different names in different animals: it is known as the **vitelline membrane** in insects, molluscs, amphibians, and birds; in tunicates and fishes it is usually called the **chorion.** In mammals the membrane of an exactly similar nature is called the **zona pellucida** (Odor, 1960). The zona pellucida thus takes the place of the

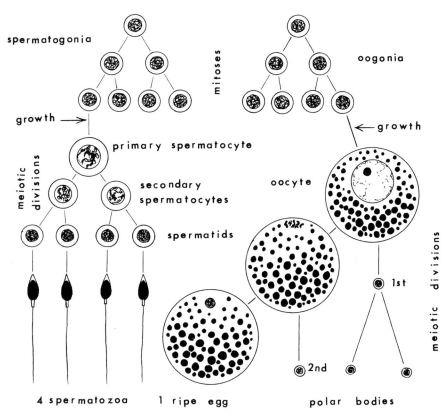

spermatogonia

oogonia

mitoses

growth →

← growth

primary spermatocyte

meiotic divisions

oocyte

secondary spermatocytes

spermatids

1st

meiotic divisions

2nd

4 spermatozoa   1 ripe egg   polar bodies

Figure 45. Comparison of gametogenesis in the male (left) and in the female (right). In both sexes the meiotic divisions produce four cells, although in the male all four cells become functional gametes (spermatozoa), whereas in the female only one cell is functional as the egg; the others are abortive.

zona radiata mentioned earlier (p. 38). The jelly coat surrounding the eggs of sea urchins belongs in the same group.

The origin of the materials forming these membranes is not quite obvious; usually the membrane is believed to be secreted by the follicle cells (Trujillo-Cenóz and Sotelo, 1959; King and Koch, 1963), but the oocyte may perhaps contribute to the formation of the membrane in some cases (Wartenberg, 1962; Okada and Waddington, 1959).

In insects a second, thicker membrane is secreted by the follicle cells on top of the vitelline membrane. This second membrane is called the **chorion.** In many insects its surface shown a complicated sculpture, which is typical for each species. The primary egg membranes usually closely adhere to the surface of the oocyte, but at a later stage a space filled with fluid may appear between the cytoplasm and the membrane; this space is called the **perivitelline space.**

In mammals an egg escaping from the ovary carries with it on the surface of the zona pellucida a layer of follicle cells known as the **corona radiata** (Fig. 47). The cells of the corona radiata are peeled off later, as the egg descends the oviduct.

The second group of egg membranes are those which are secreted by oviducts and other accessory parts of the genital organs while the egg is passing from the ovary to the exterior.

The eggs of amphibians are surrounded by a layer of jelly (Fig. 48), which protects the egg and sometimes serves to make the eggs adhere to one another and to sub-

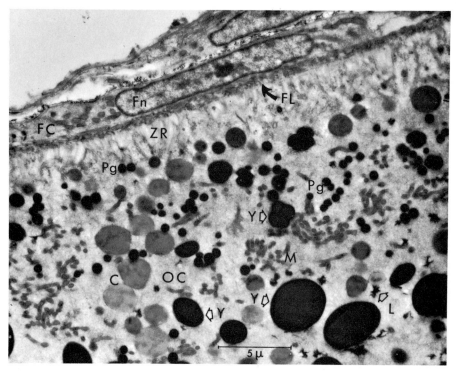

Figure 46. Low power electronmicrograph of the surface of an advanced amphibian oocyte with adjacent follicle cells. ZR, Zona radiata with interdigitating microvilli of occyte and follicle cells; FL, fibrous layer — the precursor of the vitelline membrane; FC, follicle cell; Fn, nucleus of follicle cell; OC, cytoplasm of oocyte; C, cortical granules, not yet in position at surface; Y, yolk platelets; Pg, pigment granules; M, clusters of mitochondria; L, lipochondrion.

merged objects such as water plants. This jelly is secreted as the eggs pass through the oviducts. When the amphibian egg is deposited in water, the jelly absorbs water and swells. In the oviparous sharks and rays the egg is surrounded in the oviducts (in the special parts called the **shell glands**) by a hard shell of a complicated shape. The shell is drawn out into long twisted horns which serve to entangle the eggs among seaweed.

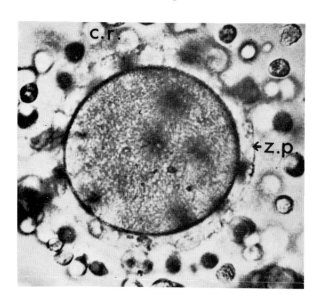

Figure 47. Mammalian egg after ovulation, surrounded by the zona pellucida (z.p.) and the corona radiata (c.r.). (After Hamilton, J. Anat., Lond. 78, 1944.)

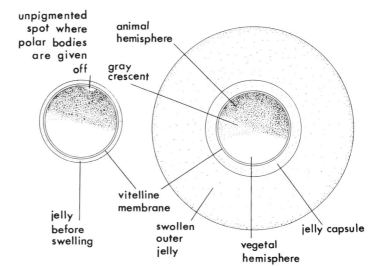

Figure 48. Frog egg as taken from the oviduct (left), and after fertilization, with swollen jelly membrane (right).

The most complicated egg membranes, however, are found in the eggs of birds, where no less than five membranes can be distinguished (Fig. 49), the innermost being the vitelline membrane—a very thin membrane covering the surface of the yellow of the egg (which is the true egg cell). This membrane actually has a double origin. An inner layer of the membrane is produced in the ovary, in the space between the oocyte and the follicle cells, as previously described. This layer is composed of very rough fibers. An outer, more finely fibrous layer is then formed on top of the first layer when the egg enters the upper portion of the fallopian tube (Bellairs, 1963). The next membrane is the white of the egg. Eighty-five per cent of the egg white is water; the rest is a mixture of several proteins, mostly albumins, which make up 94 per cent of the dry weight. A denser part of the egg white forms strands, known as the chalazae, which help to keep the egg cell in the center of the egg white. Next to the egg white are two layers of **shell membranes** which consist of fibers of keratin matted together. Over most of the surface

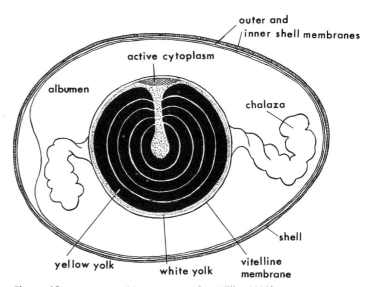

Figure 49. Diagram of hen's egg. (After Lillie, 1919b.)

of the egg the shell membranes are in contact with one another, but at the blunt end of the egg they are separated: the inner membrane adheres to the egg white, and the outer membrane adheres to the shell, leaving a space in between filled with air. The outermost membrane is the **shell,** which consists chiefly of calcium carbonate ($CaCO_3$), about 5 gm. in a hen's egg. The shell is pierced by many fine pores which are filled by an organic (protein) substance related to collagen. In an average hen's egg the pores have a diameter of 0.04 to 0.05 mm., and the total number of pores is estimated at about 7000. The membranes of a bird's egg are secreted one after another as the egg proceeds down the oviduct. The whole process takes slightly longer than 24 hours. After the egg has been released from the ovary, it quickly passes into the oviduct and descends through it for about three hours, during which most of the egg white is secreted and envelops the egg cell. The lowest portion of the oviduct is widened and is termed the uterus. Here the egg remains for 20 to 24 hours, while the remainder of the egg white and eventually the shell membranes and the shell itself are secreted.

Not only do the membranes of a bird's egg protect the egg cell, but the egg white serves also as an additional source of nourishment and is gradually used up in the course of the development of the embryo.

Chapter 4

THE DEVELOPING EGG AND THE ENVIRONMENT

In every stage of its development the embryo is a living organism, and like every living organism it requires foodstuffs for its vital processes. It assimilates its food and metabolizes organic substances to gain energy for its maintenance and for performing the processes which together constitute its development. Furthermore, the new individual has to increase in size to a variable, but usually very considerable, degree until it approximates the structure of its parents. The growth process again involves the consumption, processing, and partial combustion of a vast quantity of foodstuffs. Since the egg is a single cell and does not have any of the organs which an adult uses to procure and utilize its food supply, the supplying and utilization of food for the developing embryo have to be organized along lines that are quite special and specific for the process of embryonic development.

The primary source of nutrition in the eggs of most animals is the reserve material stored in the egg cell during its development in the ovary. (See Section 3–2.) This material is metabolized; it is partly broken down as a source of energy and partly transformed into the substances of which the various organs of the new individual are built in the course of development. In addition to materials contained in the egg, various substances may be taken up from the environment. The nature and the amount of extraneous materials used by the embryo depend very largely on the environment in which the embryo develops. A vast number of animals rather early attain a stage in which the young individual can start feeding and thus can become self-supporting. This is often greatly facilitated by the egg's developing in the first instance into a larva, instead of directly into an adult. Larvae do not necessarily have the ability to take in food. In some animals, for instance in the tunicates, no food is taken during larval life. In this latter case the advantage of a larval stage lies in the ability of the larva to move. The adult is sedentary (Fig. 1, p. 9). The larva thus serves to effect dispersal of the animals. In many parasites, especially internal parasites, larval stages serve for the infestation of new hosts.

The sea urchin egg may be taken as an example of a small egg with relatively little yolk. It develops in seawater and after a very short time (35 to 40 hours) produces a larva, called a **pluteus,** which has an alimentary canal of three parts: a foregut, opening to the exterior by a mouth, a dilated stomach, and a hindgut with anus. The pluteus swims freely in water with the aid of a ciliary band drawn out into loops along the edges of arms supported by a complicated calcareous skeleton. It feeds on minute planktonic organisms, mainly algae, which are injected into the gut by the ciliary action of the stomodeum (Fig. 50). Table 2 gives an idea of the turnover of substances during the first 40 hours of development, up to the formation of the pluteus.

Some points may be made from this table: The developing embryo takes in some water from the environment and also quite a considerable quantity of mineral sub-

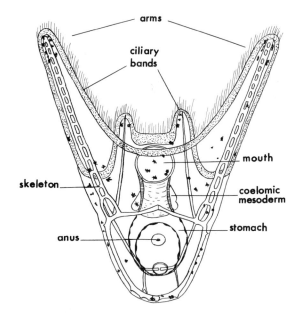

Figure 50. Pluteus—larva of a sea urchin (*Tripneustes gratilla*).

stances which are mainly utilized in building up the calcareous skeleton. The embryo loses a large part of its carbohydrates (including almost all of the glycogen) and some fat. There is only a very slight decrease in nitrogen; according to some more recent investigations (Gustafson and Hjelte, 1951), the total nitrogen of the sea urchin egg does not change during the period up to the pluteus stage.

Water is absorbed as a general rule by embryos developing in an aquatic medium. (See Table 4, p. 70.) The amount absorbed may be quite considerable, as can be seen in the case of an egg-laying dogfish, *Scyllium canicula*. Additional ash (salts) is also taken up from surrounding seawater by marine invertebrates (coelenterates, molluscs, crustaceans, and echinoderms). Among fishes, the developing embryos of the elasmobranchs absorb salts, but the embryos of some freshwater fishes, such as the Salmonidae, do not absorb salts from the surrounding medium although they take in water. The same is true of the developing embryos of the Amphibia. The animals whose eggs develop in fresh water cannot depend on the surrounding medium for the intake of salts because fresh water does not contain the necessary salts (especially Na and K ions) in sufficient quantity.

The turnover of substances in the developing egg of a frog (Table 3) will serve as

TABLE 2   Chemical Composition of Sea Urchin Embryo (In Per Cent of Wet Weight)

Substance	Unfertilized Egg	Blastula, 12 Hours	Pluteus
Water	77.3	77.3	78.8
Dry substances	22.7	22.7	21.2
Protein nitrogen	10.7	10.2	9.7
Total carbohydrates	5.43	5.46	3.4
Glycogen	5.13	4.8	traces
Fat	4.82	4.43	3.69
Ash	0.34	2.07	3.56

After Ephrussi and Rapkine, from Lehmann, 1945.

TABLE 3   Chemical Composition of the Egg and Embryo of a Frog (Weights in Micrograms)

Substance	Egg	Tadpole Soon After Hatching (Stage 20)
Total nitrogen	162	159
Extractable nitrogen (= nitrogen of active protoplasm	40	42.8
Nonprotein nitrogen (in nucleic acids ?)	4	5.0
Total carbohydrates	104	58
Glycogen	73.2	31

After Gregg, and Gregg and Ballentine, from Barth and Barth, 1954.

an example of an animal developing in fresh water. Table 3 shows that there is very little increase in the amount of active cytoplasm during the early development of the embryo, that there is somewhat more of an increase in the amount of nucleic acid nitrogen (about 25 per cent), and that simultaneously there is a very considerable decrease in the total carbohydrates, due almost exclusively to the loss of glycogen, which is thus the main source of energy for the maintenance and development of the embryo. The increase of water is considerable. It is not included in Table 3, but according to some available data, it accounts for a 75 per cent increase in the total weight of the embryo during the corresponding stages. There is also some uptake of salts, particularly calcium, from the environment.

An entirely new situation is faced by animals that have abandoned the aquatic medium and have become completely terrestrial. Some of them have attempted a compromise by returning to the water for egg laying. This is true of most of the amphibians. Other amphibians (many terrestrial frogs, some salamanders, and the Gymnophiona) lay their eggs on land but in damp places—in burrows underground, etc.—where the eggs can absorb the minimal quantities of water that are necessary for their development. Even among the reptiles there are some whose eggs take up water from the environment. This is the case with turtles. For instance, the egg of the turtle, *Malaclemys centrata,* which weighs 10.58 gm., absorbs 3.07 gm. of water during its development. This is made possible because the eggs are laid in damp sand.

In the eggs of other reptiles and of birds, there is no longer a possibility for the absorption of water from the exterior. The egg membranes have become watertight. On the other hand, the loss of water from the egg by evaporation is reduced to a minimum. The only substance that is taken from without is the oxygen necessary for oxidative processes in the egg. Otherwise, the egg has become a closed system, developing at the expense of the substances stored inside the egg itself. Such an egg, which has become self-sufficient (except for oxygen intake and CO_2 outflow) is called a **cleidoic** egg (Needham, 1931). Cleidoic means boxlike.

Independently of the vertebrates, the terrestrial arthropods have also evolved cleidoic eggs. Some of the stages of this evolution can still be traced among contemporary insects. For instance, in the eggs of grasshoppers *(Melanoplus* and *Locusta)* there is an intake of water during development. In *Melanoplus* the water content increases from 2.5 to 4.7 mg. In the eggs of other insects, however, no water can be taken in, and throughout the development of the eggs there is only a certain amount of water loss by evaporation. Such eggs are therefore as much cleidoic as the egg of a bird.

There is a fairly obvious advantage in the young individual's being more advanced in development and growth when it emerges from the egg. This advancement may be achieved by the amassing of greater and greater supplies of foodstuffs in the egg. Increase of food supplies in the egg may be correlated with the elimination of a larval stage—as in cephalopods, among the molluscs; and in elasmobranchs, reptiles, and birds, in the vertebrate phylum.

The eggs of birds present an example of a very abundant supply of the egg with food material for the nourishment of the embryo. The chicken that hatches from the egg is already essentially a bird. The principal difference, besides the small size, lies in the development of the feathers. The body of a newly hatched chicken is covered with down, which is a simplified form of a feather. As soon as the down is replaced by typical feathers, the chicken acquires all the typical features of a bird.

A very special though widespread method of increasing the chances of survival of the offspring consists in retaining the eggs in the mother's body and letting them develop there to a greater or lesser degree. The eggs are usually retained in the oviducts (which are then usually referred to as **uteri**), but in some cyprinodont fishes (*Poecilia* and *Girardinus*) the eggs begin their development while still in the ovaries. (See Needham, 1942.) Sometimes, as in some salamanders, the young hatch from the eggs inside the mother's body. Occasionally, as in the adders, the eggs are laid intact, but the young begin to hatch as soon as the eggs are laid. The bearing of young developed from eggs which have been retained in the body of the mother, but without the maternal organism providing additional nourishment for the embryo, is known as **ovoviviparity. Oviparity** is, of course, the form of reproduction in which all or most of the development of the embryo in the egg occurs after the egg has been laid.

In typical cases of ovoviviparity the embryo is nourished by the food stored in the egg. A next step in evolution may be made when the embryo absorbs some substances present in the fluids filling the oviducts. The degree to which such oviductal fluids are used for the nourishment of the embryo is very variable. In different species of the sharks and rays, all possible gradations may be found between such forms in which the embryo depends chiefly on the food supplied in the egg, and such forms in which the embryo depends chiefly on the food supplied by the mother. The proportions of these two sources of nourishment may be best estimated by comparing the weight of organic substances in the egg and in the newly born offspring. Table 4 presents a comparison of the turnover of materials in the eggs of an oviparous fish and an ovoviviparous fish (as well as a viviparous species).

TABLE 4 Chemical Composition of the Eggs and of Later Stages of Development in Three Species of Dogfish

Species		Composition of Egg and of Young at Hatching or Birth (Weight in Grams)	
		Egg	Young
Scyllium canicula (oviparous)	Total weight	1.3	2.7
	Organic substances	0.61	0.48
	Water	0.68	2.15
Mustelus vulgaris (ovoviviparous)	Total weight	3.9	60.6
	Organic substances	1.9	8.9
	Water	1.9	49.8
Mustelus laevis (viviparous)	Total weight	5.5	189.0
	Organic substances	2.8	32.0
	Water	2.6	152.0

After Ranzi, from J. Needham, 1942.

It will be seen that in the development of an egg-laying fish, the turnover of substances is very similar to that in the embryo of the sea urchin or frog: there is a decrease in organic substances, due to their combustion, and an increase in water content, which makes the young at hatching exceed the weight of the egg. In the ovoviviparous fish, however, the uptake of organic substances from the maternal body not only compensates for their loss through combustion but brings about a total increase in the amount of organic material during development. There is also a very considerable intake of water, so that the overall increase in weight of the embryo is many times that observed in a species in which the embryo depends on the egg reserves as the sole source of its organic materials. Ovoviviparity in several groups of animals seems to have been a transitional phase in the development of true **viviparity.**

True viviparity is achieved when the embryo establishes a direct connection with the maternal body, so that the nutrition can pass from the mother to the embryo without the intermediate state of being dissolved in the uterine fluid. The connection is established through a special organ, the **placenta,** which is an outgrowth of the embryo joined to parts of the maternal body especially modified for this purpose. Viviparity and placentae have been developed in several groups of the animal kingdom independently of one another. Placentae are found in the protracheates *(Peripatus)*, in the tunicates *(Salpa)*, in several elasmobranchs, and in the placental mammals. The mode of origin and the structure of the placenta are different in each case mentioned. The placenta of the mammals will be described later, in Section 10–5.

The supply of nutrition to the embryo through a placenta appears to be a much more efficient method than the absorption of nourishment from the uterine fluids. This fact may be illustrated by the third example in Table 4, that of the dogfish *Mustelus laevis*, in which a placenta is developed when the yolk sac of the embryo becomes connected to the walls of the uterus. (See Chapter 10.) In this particular case the increase in organic materials during intrauterine life is more than tenfold as compared with the roughly fourfold increase of organic substances in *Mustelus vulgaris*, taken as an example of the ovoviviparous fish.

In mammals, which have developed the placenta to a degree of perfection not found in other animals, the increase in weight of the embryo is much greater, especially since the eggs of mammals are very small as compared with the eggs of viviparous fishes.

Part Three

FERTILIZATION AND THE BEGINNING OF EMBRYOGENESIS

Chapter 5

FERTILIZATION*

Fertilization is the fusion of two **gametes** (sex cells), a male and a female one, followed by the joining together of the nuclei of the two gametes. The fusion of the gametes in the Metazoa activates the egg so that it starts to develop, and the joining together of the nuclei of the egg and the spermatozoon results in the endowment of all the cells of the developing new organism with carriers of hereditary properties derived from the maternal and paternal organisms.

It should be pointed out here that the place of fertilization in the life cycles of various organisms is not always the same. Although in the Metazoa and Metaphyta fertilization starts the egg on its way of development into an embryo, in many protozoans the zygote enters into a dormant state, which may be interrupted later, usually by changes in the environmental conditions. Similarly, the fusion of the nuclei of the two gametes is deferred in some organisms (especially fungi) and occurs many cell generations later. There are thus two essentially independent aspects to fertilization: one is the **activation** of the egg, the other is the intermingling of the paternal and maternal hereditary characters in the offspring. The latter aspect of fertilization is also known as **amphimixis** and really lies within the sphere of the science of genetics.

We will now consider the consecutive steps which lead to the fertilization of the egg and the changes in the egg occurring during and after fertilization, stressing mainly those aspects of fertilization which contribute to the transformation of the egg into a new organism.

5–1 APPROACH OF THE SPERMATOZOON TO THE EGG

After both the eggs and the spermatozoa are discharged into the surrounding (aquatic) medium, or after the spermatozoa have been introduced into the genital ducts of the female, in the case of internal fertilization, the first step is the encounter of the spermatozoon and the egg. This encounter is brought about by the swimming movements of the spermatozoa. These movements have been carefully studied in the case of the fertilization of sea urchin eggs, which occurs when both the eggs and the sperm are released into seawater. It was found that movements of the spermatozoa are entirely at random, and the spermatozoa collide with the eggs as a matter of pure chance. That this chance encounter occurs regularly in nature is partly a result of the enormous number of spermatozoa produced by the male gonads, and partly the result of the egg's being a relatively very large target, so that it can be hit fairly easily (Rothschild, 1956). This state of affairs probably holds for most animals. There are a few cases,

---

*Further references: Rothschild, 1951 and 1956; Tyler et al., 1957; Austin, 1965 and 1968; Metz and Monroy, 1967.

however, in which the spermatozoa appear to be guided to the egg by chemical substances. In the hydroid, *Campanularia*, the eggs are produced by reduced female gonangia, which are enclosed in a theca with an opening at the distal end (Fig. 51*a*). During fertilization spermatozoa have been observed to converge toward the opening of the theca (Fig. 51*b*).

Furthermore, it has been noted that the spermatozoa swim toward the mouth of a pipette filled with a solution prepared by extracting female gonangia with alcohol, drying the extract, and redissolving it in seawater. The substance contained in the solution was found to have a small molecular structure, with a molecular weight less than 5000. A weaker attractant solution could be prepared from the stems of the hydroid colony (Miller, 1966). A somewhat similar case is presented by fish eggs which are enclosed in a hard chorion (p. 62). The chorion is perforated by narrow canals called micropyles through which the sperm enters to reach the egg cytoplasm. It has been observed that spermatozoa accumulate at the opening of the micropyle, as if they are attracted to it (Yamamoto, 1961). It is noteworthy that in both cases (hydroids and fish)

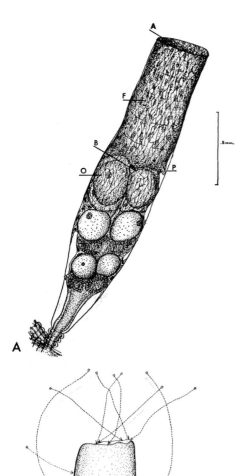

Figure 51. *A*, Gonangium of the hydroid *Campanularia*. A, aperture of theca; B, blastostyle; F, funnel of theca; P, perisarc of theca; O, ovum. *B*, Tip of theca with plotted paths of spermatozoa. (From Miller, 1966.)

the egg is not accessible from every angle and the spermatozoa are directed to a certain opening which leads to the egg.

A very fine chemical mechanism is involved in the next step. In the presence of ripe eggs, or even in the water in which ripe eggs of the same species have been lying for some time ("egg water"), the spermatozoa become "sticky" and adhere to the surface of the egg and even to each other. The mutual adhesion of the spermatozoa results in their clumping or **agglutination.** The agglutination is more easily observed with sperm of some animals (especially sea urchins) than with sperm of others (for instance starfish), and it also depends to a large extent on environmental conditions. But the sticking of sperm to eggs of the same species is found in most animals, and possibly it is always present.

The substance causing agglutination of spermatozoa was studied in detail by F. R. Lillie (1919a), who called it **fertilizin.** He believed that fertilizin is continuously emitted from mature eggs and that it is directly involved in the reaction between the egg and spermatozoon. It was found later that the main source of fertilizin, after the maturation of the egg, is not the egg itself but the layer of jelly surrounding the egg (in echinoderms). This jelly consists wholly of fertilizin which gradually dissolves in the seawater after the eggs are laid. It is not definitely known whether the jelly surrounding sea urchin eggs is primarily produced by the oocyte or by the follicle cells, the latter being the more likely explanation.

Chemically, fertilizin is a glycoprotein or mucopolysaccharide. As a protein it contains a number of amino acids, and as a polysaccharide it includes molecules of one or more monosaccharides. The monosaccharides (glucose, fucose, fructose, or galactose) are esterified by sulfuric acid, as shown in the following formula (Runnström, 1952):

$$\text{—O}_3\text{SO} \quad \text{fertilizin structural formula}$$

Both the amino acids and the monosaccharides vary from one species to another, so that it is more correct to speak of **fertilizins** rather than of one fertilizin found in different animals. The molecules of the fertilizins are quite large—the molecular weight is about 300,000—and each molecule may have more than one "active group," so that one fertilizin particle may become attached to two or more spermatozoa, thus binding them together (Fig. 52). (See Metz, 1957; Rothschild, 1956.)

The surface layer of the cytoplasm of the spermatozoon contains another substance (or substances) known as the **antifertilizin** (or antifertilizins). The antifertilizins can be extracted from the spermatozoa by heating, freezing and thawing, or acidifying the water. Their properties show that they are acid proteins with a fairly small molecule (molecular weight about 10,000).

The remarkable peculiarity of the fertilizins and antifertilizins is that they combine in a specific way; that is, the egg fertilizin of any species reacts best with the sperm antifertilizin of the same species. Reactions with other species, although possible, are very much weaker, and even so they occur only when two species are fairly nearly related to each other. The reaction between fertilizin and antifertilizin is thus very similar to reactions which take place between the substances in the serum of immunized animals (antibodies) and the foreign substances which cause the immunizations (antigens)

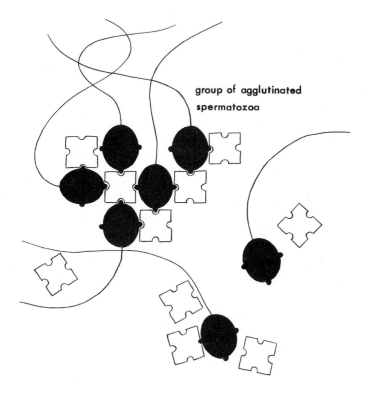

group of agglutinated
spermatozoa

Figure 52. Diagram showing binding of sea urchin spermatozoa by particles of fertilizin. (Modified after Rothschild, 1951.)

(Tyler, 1948). In both cases, the bond between the two complementary substances is supposed to depend on the spatial arrangement of the atoms in certain parts of the molecules in such a way that the shape of the surface of one molecule fits closely onto the surface of the other. The two bodies correspond to each other as a template to the model. Another often-used analogy is the correspondence of a key to a lock. When the latter analogy is used, it may be pertinent to remember that sometimes a lock may be opened by a wrong key, if it is sufficiently similar to the right one. This situation finds its counterpart in the cross reactions of fertilizins and antifertilizins of sufficiently nearly related species.

The adhesion of the spermatozoon to the surface of the egg has been interpreted as the result of the presence of fertilizin molecules in the surface layer of the egg cytoplasm. The linking of the fertilizin molecules with the antifertilizin molecules present on the surface of the spermatozoon would account for the establishment of an initial bond which would later lead to the penetration of the spermatozoon into the egg. Unfortunately, the evidence as to whether any fertilizin is present in the egg or at its surface is very contradictory. (See Metz, 1957.) No fertilizin can be extracted from sea urchin eggs. In view of this it has been suggested that the main function of the fertilizin-antifertilizin reaction is to thin out the number of spermatozoa around the egg, so that the chances of two or more spermatozoa penetrating into the egg at the same time are diminished (Runnström et al., 1959).

We have seen (Section 3–5) that the surface of the ripe egg is very seldom naked (as in coelenterates), but usually surrounded by membranes, follicle cells, or both (as in mammals). The spermatozoon has to penetrate through these before it can reach the egg.

The mechanism of penetration is chemical. The spermatozoon produces substances of an enzymatic nature, known under the general name of **sperm lysins,** which dissolve the egg membranes locally and make the path clear for the spermatozoon to reach the surface of the egg. (See Tyler, 1948.) The sperm lysins are produced presumably by the acrosome of the spermatozoon (Colwin and Colwin, 1961). This is in accord with the origin of the acrosome; we have seen that the acrosome is developed from the Golgi bodies of the spermatid. There is clear evidence that in a normal animal cell at least some of the secretory activity is performed by the Golgi bodies. The production of sperm lysins is thus one of the aspects of the secretory activity of these bodies.

The sperm lysins certainly differ from one animal group to another. In some cases the dissolution of the egg membranes may be brought about by simpler means. Thus, it is believed that the jelly coat of echinoderm eggs may be dissolved as a result of the acidification of seawater by carbon dioxide produced by the spermatozoa in the course of their respiration. In the case of eggs with very thick and resistant membranes, such as the egg membranes of fishes or insects, the sperm cannot reach the egg at all points but must penetrate through a special canal (the micropyle), or canals, left in the egg membrane. (See Section 3–1.)

In mammals, as we have seen (p. 63), the situation is complicated by the presence around the egg membrane, the zona pellucida, of a layer of follicle cells, the corona radiata, which the egg carries with it when it leaves the ovary. The spermatozoon has to pass through this layer on its way to the egg. It has been found that mammalian spermatozoa produce an enzyme, **hyaluronidase,** which dissolves the mucopolysaccharide, hyaluronic acid, the substance cementing the follicle cells together. The enzyme thus enables the spermatozoon to make a way for itself through the corona radiata. Under experimental conditions, extracts containing large amounts of hyaluronidase may remove all the follicle cells from the surface of the egg, but this does not occur during normal fertilization; the corona persists for some time after the penetration of the spermatozoon into the egg. It later disperses before the implantation of the egg, but this is not the result of the action of the spermatozoa.

The agglutination of the spermatozoa is not the only change which they undergo as a result of the egg water or the proximity of ripe eggs. Under the influence of substances diffusing from the eggs (whether the substance in question is fertilizin or some other compound has not yet been determined), the structure of the spermatozoon becomes changed, as can best be seen by using the electron microscope. The main change concerns the acrosome. The outer membrane of the acrosome breaks down, and the acrosomal granule is extruded and partially dissolved in water (Fig. 53) (possibly releasing the lysins which were referred to earlier). At the same time a structure, arising from the posterior wall of the acrosome (the axial body or acrosomal cone, where one is present; see p. 28), starts elongating and gives rise to a long (1 to 75 μ) thin filament, or possibly a tube, which has considerable rigidity and protrudes forward from the head of the spermatozoon (Figs. 53 and 54). This structure has been called the **acrosomal filament** (Dan and Wada, 1955). In a case which has been studied by electron microscopy, that of the annelid *Hydroides hexagonus*, several short acrosomal filaments take the place of a single long one (Fig. 55) (Colwin and Colwin, 1961). The spermatozoa which have developed acrosomal filaments are considered to have been activated and are then ready to penetrate into the egg. As the spermatozoon approaches the egg, the acrosomal filament is pushed through the jelly (where present) and through the vitelline membrane, the pathway for it being cleared by the action of the lysins, as previously explained (Fig. 55). Eventually, the tip of the acrosomal filament touches the surface membrane of the egg cytoplasm and triggers the next phase of the process of fertilization.

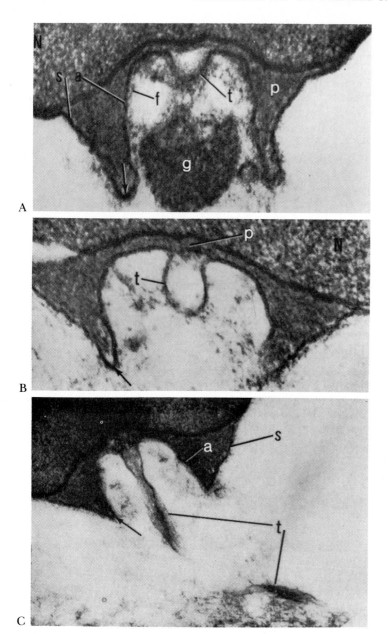

Figure 53. Three stages of activation of the spermatozoon of the enteropneust *Saccoglossus*. *A*, Acrosomal granule in the process of being released from the acrosome; *B*, formation of the acrosomal filament; *C*, acrosomal filament fully extended. a, Membrane surrounding acrosomal granule; N, nucleus of spermatozoon; p, periacrosomal material; s, spermatozoon plasma membrane; t, acrosomal filament (tubule). Arrows point to edge formed by breaking out of acrosomal granule. (From Colwin and Colwin, 1963.)

Internal fertilization, as it occurs in mammals, shows some peculiarities not observed during fertilization of typical marine invertebrates, discussed in the preceding paragraphs. The spermatozoa of mammals, though possessing acrosomes, do not develop acrosomal filaments. Spermatozoa removed from the seminiferous tubules of the testis, and even from the ductus deferens, are not capable of fertilizing eggs, though they are already motile in the latter case. Only after being injected into the vagina in the process of copulation, and as a result of being exposed to the influence of the fluids in the reproductive tract of the female, do the spermatozoa become capable of penetrating through the egg membranes and fusing with the egg. The change in the mammalian spermatozoon which makes it capable of fertilizing the egg has been called **capacitation.** The process is analogous to the change produced in the spermatozoa of marine

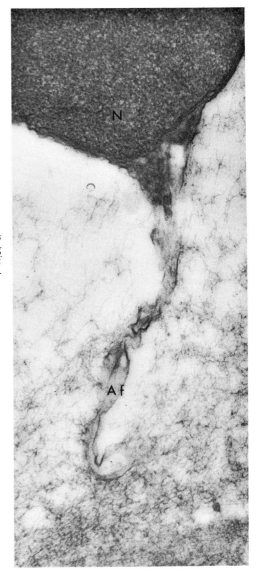

Figure 54. Tip of spermatozoon of the enteropneust *Saccoglossus kowalevskii* penetrating through the vitelline membrane of the egg (electronmicrograph). Af, Acrosomal filament; N, nucleus of spermatozoon. (Courtesy of Dr. L. H. Colwin and Dr. A. L. Colwin, 1963.)

invertebrates by "egg water." From a morphological viewpoint, as shown by electron microscopic studies, capacitation involves the breaking up of the membranes surrounding the acrosome and the release of the contents of the acrosome, consisting of sperm lysins which aid the spermatozoon in penetrating the layer of follicle cells and the zona pellucida (Bedford, 1967). No acrosomal filament is formed, and the spermatozoon appears to contact the surface of the egg by its lateral aspect. Following this action, the plasma membranes of the egg and the spermatozoon dissolve at the point of contact, and the spermatozoon is drawn into the interior of the egg. (See Austin, 1968.)

5–2 THE REACTION OF THE EGG

Immediately after the acrosomal filament of the spermatozoon touches the surface of the egg, the cytoplasm of the egg bulges forward at the point of contact, producing a

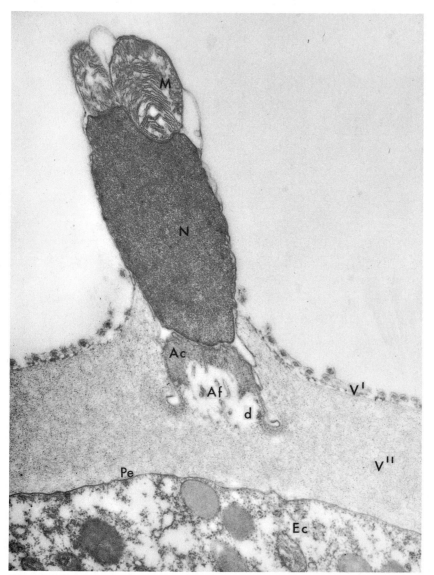

Figure 55.   Spermatozoon of the annelid *Hydroides hexagonus* penetrating through the vitelline membrane of the egg. The electronmicrograph shows the formation of multiple acrosomal filaments and the resorption of the vitelline membrane by secretions from the acrosome of the spermatozoon. Ac, Acrosome; Af, acrosomal filaments; d, area where substance of vitelline membrane has been dissolved away; Ec, egg cytoplasm; M, mitochondrial body; N, nucleus of spermatozoon; Pe, plasma membrane of egg; V', external layer of vitelline membrane; V", internal layer of vitelline membrane. (Courtesy of Dr. L. H. Colwin and Dr. A. L. Colwin, 1961.)

process of hyaline cytoplasm, the **fertilization cone.** The fertilization cone may be in the form of a more or less simple conical protrusion (Fig. 56), it may consist of several irregular pseudopodium-like processes, or in some cases it may take the form of a cytoplasmic cylinder stretching forward along the acrosomal filament (Colwin and Colwin, 1957). Whatever its shape, the fertilization cone gradually engulfs the spermatozoon and then begins to retract, carrying the spermatozoon inward, surrounded by the hyaline cytoplasm of the fertilization cone (Fig. 57).

Even before the spermatozoon thus penetrates into the interior of the egg, the egg

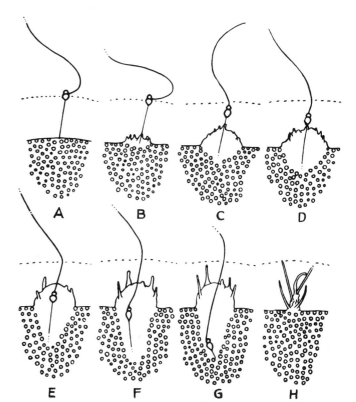

Figure 56. Sperm entry into the egg of the sea cucumber *Holothuria atra,* with formation of fertilization cone by the cytoplasm of the egg. (From Colwin and Colwin, 1957.)

shows signs that is has become profoundly changed, that it has been **activated.** By this statement we mean that the egg has become capable of starting on its way to develop into an embryo. The first observable changes in the egg, however, bear only an indirect relation to the formation of the embryo. They are changes in the surface of the egg cytoplasm which are known under the general name of the **cortical reaction** (Pasteels, 1961). The cortical reaction may differ quite considerably from one group of animals to another, and it will be useful to describe first how it occurs in fertilized eggs of sea urchins, in which it has been studied most extensively, and then to review the corresponding phenomena in other animals.

If sea urchin eggs are observed under the microscope with darkfield illumination at the time of fertilization, the color of the egg surface can be seen to change from yellow to white. The change starts from the point of attachment of the spermatozoon and gradually spreads over the whole surface of the egg (Runnström, 1952) (Fig. 58), taking 20 seconds to be completed. The structural equivalent of this color change is not quite clear, but immediately following the color change a series of transformations takes place, profoundly affecting the organization of the egg cortex and involving, in particular, the cortical granules mentioned in Section 3–3. In the mature sea urchin egg the plasmalemma is seen as being rather thick (150 to 200 Å) and double layered. Immediately after fertilization the two layers of the plasmalemma split from each other, and the outer layer separates from the surface of the egg as the **fertilization membrane** (Wolpert and Mercer, 1961; Endo, 1952; Anderson, 1968). At the same time the cortical granules start swelling rapidly and "explode." The mucopolysaccharides contained in the granules become liquefied through the absorption of water, and the resulting fluid is released into the space between the egg cytoplasm and the

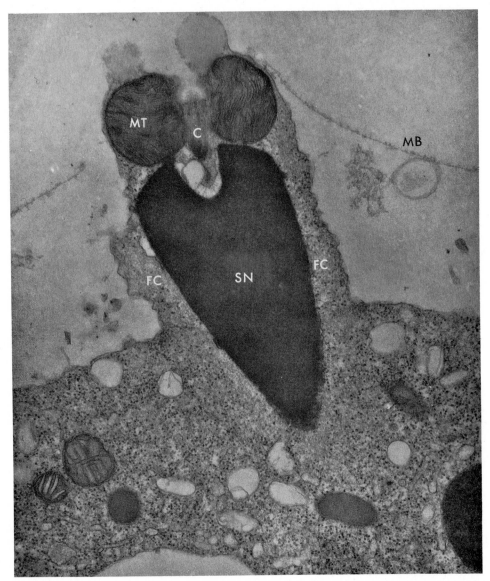

Figure 57. Spermatozoon of the sea urchin *Arbacia punctulata* being drawn onto the egg cytoplasm by the fertilization cone (electronmicrograph). FC, Fertilization cone; C, centriole of spermatozoon; MB, fertilization membrane; MT, mitochondrial body; SN, spermatozoon nucleus. (Courtesy of Dr. F. J. Longo and Professor E. Anderson.)

fertilization membrane (compare Fig. 43, p. 59 and Fig. 59, p. 85). Bits of cytoplasm lying between the outer surface of the cortical granules and the plasmalemma may become disconnected from the egg cytoplasm and adhere to the fertilization membrane (Balinsky, 1961a; Anderson, 1968). The fertilization membrane, as formed in the first minutes after fertilization, is soon reinforced by the denser lamellar substance released from the exploding cortical granules (Endo, 1952; Afzelius, 1956; Wolpert and Mercer, 1961) or possibly secreted independently from the egg cytoplasm (Anderson, 1968). The fertilization membrane thus becomes much thicker (up to 900 Å) and stronger. The liquefied component of the cortical granules fills the space (the perivitelline space) between the surface of the egg and the fertilization membrane, and by fur-

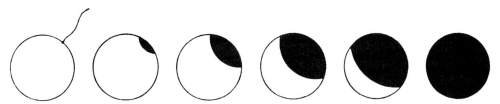

Figure 58. Spreading of cortical change (black) in a sea urchin egg after fertilization. The figure at the left represents the moment when the spermatozoon contacts the egg surface. (From Rothschild, 1956.)

ther imbibition of water it assists in lifting the membrane still further, so that the lifted membrane becomes visible even with low powers of the microscope.

In addition to strengthening the fertilization membrane and producing, or helping to produce, the fluid filling the perivitelline space, the cortical granules also give rise to a viscous **hyaline layer** which adheres closely to the surface of the egg and during cleavage helps to keep the blastomeres together. The jelly layer which surrounds the unfertilized egg persists for some minutes after fertilization but eventually dissolves without leaving a trace (Fig. 59).

Since much of the experimental work on fertilization has been done on sea urchin eggs (See Rothschild, 1951), it has often been tacitly assumed that the way fertilization occurs in sea urchins is typical of animals in general and, in particular, that the formation of a fertilization membrane is an essential part of the process of fertilization. It has now become obvious, however, that the reactions of the sea urchin egg present a very special case, differing in several important aspects from what can be observed in most other animals. The special way in which the fertilization membrane is formed in the sea urchin egg has apparently not been found in any other animal. (See Pasteels, 1961.)

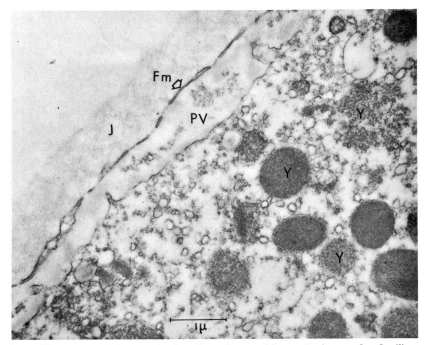

Figure 59. Electronmicrograph of surface of sea urchin egg 2 minutes after fertilization, showing absence of cortical granules and formation of the fertilization membrane. Fm, Fertilization membrane; J, jelly coat; Y, yolk; PV, perivitelline space.

The greatest similarity to sea urchin fertilization may be seen in those animals which have cortical granules, such as bony fishes and frogs. In the eggs of these animals the cortical granules are broken down after sperm attachment, their contents becoming liquefied and extruded onto the surface of the egg (Figs. 60 and 61). In contrast to those of sea urchins, eggs of bony fishes and frogs are surrounded by distinct membranes which are produced while the eggs are in the ovary (the vitelline membrane in the frog and the chorion in the bony fish; see Section 3–5). The perivitelline space— the space between the egg cytoplasm and the membrane—may be narrow in unfertilized eggs, especially in frogs, but after the breakdown of the cortical granules and the release of their contents into the perivitelline space the amount of fluid increases rapidly, and the membrane is lifted from the surface of the egg. In the fish the membrane is said to be hardened or "tanned" after fertilization, but no new membrane is developed which could be compared to the fertilization membrane of the sea urchin.

In mammals the cortical granules are released into the perivitelline space (the space between the egg and the membrane, the zona pellucida) and disappear, probably becoming dissolved (Hadek, 1963a, b). No new membrane is formed around the egg at the time of fertilization. The perivitelline space may increase slightly, owing not so much to the lifting of the membrane as to a shrinkage of the egg itself, which loses a certain amount of water to the perivitelline fluid.

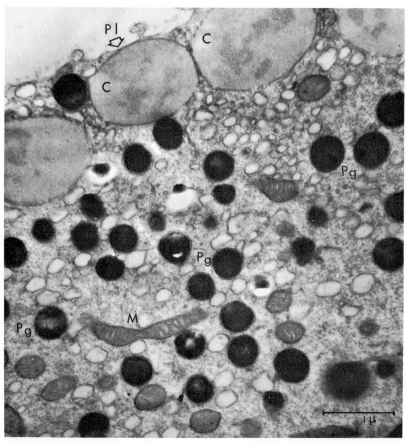

Figure 60.   Surface of ripe unfertilized frog's egg showing cortical granules (electronmicrograph). C, Cortical granule; Pg, pigment granule; M, mitochondrion; Pl, plasmalemma. The surface of the cytoplasm is extended in short processes.

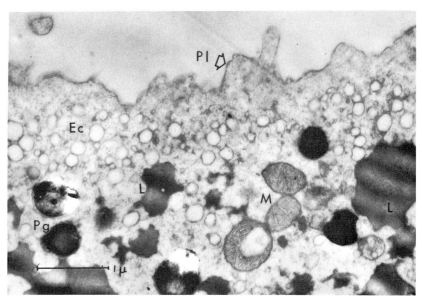

Figure 61. Surface of fertilized uncleaved frog's egg; the cortical granules have disappeared (electronmicrograph). Pl, Plasmalemma; L, lipochondrion; M, mitochondrion; Ec, vesiculated ectoplasm; Pg, pigment granules.

Very peculiar conditions prevail in lamellibranch molluscs. These animals possess cortical granules containing acid mucopolysaccharides (p. 57), but the cortical granules do not change at the time of fertilization, and there is also no change in the membranes or the perivitelline space surrounding the eggs. A gradual breakdown of the cortical granules may be observed, but this occurs both before and after fertilization and does not seem to be affected by the attachment and penetration of the spermatozoon; in addition, the contents of the granules are not released to the exterior but rather into the egg cytoplasm (Pasteels and de Harven, 1962).

Lastly, there are many animals that do not possess cortical granules. In these animals, such as urodele amphibians, insects, and others, there do not seem to occur at fertilization any morphological changes comparable to the "cortical reaction" of sea urchin eggs.

Normally, only one spermatozoon penetrates the egg in most classes of the animal kingdom (particularly coelenterates, annelids, echinoderms, bony fishes, frogs, and mammals). Fertilization performed by the penetration of only one spermatozoon is called **monospermic.** Under abnormal conditions, when the egg is adversely affected in any way or if there is an abnormally high concentration of the spermatozoa around the egg, more than one spermatozoon may establish contact and penetrate into the egg. The development resulting from such **polyspermic** fertilization is inevitably abnormal **(pathological polyspermy),** and the embryo is not viable.

It has often been said that the cortical reaction of the egg and, in particular, the lifting of the fertilization membrane are defense reactions which prevent the penetration of additional spermatozoa after the first one has established contact with the egg. This explanation does not seem to be entirely valid. First of all, the cortical reaction as described above is too slow to account for the exclusion of additional spermatozoa contacting the egg. We have seen that in sea urchin eggs it takes 20 seconds for the first detectable cortical reaction (color change) to spread over the whole surface of the egg. It has been calculated that very numerous "collisions" of additional spermatozoa with parts of the eggs which have not yet reacted could occur during this period. The elevation of the fertilization membrane takes even more time to complete ($\pm$ 60 seconds). Nevertheless, in nature and under favorable experimental conditions, practically all fertilized eggs are monospermic. Furthermore, in most animals no fertilization membrane is formed, and in many there occurs no detectable change in the cortex or the egg membranes upon attachment of the spermatozoon (apart from the formation of the fertilization cone). It appears, therefore, that a more subtle mechanism is involved in preventing pathological polyspermy. This mechanism is still to be investigated.

There are certain groups of animals, mainly those having yolky eggs, such as some molluscs, selachians, urodeles, reptiles, and birds, in which several spermatozoa enter the egg as a rule **(physiological polyspermy),** but of these only one participates fully in the development of the embryo. The rest degenerate sooner or later, although in some cases (as in birds and reptiles) the nuclei of the accessory spermatozoa may undergo several abortive divisions in the cytoplasm of the egg. The egg in these animals thus possesses some mechanism which serves to eliminate additional spermatozoa and insures that only one remains fully active.

5-3 THE ESSENCE OF ACTIVATION

If the condition of the egg before fertilization is compared with that after fertilization, it becomes clear that the system is brought from a quasi-stationary state to a condition characterized by a series of rapid changes. If it is not fertilized, the ripe egg remains quiescent for some time, eventually becomes subjected to degenerative processes, and in the end becomes necrotic. If the egg is fertilized, it goes into action, as it were: the reduction divisions, if not performed before, are brought to completion; the male and female pronuclei fuse; complicated dislocations of cytoplasmic substances of the egg may take place (Section 6–6); and the egg enters into a period of rapid divisions (cleavage).

What is the physiological mechanism of this activation of the egg, and how does the spermatozoon produce the profound change in the egg's condition? A rather logical hypothesis would be that the action of the spermatozoon rests on the introduction of some special substance into the egg which triggers off all the other reactions, of which the changes in the metabolism of the egg are visible expressions. In view of the importance of the fertilizin-antifertilizin reaction for the entry of the spermatozoon into the egg, it has been suggested by Lillie (1919a) that the activation of the egg is also due to the action of the same substances. The combining of the fertilizin of the egg with the antifertilizin of the sperm would then be the beginning of the changes in the physicochemical system of the egg—the key reaction of fertilization.

We shall not go into the further assumptions of the theory, since in any case it cannot be accepted for quite a number of reasons, one being that no plausible explanation has been given as to how the binding of fertilizin and antifertilizin is connected with the other processes involved in the activation of the egg. Antifertilizin, as we have seen, can be extracted from the spermatozoa and added to the water surrounding ripe eggs. The presence of the antifertilizin can be ascertained by the sticking together of the eggs—an exact counterpart of the agglutination of the spermatozoa by the fertilizin contained in egg water—but there is observed no trace of activation of the egg by this treatment (Metz, 1957).

A different approach to the problem would be to compare the processes of metabolism and other physicochemical properties of unfertilized and fertilized eggs. In 1908 Warburg measured the oxygen consumption of unfertilized and fertilized sea urchin eggs and found that immediately after fertilization there was a sharp increase in oxygen consumption of up to 600 per cent as compared with unfertilized eggs. It would have been tempting to suggest that the primary action of the spermatozoon consists in release or activation of the oxidative enzymes of the egg, and that the ensuring increase in oxidation provides the energy necessary for the performance of the other changes in the egg, as previously outlined, and for the development of the egg in general. It was soon discovered, however, that the increased oxidation rate in eggs is by no means a general rule: although the oxygen uptake increases in sea urchins and some annelids *(Nereis)*, it does not change appreciably in starfish and amphibians, and it actually decreases after fertilization in the eggs of the mollusc *Cumingia* and the annelid *Chaetopterus* (Needham, 1942; Rothschild, 1956).

Obviously, the increase in oxidation rate cannot be considered as the "key reaction" of fertilization. Other metabolic changes found after fertilization in some, though by no means all, animals are (1) production of considerable amounts of acid during the first few minutes after fertilization, (2) increase in the permeability of the egg membrane, and (3) increase in the activity of proteolytic enzymes.

A new opening for understanding the essence of fertilization has been provided by the studies on the mechanism of gene action and protein synthesis. It has been shown that in sea urchins the ability of an unfertilized egg to incorporate radioactively labeled precursors into proteins is very low, but that fertilization of the egg is almost immediately followed by an increase in protein synthesis. The reason for the change cannot be a lack of amino acids which are necessary for protein synthesis, as these are available both in the fertilized and the unfertilized egg. The ribosomes are also present in unfertilized eggs and can be shown to be able to participate in protein synthesis. It would be attractive to think that messenger RNA is lacking in the unfertilized egg and that after fertilization the nucleus emits new quantities of messenger RNA. It has been shown, however, that enucleated fragments of the egg may be activated to parthenogenetic development (see next section), and that in these fragments an increase of protein synthesis occurs as in normal fertilized eggs (Monroy and Tyler, 1963). As the fragments do not possess nuclei, the necessary messenger RNA, in this case, could not be produced on chromosomal DNA and thus must have been present, though inactive, in the cytoplasm of the unfertilized egg.

Before fertilization, the ribosomes in the egg cytoplasm are single. After fertilization, increasing quantities of ribosomes are joined in groups (Monroy and Tyler, 1963), which shows that they are actively engaged in protein synthesis. As both ribosomes and messenger RNA are present in the cytoplasm of the unfertilized egg, it has been concluded that either the messenger RNA or the ribosomes or both are inactivated in some way or other in the unfertilized egg. There is evidence that a protein may be responsible for this inactivation. Ribosomes isolated from unfertilized sea urchin eggs by homogenization and subsequent centrifugation have been treated with trypsin for 30 minutes. After this treatment it was found that the ribosomes were capable of supporting protein synthesis in a cell-free system containing protein precursors (the activated amino acids). Furthermore, it was found that the treated ribosomes could support protein synthesis both with messenger RNA prepared from developing embryos and without any other messenger added.

In the latter case the messenger RNA was derived from the unfertilized eggs and actually must have been attached to the ribosomes, because it was segregated from the other cellular constituents together with the ribosomes during the centrifugation of the cell homogenate (Monroy, Maggio, and Rinaldi, 1965). It follows that in the unfertilized eggs the messenger RNA was present and already attached to the ribosomes but could not act in synthesizing protein. As a proteolytic enzyme could bring the synthesizing system into action, it seems logical to conclude that a protein which was binding the system was destroyed or removed by the enzyme. In other words, the messenger RNA of unfertilized eggs is "masked" and becomes "unmasked" during fertilization (Tyler, 1967). One of the physiological changes during fertilization is an increase of proteolytic activity in the egg immediately following the penetration by the spermatozoon. (See preceding discussion.) It may well be that this burst of proteolytic activity causes the breakdown of protein attached to the messenger RNA and ribosome particles, unmasks the messenger RNA, and puts into action the protein synthesis system of the egg.

It is important to note that if this interpretation is correct, the new protein is synthesized on the information provided by the messenger RNA which is already in the egg cytoplasm before fertilization, that is, messenger RNA produced by the chromosomes of the oocyte. Although the final solution to the problem of fertilization is not yet available, it is evident that some narrowing of the possibilities has been achieved and that the questions can now be set in a more precise form, which can lead to fruitful experimental investigations in the near future.

5–4 PARTHENOGENESIS

A different approach to the solution of the problem of fertilization lies in attempting to imitate the action of the spermatozoon by some known agent. It has long been known (p. 11) that in some animals an egg can develop without fertilization, as in the aphids, phyllopods, and rotifers at some times of the year, or as in bees and wasps in which a fertilized egg produces a female individual and an unfertilized egg develops into a male. These are cases of **natural parthenogenesis** (virginal reproduction). In other animals, such as most of the echinoderms and many others, the eggs under natural conditions do not develop unless they are fertilized. It has been found, however, that certain treatments of the ripe eggs may incite them to develop, and this phenomenon is known as **artificial parthenogenesis.**

A great amount of work on artificial parthenogenesis has been done with sea urchin eggs. O. Hertwig and R. Hertwig discovered that ripe sea urchin eggs may be caused to start developing by treatment with chloroform or strychnine (Hertwig, 1896). Later, it was found that the same and even better results may be obtained by treating the eggs with a variety of substances: hypertonic or hypotonic seawater; various salts, such as the chlorides of potassium, sodium, calcium, magnesium, etc.; weak organic acids—butyric acid, lactic acid, oleic and other fatty acids; fat solvents—toluene, ether, alcohol, benzene, and acetone; and urea and sucrose. Similar results are obtained by temperature shocks—that is, by transferring the eggs for a short time to warm (32° C.) or cold (0° to 10° C.) water; by electric induction shocks; by ultraviolet light; and even by shaking the eggs in ordinary seawater (Loeb, 1913; Harvey, 1956). This long list of agents (which is by no means exhaustive) clearly shows that there is no one agent which can be recognized as the specific cause of the activation of the egg, the cause which determines the nature of the processes that are to take place. Obviously, factors determining the nature of the reaction of the egg are contained in the egg itself. The agents causing the artificial parthenogensis of the egg are instrumental only as factors triggering off this reaction of the egg, and if this is so it becomes plausible that the spermatozoon itself exercises only a similar triggering action.

Something further may be deduced, however, from the array of agents causing artificial parthenogenesis. Most of the agents used are of such a nature that they damage the cells to a greater or lesser degree, and if applied in greater intensity, or for a longer time, they can cause the death of the cells. Thus, it is reasonable to suppose that activation of the egg involves some type of sublethal damage to the egg cytoplasm. Furthermore, some of the agents used, such as fat solvents, may be expected to affect primarily the cell surface, which is known to consist partly of lipids; all other chemical agents may likewise primarily affect the surface of the cytoplasm, which would also not be immune from the action of physical agents.

We may therefore go one step further and suggest that the activation of the egg is connected somehow with damage to the cortical layer of the egg cell cytoplasm. This is the more likely because (1) the reaction of the egg is started as soon as the spermatozoon comes in contact with the surface of the egg, and (2) the first detectable reactions of the egg occur in the cortical layer and actually involve the breakdown of parts of the cortex (the cortical granules). We have seen (p. 89) that one of the essential features of fertilization is the setting in motion of the protein synthesis in the egg cytoplasm, affected possibly by a proteolytic action on the bound messenger RNA-ribosome complexes. The sublethal damage to the egg cortex may well be the direct cause of proteoltyic action in the egg cytoplasm. This is the more likely since small lysosome-like particles have been found close to the plasma membrane of eggs in several invertebrates (Smith, 1964). This does not explain, however, the mechanism by which the spermatozoon releases the proteolytic reaction.

We will conclude this section by adding some information on natural and artificial parthenogenesis in some vertebrates. In frogs, artificial parthenogenesis may be achieved by some of the methods used on echinoderm eggs, such as the use of hypertonic and hypotonic solutions, some poisons (corrosive sublimate), and electric shock. The activation achieved by these methods is incomplete, however, and the development does not go beyond abortive cleavage. A more efficient method is pricking the eggs with a fine glass needle (a method also used successfully with echinoderm eggs); for full success with this method, however, it is necessary that the needle be smeared with blood or be contaminated by cells or cellular particles from other tissues. If particles from foreign cells are thus introduced into the ripe egg cell, the cleavage is greatly improved, and a small percentage of the treated eggs may go through the whole development apparently quite normally (Bataillon, 1910). Instead of pricking the eggs with a contaminated needle, constituent parts of cells may also be introduced into the egg with a micropipette. In this way it has been possible to investigate what fractions of cell constituents are most active in causing parthenogenesis, and it has been found that "large granules" of a centrifuged tissue homogenate (this fraction includes the mitochondria) have the strongest effects, while the liquid supernatant had no more effect than did the pricking with a clean needle or a needle wetted with a buffer saline solution (Shaver, 1953 — see Table 5).

TABLE 5   Activation of Frog Eggs by Injecting Them with Tissue (Testis) Extract and Various Fractions of the Same Obtained by Centrifugation

Preparation	*Eggs Reaching Blastula Stage (%)*
Whole tissue extract	9.7
Large granules	35.2
Medium and small granules	19.5
Supernatant	0
Phosphate buffer	0.3
Pricked "dry"	0

From Shaver, 1953.

So far, no successful experiments have been reported in producing artificial parthenogenesis in birds, but in connection with other work it may be of interest to note that spontaneous parthenogenetic development has been observed in two representatives of the class, the domestic turkey and the common domestic hen. Eggs laid by female turkeys which had been isolated from males start developing in a fairly high percentage of cases (up to 41.7 per cent in some recent experiments—Olsen, 1960a); although there is still a high percentage of abnormal development, a small number of embryos reach the hatching stage. All the poults, the sex of which could be determined, were males. A few of these reached maturity and one was used, with success, to fertilize eggs of normal females (Olsen, 1960b).More recently, similar results have been obtained with hens (Sarvella, 1973).

A very peculiar case of spontaneous parthenogenesis is presented by some lizards, in particular the lizard *Lacerta saxicola armeniaca* which is found in the Caucasus. In this subspecies no males exist in nature, and the population is propagated exclusively by parthenogenetic females. Some embryos start developing as males, but these all die in early stages (Darevskii and Kulikowa, 1961).

In mammals the possibility of artificial parthenogenesis was discovered in connection with experiments on cultivating unfertilized eggs *in vitro* which had been collected from the fallopian tubes. Extensive investigations in this field were underteken by Pincus and his collaborators. Most of the experiments were done with rabbits' eggs (Pincus, 1936). It was found that if the eggs are simply kept for up to 48 hours in the ordinary tissue culture medium (blood plasma plus embryo extract), some of them become activated. The first sign of activation in this case is completion of the second meiotic division and extrusion of a second polar body. Some of the eggs even go beyond that stage and start cleaving. Chemical treatment (with butyric acid, which has been used for activating sea urchin eggs) does not seem to yield any better results. However, a temperature shock, particularly a treatment with cold, is more effective.

To allow the activated eggs to develop further, in some experiments they have been injected into the fallopian tubes of rabbit does, which were made "pseudopregnant" by mating with a sterile buck or by injection of hormones (the luteinizing hormone). In the body of the female the development progressed further, and quite a number of embryos reached the blastocyst stage (18 per cent in one experiment in which the eggs were activated by cooling for 24 hours at 10° C.—Chang, 1954). In two cases in which the fertilization of the eggs by spermatozoa seems to have been completely excluded, the parthenogenetic embryos completed intrauterine development, and one of the young was born alive (the other was stillborn—Pincus, 1939). In another case the eggs were given a cold shock *in vivo* by opening the body cavity of a rabbit doe which had unfertilized eggs in its fallopian tubes and cooling the fallopian tubes with cold water. One live young was born (Pincus and Shapiro, 1940). So far, these are the only records of living mammalian young produced by parthenogenesis. (See also Beatty, 1957.)

5–5 THE SPERMATOZOON IN THE EGG INTERIOR

There is some variation in different animals as to how much of the spermatozoon is taken into the interior of the egg. In many animals, notably in the mammals, the whole of the spermatozoon, head, middle piece, and tail penetrates into the cytoplasm and for a short time may be seen lying intact in the interior of the egg. In some animals (echinoderms), however, the tail of the spermatozoon breaks off and is left outside the vitelline membrane, and even the middle piece of the spermatozoon may be left without, so that only the head and the centrosome enter the egg *(Nereis)*. That the tail of the spermatozoon often does not enter the egg gives additional proof that its functions are purely locomotive.

The information concerning the middle piece of the spermatozoon is not unequivocal. Although the middle piece appears to enter the egg cytoplasm in most cases, there is no definite proof that any constituents of the spermatozoon except for the nucleus and the centrosome play an active part in subsequent development. The mitochondria contained in the middle piece have been observed in some cases to scatter in the cytoplasm of the egg, but it is not known how long they maintain their existence there.

The subsequent behavior of the spermatozoon, or rather its head, is dependent on the stage of maturation (reduction divisions) which the egg has reached at the time of fertilization. In the sea urchins the eggs are shed and become fertilized after the reduction divisions have been completed and both polar bodies have been extruded from the egg. This is, however, by no means a general occurrence. In vertebrates the rule is that the egg completes its first reduction division in the ovary and reaches the metaphase stage of the second meiotic division. At this stage all further progress is arrested, ovulation takes place, and the egg may become fertilized. The second reduction division is completed and the second polar body extruded only if the egg is fertilized by a spermatozoon or activated in some other way (p. 90). In ascidians, some molluscs, and annelids, the egg reaches only the metaphase of the first meiotic division when it becomes ripe and is fertilized; only then does the egg complete the first reduction division and carry out the second. Lastly, in some annelids, in nematodes, and in chaetognaths, the eggs are fertilized even before the beginning of meiotic division, while the occyte nucleus is still intact.

It follows that although in animals like the sea urchins the spermatozoon nucleus may immediately proceed to join the egg nucleus, in other cases the immediate effect of the fertilization, as far as the nuclear apparatus is concerned, is the completion of the meiotic divisions, only after which the fusion of the male and female pronuclei may take place.

When the spermatozoon first penetrates the egg cytoplasm, it moves with the acrosome (or acrosomal filament) at its front. The nucleus and the centrosome, in that order, are arrayed behind the acrosome. Soon, however, a rotation of the nucleus and the centrosome can be observed in many, though not in all, animals, the centrosome coming ahead of the nucleus and the nucleus turning through 180 degrees so that its original posterior end turns forward. The other parts of the spermatozoon, if still discernible by this stage, lose connection with the nucleus and the centrosome. Both the nucleus and the centrosome change in appearance. The nucleus, which is now referred to as the **male pronucleus,** starts swelling, and the chromatin, which is very closely packed in the spermatozoon, again becomes finely granular (Fig. 62) (Longo and Anderson, 1968). By imbibition of fluid from the surrounding cytoplasm the pronucleus becomes vesicular. The centrosome, at the same time, becomes surrounded by an aster, similar to that of the centrosome in the early stages of an ordinary mitosis. While these changes are occurring, the sperm nucleus, together with the centrosome, moves through the egg cytoplasm toward the area where fusion with the egg nucleus, the **female pronucleus,** is to take place. This area is generally near the center in holoblastic eggs having a fairly small amount of yolk, but in telolecithal eggs it is in the center of the active cytoplasm at the animal pole of the egg. As the sperm head moves inward it may be accompanied by some cortical and subcortical cytoplasm. If the latter is heavily pigmented, as in amphibian eggs, the trajectory of the sperm head may be marked by

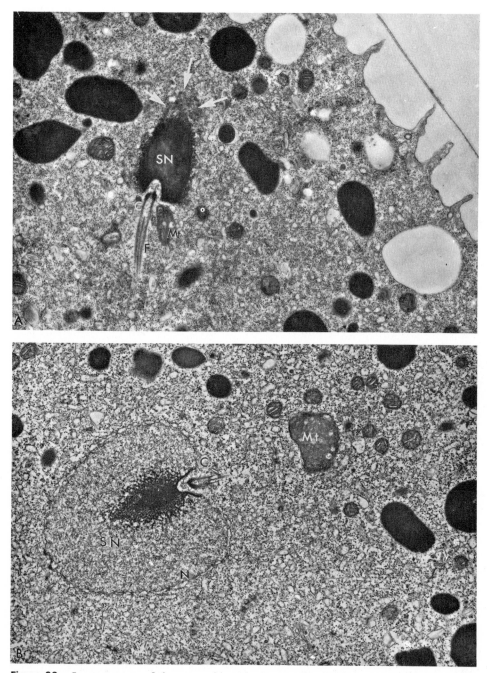

Figure 62. Spermatozoon of the sea urchin *Arbacia punctulata* in the process of transformation into the male pronucleus after penetrating into the egg cytoplasm. *A*, Beginning of dispersal of the chromatin of the spermatozoon nucleus (arrows); *B*, nucleus has almost attained the structure of an ordinary interphase nucleus, but a core of condensed chromatin still remains (electronmicrographs). C, Centriole of spermatozoon; F, remains of spermatozoon flagellum; Mt, remains of mitochondrial body of spermatozoon; N, nuclear membrane; SN, spermatozoon nucleus. (Courtesy of Dr. F. J. Longo and Professor E. Anderson.)

pigment granules trailing along its path. This is sometimes referred to as the **penetration path.**

The female pronucleus also has to traverse a greater or lesser way before it reaches the male pronucleus. At the beginning of its migration the female pronucleus is invariably at the surface of the egg, where the second meiotic division has been taking place, since it is only after the completion of the meiotic division that the nucleus of the egg may fuse with the nucleus of the spermatozoon. The haploid nucleus of the egg, after the completion of the second meiotic division, is often in the form of several vesicles known as the **karyomeres.** These fuse together to form the female pronucleus, which swells and increases in volume as it approaches the male pronucleus (Fig. 63). In the

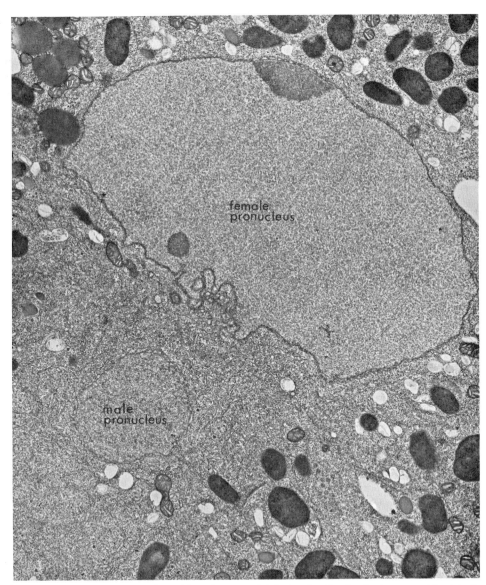

Figure 63. Meeting of the male and female pronuclei during fertilization in the sea urchin *Arbacia punctulata* (electronmicrograph). (Courtesy of Dr. F. J. Longo and Professor E. Anderson.)

last stage before they meet, the male and female pronuclei may become indistinguishable.

The actual fusion of the male and female pronuclei may differ in detail among different animals. In some animals the two pronuclei actually fuse together; that is, the nuclear membranes become broken at the point of contact, and the contents of the nuclei unite into one mass surrounded by a common nuclear membrane. At the approach of the first cleavage of fertilized eggs of sea urchins and vertebrates, the nuclear membrane dissolves, and the chromosomes of maternal and paternal origin become arranged on the equator of the achromatic spindle. In other cases, however, the male and female pronuclei do not fuse as such, but the nuclear membranes in both dissolve and the chromosomes become released. In the meantime, the centrosome of the spermatozoon has divided in two and an achromatic spindle has been formed to which the chromosomes derived from the male and the female pronuclei become attached. (It is important to note that in normal fertilization the achromatic figure of the first and subsequent cell divisions is produced by the centrosome of the spermatozoon.) Only after completion of the first division of the fertilized egg do the paternal and maternal chromosomes become enclosed by common nuclear membranes in the nuclei of the two daughter cells into which the egg has become divided (*Ascaris,* some molluscs, and annelids). In both types of fusion the chromosomes of the maternal and paternal sets retain, of course, their individuality.

Lastly, in some animals, of which the copepod *Cyclops* is a well-known example, the paternal and maternal nuclear components remain separate for some time, even after cleavage has started, so that each blastomere has a double nucleus consisting of two parts lying side by side, but each surrounded by its own nuclear membrane.

A closer union of the homologous chromosomes takes place much later in preparation for meiosis in the gonads of the new individual and also in cases of somatic conjugation of chromosomes, as in the salivary gland chromosomes of *Drosophila.*

5–6 CHANGES IN THE ORGANIZATION OF THE EGG CYTOPLASM CAUSED BY FERTILIZATION

In addition to activating the egg and providing an opportunity for amphimixis, the penetration of the spermatozoon into the egg (or the parthenogenetic activation of the egg) causes in many animals, perhaps in all, far-reaching displacements of the cytoplasmic constituents of the egg. As a result of this, the distribution of various cytoplasmic substances and inclusions in the egg at the beginning of cleavage may be considerably different from that in the unfertilized egg, and even qualitatively new areas may sometimes appear. It will be evident later that these changes in the organization of the egg at fertilization may be profoundly important for further development of the fertilized egg.

One result of the extrusion of the cortical granules is that a large part of the original outer egg cell surface becomes replaced by the inner surfaces which surrounded the cortical granules and now are everted onto the exterior (Afzelius, 1956). In view of the importance of the cell surfaces for morphogenetic processes (see further, Sections 6–7 and 9–2), this change in the composition of the surface layer of the egg may have a considerable significance, though at present it has not actually been shown to what degree the changes in the physiological properties of the fertilized egg, as compared with those of the egg before fertilization, are due to the replacement of part of its outer surface.

Displacements of cytoplasmic substances may be observed best in those eggs in which easily distinguishable kinds of cytoplasm have been evolved during the growth and maturation of the oocyte.

We have seen that in the ascidian *Styela partita* the surface of the mature egg is covered by a layer of cortical cytoplasm containing yellow granules. The position of the egg nucleus close to the animal pole stresses the polar organization of the egg, but there is no trace of a bilateral organization: all meridians of the egg are exactly alike. The spermatozoon penetrates the egg at any meridian and nearer to the vegetal pole.

As soon as the spermatozoon enters the egg, the yellow cortical cytoplasm falls into violent commotion (Fig. 64). At first, the yellow cytoplasm begins to stream down along the surface of the egg toward the vegetal pole, and for a time it accumulates as a cap on the vegetal pole. As the spermatozoon pronucleus penetrates deeper into the cytoplasm on its course toward the egg pronucleus, the yellow cytoplasm reverses its movement and streams upward, but this time on one side of the egg only, namely, on the side where the spermatozoon entered the egg. Shortly before the first cleavage, the yellow cytoplasm takes up a position just below the equator of the egg, forming there a crescentic area which later gives rise to mesoderm and may be called the **mesodermal crescent.** At the same time, a crescent of light gray cytoplasm appears subequatorially on the opposite side of the egg. The origin of this cytoplasmic substance has not been traced, but it is found to give rise later to the notochord and may therefore be denoted as the **notochordal crescent.** Thus, the cytoplasmic displacements following fertilization not only bring some kinds of cytoplasm to more strictly defined, restricted areas, but also in this case give the egg a distinct bilateral structure (Conklin, 1905). Four different kinds of cytoplasm are now present in the egg: the yellow cytoplasm on one side and the light gray cytoplasm on the other side together form a belt surrounding the egg just below the equator. Below this subequatorial zone, toward the vegetal pole, the cytoplasm contains abundant yolk granules, and in the subcortical layer, also large numbers of mitochondria (Fig. 95, p. 130) (Mancuso, 1964). This **vegetal cytoplasm** is slaty gray in color. Above the equator, in the animal hemisphere, the cytoplasm contains less yolk and few mitochondria and appears more transparent than the rest of the egg. As we will see later, during development the different kinds of cytoplasm give rise to specific parts of the embryo.

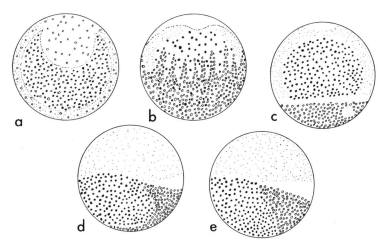

Figure 64. Displacements of cytoplasmic substances in the egg of the ascidian *Styela partita* following fertilization. *a,* Ripe unfertilized egg; *b,* egg immediately after entry of spermatozoon; *c, d,* two successive stages of the egg at the time when ♂ pronucleus moves to meet ♀ pronucleus; *e,* egg just before first division. Yolk granules are represented by black dots, yellow granules by circles. (After Conklin, from Huxley and de Beer, 1934; and from Morgan, 1927.)

Displacements of cytoplasmic substances follow fertilization in other animals also. In the sea urchin *Paracentrotus lividus*, the mature egg contains red granules in the cortical cytoplasm which are evenly distributed over its whole surface. After fertilization, streaming movements of the surface layer bring the pigmented granules into a subequatorial zone, leaving the entire animal hemisphere clear of pigment and also removing pigment from a smaller area at the vegetal pole (Fig. 65).

The structure of the amphibian egg at the time of fertilization was described on page 55. Only a few minutes after the sperm has penetrated the egg, the surface layer of protoplasm starts contracting toward the animal pole of the egg, and in the course of this contraction the pigmented protoplasm covers up the gap in the pigment layer at the animal pole, where the first polar body has been given off. Considerably later (10 minutes after insemination of a *Rana temporaria* egg at 18° C.), a much more extensive movement of the cytoplasm begins. This involves practically the whole cortical layer of the egg, which is rotated with respect to the internal mass of cytoplasm (Ancel and Vintemberger, 1948). At the future dorsal side of the egg, the cortical cytoplasm moves upward toward the animal pole, and at the opposite, future ventral, side of the egg the cortical cytoplasm moves downward toward the vegetal pole (Fig. 66). As the edge of intensely pigmented cortical cytoplasm rises above the equator on the dorsal side, it reveals the deeper-lying ring of marginal cytoplasm. The marginal cytoplasm, although bearing some pigment, is much lighter colored than the cortical cytoplasm. The result is that on the dorsal side of the egg a subequatorial, lightly pigmented, crescentic area, known as the **gray crescent,** appears.

The gray crescent presents an arrangement of cytoplasmic constituents which had not occurred in the egg previously. Owing to the sliding upward of the cortical layer, not only is the heavily pigmented surface protoplasm removed from this area, but the cortical layer of the vegetal field is drawn over it, with some of the yolky cytoplasm of that part of the egg being superimposed over the marginal cytoplasm. The superimposed layer is sufficiently thin not to conceal the pigment of the deeper-lying marginal cytoplasm. On the ventral side of the egg the reverse occurs; the marginal cytoplasm is partly covered by the layer of pigment.

As a result of these displacements, the egg acquires a very well-defined bilateral symmetry, with the median plane bisecting the gray crescent. The cytoplasm of the egg is found in this condition at the beginning of cleavage.

The new distribution of cytoplasmic substances in the egg is immediately followed by changes in the physiological properties of the surface layer of the egg. Before the appearance of the gray crescent the permeability of the cortical cytoplasm for vital dyes is the same over the whole surface of the egg, but after the rotation of the cortical cytoplasm differences become discernible. The permeability to vital dyes becomes de-

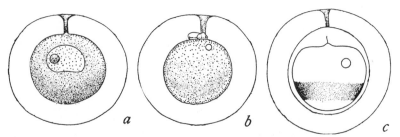

Figure 65. Displacement of pigment in the egg of the sea urchin *Paracentrotus lividus* following fertilization. *a*, Oocyte; *b*, ripe unfertilized egg; *c*, fertilized egg. (After Boveri, from Morgan, 1927.)

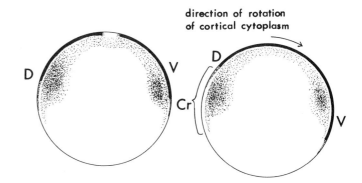

direction of rotation
of cortical cytoplasm

Figure 66. Diagrammatic median sections showing displacement of the cortical cytoplasm in the frog's egg following fertilization, with the formation of the gray crescent (Cr). D, Dorsal surface; V, ventral surface of the egg. (According to Banki, 1929, and Ancel and Vintemberger, 1948.)

creased around the animal pole and on the ventral surface of the egg, but in the gray crescent area this decrease does not take place. The permeability at the gray crescent thus becomes relatively higher than at other parts of the egg's surface (Dollander, 1953). Further physiological distinctions follow in later stages, and at the approach of gastrulation the gray crescent takes the lead in the formation of the blastopore, as will be explained in Section 7–4.

There is some experimental evidence that factors external to the egg may determine on which side of the amphibian egg the layer of cortical cytoplasm moves upward, and on which side, therefore, the gray crescent will develop. In 1885 W. Roux attempted to control the point of entry of the spermatozoon into the frog egg by applying a sperm solution on one side of the egg only, and he found that the gray crescent appeared on the side opposite to that of the point of entry into the egg. This result has been confirmed by more recent work (Ancel and Vintemberger, 1948). The arrangement of the cytoplasmic substances in the egg of an ascidian follows, as we have seen, the same course. However, if the frog's egg is stimulated to parthenogenetic development by pricking, the gray crescent appears without any relation to the point at which the egg was pricked. In the egg of the sturgeon there appears after fertilization and before the beginning of cleavage a "gray crescent" similar to that of the amphibian egg and bearing the same relationship to the later formation of the blastopore. In the sturgeon, however, the spermatozoon enters the egg through a micropyle exactly at the animal pole; it can therefore in no case determine the position of the gray crescent or the plane of symmetry of the embryo (Ginsburg, 1953).

Thus, it is obvious that the direction of rotation of the cortex of the frog's egg and the development of the gray crescent in a frog's or in a sturgeon's egg can be determined by some special properties of the cytoplasm of the egg and independently of the entry of the sperm. What these properties are is not fully understood. In this connection, it is noteworthy that if the unfertilized egg of a frog is allowed to rotate freely, it takes up a position with its main axis not strictly vertical but slightly oblique. The part of the equator which is raised in this position is that part where the gray crescent will later develop and which corresponds later to the dorsal side of the embryo. The fact that the dorsal side tends to be raised and that the opposite ventral side sinks down indicates that the cytoplasm on the dorsal side has a slightly lower specific gravity. This may mean that there is less yolk on this side (though of course the greatest concentration of yolk is toward the vegetal pole of the egg). It is doubtful, however, whether the amount of yolk is the decisive factor in determining the position of the gray crescent.

It was previously mentioned (p. 55) that the unfertilized egg of a frog may show some traces of bilateral symmetry: the ring-shaped zone of marginal cytoplasm has been found to be slightly thicker on one side, and this may well be the side on which the gray crescent develops in a parthenogenetically activated egg.

On the other hand, the possibility of determining the plane of bilateral symmetry of the egg by the point of entry of the spermatozoid shows that the inherent differences in the egg may be overridden by an external factor. In other words, any part of the marginal zone which lies opposite the point of entry of the spermatozoon may be transformed into the gray crescent. With respect to the development of its bilateral symmetry, the egg appears to have a very considerable amount of flexibility.

Chapter 6

CLEAVAGE

One of the peculiarities of sexual reproduction in animals is that the complex multicellular body of the offspring originates from a single cell—the fertilized egg. It is necessary, therefore, that the single cell be transformed into a multicellular body. This transformation takes place at the very beginning of development and is attained by means of a number of cell divisions following in rapid succession. This series of cell divisions is known as the process of **cleavage.**

Cleavage can be characterized as that period of development in which:

1. The unicellular fertilized egg is transformed by consecutive mitotic divisions into a multicellular complex.
2. No growth occurs.
3. The general shape of the embryo does not change, except for the formation of a cavity in the interior—the blastocoele.
4. Apart from transformation of cytoplasmic substances into nuclear substance, qualitative changes in the chemical composition of the egg are limited.
5. The constituent parts of the cytoplasm of the egg are not displaced to any great extent and remain on the whole in the same positions as in the egg at the beginning of cleavage.
6. The ratio of nucleus to cytoplasm, very low at the beginning of cleavage, is, at the end, brought to the level found in ordinary somatic cells.

6-1 PECULIARITIES OF CELL DIVISIONS IN CLEAVAGE

The cleavage of the fertilized egg is initiated by the division of the nucleus (the synkaryon), and as a general rule, the division of the nucleus is followed by the division of the cytoplasm, so that the egg cell divides into two daughter cells (Fig. 67). The daughter cells are termed the **cleavage cells** or **blastomeres.** The first two blastomeres divide again, thus producing four blastomeres, then 8, 16, 32, and so on. The first cleavages tend to occur simultaneously in all the blastomeres, but sooner or later the synchronization is lost, and the blastomeres divide at different times, independently of one another.

The division of the blastomeres is essentially a typical mitosis, and the chromosomes have the appearance and structure of somatic chromosomes. There is, however, one very important difference between the cell divisions in cleavage and the cell divisions in later stages of development and in the adult organism. In the later stages and in the adult, cell division is intimately connected with growth. After each division the daughter cells grow, and when they are approximately doubled in size they divide again. The cells thus maintain an average size in every type of tissue. During cleavage this is not so. The consecutive divisions of the blastomeres are not separated by periods of growth; a

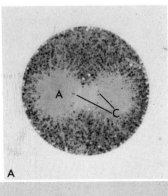

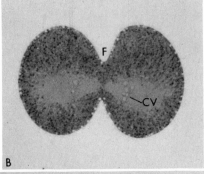

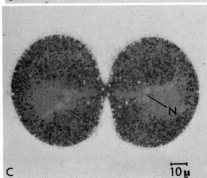

Figure 67. Three stages of the first cleavage division of an oligolecithal egg (sea urchin). A, Yolk-free cytoplasm surrounding the mitotic aster; C, chromosomes; CV, chromosomal vesicles; F, cleavage furrow; N, daughter nucleus. (Courtesy of Dr. T. E. Schroeder, 1972.)

blastomere does not increase in size before the next divisions begins. Consequently, with each division the resulting blastomeres are only half the original size. Thus, cleavage begins with one very large cell and ends with a great number of cells, each of which is no longer very much larger than the tissue cells of the adult animal. Indeed, at the end of cleavage, the cells are usually even smaller than most of the differentiated cells of an adult animal, because cellular differentiation is often accompanied by an increase in the size of an individual cell.

The nuclei of the early cleavage cells are considerably larger than they are in ordinary somatic cells of the same animal. This is, however, mainly due to the presence of larger amounts of nuclear sap and not to more chromosomal material. The amount of nucleic acid in individual nuclei may be determined by using the property of these acids of absorbing ultraviolet light at a wave-length of 2600 Å. If a very narrow beam of light of this wavelength is passed through a nucleus in a microscopic preparation, the light absorbed by the nucleus may be measured, and the actual content of the nucleic acid may then be calculated. Alternatively, the nuclei may be stained in the preparation by the

TABLE 6   Mean Amounts of Deoxyribonucleic Acid per Nucleus in Arbitrary Units in Early Development of the Sea Urchin, Lytechinus variegatus

Stage	DNA	Condition of Nucleus
1-cell	1.11 ± .06	near 4n
2-cell	1.02 ± .04	near 4n
4-cell	1.09 ± .04	near 4n
8-cell	1.08 ± .07	near 4n
16-cell micromeres	0.82 ± .06	2n < 4n
mesomeres	1.04 ± .04	near 4n
macromeres	1.02 ± .04	near 4n
28-cell micromeres	1.10 ± .05	near 4n
mesomeres	0.95 ± .06	near 4n
macromeres	1.01 ± .03	near 4n
blastula ectoderm	0.68 ± .04	2n < 4n
gastrula ectoderm	0.60 ± .03	2n +
early pluteus ectoderm	0.62 ± .03	2n +
late pluteus ectoderm	0.60 ± .03	2n +

From McMaster, 1955.

Feulgen method which is specific for deoxyribonucleic acid, and the absorption of light by the fuchsin fixed in the nucleus may be taken as an estimate of the nucleic acid present.

Since the chromosomal material in the cleavage cells is being doubled at each mitotic division, one would expect the amounts of deoxyribonucleic acid to vary from a diploid amount in cells that have just undergone mitosis to a tetraploid amount in cells that are ready to divide again. This has actually been found in the embryo of the sea urchin, *Lytechinus,* from the one-cell stage (zygote nucleus) to the larval stage (McMaster, 1955 — see Table 6). As an example, we may also quote results obtained by measurement of the deoxyribonucleic acid content in nuclei during fertilization and cleavage in an annelid, *Chaetopterus* (Alfert and Swift, 1953 — see Table 7.)

In spite of the greater volume of the cleavage nuclei, these are small in proportion to the volume of the cell. In a mature sea urchin egg before fertilization the ratio $\frac{\text{volume of the nucleus}}{\text{volume of cytoplasm}}$ is 1/550 (Brachet, 1950). At the end of cleavage (in the blastula stage) the same ratio is 1/6.

If the numbers of cells in a developing embryo at various stages are counted, the change in the rate of reproduction of cells between the period of cleavage and the later

TABLE 7   Nucleic Acid Content in Nuclei During Gametogenesis and Cleavage in Chaetopterus (Deoxyribonucleic Acid in Arbitrary Units — Average from Several Determinations)

Stage	Amount of DNA	Condition of Nucleus	
First polar body	127 ± 3	2n	
Spermatozoon	61 ± 1	1n	
Cleavage, interphase	210 ± 9	2n ——————— 4n
Cleavage, prophase	263 ± 10	4n	
Cleavage, telophase	124 ± 3	2n	

From Alfert and Swift, 1953.

development becomes very obvious. In Figure 68 the logarithm of the number of cells in a frog embryo is plotted against time. The curve shows a very distinct bend at about 40 hours, which is just at the end of cleavage (between the blastula stage and the beginning of gastrulation), if development has been proceeding at 15° C. Before this time the rate of increase is far more rapid than afterwards (Sze, 1953).

The rhythm of cleavage as measured by the time interval between two consecutive divisions is not quite the same in different animals. In the goldfish, divisions follow each other continuously at rather regular intervals of about 20 minutes. The interval is nearer to one hour in the case of the frog, though it depends very much on temperature. The eggs of mammals cleave very much more slowly, about 10 to 12 hours elapsing between consecutive cell divisions in the mouse at 37° C. (Kuhl, 1941).

6–2 CHEMICAL CHANGES DURING CLEAVAGE

Although there is no growth during the period of cleavage, chemical transformations do occur, and at least some are markedly intensified, as compared with the conditions in the unfertilized egg.

The most obvious change observed during cleavage is a steady increase of nuclear material at the expense of cytoplasm. The number of nuclei is of course doubled with every new division of the blastomeres, and this doubling is accompanied by an increase of nuclear substance, which involves an increase of deoxyribonucleic acid – the amount per nucleus of the latter remaining constant (McMaster, 1955 – see Table 6).

The increase in the chromosomal deoxyribonucleic acid, at least during the earlier phases of development, must be at the expense of materials contained in the egg. There are several possible sources of such materials. First, the nucleic acids present in the cytoplasm of the eggs should be mentioned. In sea urchin eggs there is a large amount of ribonucleic acid in the egg cytoplasm, and this gradually disappears later in development (Brachet, 1950b). When sea urchin embryos are supplied with radioactively tagged uridin, some of it is later incorporated into the DNA (Czihak et al., 1967). Such incorporation is made possible by the presence in developing sea urchin eggs of an enzyme, ribonucleotide reductase, which converts ribonucleotides into deoxyribonucleotides (Noronha, Sheys, and Buchanan, 1972). DNA may, however, be synthesized in the cleaving egg directly from low molecular weight precursors. This has been proved by supplying such precursors, labeled with radioactive atoms, to cleaving eggs of sea

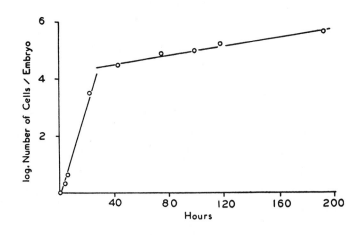

Figure 68. Increase in number of cells during early development of the frog's egg. (From Sze, 1953.)

urchins and amphibians. When cleaving sea urchin eggs were kept in seawater containing C[14] labeled glycine (which may be used for the synthesis of purine groups in the nucleic acid molecule), it was found that the racioactive carbon atoms were incorporated in large amounts into the deoxyribonucleic acid, bypassing the cytoplasmic ribonucleic acid (Abrams, 1951). Also, when C[14] labeled glycine was injected into fertilized frog eggs, some of it was incorporated into deoxyribonucleic acid (Grant, 1958).

The second important aspect of metabolism during cleavage is the synthesis of ribonucleic acids, which is believed to be very limited, although not absent altogether. In frogs ribosomal RNA apparently is not produced at all until after completion of cleavage. As the nucleolus is the site of synthesis of rRNA, this organoid is completely lacking in these animals during cleavage (Brown, 1966). It reappears in the nuclei at the onset of gastrulation simultaneously with the resumption of ribosomal RNA synthesis (Gurdon and Brown, 1965). In the sea urchin there is very little ribosomal RNA produced during cleavage, but in both the amphibians and sea urchins synthesis increases drastically at the onset of gastrulation (Fig. 69). Messenger RNA and transfer RNA, on the other hand, are synthesized during cleavage, or at least in the later stages of cleavage (Tyler and Tyler, 1966; Gurdon, 1969). Synthesis of RNA, however, does not seem to be necessary for cleavage, since eggs which are treated with actinomycin D and in which presumably DNA dependent RNA synthesis is suppressed continue cleaving normally. It is concluded, therefore, that any messenger RNA produced during cleavage remains inactive or "masked," similar to the messenger RNA in unfertilized eggs (p. 90).

It has already been noted (p. 89) that fertilization in sea urchins leads to a spectacular increase in protein synthesis, and this is continued throughout the period of cleavage. In other animals, such as amphibians, protein synthesis does not markedly change after fertilization; a certain amount of protein synthesis, however, takes place throughout the period of cleavage. The amount of active cytoplasm increases. One in-

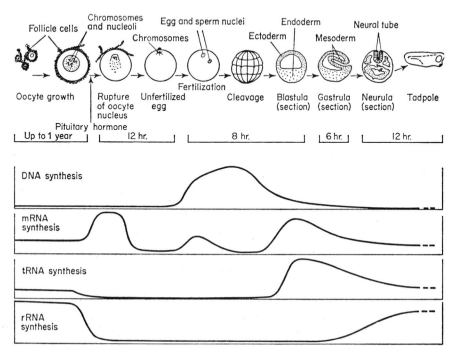

Figure 69. Changes in the synthesis of various classes of nucleic acids in oogenesis, at fertilization, and during early development in the frog embryo. (From Gurdon, 1968.)

dication of this is the steady increase of respiration throughout the period, which is generally attributed to an increase in the amount of active cytoplasm (Boell, 1945; Weber and Boell, 1955).

Much of the protein newly produced during cleavage is directly involved in the process of cell multiplication. One group of such proteins is the nuclear histones, which are needed for the chromosome replication in the same degree as additional quantities of DNA. In mid-cleavage of sea urchin embryos as much as 50 per cent of the newly synthesized protein is located in the nucleus (Kedes and Gross, 1969). The mRNA for these proteins is transcribed during cleavage, and contrary to other mRNA's, does not become masked, but is immediately used for translation into protein (Kedes and Gross, 1969). This exception is probably due to the need for rapid synthesis of large quantities of nuclear histones. Some mRNA for nuclear histones is, however, present in the egg before fertilization.

Another protein synthesized during cleavage is tubulin, the constituent protein of microtubules—the fibers of the achromatic figures appearing during the mitotic divisions of cleavage cells. Tubulin is synthesized on messenger RNA already present in the egg (Raff et al., 1972). In the course of cleavage, tubulin is synthesized in increasing quantities, presumably as a result of progressive "unmasking" of the corresponding mRNA.

A third protein synthesized during cleavage is the enzyme ribonucleotide reductase, mentioned previously, which in sea urchin embryos converts cytoplasmic ribonucleotides into deoxyribonucleotides, and thus provides a source of material for the replication of the chromosomal DNA. The messenger RNA for ribonucleotide reductase is present in the unfertilized egg, but becomes active (is unmasked) after fertilization (Noronha, Sheys, and Buchanan, 1972). A fourth protein necessary for chromosomal replication, the DNA polymerase, is already present in necessary quantities in the egg, and its quantity does not increase during early development (De Petrocelis and Monroy, 1974).

If cleaving eggs are treated with puromycin, which inhibits RNA dependent protein synthesis, cleavage stops immediately, thus showing that protein synthesis is indispensable for cleavage to take place. This is in marked contrast to cleavage being able to proceed in the presence of actinomycin D which effectively prevents the production of new RNA and particularly of new messenger RNA.

Possibly the most important of the proteins which have to be synthesized for cleavage to proceed are those which are used in the replication of the chromosomes: the nucleohistones, actually incorporated into the chromosome structure, and the ribonucleotide reductase, without which the cells are unable to use the supplies of RNA in the cytoplasm for replication of the nuclear DNA. The synthesis of tubulin seems to be less important, as asters may be formed in the cytoplasm in the presence of puromycin, but do not lead to cell division (Brachet, 1969). Presumably there is enough tubulin in the egg for aster formation, without new synthesis.

6–3 PATTERNS OF CLEAVAGE

The way in which the egg is subdivided into the daughter blastomeres is usually very regular. The plane of the first division is, as a rule, vertical; it passes through the main axis of the egg. The plane of the second division is also vertical and passes through the main axis, but it is at right angles to the first plane of cleavage. The result is that the first four blastomeres all lie side by side. The plane of the third division is at right angles of the first two planes and to the main axis of the egg. It is therefore horizontal or parallel to the equator of the egg. Of the eight blastomeres, four lie on top of the

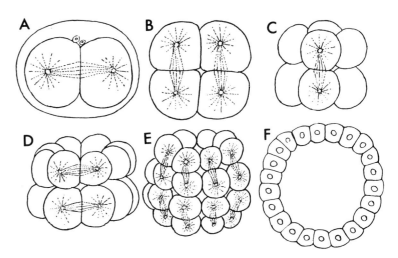

Figure 70. Radial cleavage with almost equal size of blastomeres in sea cucumber *Synapta digitata. A*, Two-cell stage; *B*, four-cell stage (viewed from animal pole); *C*, eight-cell stage, lateral view; *D*, 16-cell stage; *E*, 32-cell stage; *F*, blastula, vertical section. (After Selenka, from Korschelt, 1936.)

other four, the first four comprising the animal hemisphere of the egg, the second the vegetal hemisphere.

If each of the blastomeres of the upper tier lie over the corresponding blastomeres of the lower tier, the pattern of the blastomeres is radially symmetrical (Fig. 70). This is called the **radial type** of cleavage. In many animals, however, the upper tier of blastomeres may be shifted with respect to the lower tier, and the radially symmetrical pattern becomes distorted in various degrees. The distortion may sometimes be due to individual variation, but there are certain groups of animals in which distortion always takes place and is the result of a specific structure of the egg.

In the annelids, molluscs, nemerteans, and some of the planarians (the Polycladida), all the blastomeres of the upper tier are shifted in the same direction in relation to the blastomeres of the lower tier, so that they come to lie not over the corresponding vegetal blastomeres, but over the junction between each two of the vegetal blastomeres (Fig. 71). This arrangement comes about not as a result of secondary shifting of the blastomeres, but because of oblique positions of the mitotic spindles, so that from the

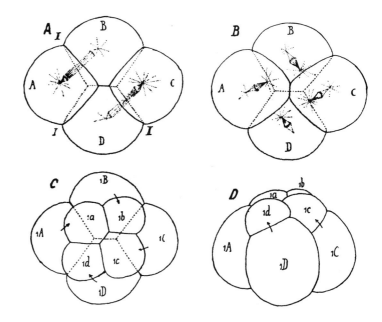

Figure 71. Spiral cleavage in the mollusc *Trochus. A*, Four-cell stage, just after second division (spindles of second division still visible); *B*, four-cell stage, but in preparation for third division (metaphase); *C*, eight-cell stage, viewed from animal pole; *D*, eight-cell stage, lateral view. (After Robert, from Korschelt, 1936.)

start the two daughter cells do not lie one above the other. The four spindles during the third cleavage are arranged in a sort of spiral. This type of cleavage is therefore called the **spiral type** of cleavage.

The turn of the spiral as seen from above may be in a clockwise direction or in a counterclockwise direction. In the first case the cleavage is called **dextral**; in the second case it is called **sinistral.** Since the cleavage planes are at right angles to the spindles, they also deviate from the horizontal position found in radial cleavage, and each cleavage plane is inclined at a certain angle. The spiral arrangement of the mitotic spindles can be traced even in the first two divisions of the egg; the spindles are oblique and not vertical as in radial cleavage. However, the resulting shifts in the position of the blastomeres are not so obvious as after the third cleavage. During the subsequent cleavages the spindles continue to be oblique, but the direction of spiraling changes in each subsequent division. Dextral spiraling alternates with sinistral, so that the spindle of each subsequent cleavage is approximately at right angles to the previous one.

Note that the type of cleavage of the egg as a whole, whether dextral or sinistral, depends on the direction of spiraling occurring during the third division of the egg.

Peculiarities of the cleavage pattern can also be introduced by differences in the size of the blastomeres. Of the four blastomeres in the four-cell stage of eggs having a spiral type of cleavage, one blastomere is often found to be larger than the other three (Fig. 72). This allows us to distinguish the individual blastomeres. The four first blastomeres are denoted by the letters A, B, C, D, the letters going in a clockwise direction (if the egg is viewed from the animal pole) and the largest blastomere being denoted by the letter D. In some animals which otherwise have an approximately radial type of cleavage, two of the first four blastomeres may be larger than the other two, thus establishing a plane of bilateral symmetry in the developing embryo. Subsequent cleavages may make the bilateral arrangement of the blastomeres still more obvious (as in tunicates and in nematodes, although in a different way). The resulting type of cleavage is referred to as the **bilateral type.** (See Figure 93, p. 129).

A very special type of cleavage showing bilateral symmetry is found in nematodes. The first division produces two unequal cells: a slightly larger cell designated as cell AB and a smaller cell P_1. The two cells divide next in mutually perpendicular planes, so that the blastomeres in the four-cell stage are placed in the form of a letter T (Fig. 73B). The transverse shaft of the T is made up of blastomeres A and B (descendants of the cell AB), and the vertical shaft is made up of the offspring of blastomere P_1. The cells are designated as EMSt (abbreviation for endoderm, mesoderm, stomodeum—which shows

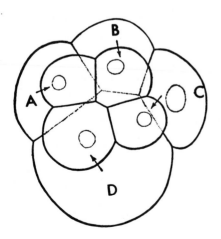

Figure 72. Spiral cleavage in the mollusc *Unio* with blastomere D distinctly larger than blastomeres A, B, and C. (After Lillie, from Kellicott, General Embryology, 1914.)

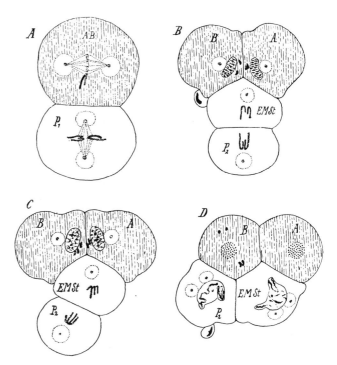

Figure 73. Cleavage of the nematode *Ascaris,* up to four-cell stage. (After Boveri, from Korschelt, 1936.)

the destiny of this cell) and as P_2. The "T" arrangement is, however, only temporary, the P_2 cell soon shifting toward the B cell. The blastomeres are then arranged in a rhomboid figure (Fig. 73D). Next, the third division enhances the bilateral symmetry of the embryo, because the blastomeres A and B each divide into a right and left daughter cell, while the other two blastomeres produce a group of four cells lying one behind the other in the median plane (Fig. 74). The blastomeres of this group are designated Mst, E, P_3, and C, respectively.

The cleavage in nematodes is also an example of **determinate cleavage** in which definite blastomeres give rise to specific parts of the embryo. Thus, blastomeres A, B, and C give rise to the skin of the animal, blastomere E gives rise to the endoderm of the alimentary tract, blastomere MSt gives rise to the mesoderm and the stomodeum, and blastomere P_3 eventually produces the reproductive cells.

From the stage of eight cells a slight asymmetry is noticeable between the right and the left halves of the embryo (Fig. 74).

The yolk, which is present in the egg at the beginning of cleavage in greater or lesser quantities, exerts a very far-reaching effect on the process of cleavage. Every mitosis involves movements of the cell components—the chromosomes, parts of the cytoplasm constituting the achromatic figure, the mitochondria, and the surface layer of the cell— the activity of which along the equator of the maternal cell leads to the eventual separation of the daughter cells. During these movements, the yolk granules or platelets behave entirely passively and are passively distributed among the daughter blastomeres. When the yolk granules or platelets become very abundant, they tend to retard and even to inhibit the process of cleavage. As a result, the blastomeres which are richer in yolk tend to divide at a slower rate and consequently remain larger than those which have less yolk. The yolk in the uncleaved egg is more concentrated toward the vegetal pole of the egg. It is therefore at the vegetal pole of the egg that cleavage is most retarded by the presence of yolk, and where the blastomeres are of the largest size.

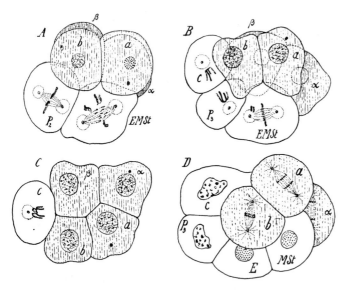

Figure 74. Cleavage of the nematode *Ascaris*, up to eight-cell stage. (After Boveri and zur Strassen, from Korschelt, 1936.)

A good example of the effect of the yolk on cleavage is provided by the frog's egg (Fig. 75). The yolk's influence may be detected even during the first division of the fertilized egg. During the anaphase of the mitotic division, a furrow appears on the surface of the egg which is to separate the two daughter blastomeres from each other. This furrow, however, does not appear simultaneously all around the circumference of the egg, but at first only at the animal pole of the egg, where there is less yolk. (It has been indicated that the first cleavage plane is vertical and therefore passes through the animal and vegetal poles of the egg.) Only gradually is the cleavage furrow prolonged along the meridians of the egg, until, cutting through the mass of yolk-laden cytoplasm, it eventually reaches the vegetal pole and thus completes the separation of the first two blastomeres. The same process is repeated during the second cleavage. During the third cleavage, when the plane of separation of the daughter blastomeres is horizontal, the

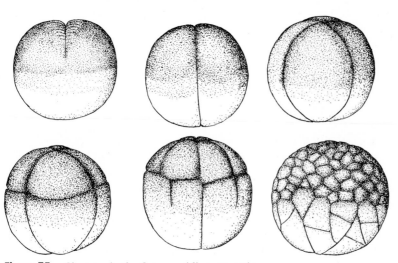

Figure 75. Cleavage in the frog, semidiagrammatic.

furrow appears simultaneously over the whole circumference of the egg, for it meets everywhere with an equal resistance from yolk.

A further accumulation of the yolk at the vegetal pole of the egg causes still greater delay in the cell fission at this pole, so that the cleavage becomes inhibited more and more. This can be clearly traced in a series of various ganoid fishes, whose eggs possess an increasing amount of yolk. In *Acipenser* the cleavage is complete, as in the amphibians, but the difference between the micromeres of the animal hemisphere and the macromeres of the vegetal hemisphere is much greater than in amphibians. In *Amia* (Fig. 76) cleavage starts at the animal pole, and the cleavage furrows reach the vegetal pole, but they are so retarded that subsequent divisions begin at the animal pole before the preceding furrows cut through the yolk at the vegetal pole. In *Lepidosteus* the cleavage starts at the animal pole as in *Amia*, but the cleavage furrows never reach the vegetal pole, so that the vegetal hemisphere of the egg remains uncleaved, resulting in what is called **incomplete cleavage.** The eggs which are completely divided into blastomeres are called **holoblastic;** those with incomplete cleavage are known as **meroblastic.** As a result of incomplete cleavage the egg is divided into a number of separate blastomeres and a residue, which is an undivided mass of cytoplasm with numerous nuclei scattered in it.

In eggs in which the yolk is segregated from the active cytoplasm (elasmobranchs, bony fishes, birds, and reptiles), the cleavage, right from the start, is distinctly recognizable as meroblastic or incomplete. At first, all the cleavage planes are vertical, and all the blastomeres lie in one plane only (Fig. 77). The cleavage furrows separate the daughter blastomeres from each other but not from the yolk, so that the central blastomeres are continuous with the yolk at their lower ends, and the blastomeres lying on the circumference are, in addition, continuous with the uncleaved cytoplasm at their outer edges. As the nuclei at the edge divide, more and more cells become cut off to join the ones lying in the center, but the new blastomeres are also in continuity with the uncleaved yolk underneath (Fig. 78).

In a later stage of cleavage, the blastomeres of the central area become separated from the underlying yolk in one of two ways: either slits may appear beneath the nucleated parts of the cells, or else the cell divisions may occur with horizontal (tangential) planes of fission. In the latter case one of the daughter cells, the upper one, becomes completely separated from its neighbors, while the lower blastomere retains the connection with the yolk mass. The marginal cells, which remain continuous with one another around their outer edges, are also continuous with the mass of yolk and hence with the lower cells resulting from tangential divisions (Figs. 78 and 79). All these blastomeres eventually lose even those furrows which partially separated them from one another and fuse into a continuous syncytium with numerous nuclei but no indication of individual cells.

The cytoplasmic cap on the animal pole of the egg is now, in its central part, subdi-

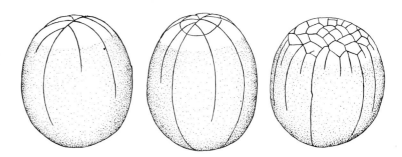

Figure 76. Cleavage in the ganoid fish *Amia*. (After Whitman and Eyklesheimer, from Korschelt, 1936.)

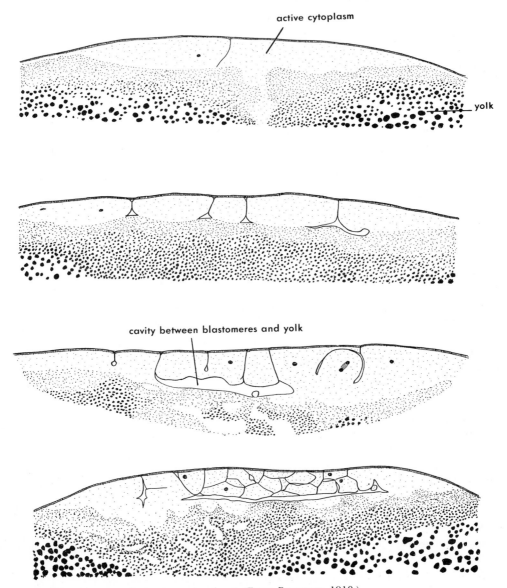

Figure 77. Cleavage of a hen's egg (sections). (From Patterson, 1910.)

vided into a number of "free" blastomeres, while around the outer margin of the mass of blastomeres and underneath, closely adhering to the yolk, is the syncytial layer which is called the **periblast.** The periblast is not destined to participate directly in the formation of the embryonic body, but it is supposed to be of some importance in breaking down the yolk and making it available for the growing embryo (Fig. 80).

Another type of incomplete cleavage is found in centrolecithal eggs. At the beginning of cleavage the nucleus (synkaryon) lies in the center of the egg surrounded by a small amount of cytoplasm. The mitotic division of the nucleus starts in this position, but at first it is not followed by a division of cytoplasm (Fig. 81). As a result, a number of nuclei are formed, all embedded in the undivided central mass of cytoplasm. After several divisions have taken place (the actual number of divisions varies in different animals), the nuclei start moving away from the center of the egg. Each nucleus, sur-

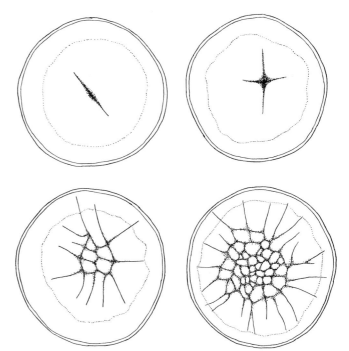

Figure 78. Cleavage of a hen's egg, surface view. (From Patterson, 1910.)

rounded by a small portion of the original central cytoplasm, travels outward toward the surface of the egg. The central mass of cytoplasm is thus broken up and disappears. When the nuclei reach the surface of the egg, the cytoplasm surrounding them fuses with the superficial layer of cytoplasm. The surface layer of the embryo is then a

Figure 79. Discoidal cleavage in the cephalopod mollusc, *Sepiola rondeletii.* The micromeres at the animal pole lie loosely at this stage and do not yet form a continuous layer. The large cells at the edges of the blastodisc are not completely separated from one another or from the uncleaved yolk. (Courtesy of Dr. G. de Leo, 1972.)

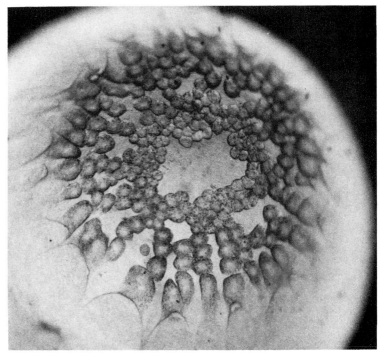

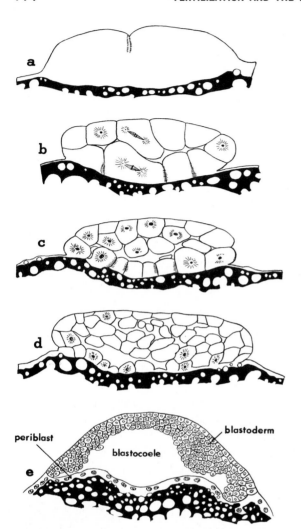

Figure 80.   Cleavage and blastula of bony fish. *a, b, c, d,* Trout (after Kopsch); *e, Muraena* (after Bocke). (From Brachet, 1935.)

syncytium with numerous nuclei embedded in an undivided layer of cytoplasm. In the next stage the cytoplasm becomes subdivided, by furrows going inward from the surface, into as many sections as there are nuclei. The sections can now be called cells, even though they are at first still connected to the yolk mass. Sooner or later the cells become completely separated from the yolk. The latter persists as a compact mass until it is gradually used up as a food reserve by the developing embryo. This type of cleavage, called **superficial cleavage,** is the usual type in insects and many other arthropods.

6–4 MORULA AND BLASTULA

The blastomeres in the early cleavage stages tend to assume a spherical shape like that of the egg before cleavage. Their mutual pressure flattens the surfaces of the blastomeres in contact with one another, but the free surfaces of each blastomere remain spherical, unless these outer surfaces are also compressed by the vitelline membrane. The whole embryo acquires, in this stage, a characteristic appearance resembling a

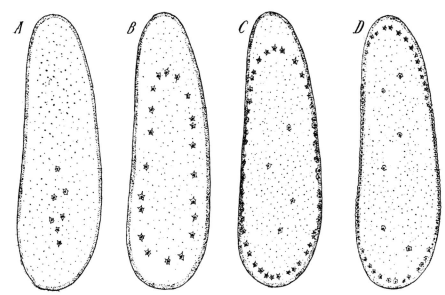

Figure 81. Superficial cleavage of a beetle (*Hydrophilus*). (After Heider, from Korschelt, 1936).

mulberry. (See Figure 70*E*, p. 107.) Because of this superficial similiarity, the embryo in this stage is called a morula (Latin for mulberry).

The arrangement of the blastomeres in the morula stage may vary in the different groups of the animal kingdom. In coelenterates it is often a massive structure, with blastomeres filling all the space that had been occupied by the uncleaved egg. Some of the blastomeres then lie externally and others in the interior. (Some embryologists apply the name morula to this type of embryo only.) More often, as the egg undergoes cleavage, the blastomeres become arranged in one layer, so that all the blastomeres participate in the external surface of the embryo. In this case a cavity soon appears which at first may be represented just by narrow crevices between the blastomeres, but which gradually increases as the cleavage goes on. This cavity is called the **blastocoele.**

As cleavage proceeds, the adhesion of the blastomeres to one another increases, and they arrange themselves into a true epithelium. In cases in which a cavity has been forming in the interior of the embryo, the epithelial layer completely encloses this cavity, and the embryo becomes a hollow sphere, the walls of which consist of an epithelial layer of cells. Such as embryo is called a **blastula.** The layer of cells is called the **blastoderm,** and the cavity is the blastocoele, as already mentioned (Fig. 82*a*).

In oligolecithal eggs with complete cleavage (echinoderms, *Amphioxus*), the blastomeres at the end of cleavage are not of exactly equal size, the blastomeres near the vegetal pole being slightly larger than those on the animal pole. When the blastula is formed, the cells arrange themselves into a simple columnar epithelium enclosing the blastocoele. Because the vegetal cells are larger than the animal cells, the blastoderm is not of an equal thickness throughout; at the vegetal pole the epithelium is thicker, and at the animal pole it is thinner. Thus, the polarity of the egg persists in the polarity of the blastula.

Animals with a larger amount of yolk, such as the frog, show a difference in the size of the cells of the blastula that may be very considerable, and the blastula still further departs from the simple form of a hollow sphere. The cells here are also arranged in a

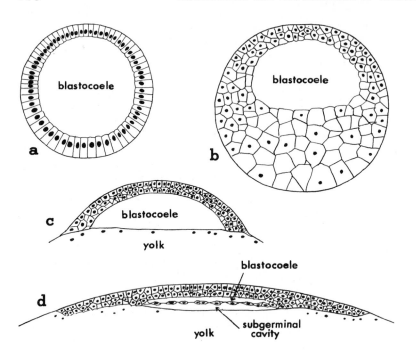

Figure 82. Diagrammatic comparison of blastulae of an echinoderm (*a*), a frog (*b*), a bony fish (*c*), and a bird (*d*).

layer surrounding the cavity in the interior, but the layer is of very uneven thickness. The layer of cells at the vegetal pole is very much thicker than at the animal pole, and the blastocoele is consequently distinctly eccentric, nearer to the animal pole of the embryo (Fig. 82*b*). Furthermore, the blastoderm is no longer a simple columnar epithelium but is two or more cells thick. The cells in the interior are rather loosely connected to one another, but at the external surface of the blastula the cells adhere to one another very firmly, because of the presence of a cementing substance joining the surfaces of adjoining cells in a narrow zone just underneath the surface of the blastoderm. (See Figure 180, p. 237.)

A process corresponding to blastula formation occurs also in animals whose eggs have incomplete cleavage. In a bony fish or a shark the early blastomeres tend to round themselves off, showing that they are only loosely bound together. Later, the blastomeres adhere to one another more firmly and thus become converted into an epithelium. Again, as in amphibians, the superficial cells are firmly joined to one another, while the cells in the interior may remain loosely connected until a later stage. The epithelium, however, cannot have the form of a sphere. Since the cleavage is restricted to the cap of cytoplasm on the animal pole of the egg, the blastoderm is developed only in the same polar region. The blastoderm therefore assumes the shape of a disc lying on the animal pole. The disc, which is called the **blastodisc,** is more or less convex and encloses, between itself and the uncleaved residue of the egg, a cavity representing the blastocoele (Fig. 82*c*).

The earlier stages of the discoidal cleavage in the eggs of reptiles and birds resemble essentially the cleavage in meroblastic eggs of the fishes. At the time when the blastoderm is being formed, however, an essential difference becomes apparent. The cavity underneath the layer of blastomeres does not extend through the whole blastoderm in reptiles and birds, as it does in fishes, but is restricted to the central part of the blastoderm. No cavity is formed under the layer of blastomeres in the periphery of the blastodisc. The blastodisc is thus subdivided into two parts: the central part, called the **area**

pellucida, and the peripheral part, the **area opaca.** The names indicate that the central part appears more transparent in the living embryo, owing to the presence of the cavity under the layer of blastomeres; the peripheral area is opaque because the blastomeres rest directly on the yolk.

Only the area pellucida furnishes material for the formation of the body of the embryo. The cells of the area opaca are concerned with the breakdown of the underlying yolk and thus indirectly supply the embryo with foodstuffs.

In tortoises the blastoderm remains a single layer, but in most reptiles, especially lizards and snakes, and also in birds, a thin second layer of cells appears rather early in development in the space between the thicker surface layer and the uncleaved yolk underneath (Figs. 82 and 83 *C, D*). The presence of this lower layer of cells has caused much controversy among embryologists as to its origin, significance, and homologies. The two layers certainly do not represent ectoderm and endoderm respectively, as will be shown in Chapter 7 on gastrulation. It has been found advisable, therefore, to refer to the two layers by the terms **epiblast** (for the upper layer of cells) and **hypoblast** (for the lower layer of cells). In fact, the upper layer gives rise to all three germinal layers of the embryo (the ectoderm, mesoderm, and endoderm). (See Chapter 7.) It has therefore been concluded (Pasteels, 1945, 1957) that the two-layered embryo of birds and reptiles is still in the blastula stage and that the lower layer, the hypoblast, corresponds to the cells lying on the floor of the blastocoele cavity in animals such as amphibians. The cavity between the epiblast and the hypoblast of birds and reptiles thus corresponds to the blastocoele of amphibians and fishes (Fig. 82*d*). The space between the hypoblast and the yolk may then be distinguished as the **subgerminal cavity.**

The origin of the cells constituting the hypoblast is very difficult to ascertain by experimental methods, since the eggs, being fertilized internally, start cleaving and may reach the blastula stage while they are still in the oviducts. In the hen, which is the most convenient animal for experimental research into bird development, the blastodisc is already two-layered at the time the egg is laid. This is not the case in certain other birds, such as the dove, but no work on hypoblast development has been done on dove eggs.

It is a recognized fact that the hypoblast first appears in the posterior part of the area pellucida and extends forward in later stages. The most plausible interpretation of hypoblast development is that its cells originate from the lower levels of loosely arranged cleavage cells near the posterior border of the area pellucida and area opaca, and that from this position the cells move forward underneath the epiblast, at the same time joining to form a continuous layer of flattened cells. When a hen's egg is laid, the hypoblast is usually not complete anteriorly, and its development may be studied by experimental methods. It has been found, beyond doubt, that the sheet of hypoblast cells moves in an anterior direction (Spratt and Haas, 1960; Vakaet, 1962).

In centrolecithal eggs having a superficial cleavage (insects), there is no cavity comparable to the blastocoele. Nevertheless, the formation of the epithelium on the surface of the egg, after the nuclei have migrated to the exterior, can be compared to the formation of the blastula. The layer of cells thus formed on the surface of the embryo is the blastoderm. Instead of surrounding a cavity, the blastoderm envelops the mass of uncleaved yolk. We may also compare this stage to an embryo whose blastocoele has been filled with yolk.

Up to the blastula stage, the developing embryo preserves the same general shape as the uncleaved egg. So far, the results achieved are the subdivision of the single cell into a multiplicity of cells and the formation of the blastocoele. In addition, the substances contained in the egg remain basically in the same position as before. The yolk remains near the vegetal pole. In pigmented eggs, such as those of amphibians, the pigment remains as before, more or less restricted to the upper hemisphere of the egg.

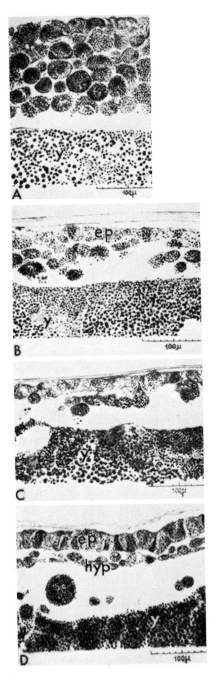

Figure 83.   Formation of the epiblast and hypoblast in the duck embryo. *A*, Late cleavage stage, all cells in one undifferentiated layer. *B, C*, Intermediate stage, loose hypoblast cells underneath the epiblast, already in the form of an epithelial layer. *D*, Epiblast in the form of a columnar epithelium, and hypoblast cells joined to form a continuous layer. There are a few loose cells above and below the hypoblast. ep, Epiblast; hyp, hypoblast; y, uncleaved yolk. (From Pasteels, 1945.)

Only a slight intermingling of cytoplasm seems to be produced by the cleavage furrows cutting through the substance of the egg.

We have pointed out that during cleavage qualitative changes in the chemical composition of the developing embryo are very limited. Few new substances, either chemically defined or microscopically detectable, have been found to appear during cleavage. It is conceivable, however, that the substances present in the egg may be redistributed in some way during cleavage and that such a redistribution may be essentially important for further development.

In this connection we will first examine whether the numerous nuclei produced during the mitotic divisions of the egg are all alike in their properties, or whether any differences may be discovered among them.

6-5 THE NUCLEI OF CLEAVAGE CELLS

In his "germ plasm theory," A. Weismann presented a hypothesis to explain both heredity and the ontogenetic development of organisms. According to Weismann (1904), every distinct part of an organism (animal or plant) is represented in the sex cell by a separate particle: a **determinant.** Thus, the sum total of determinants would represent the parts of the adult organism with all their peculiarities. The complete set of determinants is supposedly handed down from generation to generation, which would account for hereditary transmission of characters. The determinants, according to Weismann, are localized in the chromosomes of the nucleus, just as the genes of modern genetics are. However, there is a difference: the genes are not supposed to represent *parts* of the organism but rather properties which may sometimes be discernible in all the parts of the body.

During the cleavage of the egg, the various determinants, according to Weismann, become segregated into different cells. The blastomeres would receive only some of the determinants, namely, those which correspond to the fate of each blastomere. The successive segregation of the determinants would eventually result in each cell's having determinants of only one kind, and then nothing would be left to the cell but to differentiate in a specific way in accordance with the determinants present. Only the cells having the sex cells among their descendants, Weismann held, would preserve the complete set of determinants, since these would be necessary for directing the development of the next generation.

Even though Weismann's conception of the properties of determinants is not tenable from the genetic viewpoint, it is still important to know whether the difference in the fate of the blastomeres and parts of the embryo may be attributed to differences in the nuclei of the cleavage cells. Experiments now yield convincing information on this point.

Our knowledge of the development of complete embryos from one of the two daughter blastomeres of the egg (as a result of either separating the first two blastomeres or killing one of the first two blastomeres—see Section 1–3) contradicts Weismann's hypothesis about the segregation of determinants during cleavage. The evidence becomes still more conclusive because it has been found that not only are the first two blastomeres capable of developing into complete embryos, but the blastomeres in later cleavage stages sometimes possess the same ability. In the case of sea urchins, one of the first four blastomeres, and even occasionally one of the first eight, may develop into a whole embryo with all the normal parts but reduced in size. (See Figure 84—Hörstadius and Wolsky, 1936.)

The method of isolating blastomeres, however, does not permit one to test the properties of nuclei of later generations of cells. In the four-cell stage, the quantity of cytoplasm contained in one blastomere may already be too small for development to take place in an approximately normal fashion. This is true in increasing degree as cleavage proceeds and the individual blastomeres become smaller and smaller.

To further this investigation, a different method has been devised. A most elegant experiment in this field was carried out by Spemann (1928). Spemann constricted

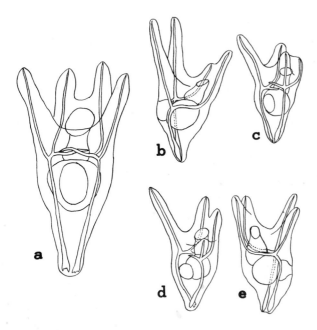

Figure 84. Results of separation of the first four blastomeres of a sea urchin's egg. *a,* Normal pluteus. *b, c, d,* and *e,* Plutei of normal structure but diminished size; each developed from one of the blastomeres of the four-cell stage. All drawn to the same scale. (Front Hörstadius and Wolsky, 1936.)

fertilized eggs of the newt *Triturus (Triton)* into two halves with a fine hair, just as they were about to begin to cleave. The constriction was not carried out completely, so that the two halves were still connected to each other by a narrow bridge of cytoplasm. The nucleus of the fertilized egg lay in one half, and the other half consisted of cytoplasm only. When the egg nucleus began to divide, the cleavage was at first restricted to that half of the egg which contained the nucleus (Fig. 85). This half divided into two, four, eight cells, and so on, while the non-nucleated half remained uncleaved. At about the stage of 16 blastomeres, one of the daughter nuclei, now much smaller than at the beginning of cleavage, passed through the cytoplasmic bridge into the half of the egg which had hitherto no nucleus (Fig. 85c). Forthwith this half also began to cleave. After both halves of the egg were thus supplied with nuclei, Spemann drew the hair loop tighter and completely separated the two halves of the egg from each other. They were then allowed to develop into embryos. In a number of cases two completely normal embryos developed from the two halves of the egg as a result of this experiment.

Of the two embryos, each started by having one half of the egg cytoplasm, but as to the nucleus they were in very different situations. While one of the embryos possessed 15/16 or even 31/32 of all the nuclear material of the egg, the other received only 1/16 or 1/32 of the nuclear material. The experiment proves conclusively that even in the 32-cell stage every nucleus has a complete set of hereditary factors necessary for the achievement of normal development. All the nuclei in this stage are completely equivalent to one another and to the nucleus of the fertilized egg. The hypothesis of an unequal division of the hereditary substance of the nucleus, of the segregation of determinants or genes to the different cleavage cells, is thus disproved. It is now assumed that every cell of the metazoan body has a complete set of nuclear factors necessary for development (a complete set of genes, in the terminology of modern genetics).

The experiments on the delayed nuclear supply to one half of an amphibian egg have been corroborated by experiments on the eggs of other animals. It will be useful to relate a corresponding experiment carried out on a very different kind of animal, the dragonfly *Platycnemis pennipes* (Seidel, 1932). In the dragonfly, cleavage is incomplete,

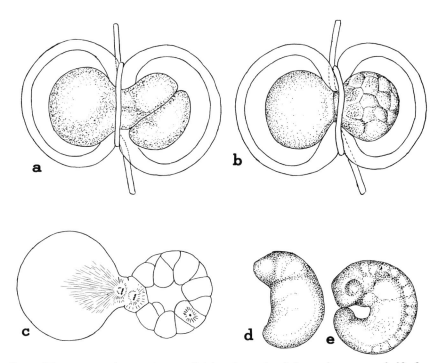

Figure 85. Spemann's experiment of delayed supply of the nucleus to one half of a newt's egg. *a*, Beginning of cleavage in nucleated half. *b*, One of the nuclei has penetrated into the uncleaved half, and a cell boundary appears between this and the half which originally contained the zygote nucleus. *c*, Same stage in section. *d*, Embryo from the half with delayed nuclear supply. *e*, Embryo from the half which contained the zygote nucleus. (From Spemann, 1938.)

and only the nucleus divides at first, the cytoplasm remaining uncleaved. The egg is elongated, and after the first division the daughter nuclei move, one into the anterior half and the other into the posterior half of the egg. When eight nuclei are available, they are spaced along the length of the egg. By further divisions nuclei are provided for all cells in the respective regions (Fig. 86*a–d*). In the stage when two nuclei are present, either of them may be killed by a short exposure to a narrow beam of ultraviolet light, which does not damage the cytoplasm to any great extent (Fig. 86*e*). The remaining nucleus continues to divide, and its daughter nuclei are distributed to all parts of the egg instead of supplying only one half of it. Completely normal embryos develop, no matter which of the two nuclei is allowed to survive (Fig. 86*f*). The two nuclei prove to be completely equivalent for development, although normally they would have supplied different parts of the embryo.

The methods used in the preceding experiments for testing the properties of the cleavage nuclei are of necessity confined to the earlier stages of cleavage. At present, a more universal method is available which allows the investigation to be extended to nuclei of cells of much more advanced embryos, and possibly it may ultimately be applied even to fully differentiated cells of an adult organism. This is the method used for transplantation of nuclei. The transplantation of nuclei from one cell to another was first carried out successfully on *Amoeba,* and the method was then applied to test the properties of nuclei in developing frog embryos (Briggs and King, 1952, 1953, 1957; King and Briggs, 1954, 1956). The method, as applied to the frog embryo, consists essentially in taking the nucleus of any cell from a developing embryo and injecting it into an enucleated uncleaved egg.

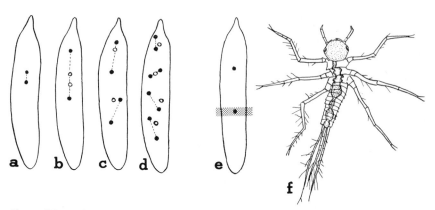

Figure 86. *a, b, c,* and *d,* Early cleavage in the dragonfly, *Platycnemis pennipes.* Hollow circles indicate the previous positions of the nuclei in each case. *e,* Killing one of the first two cleavage nuclei by ultraviolet irradiation. Irradiated area is stippled. *f,* Larva which developed from an egg treated as in *e.* (Adapted from Seidel, 1932.)

The egg receiving the nucleus must be specially prepared. The ripe eggs are removed from the oviducts and activated by pricking with a glass needle. (See Section 5 – 4.) The egg nucleus then approaches the surface of the cytoplasm in preparation for the second maturation division and is removed by a second prick with a glass needle at the exact spot where the nucleus is located. Next, a cell of an advanced embryo is separated from its neighbors and sucked into the tube of an injection pipette. The diameter of the pipette is smaller than that of the cell and as a result the surface of the cell is broken. The contents of the pipette, consisting of the nucleus and the debris of the cytoplasm, are injected deep into the enucleated egg. When the pipette is withdrawn, the egg cytoplasm tends to escape through the hole in the egg membranes, forming an extraovate protrusion. This must be cut off by a pair of glass needles to prevent further loss of egg substance (Fig. 87).

As a result of these procedures, up to 80 per cent (King and Briggs, 1956) of the eggs operated on start cleaving and producing numerous cells, the nuclei of which are derived from the injected nucleus and the cytoplasm of which is from the enucleated egg (the small amount of cytoplasm injected with the nucleus, comprising less than 1/40,000 of the volume of the egg cytoplasm, may be ignored). Not all eggs which start cleaving develop normally later, but at least a small proportion do and may proceed through all the stages of embryonic and postembryonic development up to metamorphosis (Fig. 88). In the late blastula, the stage used for some of these nuclear transplantations, there are 8000 to 16,000 cells, which means that about 13 to 14 divisions (or generations) of the original nucleus of the fertilized egg had been performed without diminishing the power of the nucleus to support every type of differentiation provided for by the specific genotype. Nuclei of cells of an early gastrula show the same properties.

In further experiments the potentialities of nuclei of cells, which are even more advanced in the process of differentiation, were tested by transplanting them into enucleated eggs. Normal tadpoles developing up to and through metamorphosis were obtained by using nuclei from the neural plate of a frog embryo (DiBerardino and King, 1967) or nuclei from already ciliated cells of the alimentary tract of a swimming tadpole (Gurdon, 1968). Although the cells of the neural plate (p. 172) were already well on the way to becoming cells of the nervous system, and the cells of the gut were already functionally differentiated, their nuclei were still capable of providing the neces-

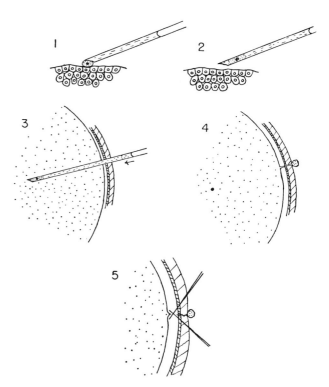

Figure 87. Diagram illustrating method of transplanting cell nuclei into enucleated eggs. (From Briggs and King, 1953.)

sary genetic information for the diffentiation of all the various tissues and cell types of an adult frog. Fairly normal embryos were also produced by implanting into enucleated eggs the nuclei from a frog adenocarcinoma (King and DiBerardino, 1965).

Finally, a modification of the original methods has made it possible to test the potentialities of nuclei of adult animals by transplanting them into enucleated eggs. When nuclei are taken directly from adult tissues and transplanted into eggs, they are not able to support development. If adult cells are cultured *in vitro*, however, where they lose part of the properties of differentiated cells and start reproducing mitotically, and if their nuclei are then transplanted into eggs, development may be initiated in a fair proportion of cases. Even so, the embryos develop abnormally, producing blastulae with mostly partial cleavage. If nuclei of cells from the more healthy parts of the partial embryos are then removed and transplanted into eggs, the development of such sec-

Figure 88. Metamorphosing tadpole developed from an egg with transplanted nucleus. (From King and Briggs, 1954.)

ond generation embryos proceeds much better; swimming larvae and occasionally even larvae which become metamorphosed into froglets develop. The tissues used successfully as sources of nuclei are kidney, lung, and skin of adult frogs. It is noteworthy that no differences in the development of tadpoles could be noted when the nuclei were taken from these three different tissues (Laskey and Gurdon, 1970).

It was noted that with progression of development of the cells from which nuclei were taken the proportion of experiments leading to completely normal tadpoles became increasingly reduced, and more and more operated eggs were arrested in their development in various early stages. On careful investigation it was found that the chromosomes in the nuclei of arrested embryos showed various defects, such as aneuploidy (chromosomes missing) or defects within chromosomes (deletions, translocations) (DiBerardino and King, 1967). These defects are due to the inability of the chromosomes in nuclei taken from advanced embryos and adult tissues to adapt themselves to the rapid reproductive rhythm of early development. The duplication of the interphase chromosomes is too slow, and many of them do not complete duplication by the time cell division sets in. The result is, of course, severe damage to and incompleteness of the chromosome set (Gurdon, 1968). Whatever the explanation for the defective development, it remains that at least a proportion of nuclei from cells well on the way to differentiation and from those that have become malignant retain their full potentialities for controlling and directing normal development.

The preceding discussion is not meant to imply that nuclei of differentiating cells are in every respect similar to the nuclei of fertilized eggs. In experiments with frogs, nuclei transplanted into the egg cytoplasm undergo a change which makes them resemble early embryonic nuclei. The nuclei of newly fertilized eggs, although small in proportion to the egg cytoplasm, are much larger than the nuclei at later stages. Accordingly, transplanted nuclei increase up to 30-fold in volume during the first 40 minutes after transplantation. The functioning of the nuclei also changes drastically. Nuclei of advanced embryos do not divide rapidly, and accordingly the synthesis of DNA, necessary for the replication of chromosomes, is slow. On the other hand, they synthesize large quantities of ribosomal RNA, and, as this kind of RNA is produced in the nucleolus, they have prominent nucleoli. Nuclei of cleavage cells synthesize DNA rapidly but do not synthesize any ribosomal RNA and therefore do not show any nucleoli. Transplanted nuclei cease to synthesize ribosomal RNA, and their nucleoli disappear. Instead, they start to synthesize DNA rapidly as do normal cleavage nuclei. By the time the embryo reaches the stage of gastrulation, DNA synthesis in the descendants of the transplanted nucleus slows down, and ribosomal RNA synthesis is renewed; nucleoli reappear. Thus, a transplanted nucleus reverts in every respect to the condition in which nuclei would be after normal fertilization. (Gurdon and Brown, 1965; a review of this line of research may be found in Gurdon and Graham, 1967, and in Gurdon, 1968b.)

Transplantation of nuclei into eggs has also been performed in insects. In earlier experiments nuclei from late cleavage embryos of *Drosophila* (blastoderms with 200 to 2000 nuclei) were transplanted into unfertilized eggs, with the result that the embryo developed to the hatching larva stage (Ilmensee, 1968). With improvement of techniques, it was possible to produce larvae that developed to late larval stages, and even to pupation in one case, using single nuclei from different body regions of the blastula and gastrula embryos. Lack of complete development is believed to be the result of damage to the egg during the operation. More advanced development at the expense of transplanted nuclei was achieved by taking parts of the defective embryos and cultivating them *in vivo*, first in the bodies of adult flies, and then after a second transplantation, in premetamorphic larvae. The grafts metamorphosed together with their hosts and developed into a variety of adult structures (Ilmensee, 1972, 1973) (Fig. 89). It is evident from these experiments that in an insect, as well as in a frog, nuclei well advanced in development possess the full potentialities for producing all parts of the adult body.

6–6 DISTRIBUTION OF CYTOPLASMIC SUBSTANCES IN THE EGG DURING CLEAVAGE

The fact that the nuclei of the developing embryo are fully equivalent to one another does not allow us to ascribe to the nucleus the origin of differences in the fate of the cells during development of the embryo. What then is the cause of the observed differences in the developmental behavior of cells? By exclusion, we may suppose that the cause of the differences lies in the cytoplasm and that the cells produced during cleavage do not receive the same cytoplasmic substances. There is direct experimental evidence in favor of this supposition.

Curiously enough, a further elaboration of Spemann's experiment of delayed nuclear supply to one half of the *Triturus* egg can be brought to bear on this problem (Spemann, 1928).

It has already been mentioned that pigment is unequally distributed over the surface of amphibian eggs (Sections 3–3 and 5–6). The animal hemisphere of the egg is more or less heavily pigmented; the region around the vegetal pole is poor in pigment or completely white; the equatorial, or rather subequatorial, marginal zone is more lightly pigmented than the animal hemisphere but more darkly pigmented than the vegetal region. On one side of the egg the marginal zone is broader than on the other, and this region is known as the **gray crescent** (from its color in the common European

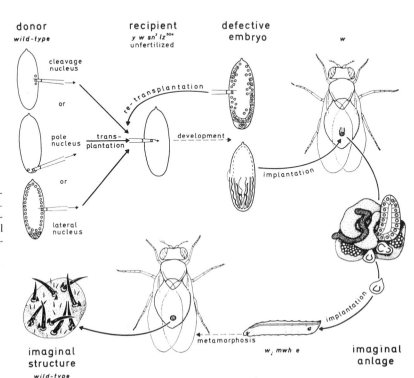

Figure 89. Nuclear transplantation, *in vivo* culture and development of adult structures with nuclei derived from the original graft in *Drosophila* (diagrammatic). (From Illmensee, 1972.)

frogs). The gray crescent indicates the future position of the dorsal lip of the blastopore and also corresponds to the **dorsal** region of the embryo (Section 7–4).

In the egg of the newt the pigment is not very abundant, but the gray crescent is readily distinguishable in the uncleaved egg. It is therefore possible to determine the position of the ligature with respect to the gray crescent (Fig. 90). If the ligature subdivides the gray crescent cytoplasm equally between the two halves of the egg, two more or less normally developed embryos are found to develop (Fig. 90b, e). If the ligature comes to lie so that all the gray crescent cytoplasm is contained in one half of the egg, whereas the other half does not get any of it, then a complete embryo is developed only from the first half. The second half remains highly abnormal; it does not develop a nervous system, a notochord, or segmented muscle, and the only parts that can be discerned are epidermis, unsegmented (lateral plate) mesoderm, and yolk endoderm. All these are parts of the ventral region of the body, thus justifying the name "belly" (Bauchstück) given by Spemann to the defective embryos (Fig. 90a, c, d).

Now it is found that the fate of the egg half, whether it develops into a complete embryo or into a "belly," does not depend on which of the two halves had retained the egg nucleus originally and which was delayed in receiving a nucleus. A half with a delayed nuclear supply will develop into a complete embryo if it contains the gray crescent cytoplasm. On the other hand, the egg half without gray crescent cytoplasm will become a "belly" even if it contained the egg nucleus right from the beginning. The difference in the time of nuclear supply and the amount of nuclear material received is completely overridden by differences in the cytoplasmic composition of the two egg halves.

The importance of cytoplasmic substances in the egg for the differentiation of parts of the embryo is further proved by a number of experiments carried out on the eggs of various animals. The following experiments will be chosen as examples.

In the egg of the mollusc *Dentalium* (Wilson, 1904), there may be distinguished three layers of cytoplasm: a layer of clear cytoplasm at the animal pole of the egg, a broad layer of granular cytoplasm in the equatorial zone (actually forming the bulk of the egg substance), and a second layer of clear cytoplasm at the vegetal pole of the egg (Figs. 41 [p. 56] and 91). When the egg begins to cleave, the clear cytoplasm at the veg-

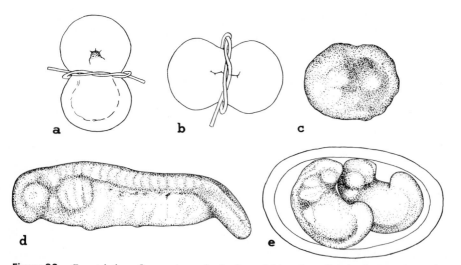

Figure 90.   Constriction of a newt's egg in the frontal (*a*) and medial (*b*) planes. After frontal constriction the ventral half developed into a "belly" (*c*), the dorsal half into a complete embryo (*d*). After medial constriction (*b*) both halves developed into complete embryos (*e*). (From Spemann, 1938.)

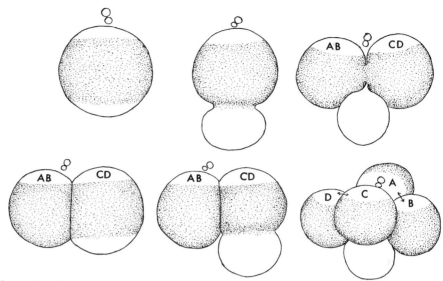

Figure 91. Cleavage of the mollusc *Dentalium*. (From Wilson, 1904.)

etal pole is pushed out in the form of a **polar lobe.** As the division of the egg draws to its conclusion, the polar lobe is rounded off and remains connected to the rest of the egg only by a narrow cytoplasmic bridge. When the division is completed, the polar lobe is found to be connected to one of the daughter cells, namely, cell CD (Fig. 91) (the cleavage of *Dentalium* is spiral), and at the end of mitosis it is fused with this cell. In this way cell CD gets all three types of cytoplasm, while the other daughter cell, cell AB, gets half of the animal pole's clear cytoplasm, half of the granular cytoplasm, but no vegetal cytoplasm. Cell CD is consequently slightly larger than cell AB.

The formation of the polar lobe is repeated when the second cleavage begins, that is, when the egg divides into four blastomeres. Of the two daughter cells of blastomere CD, the polar lobe passes to blastomere D, while blastomere C receives only the clear cytoplasm of the animal pole and the granular cytoplasm.

The two blastomeres of the two-cell stage and the four blastomeres of the four-cell stage are therefore not equivalent as to their cytoplasmic composition, for only one blastomere (CD or D, respectively) contains all the vegetal polar lobe material. The behavior of the blastomeres is also found to be different. If the first two blastomeres are separated and allowed to develop, blastomere CD is found to produce a complete larva (a **trochophore** in this case), though of diminished size. Blastomere AB also develops into a larva, but it is defective in some respects, the most significant defect being that it completely lacks the rudiment of the mesoderm normally contained in the posterior or "post-trochal" part of the larva (Fig. 92).

If the blastomeres are isolated in the four-cell stage, they are again found to differ in their development. Only blastomere D, which possesses the clear cytoplasm of the vegetal pole, develops into a complete, though smaller, trochophore larva; the other three blastomeres develop into defective larvae lacking the mesoderm rudiment. It has been concluded, therefore, that the clear cytoplasm of the vegetal pole is necessary for the development of the mesoderm.

This conclusion is further supported by the following experiment: At the time of the first or second cleavage the polar lobe containing the clear cytoplasm of the vegetal pole can be nipped off. The remainder then contains all the clear cytoplasm of the animal pole and the granular cytoplasm. It also contains all the nuclear material but no

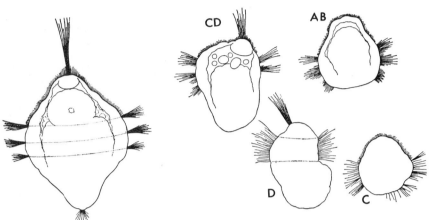

Figure 92. Larvae of *Dentalium* developing from a whole egg (left) and from separated blastomeres (right). The letters indicate from which blastomere each larva has developed (From Wilson, 1904.)

vegetal polar lobe material. The result is that the larva developed from such an egg is defective and lacks the mesoderm rudiment. This is a very clear proof that it is the cytoplasm and not the nucleus that makes blastomere D different from the other three blastomeres and able to produce a certain differentiation (the mesoderm rudiment) which the others are incapable of producing.

In the egg of the ascidian *Styela partita*, no less than four cytoplasmic substances may be distinguished because of differences in the color of these substances (Conklin, 1905). In the egg at the beginning of cleavage, the animal hemisphere consists of clear, transparent cytoplasm. The vegetal pole of the egg is distinguished by a slaty gray cytoplasm, rich in yolk. In between, slightly below the equator, two crescentic areas may be discerned on opposite sides of the egg. One crescentic area consists of light gray cytoplasm, and the other consists of yellow cytoplasm, the yellow color being due to the presence of yellow granules (Fig. 93).

During cleavage the four kinds of cytoplasm are distributed to different cells. Later, the cells containing clear cytoplasm give rise to ectoderm; the cells containing slaty gray cytoplasm become endoderm; those containing yellow cytoplasm develop into mesoderm; and those containing light gray cytoplasm become neural system and notochord. When examined with the electron microscope, the yellow color of the cytoplasm giving rise, in *Styela partita*, to the muscle tissue and mesenchyme is found to be due to the presence of lipochondria containing a yellow pigment (Berg and Humphreys, 1960) (Fig. 94). The yellow cytoplasm is also distinguished by large numbers of mitochondria. In other ascidians, such as *Ciona intestinalis*, the cytoplasm in the same position and having the same destiny is devoid of the yellow pigment but also contains dense masses of mitochondria (Berg and Humphreys, 1960; Mancuso, 1969) (Fig. 95). The greatest concentrations of both the yellow pigment and the mitochondria are in the cells which give rise to the muscles of the larva, and smaller concentrations characterize the blastomeres which produce mesenchyme. The slaty gray cytoplasm giving rise to the endoderm is filled with coarse and densely packed yolk platelets (Fig. 95). (Yolk platelets are almost completely absent in the muscle-mesenchyme cytoplasm.) The animal hemisphere "clear" cytoplasm, destined to produce skin ectoderm, contains moderate amounts of yolk and mitochondria (Mancuso, 1969). The light gray cytoplasm (chordoneuroplasm) has very little yolk and few mitochondria but has large masses of

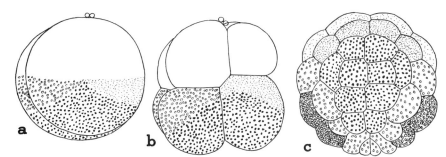

Figure 93. Three cleavage stages of the ascidian *Styela partita. a,* Two-cell stage; *b,* eight-cell stage; *c,* 64-cell stage. Large black dots represent slaty gray cytoplasm; circles represent yellow cytoplasm; loosely arranged circles represent mesenchymal cells; closely spaced circles represent muscle cells; heavy stippling represents presumptive notochord; light stippling represents presumptive neural plate. (After Conklin, 1905, redrawn from Korschelt, 1936.)

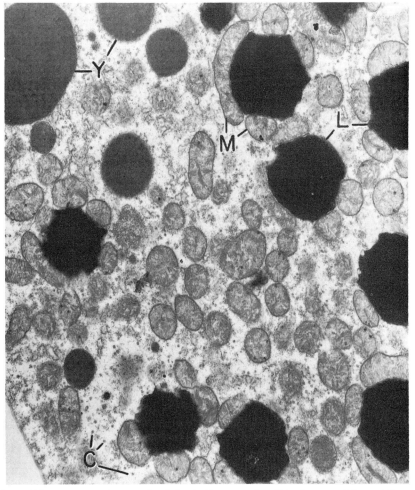

Figure 94. Electronmicrograph of the "yellow cytoplasm" in the egg of the ascidian, *Styela partita.* C, Fine granules (ribosomes?); L, lipid droplets; M, mitochondria; Y, yolk platelets. (From Berg and Humphreys, 1960.)

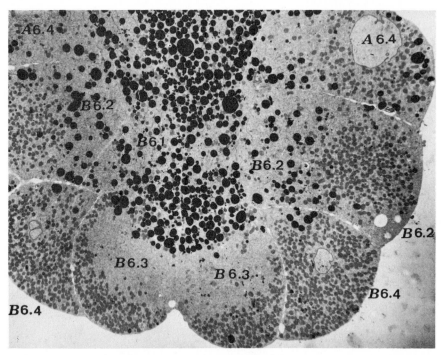

Figure 95. Low power electronmicrograph of part of the cleavage stage of the ascidian, *Ciona intestinalis,* showing different kinds of cytoplasm. Individual cells marked after Conklin's (1905) nomenclature. A6.4, Mesenchymal cells (right with nucleus); B6.1, endodermal cell with abundant yolk platelets (black); B6.2, B6.3, and B6.4, mesodermal cells, containing muscle cytoplasm (densely packed mitochondria) on outer segments and finely granular mesenchymal cytoplasm in inner segments. (Courtesy of Dr. V. Mancuso, 1969.)

fine granules (ribosomes?) and small vesicles representing endoplasmic reticulum (Berg and Humphreys, 1960).

The assumption may be made, therefore, that the development of the parts of the embryo which have been enumerated is caused by the kind of cytoplasm that is contained in the cells. This can actually be proved by centrifuging the eggs before they have begun cleaving (Conklin, 1931). In this way the cytoplasmic substances within the egg may be displaced, and when the eggs begin to cleave, the various kinds of cytoplasm are held in abnormal positions when the substance of the egg is subdivided into cells. An embryo possessing recognizable tissues and rudiments of organs, may develop from a centrifuged egg, but these tissues and organ rudiments are arranged in a chaotic way. As the distinctions between the various cytoplasmic substances can still be discerned, it is discovered that each organ and tissue develops in that position where the respective cytoplasmic substance came to lie as the result of centrifugation (Fig. 96). The outcome of these experiments supports the assumption previously made about the decisive part played by the cytoplasmic substances in differentiation.

6-7 ROLE OF THE EGG CORTEX

Distribution of visible materials in the egg and in the embryo during cleavage is, however, not always of crucial importance for the localization of parts in the developing embryo. Some substances and cellular inclusions may be displaced without disturbing the normal segregation of the embryo into its subordinate parts. This displacement

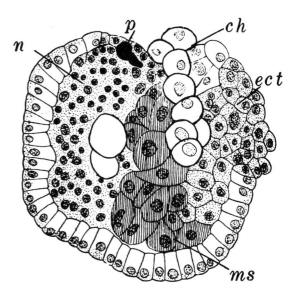

Figure 96. Displacement of organ rudiments in an embryo of *Styela partita* as a result of the centrifugation of the egg. n, Neural system; p, eye pigment; ch, notochord; ect, ectoderm; ms, muscle cells. (From Conklin, 1931.)

can most conveniently be accomplished by centrifugation of the uncleaved eggs. Centrifuging the eggs of most animals for a few minutes with moderate speeds, at an acceleration of several thousand times gravity, is sufficient to rearrange various cellular inclusions in the interior of the egg according to their specific gravities.

After sufficiently strong centrifugation, the eggs become stratified and show at least three typical layers. At the centripetal pole there is usually an accumulation of fat or lipid droplets, which are the lightest constituents of the egg cytoplasm. A layer of hyaline cytoplasm, which is the ground substance of the egg, follows. The nucleus or asters with chromosomes (if the centrifuged egg was in meiosis or mitosis) are also found in the hyaline layer. The yolk, as the most dense and heavy constituent of the egg, accumulates at the centrifugal pole.

In the eggs of some animals the vegetal pole is so much heavier, owing to the presence of yolk, that it becomes oriented centrifugally during centrifugation. In this case the yolk is not displaced from its normal site at the vegetal pole but is only more concentrated. To displace the yolk in these cases the eggs have to be fixed in a desired position, so that they cannot freely rotate. This can sometimes be done by sucking them into narrow capillaries or by embedding the eggs in gelatin prior to centrifugation.

If the main axis of the embryo lies at random to the centrifugal force, as often happens, cellular inclusions will be dislocated to different parts in individual eggs. Figure 97 shows the results of centrifuging eggs of the sea urchin *Arbacia*. The red pigment granules present in these eggs are concentrated at the centrifugal end of the egg (Fig. 97a). Some scattering of the granules occurs after the centrifuging is stopped and before cleavage progresses sufficiently to prevent further redistribution of the granules by subdividing the egg into blastomeres. It now becomes evident that the granules are concentrated in different positions in respect to the egg axis: near the vegetal pole (Fig. 97b), near the animal pole (Fig. 97c), or toward one side (Fig. 97d). Independently of the position of the granules, the invagination of the blastopore occurs at the vegetal pole, so that the region of the blastopore will contain the pigment granules in some embryos but not in others (Fig. 97e, f) (Morgan, 1927).

It has often been found that the pattern of cleavage may be highly independent of the distribution of cytoplasmic substances inside the egg. We have already described

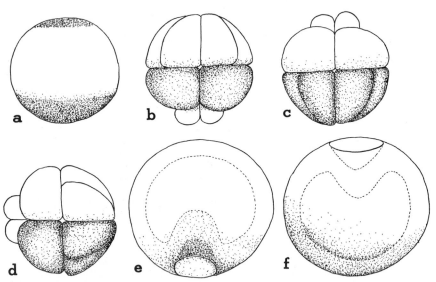

Figure 97. Centrifugation of the egg of the sea urchin *Arbacia. a,* Egg immediately after centrifugation: red pigment (stippling) thrown to centrifugal pole, oil droplets assembled at centripetal pole. Drawings show position of red pigment with respect to the main axis during cleavage (*b, c, d*) and at the beginning of gastrulation (*e, f*). (From Morgan, 1927.)

the cleavage of the mollusc *Dentalium,* in which a polar lobe appears at the vegetal pole during the first and second divisions of the egg and contains the cytoplasm necessary for the development of the mesoderm in the larva. A similar polar lobe is observed during cleavage of another mollusc, *Ilyanassa.* The polar lobe in this species is normally filled with yolky cytoplasm, while at the animal pole the egg cytoplasm is fairly free from yolk. Eggs of *Ilyanassa* have been centrifuged "in reverse," that is, with the vegetal pole fixed in position, facing the axis of the centrifuge. As a result, the heavy yolk is thrown into the animal hemisphere (still marked by the position of the polar bodies), and the hyaline cytoplasm and lipid droplets are concentrated at the vegetal pole. Nevertheless, the polar lobe appears at the vegetal pole when the egg starts cleaving, although the lobe now contains mainly hyaline cytoplasm and lipid instead of the yolk granules (Fig. 98). Obviously, the formation of the polar lobe is not dependent on the yolk normally located at the vegetal pole, but on something that is not displaced by the centrifugal force (Morgan, 1927).

What can this something be? There are two possibilities. The first is that in the cytoplasm some fixed network exists with sufficiently broad meshes to allow for the free movement of yolk granules and other inclusions without itself being torn or distorted. The existence of such a network cannot, however, be supported by any observation of the physical state or ultramicroscopic structure of cytoplasm. There remains the other alternative that the fixed system which is not displaced by centrifugation is the cortical layer of cytoplasm, or the **cortex** of the egg. This is in conformity with direct observation, for the cortical granules (see Section 3–3) are not displaced by centrifugation (Harvey, 1946). The layer of cytoplasm in which they are embedded is therefore sufficiently viscous to resist the forces usually generated in centrifugation experiments.

Since the immovable cortex of the egg appears to determine the point at which invagination begins in centrifuged *Arbacia* eggs, as well as the position of the vegetal polar lobe in *Ilyanassa,* the further suggestion may be made that the cortex is the actual carrier of the polarity of the egg, or that the polarity of the egg is ingrained in its cor-

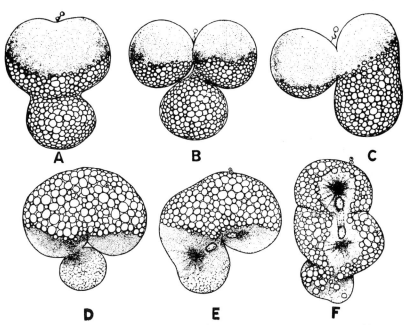

Figure 98. *A, B, C,* Normal cleavage of the mollusc *Ilyanassa. D, E, F,* Cleavage stages after centrifugation "in the reverse," with the heavy yolk displaced toward the animal pole. (From Morgan, 1927.)

tex. If this were the case, the distribution of substances in the interior of the egg might be expected to be controlled or determined by the egg cortex. This suggestion finds support in some further results of centrifugation experiments, namely, the fact that cell constituents tend to return to their normal positions after the cessation of centrifuging. A scattering or mixing up of the strata into which the egg contents had been arranged by centrifugation could be the result of random movement of particles. This explanation, however, does not apply to cases in which after centrifugation certain particles not only move from the position to which they were brought by the centrifugal force but take up a very definite location in the egg.

We have seen that in the egg of the ascidian *Styela* different kinds of egg cytoplasm may be displaced by centrifugation. Immediately after centrifugation the eggs show a clear stratification, as can be seen in Figure 99a. The yellow granules, which go into the formation of mesoderm (p. 128), are displaced to the centripetal pole; the yolk is displaced to the centrifugal pole; and the hyaline cytoplasm remains as a layer in between. When the eggs are left to themselves after centrifugation, the egg substances start flowing and rearranging themselves in the interior of the egg. If the cell divisions set in sufficiently soon, this rearrangement is stopped by partitions appearing between the cleavage cells, and the result is the formation of the abnormal embryos dealt with earlier. If, however, the eggs are centrifuged well in advance of the beginning of cleavage, or if the first divisions of the egg are delayed, the redistribution of the egg contents may proceed so far that normal conditions are attained. Figure 99b shows the end result of such a redistribution: the yolk is at the vegetal pole; the animal hemisphere is filled with hyaline cytoplasm; and the yellow granules take up a subequatorial area on one side of the egg, corresponding to the mesodermal "yellow crescent" of normal development (Conklin, 1931).

Another and perhaps even more impressive example is that of the egg of the mollusc *Aplysia limacina.* In the eggs of this animal there are granules containing ascorbic

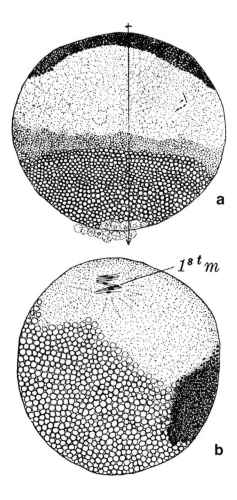

Figure 99. Redistribution of cytoplasmic inclusions in the egg of *Styela partita* following centrifugation. *a*, Egg immediately after centrifugation; *b*, centrifuged egg 22 hours later (the yellow granules, shown black, have taken up a subequatorial position on one side of the egg). 1st m. Spindle of first meiotic division. (From Conklin, 1931.)

acid (vitamin C), probably connected with the Golgi bodies. In immature oocytes the ascorbic acid granules are uniformly distributed throughout the egg cytoplasm. During maturation the granules accumulate in a ring lying inside the cortical cytoplasm and somewhat above the equator. By centrifugation the granules are concentrated at the centrifugal pole (Fig. 100), but after cessation of centrifugation the ascorbic acid granules soon start moving and again take up their normal position as a supraequatorial ring (Peltrera, 1940).

The most plausible explanation of the two preceding experiments is that the displaced cytoplasmic particles tend to return to the proximity of certain regions of the egg cortex which had remained in their respective positions all the time that the egg was being centrifuged. It follows that in normal development the cytoplasmic substances in the egg are distributed in relation to local differences in the egg cortex, and that it is the egg cortex, therefore, that foreshadows, in some way, the pattern of future development of the embryo.

6–8 THE MORPHOGENETIC GRADIENTS IN THE EGG CYTOPLASM

While we have learned that the cytoplasm of the egg, and particularly its cortex, starts the chain of reactions which eventually leads to the differentiation of parts of the

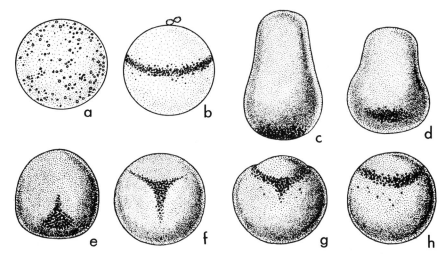

Figure 100. Redistribution of vitamin C granules in eggs of *Aplysia limacina* during matura-tion and following centrifugation. *a,* Immature egg (oocyte); *b,* mature egg; *c,* egg immediately after centrifugation. *d–h,* Consecutive stages of recovery of the centrifuged egg, with the vitamin C granules returning to their normal position. (After Peltrera, from Raven, 1958.)

embryo, it should not be assumed that the structure of the future embryo is rigidly de-termined by local differences in the cytoplasm of the egg. The local peculiarities of the egg cytoplasm are only some of the factors necessary for the formation of organ rudi-ments. That this is so can be shown by examining the development of the sea urchin, for instance.

In the 16-cell stage of the sea urchin, the blastomeres are of three different sizes. The animal hemisphere consists of eight blastomeres of medium size, the **mesomeres,** which are destined to produce most of the ectoderm of the larva. The vegetal hemi-sphere consists of four very large blastomeres, the **macromeres,** and of four very small blastomeres, the **micromeres,** which lie at the vegetal pole of the egg. The macromeres also contain some material for the ectoderm and all the material for the endoderm. In a subsequent cleavage the future ectoderm and endoderm are segregated from each other into an upper tier of macromeres and a lower tier of macromeres. The upper tier, lying immediately under the equator of the embryo, contains the ectodermal mate-rial; the lower tier, lying nearer to the vegetal pole, contains the endodermal material. The micromeres develop into mesenchyme, which later produces the larval skeleton consisting of calcareous spicules (Hörstadius, 1935). In the sea urchin *Paracentrotus livi-dus,* the cytoplasm of the macromeres possesses a surface layer of red pigment gran-ules, making the macromeres easily distinguishable from the other cleavage cells. The red pigment is already present in the egg at the beginning of cleavage as a broad sub-equatorial zone (Fig. 101).

In the blastula stage, the cytoplasmic substances are found in the same arrange-ment as at the beginning of cleavage. Subsequently, the descendants of the micromeres migrate into the blastocoele where they develop the skeletal spicules; the descendants of the lower tier of macromeres invaginate, forming a pocket-like cavity—the gut; and the ectoderm produces a ciliary band, serving for locomotion, and a tuft of rigid cilia at the former animal pole. The ectoderm also sinks inward to produce a stomodeum, coming into communication with the endodermal gut at its anterior end, while the original opening of the pocket-like invagination (the blastopore) becomes the anal opening (Fig. 102).

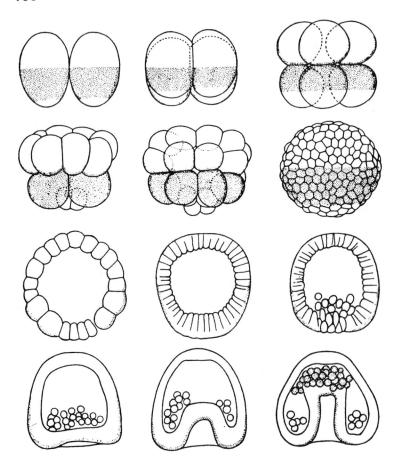

Figure 101. Cleavage and gastrulation of the sea urchin *Paracentrotus lividus*. (After Boveri, from Spemann, 1936.)

The embryo of the sea urchin develops into a larva called a **pluteus** (Fig. 50, p. 68). If the blastomeres of the sea urchin are separated in the two-cell stage or in the four-cell stage, each of them develops into a complete pluteus of diminished size. This may be related to the fact that the first two cleavage planes are meridional, passing through the main axis of the egg. All of the first four blastomeres therefore get equal portions of the three cytoplasmic regions of the egg (ectodermal, endodermal, and mesenchymal). A different result is observed if the egg is separated into the animal and vegetal halves after the third cleavage, the third cleavage plane being equatorial. In this case both halves produce, as a rule, defective embryos (Fig. 103). The animal half differentiates as an ectodermal vesicle; it does not produce a gut. Even purely ectodermal structures are abnormally differentiated: the ciliary band is not developed, whereas the tuft of long cilia on the animal pole develops excessively, growing over a much greater surface than it normally does (Fig. 103a). The vegetal half is differentiated into an ovoid embryo with a disproportionately large endodermal gut but without a mouth. There may be a few irregular skeletal spicules but no arms and no ciliary band (Fig. 103b).

In later cleavage stages, it is possible to cut the morula transversely below the equator. In such cases the vegetal part develops still more abnormally; it produces a large endodermal gut and a small ectodermal vesicle (Fig. 103c). The gut does not lie inside the ectodermal vesicle but is evaginated to the exterior (turned inside out), owing to interference with the normal processes of gastrulation. This phenomenon is known as

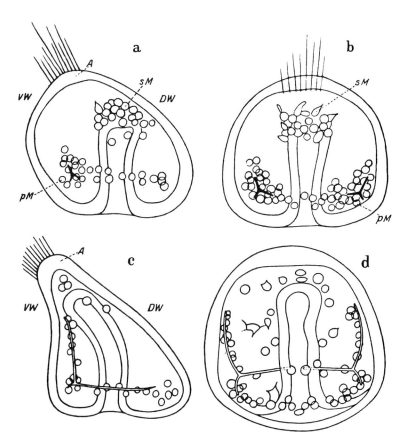

Figure 102. Early larvae of the sea urchin *Parechinus* showing development of skeleton. pM, Primary mesenchyme; sM, secondary mesenchyme; A, apical tuft of ciliae; VW and DW, ventral and dorsal body walls. (After Schmidt, from Spemann, 1936.)

exogastrulation. The mesenchyme cells migrate into the interior, but they usually produce no skeletal spicules or only very small and abnormal ones (Hörstadius, 1928, 1935).

The result of this experiment is obviously due to each of the two halves lacking some parts contained in the other half. However, each half does not simply produce what would normally have been the fate of the respective part. Instead, the differentiation of each half seems to be "exaggerated" as compared with its normal fate. This is especially clear in the case of the increased animal tuft of cilia.

The same "exaggeration" of the ectodermal or endodermal differentiations can be achieved by exposing the developing eggs to certain chemical substances, even without

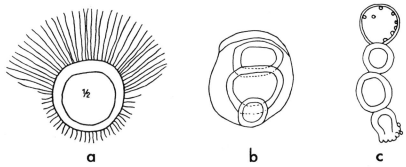

Figure 103. Animalized (*a*) and vegetalized (*b*, *c*) larvae of a sea urchin. (After Hörstadius, from Lehmann, 1945.)

removing any parts of the egg or embryo. If the fertilized sea urchin egg is exposed to seawater containing some lithium salts in solution, the embryo develops just as if it were only the isolated vegetal half. The gut is increased and tends to exogastrulate instead of invaginating toward the interior (Herbst, 1893). The skeleton is absent or abnormal, and the ectoderm is represented by an epithelial vesicle and fails to differentiate further. The increase in the size of the gut occurs at the expense of the ectoderm, and in extreme cases, most of the embryo differentiates as an enormous everted gut, with the ectodermal vesicle reduced to a tiny appendage. The opposite effect is achieved if, before fertilization, the egg is exposed to artificial seawater lacking calcium ions but with sodium thiocyanate (NaSCN) added to it. In this case the gut is diminished or completely absent, the ciliary bands in the ectoderm fail to develop, and the tuft of stiff cilia at the animal pole is increased in size (Lindahl, 1933).

It appears that all these phenomena may be accounted for by assuming that in the sea urchin's egg two factors or principles exist which are mutually antagonistic and yet interact with each other at the same time, and that normal development is dependent on a certain equilibrium between the two principles. Each has its center of activity at one of the poles of the egg. The activity diminishes away from the center, producing a **gradient** of activity. The two gradients of activity decline in opposite directions: the one from the animal pole, the other from the vegetal pole. Taking into account this type of distribution of activities, the factors or principles themselves have been called **gradients.** The two gradients are therefore the **animal gradient,** with a center of activity at the animal pole, and the **vegetal gradient,** with a center of activity at the vegetal pole.

According to this theory (originally suggested by Boveri in 1910, and developed in application to the sea urchin egg by Runnström, 1928, and Hörstadius, 1928), normal development depends on the presence of both gradients and on an equilibrium between them. If the animal gradient is weakened or suppressed, the vegetal gradient becomes preponderant, and the embryo is **vegetalized;** i.e., it develops, in excess, parts pertaining to the vegetal gradient, such as the gut. If the vegetal gradient is weakened or suppressed, the animal gradient becomes preponderant, and the embryo is **animalized;** i.e., it develops, in excess, parts pertaining to the animal gradient, such as the tuft of stiff cilia at the animal pole. Other structures of the embryo, such as the ciliary bands and the skeleton, can develop only if both gradients are active, and the development is the more nearly normal the more the two gradients approach a correct equilibrium.

The equilibrium between the two gradients may be upset in various ways. The animal gradient may be weakened, with concomitant vegetalization of the embryo, by removing its center of activity (the blastomeres at the animal pole of the egg) or by the action of lithium salts. The vegetal gradient may be weakened by removing its center of activity (the vegetal part of the egg) or by the action of sodium thiocyanate.

The gradient concept aptly covers the results of various experiments on sea urchins' eggs. In addition, we will consider the following experiments. It is possible to separate from one another the three groups of blastomeres in the 16-cell stage—the mesomeres, the macromeres, and the micromeres—and then to recombine them at will (Hörstadius, 1935). In isolation the three groups develop as follows:

1. Isolated mesomeres develop into a vesicle with a tuft of cilia (as previously mentioned) owing to a preponderance of the animal gradient and animalization.
2. Isolated macromeres plus micromeres develop into an extremely vegetalized embryo with evaginated gut, the result of a preponderance of the vegetal gradient.
3. Mesomeres plus macromeres develop into a practically normal embryo; the macromeres bear a sufficiently strong vegetal gradient to counterbalance the

animal gradient. Mesenchyme and skeleton develop in such embryos in spite of the absence of micromeres.

4. Micromeres alone are not capable of any development, since they do not keep together but fall apart.

5. Mesomeres (ectoderm) plus micromeres (mesenchyme) develop into a complete and more or less normal pluteus. This combination is especially illuminating since the gut of such embryos develops in spite of the absence of the macromeres, which should normally have supplied the material for the gut. However, the embryo possesses the two gradients, the animal gradient borne by the mesomeres and the vegetal gradient borne by the micromeres, and the possession of the two gradients seems to create the necessary conditions for normal development.

What has been said is sufficient to show that the presence of different cytoplasmic substances in the egg does not necessarily mean that there is a direct relation between these substances and certain specific organs, in the sense that the cells containing the particular kind of cytoplasm develop directly into the corresponding organ. In the last experiment mentioned, the absence of the cytoplasm with red pigment did not prevent the development of the gut because the necessary conditions for gut development were provided for in another way.

Physicochemical Nature of the Animal-Vegetal Gradient System in Sea Urchin Eggs. The existence of animal and vegetal gradients is proved not only by the patterns of morphogenetic processes occurring after certain interventions in the normal development of the egg, but also by the direct demonstration of the gradients as peculiar physicochemical states of the cells of the developing embryo. One way of proving their existence is to study the reduction of dyes by the embryo under conditions of anaerobiosis. Sea urchin embryos (or embryos of other animals for that matter), slightly stained with the vital dye Janus green (diethylsafraninazodimethylaniline), are placed in a small chamber sealed off with petroleum jelly. After some time all the free oxygen contained in the chamber is used up as result of the respiration of the embryos. The embryos then respire using Janus green as an acceptor of hydrogen, which is first reduced to a red dye, diethylsafranin, and further to a colorless substance, leucosafranin. The light grayish blue color of Janus green first changes to red, and then, as a second step in the reduction of the dye, the color disappears completely.

The reduction of Janus green, shown by the color change, does not occur simultaneously in the whole embryo but follows a characteristic sequence. In the late blastula and early gastrula of the sea urchin, the change in color is first noticeable at the vegetal pole, at the point where the primary mesenchyme is given off (Fig. 104). Then the color change spreads, involving the whole vegetal hemisphere, and reaches the equator. At this stage a spot of changed color appears at the animal pole and also gradually increases, so that the bluish color remains only in the form of a ring lying well above the equator in the animal hemisphere. Even this ring eventually disappears. The red color then begins to fade in the same sequence, starting first from the vegetal pole and then from the animal pole (Child, 1936; Hörstadius, 1952).

The order in which the dye is reduced in the embryo shows such a nice correspondence to the postulated animal and vegetal gradients that this alone would justify our mentioning these experiments. However, the connection goes much further. When either the animal or vegetal gradient is suppressed, the corresponding gradient of reduction also disappears. In isolated animal halves of sea urchins' eggs, the animal tendencies of development are predominant, and it is found that such animalized embryos start reducing Janus green at the animal pole only; there is no center of reduction at

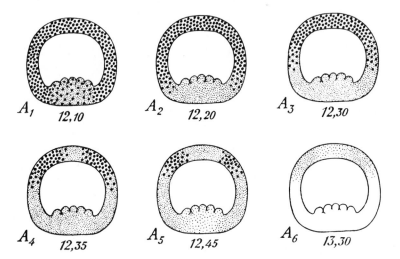

Figure 104. Reduction of Janus green in a late blastula of a sea urchin. Large dots represent blue color; small dots, red; no dots show a complete fading of the stain. Figures indicate time of observation. (From Hörstadius, 1952.)

the vegetal pole (Fig. 105). In isolated vegetal halves the animal tendencies of development are suppressed and the vegetal tendencies are supreme. Correspondingly, the only center of reduction is at the vegetal pole; no center of reduction appears at the animal pole (Fig. 106). Embryos animalized or vegetalized chemically show reduction gradients similar to isolated animal and vegetal halves of the egg (Hörstadius, 1955).

We have seen that the micromeres are carriers of the highest point of the vegetal gradient, and they preserve this property after transplantation. Accordingly, the micromeres can also serve as the center of a reduction gradient. If a group of four micromeres is implanted laterally and the embryo is tested for reduction of Janus green, it can be seen that in addition to the two normal gradients, one from the vegetal pole and one from the animal pole, a third gradient appears, having the implanted micromeres as its center but spreading out from these to the adjacent cells. In short, every modification

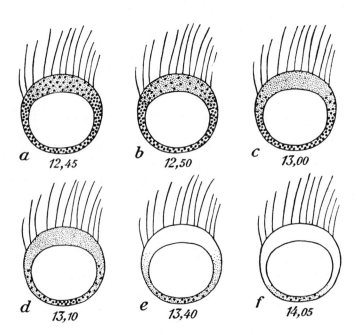

Figure 105. Reduction of Janus green in animalized sea urchin embryo (isolated animal half). (From Hörstadius, 1952.)

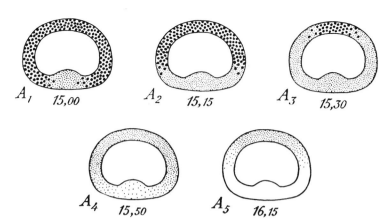

Figure 106. Reduction of Janus green in vegetalized sea urchin embryo (isolated vegetal half). (From Hörstadius, 1952.)

of the gradient system of the embryo that is postulated from the morphogenetic behavior of the embryo is reflected in the gradients of reduction of Janus green (Hörstadius, 1952).

If the micromeres are able to establish a (vegetal) gradient in the adjoining parts of the blastoderm, we would expect that they do so by producing some substance which diffuses into the nearby cells. To prove the existence of such a transmission, advantage was taken of an unusual peculiarity of the micromeres. After the fourth division of the egg, the micromeres lag behind in cleavage. While the mesomeres and macromeres continue rapid division and accordingly synthesize DNA, the micromeres pause with the next divisions and start synthesizing RNA instead. This synthesis can be shown very clearly if the embryo is supplied with radioactively tagged uridin; only the micromeres take up the label (Czihak, 1965) (Fig. 107). If the micromeres are then transplanted into an isolated animal half of a 16- or 32-cell embryo, radioactivity can be detected as being widely spread in the *cytoplasm* of the host half-embryo (Czihak and Hörstadius, 1970). The RNA synthesized by the micromeres thus diffused into the cytoplasm of the adjoining cells. From previous experiments (p. 138) it is known that micromeres implanted in an animal half-embryo cause the development of an endodermal gut from cells which normally would have produced only ectoderm. The foregoing experiment does not prove, in itself, that the RNA diffusing from the micromeres is responsible for the newly established vegetal gradient in the animal hemisphere cells. This explanation, however, is shown to be more likely by another modification of the labeling experiment. If a 16-cell embryo, instead of being provided with uridin, is sup-

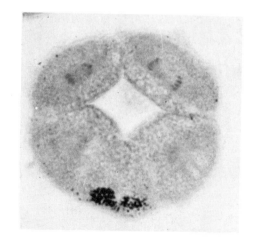

Figure 107. Autoradiograph of a sea urchin embryo supplied with tritiated uridine in the 16-cell stage. Black dots in the emulsion show that the micromeres have taken up the radioactive label. (Courtesy of Professor G. Czihak, 1965.)

plied with 8-azaguanin, an analogue or uridin, this substance is incorporated into the RNA by the micromeres. An abnormal RNA is produced. After the treatment the progeny of the micromeres—the primary mesenchyme—develop quite normally, but the gut completely fails to develop. This experiment has been interpreted as proving that the RNA diffusing from the micromeres is essential for gut development and that if this RNA is of an abnormal composition, it cannot perform its function (Czihak, 1965).

From the setup of the preceding experiments it appears highly probable that the spreading of substances from the micromeres into the other parts of the blastoderm occurs directly from cell to cell and not through the fluids surrounding the cells. This point will be further discussed in the section dealing with gradients in regeneration (Section 19–7).

The animal and vegetal gradients can be considered to be definite metabolic processes or systems of metabolic reactions which involve oxidation and which have their points of highest intensity at the animal and vegetal poles respectively. The nature of these reactions is probably very complex and is not, as yet, fully understood, but some indications concerning these reactions may be deduced from a comparison of the chemical agents which may cause vegetalization or animalization.

Apart from the use of lithium ions, vegetalization may be caused by sodium azide (Child, 1948) and dinitrophenol. Both of these substances belong to a group of enzyme poisons; the azide is known to inactivate the cytochrome oxidase system, and the dinitrophenol disturbs respiration by preventing oxidative phosphorylation, that is, the formation of energy-rich bonds between phosphoric acid and adenosine diphosphate (resulting in the formation of adenosine triphosphate). This might mean that vegetalization is based essentially on a disturbance of oxidative processes in the embryo (Lindahl, 1936) and, perhaps even more specifically, a disturbance of the processes of phosphorylation (Hörstadius, 1953b).

That the action of lithium ions is along the same lines may be deduced from a number of observations of which we shall single out the following. Lithium salt treatment suppresses the rise of oxygen consumption which occurs normally at the beginning of gastrulation, that is, at the same time as the morphological effects of vegetalization begin to be apparent (Lindahl, 1936). Furthermore, inorganic phosphate accumulates in the seawater during the development of lithium ion-treated embryos—again a hint that these embryos are incapable of utilizing the energy of oxidation for phosphorylation and synthesis of adenosine triphosphate.

The whole sequence of reactions involved in vegetalization must, of course, be far more complicated. Thus, it has been found that the relative amounts of various amino acids change as a result of vegetalization in lithium ion-treated embryos (Gustafson and Hjelte, 1951). Furthermore, if a modification of the respiratory system of the embryo is the essential feature of vegetalization, it remains to be discovered why the processes of morphogenesis at the animal pole of the embryo (essentially the ectodermal organs) are more severely affected than those of endodermal parts developing at the vegetal pole.

With regard to animalizing agents, the first of those previously mentioned, sodium thiocyanate, has been found to be rather unreliable. Some batches of eggs treated by this chemical do not react at all. In other batches individual embryos become animalized to greatly varying degrees. Subsequently, many other substances have been found whose animalizing action is much more predictable. These belong to several distinct groups:

1. Some metals: zinc, mercury.
2. Some proteolytic enzymes: trypsin, chymotrypsin.
3. Many acidic dyes: in particular some possessing sulfonic (HSO_3) groups in their

molecules, such as Evans blue, chlorazol sky blue, trypan blue, and Congo red; and others possessing carboxyl groups (COOH), such as uranin and rose bengal.

4. Some other sulfonated organic compounds: germanin and others.

5. Some anionic (acid) detergents.

Animalization may be achieved by extreme dilutions of a chemical; for instance, 1:50,000 in the case of Evans blue.

The common property of most, if not all, of the preceding agents seems to be their ability to attack proteins or to form compounds with proteins, especially basic proteins. In the case of the many sulfonated and carboxylated dyestuffs, it is fairly certain that their acidic groups become attached to the functional side groups of protein molecules. The two metals previously mentioned, zinc and mercury, also become easily bonded to side groups of protein molecules. The case of mercury is very remarkable: mercuric chloride, a powerful poison used therefore as a fixative for proteins, causes animalization when applied for a very short time in great dilutions (1:90,000). By blocking the active side groups of the protein molecules, the agents referred to probably "immobilize" these molecules, preventing them from interacting normally with other cell constituents. Even the steric configurations of protein molecules may thus be altered, since they are based on lateral bonding between parts of the polypeptide chains. A far-reaching change in the properties of proteins may thus be brought about (Lallier, 1956, 1957).

Fortunately, in the case of animalizing agents, their point of attack can be clearly demonstrated. The dyes which have been used for this purpose stain the cells of the embryo, and it can be seen that the first cells to be stained are the primary mesenchyme cells at the vegetal pole. With higher concentrations of the dye or with a longer duration of the treatment, the adjoining presumptive endodermal cells show the staining too. The zinc taken into the cells of embryos may be made visible because it yields a pink coloration with dithizone. In embryos first treated with zinc and later immersed in a dithizone solution, the pink color is detectable in the same position as in the case of animalizing dyes (Lallier, 1956, 1957). It is thus quite clear that animalization is due to damage to the center of the vegetal gradient.

While the treatment of embryos with various chemicals of simple and known composition (and, in part, of known mode of action) is useful in establishing the existence of gradients, it does not reveal the chemical mechanism of the gradients under normal conditions. A more direct approach is here indicated. Such an approach was made in attempts to isolate animalizing and vegetalizing substances from the embryo itself. By homogenizing sea urchin unfertilized eggs (Josefson and Hörstadius, 1969) or early cleavage stages (Hörstadius and Josefson, 1972) and by separation of the materials by centrifugation and column chromatography, some fractions were obtained which had a distinct animalizing action on developing embryos. Several fractions with a slightly different effect could be detected. Less success was obtained with vegetalizing substances, though some vegetalizing activity was also recorded. The chemical nature of the fractions has not, as yet, been determined, but the absorption curve of one fraction in ultraviolet light resembles that of the amino acid tryptophan, while the absorption curve of another suggests nucleotide structure. Further characterization of the substances is to be expected.

6–9 MANIFESTATION OF MATERNAL GENES DURING THE EARLY STAGES OF DEVELOPMENT

In the previous sections evidence has been presented that (1) the differences between the various parts of the embryo are not due to differences in their nuclei but to differences in the cytoplasm. From this it may be inferred that (2) the nuclei of the cleavage cells and the genes contained therein do not control the earliest stages of development of the animal egg.

This does not mean, however, that genetic factors have nothing to do with the early stages of development. The genetic factors, the genes, do play a part during this period, but in a very special way. It is not the genes contained in the nuclei of the blastomeres but the genes in the cells of the maternal body that determine the peculiarities of the egg and its early development. The best-known example of this type of gene action is concerned with the inheritance of coiling in gastropod molluscs.

In the snails, as a rule, the shell is coiled spirally, and the internal organization of the animal shows a corresponding dislocation, some of the organs (heart, kidney, and gills) being twisted around through an angle of nearly 180°. The direction of coiling of the shell is clockwise, if viewed from the apex of the shell. This type of coiling is called **dextral.** As an exception, the coiling may be counterclockwise, or **sinistral.** Individuals having a sinistral coiling of the shell have the internal organs dislocated in the opposite direction from that in dextral individuals. In short, dextral and sinistral individuals are in every detail of organization mirror images of each other.

The eggs of gastropod molluscs show a spiral type of cleavage (Section 6–3), and it has been noted that the type of cleavage of the egg has a definite relation to the type of coiling of the adult. If the cleavage is dextral—that is, if the cleavage spindles show a clockwise spiraling—the adult snail also has a dextrally coiled shell. If the cleavage is sinistral, the coiling of the shell is likewise sinistral (Fig. 108). The connection between the type of cleavage and the coiling of the shell is established through the position of blastomere D in respect to the other blastomeres. Since it is from blastomere D that the mesoderm develops, its situation is reflected in the position of the internal organs, and the coiling of the shell is one of the secondary expressions of this asymmetry.

There are some species of gastropods in which all the individuals are sinistral, but the main interest attaches to a species in which sinistral individuals occur as a mutation among a population of normal dextral animals. Such a mutant was discovered in the freshwater snail *Limnaea peregra* (Boycott, Diver, Garstang, and Turner, 1930). Breeding and crossbreeding of dextral and sinistral snails showed that the difference between the two forms is dependent on a pair of allelomorphic genes, the gene for sinistrality being recessive (l), and the gene for the normal dextral coiling being dominant (L). The two genes are inherited according to Mendelian laws, but the action of any

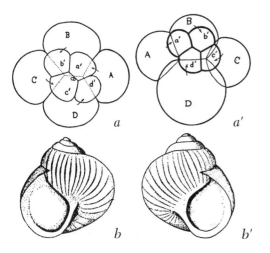

Figure 108. Correlation of sinistral (*a*) and dextral (*a'*) cleavage with sinistral (*b*) and dextral (*b'*) coiling of the shell in gastropods. (After Conklin, from Morgan, 1927.)

genetic combination is manifested only in the next generation after the one in which a given genotype is found. Thus, if the eggs of a homozygous sinistral individual are fertilized by the sperm of a dextral individual, the eggs cleave sinistrally, and all the snails of this F_1 generation show a sinistral coiling of the shell. The genes of the sperm do not manifest themselves, although the genotype of the F_1 generation is Ll.

If a second generation (F_2) is bred from such sinistral individuals, it is all dextral, instead of showing segregation as would be expected in normal Mendelian inheritance. In fact, segregation does take place in the F_2 generation as far as the genes are concerned, but the new genic combinations fail to manifest themselves, since the coiling is determined by the genotype of the mother. The genotype being Ll, the gene for dextrality dominates and is responsible for the exclusively dextral coiling of the second generation. Only in the third generation (F_3) does segregation in the proportion of 3 to 1 become apparent, and then not as segregation among individuals in each brood, but as a segregation of broods. Each brood—that is, the offspring of an individual of the F_2 generation—shows a coiling that is determined by the maternal genotype. Since the individuals of the F_2 generation have the genotypes 1 LL, 2 Ll, and 1 ll, three quarters of them, on the average, produce eggs developing into dextral snails, and one quarter produce eggs developing into sinistral individuals. Of the dextral broods, one third breed true, producing only dextral offspring, and two thirds show further segregation among the broods of the next generation.

It is easy to understand that the results of a reciprocal cross (i.e., fertilization of the eggs of a homozygous dextral individual by the sperm of a sinistral individual) will lead to a somewhat different type of pedigree: the F_1 generation will be all dextral (with genotype Ll), and the F_2 generation again will be all dextral (with genotypes LL, Ll, lL, and ll). The F_3 generation will show segregation among broods, just as in the cross examined first.

The whole case becomes clear if it is realized that the type of cleavage depends on the organization of the egg which is established before the maturation divisions of the oocyte nucleus. The type of cleavage is therefore under the influence of the genotype of the parent producing the eggs (the mother). The haploid state of the egg nucleus continues for only a very short time and cannot materially affect the organization of the egg. The sperm enters the egg after this organization is already established. Similarly, we should expect that the elaboration of cytoplasmic organ-forming substances is under the control of maternal genes, even though this cannot be proved because no mutants are known which produce a difference in these substances.

It is conceivable that maternal genes might produce conditions in the egg leading to morphogenetic processes, which take place at a later stage of development, without visibly modifying the cleavage pattern. Several cases of such an influence are actually known.

In the axolotl (*Ambystoma mexicanum*) a lethal gene "o" has been discovered which has a maternal effect. The gene is recessive, so that heterozygote individuals (the + o individuals) are completely normal. Homozygote animals (oo), produced by a cross between two heterozygote parents, develop normally in the early stages but in later life show a slight retardation of growth. Their regeneration capacity is severely reduced, so that amputated legs are not restored as would occur in normal axolotls. The hymozygote males are sterile; their testes are underdeveloped and spermatogenesis does not go beyond the spermatogonial stages. The homozygote females produce eggs capable of fertilization and normal cleavage, but at the onset of gastrulation the development is retarded. The embryos start gastrulation but usually do not go beyond the crescentic blastopore stage (p. 162) and then die. Occasionally gastrulation is completed but the affected embryos never enter the stage of organogenesis. This abnormal course of development is in no way affected by the genotype of the spermatozoon; the spermatozoon may carry the normal + gene or a mutant o gene (if the male is a heterozygote). The arrest in development is in both cases exactly the same, showing that the particular type of abnormal early development does not depend on the genes present in the cells of the developing embryo, but exclusively on the genes of the mother (maternal effect) (Fig. 109).

The effect of the maternal genotype oo may be neutralized by injecting into freshly fertilized eggs or into eggs at the beginning of cleavage (two blastomeres) the cytoplasm from a normal, ripe egg. As little as 1 to 5 per cent normal cytoplasm improves development and causes most of the embryos to complete gastrulation, enter organogenesis, and sometimes produce swimming larvae. This experiment proves that the lethal embryos lack something which is present in the cytoplasm of the normal embryo. This "something," which has been termed the "corrective factor," must be the result of the activity in the mother of the unmutated + gene. In the homozygous female the o gene is unable to produce the corrective factor; the maternal lethal effect is thus due to a specific deficiency (Briggs and Cassens, 1966).

Further experiments have shown that in the oocyte a high concentration of the corrective factor is contained in the nuclear sap (0.2 to 0.5 per cent is sufficient to improve development) but that the cytoplasm of the oocyte when injected into a lethal egg has hardly any effect. It is only when the nuclear membrane of the oocyte ruptures in the process of maturation and the nuclear sap mixes with the cytoplasm that the cytoplasm acquires the ability to correct development of the lethal embryos. Centrifugation of the cytoplasm of normal eggs showed that the corrective agent remains in the supernatant after sedimentation of all cellular inclusions (yolk, pigment, mitochondria, and ribosomes) but that it sediments after further prolonged high speed centrifugation. It is thus a macromolecular substance which is not bound to any cellular organoids. The substance is inactivated by heating between 50 and 55° C. or by treatment

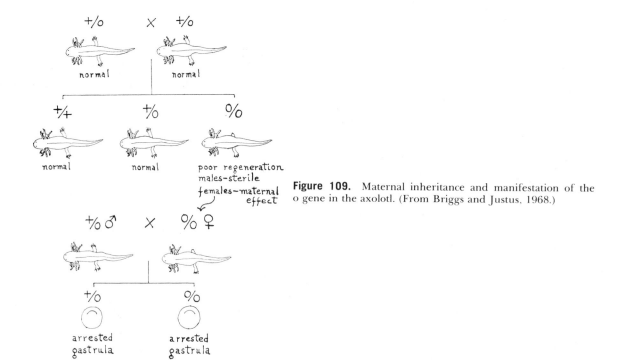

Figure 109. Maternal inheritance and manifestation of the o gene in the axolotl. (From Briggs and Justus, 1968.)

with trypsin, and it may be sedimented by ammonium sulfate. Thus, it shows character-istics of a protein (Briggs and Justus, 1967).

From this work it emerges that the normal allele of the gene o is responsible for the accumulation in the nuclear sap of immature oocytes of a protein-like substance, which, upon the nuclear membrane's breaking up during maturation, enters the cyto-plasm and is indispensable if the embryo is to go through gastrulation and enter the phase of organogenesis.

The corrective substance is, of course, not the only substance produced under the influence of the maternal genes while the egg is developing in the ovary. Many, if not all, cytoplasmic substances in the egg mentioned in Section 3–3 must have a similar origin. The number of different chemical substances in an oocyte is very great. Using very refined methods of study it has been estimated that perhaps as many as 10,000 dif-ferent genes are producing mRNA's in the oocyte of the frog *Xenopus laevis* (Davidson and Hough, 1971). It would be expected that most of the mRNA's, or at least a large proportion of them, would be directing the synthesis of proteins in the oocytes or during cleavage. The cytoplasmic substances determining the properties of blastomeres during cleavage may thus be an aftereffect of the action of genes at a previous stage of the re-productive cycle.

Part Four

GASTRULATION AND THE FORMATION OF THE PRIMARY ORGAN RUDIMENTS

The process of gastrulation is one of displacement of parts of the early embryo. As a result, the endodermal and mesodermal organ rudiments are removed from the surface of the embryo, where the presumptive material for these rudiments is to be found in the blastula stage, and brought into the interior of the embryo, where the respective organs are found in the differentiated animal. At the same time the single layer of cells, the blastoderm, gives rise to three germinal layers—the ectoderm, the endoderm, and the mesoderm.

The most prominent features of gastrulation are:

1. A rearrangement of cells of the embryo by means of **morphogenetic movements.**
2. The rhythm of cellular divisions is slowed down.
3. Growth, if any, is insignificant.
4. The type of metabolism changes; oxidation is intensified.
5. The nuclei become more active in controlling the activities of the embryonic cells. The influence of the paternal chromosomes becomes evident during gastrulation.
6. Proteins of many new kinds, that were not present in the egg, begin to be synthesized.

Chapter 7

MORPHOLOGICAL ASPECTS OF GASTRULATION AND PRIMARY ORGAN FORMATION

7–1 FATE MAPS

A correct interpretation of gastrulation is impossible without a knowledge of the position which the presumptive germinal layers occupy in the blastula. This position may be ascertained in various ways. A chart, showing the fate of each part of an early embryo, in particular, a blastula, is called a **fate map.**

In tracing the fate of various parts of the blastoderm, it is sometimes possible to make use of the peculiarities of the cytoplasm in certain parts of the egg, such as the presence of pigment granules. In the developing amphibian egg, for instance, one may trace in which part of the differentiated embryo the black pigment comes to lie. Originally this pigment is restricted to the animal hemisphere of the egg. However, peculiarities of pigmentation are seldom sufficient to make it possible to reconstruct the fate map in any great detail. Recourse must be had to artificial marking of parts of the blastoderm.

A satisfactory method of marking was devised for this purpose by Vogt (1925). The method consists in soaking a piece of agar in a vital dye (Nile blue sulfate, neutral red, Bismarck brown) and then applying the piece of agar to the surface of the embryo in the necessary position. The dye diffuses from the agar, and in a matter of minutes the cells of the embryo to which the agar has been applied take up sufficient dye to produce a stain on the surface of the embryo. This marking can be done without removing the vitelline membrane, since it is permeable to the vital dyes, and thus the embryo continues to develop normally. The presence of the stain does not change the normal development of the embryo, and the position of the stained cells in the differentiated embryo clearly shows the fate of the stained area. It has been established by trial that the vital dye remains, on the whole, restricted to the cells which had originally taken up the dye and to their descendants. The diffusion of the stain, if the staining has been done correctly, is negligible and does not interfere with interpreting the results. Several stain marks may be made on the surface of the same embryo, using different colors (red, blue, brown). In this way one experiment may disclose the fate of many parts of the early embryo at the same time.

It was later found that cellophane can also be used as a stain carrier instead of agar and that it is actually a more convenient one, as cellophane comes in thin sheets from which it is easy to cut out pieces of the desired size and shape. All vital staining is now done with cellophane as the stain carrier.

Independently of vital staining, another method has also been devised for marking cells of a developing embryo. This consists in applying tiny particles of carbon to the surface of the embryo. Carbon particles stick to the surface of the cells and can thus be used as markers enabling the investigator to follow the movements of the cells and to draw up fate maps (Spratt, 1946).

The vital stain marking method was first applied to the reconstruction of the fate map in the amphibian embryo, and the original investigations have subsequently been checked by many embryologists. It is most advantageous, therefore, that the fate map of an amphibian embryo, such as the embryo of a newt *(Triturus)* or axolotl *(Ambystoma),* be described first.

The Fate Map of an Amphibian Embryo.   The whole surface of the blastula of an amphibian may be roughly divided into three main regions (Vogt, 1925): (1) a large area on and around the animal pole, (2) an intermediate zone, also known as the **marginal zone,** extending all around the equator of the blastula, and (3) the area on and around the vegetal pole (Fig. 110). These three main regions coincide approximately with areas which differ in their pigmentation. The whole of the animal region is deeply pigmented. The marginal zone, which is much broader on one side of the embryo, is pigmented but not so deeply as the region around the animal pole. The vegetal region is more or less devoid of pigment.

Inside each region we find areas corresponding to the future organs of the animal. The **animal region** consists of two main areas: the area whose fate is to develop into the nervous system of the embryo, and the area which is to become the skin epidermis of the embryo. The material for the sense organs is also contained in these two areas. Inside the nervous system area, a small subarea may be traced which is to participate in the formation of the eyes of the embryo; inside the epidermis area, the material for the nose, the ears, and the ectodermal part of the mouth may similarly be traced. In the intermediate or **marginal zone,** we find the material for the notochord occupying a large area on the dorsal side of the blastula. This is followed by an area lying nearer to the vegetal pole and containing the material for the prechordal connective tissue; this area is known as the prechordal plate. Farther down, toward the vegetal pole but still inside the marginal zone, lies the material for the anterior parts of the alimentary canal: the endodermal lining of the mouth, gill region, and pharynx. The parts of the marginal zone on both sides of the notochordal area are taken up by the material for the segmental muscles of the body. The lateral and ventral parts of the marginal zone give rise to the

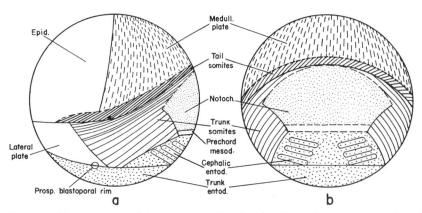

Figure 110.   Fate maps of the early gastrula of the axolotl. *a,* Lateral view; *b,* dorsal view. (After Pasteels, from Willier, Weiss, and Hamburger, 1955.)

mesodermal lining of the body cavity, the kidneys, etc. (ventrolateral mesoderm). The **vegetal region** is composed of cells which are later found in the midgut and hindgut.

It will be noticed that the areas destined to develop into the organs of the mid-dorsal part of the animal lie on one side of the blastula, on that side where the marginal zone is taken up by the notochord area. Nearer to the animal pole the area of the neural system is situated. This side of the blastula corresponds therefore to the dorsal side of the embryo. Similarly, parts of the head (eye, nose, ears) develop from areas near the animal pole of the blastula, which therefore corresponds to the anterior end of the embryo. Since the materials of the egg have not been displaced to any great extent during the cleavage, it may be inferred that the animal pole of the fertilized egg corresponds to the anterior end of the embryo. The side where the marginal zone is the broadest is the dorsal side; the opposite side may be considered as ventral, and the vegetal pole as the posterior. We find, however, that the foregut area (pharynx, part of the epithelium of the mouth) is also situated on what we have agreed to call the dorsal side of the egg. The explanation of this fact will be found later (Section 7–4).

On the whole, however, the location of the areas destined to develop into most of the organs does not seem to have anything in common with the position of the same organs in the adult animal. Organs which later are situated in the interior of the animals' body—such as the notochord, the gut, or the brain—are represented by areas laid out on the surface of the blastula. The cells destined to cover the whole of the animal's body, such as the epidermis of the skin, occupy only a limited area on the surface of the blastula. Parts of the three germinal layers—ectoderm, endoderm, and mesoderm—are all located on the surface of the embryo, instead of being in a concentric arrangement, with the ectoderm on the outside, the endoderm in the middle, and the mesoderm in between (p. 5). It is obvious that a far-reaching displacement or reshuffling of the parts of the blastoderm must take place before each cell can arrive at its final position.

This displacement of parts of the blastoderm, which are eventually rearranged in a system of concentric layers of cells, is the essence of the process of **gastrulation.** We shall now, for a time, leave the amphibians and consider the process of gastrulation in an animal with a more simple organization, a representative of lower chordates, the *Amphioxus*. Later, we will describe gastrulation as it occurs in amphibians and, lastly, in the meroblastic eggs of the fishes and birds.

7–2 GASTRULATION IN *AMPHIOXUS*

In *Amphioxus*, there are differences in the various regions of the egg cytoplasm which permitted Conklin (1932) to trace these regions into the later stages of development and thus to reconstruct a fate map, at least in rough outlines.

At the beginning of cleavage three regions can be distinguished in the *Amphioxus* egg. Near the vegetal pole a mass of cytoplasm is found which contains the greatest amount of yolk (although yolk in this case is not abundant and the yolk granules are relatively very small). The animal hemisphere of the egg consists of cytoplasm that has less yolk and is consequently more transparent. On one side of the egg, in a position roughly corresponding to the marginal zone of the amphibian egg, there is a special type of cytoplasm; it does not contain much yolk, but it can be distinguished from the animal cytoplasm by its ability to be deeply stained by basic dyes. The mass of cytoplasm of this kind has a crescentic shape, the attentuated ends of the crescent being drawn out along the equator of the egg about halfway around (Fig. 111).

During the period of cleavage the three regions become subdivided into blasto-

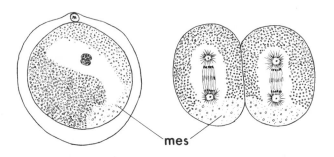

mes

Figure 111. Medial section of uncleaved egg of *Amphioxus* (left), and an equatorial section of the two-cell stage (right), showing the position of the mesodermal crescent. (After Conklin, 1932.)

meres without the cytoplasmic substances having been displaced to any great extent. The distinctions which could be traced in the cytoplasm of the egg now become accentuated by further distinctions in the size and shape of the blastomeres. The vegetal material is now contained in a number of rather large cells taking up the position on and around the vegetal pole of the blastula. The animal hemisphere is made up of cells containing the clear cytoplasm. The cells are columnar and form a very closely packed columnar epithelium. The cells containing the basophilic cytoplasm are clearly discernible even as to shape. They are the smallest cells in the blastula, even smaller than the animal cells, and they are rather loosely packed, the external surfaces bulging out, as is usually found in the earlier cleavage stages (Fig. 112).

The fate of the three regions is the following: The clear cytoplasm that later becomes the animal hemisphere of the blastula develops mainly into skin epidermis. The granular cytoplasm, which takes up the region around the vegetal pole of the blastula, develops into the lining of the alimentary canal. The crescent of basophilic cytoplasm is the material which gives rise to the muscles and the lining of the body cavity and thus represents a mesodermal area.

More recently, the method of local vital staining has been applied to the study of *Amphioxus* development (Tung, Wu, and Tung, 1962a). It was found that the presumptive mesodermal area is not restricted to one half of the egg only but reaches farther around the equator. On the opposite side a zone giving rise to the notochord could be detected, and above that a crescentic area which develops into the nervous system (Fig. 113). The similarity between this fate map and that of the amphibians is practically complete. There is also a striking similarity to the distribution of different kinds of cytoplasm in the ascidian embryo. (See Section 5–6.)

As a result of cleavage in *Amphioxus*, a blastula is formed which has a large blastocoele and a blastoderm consisting of a single layer of columnar cells (Fig. 115*A*). The

ect

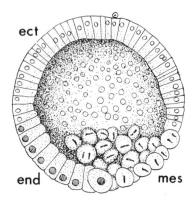

end   mes

Figure 112. Blastula of *Amphioxus* showing distinction between the cells of the mesodermal crescent (mes) and the cells of the presumptive endoderm (end) and ectoderm (ect). (After Conklin, 1932.)

cells at the vegetal pole are somewhat larger than at the animal pole, and the blastoderm therefore is thicker. The various cytoplasmic substances present in the egg suffer no appreciable displacement during cleavage, and the fate map presented in Figure 113 can equally well be applied to the blastula. The necessity of displacement of the parts so as to put them in the positions where they are situated in the adult animal is amply evident.

Gastrulation is initiated when the blastoderm at the vegetal pole becomes flat and subsequently bends inward, so that the whole embryo, instead of being spherical, becomes converted into a cup-shaped structure with a large cavity in open communication with the exterior on the side that was originally the vegetal pole of the embryó. The cup has a double wall, an external one and an internal one, the latter lining the newly formed cavity. The external and internal epithelial layers are continuous with each other over the rim of the cup-shaped embryo. In this stage there is still a space between the external and internal walls representing the remnants of what was the blastocoele of the blastula (Figs. 114 and 115).

The external lining consists of presumptive epidermis and presumptive nervous system. In other words, it consists of parts which have been classified as ectoderm. The internal lining consists mainly of the presumptive gut material, that is, of endoderm. The presumptive material of the notochord and the mesodermal crescent at first lie on the rim of the cup but very soon they shift inward so as to occupy a position on the internal wall of the cup. In this way the endoderm, the mesoderm, and the notochord

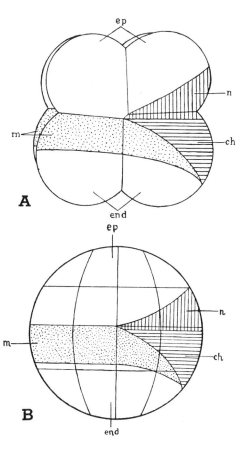

Figure 113. Fate maps of *Amphioxus* (*Branchiostoma*) *belcheri* at the eight-cell stage (*A*) and at the 32-cell stage (*B*). ch. Presumptive notochord; end, presumptive endoderm; ep, presumptive epidermis; m, presumptive mesoderm; n, presumptive neural system. (From Tung, Wu, and Tung, 1962a.)

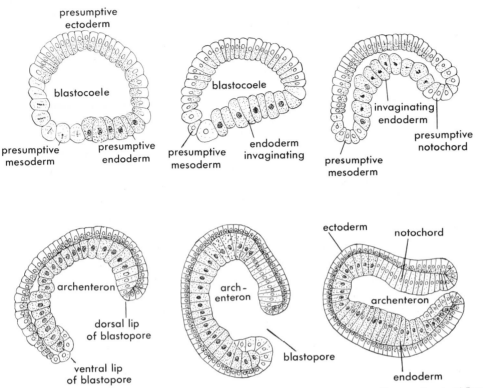

Figure 114. Gastrulation of *Amphioxus*. A series of consecutive stages (median sections). (After Conklin, 1932.)

disappear from the surface of the embryo into the interior where they belong. The external surface of the embryo now consists of ectoderm.

The embryo in this stage of development is called a **gastrula.** The movements of infolding or inward bending of the endoderm and mesoderm are known as **invagination.** The cavity arising through the invagination of the endoderm and mesoderm is called the **primary gut** or **archenteron.** The opening of the archenteron to the exterior is called the **blastopore.** At the same time, the blastopore denotes the pathway by which the endoderm and mesoderm pass into the interior of the embryo. The blastopore, being the opening leading into the primary gut, has been likened to a mouth; its rims, therefore, are usually referred to as the **lips** of the blastopore. We may distinguish the dorsal lip, the ventral lip, and the lateral lips of the blastopore, respectively.

The blastopore is very broad in the initial stage of gastrulation, but soon the lips of the blastopore begin to contract, so that the opening which leads into the archenteron becomes smaller and is eventually reduced to an insignificant fraction of the original orifice. This contraction of the lips of the blastopore is connected with the disappearance of the mesodermal crescent material and the presumptive notochord from the rim of the cup-shaped embryo. As more material is shifted into the interior of the gastrula, the remnants of the blastocoele become completely obliterated by the two walls of the embryo coming in contact with each other.

As the presumptive notochord and the mesodermal crescent shift into the interior of the gastrula, they also change their position relative to each other (Fig. 116). In the blastula, these two areas lie on opposite sides of the embryo. Now the lateral horns of the mesodermal crescent converge toward the dorsal side of the embryo and come to lie on

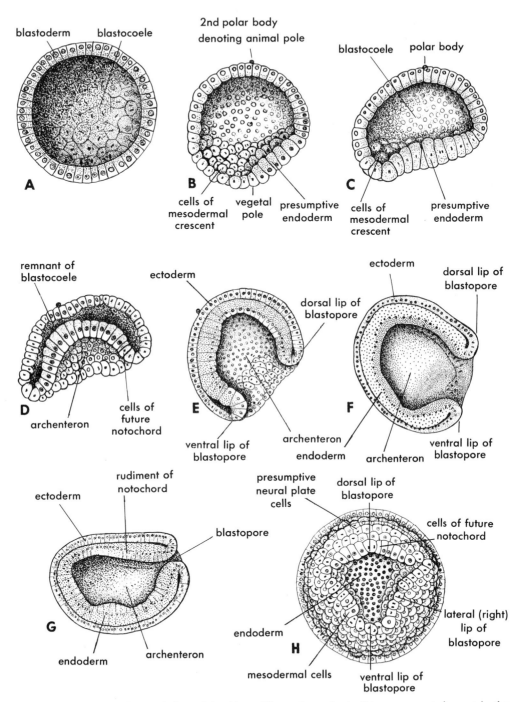

Figure 115. Stages of gastrulation of *Amphioxus*. The embryos in *A–G* are represented as cut in the median plane. *A*, Blastula; *B* and *C*, beginning of invagination; *D*, invagination advanced, the embryo attaining the structure of a double-walled cup with a broad opening to the exterior; *E* and *F*, constriction of the blastopore; *G*, completed gastrula; *H*, middle gastrula, whole, viewed from side of blastopore. (From Conklin, 1932.)

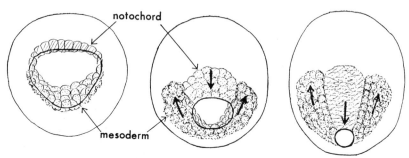

Figure 116. Change in the relative position of the presumptive mesoderm and the presumptive notochord during closure of the blastopore in *Amphioxus* (diagrammatic). (After Conklin, 1932.)

both sides of the presumptive notochord. In the next stage that follows the contraction of the rim of the blastopore, the embryo becomes elongated in the anteroposterior direction, all the various presumptive areas participating in this elongation. The elongation of the notochordal and the mesodermal material brings them into still closer contact with each other, the notochordal material shifting backward, in between the two horns of the mesodermal crescent material. As a result of these movements, the notochordal material becomes stretched into a longitudinal band of cells lying medially in the dorsal inner wall of the gastrula and flanked on both sides by bands of mesodermal cells similarly stretched in a longitudinal direction. The remainder of the lateral, ventral, and anterior parts of the inner wall of the gastrula consists of endodermal cells.

The external wall of the gastrula similarly takes part in the elongation of the embryo. One of the results of this is that the presumptive material of the nervous system becomes stretched into a longitudinal band of cells lying mediodorsally over the notochordal material but being somewhat broader than the latter.

7–3 FORMATION OF THE PRIMARY ORGAN RUDIMENTS IN *AMPHIOXUS*

Immediately after the germinal layers have taken up their positions in the inside and on the surface of the gastrula, the next phase of development sets in. The sheets of epithelium which were representative of the various parts of the future animal become broken up into discrete cell masses of diverse shape which can be called the **primary organ rudiments.** The term "primary" indicates that the structures in question are not final; they are actually complex in nature, and a further subdivision of the cell masses, or at least of most of them, is necessary before every single organ and structure of the adult animal appears as such.

In the interior of the embryo the presumptive materials of the notochord, the mesoderm, and the gut become separated from one another by crevices appearing along the boundary lines of each (Fig. 117). The strip of cells lying mid-dorsally rounds itself off forthwith and becomes a cylindrical cord of cells, the **notochord,** which still differs from the same organ of the adult *Amphioxus* in the structure of its cells.

As the presumptive material of the mesoderm becomes separated from both the notochord and the endoderm, it breaks up into a series of roughly cuboidal masses of cells lying on each side, one behind the other, along the length of the animal's body. These blocks of mesodermal cells are called the **mesodermal segments.** In connection with their formation, one detail should be noted. Just before the mesodermal cells become separated from the endoderm and notochord, a longitudinal groove appears on

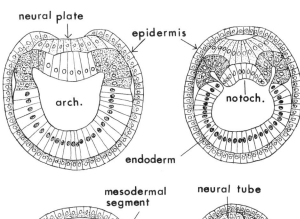

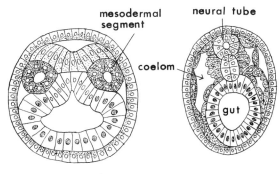

Figure 117. Formation of the primary organ rudiments in *Amphioxus* (transverse sections). (After Hatchek, from Korschelt, 1936.)

the inner surface of the mesodermal bands, that is, on the surface toward the archenteron. The groove becomes drawn out into each of the mesodermal segments, so that a pocket-like invagination is formed in each. At the same time, crevices cut in between adjoining segments from the outside, separating them from each other (Figs. 118 and 119). When the mesodermal segments eventually become separated from the endoderm and notochord, the pocket-like invaginations in each become completely closed off from the cavity of the archenteron and become small cavities inside the mesodermal segments. In a later stage of development, these cavities expand and become the secondary body cavity or **coelom** of the adult animal. The derivation of the cavities of the mesodermal segments from the archenteric cavity is distinct in the case of the host anterior segments, but the posterior mesodermal segments are solid masses of cells at first and acquire cavities later as a result of the separation of the cells in the middle.

When the notochord and the mesodermal segments dissociate themselves from the endodermal material, the free edges of the endoderm approximate each other and

Figure 118. Embryo of *Amphioxus* in lateral view showing mesodermal segments overlying the notochord and neural plate. (After Conklin, 1932.)

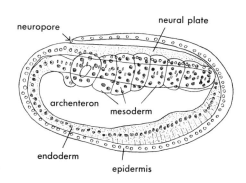

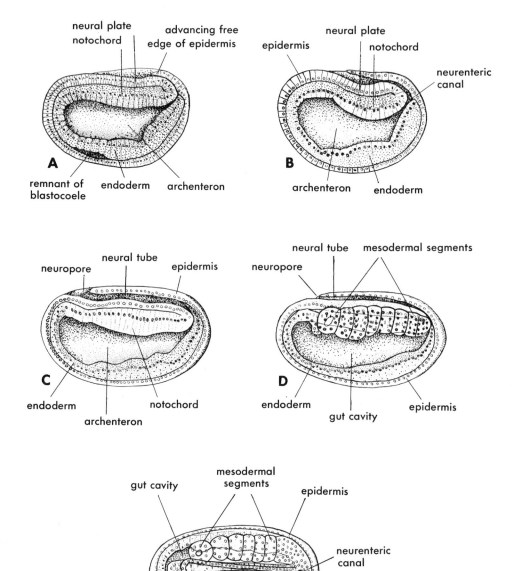

Figure 119.   Stages of neurulation of *Amphioxus.* In *A*, *B*, and *C* the embryos are represented as being cut in the median plane. *A*, Earliest stage of neurulation; *C*, almost completed neurula; *D*, slightly later stage than *C* but cut paramedially so that the right row of mesodermal segment can be seen; *E*, completed neurula whole, seen from the dorsal side. The transparency of the embryo allows one to see at the same time the various parts superimposed over one another (neural tube, notochord, mesodermal segments, and gut). (From Conklin, 1932.)

fuse along the dorsal midline. The endoderm thus becomes converted into a closed sac. The cavity of this sac becomes the lumen of the alimentary canal.

In the ectoderm the presumptive nervous system material becomes separated from the surrounding presumptive epidermis in the form of an elongated plate, the **neural plate.** The neural plate sinks below the level of the remainder of the ectoderm and is then covered by the free edges of the epidermal epithelium (Fig. 117). The neural plate forthwith rolls itself into a tube, the lateral edges folding themselves upward and fusing along the midline. The neural plate is thus transformed into the **neural tube,** which becomes the spinal cord of the animal. The neural tube does not close completely at the anterior end but leaves an opening, the **neuropore,** which remains patent until the later stages of development.

The epidermal epithelium which covers the neural plate by shifting over its surface is derived from the sides and from the area posterior to the neural plate. Because the hind end of the neural plate borders on the blastopore, the posterior portion of epidermis covering the neural plate is derived from the region below (ventral to) the blastopore. As the epidermis shifts forward over the surface of the neural plate, it also passes over the blastopore (Fig. 119). The blastopore becomes cut off from the exterior and opens into a space lined by the walls of the neural tube. The canal thus formed, which connects the blastopore, and therefore the archenteron, with the cavity of the neural tube, is called the **neurenteric canal.** The canal persists for only a short time, until the cavity of the neural tube (later becoming the central canal of the spinal cord) becomes completely separated from the cavity of the archenteron, which in turn becomes the cavity of the alimentary canal. The cavity of the alimentary canal later acquires communication with the exterior by means of the oral and anal openings which break through the body wall, but this does not take place until a very much later stage of development.

After the formation of the spinal cord, the notochord, the gut, and the mesodermal segments, the early embryo of *Amphioxus* possesses most of the essential features of the organization of chordate animals. The features which are still lacking are a mouth, an anus, and the gill clefts. The gill clefts, as well as the mouth, are developed in all chordates at a comparatively late stage. We will see that a stage of development in which an embryo possesses a spinal cord, notochord, gut, and segmented mesoderm recurs with great tenacity in all vertebrates, and that the subsequent stages of development in all classes of vertebrates are much more similar than are the processes of cleavage and gastrulation.

7–4 GASTRULATION IN AMPHIBIANS

The purpose which gastrulation must achieve in the amphibians is the same as in *Amphioxus:* the areas of the blastoderm which are destined to become gut, notochord, muscle, etc. have to be brought into the interior of the embryo. However, in the amphibians this cannot be done by the bending inward of the vegetal region of the blastoderm, as in *Amphioxus,* because the vegetal wall of the blastula is far too thick and overladen with yolk and is therefore rather passive in its behavior. The processes leading to the disappearance of the material for the internal organs from the surface are therefore carried out mainly by the more active cells of the marginal zone. A correct understanding of amphibian gastrulation has been achieved by the method of localized vital staining—by applying marks of vital stain on the surface of the embryo and tracing the movements of the stained areas in the course of gastrulation (Vogt, 1929).

The first trace of gastrulation which can be observed in an egg of a newt or a frog is the formation of a depression or groove on the dorsal side of the embryo, at the boundary between the gray crescent (pp. 98 and 125) and the vegetal region (Figs. 120 and 125). If the embryo is studied in sections at this stage, one may observe that the cells at the bottom of the groove are streaming into the interior of the embryo toward

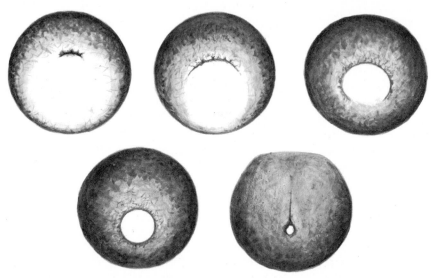

Figure 120. Changes in shape of the blastopore and closure of the blastopore during gastrulation in a frog.

the blastocoele cavity (Fig. 121). In the next phase of gastrulation, the groove begins to spread transversely, and its lateral ends are prolonged all along the boundary between the marginal zone and the vegetal region until they meet at the opposite, ventral, side of the embryo, thus encircling the vegetal region (Fig. 120). The groove represents the blastopore. It is produced by the invagination of the endodermal and mesodermal material into the interior of the embryo.

The invagination in this case is not quite similar to the invagination in *Amphioxus*. The original groove may be said to be the result of the bending inward of a portion of the blastoderm and the formation of a pocket-like depression. Once the depression has been formed, however, the further invagination can be better described as the rolling of the superficial cells over the rim of the blastopore and into its interior while new portions of the blastoderm approach the rim in their stead. While the superficial material is rolling over the rim of the blastopore, the rim itself does not remain stationary

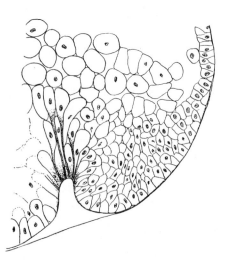

Figure 121. Median section through the blastopore region of an early gastrula of a newt, showing the cells at the bottom of the pit streaming into the interior. (After Vogt, 1929.)

but gradually shifts over the surface of the vegetal region, moving away from the animal pole and toward the vegetal pole. By the time the blastopore has the form of a ring, the cells of the vegetal region may still be seen filling the space enclosed by the edge of the blastopore. These cells are then called the **yolk plug** (Figs. 124, 125, and 126). The rim of the blastopore, however, continues to contract and at last covers the yolk plug altogether. In this way the material of the vegetal region disappears eventually into the interior of the embryo. When the blastopore contracts to such a degree that the yolk plug disappears from view, the blastopore is said to be "closed." This term is not quite exact, however, since a narrow canal leading into the interior still persists (Fig. 124).

If the blastoderm above the rim of the blastopore is marked with spots of vital stain, it can easily be seen that the stained areas stretch toward the lips of the blastopore, approach its rim, roll over the edge, and disappear inside (Fig. 122). Once inside, the stained material does not come to rest but continues its movements in the interior of the embryo, but this time it moves away from the blastopore in the opposite direction from that which it followed while it was still on the surface of the embryo. Also, the movement of the superficial material can be seen to be most rapid and extensive on the dorsal meridian of the embryo. More material is invaginated over the dorsal lip of the blastopore than over the lateral lips and the least over the ventral lip of the blastopore. This accounts for the varying breadth of the marginal zone around the circumference of the egg. In fact, the upper limit of the marginal zone is none other than the limit to which the blastoderm is invaginated during the process of gastrulation. The vegetal region is similarly that part of the blastoderm which is enclosed by the rim of the blastopore. The animal region is the part of the blastoderm that does not pass into the interior by way of the blastopore. The extent of the invagination can be clearly seen if a series of vital stain spots is made along the mid-dorsal meridian of the embryo, as shown in Figure 123.

On the dorsal side of the embryo, where the groove first appeared and where the streaming of cells into the interior of the embryo is the most active, a cavity is soon formed leading from the groove on the surface into the interior of the embryo. This cavity is lined on all sides by the invaginated cells and represents the archenteron (Figs. 124, 125, and 126). The archenteron is at first a narrow slitlike cavity, but later, as more material becomes invaginated, it expands at its anterior end and in so doing encroaches on the blastocoele. The latter is eventually obliterated (Fig. 124). While the rim of the blastopore moves over the surface of the vegetal region, the vegetal region is drawn into the interior and at the same time is caused to rotate, so that after the end of gastrulation it comes to lie in the ventral part of the archenteron, its originally exterior surface facing its cavity (Fig. 124). The opposite, dorsal, wall of the archenteron consists of cells of the marginal zone which have rolled into the interior over the dorsal lip of the blastopore.

We can now trace the displacements and movements of each area of the blastula individually.

The Ectoderm.   The material of the animal region, including the presumptive epidermis and the presumptive nervous system, greatly increases its surface during gastrulation, and at the end of gastrulation it covers the whole embryo after the mesoderm and the endoderm have disappeared inside. The **expansion** of the ectoderm is an active process, and the increase of surface area proceeds at the expense of a thinning out of the epithelial layer. The presumptive epidermis expands in all directions, but in the case of the presumptive nervous system the expansion is mainly in the longitudinal direction, i.e., toward the blastopore. In the transverse direction, on the other hand, the presumptive nervous system area contracts, and the material of the lateral horns of

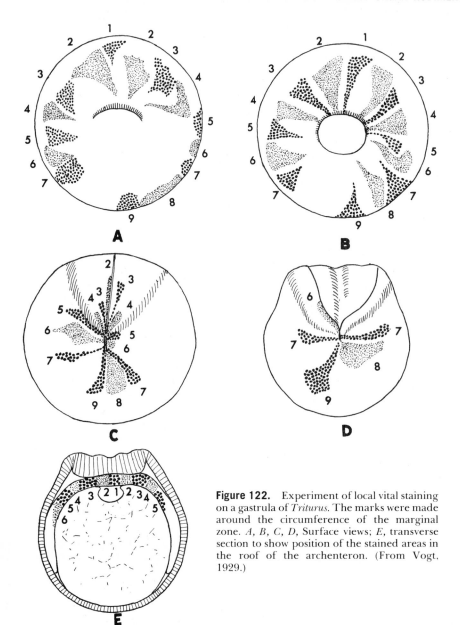

Figure 122. Experiment of local vital staining on a gastrula of *Triturus*. The marks were made around the circumference of the marginal zone. *A*, *B*, *C*, *D*, Surface views; *E*, transverse section to show position of the stained areas in the roof of the archenteron. (From Vogt, 1929.)

the crescentic area is drawn in toward the dorsal side of the embryo. As a result, the whole nervous system area changes its shape and becomes oval, elongated in an antero-posterior direction. The spreading of the ectoderm to cover other parts of the embryo is a case of so-called **epiboly.**

The Notochord.   The notochord rolls over the dorsal lip of the blastopore into the interior of the embryo and becomes stretched along the dorsal side of the archenteron, forming the mid-dorsal strip of the archenteron roof. As it does so, the presumptive notochord undergoes a very considerable elongation in the longitudinal direction and a corresponding contraction in the transverse direction. The notochordal material be-

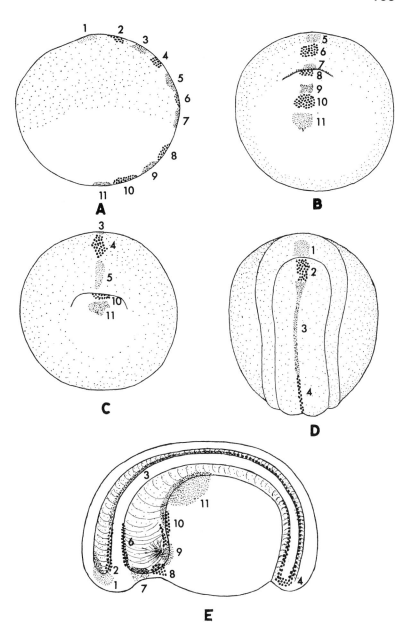

Figure 123. Experiment of local vital staining on a gastrula of *Triturus.* The marks were made along the dorsal median surface. *A, B, C, D,* Surface views; *E,* the embryo dissected in the medial plane to show the position of the stained areas in the inferior. (From Vogt, 1929.)

comes concentrated on the dorsal side of the embryo, as is the presumptive neural system, but to a much greater degree.

The Prechordal Plate.   The prechordal plate, which in the blastula lies just below the presumptive notochord, is the first mesoderm to invaginate and does so by rolling over the dorsal lip of the blastopore and becoming a part of the archenteron roof in front of the anterior end of the notochordal material.

The Mesoderm.   Of all parts of the blastula, the mesodermal area undergoes the most complicated movements. Most of the mesoderm invaginates into the interior by rolling over the lateral and ventral lips of the blastopore. Once inside the embryo, the mesoderm moves in an anterior direction as a sheet of cells, penetrating between the

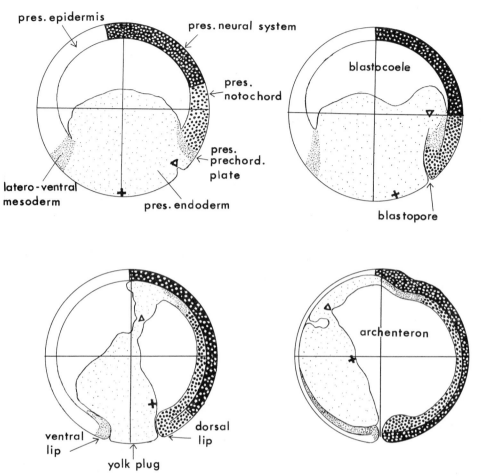

Figure 124. Diagrammatic median sections of amphibian embryos from the beginning to the completion of gastrulation. The triangle marks the cells which start invagination. The cross marks the position of the cells which were at the vegetal pole when gastrulation began. (After Vogt, 1929.)

ectoderm on the outside and the endoderm on the inside. The mesoderm in Urodela detaches itself from the endoderm and moves forward between the ectoderm and the endoderm, having a free edge anteriorly but preserving an uninterrupted connection with the notochordal material on the dorsal side of the embryo (Fig. 127). The notochordal and the mesodermal materials in this stage are in the form of one continuous epithelial sheet, the **chordomesodermal mantle.** In the Anura, the mesoderm does not split off from the adjoining endoderm until gastrulation is nearly finished. The result is, however, the same: The formation of the chordomesodermal mantle lying between the ectoderm and the endoderm. As the mesoderm moves from the posterior end of the embryo (represented by the blastopore) toward the anterior end, there remains at the anterior end a region which the mesodermal mantle has not yet reached. This region, which is free of mesoderm, diminishes as gastrulation proceeds but does not disappear completely. It is in this mesoderm-free region at the anterior end of the embryo that the mouth is later formed.

The concentration toward the dorsal side of the embryo, noted in respect to the notochordal and nervous system material, is also very distinct in the case of the meso-

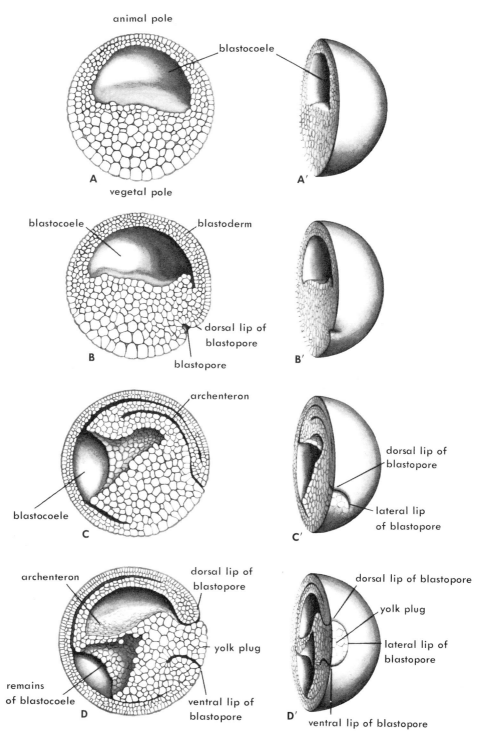

Figure 125. Four stages of development of a frog embryo; *A, A,'* late blastula stage; *B, B,'* beginning of gastrulation; *C, C,'* middle gastrula stage; *D, D,'* late gastrula stage (semidiagrammatic). Drawings on the left represent the embryos cut in the median plane; drawings on the right represent the same embryos viewed at an angle from the dorsal side (*A, B, C*) or from the posterior end (*D*).

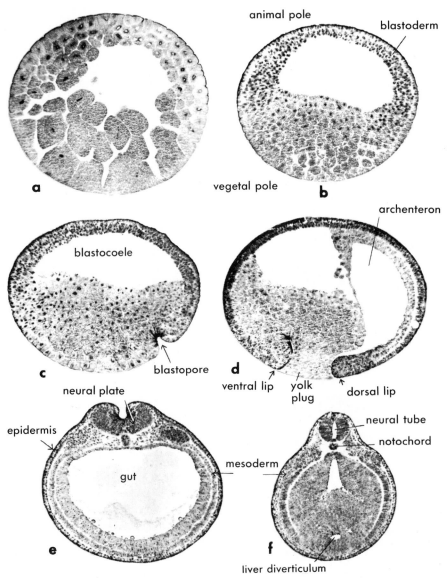

Figure 126. Blastula, gastrulation, and formation of primary organ rudiments in the frog. *a*, Late cleavage stage showing difference in the size of the blastomeres; *b*, late blastula; *c*, early gastrula; *d*, middle gastrula; *e*, neurula, transverse section; *f*, transverse section after completion of neurulation.

dermal mantle. The movement of the mesoderm inside the embryo does not follow the same path as on the surface. If the trajectories of all parts of the presumptive mesoderm are traced, they are seen to converge toward the dorsal side where they represent the movement of the mesoderm after its invagination (Fig. 128). As a result, the mesodermal material becomes concentrated toward the dorsal side. The mesodermal layer is thickest in the roof of the archenteron, where the mesoderm adjoins the notochord; it is thinned out in the lateral part and still more so in the ventral part of the embryo. The mesoderm continues to invaginate by rolling over the rim of the blastopore, even after the endoderm has come to lie inside and the yolk plug has disappeared from the

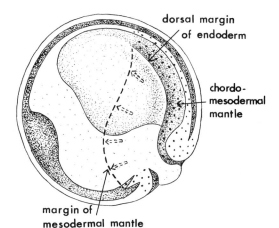

Figure 127. Gastrula of *Triturus* cut in the median plane to show separation of the chordomesodermal mantle from the endoderm. (After Vogt, from Hamburger, 1947.)

surface. The invagination of the mesoderm may therefore be considered as retarded when compared with the invagination of the endoderm. The degree of retardation varies in different animals. It is the least in frogs, greater in the urodeles, and greatest in the lamprey whose gastrulation is, otherwise, similar to that of the amphibians (Pasteels, 1940). As a result of the retardation, the presumptive mesoderm of the tail region and sometimes also the presumptive mesoderm of the posterior trunk region may still be on the surface of the embryo when the blastopore is "closed."

The Endoderm. The presumptive endoderm is found, in the blastula, partly in the marginal zone and partly in the vegetal region. The two parts of the presumptive endoderm invaginate in different manners; the part lying in the marginal zone is mainly absorbed into the original pitlike invagination of the blastopore, and the vegetal region disappears from the surface when it becomes covered up by the contracting rim of the blastopore.

The blastopore first appears as a pit in the endodermal area, between the marginal zone endoderm and the vegetal endoderm. The endodermal cells lying at the bottom of the pit are later found in the duodenal region of the embryo. In the early gastrula the presumptive endoderm of the pharyngeal and oral region lies on the anterior slope of the pit and is therefore invaginated as part of the dorsal lip of the blastopore. This

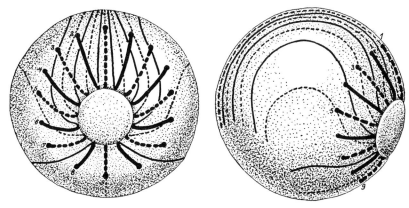

Figure 128. Trajectories of the movements of parts of the marginal zone during gastrulation in amphibians. Thick lines show movements of cells on the surface of the embryo; thin lines show movement of invaginated cells. (From Vogt, 1929.)

material later forms the most anterior part of the advancing archenteron. In the later stages of gastrulation, the oral and pharyngeal endoderm expands so as to form the spacious **foregut,** whose lateral, ventral, and anterior walls then consist of a rather thin layer of endoderm. Only part of the dorsal wall of the foregut is taken up by the prechordal plate and the anterior tip of the notochord.

The endoderm of the vegetal region passes into the interior of the embryo more or less passively and there comes to lie in the floor of the archenteron. It is not pushed forward as far as the marginal zone endoderm, but remains confined to the middle and posterior parts. The layer of endoderm in the floor of the archenteron is very thick, and the cavity of the archenteron is therefore reduced posteriorly to a rather narrow canal underneath the chordomesodermal mantle. This narrow part of the archenteron later becomes the **midgut** of the embryo. The lateral endodermal walls of the midgut are much thinner, and dorsally they end at a free edge after the mesoderm has been split off from the endoderm.

7–5 FORMATION OF THE PRIMARY ORGAN RUDIMENTS IN AMPHIBIANS

At the time of the closure of the blastopore, the separation of organ rudiments in amphibians has gone somewhat further than in *Amphioxus;* the chordomesodermal mantle has, to a great extent, separated itself from the endoderm, and the mesoderm has attained its definitive position between the endoderm and the ectoderm. To make the separation complete, the notochord becomes split at its anterior end from the prechordal plate, and the prechordal plate itself separates from the adjoining endoderm. The endoderm forthwith closes under the prechordal plate. The foregut is then surrounded on all sides by endodermal cells (Figs. 126 and 129). In the midgut region the free edges of the endoderm approach each other in the midline and fuse, thus completely enclosing the midgut. The chordomesodermal mantle becomes subdivided into its two components: the notochord in the middle and the mesoderm on both sides. As the notochord is separated from the mesoderm, the strand of notochordal cells becomes converted into a cylindrical body, which is round in cross section. The mesoderm, at the same time, becomes subdivided into a series of mesodermal segments. Contrary to what is found in *Amphioxus*, however, not the whole of the mesoderm becomes segmented but only its dorsal part, the part adjoining the notochord. The mesodermal layer has been thickened here owing to the convergence of the mesoderm toward the dorsal side of the embryo during gastrulation. This thickened part now becomes segmented by a series of transverse crevices cutting into the mass of mesodermal cells and separating them (Fig. 130). The thinned out lateral and ventral parts of the mesodermal mantle on each side do not become subdivided into segments, and these parts of the mesoderm are known as the **lateral plates.** The mesodermal segments in amphibians, called **somites,** are not fully homologous to the primary mesodermal segments of *Amphioxus.* Each somite is separated from the adjoining somites but remains connected to the dorsal edge of the lateral plate. The strand of cells connecting the somite to the lateral plate is known as the **stalk** of the somite, and these cells, as will be shown later, give rise to the excretory system of the embryo.

At about the same time that the segmentation of the dorsal part of the mesoderm is occurring, the lateral plate mesoderm becomes split into two layers, an external layer applied to the ectoderm known as the **parietal layer,** and an internal layer applied to the endoderm, the **visceral layer.** The narrow cavity between the two layers is the **coelomic cavity** or **coelom.** In later stages the cavity expands and becomes the body cavity

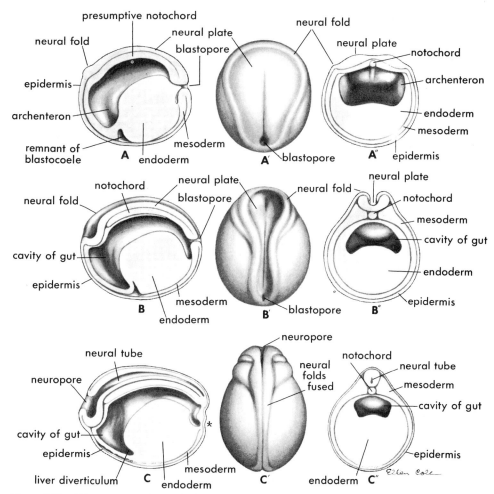

Figure 129.   Three stages of neurulation in a frog embryo. The drawings in the middle show whole embryos in dorsal view. The drawings on the left show the right halves of embryos cut in the median plane. The drawings on the right show the anterior halves of embryos cut transversely. *A, A', A''*, Very early neurula; *B, B', B''*, middle neurula; *C, C', C''*, late neurula with neural tube almost completely closed. *C* shows the blastopore closed; the asterisk indicates the point at which the anal opening will break through.

Figure 130. Segmentation of the dorsal mesoderm in an embryo of a salamander (*Ambystoma punctatum*). (Modified from Adelmann, 1932.)

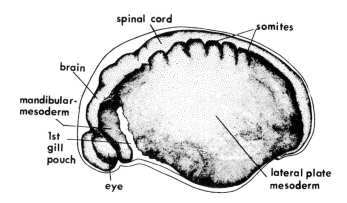

of the adult animal. (See Figure 323, p. 383.) Small cavities also appear in the somites, but these are obliterated later and disappear without a trace. There is never any connection between the coelomic cavities in the mesoderm of the amphibians and the archenteric cavity.

The presumptive nervous system has been traced previously to the stage when it forms an approximately oval-shaped area on the dorsal side of the embryo. The neural area covers the areas of the notochord and the somites in the middle and posterior parts of the embryo and the prechordal plate in front. The anterior end of the neural area is underlain by part of the endodermal lining of the foregut. After the closure of the blastopore, the presumptive area of the nervous system becomes differentiated from the rest of the ectoderm in the form of the **neural plate** (Fig. 129). The ectodermal epithelium of the neural plate becomes thickened by the further concentration of the epithelium toward the dorsal side of the embryo. At the same time, the cells of the neural plate change in shape; they become elongated and arrange themselves into a columnar epithelium. Thus, they are different from the cells of the epidermis, which remain more or less cuboidal and arranged as a stratified epithelium usually two cells thick.

Superficially the neural plate becomes visible because of a concentration of pigment at the edges of the plate. Soon, however, the edges of the neural plate become thickened and raised above the general level as **neural folds.** A shallow groove may be seen at the same time along the midline of the neural plate, separating it into right and left halves. The neural plate continues to contract in a transverse direction, especially in its posterior parts. The neural folds become higher, so that eventually the neural plate is converted into a longitudinal depression, bordered laterally and in front by the neural folds (Fig. 129). Subsequently, the folds make contact in the midline and fuse, beginning from the point corresponding to the occipital region of the embryo and progressing forward and backward. In this way the neural plate is transformed into the neural tube (Fig. 126 and 131). The cavity of the neural tube is broadest in the anterior part of the tube, which later develops into the brain with the brain cavities. In the posterior part of the neural tube, the cavity is much narrower. It later becomes the central canal of the spinal cord. Sometimes the central canal is not found in the posterior part of the neural tube at the time of the closure of the neural folds, and in this case the tube is hollowed out later by the separation of the cells in the middle of the organ.

The formation and closure of the neural plate is known as **neurulation,** and the embryo, in the stages when it possesses a neural plate, is called a **neurula.**

In frogs, the neural folds reach posteriorly to the level of the blastopore or even slightly beyond it. As a result the blastopore, now reduced to a narrow canal, is surrounded by the neural folds. When the folds fuse, the external end of this canal is found in the floor of the neural tube. In most frogs, by the time the neural tube is formed, the blastopore becomes completely occluded by the fusion of its edges. In some amphibians, however, particularly the clawed toad, *Xenopus*, the canal remains open and becomes the neurenteric canal, which for some time provides a direct communication between the posterior end of the archenteron and the neural tube (Fig. 132). (Compare the same formation in *Amphioxus*, Figure 119 and p. 160.) The neurenteric canal is later interrupted and disappears without a trace.

After the blastopore ceases to provide an opening for the gut to the exterior, a new opening breaks through at the posterior end of the embryo, slightly below the position of the blastopore. This opening becomes the anus (Tahara and Nakamura, 1961). In urodeles, however, the blastopore remains outside the neural folds and becomes directly transformed into the anal opening.

After the neural folds have fused in the midline, the neural tube separates itself

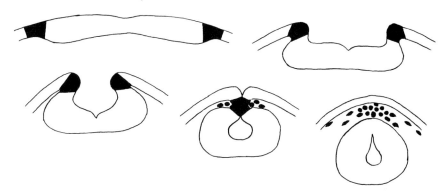

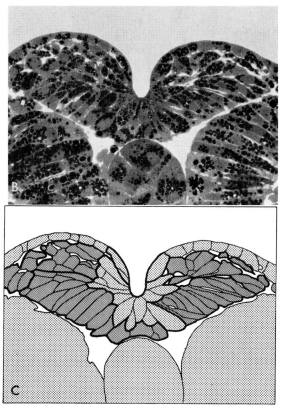

Figure 131. *A*, Stages in the formation of the neural plate and neural tube in amphibians. Transverse sections (diagrammatic). Neural crest cells are shown in black. *B*, Cross-section of neural plate and adjacent mesoderm in the trunk region of a *Xenopus laevis* embryo (stage 18½). Neural groove is U-shaped. *C*, Diagram of cross-section shown in *B*, Cellular outlines illustrate kinds of cellular shape changes that accompany neurulation in this embryo. (*B* and *C*, Courtesy of Dr. T. E. Schroeder.)

completely from the overlying epidermis. The free edges of the epidermis fuse, so that the epidermis becomes continuous over the back of the embryo. A certain number of cells, however, do not become included either in the neural tube or in the epidermis. These cells can be traced back to strips running all along the crest of the neural folds (Fig. 131). After the neural tube has become separated from the epidermis, these cells are found as an irregular flattened mass between the neural tube and the overlying epidermis. This mass of cells is the **neural crest.** It may be continuous across the midline at first, but soon the mass of cells separates into a right and a left half, lying dorsolaterally to the neural tube. The neural crest cells have a special part to play in the development of the embryo, as will be indicated later (Section 12–3).

During neurulation, the embryo already begins to stretch in length. At the same

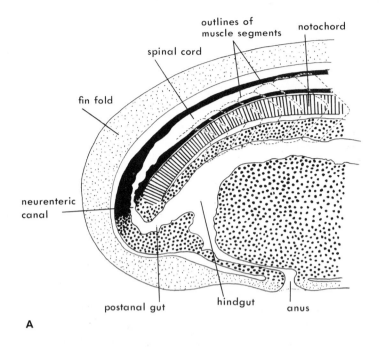

A

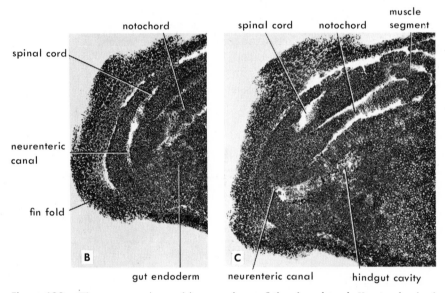

Figure 132.   The neurenteric canal in an embryo of the clawed toad, *Xenopus laevis. A,*
Reconstruction of a median section through the embryo, showing the continuity of the gut
cavity and of the central canal of the neural tube at the posterior end of the embryo. *B* and
C, Photographs of two of the sections used for the reconstruction; owing to a slight
curvature of the neurenteric canal, only part of it is visible in any one section.

time, it is flattened from side to side and also becomes smaller in a dorsoventral direction,
but the volume of the embryo does not change appreciably. The stretching is greatest in
the posterior part of the embryo. As the neural tube takes part in this general stretching, it becomes still more attenuated in its posterior parts, so that the difference between
the presumptive brain and the presumptive spinal cord is increased. The part of the

embryo above the blastopore becomes elongated beyond the blastopore, and this elongation becomes the rudiment of the tail, known as the tailbud. (See Figure 233, p. 303.) The pharyngeal pouches are formed as lateral outpushings of the foregut, and in later development they give rise to the gill clefts. The mouth later breaks through at the anterior end of the foregut.

7–6 GASTRULATION AND FORMATION OF THE PRIMARY ORGAN RUDIMENTS IN FISHES

In fishes the yolk formed in the oocytes during oogenesis initially does not differ substantially from the yolk in amphibian oocytes and eggs. It consists of yolk platelets lying in the cytoplasm and intermingled with other cytoplasmic components (mitochondria and the like). The cores of the yolk platelets, similar to those of amphibians, may show a crystalline structure, although this structure is lost by the time the eggs are ripe. In the immature egg the surface is formed by yolk-free cytoplasm containing numerous cortical granules. After fertilization, however, the cytoplasm moves toward the animal pole of the egg, and there it forms a cytoplasmic "cap." This is particularly the case in bony fishes. Active cytoplasm is withdrawn both from the surface and from the interior of the egg. As a result, the yolk platelets are pressed closer together. The process is a gradual one and occurs earlier or later in different fishes. In the more primitive fishes the condensation of the yolk does not go too far; cleavage remains complete, and development follows essentially the pattern found in amphibians (Dipnoi: see Pasteels, 1962).

In the Teleostei, however, the yolk platelets are pressed together in the central parts of the egg to such an extent that the cytoplasm is completely squeezed out from between them. The fertilized egg then consists of the cytoplasmic cap on the animal pole containing the nucleus or nuclei, the central mass of yolk, and a thin layer of cytoplasm surrounding the yolk. Cleavage occurs only in the cytoplasmic cap on the animal pole; the yolk and the layer of cytoplasm surrounding it remain uncleaved. Even the cytoplasmic cap of the animal pole is not completely subdivided into cells; the deeper parts of the cytoplasm are provided with nuclei in the process of cleavage, but the mass of cytoplasm fails to divide and remains in the form of a syncytial layer adjoining the uncleaved yolk. This layer is called the **periblast.** The cellular mass on top of the periblast represents the **blastoderm.** The superficially lying cells of the blastoderm adhere more firmly to one another than the cells lying in the interior, thus forming a **"covering layer"** (Fig. 133).

Only the blastoderm is responsible for the formation of the body of the embryo.

Figure 133. Diagram of blastodisc of a bony fish. DB, Deep blastomeres; EL, external cell layer; P, periblast; PN, nuclei of periblast; SC, segmentation cavity (blastocoele); Y, yolk; YCL, cytoplasmic layer on surface of yolk. (From Lentz and Trinkaus, 1967.)

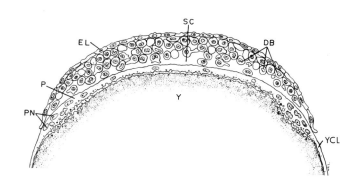

The other parts—the periblast, the yolk, and the surface layer of cytoplasm surrounding the yolk (where it is not covered by the periblast and the blastoderm)—do not contribute directly to the construction of the body of the embryo and, for this reason, are designated as **extraembryonic.**

The three germinal layers are formed within the blastoderm. How this occurs is a subject of long-standing controversy. It has been claimed (Pasteels, 1940; Oppenheimer, 1947) that the endomesoderm arises from the edge of the blastodisc, and that therefore the edges of the blastodisc correspond to the lips of the blastopore, as found in amphibians, cyclostomes, and the more primitive fishes. Experiments have furnished evidence against this view by showing that there is no significant sinking in of the superficial material into the interior and that the endomesodermal layer is formed by a rearrangment of the deep-lying cells of the blastodisc (Ballard, 1966a, b, c). Whatever the mechanism of its formation, the endomesodermal layer is found along the edges of the blastodisc, where it is continuous with the superficial (ectodermal) layer (Fig. 134). Furthermore, the endomesodermal layer is best developed at what will be the posterior edge of the blastodisc. Here the blastodisc is thickened, forming an **embryonic shield,** and it is here that the primary organ rudiments are formed: the neural plate and tube, the notochord, and laterally, the somites. The endoderm is separated from the mesoderm rather late and is then represented as an elongated plate of cells stretched longitudinally underneath the notochord. It does not contain a cavity until a much later stage. After the separation of the endoderm, the remaining layer becomes equivalent to the chordomesodermal mantle.

In fishes the formation of the primary organ rudiments starts with the most anterior parts and progresses backward. The lateral edges of the blastodisc are gradually involved in the body formation; they are drawn toward the midline and contribute to the formation of the more posterior parts of the body and the tail (Fig. 135).

Even before the primary organs of the embryo are laid down, the extraembryonic yolk is brought into the orbit of development of the new organism by the blastodisc's spreading over the surface of the yolk and eventually surrounding it completely.

Immediately after fertilization the cytoplasmic cap rises quite steeply over the surface of the yolk, acquiring a very nearly hemispherical shape. During cleavage the mass of blastomeres is still markedly elevated, but at the end of cleavage it spreads out over

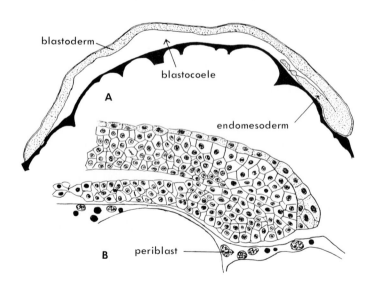

Figure 134. Early stage of gastrulation in a trout (median section). *A*, View of the whole blastodisc under low magnification; *B*, part adjoining the posterior edge of blastodisc at higher magnification.

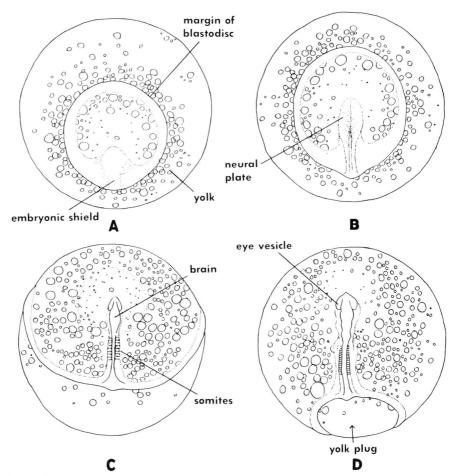

Figure 135. Stages of development of the trout. *A*, Blastodisc in the stage of gastrulation; *B*, neural plate stage; *C*, formation of the brain, eye vesicles, and somites; *D*, overgrowth of yolk almost completed.

the yolk and becomes flatter; it can then be likened to an inverted saucer and is very well characterized by the term "embryonic disc." During the formation of the germinal layers the disc spreads over a wider area covering a greater part of the surface; this continues during organ formation. As a result, an ever increasing part of the yolk is covered by the blastodisc (Fig. 135). It has been shown that the spreading of the embryonic disc occurs at the expense of a thinning of the cellular layer, without an appreciable increase in its mass. It is therefore a process of **epiboly,** similar to the spreading out of the animal hemisphere of the amphibian embryo during gastrulation.

The advancing edge of the disc consists of three layers: the periblast, the mesoderm, and the ectoderm. No endoderm is present at the edge of the blastodisc, except in a small midposterior portion; thus, endoderm does not participate in the overgrowth of the yolk.

If the cellular part of a fish blastula is removed, the remaining periblast nevertheless spreads out over the yolk (Trinkaus, 1951). It may therefore be concluded that this syncytial layer is essential for the overgrowth of the yolk and that it provides a suitable substrate over which the germinal layers (mesoderm and ectoderm) may also expand and shift over the surface of the yolk. Eventually the edges of the blastodisc, having surrounded the entire yolk, converge and close at the posterior end of the embryo, similar to the closure of the blastopore in an amphibian or *Amphioxus*. Just before the yolk is completely closed in, a yolk plug may be seen projecting between the constricting edges of the blastodisc. This yolk plug looks very much like that of the amphibians but differs from it insofar as it consists only of nonsegmented yolk, whereas the yolk plug in amphibians consists of cells of the vegetal region. The yolk is thus enclosed in an envelope consisting of periblast, mesoderm, and ectoderm.

The process of neurulation in the bony fishes differs considerably from that in amphibians. Although a neural plate is found, the neural folds are only slightly indicated. The neural plate does not roll into a tube but narrows gradually, at the same time sinking deeper and deeper into the underlying tissues. Eventually it separates from the epidermis, which becomes continuous over the dorsal surface of the embryo. No cavity appears in the nervous system rudiment while it is being separated from the remainder of the ectoderm. The ventricles of the brain and the central canal of the spinal cord are formed later by the separation of cells in the middle of these organs.

Gastrulation and the formation of the primary organ rudiments in elasmobranchs follow the same pattern as in the bony fishes—with the difference, however, that an archenteron is formed at the posterior edge of the blastodisc. Likewise, the rudiment of the nervous system is not solid, as in bony fishes, but the neural plate is rolled into a tube having a distinct cavity right from the start.

7–7   GASTRULATION AND THE FORMATION OF THE PRIMARY ORGAN RUDIMENTS IN BIRDS*

Numerous experiments have been conducted to study the fate map and the gastrulation movements in the bird embryo. The method of local vital staining, as applied with success in the study of amphibian gastrulation, was used in the earlier work on bird embryos (Wetzel, 1929; Pasteels, 1940, and others). The results have been disputed, since vital stains tend to diffuse and thus could give misleading results. An alternative method has been to mark portions of the blastoderm with fine carmine or carbon (charcoal) particles which stick to the surface of the cells and are carried around by the cells during their movements (Spratt, 1946, and others). The particles, however, do not become constituents of the cells and could possibly be shed by the cells on their way, at least in some cases. The uncertainties in experimental results have been augmented by the fact that the carbon markings were made on blastoderms cultivated *in vitro*, removed from the egg and placed, usually upside down, on the surface of agar gel or coagulated albumen. The expansion of the blastoderm, as in normal development, does not proceed on agar blocks. The most reliable method was found to be to mark parts of the blastoderm with radioactive tracers (tritiated thymidine). The method is somewhat complicated. First, an explanted blastoderm is immersed in a medium containing tritiated thymidine. In three to eight hours the tritiated thymidine is incorporated into the chromosomal DNA of the growing and reproducing cells of the embryo. The embryo labeled with tritiated thymidine now serves as a donor. To establish the fate and the movements of a particular part of the embryo, a recipient embryo is chosen which is in the same stage of development as that attained in the meantime by the donor. A small area of the recipient embryo is then excised and replaced by a corresponding piece from the donor embryo (Fig. 136). Healing usually occurs quickly, and the develop-

---

*The subject has been reviewed by Bellairs (1972).

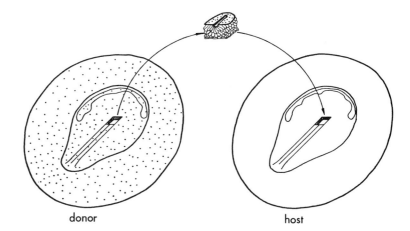

Figure 136. Diagram explaining the method of testing the fate of a circumscribed part of a chick blastoderm by implanting a corresponding part from a tritiated thymidine-labeled donor (stippled) into an unlabeled host. (From Nicolet, 1970a.)

donor   host

ment is not disturbed. The chromosomal DNA does not pass out of the nuclei of the labeled cells but remains in the chromosomes of their descendants, though it is gradually diluted with each subsequent replication of the chromosomes. In spite of this the radioactivity remains for a considerable time, and if the embryos are later fixed and sectioned, autoradiograms can be prepared by coating the sections with photographic emulsion. Silver grains in the developed emulsion provide unequivocal evidence that particular cells or groups of cells are descendants of the original graft.

Unfortunately, the work has to be done on embryos cultivated *in vitro*. In addition, having to label the donor embryos for several hours before the operation precludes the study of blastoderm development immediately after the eggs are laid, but only after the first stages of development had already passed.

Nevertheless, by the use of tritiated thymidine marking, fate maps of bird embryos have been obtained which are far superior in detail and reliability to any which have been based on vital staining or carbon particle marking. The fate map in Figure 137 is based on this work (Rosenquist, 1966; Nicolet, 1970a). The map refers to the epiblast only. Roughly in the center of the area pellucida lies a small area which will produce the notochord. Posterior to this, in the median plane of the embryo, lies an elongated oval area which is the presumptive endoderm. The part nearer to the notochordal area is incorporated in the body of the embryo proper, and it produces the gut with its subordinate parts. Further toward the posterior edge of the area pellucida lies the extraembryonic endoderm which will form part of the lining of the yolk sac. (See Section 10–1, p. 254.) To the right and left of the presumptive notochord and endoderm and also posterior to the presumptive extraembryonic endoderm lie the various subdivisions of the presumptive mesoderm. Nearer to the notochordal area is situated a not very precisely defined section producing the prechordal (head) mesoderm. Posterior and lateral to that section are the presumptive somite areas, and posterior to these is the presumptive lateral plate mesoderm; the right and left lateral mesoderm areas are in continuity with each other posterior to the endodermal area. Toward the posterior edge of the area pellucida a fairly large crescentic area is situated which will give rise to extraembryonic mesoderm, particularly to the extraembryonic blood vessel system (area vasculosa, p. 254). The presumptive ectoderm occupies the anterior and lateral parts of the area pellucida. A roughly semicircular area anterior to the notochordal material is the presumptive neural system. In a band surrounding this is the presumptive embryonic epidermis (including material for the various epidermal derivatives).

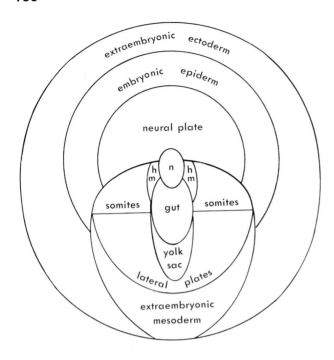

Figure 137. Fate map of a bird embryo just before the beginning of gastrulation. (After experiments of Rosenquist, 1966, and Nicolet, 1970a.) It is assumed that if there is a large scale forward thrust in the epiblast from the posterior margin, this thrust has already occurred at a previous stage. hm, Presumptive head mesoderm; n, presumptive notochord.

Still further outward, forming almost a complete ring at the outer edge of the area pellucida, lies the extraembryonic ectoderm—ectoderm which will not take part in building the body of the embryo proper. This it has in common with the entire area opaca, as has been indicated previously.

The presumptive areas of the blastoderm take up their positions in the embryo by means of a series of morphogenetic movements. One of these, previously mentioned (Section 6–4, p. 117), is the forward movement in the hypoblast (Spratt and Haas, 1960). The movement starts very early in the chick, even before the egg is laid, and contributes to the formation of the hypoblast, if it is not actually the means by which the hypoblast is formed. Even after the hypoblast is complete, the forward streaming continues; from the region at the posterior end of the area pellucida the hypoblast cells flow in a "fountain-like" fashion spreading out toward the anterior and lateral edges of the area pellucida (Fig. 138). Some of the hypoblast cells are embedded in the thick-

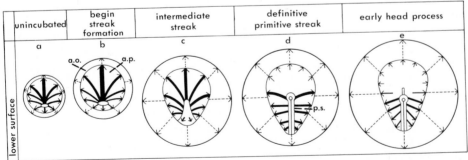

Figure 138. Diagram showing morphogenetic movements in the hypoblast of chicken embryo in stages leading to the formation of the primitive streak. ao, Area opaca; ap, area pellucida; ps, primitive streak. (From Spratt and Haas, 1960.)

ened border between the area opaca and the area pellucida at the anterior edge of the latter.

Several hours after the beginning of incubation of the laid egg, a thickening can be noticed at what will be the posterior edge of the area pellucida. This marks the beginning of the processes that lead to gastrulation and the formation of the three germinal layers in the bird embryo. Since the organ-forming part of the blastoderm in birds occupies only a limited area inside the blastodisc (only the central part of the area pellucida), the invagination of the endoderm and mesoderm takes place in a special region inside the area pellucida. This region is a median strip of blastoderm, starting from the thickening, already mentioned, at the posterior edge of the area pellucida and eventually extending to about three fifths or three quarters of the entire length of the area. This strip of blastoderm becomes thickened during the stages of gastrulation and is known as the **primitive streak.** Along the middle of the primitive streak, when it is fully developed runs a narrow furrow, the **primitive groove.** At the anterior end of the primitive streak there is a thickening, the **primitive knot** or **Hensen's node** (Fig. 139*B*). The center of Hensen's node is excavated to form a funnel-shaped depression. In some birds (the duck, for instance) this depression extends forward into a narrow canal which will be referred to later.

The thickening of the blastoderm is brought about by a convergence of its surface layer, the epiblast, toward the midline in the posterior half of the area pellucida. As the cells nearest the midline become concentrated to form the early primitive streak, the parts of the epiblast lying farther out laterally and anterolaterally swing in a curve backward and inward to take the place of parts of the blastoderm shifting toward the midline (Fig. 140). The primitive streak first becomes visible in the hindmost part of the area pellucida and is designated the **short primitive streak** (Fig. 139*A*).

The primitive streak now elongates by the concentration of more and more material from the sides toward the midline in front of the original short primitive streak. The early primitive streak is broad and its edges are somewhat vaguely indicated. In later stages it contracts in a transverse direction, becoming narrower and quite sharply delimited, and is called the **definitive primitive streak** (Fig. 139*B*). In the process of this transformation, the primitive streak elongates and its anterior end is pushed even farther forward, though the amount of this forward thrust has been a matter of controversy among embryologists. (See Waddington, 1952.)

The movements in the blastoderm leading to the final placement of cells in the hypoblast and to the formation of the primitive streak in the epiblast may be called **pregastrulation movements,** to distinguish them from the gastrulation movements proper.

At the stage of the short primitive streak, the cells of the blastoderm already begin to invaginate into the space between the epiblast and the hypoblast. (See Section 6–4.) The process of invagination is different from that in *Amphioxus* and in the amphibians. There is no infolding of the epithelial layer of the blastoderm. In the birds, the cells seem to be moving downward singly, even if there are many cells moving in the same direction. The regular epithelial arrangement of cells found in the epiblast is lost (Figs. 141, 142, and 145). This type of gastrulation movement bears the name of **immigration.**

Soon the migrating cells reach the hypoblast and establish intimate contact with the cells of this layer. Henceforth the entire primitive streak is a mass of moving cells. The direction of movement is mainly downward from the surface of the blastoderm toward the hypoblast, but the mass of migrating cells also spreads out sideways and forward from the anterior end of the primitive streak.

Although the cells of the chick blastoderm move individually, their movements are

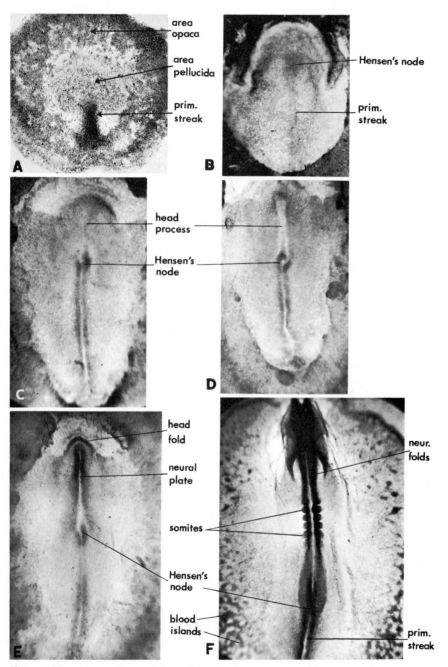

Figure 139. Gastrulation and formation of primary organ rudiments in a chick embryo (surface views). *A*, Short primitive streak (stage 3) (from Hamburger and Hamilton, 1951); *B*, definitive primitive streak (stage 4); *C*, head process stage (stage 5); *D*, retreating primitive streak (stage 6); *E*, neural plate (stage 6 +); *F*, closure of neural tube and formation of somites (stage 9 −). (Numbers of stages after Hamburger and Hamilton, 1951.)

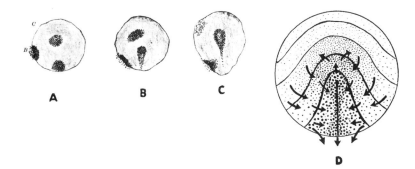

Figure 140. Movements in the epiblast during the formation of the primitive streak in a chick embryo as shown by displacement of vitally stained areas (*A, B, C*). *D,* Diagram of movements based on this and similar experiments. (From Wetzel, 1929.)

obviously directed by common causes, and therefore the whole mass of cells moves in a coordinated fashion. The formation of depressions on the surface of the blastoderm, the furrow along the midline of the primitive streak, and the funnel-shaped depression in the center of Hensen's node are due to this mass movement of cells from the surface of the blastoderm into the interior.

As the cells of the epiblast migrate into the interior, whole areas of the blastoderm disappear from the surface. They are replaced, however, by the adjoining areas moving toward the midline and taking their places in the primitive streak. The newly arrived cells in their turn migrate down into the interior. Thus the primitive streak persists, although the cells of which it is made do not stay in the same place but are constantly replaced. The first areas to start invaginating are the presumptive endoderm, the notochord, and the presumptive head mesoderm.

The presumptive notochordal cells become concentrated in the definitive primitive streak in the deeper parts of Hensen's node. They then start moving as a dense mass in the midline straight forward from Hensen's node underneath the surface of the epiblast (Fig. 143).

The mass of notochordal cells can be distinguished from the surrounding (mainly mesodermal) cells and is called the **head process,** or more correctly, the **notochordal process** (Fig. 139C). The narrow canal mentioned previously which is found in some

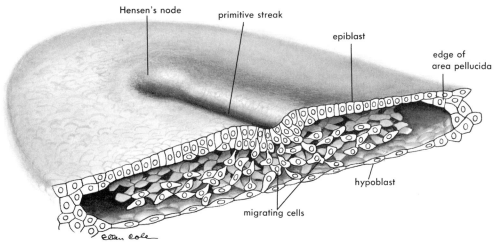

Figure 141. Anterior half of the area pellucida of a chick embryo cut transversely to show the migration of mesodermal and endodermal cells from the primitive streak.

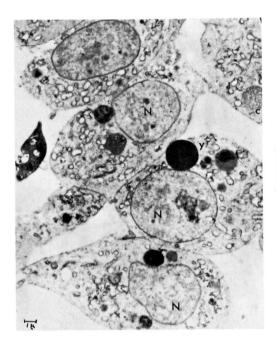

Figure 142. Cells in the zone of migration underneath the primitive streak of a chick embryo (electronmicrograph). N, Nuclei of mesenchyme cells; y, yolk granule. (From Balinsky and Walther, 1961.)

birds (e.g., the duck) penetrates into this notochordal process. Since formation of the canal is due to the movement of the invaginating cells of the gastrula, the cavity of the canal must be recognized as corresponding to a part of the archenteron, even though the lining of the canal consists exclusively of presumptive notochordal cells. There is no other cavity in the development of birds that could be considered as a homologue of the archenteron. The invagination of the mesoderm or the endoderm does not lead to the appearance of a cavity. We have already seen that in other vertebrates the archenteron may also be completely absent (pp. 175–178).

The endoderm starts invaginating at an early stage, even before the primitive streak is fully formed. Cells of the presumptive endoderm penetrate into the hypoblast and push outward and forward the cells originally composing this layer. As a result, an extensive part of the hypoblast around and in front of the Hensen's node becomes replaced by cells derived from the primitive streak (Hunt, 1937; Rosenquist, 1966; Nicolet, 1970a). This area shifts forward and later gives rise to the foregut. The endoderm lying in the more posterior part of the primitive streak after invagination moves farther laterally, and even in a laterocaudal direction, replacing the original hypoblast and later giving rise to part of the lining of the yolk sac (p. 254).

As the notochordal and endodermal materials disappear from the surface, the presumptive somite areas converge medially and enter the primitive streak at its anter-

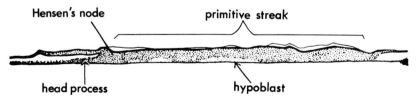

Figure 143. Longitudinal medial section of a chick embryo in the head process stage. (After Wetzel, 1929.)

ior end, behind Hensen's node. From this part of the streak somite material invaginates into the interior. After passing into the interior, the cells of the presumptive somites migrate outward and forward and become distributed in strips on each side of the no-tochordal process (Fig. 144). The presumptive lateral plate mesoderm likewise enters the primitive streak after disappearance of the endodermal area from the surface. The part of the streak occupied by lateral plate mesoderm is immediately posterior to the position in the streak of the somite mesoderm and forms the middle section of the primitive streak throughout the greater part of the period of gastrulation. Having migrated inward the material of the lateral plates shifts laterally and forward, arranging itself to each side of the somites (Fig. 144). The material for the heart and kidneys moves in together with the lateral plate mesoderm.

The posterior part of the primitive streak (about two fifths of its length, when the streak is fully developed) contains only material for extraembryonic mesoderm.

After its first appearance, the primitive streak increases somewhat in length, owing mainly to an elongation of its posterior half. The adjoining part of the area pellucida is also involved in the stretching, so that the area pellucida loses its circular shape and becomes more or less pear-shaped, the attenuated end being directed posteriorly. The elongation of the primitive streak is, however, only temporary. As the cells destined to become notochord, endoderm, and mesoderm migrate into the interior of the embryo, the material of which the primitive streak consists becomes gradually exhausted. The influx of cells from the sides becomes retarded and can no longer compensate for the expenditure of cells due to immigration. The whole primitive streak begins to shrink, the anterior end receding backward, while the posterior end remains more or less stationary (Fig. 139D, E). It has been ascertained by marking with carbon particles and by tritiated thymidine marking that Hensen's node is carried backward bodily during this recession of the primitive streak (Spratt, 1947; Rosenquist, 1966). The presumptive notochordal cells contained in Hensen's node one after another continue migrating downward and forward, so that the notochordal process is prolonged backward, owing to continual apposition of new cells. A part of the presumptive endodermal area remains just posterior to Hensen's node. As the node retreats, cells from this area continue moving inward and form a strip of definitive endoderm underneath the notochord.

Figure 144. Diagram showing morphogenetic movements during gastrulation in the bird embryo. Continuous lines (on left side) show movements in the epiblast. Broken lines (on right side) show movements of the cells which have immigrated into the interior through the primitive streak. (Based on Rosenquist, 1966.)

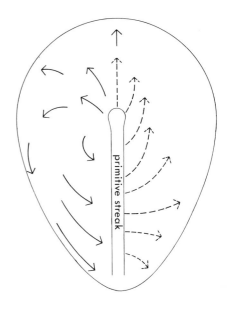

This endodermal strip gives rise to the middle and posterior sections of the embryonic gut. In the same way, the strip of somite mesoderm and the sheets of lateral plate mesoderm extend backward from parts of the primitive streak which retain their properties although the whole structure is reduced in scale. By the end of gastrulation, the primitive streak has shrunk almost to nothing (Fig. 139F); the residue becomes partly incorporated into the tailbud, which is formed at the posterior end of the body of the embryo, and partly into the cloacal region of the embryo.

As the primitive streak recedes backward, the neural system area, which lies just in front of Hensen's node and even forms the most anterior half of the node itself, stretches backward, while the lateral horns of the area are drawn in toward the midline. As a result, a strip of presumptive neural plate is left in the wake of the receding primitive streak has shrunk almost to nothing (Fig. 139F); the residue becomes partly where Hensen's node was located originally. This part will become the prechordal part of the brain.

The primitive streak, being the pathway by which the presumptive internal organs are invaginated into the interior of the embryo, must be considered as the blastopore, even if it does not lead into an archenteron. The recession and disappearance of the primitive streak correspond therefore to the closure of the blastopore. The remnant of the blastopore, as in the amphibians, is to be found in association with the anus (or cloaca).

We have seen that the primitive streak, while it is active, is a mass of cells continuous with both the epiblast and the hypoblast, as well as with the sheets of mesodermal cells migrating in between the two into the blastocoele (Fig. 145a). Cells from right and left, and also to a certain extent from above and below, intermingle within the primitive streak. There is unequivocal evidence (Rosenquist, 1966) that cells situated originally in areas on one side of the streak, after passing through the streak, may come to lie on the opposite side of the embryo. It has also been found (Fraser, 1954) that some cells from the hypoblast find their way into the primitive streak and Hensen's node, and eventually into the notochordal rudiment. In the part of the embryo from which the primitive streak has receded, the continuity of all three layers no longer occurs. The sheet of mesodermal cells, together with the notochordal process, is separated from the overlying epithelium, which no longer contains presumptive mesoderm or endoderm and from then on is pure ectoderm. Likewise, the chordomesodermal mantle has been split off from the endodermal layer underneath.

The hypoblast of the early embryo, without losing its continuity, has been completely transformed by the immigration of endodermal cells from the epiblast. A median strip has been replaced by cells from the epiblast, and these are the only cells from which the gut and its derivatives are formed. The original hypoblast cells contribute to the development of extraembryonic parts, such as the yolk sac, and even the latter is lined in part by cells migrating from the primitive streak.

The fact that the gut of the embryo is derived from the epiblast, and not the hypoblast, justifies our claim previously made (p. 117) that the epiblast cannot be equated with ectoderm, nor the hypoblast with endoderm.

Neurulation and the formation of the other primary organ rudiments proceed in an anteroposterior direction, just as in the fishes. The neural plate appears in the brain region while the gastrulation movements are still in full swing. (See Figure 139E.) As Hensen's node recedes, farther and farther parts of the neural plate become differentiated, and the anterior parts of the neural plate proceed to close into a tube, the neural tube. The neural tube is formed in a typical way, with a large cavity enclosed by the walls of the tube (Fig. 145b, c). The notochord becomes separated from the adjoining

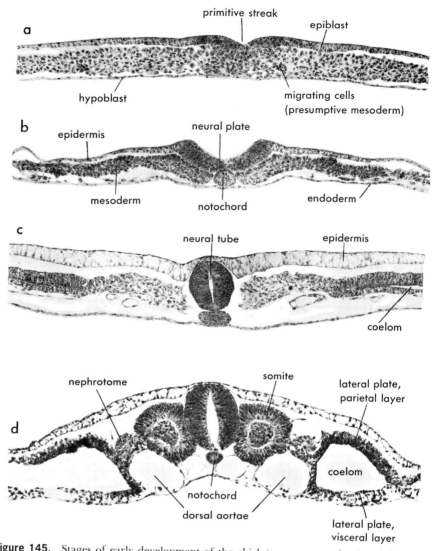

Figure 145. Stages of early development of the chick (transverse sections). *a,* Primitive streak; *b,* neural plate stage; *c,* neural tube stage; *d,* primary organ rudiments.

sheets of mesoderm, and the dorsal mesoderm becomes subdivided into segments—the somites (Fig. 145*d*).

The presumptive alimentary canal of the embryo is represented by a narrow median strip of the endodermal layer. Folds directed downward and inward arise on both sides of this strip; the folds approximate each other and fuse, enclosing a cavity which will be the lumen of the alimentary canal. (See also Section 14–1, p. 444.) This process however, does not take place along the whole length of the embryo but only at its anterior and posterior ends. In the middle part, although a groove in the endodermal epithelium may be present, connecting the anterior and posterior portions of the alimentary canal, this groove continues to be open toward the underlying yolk. The fate of this opening will be considered later in another connection (Section 14–1).

The sheet of mesoderm invaginated through the primitive streak at first consists of loosely lying cells. At the stage when the segmentation of the mesoderm begins, or shortly before, the mesodermal cells reunite into an epithelial arrangement. After the somites and the lateral plates have been formed, the mesoderm of the lateral plate splits into two layers: the external or parietal layer and the internal or visceral layer. The cavity between the two layers is the coelom (Fig. 145c, d). Small coelomic cavities also appear in the somites, but these cavities, as in the amphibians, soon disappear again.

Chapter 8

DIVERSIFICATION OF EMBRYONIC PARTS AND ITS CONTROL DURING GASTRULATION AND PRIMARY ORGAN FORMATION

Development of the three germinal layers in gastrulation and subsequent formation of primary organ rudiments are the outward expressions of profound changes in the properties of different groups of cells in the embryo during this period. These changes are based on intricate chemical processes in the cells which affect both their metabolism and their fine structure. Furthermore, the changes in properties in different regions of the embryo proceed in an organized and strictly controlled way, so that in spite of increasing diversification the embryo achieves a higher level of integration as a whole.

8–1 GENERAL METABOLISM DURING GASTRULATION

Throughout gastrulation, the volume of the embryo does not change appreciably. Every expansion in one direction occurs at the expense of a contraction in another direction or directions. What has been said about the absence of growth during cleavage applies in the same way to the period of gastrulation. Division of cells by mitosis continues, however, throughout gastrulation, and thus there is an increase of nuclear material at the expense of the cytoplasmic substances. Breakdown and assimilation of reserve materials are also proceeding, but here a new feature is observed that makes the metabolism of the gastrula different from the metabolism of a blastula.

The morphogenetic movements during gastrulation could be expected to cause an increased expenditure of energy and consequently increased oxidation. This is what is actually found: the oxygen consumption during gastrulation shows a further increase as compared with the cleavage stages and with the blastula. The oxygen consumption of the frog embryo is shown in Table 8. A similar sharp increase in total oxygen consumption is also observed in sea urchin eggs.

One of the substances particularly involved in the supply of energy during gastrulation in amphibians is glycogen. It has been discovered that the amount of glycogen becomes considerably diminished in the invaginating cells of the dorsal lip of the blastopore. This was first discovered by histochemical methods, by using a specific stain for glycogen on sections of gastrulating embryos (Woerdeman, 1933). Later, by methods of

TABLE 8   Oxygen Consumption during Cleavage and Gastrulation in the Frog

Stage	Oxygen Consumption (In Microliters per Milligram of Dry Weight per Hour)
Cleavage (stage 6 +)	0.054
Early blastula (stage 8 +)	0.069
Late blastula (stage 9)	0.108
Midgastrula (stage 11 −)	0.141
Late gastrula (stage 12 +)	0.162

After Cohen, from Barth and Barth, 1954.

chemical analysis, the exact amounts of glycogen consumed were determined; it was found that in the dorsal lip of the blastopore 31 per cent of the glycogen is lost during gastrulation, whereas in other parts of the embryo only from 1 per cent to 9 per cent is lost during the same time (Brachet, 1950b).

Rapid breakdown of glycogen in the dorsal lip suggests particularly active respiration in this area. Direct measurements of respiration, however, showed that the dorsal lip region is by no means the part of the gastrula which respires at the highest rate. In order to compare respiration in different parts of the embryo, frog gastrulae were cut into several regions, as indicated in Figure 146. Two big pieces were made of the vegetal hemisphere (a dorsal and a ventral one) and four pieces of the marginal zone and the animal hemisphere (pieces 1 to 4), starting with the dorsal lip of the blastopore. The oxygen consumption was determined for each piece as well as the dry weight, total nitrogen, and extractable nitrogen—the latter, as was indicated previously (see Chapter 4), is equivalent to the nitrogen of the active cytoplasm (total nitrogen less the nitrogen contained in the yolk).

When the oxygen consumption is related to dry weight of the fragments, it is seen that there is a great difference in the oxygen consumption of various parts of the embryo (Table 9).

The highest oxygen consumption, as related to dry weight or to total nitrogen— that is, to the whole mass of the embryonic tissue— is found at the animal pole of the gastrula, and the lowest oxygen consumption is at the vegetal pole. The dorsal side, including the dorsal lip, has a distinctly higher oxygen consumption than the ventral side. This would account for the difference in the breakdown of glycogen between the dorsal and the ventral lips. (See preceding paragraphs.) We must realize, however, that not all parts of the cells of a frog gastrula respire; the yolk presumably does not respire at all, whereas it contributes to the dry weight and to the total nitrogen of parts of the embryo.

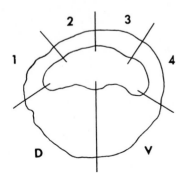

Figure 146. Diagram showing the pieces into which frog gastrulae were cut to study respiration of the different parts. The dorsal side, with the blastopore, is on the left. (After Sze, from Barth and Barth, 1954.)

TABLE 9   Oxygen Consumption in Different Parts of the Frog Gastrula in Microliters of Oxygen Consumed per Hour per Unit of Reference

Per Unit of	Vegetal Dorsal	Dorsal Lip Piece 1	Animal Pole Piece 2	Piece 3	Ventral Lip Piece 4	Vegetal Ventral
Dry weight	1.2	3.7	4.4	3.9	2.7	0.9
Total nitrogen	12.0	37.0	53.0	46.0	30.0	8.2
Extractable nitrogen (nitrogen of active cytoplasm)	11.0	12.0	13.0	13.0	13.0	11.0

From Barth and Barth, 1953.

It would be desirable to eliminate the yolk from calculations of embryonic respiration, and this was done by calculating the oxygen consumption per unit of extractable nitrogen (= nitrogen of active cytoplasm). The data are given in the last line of Table 9, and they show that equal amounts of active cytoplasm have practically the same respiration rate in all parts of the gastrula. In other words, the observed differences in oxygen consumption are due to different amounts of active cytoplasm in relation to yolk, that is, to the gradient of yolk distribution (dealt with in Section 3–2), and not to a local specifically higher rate.

The second peculiarity of the metabolism during the gastrulation period is a sharp increase in protein turnover, and particularly in protein syntheses. In a newt embryo (*Triturus*), the rate of synthesis of protein, as measured by the uptake of radioactive precursors, increases fivefold between the beginning of gastrulation and the late tailbud stage when gastrulation is completed and most of the organ rudiments are formed (Tiedemann and Tiedemann, 1954). In the same way, the rate of radioactive precursor intake into proteins in the sea urchin embryo increases roughly threefold between the earliest gastrula, with primary mesenchyme migrating into the blastocoele, and the middle gastrula (Monroy and Vittorelli, 1962).

The source of materials for the protein synthesis is mainly the protein yolk, contained in the eggs of most animals. The breakdown of yolk granules has been investigated in amphibian embryos both electron microscopically and chemically. With the electron microscope, it can be seen that in amphibian embryos the first change in the yolk platelets consists in the disappearance of the amorphous or granular peripheral layer which contains, besides protein, considerable quantities of ribonucleic acid. The disappearing material goes into solution in the cytoplasm and becomes available for synthetic processes. The solubilization of the peripheral layer occurs in the invaginating chordomesoderm during gastrulation, in the neural plate during late gastrulation and early neurulation, and still later in the epidermis (Karasaki, 1963b). The solubilization of the crystalline core ("main body") of the yolk platelets occurs considerably later, and in endoderm, it is delayed till just preceding the stage when the larvae start feeding.

Chemically, the solubilization of the yolk platelets can be recorded either spectrophotometrically by the decrease of light absorption of the yolk platelets in a microscopic preparation (Vahs, 1962) or by separating the yolk platelets from homogenates of embryos and measuring their protein content (Rounds and Flickinger, 1958). By both methods it was shown that there is a rapid decrease of yolk platelet protein in the invaginating chordomesoderm starting from the beginning of gastrulation and a slower decrease in the ectoderm in neurulation stages.

8–2   GENE ACTIVITY DURING GASTRULATION

Fragments of sea urchin eggs not containing the nucleus may be treated with agents which cause parthenogenetic activation of the egg (viz., hypertonic seawater). It has been found (Harvey, 1936) that the non-nucleated fragments begin to cleave and may reach the morula stage. The embryos, however, do not form a regular blastula and cannot gastrulate, which shows that the presence of nuclei is necessary for development to go beyond cleavage.

During cleavage the main activity of the chromosomes is to duplicate themselves at each cell division. At the completion of cleavage the genes contained in the chromosomes enter a period of new activity, which is the production of large quantities of RNA, particularly messenger RNA. This production can be recorded and measured in the embryo with radioactively labeled uridine which is taken up into ribonucleic acid and not into the deoxyribonucleic acid of the genes themselves. After labeling the cells, the ribonucleic acids can be separated chemically. The product initially obtained is a mixture of different RNA's; in addition to the messenger RNA, transfer RNA and ribosomal RNA are also present. The different kinds of RNA have different molecular weights, and as result they are sedimented at different rates when subjected to ultracentrifugation. Certain fractions isolated in this way represent messenger RNA. The actual amounts can then be estimated by measuring the radioactivity of the sample in an apparatus called a "scintillation counter" (which records the number of particles emitted by the radioactive sample).

Using this kind of method, it has been found that although certain amounts of RNA are produced in the eggs after fertilization and during cleavage (see p. 105 and Fig. 69), at the approach to gastrulation the production of mRNA increases sharply (Brown and Littna, 1966; Bachvarova et al., 1966). The increase is quite sudden, more than tenfold within one hour in the frog *Xenopus laevis*.

It would be interesting to determine whether or not the mRNA synthesized just before the onset of gastrulation is the same as the mRNA present in the egg cytoplasm or the mRNA produced immediately after fertilization and during earlier cleavage stages. The method of DNA-RNA hybridization has been used in attempts to answer this question. By raising the temperature (to near boiling point) the two strands of nucleotides, of which a chromosomal DNA molecule consists, can be made to separate. If the temperature is lowered, the single strands join together again. Since each base can pair with only the complementary base (thymine with adenine, and guanine with cytosine), and since a whole sequence of bases in one strand must correspond with the sequence in the other strand, pairing occurs only between corresponding or similar sections of DNA molecules. Because the messenger RNA is complementary in its base structure to the DNA strand on which it has been modeled, a messenger RNA molecule may also pair with a DNA single strand, if a sample of mRNA is mixed with a DNA sample when the latter is subjected to heating (to separate the two strands of the double helix) and then is cooled. In this way, the RNA molecules will "hybridize" with corresponding sections of DNA. Once such a "hybridization" takes place, other RNA molecules, identical to the first ones, cannot find a suitable place on the DNA molecule and thus cannot attach themselves to it. If a sufficient amount of mRNA is used, and adequate time is allowed for the hybridization to take place, the sites on the purified single-stranded DNA may be saturated with corresponding mRNA's. If a second sample of mRNA is then added, only those mRNA molecules that are different from the mRNA molecules in the first sample will become attached to the DNA. For these molecules, corresponding sites on the DNA strand will still be free, whereas the mRNA molecules identical to ones from the first sample will find their places already occupied

(Fig. 147). Consequently, the proportion of molecules which fails to hybridize is an indication of the degree of similarity between the first and the second samples of mRNA.

An alternative method, which is actually more often used, is to add both samples of mRNA to the same preparation of single-stranded DNA at the same time. If there are similar mRNA molecules in the two samples, they will compete, and the proportion of such molecules hybridizing with the DNA will be reduced, whereas the hybridization of the mRNA molecules which in each sample are different from those in the other sample would proceed unimpeded. Thus again, a lowering of the proportion of hybridizing molecules is an indication of the presence of the same kinds of mRNA molecules in the samples being compared. Both variations of the method yield essentially similar results.

Using the first variation on sea urchin embryos, it was found that previous hybridization with blastula mRNA reduces subsequent binding of mRNA from late gastrula (to be exact the "prism" stage) less than previous hybridization with late gastrula mRNA. The difference, which amounted to about 40 per cent, was due to the fact that some

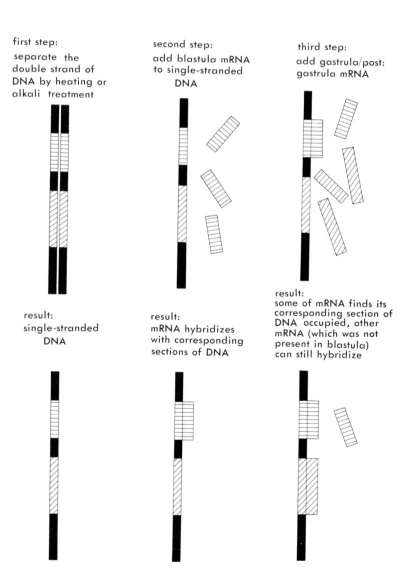

Figure 147. Diagrammatic presentation of a DNA-RNA hybridization experiment showing that some messenger RNA is present in a sea urchin gastrula, which was absent in the blastula stage. For the sake of simplicity, only one kind of mRNA is shown as being present in both the blastula and gastrula, and one kind of mRNA as present in the gastrula only. Also, the strands of DNA are shown as lying parallel to each other, instead of being spirally twisted around each other, as they are in reality.

first step:
separate the double strand of DNA by heating or alkali treatment

result:
single-stranded DNA

second step:
add blastula mRNA to single-stranded DNA

result:
mRNA hybridizes with corresponding sections of DNA

third step:
add gastrula/post: gastrula mRNA

result:
some of mRNA finds its corresponding section of DNA occupied, other mRNA (which was not present in blastula) can still hybridize

mRNA molecules present in late gastrula (prism) mRNA were not identical to blastula mRNA molecules and thus were able to find sites on the DNA molecules not occupied by bound blastula mRNA molecules (Whiteley et al., 1966). These results prove that a proportion of genes which were inactive up to the blastula stage had become active in the gastrula stage. This experiment is not entirely satisfactory inasmuch as it does not differentiate between repetitive sequences and single copy sequences of DNA (and RNA). (See p. 484.) It is highly likely, however, that the latter are also among the sequences that are being transcribed in the gastrula and not in the cleaving embryo.

Indirect proof that new kinds of messenger RNA are produced in the gastrula stage is provided by the fact that when protein synthesis becomes enhanced at this stage, the new proteins are qualitatively different from those present in the egg. This has been proved in sea urchins and in amphibians by applying the extremely sensitive methods of immunology. (See also Section 16–4.) It has been found that the gastrula contains antigens, capable of causing the formation of antibodies, which were not present before, besides containing antigens already present in the egg at the beginning of development (Perlman and Gustafson, 1948; Clayton, 1951).

The new activities of genes at the beginning of gastrulation are not restricted to producing new kinds of mRNA (which enables new kinds of proteins to be synthesized). The other kinds of RNA, transfer RNA and ribosomal RNA, must also be supplied in adequate quantities if development is to proceed normally, and we have already seen that these RNA's, which have a supporting function in protein synthesis, are likewise transcribed from nuclear genes at the start of gastrulation. What happens if one of these supporting factors is lacking has been shown by studies on a peculiar mutation in the frog *Xenopus laevis*. In this mutation the section of DNA which codes for ribosomal RNA is missing. As a result, individuals carrying this character in a homozygous state are unable to synthesize rRNA and are devoid of nucleoli (the "anucleolate" mutation). In the absence of rRNA there can be no ribosomes and no protein synthesis. The mutation can be perpetuated in the heterozygous state, in which the nonmutant chromosome set provides the locus for rRNA transcription and thus for protein synthesis. Homozygous individuals, produced by mating two heterozygous parents, start their lives normally, since enough ribosomes are supplied in the egg by the heterozygous genotype of the female to take the embryo through cleavage and early gastrulation stages. The production of new rRNA and thus of new ribosomes, however, which normally occurs during gastrulation, cannot take place. Protein synthesis as a result slows to a halt, and the embryos die in the early tadpole stage (Brown and Gurdon, 1964).

8–3 INVOLVEMENT OF PARENTAL GENES IN THE CONTROL OF DEVELOPMENT

Apart from the foregoing data, which characterize the activity of the nuclear DNA in general terms, there are many indirect indications that nuclear genetic factors (the genes), after having been dormant throughout the period of cleavage, begin to manifest themselves during gastrulation and in ever-increasing measure control the processes of development from this stage onward. Evidence to this effect may be derived from certain hybridization experiments. These must be considered here in greater detail. If the egg of one species is fertilized by a spermatozoon of another species, the nuclei of the two gametes may fuse and together may participate in cleavage and development. Each cleavage cell would then consist of maternal cytoplasm (because the contribution of the spermatozoon to the cytoplasm of the fertilized egg is negligible), the maternal nuclear half, and the paternal nuclear half. Embryos of this composition are **true hybrids.** The eggs, however, may be treated just after fertilization, before the nuclei of the egg and the sperm have time to fuse, in order to remove the egg nucleus. This can be achieved, for instance, by puncturing the egg and allowing a small quantity of cytoplasm together with the egg nucleus to flow out of the egg. If the egg is fertilized with sperm of another species, an embryo will develop in which the cytoplasm is maternal, but the nucleus is paternal (Baltzer, 1940). Such embryos will be referred to as **hybrid andromerogones.** (The term **merogony** is applicable to the fertilization of a part or fragment of the egg; **androgenesis** is the development of an embryo having only paternal nuclei.) **Andromerogones** are embryos in which an enucleated egg or egg fragment is fertilized by sperm of the same species. As a rule, andromerogones are haploid.

The results of hybridization depend primarily on the relationship between the two parents. The penetration of the egg by foreign sperm is much more easily achieved than the subsequent cooperation with the maternal cytoplasm. Cases are known in which eggs were "fertilized" by sperm belonging to animals of a different class and even of a different phylum (e.g., fertilization of the sea urchin *Sphaerechinus* with sperm of the sea lily *Antedon;* fertilization of the sea urchin *Strongylocentrotus* with sperm of the mollusc *Mytilus*).

In these cases, the sperm penetrates the egg and activates it but does not play any part in the subsequent development. This is tantamount to a parthenogenetic development of the egg, caused by the penetration of foreign sperm. (See Morgan, 1927.) If the animals are more closely related, as for instance in the case of a toad's eggs being fertilized by frog's sperm *(Bufo vulgaris×Rana temporaria)*, the sperm nucleus fuses with the nucleus of the egg. The development (cleavage) begins, but the embryos die before the beginning of gastrulation owing to incompatibility between the sperm nucleus and the egg nucleus and cytoplasm. If the animals are still more closely related, such as species of the same genus, then the hybrids may be fully viable throughout their entire development.

In the cases of rather closely related species, producing viable hybrids, the influence of the nucleus can best be analyzed. Much of the work has been done on sea urchin development. If two species of sea urchin, differing in rate of cleavage, are crossed, the cleavage of the hybrid is always the same as the cleavage of the maternal species. In the sea urchin *Dendraster eccentricus,* the egg cleaves at the rate of 29 to 30 minutes between successive divisions. In *Strongylocentrotus franciscanus,* the interval between successive cleavages is 47 minutes. The hybrid has the maternal rate of cleavage. Also, if fragments of the egg devoid of a nucleus are fertilized, the resulting hybrid andromerogones cleave with a rhythm that is the same as in the maternal species (A. R. Moore, 1933).

The rate of cleavage cannot be influenced by the sperm nucleus. On the other hand, the characters of the larva—the pluteus—are clearly intermediate between the two parents; at the larval stage the sperm nucleus has been able to exert its influence. In a special case, the influence of the sperm nucleus has been discovered as early as the beginning of gastrulation. In the sea urchin *Lytechinus,* the mesenchyme cells produced from the micromeres (Section 6–8) migrate into the blastocoele slightly earlier than the beginning of invagination of the archenteron. In the sea urchin *Cidaris,* the mesenchyme cells do not separate from the epithelium until after the archenteron begins to invaginate; they are consequently given off from the inner end of the archenteron. Furthermore, the cleavage of *Lytechinus* proceeds at a much greater pace than in *Cidaris;* the blastula stage is reached after 5.5 hours in the *Lytechinus* embryos, but only after 16 hours in *Cidaris* embryos. The beginning of gastrulation is correspondingly delayed in the second species. When the *Cidaris* eggs were fertilized by *Lytechinus* sperm, the early stages of development up to the gastrula stage proceeded exactly as in the maternal species, *Cidaris*. The rate of cleavage corresponded to that typical for *Cidaris*. As the gastrulation approached, the mesenchyme cells began to be separated from the blastoderm just as the invagination of the archenteron first became visible. The stage at which the mesenchyme cells migrated into the blastocoele was therefore intermediate between the two species, thus proving that at the beginning of gastrulation the paternal nucleus was already able to manifest itself (Tennent, 1914).

In hybridization experiments performed on amphibians, it is a rule without exception that the cleavage rate is strictly maternal. This also holds true in cases of hybrid andromerogones. Therefore, one may conclude that the cytoplasm alone determines the rate of cleavage. The pigmentation of the egg and early developmental stages is, of course, always maternal, because the egg pigment is synthesized before maturation. Differences in the pigmentation of the larvae become evident only much later when the melanophores, derived from the neural crest, become differentiated.

During subsequent development the pigmentation depends on the nuclear factors. A hybrid andromerogone was produced by using eggs of the black race of the axolotl (*Ambystoma mexicanum*) and sperm of the white race. In the embryo, the cytoplasm therefore derived from the black race and the nucleus from the white race. The animals were all white, thus showing that the nucleus dominated over the cytoplasm (Baltzer, 1941).

No crosses are known in amphibians in which morphological differences between the two parent species could be discovered in the early stages—the gastrulation or the neurulation stages. It has been found, however, that the end of cleavage and the beginning of gastrulation is the stage when the incompatibility of the sperm nucleus and the egg nucleus and cytoplasm first manifests itself. In many hybrid combinations the hybrid goes through the early stages of cleavage more or less normally, but the development stops in the blastula or early gastrula stage (*Rana esculenta* × *Bufo vulgaris*; *R. esculenta* × *R. temporaria*; *B. vulgaris* × *R. temporaria*; *Triturus palmatus* × *Salamandra maculosa*; *R. pipiens* × *R. sylvatica*; *R. pipiens*× *R. clamitans*; and others) (Morgan, 1927; J. A. Moore, 1946; A. B. C. Moore, 1950). If the true hybrid between two species is viable, the hybrid andromerogone may not be. The andromerogone embryos, even without hybridization, are always weaker than normal embryos owing to the haploid state of their nuclei. However, the hybrid andromerogones are much inferior in their capacity for development. The hybrid combination *R. pipiens* × *R. capito* is fully viable up to the adult stage when the chromosomes of both parental species are present. The andromerogone of *R. pipiens* fertilized with sperm of the same species develops into a tadpole. The hybrid andromerogone (*R. pipiens* × *R. capito*) develops normally through cleavage and gastrulation stages but dies in the early neurula stage (Ting, 1951). The paternal chromosomes in the presence of maternal chromosomes may function normally and transmit the paternal characters to the diploid hybrid, but left alone with the foreign cytoplasm they prove to be inadequate for supporting development. Further cases are known in which the andromerogonic hybrid dies during gastrulation or shortly after (*R. pipiens* × *R. palustris*; *Triturus alpestris* × *T. palmatus*; *T. palmatus* × *T. cristatus*) (Hadorn, 1932; de Roche, 1937; A. B. C. Moore, 1950).

There is some direct evidence that in hybrids the paternal chromosomes cause the synthesis of proteins of their own species about the time of the beginning of gastrulation. The evidence was obtained by immunological methods in the following experiment. Eggs of the sea urchin *Paracentrotus lividus* were fertilized with sperm of *Psammechinus microtuberculatus*. Using the blood plasm of animals immunized against proteins of *Psammechinus microtuberculatus*, it was found that the fertilized eggs showed no trace of these specific antigens. Twenty-four hours after fertilization, however, when the hybrid embryos were in the mesenchyme blastula stage, antigens of the paternal species, *Psammechinus microtuberculatus,* could be clearly detected. Since the paternal antigens could not be discovered earlier, it is evident that these antigens (predominantly proteins) had been built up in the cytoplasm, which is derived mainly from the egg, under the influence of the sperm chromosomes (Harding, Harding, and Perelman, 1954). The death of nonviable hybrids and hybrid andromerogones is preceded by an arrest of at least some of the physiological processes which characterize the onset of gastrulation, such as the increase in oxygen consumption, glycolysis, and the increase in the amount of ribonucleic acid (Brachet, 1950a); the latter is especially suggestive, since it involves a synthetic process. We have seen that besides the synthesis of ribonucleic acids the embryo during gastrulation begins synthesizing new proteins, manifested as new antigens when investigated by serological methods. It is highly probable that in cases of incompatibility between the foreign paternal nuclei and the maternal cytoplasm these synthetic processes cannot go on, or cannot go on satisfactorily, hence the arrest of development and death.

If, on the other hand, the nuclei and the cytoplasm are compatible, the synthesis of new proteins and nucleic acids (and probably also other cytoplasmic substances) goes on progressively. Since some of the newly synthesized substances are different from the ones present before, rapid changes in the constitution of the embryo are inaugurated, changes that involve not only the position and arrangement of cells, as in the gastrulation movements, but also their inherent physiological properties. This leads to the next period of development in which the various parts of the embryo become progressively differentiated from one another: the period of **organogenesis.**

8-4 DETERMINATION OF THE PRIMARY ORGAN RUDIMENTS*

The activation of genes at the beginning of gastrulation and the resulting synthesis of new kinds of proteins is an important step in development, but in itself it would not be sufficient for initiating the formation of different organs and organ systems of the embryo. In order for this to occur, the new proteins must be placed in different parts of the embryo, so that these different parts may each undergo a particular type of differentiation. Displacing the masses of cells of the embryo in gastrulation is a means of attaining the same result, some groups of cells remaining on the surface, as ectoderm, and some sinking into the interior, as mesoderm and endoderm. Different positions in the embryo become a factor in guiding the cells to different goals. Also, the change in the relative positions of cells brings some of the groups of cells into the proximity of other groups, which were distant from one another in their original positions, prior to gastrulation. This change in position provides possibilities for new interactions between parts of the embryo. Such new interactions have been studied in great detail in the development of vertebrate embryos, and the following sections will be devoted to the results of these studies.

The experiments of local vital staining and similar methods, allowing the experimenter to reconstruct fate maps, do not furnish any information as to whether or not the cells in the various areas actually have different potentialities for development. The only information that these experiments give concerns the eventual destination of each part of the early embryo in normal development. Certain distinctions between cells

---

*A review of this subject has been published by Saxén and Toivonen (1962).

belonging to the various areas are apparent, such as differences in yolk content, pigmentation, and so on. There is, however, no *a priori* reason to believe that such differences are actually connected with the future development of each part.

To find out whether the presumptive areas of a blastula are actually preformed for a specific part in the future development, methods other than observing normal development are necessary. One of the methods used is the method of **transplantation,** already mentioned in another connection (Section 7–7). Small pieces may be cut out of embryos in various stages of development and inserted into suitably prepared wounds of the same or another embryo. In the case of transplantation of a piece of an embryo to another place in the same embryo, the transplantation is said to be **autoplastic.** If the transplantation is from one individual to another of the same species, the transplantation is **homoplastic.** If the transplantation is to an individual of another species belonging to the same genus, the transplantation is **heteroplastic.** A transplantation to an individual more distantly related than species of one genus is called **xenoplastic.** The animal (embryo) from which a part is taken is referred to as the **donor;** the animal to which the part is transplanted is called the **host.**

In adult animals, especially highly organized animals such as vertebrates, transplantation is not easy, and only autoplastic and homoplastic transplantations are usually successful. In embryos, however, and in lower invertebrates (such as the coelenterates), the grafts may heal successfully even after xenoplastic transplantations. Successful transplantations have been carried out between embryos of frogs and salamanders and between mammals and birds.

It is usually important to know later which tissues and cells are derived from the graft and which from the host. The distinction is easy if donor and host are sufficiently different from each other. The cells may be distinguished by their size, staining properties, and other factors. Sometimes differences may be found even between cells belonging to closely related animals. Differences in pigmentation may be very useful. The eggs of *Triturus (Triton) cristatus,* for instance, are entirely devoid of pigment, whereas eggs of other species of *Triturus (T. taeniatus, T. alpestris)* have fine granules of black pigment in their cytoplasm. These pigment granules are distributed to all the cells during cleavage. If tissues of *T. taeniatus* and *T. cristatus* are joined in transplantation, the cells of *T. taeniatus* can be distinguished both macroscopically and in sections by the presence of pigment granules. If no natural differences can be discovered between the two animals (embryos) used for transplantation, artificial differences may be introduced by staining one of the embryos with vital dyes. The presence or absence of the dye will indicate the position of the graft. In some cases the vital staining may be preserved in sections of the operated embryo by special methods.

Heteroplastic transplantation between *Triturus cristatus* and *Triturus taeniatus* embryos has been used to investigate the determination of the presumptive ectodermal areas—the epidermis area and the neural system area (Spemann, 1921). A small piece of the epidermis area of an early gastrula was transplanted into the neural system area of another embryo in the same stage of development (Fig. 148). The result was that the grafted material developed in conformity with its new surroundings and became first a part of the neural plate of the host (Fig. 148c) and later a part of the neural tube (Fig. 149). There was every reason to conclude that it was differentiating as nervous tissue. In the case of a reverse transplantation, presumptive neural material differentiates into skin epidermis in conformity with its new surroundings (Fig. 148d). Presumptive ectoderm (either from the epidermis area or from the neural system area) was also transplanted into the marginal zone of an early gastrula. In this case, the graft was drawn into the blastopore in the course of gastrulation and was later found in different places in the interior of the host. In every case, the graft differentiated in conformity with its surroundings and took part in the development of various organs of the host: the noto-

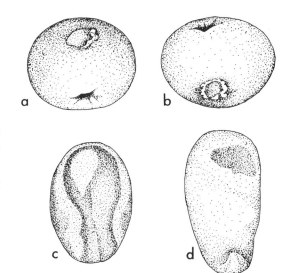

Figure 148. Exchange of presumptive epidermis and presumptive neural plate in the early gastrula stage of two *Triturus taeniatus* embryos differing slightly in density of pigmentation. *a, b,* Embryos immediately after the operation. *c, d,* The same embryos in the neural plate stage. In *c* the graft forms part of the neural plate. In *d* the graft forms part of the anteroventral epidermis. (From Spemann, 1938.)

chord, somites, lateral plate mesoderm, kidney tubules, and wall of the gut (Mangold, 1923).

It is thus evident that the fate of presumptive nervous system and presumptive epidermis is not fixed at the time of the stages used in the preceding experiments. Besides having a definite normal fate—called the **prospective significance**—the parts tested possess abilities to develop in various other ways, under experimental conditions. This ability of the parts of an early embryo to develop in more than one way is called the **prospective potency** of these parts.

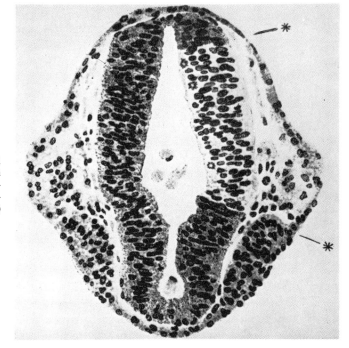

Figure 149. Section through anterior head region of a *Triturus taeniatus* embryo in which part of the brain (shown by asterisks) is developed from grafted *Triturus cristatus* presumptive epidermis. (From Spemann, 1938.)

The prospective potency of the neural area in the early gastrula stage is therefore shown to include not only epidermis but also mesodermal and endodermal tissues. The prospective potency of the epidermis area includes nervous system and mesodermal and endodermal tissues as well. The two prospective potencies are practically identical, in spite of the different prospective significance.

The potencies of the marginal zone and the vegetal region could not be tested as easily as those of the presumptive ectoderm, because they both tended to invaginate into the interior from every position in which they were placed. However, in exceptional cases parts of the transplanted marginal zone may remain on the surface at the end of gastrulation, and then they also conform to their surroundings in their development and differentiate as epidermis or neural system.

An entirely different result is observed if the transplantation of pieces of ectoderm is performed at the end of the period of gastrulation. A transplanted piece of presumptive neural system area will differentiate as brain or spinal cord in whatever part of the embryo it has been placed. Usually in this case, the transplanted tissue sinks from the surface of the embryo—this would correspond to neurulation in normal development—and develops into a vesicle with thickened walls. Sometimes definite parts of the neural system can be recognized, such as one or other of the brain vesicles or an eye (Spemann, 1919). The presumptive epidermis of a completed gastrula likewise loses the ability to differentiate as nervous system. If it is transplanted into the nervous system area it does not conform to its surroundings and may interfere with the closing of the neural tube, but even if it does become enclosed inside the neural tube it differentiates into epidermis and not into nervous tissue.

Obviously some change has occurred between the early gastrula and the completion of gastrulation. The prospective potencies of presumptive neural tissue and epidermis have been narrowed down and have become the same as their prospective significance, as far as these experiments go. This narrowing down of the prospective potency, which is equivalent to fixing the fate of a part of the embryo, is known as **determination.** After the process of determination has taken place the respective parts are said to be **determined.** The experiments described so far show that the neural system as a whole is determined by the end of gastrulation. They do not show, however, whether *every part* of the neural system is determined at the same time. The determination of parts inside the primary organ rudiments will be dealt with later (Chapters 12, 13, and 14).

The experiments just described also show that the determination of parts of the ectoderm does not take place from causes inherent in the ectoderm itself. The differentiation of parts of the ectoderm is dependent on the surroundings in which the ectodermal cells find themselves. Further experiments have proved that the condition which is decisive in determining the development of the neural plate is contact of the ectoderm with the roof of the archenteron—i.e., with the sheet of cells representing mainly the presumptive notochord and somites—which shifts forward underneath the ectoderm during gastrulation. Parts of the ectoderm in contact with the archenteron roof differentiate into neural plate and nervous tissue.

One of the most spectacular experiments showing the role of the archenteron roof was performed by Spemann's pupil, Hilde Mangold (Spemann and H. Mangold, 1924). H. Mangold transplanted heteroplastically (from *Triturus cristatus* to *Triturus taeniatus*) a piece of the dorsal lip of the blastopore of an early gastrula. The graft was placed near the lateral lip of the blastopore of the host embryo (also an early gastrula), and subsequently most of the graft invaginated into the interior, leaving, however, a narrow strip of tissue on the surface. When the host embryo developed further, it was found that an additional whole system of organs appeared on the side where the graft had been placed. The additional organs together comprised an almost complete secondary embryo. This secondary embryo lacked the anterior part of the head. The posterior part of the head was present, however, as indicated by a pair of ear rudiments. There was a neural tube flanked by somites and a tailbud at the posterior end (Fig. 150c). A microscopic examination revealed the presence of a notochord, kidney tubules, and an additional lumen in the endoderm representing the gut of the secondary embryo. All the parts of the secondary embryo were found to lie at about the same level as corresponding parts of the host. Since heteroplastic transplantation was used in this experiment, it was possible to determine which parts of the secondary embryo were developed from the graft and which were developed from the cells of the host (Fig. 150d). It was found that the notochord of the secondary embryo consisted exclusively of graft cells; the somites consisted partly of graft and partly of host cells. A small number of graft cells were present in the neural tube; these were certainly the cells that had not invaginated. The bulk of the neural tube, part of the somites, the kidney tubules, and the ear rudiments of the secondary embryo consisted of host cells. The additional lumen in the endoderm was also surrounded by host cells. All these latter parts would not have de-

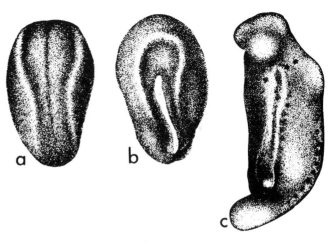

Figure 150. Induction of secondary embryo in *Triturus* by means of a transplanted piece of blastopore lip. *a* and *b*, Operated embryo in the neurula stage; *a*, dorsal view with host neural plate; *b*, lateral view with induced neural plate. *c*, Same embryo in the tailbud stage with an induced secondary embryo on the side. *d*, Same embryo, transverse section. (After Spemann and H. Mangold, from Spemann, 1938.)

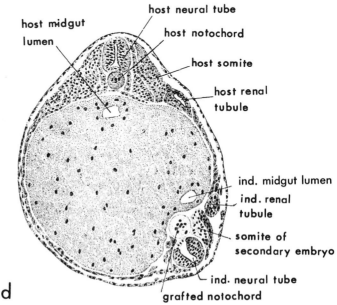

veloped if the graft had not been present. It follows that their development was due to some influence of the graft. This influence, causing the development of certain organs of the embryo, is called **embryonic induction.** The part which is the source of the influence is called the **inductor.**

From the experiment just described, it may be concluded that the induction of the neural system is due to the activity of the underlying tissues, the presumptive notochord and the presumptive somites in particular. The issue is somewhat complicated by the participation of both graft and host cells in some of the organs of the secondary embryo. This was later shown not to be essential; in other experiments the whole of the induced neural tube was developed exclusively from host tissue. Furthermore, the anterior parts of the brain could also be induced by the roof of the archenteron, so that the induced nervous system might be complete.

In special experiments, it has been ascertained which parts of the gastrula are capable of inducing the neural system. Pieces were taken from all parts of the early gastrula and each slipped into the blastocoele of an embryo in the early gastrula stage. Only grafts taken from the dorsal lip of the blastopore and the adjoining parts of the marginal zone were found to be able to induce. The area capable of induction coincided, in fact, almost exactly with the presumptive areas of notochord, somites, and prechordal plate (Bautzmann, 1926).

Likewise, it has been ascertained which embryonic tissues are able to react to the induction by developing as nervous system. The tissue in question is the ectoderm or presumptive ectoderm of the gastrula. The reactive ability is highest in the early gastrula stage, is still high in the midgastrula stage, begins to decline in the late gastrula, and fades away with the beginning of neurulation. This can be shown by transplanting the inductor underneath the ectoderm in successive stages, and the result may be expressed as a percentage of successful inductions (Machemer, 1932).

In later stages, not only does the percentage of inductions diminish, but the volume and degree of differentiation are also reduced. Complete neural systems or large brain vesicles may be induced by transplantation of inductors in the early gastrula stage, but in the early neurula inductions are very weak; they may consist of only a few cells differentiating as nervous tissue. The presumptive ectoderm of the blastula is not able to react to the inductive stimulus; if the inductor is transplanted in the blastula stage, the ectoderm reacts by the formation of a neural plate only after the gastrulation is completed. In all cases, the induced neural system develops simultaneously with the neural system of the host.

All this shows that the reacting cells must be in a particular state to be able to differentiate into nervous system under the influence of the inductor. This particular state of reactivity is referred to as **competence.** The competence for neural induction is restricted to the ectoderm and is present during a short period only. If during this period the presumptive ectoderm is not stimulated to differentiate as neural tissue, it differentiates as epidermis. No special stimulation (induction) is necessary to cause the latter differentiation, though it has been found that certain conditions must be satisfied if the epidermis is to develop progressively and to produce normal skin epithelium. The condition is the presence of underlying connective tissue. Without underlying connective tissue, epidermis quickly degenerates. The cells lose their polarity; they do not remain arranged as an epithelial layer but form a spongy mass consisting of irregular strands of cells, and eventually they perish.

8–5 SPEMANN'S PRIMARY ORGANIZER

Hilde Mangold's experiment revealed the fact that the transplanted dorsal lip of the blastopore exercises its influence not only on the ectoderm, by inducing the neural plate, but also on the mesoderm and endoderm. This has also been corroborated by further experiments. Extensive inductions can be produced in the mesoderm by transplanting pieces of the dorsal lip of the blastopore. The organs developed from the graft are usually supplemented by parts produced from the host tissue as a result of the induction, so that a more or less complete whole is developed. Together with the parts developed from ectoderm, a secondary embryo of various degrees of completeness arises. All the parts of the secondary embryo may be in harmonious relationship with one another, both in size and position.

It is the ability of the dorsal lip of the blastopore to cause, when transplanted, the development of a complete whole that led to the dorsal lip of the blastopore being called the **organizer,** or **primary organizer** (Spemann, 1938). The organizer of the gastrula is that region of the egg that can be distinguished in the stages before the beginning of cleavage, as the **gray crescent.** Experimental evidence is now available that already at this early stage the gray crescent possesses the special properties of the organizer; furthermore, it has been shown that the properties of the organizer are inherent in the cortex of the egg.

A. S. G. Curtis has succeeded in performing the remarkable experiment of transplanting parts of the cortex in the fertilized egg of the frog *Xenopus laevis* at the beginning of cleavage. The gray crescent cortex was excised and transplanted into a ventral position on another embryo. The graft healed in its new position, and cleavage occurred fairly normally, but after gastrulation a secondary embryo with neural tube and notochord was found to have developed at the site of the operation. In another experiment the cortex in the gray crescent area was removed—the result: failure of the embryo to develop any embryonic organ rudiments (Curtis, 1960, 1963).

We can thus link up the phenomena occurring during gastrulation with factors present at the beginning of development; that is, we can trace the dorsal lip of the blastopore with its peculiarities to a specific part of the cytoplasm of the fertilized egg. The gastrulation movements and the determination of parts of the embryo during and after the gastrulation are thus foreshadowed by the organization of the egg before cleavage.

The inductions produced by a transplanted organizer or its part may vary in degree of complexity. When the induction is weak and the quantity of induced neural tissue is small, it is sometimes impossible to compare the induced tissue with any part of the normal nervous system. If, however, the inductor is powerful, and the reaction of the competent tissue is good, large organs or groups of organs are developed. In this case, the induced parts can be recognized as representing certain organs of the normal embryo: specific parts of the brain, such as the forebrain, midbrain, and medulla, the spinal cord, eyes, nose rudiments, ear vesicles, somites, pronephric tubules, and tailbuds. The various parts are always in a more or less harmonious relationship, so that parts belonging to the head are not mixed at random with parts belonging to the posterior trunk and tail region. In other words, the inductions are regionally specific.

The regional specificity may be imposed on the induced organs by the inductor. The anterior part of the archenteric roof induces head organs and may be called a **head inductor;** the posterior part of the archenteric roof induces trunk organs and tailbuds and may be called the **trunk** or **spinocaudal inductor.** The head inductor may be further differentiated into the **archencephalic** inductor, inducing forebrain, eyes and nose rudiments, and the **deuterencephalic** inductor, inducing hindbrain and ear vesicles (Lehmann, 1945) (Fig. 151).

During gastrulation, the anterior part of the archenteric roof invaginates over the dorsal lip of the blastopore earlier than the posterior part of the archenteric roof. The dorsal blastopore lip of the early gastrula therefore contains the archencephalic and deuterencephalic organizer (inductor), and the dorsal blastopore lip of the late gastrula

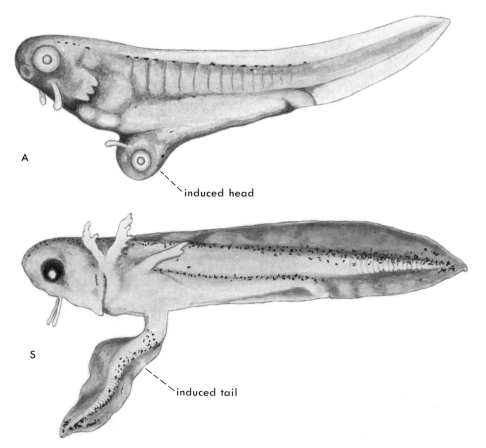

A

induced head

S

induced tail

Figure 151. Archencephalic (A) and spinocaudal (S) induction in newt (*Triturus*) embryos. (Redrawn from Mangold, 1932, and Tiedemann, 1963.)

contains the spinocaudal organizer (inductor). The inductions produced by the dorsal lip of the blastopore taken from the early and the late gastrula differ in accordance with expectation: the first tends to produce head organs, and the second tends to produce trunk and tail organs (Spemann, 1931).

In the case illustrated in Figure 150 the induction could be classified as combined deuterencephalic and spinocaudal, but lacking archencephalic parts.

Although the "primary organizer" was first discovered in urodele amphibians, it was soon found that the dorsal lip of the blastopore and the roof of the archenteron of other vertebrates have the same function in development. In particular, the chordomesoderm in all vertebrates was found to induce the nervous system and the sense organs.

In frogs, the induction of a secondary embryo can be performed by the dorsal lip of the blastopore transplanted into the blastocoele of a young gastrula, in much the same way as in newts and salamanders (Fig. 152; Schotté, 1930). The same applies to the cyclostomes, particularly the lampreys, whose cleavage is holoblastic and whose blastula and gastrula resemble those of amphibians (Fig. 153a; Bytinski-Saltz, 1937; Yamada, 1938). In bony fishes, inductions of secondary well-developed embryos were produced by transplanting the posterior edge of the blastodisc, which corresponds to the dorsal lip of the blastopore, into the blastocoele of another embryo (Fig. 153b and c; Oppenheimer, 1936, in *Fundulus* and *Perca*), or by transplanting the chordomesoderm and endoderm (Luther, 1935, in *Salmo*). Neural inductions were also obtained by trans-

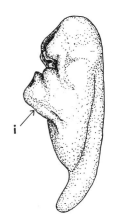

Figure 152. Induction of a secondary embryo (i) by means of a grafted primary organizer in a frog. (After Spemann and Schotté, 1932.)

planting the dorsal lip of the blastopore in the sturgeon (Ginsburg and Dettlaff, 1944). (See also Oppenheimer, 1947).

In the birds, the existence of the primary organizer was established by experiments by Waddington and his collaborators. The inducing part was found to be the anterior half of the primitive streak, which for other reasons we have recognized as corresponding to the lips of the blastopore in amphibians. To test the inducing ability of the primitive streak, whole blastoderms were removed from the egg in early gastrulation or pregastrulation stages and cultivated *in vitro* on a blood plasma clot. Parts of the primitive streak from another embryo were then inserted between the epiblast and hypoblast. Very good inductions of secondary embryos were obtained in this way (Fig. 154; Waddington, 1933; Waddington and Schmidt, 1933; see also Waddington, 1952). Hensen's node does not differ in inducing ability from the part of the primitive streak which immediately follows, but the posterior third of the primitive streak cannot induce neural differentiation. We have seen that cells migrating through the posterior part of the primitive streak give rise to extraembryonic mesoderm (p. 185). This part of the streak does not contain materials which have inducing power in an amphibian embryo (notochord, somitic mesoderm), so that its lack of inducing ability is not surprising. In later stages neural induction may be caused by transplantation of the head process—the notochordal rudiment.

Much light on the relation of the hypoblast to the epiblast of the avian blastoderm has been thrown by experiments performed by Waddington and his collaborators (Wad-

Figure 153. Induction of secondary embryos by grafting of the primary organizer in the lamprey (*a*) and the perch (*b, c*). p, Primary embryo; i, induced secondary embryo. (*a*, From Yamada, 1938; *b, c,* from Oppenheimer, 1936.)

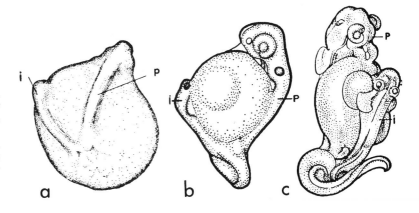

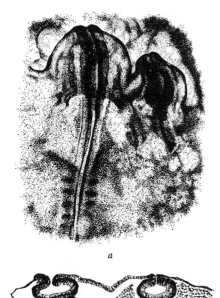

a

b

Figure 154. Induction of a secondary embryo by means of a grafted primitive streak in a bird. *a*, Surface view; the secondary embryo is on the right. *b*, Section through the same embryo, showing the host axial system (right) and the induced neural tube lying above the neural tube and mesoderm developed from the graft (left). (After Waddington and Schmidt, from Waddington, 1952.)

dington, 1933, 1952). After excision of the whole blastoderm it is possible to separate the hypoblast from the epiblast and to join them together so that the anteroposterior (longitudinal) axis of the epiblast is placed at an angle of from 90 to 180 degrees to the anteroposterior axis of the hypoblast. The preparation is then cultivated *in vitro* on a blood plasma clot. It was found that, as a result of the discrepancy in the orientation of the epiblast and hypoblast, the primitive streak developing in the epiblast became curved, so as to coincide in part with the orientation of the hypoblast. When the rotation of the epiblast was 180 degrees, a completely new primitive streak appeared in some cases, with its anteroposterior axis orientated in accord with the position of the hypoblast, and thus pointing in the opposite direction to the one in which it should have been pointing normally. The hypoblast is thus capable of imparting its polarity to the epiblast and causing the complicated system of movements in the epiblast already described (Section 7–7). The primitive streak in birds is thus dependent on the underlying hypoblast for its formation. This process of induction is something for which no analogue has been found so far in other vertebrates.

No experiments on the primary organizer have been done on reptiles, but one would expect that the archenteron has the same inducing activity in this class as in other vertebrates.

Because of technical difficulties, only inadequate information is available on the primary organizer and neural induction in mammals. A successful neural induction has been performed in a rabbit embryo by cultivating the early blastodisc on a plasma clot and implanting the primitive streak of the chick as an inductor (Fig. 155; Waddington, 1934). The experiment proves that the tissues of a mammalian gastrula have the competence for neural induction. It is practically certain that the primitive streak and the cells migrating from the streak (chordomesoderm) are the source of the inducing stimuli. This suggestion may be supported by the fact that the anterior end of a somewhat

Figure 155. Induction of a neural plate in a rabbit embryo by means of a grafted piece of primitive streak of the chick. The induced neural plate (I.n.p.) is joined to the host neural plate (H.n.p.). G., Cyst developed from the graft. (From Waddington, 1934.)

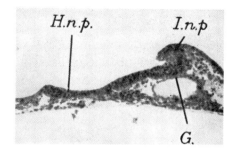

later rabbit embryo (with two pairs of somites) induced a neural plate in a chick embryo when placed under a chick blastoderm (Waddington, 1936).

More recently, it has been proved that the primary organizer and neural induction exist also in the lower chordates, and particularly in *Amphioxus*. Tung, Wu, and Tung (1962b) transplanted pieces of tissue from the inner surface of the dorsal blastopore lip of an early gastrula of *Amphioxus* into the blastocoele of another embryo in the same stage (Fig. 156*A*). They observed that a secondary embryo developed in the ventral region of the host with a notochord and mesoderm produced by the graft and a neural tube induced from host tissue (Fig. 156*B*).

In ascidians, the development of the brain is dependent on the notochordal rudiment and also on the anterior endoderm, but the relation between these parts is of a somewhat different nature: the ability to react to the induction is present only in the cells of the presumptive brain area. Other cells, such as cells of presumptive epidermis, cannot be converted to nerve cells by the action of the notochordal rudiment. In this way the "organizing" action of the notochordal rudiment is much more restricted than in amphibians and other chordates (Reverberi, Ortolani, and Farinella-Ferruzza, 1960).

We will now consider the nature of the inducing stimuli. Most of the work has been done with amphibian material as the reacting system.

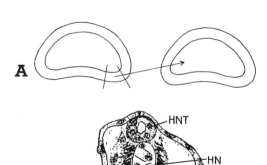

Figure 156. Neural induction in *Amphioxus* (Branchiostoma) *belcheri. A,* Diagram of operation: cells lying inside dorsal lip of the blastopore transplanted into the blastocoele of an embryo in the same stage of development. *B,* Transverse section of the host embryo with a secondary neural tube (NT) induced by the grafted notochord material (GN); HN, host notochord; HNT, host neural tube; sec. emb, secondary embryo. (From Tung, Wu and Tung, 1962b.)

0.05 mm.

8–6   ANALYSIS OF THE NATURE OF INDUCTION

Induction of the neural plate by the underlying chordomesoderm suggests the problem of analyzing the nature of the influence exerted by the inductor on the reacting ectoderm. Spemann gave the active region the name of organizer, meaning that this is the part which organizes the process of development. He wanted to suggest that the term "organizer" is not merely a metaphor and to say that the action of the organizer "is not a common chemical reaction, but that these processes of development, like all vital processes, are comparable, in the way they are connected, to nothing we know so much as to those vital processes of which we have the most intimate knowledge, viz., the psychical ones" (Spemann, 1938, p. 372).

Spemann's opinion has not, however, been borne out by subsequent experiments, or at least it has not been supported in its original form. It has been found that the nature of the inducing agent can be analyzed by physicochemical methods.

In the study of the nature of induction, the first question that had to be answered was whether some vital activity of the organizer is essential for neural plate induction, that is, whether the organizer can exercise its action only in the living state. To answer this question, attempts were made to test whether the organizer could retain its inducing power after its vital activities had been stopped. The organizer, the dorsal lip of the blastopore, was killed by various means—by boiling, by treating it with alcohol or petrol ether, by freezing, or by desiccation—and was then implanted into a living embryo in an appropriate stage of development (the early gastrula stage). It was found that a killed organizer can still induce (Bautzmann, Holtfreter, Spemann, and Mangold, 1932). The vital activity of the organizer is therefore not essential for induction. These experiments made it very likely that the roof of the archenteron produces its effect by liberating some chemical substance which is the immediate cause of induction. It is plausible that such a substance could be liberated even from dead tissue.

If chemical substances liberated by the inductor were responsible for its effect on the reacting cells, it should be possible to observe induction under conditions which would exclude the immediate contact between inductor and reacting tissue which occurs in normal development. For many years, all attempts to prevent immediate contact between the neural inductor and reacting presumptive ectoderm led either to an absence of induction or produced very dubious results. Eventually, however, clear-cut neural induction was obtained by interposing, *in vitro*, a 20 μ-thick Millipore filter between a piece of organizer material (dorsal blastopore lip) and a piece of presumptive early gastrula ectoderm of *Triturus* (Saxén, 1961). The presumptive ectoderm exposed to the treatment later became rounded up, forming a vesicle with distinct brain parts, including an eye with lens in one case. The average pore size of the filter was 0.8 μ, which of course allows very large molecules, including protein molecules, to pass through.

Some embryologists have claimed that under conditions of normal development the inductor, the roof of the archenteron, must be in close contact with the reacting system, the ectoderm. It has even been suggested that temporary intercellular bridges may be established between cells of the archenteron roof and those of the ectoderm. Electron microscopic study has shown, however, that between the roof of the archenteron and the ectoderm there is always an "interspace," and that the cells of the two layers do not approach each other anywhere closer than 500 Å (Kelley, 1969) (Fig. 157). This is several times the width of the intercellular space between epithelial cells. Thus, this gap of 500 Å has to be traversed by the inducing substance.

Experiments with Abnormal Inductors and Induction by Substances of Known Chemical Composition.   After the first indications were obtained that the action of the inductor is mediated by chemical means, the obvious thing to do was to find out what kind of substance or substances were involved. This task turned out to be quite a difficult one, largely because of the small size of the embryo and the very minute quantities of the substances operating in the induction. The problem was therefore approached in an indirect way: first, by trying to find sources of the "inducing substance" other than the

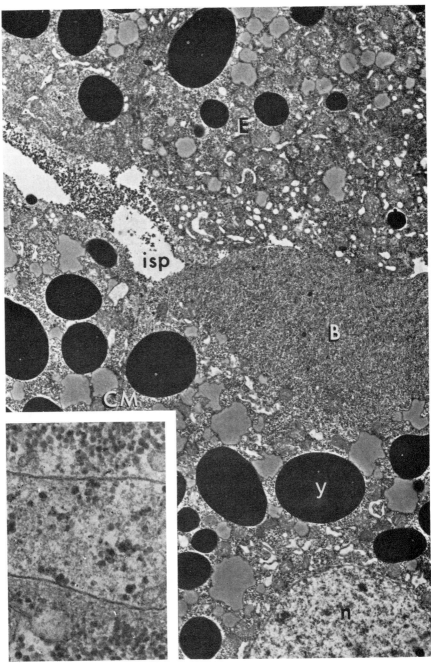

Figure 157. Contact zone between a roof of archenteron cell (below left) and an ectodermal cell (above right) in *Xenopus* (electronmicrograph). Inset: Interspace between the roof of archenteron and ectoderm at high magnification. B, Cytoplasmic "bleb." protrusion of chordomesodermal cell; CM, chordomesodermal cell; E, ectodermal cell; isp, interspace between roof of archenteron and ectoderm; n, nucleus; y, yolk platelet. (Courtesy of Dr. R. O. Kelley, 1969.)

roof of the archenteron, and second, by trying out various chemicals for their ability to replace the natural induction.

A large series of experiments, carried out by many researchers, investigated whether the inducing substances can emanate only from the roof of the archenteron, or whether perhaps other tissues can also produce similar substances. It was shown experimentally that the second alternative is the correct one. First, it was found that the neural plate, once it has begun to differentiate, may be a source of an inductive stimulus for neural differentiation (Mangold and Spemann, 1927). The inductive ability is not restricted to the early stages of neural development but is also present in the tissues which are derived from the neural plate, that is, in the tissues of the brain and spinal cord in the larva and in the adult animal. If the roof of the archenteron is the normal inductor of the neural plate, the nervous tissue may be called an abnormal inductor of the same. The notochord and the dorsal mesoderm (the muscles derived from the somites) also preserve their ability to induce long after the time when neural plate induction takes place in the normal embryo. It has been found, furthermore, that almost any tissue of the adult animal can induce a neural plate, and other organ rudiments besides, if it is placed under the presumptive ectoderm of an embryo in the early gastrula stage. Liver, kidney, and muscles have been found to be very good inductors, but gut, skin, and various other tissues have also been found to be effective (Holtfreter, 1934b).

In earlier experiments on organizer action, the inductor (the dorsal lip of the blastopore) was transplanted into a wound made in the blastoderm of the host embryo. The graft healed in and, at first, became part of the blastoderm. It later invaginated, owing to its inherent ability for that kind of morphogenetic movement, and it came to lie underneath the reacting presumptive ectoderm. This method obviously could not be used with "abnormal inductors," which do not have the ability to invaginate. Instead, a different method was used: the piece of tissue to be tested, or a piece of coagulated extract, was pushed into the blastocoele of a late blastula or early gastrula through a cut made in the animal hemisphere of the host embryo (Fig. 158). The cut heals very quickly and later, when the gastrulation proceeds, the graft becomes wedged in between the invaginating endomesoderm and the ectoderm, usually somewhere in the liver region. The graft is pressed against the ectoderm and thus is in a good position to exercise its ability to induce.

Further methods were devised to eliminate the possible interaction of the graft with the primary organizer of the host embryo. The "sandwich method" consists in cutting out pieces of presumptive ectoderm from the animal hemisphere of a blastula or beginning gastrula and then placing the inductor between two flaps of ectoderm. The edges of the pieces of ectoderm soon grow together, and the inductor is thus enclosed in a vesicle consisting of reacting material. A further problem is posed when fluid extracts are to be tested for their possible inducing ability. In this case it is undesirable for the presumptive ectoderm to curl up and form vesicles because the distal surface of ectoderm, which covers such ectodermal vesicles from outside, is apparently less permeable than the inner surface, which comes into contact with the inductor in normal development. To prevent them from curling, the pieces of ectoderm are kept flat by holding them between two layers of nylon (Yamada and Takata, 1961) or by pressing them to the surface of a coverslip by a piece of silk or filter paper (Tiedemann, 1962). The four main methods of testing inductors are represented diagrammatically in Figure 158.

The "sandwich method" has been used to investigate the temporal relationships in induction. Inducing tissue was wrapped in a layer of early gastrula ectoderm, and after varying periods of time the inducer was removed, and the ectoderm left to differentiate. When ectoderm of *Triturus* was used for the experiment and the inductor was

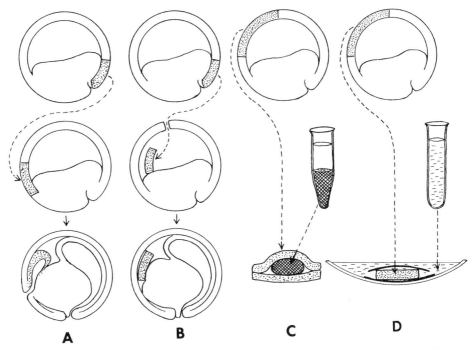

Figure 158. Four methods of exposing gastrula ectoderm to induction (diagrammatic). *A*, Transplantation of part of dorsal lip (stippled) into the ventral marginal zone, where it invaginates and forms a secondary archenteron. *B*, Transplantation of part of dorsal lip (stippled) into the blastocoele through a slit at the animal pole. Graft is eventually pressed against the ventral ectoderm. *C*, "Sandwich experiment": Inductor (such as a centrifuged tissue homogenate) is placed between two pieces of presumptive gastrula ventral ectoderm. *D*, Cultivation of a piece of presumptive ventral ectoderm of a gastrula in a watch glass filled with fluid containing inducing substance in solution. The reacting ectoderm is held between two layers of supporting material (silk) to prevent it from curling.

removed after one hour, only very weak inductions were produced. After three hours, or still better after four hours, well-developed and large inductions were obtained (Johnen, 1956, 1964; Toivonen, 1961; Suzuki, 1968b). The claim that in *Ambystoma mexicanum* inductions can be recorded after only five minutes of contact (Johnen, 1956) seems to be somewhat doubtful, especially since *Ambystoma* tissues react very easily to unspecific stimulation (via sublethal cytolysis?). (See further, p. 213.)

It is of special interest that inductions can be performed not only by tissues of the same or closely related species but also by tissues of various animals, even of animals belonging to different phyla of the animal kingdom. Using the newt as the host, the tissues of the following animals were found to be effective as inductors: *Hydra*, insects, fishes, reptiles, birds, and mammals (mouse, guinea pig, man). This lack of specificity applied to both normal and abnormal inductors. The conclusion from these experiments must be either that the inductive substance is very widely distributed in animal tissues or that various substances may act as inducing agents.

Just as the normal inductor can act after being killed, the abnormal inductors can also induce after being killed by boiling or by immersion in alcohol or ether. A rather remarkable discovery made in this connection is that some tissues can induce better when they have been killed than when they are used in the fresh state. Some tissues that do not induce in the living state may induce when dead. This is the case with parts of the amphibian gastrula and blastula other than the dorsal blastopore lip region. Presumptive ectoderm and endoderm if killed by boiling or in other ways and implanted into the blastocoele will induce neural plates (Holtfreter, 1934a). Whether the inducing agent in this case is the same as the one emerging from the archenteron roof cannot be affirmed without further proof.

The next stage in induction research was obviously to try to isolate the active substance from the inducing tissues. Such attempts date back to the early 1930's, when active preparations were made (from "abnormal inductors," such as liver or muscle tissue) by extraction with ether (Waddington, Needham, et al., 1934) or by boiling in water (Wehmeier, 1934; Fig. 159). Chemicals of known composition, in particular certain carcinogens, were used with some success. Sufficient evidence was not obtained that the inducing preparations were identical, or even similar, to the active substances of normal development. Thus, for instance, it was accidentally discovered that isolated gastrula ectoderm may be stimulated to produce neural tissue (in the form of neural tubes) by keeping it in a weak solution of the dye methylene blue, which obviously has nothing to do with the development of the normal embryo (Waddington, Needham, and Brachet, 1936). Lastly, it was found (Holtfreter, 1947) that isolated presumptive ectoderm may be caused to develop into nervous tissue by a short exposure to saline solution of a composition which is not quite favorable to the cells. This is the case if the standard solution is more acidic or more alkaline than it should normally be—if the pH is lower than 5.0 or higher than 9.2.

Continued development in strongly acid or alkaline solutions is of course impossible; embryonic tissue, exposed to these solutions, very soon shows signs of damage. First, the connections between the cells are broken; embryonic epithelia fall apart. The cytoplasm of the cells treated with acids or alkalis shows signs of liquefaction, which are noticeable because of an increased brownian movement in the interior of the cells, and the amount of the hyaline cytoplasm at the periphery of the cells increases. If the acid or alkaline treatment is continued, cytolysis sets in, the cell membrane breaks down and the cells disintegrate. (See Figure 160.) If the duration of the treatment is short, and

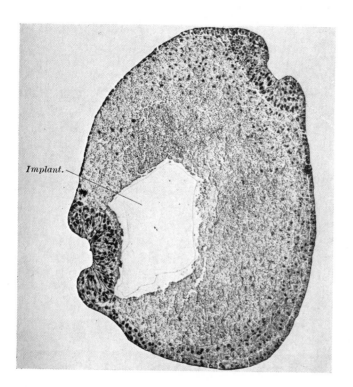

Implant.

Figure 159. Induction of a secondary neural plate in an axolotl embryo by means of an implanted piece of agar containing substances extracted from muscle by boiling the muscle in water. (From Wehmeier, 1934.)

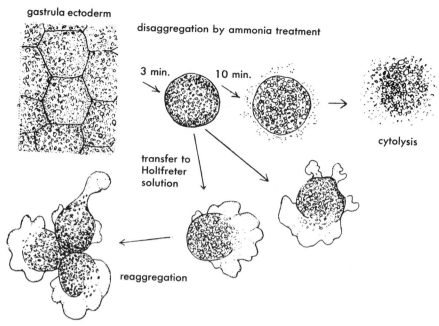

gastrula ectoderm

disaggregation by ammonia treatment

3 min.   10 min.

cytolysis

transfer to
Holtfreter
solution

reaggregation

Figure 160. Treatment of amphibian embryonic cells with ammonia, with and without return of the cells to a normal medium. (From Karaskaki, 1957.)

the cells are soon transferred to a normal solution, the disintegration of the cells is checked. For a time they may show active ameboid movement, but later the cells tend to clump together—reaggregate—and may resume normal development except that the end result of this development is changed: neural differentiation is observed in a great proportion of the aggregates. Complete dispersal of the tissues into individual cells is not necessary for the change in their differentiation. The main effect appears to be a sublethal damage to the cells from which they can recover. Some agents such as urea, causing damage to the cells without changing the pH of the medium, may have the same effect as acids and alkalis (Karasaki, 1957).

The neural induction by cytolyzing agents is obviously a completely different way of controlling differentiation from the normal processes of induction by the roof of the archenteron. The action of the roof of the archenteron does not cause a sharp swing of the pH to the acid or the alkaline side; neither does it cause cytolysis in the overlying ectoderm. Even if it could be assumed that some specific inducing substances pass from the roof of the archenteron (or from pieces of adult tissues) into the ectoderm and thus change it qualitatively, no similar mechanism can be at work when solutions of ammonia or hydrochloric acid are used in cytolysis experiments. It has been suggested that the cytolysis "unmasks" some substance present in the ectodermal cells themselves, and that this substance, once released or activated, changes the presumptive epidermal cells into neural cells. Whether this interpretation is correct or not, it causes us to regard with extreme caution any experiment in which neural differentiation is caused by treatment with various chemical substances. These substances may have nothing to do with normal induction but cause neural differentiation through their toxic action, which causes sublethal cytolysis of cells.

Emission of Inducing Substances by the Natural Inductors. We have seen that extracts of inducing substances made by chemical methods cannot be entirely relied on, since a variety of substances may produce inductions. It was therefore a great advance when Niu and Twitty discovered that tissues capable of induction themselves release inducing substances into the surrounding medium. It was only necessary to find a means of accumulating these substances in sufficient concentrations and to discover the best way to apply them to cells competent to react to the induction. The first part of the problem was solved by cultivating inductor tissues, such as the dorsal lip of the amphibian blastopore, parts of the neural plate, or rudiments of the notochord and somites, in small quantities of saline by the hanging drop method (a drop of culture medium with the tissue to be cultivated is placed on a coverslip, and this is inverted over a hollow on a slide; the preparation is then sealed to prevent desiccation). The saline solution used was a specially modified one, now known as the "Niu and Twitty solution" (Niu and Twitty, 1953). The inducing substances accumulate in the medium gradually; the active concentration is reached only after a week to 10 days. The fragment of early gastrula ectoderm is then placed in the same drop of medium and cultivated for an additional one to three weeks. It was found important for the success of the experiment that the fragment of ectoderm be very small, containing approximately 15 to 20 cells. If the fragment is large, it rounds off and forms a continuous epithelial layer on the surface, which prevents substances from without from penetrating into the cells. With a very small fragment this does not occur, and there is a sufficient surface of unprotected cytoplasm through which the inducing substances may be absorbed.

In successful cases, and these may be up to 90 percent in some experimental series, the cells of the ectodermal explant behave as if they were a part of a normal neural plate. Many of the cells scatter in the medium and become transformed into typical melanophores. Others remain in a clump and differentiate as nerve cells. The latter may produce nerve processes, which radiate into the surrounding medium (Fig. 161). The presence of the inductor tissues is no longer necessary for the success of the induction once the medium has been "conditioned"—i.e., has accumulated the necessary amount of inducing substances. Pieces of ectoderm cultivated in drops withdrawn from the original cultures and containing no inductor cells differentiate into nerve cells and melanophores just as readily as those which are kept in a drop with a piece of inductor (Niu, 1956).

The action of "conditioned" saline must be essentially different from the shock treatment by media with very high or very low pH which was previously referred to. It was found by direct measurement that the saline after cultivation in it of inductor tissues was still very nearly neutral (it has a pH of 7.8), and neither was any trace of suble-

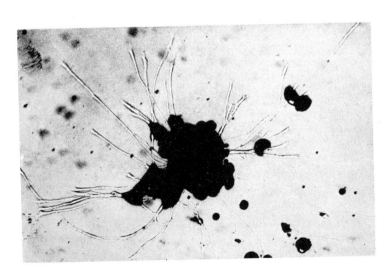

Figure 161. Neural tissue with radiating nerve fibers and melanophores developed from a piece of young *Triturus rivularis* ectoderm cultivated in conditioned medium. (From Niu, 1956.)

thal injury or cytolysis apparent. The action was obviously due to the presence of a specific substance or substances in the medium, and there appears to be no reason why these substances may not be the "natural" inducing substances, that is, the substances through the mediation of which the neural plate and other organs are induced in normal development.

As far as further insight into the chemical nature of the inducing substances is concerned, Niu and Twitty's experiments have the same drawbacks as the experiments with the normal embryos: the quantities of substances produced are too minute for an elaborate chemical analysis.

Chemical Analyses of Inducing Substances.   As has already been indicated, it is not very difficult to obtain extracts from tissues showing inductive action when brought into contact with competent embryonic cells, and such extracts may be prepared in a variety of ways. Perhaps the simplest way is to homogenize the tissue and to sediment the larger particles by centrifugation. However, chemical extraction by a variety of solvents has also been tried out. Adult tissues, as a source of inducing substances, enable the investigator to make use of a number of methods of purification which are normally applied in biochemical investigations, since one may start with sufficiently large quantities of material.

Two schools have been especially successful in purifying extracts from tissues capable of imitating the action of the natural "primary organizer": namely, the school of Yamada in Japan and the school of Tiedemann in Germany. Yamada and his collaborators used guinea pig liver and guinea pig bone marrow as their main sources of inducing substance. The tissues in their experiments were homogenized, and separation of fractions was achieved by ultracentrifugation, by sedimentation of nucleoproteins with streptomycin sulfate and of proteins with acidified solutions and with ammonium sulfate, and eventually by chromatographic separation on a diethylaminoethyl cellulose chromatographic column (Yamada and Takata, 1961; Hayashi, 1956; Yamada, 1958).

Tiedemann and collaborators used nine-day-old chicken embryos as starting material. The homogenates were extracted either with phenol, to obtain an active protein fraction, or by deoxycholic acid followed by sedimentation with ammonium sulfate, which separates a nucleoprotein fraction in the sediment and protein fraction in the supernatant, both of which were found to possess inducing ability. The last stage of the purification was also carried out on a diethylaminoethyl cellulose or a carboxymethyl cellulose chromatographic column (Tiedemann, Becker, and Tiedemann, 1961; Tiedemann, 1961, 1962, 1968).

The extracts prepared by workers of both schools contain macromolecular substances, some of which show all the chemical and spectrophotometric reactions of proteins; others are ribonucleoproteins. The inducing power of these preparations is no weaker than that of the natural organizer. Another similarity is that substances obtained by certain modifications of the preparation procedures have more specific inducing properties, such as those of archencephalic, deuterencephalic, and spinocaudal inductors (p. 203). A particular protein fraction, prepared from chicken embryos with spinocaudal inducing properties, could be further purified by dextran gel electrophoresis. In this way, from 1 kg. of chick embryos 0.6 to 0.8 mg. of presumably pure substance was obtained with a molecular weight of about 25,000. The preparation proved to be highly active and could induce in about one third of the cases, even if diluted in the proportion of $1:12,000$ by mixing with inert noninducing globulin. Each embryo in this experiment received only 2.10^{-6} mg. of inducing protein. In pure or only slightly diluted form it produced large inductions in 100 per cent of the cases (Tiedemann et al., 1965). Fractions containing only deoxyribonucleoproteins were completely inactive (Tiedemann, 1961).

It seems somewhat disconcerting that inducing activity should be present in two different groups of chemical compounds such as proteins on the one hand and ribonucleoproteins on the other. Furthermore, seeing that synthetic processes in cells are controlled by different kinds of ribonucleic acids, it would be of great importance to know whether the participation of ribonucleoproteins in induction may mean that the ribonucleic acids involved are instrumental in bringing some kind of "information" from the inductor cells to the reacting cells.

Fortunately, some further experiments provide a clear and satisfactory answer to these problems. It has been convincingly shown by various investigators that the degradation of ribonucleic acid in inducing preparations by ribonuclease does not reduce the inducing ability of such preparations. In one experiment, for instance, a very active preparation was obtained from a liver homogenate by ultracentrifugation. The microsome fraction was used, which contains most of the cytoplasmic ribonucleic acid. The experimental lot of this preparation was incubated for 3.5 hours with ribonuclease. After ribonuclease treatment the substance was precipitated, and its inducing ability (as well as that of untreated controls) was tested by placing pieces of precipitate between two layers of gastrula presumptive ectoderm. The preparations were cultivated as explants in a saline solution. Table 10 gives the results of the experiment.

The data in Table 10 show that although the ribonucleic acid content of the inductor was reduced to less than 1 per cent of the original quantity, the inducing power of the preparation was the same as in controls, not only with respect to the total number of inductions but also with respect to the type of induction observed. This result is supported by many similar experiments, and it appears therefore that ribonucleic acid cannot be the essential part in the chemical composition of an inductor.

In contrast to the action of ribonuclease, treatment of tissues or tissue extracts with proteolytic enzymes—pepsin, trypsin, or chymotrypsin—destroys their ability to induce. Figure 162 shows in diagrammatic form the results of treating a liver extract with trypsin for varying times (Hayashi, 1958). It was found that the preparation completely lost its ability to induce neural structures after treatment for 120 minutes and only caused weak atypical inductions. Figure 163a and b shows the induction in the control compared with the lack of neural induction after treatment of the liver extract with trypsin (120 mintues). There seems to be no doubt that the integrity of the proteins is essential for inductions by tissues and tissue extracts. Exactly similar results have been obtained by workers of the Tiedemann school. Likewise, the "conditioned medium" in the experiments of Twitty and Niu is not substantially affected by ribonuclease treatment, but addition of trypsin or chymotrypsin prevents induction completely (Niu, 1956).

The general conclusion from the preceding experiments is that the inducing prin-

TABLE 10   **Inducing Effect of Liver Ribonucleoprotein Treated with Ribonuclease**

	Control	*Experimental*
Ribonucleic acid content in mg. of RNA phosphorus per mg. protein nitrogen	148.6	0.84
Number of explants	49	52
Total neural inductions	100%	100%
Archencephalic inductions	14%	15%
Deuterencephalic inductions	82%	77%
Atypical inductions	18%	17%

After Hayashi, from Yamada, 1958.

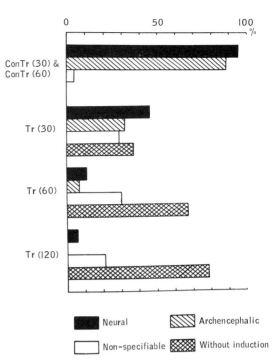

Figure 162. Progressive suppression of the inducing power of a liver preparation by treatment with trypsin. Figures on left indicate duration of trypsin treatment. Control is at top of graph. (From Hayashi, 1958.)

ciple in both types of experiment is of a protein nature. The protein may or may not be coupled with ribonucleic acid, but the inducing action is due to protein alone.

There is hardly reason to believe that the treatment with purified extracts of inducing tissues, any more than the "conditioned medium," acts in some roundabout way by "unmasking" an inducing substance present in the reacting cells. There is no trace of sublethal damage, and the specificity of the preparations and their activity in minute quantities are evidence in the same direction. Rather, one feels that the proteins contained in the tissue of advanced embryos (nine-day chicken embryos) and adults have some properties in common with the natural inducing agents liberated from the roof of the archenteron, if they are not actually identical to these agents. As a further corroboration of this view, we may mention that an extract obtained from gastrulae and neurulae of *Triturus*, and presumably containing the "natural" inducing substance, is

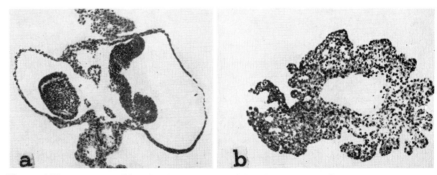

Figure 163. Results of implanting into a vesicle of *Triturus pyrrhogaster* early gastrula blastoderm of a pentose-nucleoprotein liver preparation without enzyme treatment (*a*) and after treatment with trypsin for 120 minutes (*b*). There is archencephalic induction in *a*, lack of neural induction in *b*. (From Hayashi, 1958.)

inactivated by trypsin treatment, similarly to the inactivation of inducing extracts dealt with in the foregoing pages (Tiedemann, Becker, and Tiedemann, 1961).

8–7 MECHANISM OF ACTION OF THE INDUCING SUBSTANCES

The last question which may be considered in connection with the physicochemical nature of the processes of induction concerns the way in which the inducing substances control the differentiation of cells. In principle it may be assumed that the inducing substances penetrate into the cells and, by interfering in the metabolic mechanisms in the interior of the cells, change their physicochemical composition. Although this mechanism of action seems to be fairly plausible, it is not self-evident.

Radioactive tracers have been used to learn whether, in the process of embryonic induction, substances pass from the inducing tissue into the reacting tissue. Labeled amino acids, methionine S^{35} and glycine C^{14}, and a nucleic acid precursor, orotic acid C^{14}, were used in some experiments (Ficq, 1954; Sirlin, Brahma, and Waddington, 1956; Waddington and Mulherkar, 1957). These compounds are taken up into the tissues of the embryo. Next, inducing parts—roof of the archenteron or parts of brain rudiments—are excised from the embryo containing radioactive substances and are transplanted into a normal embryo where the graft may induce neural plates or other structures from the host tissue. The embryo is then fixed and cut into sections, and autoradiograms are prepared by the method mentioned on page 42.

It has been found that the radioactive atoms do not remain restricted to the cells of the grafts, but that they become fairly widely dispersed in the host embryo. A high radioactivity is shown by induced neural plates, which may be the result of the passage of inducing substances from the graft into the host tissue. However, the host neural plates and tubes are also strongly radioactive, and radioactivity can also be discovered in the mesodermal and endodermal tissues of the host. These results are consistent with the assumption that the radioactive atoms are carried around in the host tissue by simple diffusion.

Furthermore, it cannot be claimed with certainty that the diffusing substance was actually the macromolecular material which can reasonably be assumed to be the inducing agent. It could well be that the radioactive atoms were carried around as components of small molecules (single aminio acids, mononucleotides, and even smaller organic and inorganic molecules).

The preceding criticism does not apply to another set of experiments in which use was made of immunological methods to trace the passage of substances from the inductor into the reacting tissue. Antibodies were prepared against malignant tumor tissue used in one experiment (Vainio, Saxén, Toivonen, and Rapola, 1962) and against a purified protein preparation isolated from guinea pig bone marrow in another experiment (Yamada, 1962). The antibodies in the antiserum were later coupled to a fluorescent dye, rhodamine B.200. *Triturus* ectoderm was exposed to the action of the inductor (tumor tissue in one case, protein preparation in the other) for a few hours, then fixed, and cut into sections. The sections were then treated with the antiserum coupled with the fluorescent dye. Owing to antigen-antibody binding, the antibody molecules with the attached dye became localized in the section in positions which indicated the distribution of the antigen molecules (the molecules of the inductor). The fluorescent dye made these locations easily discernible. Distinct fluorescence was found in the reacting tissue, thus showing that antigen molecules had actually penetrated into it. Since the specificity of the antibody-antigen reaction would have disappeared if the antigen molecules were degraded to any great extent, it is evident that the macromolecules of the materials used as inductors penetrated into the reacting cells without being split up into small components.

This type of experiment would have been convincing if highly purified inducing substances had been used. Unfortunately, this was not the case, and it is possible that proteins or other antigens which were detected by the antiserum molecules had nothing to do with the process of induction.

Some evidence on the mechanism of induction has been provided by the use of metabolic poisons coupled with induction experiments. If the induction is mediated by the diffusion of some specific proteins (or nucleoproteins) from the inducing to the reacting tissue, then it may be postulated that a specific protein has been manufactured in the inductor at some stage and that a corresponding mRNA had been transcribed from a locus on the DNA. Secondly, if the reacting tissue acquires new properties, this, in all probability, means that some new substances have been produced in the reacting cells or that existing substances have been modified. In both cases, it would be suspected that new proteins are involved, a circumstance which again presupposes transcription of a new kind of mRNA and translation into a previously absent new protein. That the change is caused simply by the presence in the reacting cells of the inducing substance, which had passed into them from the inductor, is highly improbable, especially in view of the possibility of neural induction by way of sublethal cytolysis. The synthetic mechanisms of the reacting cells are obviously involved in the transformation. Consequently, the question is, What would happen to the process of induction if transcription or translation were prevented in the inductor, or in the reacting tissue, or in both?

In some experiments (Tiedemann et al., 1967; Suzuki, 1968a) the dorsal blastopore lips of young *Triturus* gastrulae were cultivated for up to 20 hours in a medium containing actinomycin D or puromycin, and then the explant was confronted with reacting ectoderm. Mesodermal induction (Suzuki, 1968a) or both mesodermal and neural inductions (Tiedemann et al., 1967) were obtained. These results prove that neither transcription of mRNA nor synthesis of new proteins occurs in the inductor at the time of its action; that is, the inducing substance is already present at the beginning of gastrulation (which could be expected, since a killed inductor can induce).

If, on the other hand, the dorsal blastopore lip was explanted with the adjacent ectoderm and cultivated in a medium containing sufficient quantities of actionomycin D to inhibit RNA synthesis completely, induction could not take place. After two to seven days of cultivation it was found that while some differentiation of muscle and notochord occurred, there was no differentiation of nervous tissue (Tiedemann, 1968). Since no new mRNA could be produced under the circumstances, it follows that the mRNA for notochord and muscle differentiation was already present at the time of explantation—at the beginning of gastrulation. For the ectoderm to start neural differentiation, however, it was necessary for the reacting cells to produce new mRNA by transcription from the DNA in the reacting cells. Such transcription was made impossible by the presence of actinomycin D, and the result was that no neural induction could be detected.

Lastly, *Triturus* gastrula ectoderm, which was exposed to an inductor in a "sandwich" experiment, was separated from the inductor and treated with actinomycin D for six hours, after having been disaggregated (p. 230) to allow better penetration of the poison into the cells. In controls, the disaggregated cells fused together again and differentiated into neural and mesodermal tissues, as would be expected in view of the inductor which had been used. Actinomycin-treated explants also reaggregated and remained viable for days, but only atypical epidermal differentiation took place

(Toivonen et al., 1964). Together with the previous experiment, this experiment shows that even though an "inducing substance" may have penetrated into reactive cells, their transformation into differentiated neural and mesodermal cells cannot occur without active participation of the nuclei of reacting cells, and particularly without an RNA (presumably mRNA) being produced as a crucial link in the process of induction.

8–8 GRADIENTS IN THE DETERMINATION OF THE PRIMARY ORGAN RUDIMENTS IN VERTEBRATES

The Dorsoventral and Anteroposterior Gradients.   The phenomenon of regional specificity raises the question of whether the gray crescent does not contain, after all, two or even more different specific substances, responsible for the development of the head and the trunk regions (or the archencephalic, deuterencephalic, and spinocaudal regions), with their separate inductors or organizers. Certain results achieved with abnormal inductors have been adduced in favor of this supposition.

When different tissues of various adult animals are tested as inductors, it is found that they do not all act exactly alike; some of them act preponderantly as head inductors, and others as trunk inductors. Some induce only ectodermal parts (neural structures, sense organs); others also induce mesodermal organs and tissues (Toivonen, 1949, 1950, 1953). For instance, the liver of the guinea pig, treated with alcohol, acts as an archencephalic inductor; that is, it induces large brain vesicles, bearing a resemblance to the telencephalon, diencephalon, and mesencephalon (p. 203), sometimes with eyes. It also induces noses and balancers. (See p. 204 and Figure 151A.) The kidney of the adder is a rare case of a predominantly deuterencephalic inductor; it induces brain parts resembling the hindbrain and also ear vesicles. The kidney of the guinea pig treated with alcohol is a spinocaudal inductor; it induces mainly spinal cord, notochord, bands of muscle segments arranged as they are normally in the trunk region of the embryo, and often also complete tails and fin folds around their edges. Lastly, alcohol-treated bone marrow of the guinea pig induces almost exclusively mesodermal parts of the trunk and tail: notochord, rows of somites, nephric tubules, and limb-buds (Fig. 151B).

The inducing properties of tissues may change depending on the type of treatment. In particular, it was found that heating (or boiling) the tissues reduces their ability to induce mesodermal and spinocaudal structures, whereas archencephalic inductions are still easily obtained. This and similar results have led to the conclusion that the regional nature of inductions produced by adult tissues is the result of the interaction of two factors contained in various proportions in the different tissues (Yamada, 1950; Nieuwkoop et al., 1952; Toivonen and Saxén, 1955; Toivonen, 1958). One factor is the "neuralizing agent" of Toivonen and Saxén, called "dorsalizing agent" by Yamada and "activating agent" by Nieuwkoop et al. When it is present alone it causes archencephalic inductions. It is the active principle of the alcohol-treated liver. The second principle is Toivonen and Saxén's "mesodermalizing agent," which is the same as the "caudalizing agent" of Yamada and the "transforming agent" of Nieuwkoop et al. If present alone, this second factor induces only mesodermal parts: notochord, muscle, kidney, and limb-bud (Fig. 164). This is the active principle of the alcohol-treated guinea pig bone marrow. A deuterencephalic induction is the result of the presence of a small amount of the mesodermalizing factor in addition to the neuralizing factor. A large amount of the mesodermalizing factor added to the neuralizing factor produces a spinocaudal induction.

That deuterencephalic structures are a result of a certain balance between neuralizing and mesodermalizing inducing substances has been shown by the following ingen-

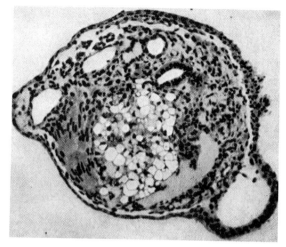

Figure 164. A purely mesodermal induction (notochord, somites, pronephros) in isolated ectoderm caused by the non-nucleoprotein fraction of bone marrow. (From Yamada, 1958.)

ious experiment (Toivonen and Saxén, 1955). Two pellets, one prepared from guinea pig liver and another prepared from guinea pig bone marrow, were implanted simultaneously into the blastocoele of a young gastrula (Fig. 165). The liver preparation alone would have induced exclusively archencephalic structures; the bone marrow alone would have induced trunk mesoderm. Together, however, they induced organs belonging to all levels, including deuterencephalic and spinocaudal structures: medulla, ear vesicles, and spinal cord.

A counterpart of the experiment of the Finnish embryologists was performed by the Tiedemann school (Tiedemann et al., 1963). By a modification of the method of extraction and purification of inducing substances derived from chick embryos, the investigators obtained a preparation which acted as a deuterencephalic inductor, inducing mainly hindbrain and ear vesicles. The preparation was then separated by chromatography on diethylaminoethyl cellulose into several fractions. Two active fractions were obtained, one which produced archencephalic inductions, and another which produced spinocaudal inductions. Deuterencephalic inductions disappeared almost completely (Fig. 166)!

The two factors are actually substances with different chemical properties. The neuralizing substance is thermostable and soluble in organic solvents. The mesodermalizing substance is insoluble in organic solvents and highly thermolabile. The result of this latter property is that spinocaudal and deuterencephalic inductors upon heat treatment induce archencephalic structures; in the fresh state they contain both sub-

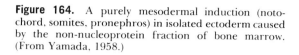

Figure 165. Simultaneous implantation of an archencephalic (liver—L) and a mesodermalizing (bone marrow—Bm) inductor into a newt's gastrula. (From Toivonen and Saxén, 1955.)

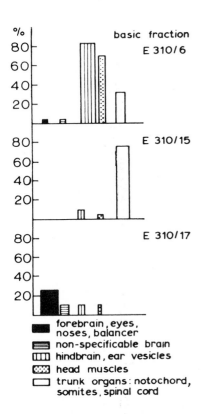

Figure 166. Results of chromatographic separation of a chick embryo extract having a deuterencephalic inducing capacity (basic fraction) into a spinocaudal inducing fraction (E 310/15) and an archencephalic inducing fraction (E 310/17). (From Tiedemann, Becker, and Tiedemann, 1963.)

stances, but the mesodermalizing substance is destroyed by heat, whereas the neuralizing substance remains unchanged (at least after a short heat treatment—prolonged heat treatment inactivates the neuralizing substance as well). By graded heat treatment of a tissue its inducing properties may be changed gradually, as the ratio between the mesodermalizing and the neuralizing substances changes in favor of the latter (Fig 167).

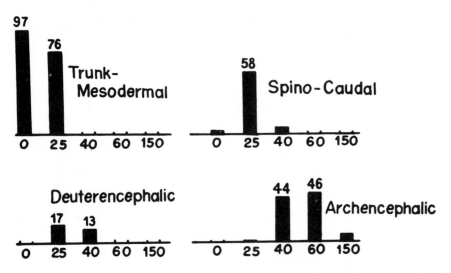

Figure 167. Change in the percentage of different types of inductions as a result of treating the inductor (slices of bone marrow) with heat for varying times (seconds) indicated below the lines on the diagram. (From Yamada, 1958.)

It has been a matter of controversy for some time whether the neuralizing agent and the mesodermalizing agent are two distinct chemical substances, two aspects of the action of one substance, or two states of one substance, a more labile state (the mesodermalizing agent) and a more stable state (the neuralizing agent). Experiments have now resolved this problem. It has been found that the neuralizing substance of the guinea pig liver can be isolated in the form of a ribonucleoprotein (Hayashi, 1956), either by sedimentation with streptomycin sulfate or by ultracentrifugation, as a result of which the microsomal fraction, containing the cytoplasmic nucleoprotein, shows the strongest archencephalic inductive action. The mesodermalizing substance of guinea pig bone marrow, on the other hand, is not sedimented by streptomycin sulfate and is not contained in the microsome fraction after ultracentrifugation (Yamada, 1958).

The mesodermalizing substance is therefore a protein which does not tend to be coupled with nucleic acid. The two agents are thus distinctly different and separate substances, one possibly bound to the microsomes, the other not. There is no contradiction between the statement that the neuralizing substance can be obtained in the form of a ribonucleoprotein, and our previous finding that the active principle of the inducing substance is a protein and not a nucleic acid: if the active nucleoprotein prepared from liver is treated with ribonuclease, the inducing action is retained; if it is treated with proteolytic enzymes, the preparation loses its inducing ability (Hayashi, 1955, 1958).

The results obtained with abnormal inductors have been used to interpret normal development. It has been assumed (Toivonen and Saxén, 1955) that the same two substances are active as inducing substances of the archenteron roof, and that the regional differentiations in the normal embryo are controlled by a balance between the two substances distributed in the form of gradients. The gradient of the neuralizing substance is highest middorsally and declines toward the lateral and ventral parts of the embryo. The mesodermalizing substance is most highly concentrated at the posterior dorsal end of the embryo and forms a declining gradient both in the anterior direction and in a lateral direction. This distribution of substances as originally contained in the archenteron roof is transmitted to the overlying ectoderm.

The anterior part of the archenteron roof (including prechordal plate) emits only the neuralizing substance and induces the archencephalon. No notochord and no somites are developed at this level. At a slightly posterior level, the notochord and somites are already present. There is some admixture of the mesodermalizing substance with the neuralizing substance, and as a result deuterencephalic structures (medulla and ear vesicles) are induced in the ectoderm. Still farther back, large amounts of the mesodermalizing substance are available, mesoderm develops into notochord and large masses of muscle, and the neural structure induced is a spinal cord (Fig. 168). Each level of the embryo has a different concentration of the neuralizing and mesodermalizing agents.

The concept of two mutually permeating gradients reminds us at once of the animal and vegetal gradients of the sea urchin egg, though the relationships of the gradients to the main axes of the embryo (and egg) are different.

Gradients, in the form of grading or fading out of certain properties, are obvious in the early development of amphibia and other vertebrates. The dorsoventral gradient (the gradient of the neuralizing agent) can be traced both in the ectodermal and in the mesodermal layers. In the ectoderm the highest level of the gradient is associated with the development of the neural plate and neural tube. The next highest level of the gradient causes the differentiation of the neural crest. There is ample evidence that the development of the neural crest is the result of a weak induction of the same nature as the induction of neural tissue. When neural plates are induced, the cells on the periphery of the region exposed to the inductor, where the inductive stimulus is fading away,

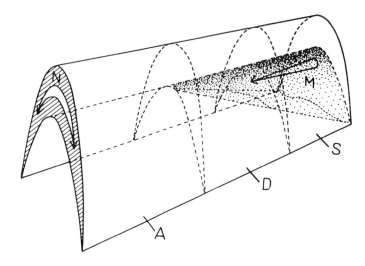

Figure 168. Diagram of the neuralizing gradient (N) and the mesodermalizing gradient (M). A, Archencephalic level; D, deuterencephalic level; S, spinocaudal level. (From Toivonen and Saxén, 1955.)

develop as neural crest. Sometimes, if the inductor is too weak to induce a neural plate, neural crest cells can still be induced. Usually the presence of neural crest cells on the site of induction can be easily noticed, because some of them differentiate as pigment cells. Possibly a still lower level of the gradient is responsible for the differentiation of the various placodes, including nose and ear rudiments, the rudiments of cranial ganglia and of the lateral line sense organ system. (See Sections 12–1 and 12–4.)

Where the gradient of the neuralizing agent fades away, ectoderm differentiates as skin epidermis. In the chordomesodermal mantle, the dorsoventral gradient may be held responsible for the segregation of the notochord and the mesoderm in the gastrula of an amphibian. Those parts of the marginal zone in which the specific cytoplasmic substance is most highly concentrated become notochord, and those parts in which the concentration of the substance is lower become mesoderm (the somites). Lastly, those parts in which the concentration of the cytoplasmic substance is minimal become unsegmented lateral-plate mesoderm (Yamada, 1937; Dalcq and Pasteels, 1937). The gradient concept does not contradict our previous statement that the action of the primary organizer may include the induction of somites and other mesodermal parts. This induction may be due to the presumptive notochord, as the highest level in the gradient system, establishing a new gradient in the surroundings into which it has been transplanted. This may involve a diffusion of the specific substance from higher levels of concentration into the surroundings where the concentration is lower.

Besides the gradient which has its highest level in the dorsomedian strip of tissues diminishing toward both sides, there is a second gradient, with a center of highest concentration at the anterior end of the embryo, which lessens in a posterior direction. This anteroposterior gradient may be considered to be another manifestation of the caudocranial mesodermalizing gradient; its reverse side, as it were. This gradient is responsible for the differentiation of parts of the nervous system, the most anterior part of the head being the part where the gradient is at its highest.

Control of Development by Influencing the Gradient System.   The gradient unites the parts of a developing embryo into one whole, into one morphogenetic system. Any factor that affects the gradient will therefore affect the whole morphogenetic system, no matter how simple or even elementary the factor itself may be. It is possible, for instance, to affect a morphogenetic system by depressing the level of the gradient at its highest point. One way of achieving this is to expose the embryo to certain chemical substances, such as magnesium chloride or lithium chloride. Another way of depress-

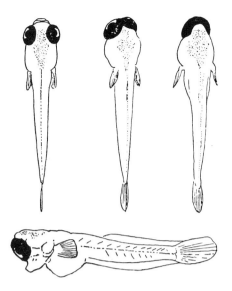

Figure 169. Cyclopia in fish larvae, caused by magnesium chloride treatment. Top left, normal control. (After Stockard, from Huxley and de Beer, 1934.)

ing the high level of the gradient is to remove part of the archenteron roof underlying the anterior end of the nervous system. The result in both cases is about the same: the most anterior parts of the nervous system fail to develop normally. The defects can best be traced in the structure of the eyes. The injured embryos develop a defect known as **cyclopia,** that is, the appearance of one median eye instead of two lateral ones (Fig. 169). The median eye is really the result of fusion in the midline of the two eye-forming areas, and every intermediate state may be found in embryos in which the injury has not been very severe. If large parts of the underlying chordomesoderm have been removed, or if the action of the chemical agent has been very strong, the defects in the structure of the eyes are still greater, the single eye being reduced in size and failing to develop altogether in the extreme cases. The eyes are not the only organs affected in cyclopic embryos; other parts of the head are changed and reduced more or less in correspondence with the defects of the eyes (Adelmann, 1936). The parts which

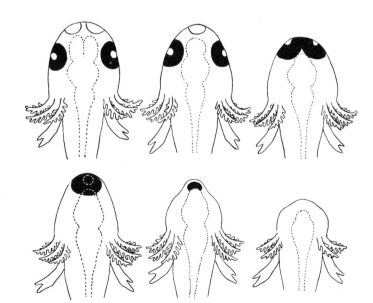

Figure 170. Increasing degrees of cyclopic deficiencies in newt larvae (diagrammatic.) Upper left, normal larva.

may be involved are the brain, the nose, the ear, the mouth, and the gill clefts. The defects spread from the front end backward as the injury is more and more pronounced. The nose rudiments and the forebrain are the first to be affected, so that the olfactory sacs become unpaired and finally disappear. Then the eyes and the mouth follow, the mouth becoming narrower and eventually disappearing. After the eye and mouth follow the posterior parts of the brain, the gill clefts, and the ear vesicles. The gill clefts fuse in the midventral line, before disappearing one by one. In extreme cases practically the whole head is lacking (Fig. 170).

It must be understood that the complete presumptive ectoderm remains in place during the experiments just described; none of it is removed, and neither do any of its parts become necrotic. All cells are there, but their development is changed, as a result of an interference with the gradient system. The experiment with the lithium chloride or magnesium chloride reveals a further important phenomenon: although the whole embryo may be exposed to the chemical substance, it is only the topmost levels of the gradient system that are affected. This is a very common result, namely, that the highest point of a gradient is more easily damaged than the other parts. In fact, a gradient can often be discovered by exposing the embryo to any mild injurious factor, such as weak poisons, abnormal temperature, or ultraviolet radiation. The effect will be observed first, and sometimes only, at the highest point of the gradient if the intensity of the injurious factor has been chosen correctly. In vertebrates in stages following gastrulation, such a sensitive region is invariably the anterior part of the head.

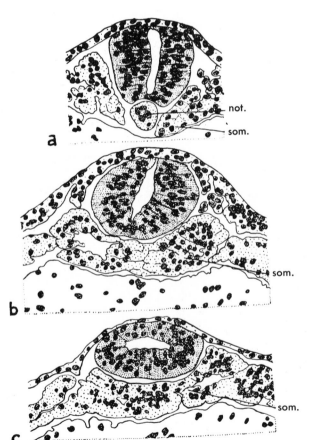

Figure 171. Transformation of presumptive notochord into somite mesoderm by means of the action of lithium in *Triturus*. *a*, Untreated control; *b*, *c*, lithium-treated embryos. not., Notochord; som., somites. (From Lehmann, 1945.)

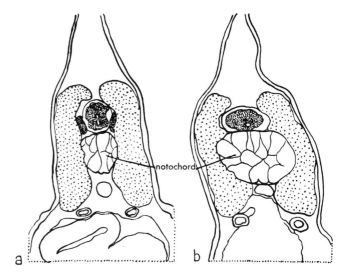

Figure 172. Raising the dorsoventral gradient in a frog embryo with thiocyanate. *a*, Normal embryo; *b*, embryo with increased notochord after thiocyanate treatment. (After Ranzi and Tamini, from Lehmann, 1945.)

Lithium can also affect the gradient in the chordomesodermal system. Treatment of the embryo during gastrulation with a weak solution of lithium chloride suppresses the development of the notochord, which has been postulated as representing the center of highest activity of the mesodermal gradient. If the development of the notochord is suppressed, the presumptive notochordal cells differentiate according to the next highest level of the gradient and develop into somite tissue (Fig. 171). The right and left rows of somites are then continuous with each other across the midline, underneath the neural tube (Lehmann, 1937, 1945).

It has been reported that the opposite effect may be produced by treating frog embryos in the blastula stage with a solution of sodium thiocyanate (NaSCN). The result is the development of embryos in which the notochord is much larger and thicker than in control animals (Fig. 172). This means that a greater than normal area has shown a differentiation characteristic of the highest level of the gradient. This action may be called raising the level of the gradient (Ranzi and Tamini, 1939).

Raising the dorsoventral gradient in the ectoderm should lead to an increase in the size of the neural system. In fact, embryos with excessively broad and massive brains have been observed after sodium thiocyanate treatment in fishes (Huxley and de Beer, 1934) and frogs (Ranzi, 1957; Fig. 173). The opposite effect of Li and NaSCN on amphibian regional differentiation has been confirmed in some experiments performed on explants *in vitro*. It was found that two- to four-hour exposure of early gastrula ectoderm of *Ambystoma* to either of these ions causes the ectoderm to differentiate (in the absence of the normal inductor—the chordomesoderm). This is the same sort of abnormal induction that was mentioned on page 212: induction with methylene blue or with urea. The point of interest in the present context is that the action of NaSCN leads to the development of archencephalic structures, and the influence of Li causes the differentiation of notochord, mesoderm, and endoderm (Johnen, 1970). The analogy between the gradient system of amphibians and that of sea urchins is thus greatly enhanced.

Cyclopia is occasionally observed in natural conditions both in man and in domestic animals (Adelmann, 1936). It is therefore significant that the same effect can be produced experimentally. In this way we get an inkling of what goes on "behind the scenes" when a cyclopic monster is born. There must have been some injury to the

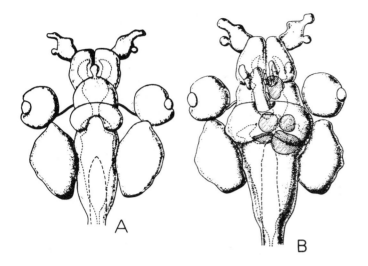

Figure 173. Reconstruction of the brain and sensory organs in a normal frog tadpole (*A*) and in a tadpole after treatment with thiocyanate (*B*). (From Ranzi, 1957.)

gradient system concerned with the formation of the head in the embryo. The injurious effect may be produced by both hereditary and environmental factors. Cyclopia or similar defects may be the result of some toxicosis of the mother occurring at an early stage of pregnancy. This is known to be the case if a woman contracts the illness known as German measles (rubella) during the early weeks of pregnancy. The illness is not a serious one for the mother, but the toxin poisons the embryo, producing defects of the anterior part of the head (Glatthaar and Töndury, 1950). On the other hand, a gene has been found in guinea pigs (Wright and Wagner, 1934) which produces, in varying degrees, defects of a cyclopic nature, from the underdevelopment of the nose and mouth to almost complete absence of the head.

Chapter 9

CREATION OF FORM DURING GASTRULATION AND IN SUBSEQUENT DEVELOPMENT

9–1 MORPHOGENETIC MOVEMENTS

During cleavage and up to the blastula stage the embryo retains roughly the shape of the egg from which it started developing. It is spherical in most cases, sometimes oval if the egg is oval, or elongated, as in the case of many insects and Cephalopods among the molluscs, which have elongated eggs. The internal structure of the blastula is also simple, consisting, as it does, of one layer of cells around a cavity. (Special cases are present in some animals with large yolky eggs, such as birds—see p. 116.)

With the onset of gastrulation the embryo's structure changes. Through the development of the archenteron and then the separation of the mesoderm, the embryo acquires a more complicated internal structure, departing from the geometric simplicity of the blastula of animals with oligolecithal eggs. The structure becomes further complicated with the formation of the primary organ rudiments, which closely follows gastrulation. Gradually the external shape of the embryo starts to change. As primary organ rudiments give rise to secondary and tertiary organ rudiments, the internal structure and the external shape of the embryo approach those of the adult animal. (The embryo or larva may leave the egg long before the final goal is achieved.)

Thus, the new organism not only acquires a diversity of parts that did not exist in the egg but also acquires the **form** typical of an animal of its species. By **form** we mean not only the external shape of the animal but also the structural organization, that is, the body's being composed of a number of parts, placed in a typical disposition inside and on the surface of the animal's body.

The new elements of the embryo's organization appearing in gastrulation (the archenteron, etc.) are produced by movements and changes in the shape of cells and groups of cells of the embryo. These movements and changes in shape are also involved in subsequent processes of form creation. The movements which we are referring to are very different from the movements of parts of an adult animal. Whereas the movement of parts of an adult are usually of a reversible nature, the gastrulation movements are irreversible; each part remains in the position into which it was brought by the preceding movement. As a result of the movements, the structure of the embryo is changed, or in other words, new structural elements are created, such as the archenteron, the neural tube, and the notochord, in place of the simple layer of cells found in the blastula stage. The movements have created new shapes, new forms. They have therefore been designated as the **morphogenetic movements** (Vogt, 1925; Trinkaus, 1969).

The morphogenetic movements appear to be movements of large parts of the whole embryo, which stretch, fold, contract, or expand. The question arises, How are these movements achieved? They cannot be ascribed to contractility in the narrow sense of muscle contractility. Neither can they be interpreted as an ameboid movement of the embryo as a whole, because each moving part consists of numerous cells, and the movement of the whole, we should expect, would be an integrated result of the movements of the individual cells.

That gastrulation cannot be a function of the embryo as a whole has been proved by investigating isolated parts of the young gastrula. We have seen that the presumptive ectoderm contributes to gastrulation by expanding its surface. The expansion is an active process depending on the presumptive ectoderm itself. If large pieces of the animal region of an amphibian blastula or early gastrula are cut out and cultivated in a suitable medium, the presumptive ectoderm rounds itself up into a vesicle, and later the epithelium of this vesicle increases its surface greatly and is thrown into a series of irregular folds as it does so. Also, if presumptive ectoderm is combined with cells of presumptive endoderm and mesoderm in such a proportion that the ectoderm is far in excess of the endodermal and mesodermal parts which it has to cover, the ectoderm tends to form folds (Spemann, 1921). These experiments show that the expansion of the ectoderm is active and proceeds independently of the other movements involved in gastrulation.

Similarly, the movements of invagination can be performed by parts of the blastoderm independently of their surroundings. The most suitable method of testing this is to transplant small pieces of the dorsal lip of the blastopore into some other part of the embryo, the animal region or the ventral part of the marginal zone. The transplanted piece will invaginate and form an archenteric cavity which may be completely independent of the archenteric cavity of the host embryo. The mechanism by which the invagination is achieved is the change in shape and the movements of the cells of the marginal zone.

The change in the shape of the cells is not due to forces exercised by the embryo as a whole but is performed by the cells themselves. This property is best seen if the cells are isolated by tearing the embryo to pieces; the cells preserve their shape, and moreover, the infolding of the surface layer may even be facilitated by releasing a piece of it from its connections with the surrounding parts (Fig. 174—Holtfreter, 1943a).

9–2 SELECTIVE AFFINITIES OF CELLS AS A DETERMINING FACTOR IN CELLULAR REARRANGEMENTS

We may ask what is the nature of the forces which control the rearrangement of cells resulting from morphogenetic movements in gastrulation and also in other morphogenetic processes; we may ask whether the new positions of the cells are the result of their own special properties, or whether the organization of the system as a whole (i.e., the whole embryo) is decisive in bringing each kind of cell into its appropriate position. These questions have been investigated by separating the cells of the embryo and allowing them to interact *in vitro* (Townes and Holtfreter, 1955). The cells of amphibian embryos placed in water, which had been made alkaline (*p*H 9.6 to 9.8) by adding potassium hydroxide, lose cohesion, and the embryo or parts of it treated with alkali disaggregate into a mass of disconnected single cells. Masses of cells prepared in this fashion from different embryonic tissues (gastrula ectoderm, mesoderm or endoderm, neurula epidermis, neural plate, and others) may then be mixed together in

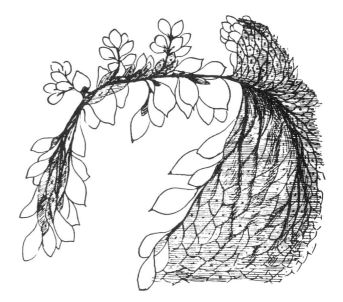

Figure 174. A fraction of an amphibian blastoporal lip with groups of cells held together by the "surface coat." (From Holtfreter, 1943a).

various combinations and returned to a fresh solution of pH 8.0, in which the cells return to their normal state and again join together (Fig. 175).

After clumps of cells are formed, the cells begin to sort themselves out according to their properties. The ectodermal and epidermal cells combined with mesodermal or neural plate cells always ended up by assembling on the surface of the aggregate. The mesodermal and neural plate cells combined either with epidermis or with endoderm disappeared from the surface and were to be found forming solid masses in the interior (Fig. 176). The masses of neural plate cells became hollowed out later and formed structures resembling brain vesicles. Masses of mesodermal cells arranged themselves around "coelomic cavities" (p. 000). If epidermis, mesoderm, and endoderm were present in an aggregate, the epidermis cells concentrated on the exterior, and mesodermal cells took up a position between the epidermis and the endoderm; the latter, rather unexpectedly, also occupied part of the external surface of the aggregate. This is, however, not unlike the position of the endoderm in the case of exogastrulation. In this way, arrangements of cells similar to those produced at the end of normal gastrulation and neurulation are brought about by the activities of individual cells beginning from a completely abnormal starting point.

In later work, more refined methods have been used for disaggregating cells of embryonic tissues. The result is achieved either by a short treatment with trypsin or by placing the tissues in a medium devoid of Ca and Mg ions and with EDTA (ethylenediaminetetra-acetic acid, a chelating agent) added. Use of the latter method is based on the discovery that Ca and Mg ions play an important part in holding cells together; EDTA binds these ions, removing them from intercellular connections.

Figure 175. Separated cells of presumptive ectoderm and endoderm joining together into a spherical mass. (From Holtfreter, 1939b.)

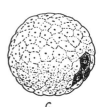

a *b* *c*

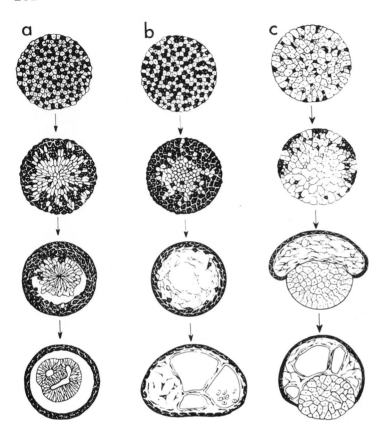

Figure 176. Rearrangement of disaggregated and reaggregated embryonic cells. *a*, Combined epidermal and neural plate cells. *b*, Combined epidermal and mesodermal cells. *c*, Combined epidermal, mesodermal, and endodermal cells. (From Townes and Holtfreter, 1955.)

The sorting out of mixtures of different kinds of cells occurs not only in cells taken from embryos in stages of gastrulation and neurulation, but also with cells taken from older embryos. Cells of the heart rudiment, procartilage, the liver, the pigmented retina, and others sort themselves out, after having been combined at random, and take up positions either on the surface or in the interior of the clump. Three different tissues may be combined, and they will rearrange themselves into three concentric layers. Some of these combinations copy the normal arrangement, as when muscle cells surround cartilages. In other cases, when the cells are of such a nature that they do not normally occur together, entirely new combinations are produced. Thus, when limb-bud procartilage, heart muscle, and pigmented retina are combined, procartilage is located in the interior and is surrounded by retina, with heart cells forming the outer layer (Steinberg, 1964).

The behavior of the cells under these conditions—and by inference also in normal development—is presumably due to their **selective affinities** (Holtfreter, 1939a; Weiss, 1947; Steinberg, 1963). When cells touch one another as a result of their random movement, they tend to remain in contact (Abercrombie and Heaysman, 1953). It is supposed that the forces holding the cells together vary, depending on the kind of cells involved. According to its special properties, a cell may either adhere to another cell and stay in contact with it or move away from a position in which the surroundings are not of such a nature that they keep the cell bound. Generally speaking, the result of these selective affinities would lead to like cells being brought together. This does not necessarily mean that the bonding between such cells is of a strictly specific nature; to allow similar cells to sort themselves out from among cells with differing properties, it would suffice that the adhesive properties of the former, at any given time, be equal in grade or intensity (Curtis, 1962; Steinberg, 1963).

A cell capable of adhering to another cell or cells may be said to possess a certain amount of free energy, the energy that would be used to establish and maintain a contact. General principles require that a system—in our case an aggregate of cells—reach an equilibrium when its free energy is at a minimum or when the adhesiveness of the cells is used to a maximal extent or is "maximalized" (Steinberg, 1963, 1964, 1970). As cells on the surface of the aggregate do not use their adhesiveness to a maximal capacity (their outer surfaces not being in contact with other cells), the requirements of a system with a minimum of free energy dictate that the aggregate have the least possible surface area. A body with the least possible surface for a given volume is, of course, a sphere, and we see that the aggregates generally assume a spherical shape (Fig. 175).

If cells with different degrees of adhesiveness are available, the least free energy requirement dictates that only those cells should stay on the surface whose adhesiveness is least; the cells with greater adhesiveness will withdraw into the interior. In the interior these cells will maximalize their adhesiveness if they appose their surfaces to cells of the same kind, rather than to cells whose adhesiveness is lower. As a result, whenever such highly adhesive cells meet, they will become drawn to each other and eventually will form large clumps within the aggregate. There seems to be no attraction of like cells from a distance (no evidence for a chemotaxis among cells), and the ability of individual cells to move within the aggregate, once it is formed, is limited to 10 to 30 μ. This was established by direct observation of pigmented retina cells which were mixed with transparent cells of the heart (Trinkaus, 1969). However, some contacts could be established as a result of even this limited mobility. The sphere of activity of cells could perhaps be extended still further by filopodia-like extensions of the cytoplasm. The facilities for contacts between like cells of an aggregate must be fairly great, however, since the eventual separation and stratification of unlike cells usually are exceedingly clear-cut.

A complication of the processes governing the arrangement of cells in aggregates is brought about if only part of the surface of the cells involved is adhesive and another part is nonadhesive. In the amphibian neurula the free outer surface of epidermal cells becomes nonadhesive, and this may be the case in some other epithelial cells as well. When such cells participate in the aggregate they end up forming a layer on the outer surface. If the number of such partially nonadhesive cells were large in relation to the volume of the whole, the surface area would no longer need to be the least possible one. In fact, the surface may be thrown into folds to accommodate all the partially nonadhesive cells.

There is one very obvious discrepancy between the results of reaggregation of isolated embryonic cells and the processes of gastrulation in intact normal embryos. In reaggregates, the endoderm in combination with mesoderm becomes arranged outside, and the mesoderm takes up a position inside the endoderm, while the reverse happens in normal development. Even if cells of all three germ layers are present in the reaggregate, the endoderm does not take up a position inside but forms part of the surface on one side of the newly formed complex (Fig. 176c).

The reason for this discrepancy is that in the stage preceding gastrulation (the blastula stage) the cells are in the form of an epithelium (the blastoderm) which is more or less distinctly polarized. As in typical epithelia, it has a distal surface, in contact with the external environment, which is different from the proximal surface, facing the inner milieu of the embryo. It has been noticed (Gudernatsch, 1913) that when an epithelial layer is bent, or when cells escape from an epithelium as ameboid or mesenchyme cells, the direction of movement is nearly always in the proximal direction, toward the internal milieu of the embryo. It is the proximal end of the cell that is more likely to start forming pseudopods and embarking on ameboid locomotion, and this is why in an intact embryo the movement of both mesoderm and endoderm is inward and not outward. Under experimental conditions, especially if ectoderm is prevented in some way from covering the endoderm (as when extensive parts of the ectoderm and mesoderm are removed by an operation), the endoderm flows to the exterior, and once on the surface it remains there.

9–3 MORPHOGENETIC MOVEMENTS IN EPITHELIA

During the early stages of cleavage the blastomeres are rather loosely connected to one another, but before gastrulation and the morphogenetic movements set in, the cleavage cells in most animals join to form more closely knit structures in the form of layers of cells—the **epithelia.** Thus, the first morphogenetic movements involve epithelia, and also at later stages of development many organ rudiments are formed by epithelial layers. It will therefore be convenient at this stage to deal in greater detail with the processes of morphogenesis in the epithelial layers of the embryo.

Thinning Out of Epithelial Layers.   This process is a transformation which has an important role in gastrulation in the form of **epiboly,** the "overgrowth" of the surface of the embryo by presumptive ectoderm. In fact, growth (increase in living mass) is not involved, and the process is one of stretching of the epithelial layer, so that increase in surface proceeds at the expense of a thinning out of the epithelium. Further expansion of the ectoderm occurs at the stage of neurulation, when the neural plate ectoderm disappears from the surface and the entire exterior of the embryo has to be covered by epidermis. The presumptive ectoderm of an amphibian blastula consists of several layers of cells; the number of cell layers becomes reduced during gastrulation and still further reduced during neurulation. In the process, some of the cells originally lying in the interior become intercalated between the superficially lying cells. The whole change is thus brought about by a rearrangement of the cells of the epithelium. After neurulation the epidermis in amphibians, as well as in many other vertebrates, consists of two layers of cells. In invertebrates the epidermis is usually a single layer of columnar cells. Once the epithelium is reduced to two or even one layer of cells, a further thinning out may be achieved by the flattening of the cells constituting the layer.

Local Thickenings of the Epithelial Layer.   Local thickenings may be multiple, leading to the formation of a number of similar organs or structures within an epithelial layer; there may be a pair or several pairs of similar thickenings, as in the development of paired organs; or there may be only one organ rudiment starting its development as a thickening of the epithelium. The formation of a neural plate takes place in this way. The thickening is brought about by the epithelial layer flowing, as it were, toward the region in which the organ rudiment is to appear, in this case the mid-dorsal region of the embryo. As the thickening becomes discernible, the epithelial cells may become elongated in a direction perpendicular to the surface of the epithelial layer.

This elongation of cells is very distinct in the neural plate. The cells become columnar, and in the case of the urodele amphibians, at least, single cells stretch from the external to the internal surface of the neural plate, in spite of its increased thickness. This means that the formation of the neural plate involves a rearrangement of the cells of the epithelial layer, because the ectoderm of the late gastrula in amphibians consists of several layers of cells, whereas in the neural plate a single layer of columnar cells is found. It has been suggested that the elongation of individual cells of the epithelium is the mechanism by which the thickening of the layer is achieved (Holtfreter, 1947a).

In the case of multiple organs, more than one thickening may develop simultaneously in the same sheet of epithelium; a clear-cut example of this may be seen in the development of the hair follicles of mammals. Before the appearance of the rudiments of the hair follicles, the epidermis is a two-layered sheet of epithelium, with cells more or less flattened and of uniform thickness. Then, rather suddenly, numerous concentrations become visible, scattered at more or less regular intervals throughout the epithelial layer (Fig. 177). In each concentration the cells lie closer together, and the individual cells, especially in the lower layer of epithelium, are less flat than before and approach the cuboidal or even the columnar shape. The concentration of cells is brought about by the cells coming closer together and not by a local increase in the proliferation of cells. This assertion has been proved by two different methods. Counting mitoses in the early hair follicles and the epidermis in between did not show greater numbers of mitoses in the follicles (Balinsky, 1950). Providing the epidermis with radioactive precursor (tritiated thymidine) showed that nuclei taking up thymidine (a reliable sign of the cells preparing for mitosis) were evenly distributed at the time of hair follicle formation (Wessells and Roessner, 1965). After the epidermal thickenings have appeared, however, growth and cell proliferation set in. As a result, the number of layers of cells in the epidermal thickenings increases. What was originally a thickened area becomes a rounded peg intruding into the dermis.

Separation of Epithelial Layers. Epithelial layers or masses of cells may be subdivided into parts by the appearance of crevices between groups of cells, the cells on the opposite sides of a crevice losing connection with one another. The crevices may appear at any spot, but perhaps the most common occurrence is for the crevices to be either parallel to the surface of the epithelium or perpendicular to it (Fig. 181*E* and *F*). In the first case, the epithelial layer is split into two layers lying one on top of the other. The original crevice may be increased by the secretion of fluid into it and may become a more or less spacious cavity (Fig. 181*F*). This is the case with the formation of the parietal and visceral layers of the lateral plate mesoderm and the coelomic cavity between them. An instance of crevices appearing perpendicular to the surface of the epithelium is found in the development of the mesodermal somites. The upper edge of the mesodermal mantle becomes subdivided into segments by crevices running between cells perpendicular to the surface of the epithelium and at the same time perpendicular to the main axis of the body. (See Figure 130.) The masses of cells thus formed (the somites) are of approximately the same size. This suggests that each somite may possibly be formed around a center of attraction whose force is limited, so that only a certain

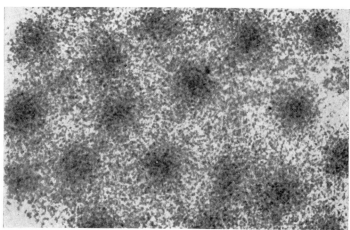

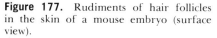

Figure 177. Rudiments of hair follicles in the skin of a mouse embryo (surface view).

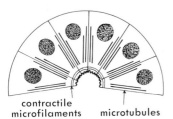

contractile
microfilaments   microtubules

Figure 178.   Folding in an epithelium as a result of the change of shape of cells (diagram).

number of cells can be kept together, the cells losing connection with one another where the attraction of the center is too weak. A rhythmic pattern of differentiation could thus be produced.

Splitting of epithelial masses often follows the formation of local thickenings. Thus, the neural tube eventually loses its connection with the epidermis.

Folding of the Epithelial Layer.   The folding of epithelial layers is perhaps the most important form of morphogenetic movement. The invagination of the archenteron as it occurs in animals such as the sea urchin or *Amphioxus* and also in the early stages of gastrulation in the amphibians is obviously a case of infolding of an epithelium. The infolding is generally accepted to be the result of a change in the shape of the cells.

If cells of a columnar epithelium contract at their external ends and expand at their internal ends, so that instead of being prismatic they become pyramidal, the whole epithelial layer must inevitably change its shape and become bent in toward the interior (Fig. 178). This is what actually takes place in the blastopore region; the external parts of the cells here become attenuated, and the opposite ends of the cells expand. The cells acquire a bottle shape, with the neck of the bottle keeping the cells in touch with the surface, while the bodies of the cells are drawn away from the surface. The area occupied by a given group of cells on the surface of the embryo contracts to a small fraction of what it had been originally. (See Figure 121.) The contraction at the external surface during the formation of the blastopore in the frog embryo is so violent that the plasmalemma of the cells is thrown into deep folds, and the deeper sections of the folds are nipped off as vesicles which sink into the cytoplasm (Balinsky, 1961b – Fig. 179).

To understand the final result one has to consider, however, one further factor: namely, that the cells are held together at their external (superficial) ends. As has been mentioned in Section 6–4, the cells at the surface of the blastoderm are more firmly held together than the cells which do not reach the surface. It has been claimed (Holtfreter, 1943a) that the adhesion of cells in the surface layer is due to the presence of an extracellular continuous cuticular membrane, the "surface coat," to which all the cells reaching the surface are firmly connected. With the aid of the electron microscope it can be shown (Balinsky, 1961b) that no such membrane exists, but that the cells reaching the surface adhere to one another very closely just at their distal ends and are probably joined by some cementing substance, which is not easily broken (Fig. 180). As a result, when numerous cells in the blastopore region start moving inward, they cause the surface of the embryo to be pulled inward with the formation of a pocket – the rudiment of the archenteron. Similarly, when the movement of the cells is

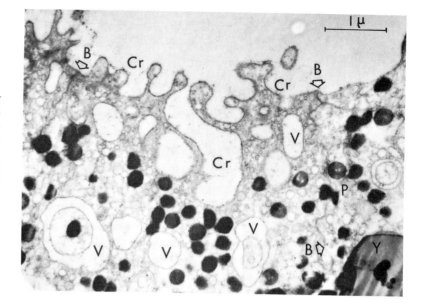

Figure 179. External surface of cells invaginating in the amphibian blastopore (electron-micrograph). Cr, Crypts formed at the surface of the cells; V, vesicles pinched off at the bottom of the crypts; P, pigment granules; B, cell boundaries; Y, yolk platelet.

a general expansion or concentration, the cells, being joined together at their external ends, move in concord, as if borne by a common force (Holtfreter, 1943a).

The folding of epithelial layers can occur either in the form of linear folds, resulting in the formation of grooves, or as approximately round depressions, forming pock-

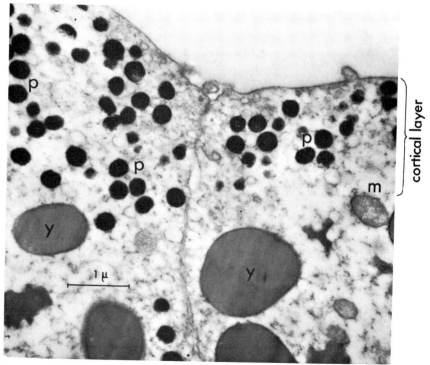

Figure 180. Surface of the blastoderm in an early gastrula of the frog *Phrynobatrachus natalensis*. The electron micrograph shows the junction of two cells and the cell boundary reaching right to the external surface. Y, Yolk; m, mitochondria; p, pigment granules.

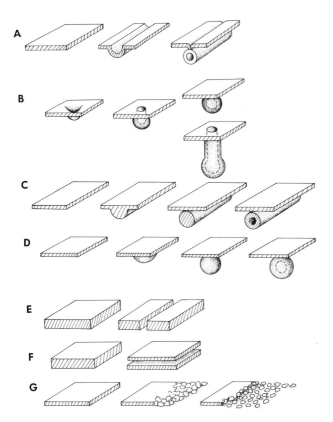

Figure 181. Different types of transformations occurring in embryonic epithelium in the course of development. *A,* Groove giving rise to a tube. *B,* Pit giving rise either to a vesicle or to a tubule open to the exterior. *C,* Development of a tube from a solid elongated thickening of the epithelium. *D,* Development of a vesicle from a solid circumscribed thickening of the epithelium. *E,* Splitting of epithelium perpendicular to surface into sections. *F,* Splitting of epithelium tangentially into two layers with a space in between. *G,* Breaking up of epithelium to form separated mesenchyme-like cells.

ets (Fig. 181*A* and *B*). The direction of folding is, as a rule, toward the originally proximal surface of the epithelial layer. In the ectoderm, and later in the epidermis, the folding is therefore directed inward. In the invaginated parts of the embryo, as in the whole of the endoderm, the folding is directed outward, that is, away from the lumen of the alimentary canal.

If a linear fold becomes closed in and separated from the original epithelial layer, the resulting structure is a hollow tube (Fig. 181*A*). If a pocket-like depression becomes closed in and separated from the original epithelial layer, the result is a hollow vesicle (Fig. 181*B*). An example of the first formation is the neural tube and of the second formation is the rudiment of the inner ear—the ear vesicle. (See Figure 310.) Most of the glands are developed as pocket-like invaginations of the epithelium, an indication of how common this type of morphogenetic process is.

Thickenings Followed by Their Excavation to Form Tubes or Vesicles.   By no means are tubes or vesicles always produced by infolding of the epithelial layer. Quite often the initial stage is a solid thickening of the epithelium (as discussed previously). Such a solid thickening may acquire an internal cavity secondarily by the separation of cells in the middle (Fig. 181*C* and *D*). This cavity may or may not be connected with the external surface of the epithelium. It is remarkable that one and the same organ in different animals may develop in different ways: as an infolding in some and as a solid thickening becoming excavated later in others. This is the case with the development of the neural tube in various vertebrates. It is formed by infolding in the elasmobranchs, in the urodele amphibians, and in the amniotes; it develops as a longitudinal thickening that later separates itself from the epidermis and acquires a central cavity in the Myxinoidea and in the bony fishes. In the Anura a sort of intermediate state is found, the groove being very shallow and the neural tube being, in part, solid at the beginning. Other organs such as the eye, the ear, and the lens may develop either as hollow pocket-like invaginations or as solid masses of cells. The underlying mechanism of both formation must therefore be similar in nature.

If the thickening of an epithelial layer is directed outward, it takes the form of an "outgrowth." The formation of an outgrowth is probably never caused by a local increase in growth rate, that is, increased multiplication or increase in size of the cells. Much as in all other cases of organ formation, the outgrowths are caused by the concentration of cells and their rearrangement. Such outgrowths are formed on the surface of the chorion in mammals (the chorionic villi—Section 10–5) or on the gills in amphibians and fishes (the gill filaments). The outgrowths may be simple and finger-like or may develop secondary outgrowths when the whole becomes dendritic or plumose. The tips of the epithelial outgrowths may be solid, but once they have appeared, they become hollowed out, so that connective tissue and blood vessels may penetrate into them. Blood capillaries develop in a similar way: first as solid strands of endothelial cells, pushing forward from existing blood vessels and being hollowed out later. The cavity in this case is the lumen of the capillary. The speed with which the solid outgrowths are hollowed out differs from case to case, and no strict boundary can be drawn between these and outwardly directed folds or outpushings which are hollow from the start.

Breaking Up of Epithelial Layers to Produce Mesenchyme. The breaking up of epithelial layers is a very important morphogenetic process, even if it does not lead directly to the formation of organ rudiments, because it furnishes the material from which other organ rudiments may develop. Transformation of epithelial cells into mesenchyme-type cells may start very early in development and may be involved in the formation of germ layers during gastrulation. An epithelial layer or a portion of it may sometimes break up into mesenchyme altogether, so that the epithelium as such disappears or a gap is left in the previously continuous layer (Fig. 181G). An example of this process is found in the development of the neural crest. A part of the ectodermal epithelium forming the edge of the neural folds splits up into disconnected cells, leaving a gap between the neural tube and the epidermis. (Sometimes, however, the neural crest cells are given off from the dorsal wall of the neural tube after its separation from the epidermis.) The other possibility is for individual cells to slip out from an epithelial layer, so that the latter remains intact though depleted of part of its cells.

Invagination of the endoderm and mesoderm into the interior of the embryo, as we have seen, cannot always be interpreted as the infolding of an epithelial layer. It often takes the form of immigration of individual cells into the blastocoele, as in the case of the primitive streak of amniotes. A study of the cells in the primitive streak of birds with the electron microscope shows that the behavior of migrating cells in this case is very similar to the behavior of cells in the amphibian blastopore (Balinsky and Walther, 1961). The cells of the epiblast of the chick are roughly cuboidal (or rather, low columnar). The cells participating in the downward migration start by changing their shape. They elongate in a vertical direction, their inner ends expand, and their outer ends become narrowed like necks of bottles reaching to the surface of the epiblast (Fig. 182).

Contrary to what happens in the amphibian blastopore, however, not all the cells in the primitive streak undergo the change in shape at the same time; while some cells become bottle-shaped, others retain the original broad columnar form. Eventually the external ends of the bottle-shaped cells lose contact with the surface, and the cells slip downward, emerging on the under (inner) surface of the epiblast. Presumably the migrating cells are not held as firmly together at their external ends as the cells of the amphibian blastopore, so that instead of pulling the surface inward, they detach themselves from the surface without causing the formation of a pocket (the archenteron). It

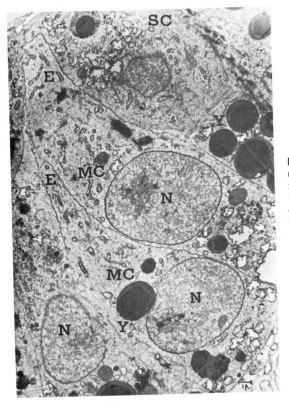

Figure 182. Cells in the primitive streak of a chick embryo (electronmicrograph). Some of the cells are withdrawing from the surface of the streak, their distal ends attenuated and their proximal ends expanding. MC, Migrating cells with their distal ends (E) attenuated; SC, stationary cell, cuboidal in shape; N, nuclei; Y, yolk granules. (From Balinsky and Walther, 1961.)

follows that the presence or absence of an archenteric cavity may be accounted for by a different degree of mutual adhesion of cells involved in the inward movement at the time of gastrulation.

The mechanism of the migration of mesenchyme cells has also been studied in the early gastrulae of sea urchin embryos (Gustafson and Kinnander, 1956; Gustafson and Wolpert, 1961). Cinemicrography was used in these studies. Gastrulae of sea urchins were photographed at 1- to 18-second intervals. When the film, taken in this way, is projected at normal speed, the movements of the cells of the gastrula become accelerated and thus are made more readily visible. Before gastrulation, the blastoderm of a sea urchin embryo is a regular columnar epithelium, somewhat thickened (the cells being larger and taller) at the vegetal pole. The cells of the primary mesenchyme start "bubbling" or "pulsating" on their inner surface (the surface facing the blastocoele); that is, the cytoplasm produces rounded protrusions or broad pseudopods which may collapse and then be formed again. Of course, the bulging of the cytoplasm on the inner surface cannot proceed without the cytoplasm being withdrawn from those parts of the cells which are nearer to the exterior. It appears that the adhesion of these cells to other cells is lost or reduced. Eventually, the mesenchyme cells are released into the blastocoele, where they become spherical in shape—an indication of complete or almost complete loss of adhesion between them (Fig. 183).

9–4 MECHANISM OF CHANGES IN THE SHAPE OF CELLS DURING MORPHOGENESIS

Electron microscopic studies of cells undergoing changes in shape have shown that a peculiar group of cytoplasmic structures are involved in these processes. They are the

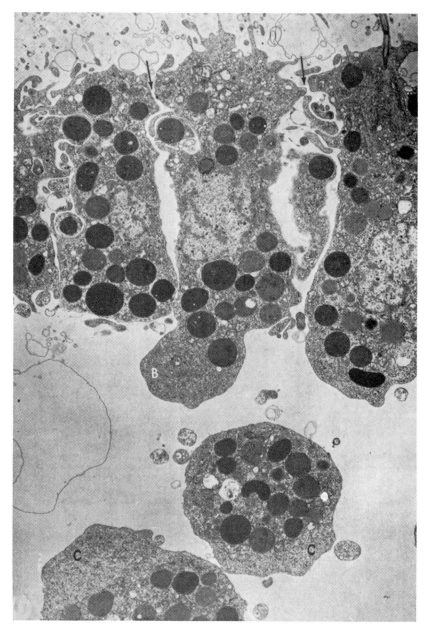

Figure 183.   Disengagement of primary mesenchyme cells from the blastoderm in a sea urchin embryo (electronmicrograph). Arrows show lack of cohesion of cells in the blastoderm. B, "Bleb" or pseudopod of a cell about to move into the blastocoele. C, Remnants of similar pseudopods in cells already lying free in the blastocoele. (From Gibbins, Tilney, and Porter, 1969.)

microtubules and **microfilaments.** (See Wessells, 1971.) The microtubules are seen as long, straight, unbranched rods, 200 to 250 Å in diameter, which appear hollow in cross sections. In cells not undergoing changes in shape the microtubules may be observed to traverse the cytoplasm at random. In cells which are elongating in a certain direction, however, the microtubules are arranged in parallel bundles running in the direction in which the cell is elongating (Fig. 184). Bundles of such microtubules are prominent in

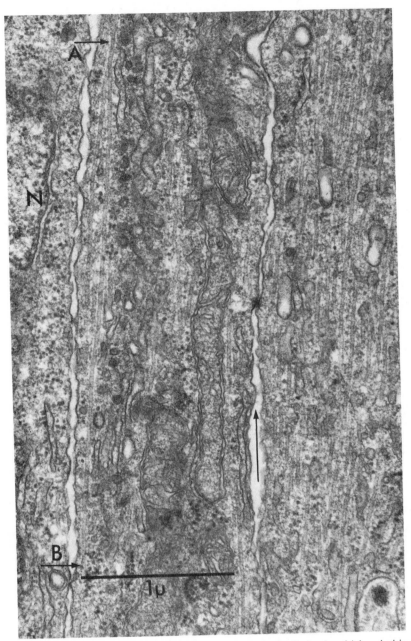

Figure 184. Electronmicrograph of "neck" region of cells in the chick primitive streak showing microtubules. A and B with arrows indicate the extent of one microtubule in the section. N, Nucleus. (From Granholm, N. H.: Cytoplasmic microtubules and the mechanism of avian gastrulation. Developmental Biology, Vol. 23, 1970. Courtesy of Dr. John R. Baker of Iowa State University.)

the narrowed "necks" of the bottle-shaped cells of the amphibian blastopore (Perry and Waddington, 1966), in the cells immigrating from the primitive streak in birds (Granholm and Baker, 1970), and in the cells of the neural plate (Waddington and Perry, 1966). Bunches of microtubules may be seen along the length of outgrowing processes of nerve cells, and they are also involved in mitosis, as the fibers of the achromatic spindle.

It was originally with the spindle fibers of mitosis that the microtubules were found to be very sensitive to certain poisons, namely, colchicine (an alkaloid extracted from plants of the lily family) and also vinblastine sulfate. In cells treated with these poisons the microtubules disintegrate, and as a result the movements of the mitotic chromosomes stop, and mitosis is arrested in metaphase. When applied to growing nerve cell processes, colchicine causes the collapse of the axon, which shortens, loses its rectilinear shape, and may eventually be completely withdrawn into the cell body (Wessells et al., 1971). When the chick primitive streak is treated with colchicine, the elongating cells shorten, and the primitive groove straightens out (Granholm and Baker, 1970). The same happens with the elongating cells of the neural plate (Karfunkel, 1970). These experiments show that the microtubules are involved in changes in the shape of the cells and particularly in the elongation of cells and the maintenance of the elongated form. The collapse of the outgrowing nerve cell processes suggests that the microtubules have a supporting or skeletal function. This explanation should not, however, be understood in too narrow a sense. Although the microtubules appear to be able to increase in length, either by growth at the ends or by intussusception along their whole length, and thus to provide continuous support to elongating parts of the cells, it would be premature to conclude that the microtubules actually "push" the protruding or elongating parts. Rather it may be suggested that the microtubules in some way lead the flow of cytoplasm in the direction which is indicated by the oriented arrangement of the microtubule bundles. This appears to be the role of microtubules in mitosis, in which chromosomes attached to the spindle fibers are drawn to the opposite poles of the achromatic figure.

The poisons which disrupt microtubules and prevent elongation of cells do not affect, to any notable extent, other processes in the cell; in particular, they do not prevent synthetic processes from occurring. Their influence on microtubules thus appears to be a direct one.

The other cell organoids previously mentioned as taking part in changes in cell form, the microfilaments, are much finer threadlike bodies, which are 30 to 60 Å in diameter. They appear to be contractile elements, and accordingly bundles of microfilaments are found in positions where their contraction could cause a narrowing or constriction of parts of the cells. In cells undergoing cleavage (the egg during its first division and subsequently during the division of blastomeres), the microfilaments appear in the form of a "contractile ring" just underneath the cleavage furrow. As the "ring" narrows, the cleavage furrow cuts deeper and eventually separates the two daughter blastomeres. Colchicine has no effect on microfilaments, but these organoids are highly sensitive to another poison, cytochalasin B, a cyclic organic compound which is extracted from a mold, *Helminthosporium dematoideum*. Cytochalasin B reversibly destroys the microfilaments, transforming bundles of these fibers into granular amorphous cytoplasm. When cleaving eggs are treated with this poison, the contractile ring consisting of the microfilaments is destroyed, cleavage stops immediately, and the cleavage furrows are reversed, that is, they become smoothed out again (Schroeder, 1972).

Systems of microfilaments have been found to participate in changes in the shape of cells which are instrumental in producing folding of epithelial layers. In neural plate cells, in addition to the microtubules, which are responsible for elongation of the cells, there are bundles of microfilaments which are arranged parallel to the distal surfaces of the cells (Balinsky, 1961b; Baker and Schroeder, 1967). If sections of the neural plate are made parallel to its surface, in each cell the microfilaments can be seen to be arranged in a ring around the cell's distal end (Fig. 185). The bundle of filaments works as a "purse string," literally pulling together the distal part of the cell and transforming it into a narrowed "neck." A similar mechanism must be involved in the narrowing of the outer ends of cells and in their transformation into bottle-shaped cells in

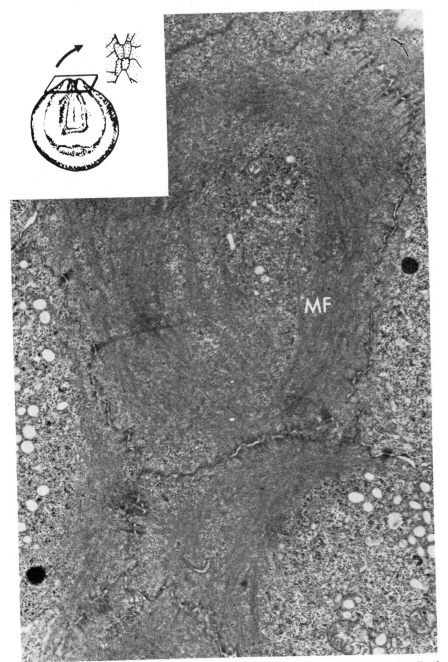

Figure 185. Electronmicrograph showing ring of microfilaments (MF) in the distal part of a *Xenopus* neural plate cell (tangential section). Plane of section indicated in inset. (Courtesy of Dr. P. C. Baker and Dr. T. E. Schroeder, 1967.)

the blastopore during gastrulation. This conclusion is supported by the fact that cytochalazin B stops invagination of the archenteron in sea urchins and in the worm *Urechis* (Wessells et al., 1971).

Most convincing experiments regarding the role of microfilaments in morphogenesis have been performed on oviducal glands in chickens (Wrenn, 1971; Wessells, 1971).

In immature birds the oviducts are simple tubes consisting of cuboidal epithelium. At sexual maturity, in response to estrogen, numerous oviducal glands are formed as simple outpocketings of the oviducal epithelium. The epithelium of the oviduct, however, is already responsive to estrogen in 7- to 12-day-old chicks. If they are injected with estrogen, their oviducts start developing glands within 24 hours. Thus, the change in shape of the cells and the resulting morphogenetic process can be studied under accurately controlled conditions, especially with respect to its timing. Even before the cells in the gland rudiment start changing in shape, a layer of microfilaments appears close to the luminal surface of the cells (Fig. 186A). Contraction of the microfilaments pulls the luminal surface of the cells together. The cells become cone-shaped, with the narrow ends facing the cavity of the oviduct, and the gland becomes discernible as a pocket-like depression in the oviducal wall. (Compare the explanation of epithelial folding, p. 236 and Figure 178.) To prove that the microfilaments are, in fact, causing the invagination of the gland, the oviduct is excised and kept *in vitro* in a medium containing cytochalazin B. As a result, the microfilaments disintegrate and become a granular mass (Fig. 186B), the cells change back to the cuboidal shape, and the depression which was to be the oviducal gland becomes smoothed out. The action of the cytochalazin B is reversi-

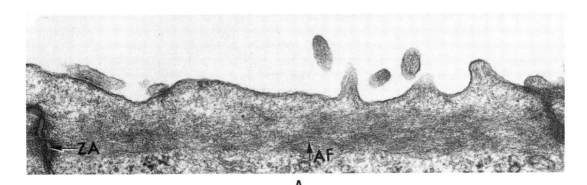

A

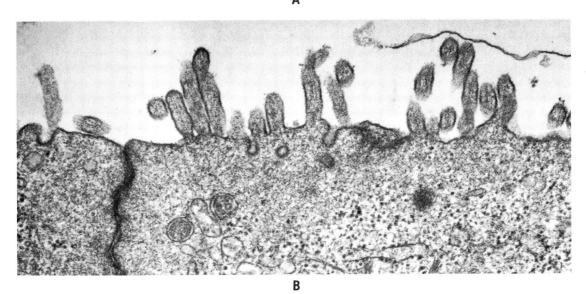

B

Figure 186. *A*, Chicken oviduct 24 hours after estrogen injection. *B*, Same, but treated with cytochalazin B; the microfilaments present in the first electronmicrograph (AF) have disappeared. ZA, Desmosome (zonula adhaerens). (From Wrenn, 1971.)

ble. If the oviducts are again placed in cytochalazin-free medium, the bundles of micro-filaments reappear. Reversal of cytochalazin B action can occur while protein synthesis in the affected cells is suppressed by another metabolic poison (cycloheximide). Thus, the formation of the microfilaments does not involve a synthesis of new proteins but is the result of a reorganization of the existing materials in the competent cells (Wessells et al., 1971).

As with hair rudiments, once the oviducal glands take shape, their further development depends on the processes of growth and cell division setting in. This phase of development can then be suppressed by exposing the gland rudiments to a chemical which blocks DNA synthesis (hydroxyurea) (Wrenn, 1971).

9–5 MORPHOGENETIC MOVEMENTS IN MESENCHYME

Immigration of mesodermal cells from the primitive streak in amniotes and migration of mesenchme cells in sea urchin gastrulation can be regarded as examples of an important step in morphogenesis: the creation of mesenchyme-type tissues, which are a new structural element of an animal's body. The mesenchyme cells are not linked together, as are epithelial cells. They are free to move, and as a result the morphogenetic processes in mesenchyme are somewhat different from those occurring in epithelial cell masses. It is surprising, however, that some of the cellular mechanisms involved in the change in shape of epithelial cells recur in mesenchyme-type cells as well.

The movements of mesenchyme cells have often been described as "ameboid movements," but this statement must be qualified. The better known type of movement of free-living amebae involves streaming of cytoplasm in the interior of the ameba's body. When pseudopods are formed at the progressing end of the ameba, internal cytoplasm flows into the pseudopod, and enlargement of the pseudopod is supported or accompanied by this forward streaming of cytoplasm. When mesenchyme cells (fibroblasts and similar types of cells) are observed *in vitro*, they are found to move by a completely different mechanism. No internal streaming of cytoplasm can be observed. Instead, the cytoplasm at one end of the cell becomes extended into a very thin edge which is seen to be in continuous undulating movement (Fig. 187). This edge has been described as the "ruffled membrane." The ruffled membrane intermittently makes and breaks contact

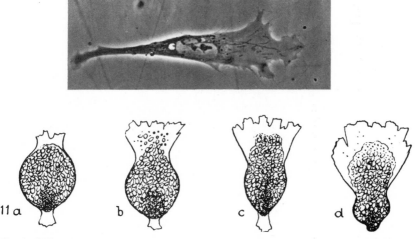

Figure 187. Embryonic cells moving *in vitro. Top,* Glial cell from a nerve ganglion of 8-day-old chick embryo. (From Spooner, Yamada, and Wessells, 1971.) *Bottom,* Ectodermal cell from an amphibian gastrula. (From Holtfreter, 1947.)

with the substrate over which the cell is moving and progressing forward. The rest of the cell trails behind, forming an attenuated "tail" at its posterior end.

The same elements that are involved in the change in shape of epithelial cells are found in migrating mesenchyme cells (Wessells, 1971; Ludueña and Wessells, 1973). The electron microscope shows a mass of microfilaments in the ruffled membrane arranged in the form of a lattice, with the filaments running in different directions. Where the edge of the membrane protrudes, the microfilaments converge to the protruding point (Fig. 188). The trailing "tail" of the cell is supported by longitudinally arranged microtubules. If the cells are treated with cytochalazin B, the movements of the ruffled membrane stop immediately, and the microfilaments are found to have been disrupted and converted into granular cytoplasm (Spooner et al., 1971). Colchicine, on the other hand, causes the distintegration of the microtubules, the result being that the "tail" of the cell collapses and is withdrawn into the cell body. Although the movement of the ruffled membrane continues, the cell seems to have become depolarized and cannot progress in one direction, as if the "tail" has been a sort of rudder keeping the cell on course.

The preceding description of the movement of cells refers to observations and experiments on cells cultivated *in vitro*. The mechanism of cell movements in the organism need not necessarily be the same. *In vitro* the cells move over a hard flat surface (the bottom of the culture dish), and their movements are thus dominated by cell-to-substrate contacts. In the organism, and especially in the case of an early embryo, the contacts are cell-to-cell. The study of movements of individual cells in the intact organism involves considerable difficulties, since the object being observed must be sufficiently

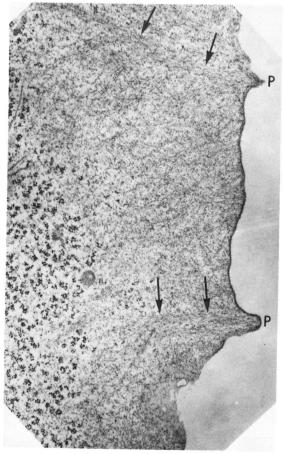

Figure 188. Edge of ruffled membrane of a glial cell from a nerve ganglion of 8-day-old chick embryo (electronmicrograph), showing system of microfilaments (arrows). P, Protuberances at the edge of the membrane. (From Ludueña and Wessells, 1973.)

small and sufficiently transparent for the cells inside the body to be accessible for microscopic observation. For this reason very few reliable studies of this kind are available at present.

Observations of the movements of primary mesenchyme cells during sea urchin gastrulation have been mentioned previously (p. 240). After becoming free of the epithelial blastoderm, the primary mesenchyme cells start sending out long thin pseudopods, rather than ruffled membranes (Fig. 189). These pseudopods may come into contact with the inner surface of the ectodermal cells and may make firm connection with them. If this occurs, the pseudopod contracts and pulls the mesenchyme cell away from the original site of its immigration into the blastocoele. If the pseudopod fails to make a firm contact or if it does not reach an ectodermal cell but is surrounded by the blastocoele fluid, the pseudopod is withdrawn, and new pseudopods are formed. By this mechanism of "trial and error" the mesenchyme cells gradually disperse from their point of origin and take up positions on the inner surface of the ectodermal layer. The pseudopods, being elongations of the cell cytoplasm, are supported by bundles of microtubules (Fig. 190).

Another case in which the transparency of the early embryo was used for observing movements of cells is the blastodisc of a bony fish (Trinkaus, 1973). The surface cells of the blastodisc form an immobilized continuous layer, but the deeper-lying cells in the blastula and gastrula stages are in a state of continuous movement and can be

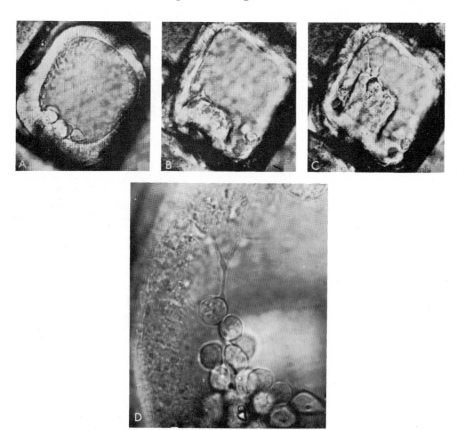

Figure 189.   Stages of gastrulation in the sea urchin embryo, photographed in the living state. *A*, Immigration of primary mesenchyme. *B*, Beginning of invagination of the endoderm. *C*, Invagination of the endoderm with the formation of the archenteron. *D*, Mesodermal cells move away from the site of the blastopore after establishing contacts with the inner surface of the ectoderm. (Courtesy of Dr. T. Gustafson.)

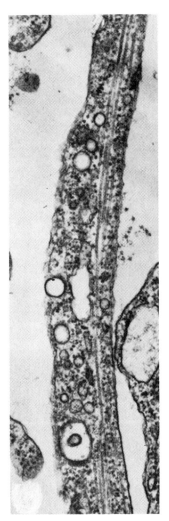

Figure 190. Microtubules in an elongated pseudopod of a primary mesenchyme cell of a sea urchin embryo. (From Gibbins, Tilney, and Porter, 1969.)

photographed on cine film through the surface layer. The locomotion of the deep cells is more similar to the classic type of ameboid movement. The movement starts by the cell's forming a broad rounded "bleb" on its surface which appears to consist of hyaline cytoplasm, without the coarser cell inclusions (mitochondria, yolk granules, etc.). The bleb may be extended into a tonguelike lobopodium. As the lobopodium is formed and extended, more cytoplasm flows into it. The bleb or lobopodium (the terms are interchangeable) can sometimes be seen to flatten and attach itself to adjoining cells. When it subsequently contracts, it pulls the cell forward (Fig. 191). The cell may, however, progress forward as a whole, without a visible contraction or withdrawal of the lobopodium. Amphibian early embryonic cells cultivated *in vitro* have also been observed to move by means of broad lobopodia (Fig. 192).

Microfilaments have not, as yet, been seen in the blebs or lobopodia of early embryonic cells, but there is some evidence that they may be present. This can be concluded from the fact that gastrulation in sea urchins is suppressed by cytochalazin B (Wessells et al., 1971). Also, movements of cells of four- to seven-day-old chick embryos are suppressed by the same drug (Steinberg and Wiseman, 1972). The latter observation was made in the course of an experiment designed to test whether active movements of cells

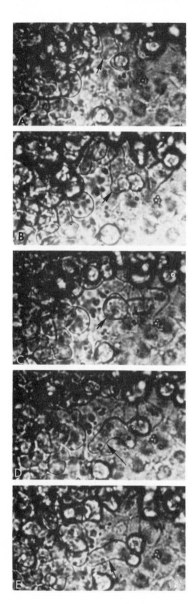

Figure 191.   Five stages in the translocation of a deep cell in a fish blastodisc, photographed *in vivo*. The same cell is indicated by an arrow in each photograph. (Courtesy of Dr. J. P. Trinkaus, 1973.)

are the means by which the reaggregation and "sorting out" of different kinds of cells are achieved after embryonic tissues have been disaggregated. It was found that cytochalazin B does not prevent the cells from adhering to one another on contact but that the aggregates formed do not attain the regular spherical shape, and the sorting out of different kinds of cells in the interior of the aggregates does not take place. These results show that the active movement of cells is directly involved in their taking up the correct positions in the aggregates (and thus also in the normal embryo!). At the same time, the results indicate that the movement of embryonic cells, as well as the movement of fibroblasts in tissue culture, depends on the presence and activity of microfilaments.

  To become a morphogenetic factor, the movement of cells must be directed in definite pathways. By moving in an organized way, the mesenchyme-type cells form accumulations in specific parts of the embryo, which become rudiments of organs (e.g., ru-

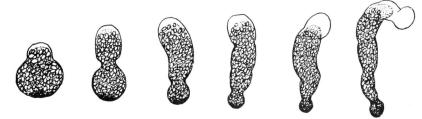

Figure 192.   Amphibian (*Ambystoma*) gastrula cell moving *in vitro* with formation of a broad pseudopod of hyaline cytoplasm. (From Holtfreter, 1946.)

diments of many skeletal parts—Fig. 193), or the cells accumulate around epithelial organ rudiments, forming connective tissue or skeletal capsules of internal organs.

The properties of the substratum along which the cells migrate are among the major factors affecting the direction of migration. As has already been indicated, the cells cannot move through space filled only with fluid. In the parts of the embryo occupied by mesenchyme, however, the intercellular spaces are filled with a colloidal solution which is in part gelated, so that fibers of molecules are stretched through the space in various directions. These fibers serve as a substratum for the migration of mesenchyme cells. If the fibers of the intercelluar substratum do not show any orientation, the migration of mesenchyme cells is disorganized, and the cells become scattered evenly in space. If the fibers of the intercellular substratum become oriented in any particular direction, the mesenchyme cells migrate along the oriented bundles of fibers. This can be proved experimentally with cells in tissue cultures. The fibers of the plasma clot on which the cells are cultivated may be caused to be stretched in a certain direction by various means, for instance by allowing the clot to set under mechanical tension.

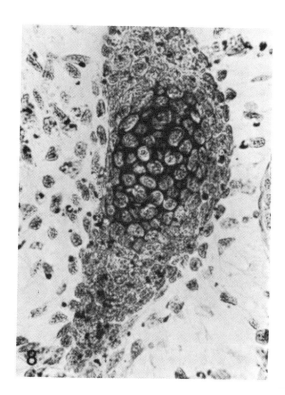

Figure 193.   Embryonic mesenchyme aggregating in a tissue culture to form a nodule of cartilage. (From Holtfreter, J.: Mesenchyme and epithelia in inductive and morphogenetic processes. In Fleischmajer, R., and Billingham, R. E. (Eds.): Epithelial-Mesenchymal Interactions, p. 21. © 1968, The Williams & Wilkins Company, Baltimore.)

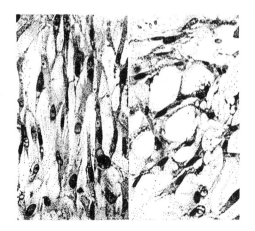

Figure 194. Cells growing in tissue culture. *Left,* micelles in substrate are oriented by allowing the blood plasma to clot under stress. *Right,* no orientation of micelles in substrate. (After Weiss, from Kühn, 1955.)

The cells will then follow the direction of the plasma clot fibers in their migration (Fig. 194 — Weiss, 1929; Doljanski and Roulet, 1934).

If the mesenchyme cells aggregate around an epithelial vesicle, this aggregation may be the result of the orienting influence of the epithelium on the fibers of the intercellular substrate. If these fibers are stretched perpendicularly to the surface of the epithelial vesicle, the migrating mesenchyme cells will converge toward the vesicle. Having reached the surface of the epithelium, the cells of the mesenchyme may be held there by intercellular bonds whose nature is not well understood but whose existence can hardly be doubted (Weiss, 1947).

The influences of adjoining parts on the movements of migrating cells is not the only form of interaction of parts playing an important role in the development of organ rudiments. Influences of a nature similar to the induction of the neural plate by the roof of the archenteron have been discovered in the development of organ rudiments. Such influences may modify the movements of cells, but not directly. The direct result is a modification of the properties of cells, and consequently the movements of cells are also altered. This can be illustrated by the development of the neural crest which differentiates together with the neural plate as a result of an induction from the roof of the archenteron. Once induced, the neural crest cells acquire new properties. They become mesenchyme cells and start migrating away from their source of origin, while both the cells of the epidermis and the cells of the neural tube remain relatively stationary.

Chapter 10

EMBRYONIC ADAPTATIONS AND THE DEVELOPMENT OF MAMMALS

In this chapter we will consider special organs or parts of the embryo which are not precursors of any of the organs of the adult or the larva but serve to satisfy the requirements of the embryo with respect to food or oxygen supply or to its protection.

The yolk in the egg cannot be considered as part of the embryo in this sense, although it fulfills the purpose of supplying nourishment for the embryo. However, in animals in which the amount of yolk is very large, special organs may be evolved to store the yolk and to utilize it.

Large amounts of yolk, which does not participate in the process of cleavage, are found in the eggs of most fishes. The organs of the embryo develop at the expense of the cells of the blastodisc, which lies at the animal pole of the egg. The yolk is eventually enclosed by the spreading edges of the blastodisc, as described in Section 7–6 (p. 175). The cell layers spreading over the yolk are the ectoderm, the mesoderm, and the periblast—the syncytial layer formed on the surface of the yolk underneath the blastoderm. The endoderm in the bony fishes does not follow the movements of the blastopore rim, so that the yolk is not enclosed by the endodermal layer.

In the meantime, the organs of the embryo are developed, and the body of the embryo raises itself away from the surface of the mass of yolk. Thus, a partial separation is introduced between the body of the embryo proper and the **yolk sac** enclosing the mass of yolk. The broad connection between the body of the embryo and the yolk sac becomes constricted later, so that the yolk sac remains connected to the embryo proper by the stalk of the yolk sac (Fig. 195). A system of blood vessels, which enter the heart by

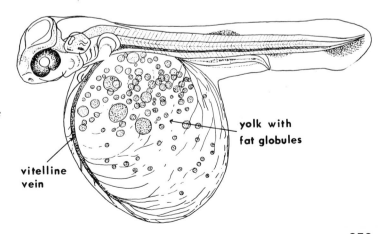

Figure 195. Trout embryo with large yolk sac.

yolk with fat globules

vitelline vein

means of a pair of vitelline veins (also known as the omphalomesenteric veins), is developed in the walls of the yolk sac. The yolk is broken down owing mainly to the activity of the periblast, and it becomes transportable through the blood stream to the embryo. As the yolk is used up, the yolk sac diminishes in size, while the embryo grows. Hatching takes place in the meantime, the embryo becoming a free-swimming larva. The yolk sac is eventually retracted into the ventral body wall of the larva.

10–1 THE EXTRAEMBRYONIC STRUCTURES IN REPTILES AND BIRDS

In reptiles and birds, as in fishes, the yolk is eventually enclosed in a yolk sac, but the way this is done differs from that found in fishes. The periphery of the blastodisc in reptiles and birds is differentiated as the area opaca, and the external edge of this spreads over the surface of the yolk and eventually covers it. The cells of which the area opaca consists, beginning with the part nearest to the embryo and progressing outward, split into three layers which are continuous with the germinal layers developed in the area pellucida: the ectoderm, the mesoderm, and the endoderm, the latter adhering to the surface of the yolk.

During the second and third days of incubation, in the case of the chick, a network of blood vessels is developed in the inner part of the area opaca, which from this time onward becomes the **area vasculosa.** The outer part of the area opaca becomes the **area vitellina.** The development of the blood vessels in the area vasculosa is intimately connected with the differentiation of the first blood cells. This happens the following way. First of all, groups of densely packed mesodermal cells appear in the area opaca all around the sides and the posterior edge of the area pellucida. These groups of cells are known as the **blood islands** (Fig. 196). Next, the cells of the blood islands become differentiated into two kinds: the cells on the periphery join to form a thin epithelial layer, the endothelium of the future blood vessels; the central cells, on the contrary, become separated from one another and are differentiated as blood corpuscles. Thus from the beginning, the blood corpuscles lie inside the blood vessels.

Meanwhile, the adjacent blood vessels establish communications among themselves, and the whole is transformed into a very irregular network. All the vessels of the

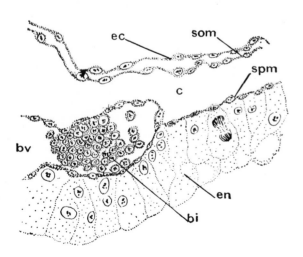

Figure 196. Section through area vasculosa of a chick embryo, showing a blood island. c, Coelom; ec, extraembryonic epidermis; som, parietal layer of mesoderm; spm, visceral layer of mesoderm; en, endoderm; bi, blood island; bv, blood vessel. (From Wieman, 1949.)

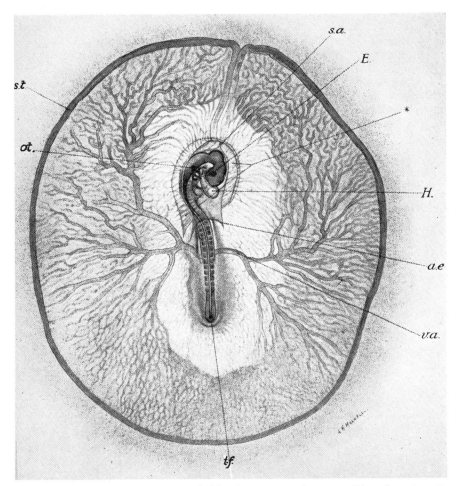

Figure 197. Chicken embryo surrounded by the area vasculosa. a.e., Edge of amnion; E.,
eye; H., heart; ot., ear vesicle; s.a., suture along the line of fusion of amniotic folds; s.t.,
terminal sinus; t.f., tail fold; v.a., vitelline artery. The asterisk indicates a depression in which
lies the head of the embryo. (From Kerr, 1919.)

area vasculosa are joined together on the periphery by the terminal sinus, which is the
boundary line between the area vasculosa and the area vitellina (Fig. 197). The network
of the area vasculosa becomes prolonged into the area pellucida and eventually establishes connection with the embryo proper. A connection with the blood system of the
embryo is formed at two points: with the venous system, by means of right and left vitelline veins, joining in an unpaired ductus venosus which enters the sinus venosus of
the heart; and with the arterial system by means of right and left vitelline arteries,
which branch off from the dorsal aorta (Figs. 198 and 271).

Before the middle of the second day of incubation the heart of the embryo begins
to beat, and between the thirty-eighth and fortieth hours of incubation the blood starts
circulating through the network of the yolk sac. The wall of the yolk sac develops outgrowths on its inner surface, that is, on the surface facing the yolk. The outgrowths,
supplied by blood vessels from the area vasculosa network, penetrate deep into the yolk
and thus facilitate its absorption. The yolk, however, is not completely absorbed during
embryonic life. Shortly before hatching, the yolk sac is retracted into the abdominal
cavity of the embryo, and the walls of the abdominal cavity close behind it.

As in the fishes, the body of the embryo in birds and reptiles becomes separated

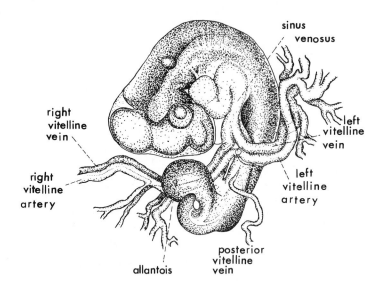

Figure 198. Three-day-old chick embryo viewed from the ventral side to show the origin of the vitelline vessels and the allantoic vesicle.

from the yolk sac. This is achieved by the formation of folds which appear all around the body of the embryo. The folds involve all three germ layers and are directed downward and inward, undercutting the body of the embryo proper. They are known as the **body folds.**

Although the body folds eventually surround the whole embryo, the various sections of the fold do not appear simultaneously. The first to appear is the part of the fold just in front of the head. This section, which may be referred to as the **head fold,** undercuts the head and anterior part of the trunk of the embryo, so that these parts project freely over the surface of the yolk sac. The lateral and the posterior parts of the body fold develop soon after, the latter undercutting the tail end and the posterior part of the trunk which now also project freely over the surface of the yolk sac, though this posterior part of the embryo is initially much shorter than the anterior part, separated from the yolk by the head fold. The body folds gradually contract underneath the embryo, and eventually only a rather narrow stalk, the **umbilical cord,** connects the embryo with the yolk sac and other extraembryonic parts. The cord includes all the germ layers: an outer ectodermal lining, a double sheath of mesoderm, and an endodermal canal connecting the cavity of the gut with the cavity of the yolk sac. In addition, it contains the blood vessels going out to the yolk sac and other structures which will be referred to later.

Besides the yolk sac, the reptiles and birds develop three other extraembryonic organs, known as the **embryonic membranes.** These are the **amnion,** the **chorion,** and the **allantois.** The amnion and the chorion are developed together as upwardly projecting folds, the **amniotic folds,** appearing on the area pellucida just outside the body folds and eventually closing over the dorsal surface of the embryo (Fig. 199). The rudiment of the amnion and chorion first appears as a transverse fold anterior to the head of the embryo. The fold bends backward over the anterior end of the head and covers it, as with a hood. Subsequently, the lateral ends of the fold are prolonged backward along both sides of the embryo. The lateral folds approximate one another over the body of the embryo and fuse from the front end backward, so that more and more of the embryo becomes covered by the folds. Eventually a fold also develops behind the embryo, and the free edges of the folds fuse, thus completely enclosing the body of the embryo in a cavity—the **amniotic cavity.** The amniotic cavity is at first a narrow slit

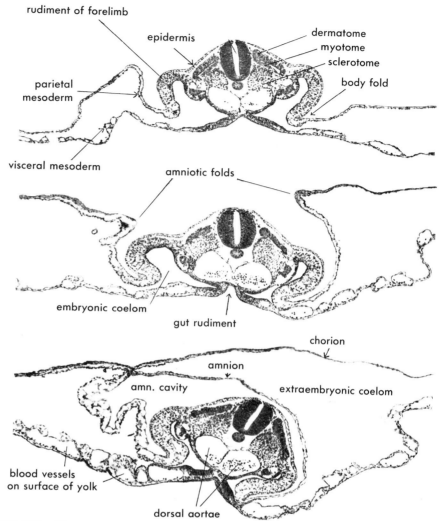

Figure 199.   Three stages in the development of the amniotic cavity in the chick (transverse sections.)

between the embryo and the inner wall of the amniotic fold, but soon a fluid is secreted into the cavity, which distends it, so that the embryo floats freely in the cavity, connected to the extraembryonic parts only by the umbilical cord.

When the amniotic fold is first formed in front of the head of the embryo, it consists only of part of the extraembryonic ectodermal layer. The lateral folds, however, from the start, consist of the ectoderm and the somatic layer of the extraembryonic mesoderm. Mesoderm also secondarily penetrates the anterior part of the fold, the composition of the fold therefore becoming uniform throughout.

The amniotic fold has two surfaces running out into an acute edge—the inner surface, facing the body of the embryo, and the outer surface, facing away from the embryo. The fusion of the amniotic folds of opposite sides occurs at the expense of their acute edges. The surface of the fold facing the embryo now lines the amniotic cavity and becomes the **amnion.** The surface of the fold facing away from the embryo, after the fusion of the folds, becomes continuous over the dorsal surface of the embryo (and over the amniotic cavity) and is called the **chorion.** Some embryologists prefer to call this membrane, as found in birds and reptiles, the **serosa,** reserving the term **chorion** for the outer embryonic membrane of mammals.

From what has been said of the composition of the amniotic folds, it is clear that the amnion consists of a layer of extraembryonic ectoderm on the inside (facing the amniotic cavity) and a layer of extraembryonic somatic mesoderm on the outside. The chorion consists of a layer of extraembryonic ectoderm on the outside and a layer of extraembryonic mesoderm on the inside. These layers are, of course, in continuity with the ectoderm and somatic mesoderm covering the yolk sac. In between the amnion and the chorion is the extraembryonic coelomic cavity, which is continuous with the coelomic cavity in the embryo proper.

The advantages that the embryo obtains from the development of the amnion, the chorion, and the amniotic cavity are clear enough: the embryo becomes immersed in a container filled with fluid and thus can accomplish its development in a fluid medium, although the egg is "on dry land." It appears therefore that the primary function of the amnion is to protect the embryo from the danger of desiccation Besides this, the fluid of the amniotic cavity is an efficient shock absorber and protects the embryo from possible concussions. Lastly, it isolates the embryo from the shell of the egg and thus protects it from adhesion to the shell or from friction against it. At the same time, the formation of the amniotic cavity has a slightly negative effect: it removes the embryo from the surface of the egg and thus from the source from which it could obtain oxygen.

The third embryonic membrane, the **allantois,** is very different in nature from the first two membranes. Originally the allantois is none other than a urinary bladder. A cleidoic egg, such as that of a bird or reptile, has no means of disposing of the waste products of protein breakdown by removing them from the egg. The problem of waste disposal is solved by excreting the protein breakdown products in the form of uric acid, which is deposited as water-insoluble crystals and stored inside the egg up to the time of hatching. The storage place is the urinary bladder. As the storage must go on throughout the time of incubation, the urinary bladder has to increase to enormous proportions.

The allantois appears as a ventral outgrowth of the endodermal hindgut. It thus corresponds exactly in nature to the urinary bladder of the amphibia. The outgrowth consists of endoderm with a layer of visceral mesoderm covering it from the outside (Fig. 200). The allantois, however, grows very rapidly and soon penetrates into the extraembryonic coelom, into the space between the yolk sac, the amnion, and the chorion. The distal part of the allantois expands and remains connected to the hindgut of the embryo by means of a narrow neck. When the body folds contract, separating the embryo from the extraembryonic parts, the neck of the allantois is enclosed together with the stalk of the yolk sac in the umbilical cord. The distal part of the allantois becomes flattened and penetrates between the amnion and the yolk sac on one side and the chorion on the other side (Fig. 201). By the middle of the incubation period, the allantois spreads over the complete surface of the egg underneath the chorion. Thus, the allantois comes into a position where it can take up a second function in addition to that of serving as a reservoir for storing uric acid.

The new function of the allantois is to supply the embryo with oxygen. A network of blood vessels develops on the external surface of the allantois, and this network is in communication with the embryo proper by means of blood vessels running along the stalk of the allantois and through the umbilical cord. The blood flows to the allantois from the dorsal aorta through the right and the left umbilical arteries. These arteries leave the dorsal aorta at a point which is much more caudal than the starting point of the vitelline arteries. The returning blood flows to the heart through a pair of umbilical veins. These veins originally enter the right and left ducts of Cuvier respectively. Soon,

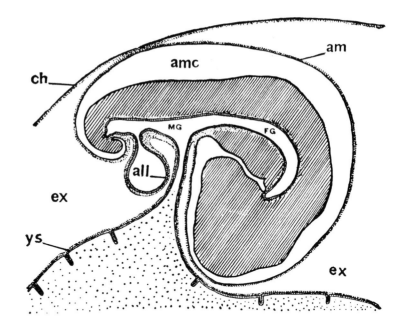

Figure 200. Early stage of development of the allantois in a four-day-old chick embryo. all, Allantois; am, amnion; amc, amniotic cavity; ch, chorion; ex, extraembryonic coelom; FG, foregut; MG, midgut; ys, yolk sac. (From Wieman, 1949.)

however, this arrangement of blood vessels changes. The right umbilical artery and the right umbilical vein disappear, and the left umbilical vein acquires a new central connection. It joins up with the left hepatic vein, whereupon the channel to the duct of Cuvier is closed, and the whole blood flow from the allantois passes into the left hepatic vein.

The allantoic circulation continues until the chick breaks the eggshell and begins to breathe the surrounding air. Then the umbilical vessels close, the circulation ceases, and the allantois dries up and becomes separated from the body of the embryo.

10–2 MAMMALIAN EGGS

The development of the mammals is greatly modified as compared with that of other vertebrates, and this is due to adaptations found in this class. There is good evi-

Figure 201. Later stage of development of the allantois in a chick embryo (diagrammatic), showing expansion of the allantois (indicated by arrows) in the extraembryonic coelom.

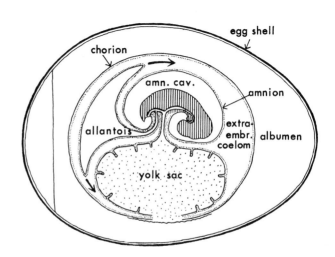

dence that the mammals are derived from ancestors which were very closely related to the early reptiles or which can actually be classified as reptiles. As such, they must have had large yolk-laden eggs, and their ontogenesis must have been much like that of reptiles. At some stage in their evolution some of the mammals became viviparous, the embryo receiving an adequate supply of nutrition from the mother while it was retained in the uterus, and the yolk supply of the egg became superfluous. Consequently, the yolk supply became progressively reduced and eventually disappeared altogether. In this respect, however, there is a distinct gradation in the three subdivisions of the class.

In the subclass Prototheria the eggs are laid, and they develop outside the maternal body (*Ornithorhynchus*), or though developing within the maternal body, they are not supplied with nourishment (*Echidna*), this adaptation not going further than the state of ovoviviparity (Chapter 4). The eggs therefore have a large amount of yolk, they are meroblastic, and subsequent development follows essentially a reptilian pattern (Flynn and Hill, 1939).

In the subclass Metatheria (the marsupials), the developing embryos receive nourishment from the mother in the uterus, although the adaptations for this mode of providing for the embryo have not advanced as far as in the next group, and the young are born poorly developed. The yolk has already become superfluous, and its quantity in the egg is insignificant. Even so, it is not used by the embryo as in the reptiles and birds, but the mass of yolk is ejected at the beginning of cleavage as the egg becomes divided into the first two blastomeres (Hill, 1918).

In the most advanced subclass of the mammals, the Eutheria (placental mammals), the egg is devoid of yolk right from the start.

Because the eggs and embryos of mammals are concealed during development inside the maternal body, the study of mammalian development presents particular difficulties as compared with the study of the development of egg-laying animals. In spite of the particular interest that mammalian development has for man, studies on mammalian eggs and embryos have lagged behind those made on lower animals. This applies to an even greater degree to the study of the embryonic development of man. In other mammals, eggs and embryos can be obtained by sacrificing fertilized and pregnant females, but in man, investigators were limited to accidental findings obtained as a result of operations or autopsies on women. Experimental studies on embryonic development, which have so greatly advanced the understanding of ontogenesis, have been particularly difficult to carry out with mammals. To advance the study of mammalian development, means had to be found to remove the eggs and embryos from the maternal body and to keep them alive and developing outside of the body, *in vitro*. Such methods have been elaborated in comparatively recent years, and as a result mammalian embryology is at present making rapid and astonishing advances.

The first step in this modern approach to the study of mammalian development is to obtain healthy unfertilized or newly fertilized eggs. To do this one must know when the ripe eggs are being released from the ovary, that is, when ovulation is taking place. In most mammals ovulation is a periodic process, which is determined by a delicate balance of hormones in the female's body. Essentially, two hormones produced by the hypophysis are responsible for the control of ovulation: the follicle-stimulating hormone (FSH) and the luteinizing hormone (LH). The follicle-stimulating hormone causes the final growth of the unripe egg follicles in the ovary. The luteinizing hormone, working in conjunction with the follicle-stimulating hormone, causes the eggs to start maturation (the formation of the first polar body) and also causes the rupture of the follicle with the release of the egg or eggs into the coelom, where they immediately enter the ostium of the oviduct. A periodic rhythm in the release of the hormones, which is due to a complicated feedback mechanism, subject to a certain extent to the external environment, and mediated at least in part by the central nervous system, determines the time of the ovulation. These determining factors apply to most mammals and to man in particular. There are a few species of mammals, however, in which the time of ovulation is determined by the occurrence of coitus. This does not contradict the previous explanation of the causes of ovulation: the release of the two hormones is, in such cases, dependent on the nervous excitation accompanying coitus. The rabbit is one of the mammals in which this mechanism is present, and for this reason the rabbit has become one of the favorite animals for the study of mammalian development. Ovulation in this case occurs 10½ hours after coitus. A sterilized "vasectomized" buck (with interrupted ductus deferens) may be used for mounting the doe, if it is desired to obtain ovulated but unfertilized eggs. The female is then sacrificed or laparotomized, and the eggs are flushed out of the oviducts.

In mice, which are also often used for embryological studies, ovulation is periodic (every four days) and independent of coitus. In these animals, however, the time of ovulation can be determined fairly accurately, since part of the seminal fluid of the male coagulates after coitus and forms a fertilization plug in the vagina of the female. The plug is easily visible from without. The male will mate only with a female that is ovulating.

Once the ova are obtained, they have to be placed in a suitable medium. Advantage has been taken of experience gained to date with tissue culturing *in vitro*, and some of the media developed for cultivating tissue cells have been used for keeping the eggs alive and developing. Cleavage of the fertilized eggs could be observed in some of the synthetic, chemically defined solutions. For more extended observations, however, it has been necessary to make use of natural body fluids, such as blood serum, at least in the experiments carried out so far.

As previously mentioned, unfertilized eggs can be obtained by mating the female with a sterile buck. The same result can be achieved both in the rabbit and in other mammals by a hormonal treatment, that is, by injecting into the female a large dose of gonadotropic hormones (the follicle-stimulating hormone and the luteinizing hormone or pregnant mare's serum, which contains a hormone with a similar action). The ovulated eggs can then be retrieved as before and fertilized by sperm *in vitro*. Such *in vitro* fertilization has been performed with a number of mammals (rabbits [Fig. 202], mice, hamsters) and also with human eggs. Experiments with human eggs are naturally of particular interest and of far-reaching importance.

In the case of humans, eggs and early embryos have occasionally been obtained during operations carried out for medical purposes. Such rare cases could not, however, form a basis for systematic research. Advantage was taken of the discovery that advanced follicular oocytes removed from ovaries of animals could be made to mature *in vitro* by a suitable hormonal treatment. Luteinizing hormone, in combination with follicle-stimulating hormone, causes explanted follicular oocytes to start the first meiotic division—the stage in which mammalian eggs normally ovulate and are then ready for fertilization. The discovery that oocytes can mature *in vitro* was used by Edwards, Steptoe, and their collaborators to develop the following procedure for obtaining human eggs for research purposes. (See Edwards and Fowler, 1970.)

Human female patients are given heavy doses of gonadotropic hormones to cause simultaneous ripening of a number of ovarian follicles. A few hours before the follicles are to rupture, instruments, consisting of a small telescope and an aspirating device, are introduced into the body cavity through a small incision in the body wall, and the swollen follicles, visible on the surface of the ovary, are sucked in by the aspirating device. After being placed in a dish *in vitro*, the eggs contained in the follicles mature,

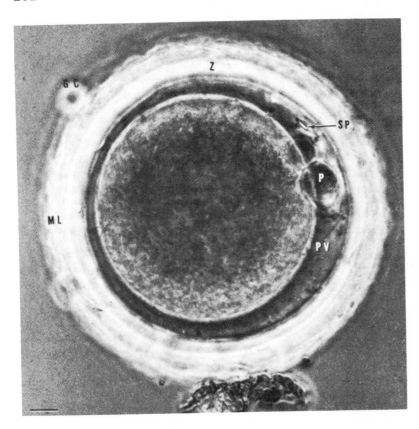

Figure 202. Fertilized egg of rabbit. GC, Adhering follicle cell; ML, mucous layer; P, polar body; PV, perivitelline space; SP, spermatozoon; Z, zona pellucida. (From Hadek, 1969.)

and most of the follicle cells fall away from the ovum, as in normal ovulation. The eggs can then be fertilized. Three or four usable eggs could be obtained in this way from one volunteer patient.

Continued development of a mammalian embryo *in vitro* is fraught with formidable difficulties, because of the embryo's dependence (after completion of cleavage) on the uterine environment for nourishment, gas exchange, etc. The nature of these difficulties, and some of the attempts to overcome them, will be dealt with later (Section 10–4). The most obvious way to insure further development of explanted eggs and embryos, however, is to return them to their customary environment, that is, to introduce them into the uterus of a female. This is, in fact, possible, provided that the uterus is in a condition to receive the embryos.

In most mammals the tissues of the uterus undergo periodic changes, which are correlated with other sexual functions: ovulation and mating. The uterine wall becomes richly supplied with blood; it swells and as a whole becomes thicker, and the uterine glands proliferate and start secreting. These changes normally occur concurrently with the processes in the ovary, which lead to the ripening of the egg follicle and subsequently to ovulation, and are mediated by hormones: estrogen, secreted by follicle cells, and subsequently progesterone, produced by the corpus luteum, which develops from remaining parts of the follicle after ovulation. If mating occurs and eggs are fertilized, the developing embryos enter the uterus and become attached to the uterine wall, and normal pregnancy ensues. If, however, the ovulated eggs are not fertilized, the female may enter a state of **pseudopregnancy,** in which the uterus remains for some time in a condition of readiness for the reception of the embryo. The condition may continue for the length of the normal pregnancy in some animals (bitch), but in others the uterus reverts to normal much earlier. It has been established that in most animals pseudo-pregnancy is a direct result of ovulation, but in rats and mice it is caused by the act of copulation, which may be substituted even by stimulation of the cervix of the uterus with a glass rod. On the chemical plane, pseudopregnancy is produced through the extended secretory activity of the corpora lutea, mediated in the case of rats and mice by the nervous system and the hypophysis.

The condition of pseudopregnancy is used for reimplantation of embryos after a sojourn *in vitro*. Reimplantation can be performed while the embryos are in the early cleavage stages or when they have completed cleavage and have reached the stage of blastocyst, which corresponds more or less to the blastula stage in lower animals. (See further, p. 266.) The embryo introduced into the uterus of a pseudopregnant female establishes connection to the uterine wall, and a normal pregnancy ensues. The embryos of many species of mammals, after having been treated *in vitro*, or after having been produced by fertilization *in vitro*, have been brought to term in this way, and healthy offspring have been produced. In animal experiments, the original female from which the eggs are taken is usually sacrificed, so that the female carrying the implanted eggs is a "foster mother."

In humans, eggs obtained by the method previously described, fertilized *in vitro* by sperm of the husband, and then implanted into the uterus of the same patient could be a means of allowing a woman to bear her own child, in cases in which normal conception is prevented by her oviducts being impassable due to a pathological process. Although attempts in this direction are being made, no successful births have been reported so far.

Once it became feasible to obtain mammalian eggs and to keep them *in vitro*, the way was open not only for observations of their early development but also for carrying out of a variety of experiments on the eggs and embryos. One of the approaches which will be mentioned here involves attempts at freezing mammalian eggs and early embryos. It is known that many lower (invertebrate) animals can be frozen, kept practically indefinitely in a frozen state, and then thawed and brought back to life. So far, to do the same with warm-blooded vertebrates has not been possible, but spermatozoa have been shown to be capable of being placed in deep freeze and to be able to regain their fertilizing ability fully after being thawed. It has been shown that mammalian eggs and early embryos can similarly be frozen and thawed. To prevent the eggs from being damaged by the formation of ice crystals in the cytoplasm, freezing must be done slowly, and the medium must be suitably adjusted (dimethyl sulfate added). Under these conditions, mouse embryos in the two- to eight-cell stages were frozen at temperatures of $-196°$ and $-269°$ C. for up to eight days, and after thawing they resumed development. Some of the embryos treated in this way were reimplanted into uteri of foster mothers, and as a result 13 litters were born. The animals thus developed from frozen embryos were perfectly normal (Whittingham, Leibo, and Mazur, 1972).

10–3 CLEAVAGE, BLASTOCYST FORMATION, AND DEVELOPMENT OF GERMINAL LAYERS IN MAMMALS

With the disappearance of the yolk, mammalian eggs have reverted to complete cleavage, but subsequent development bears ample evidence of the former presence of yolk, and in many respects the morphogenetic processes resemble those in meroblastic eggs with a discoidal type of cleavage.

Although the cleavage is complete and all the blastomeres are of more or less equal size (Fig. 203), it is by no means as regular as in oligolecithal eggs of the invertebrates and lower chordates. Synchronization of the mitoses in the blastomeres is lost very early. Even the first two blastomeres may cleave at different rates; consequently, a three-cell stage is found, and subsequently stages of five, six, seven blastomeres, and so forth (Fig. 204). Also, the overall speed of the cleavage is much lower than in many other animals: several hours elapse between successive divisions in the same line of cells, as compared with the cleavage rate of 20 to 40 minutes observed in fishes, frogs, and echinoderms. This slower rate is particularly significant, as the cleavage in mammals pro-

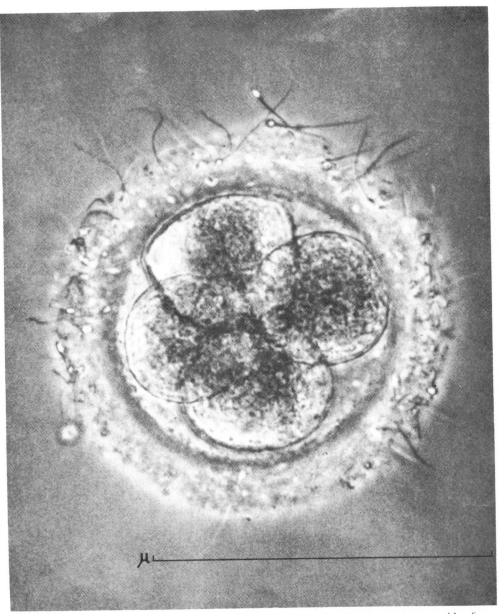

Figure 203.   Human egg, cleaving; stage of four blastomeres. Numerous spermatozoa outside of zona pellucida. (From Edwards and Fowler, 1970.)

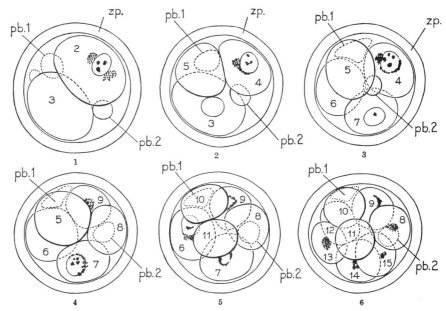

Figure 204. Cleavage of the egg of a monkey, *Macacus rhesus*. pb.1 and pb.2, First and second polar bodies; zp., zona pellucida. (From Lewis and Hartmann, 1933.)

ceeds at their body temperatures which are higher than the ambient temperatures at which the development of the animals just mentioned takes place.

The result of the cleavage is a solid mass of cells, a **morula,** in which some cells are superficial and others lie inside, completely cut off from the surface by the enveloping cells (Fig. 205*A*). In due course the superficial cells join to form a distinct epithelial layer. This layer gives rise to most of the extraembryonic parts (the embryonic membranes), serves to attach the embryo to the uterine wall, and mediates in the supply of nourishment to the embryo from the maternal body via the placenta. The outer layer of the mammalian embryo is known as the **trophoblast** (the term comes from the Greek word *trophe,* meaning nourishment). The cells lying in the interior are known as the **inner cell mass,** and it is these cells that provide material for the formation of the embryo proper. They may therefore be referred to as the **formative cells.** The two kinds of cells are distinguishable rather early by certain physiological properties, which may be shown

Figure 205. Morula (*A*) and early blastocyst (*B*) of a bat, showing differentiation into an inner cell mass and the trophoblast. (After van Benden, from Brachet, 1935.)

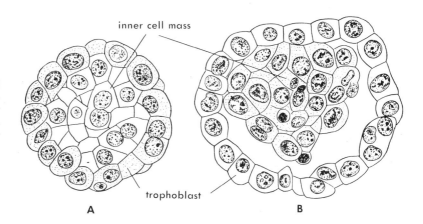

inner cell mass

trophoblast

A   B

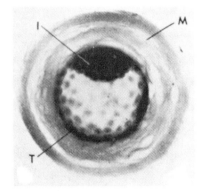

Figure 206.   Blastocyst of rabbit. I, Inner cell mass; M, egg membrane; T, trophoblast. (From Witschi, 1956.)

by the use of specific staining methods. Cells of the inner cell mass have been found (in the rat) to be more basophilic than trophoblast cells, suggesting a higher content of cytoplasmic nucleic acids (Jones-Seaton, 1950), and have also been found to contain the enzyme alkaline phosphatase, which is lacking in the trophoblast cells (Dalcq, 1954).

Sooner or later a cavity appears inside the compact mass of cells of the morula. The cavity is formed of crevices which appear between the inner cell mass and the cells of the trophoblast. Fluid is imbibed into this cavity, so that it enlarges, the whole embryo becoming bloated to the same degree. The trophoblast becomes lifted off the inner cell mass on most of its inner surface, and it remains attached on one side only. This side corresponds later to the dorsal side of the embryo (Figs. 205*B*, 206, and 207).

A mammalian embryo in this stage is called a **blastocyst.** The cavity of the blastocyst may be compared to the blastocoele, but the embryo as a whole differs essentially from a blastula, since its cells are already differentiated into two types: the inner cell mass and the cells of the trophoblast.

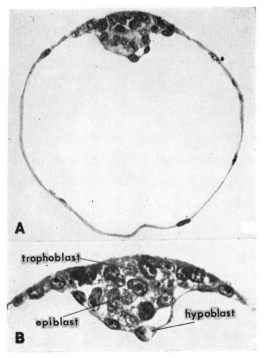

Figure 207.   Advanced blastocyst of monkey. *A*, General view (section); *B*, dorsal part of same under higher magnification. (From Heuser and Streeter, 1941.)

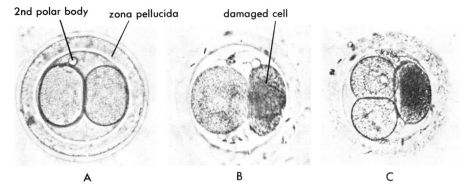

Figure 208.   The killing of one of the first two blastomeres in the rabbit. *A*, Normal two-cell stage in the cleavage of the rabbit egg. *B*, One of the two cells (on right) killed by pricking with a glass needle; the dead cell is slightly deformed. *C*, Operated embryo at a later stage; the surviving blastomere (on left) has divided in two. (Courtesy of Professor F. Seidel, 1960.)

There is experimental evidence that the properties of the inner mass cells and those of the trophoblast are already distinctly different at this stage. Three and a half-day-old mouse embryos were cut in pieces, so that some pieces contained only inner mass cells and others consisted of trophoblast cells (Gardner, 1972). It was found that clumps of inner mass cells fused together in larger masses, while clumps of trophoblast cells formed vesicles which did not adhere to one another. When introduced into uteri of pseudopregnant mice, the inner mass cells did not react with the walls of the uterus; they did not become implanted and thus did not develop further. On the other hand, the vesicles formed from trophoblast, when introduced into uteri, established connection with the uterine wall as in the development of normal embryos but later failed to proliferate. This experiment clearly reveals the roles of the two components of the early mammalian embryo.

Some embryologists have claimed that the two kinds of cells in the blastocyst may be traced back to the first cleavage of the mammalian egg, meaning that one of the first two blastomeres gives rise to the inner cell mass and the other to the trophoblast (Heuser and Streeter, 1928). This view has been definitely rejected by experimental work.

Duplicating an experiment performed long before on the frog embryo by W. Roux (p. 12), Seidel killed one of the first two rabbit blastomeres by pricking it with a fine glass needle (Seidel, 1952, 1960). The surviving blastomere continued to cleave (Fig. 208). The embryo was then introduced into the uterus of another female. A number of completely normal embryos developed and some of them were born and remained alive till adulthood (Fig. 209). Even at the four-cell stage one of the four blas-

Figure 209.   Rabbit developed from one of the first two blastomeres (on right) 2 months after birth together with its foster mother (the female, into the uterus of which the operated embryo was implanted). (Courtesy of Professor F. Seidel, 1960.)

tomeres could give rise to a normal embryo after the other three blastomeres were killed. The same results were obtained in experiments on mouse embryos (Tarkowski, 1959).

In experiments on both rabbits and mice not all blastomeres (either one of the first two or one of the four) produced whole normal embryos. In a certain proportion of cases abnormal blastocysts were formed which consisted of only extraembryonic parts: the trophoblast, embryonic membranes, and sometimes extraembryonic endoderm. These defective embryos appeared more frequently when development proceeded from a blastomere at the four-cell stage than from one of the first two blastomeres. Two interpretations were given to these results. One suggestion was that there is some sort of "organization center" in the egg and that only those blastomeres containing this organization center or a part of it give rise to blastocysts with inner cell mass and subsequently to normal embryos. The blastomeres that are devoid of the organization center produce simple purely trophoblastic vesicles (Seidel, 1960). The other interpretation (Tarkowski and Wroblewska, 1967) was that simple vesicles are defective structures resulting from insufficient cellular material in the isolated part.

It must be pointed out that incomplete fetuses consisting of embryonic membranes only and devoid of the embryo proper occur sometimes in both humans and domestic animals and also that the formation of the embryo within the blastocyst is particularly sensitive to an inadequate environment. (See further, p. 283.) On the other hand, it is known that sometimes a single egg of a mammal may produce two or more embryos. This happens occasionally in humans, and then "identical twins" are born—identical because, being derived from the same fertilized egg, they possess an exactly similar hereditary endowment. It is not known, however, at what stage the doubling occurs in identical twins. In fact, it is likely that the doubling of the embryo occurs at the primitive streak stage and involves the separation of formative cells into two lots. Twinning in this form does not therefore shed any light on the organization of the egg and the potentialities of its parts.

The development of a whole animal from one of the first two or one of the first four blastomeres of a mammalian egg is beautifully complemented by another experiment, first performed by Tarkowski (1961). (See also Mintz, 1962, 1964.) The experiment consisted in fusing two cleavage stage embryos of the mouse to form one single animal. The fusion was performed in the eight-blastomere stage. The zona pellucida was ruptured and removed by sucking the embryo into a narrow gage pipette. Two or more embryos were then pressed together in a very small drop of culture medium immersed in liquid paraffin. This caused them to fuse into a single morula, which was cultivated *in vitro* for 24 to 40 hours until a blastocyst was formed. The blastocyst was then introduced into the uterus of a foster mother, and the embryo was allowed to develop to term. Living young were born. Another method for removing mouse egg membranes is to treat the egg with pronase, a proteolytic enzyme preparation. After the removal of the membranes the blastomeres are sticky, and two or more embryos can be made to fuse just by bringing them into contact with one another (Mintz, 1962.)

It has not been feasible so far to trace the fate of individual blastomeres in the process of fusion of the embryos. The possibility cannot, therefore, be altogether excluded that during fusion like goes to like and that blastomeres destined to form trophoblast and inner cell mass sort themselves out, similar to the sorting out of dissociated cells of the early amphibian embryo (p. 231). On the other hand, the evidence is strongly in favor of the view that the inner mass cells are completely devoid of any determination in respect to developing the various tissues and organs of the embryo. This is clearly shown by some further modifications of the original experiment.

The method of fusing two mammalian embryos into one opens up possibilities of studying a variety of problems involving development. It allows, for instance, the fusion of two embryos of different sexes. (The genetic sex is, of course, determined at the time of fertilization!) In fact, intersexes, individuals with both male and female gonads, have been produced in this way (Tarkowski, 1961; Mintz, 1968). Embryos with different genotypes can be joined together to produce "chimaeras" (the term comes from Greek mythology and refers to monsters having parts of different animals). The term "allophenic" has also been coined to describe animals made up of cells with different genotypes (Mintz and Silvers, 1967). The "chimaeras" or allophenic organisms can also be used for the study of gene expression (Mintz, 1964), immunological tolerance (Mintz and Silvers, 1967), etc.

In connection with the problem of early determination in mammalian eggs, the following experiment is of particular interest. Fusion was performed of eggs derived from differently colored strains of mice: albinos and colored mice. It was found that individuals resulting from the fusion developed a fairly regular pattern of areas with white and pigmented hair (Mintz, 1967) (Fig. 210). This result shows that the cells derived from the two eggs were thoroughly mixed at the time of the formation of the embryo, making it quite impossible for individual blastomeres to have been determined at an early stage. That the white and pigmented areas were in the form of transverse stripes is an obvious result of the paths of melanocyte migration from the dorsal toward the ventral side of the embryo. (See Section 12–3.)

A further step in the progressive differentiation in the mammalian blastocyst is the appearance of a layer of very flat cells on the interior surface of the inner cell mass, that is, on the surface facing the cavity. This layer of flat cells corresponds to the lower layer of cells of the chick blastoderm, the **hypoblast** (Fig. 207). The cells represent the presumptive endoderm, or at least part of it.

The origin of the hypoblast (endoderm) cells is a subject of some controversy. The more generally accepted view is that the endodermal cells are split off from the inner cell mass. There are indications, however, that the endodermal cells in some mammals, such as the elephant shrew (van der Horst, 1942), or even in all mammals (Dalcq, 1954) are derived from the cells of the trophoblast. Some of these, near the edge of the inner cell mass, migrate inward along the internal surface of the inner cell mass and arrange

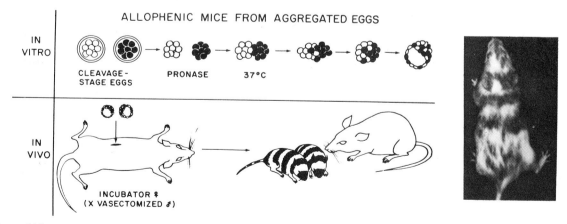

Figure 210. Production of mice consisting of "mixed up" cells from two eggs, one of an albino strain and another from a colored strain. *Left,* Diagram of the experimental procedures. *Right,* Actual photograph of a mouse developed in this way, with white and colored stripes across its back. (From Mintz, 1967.)

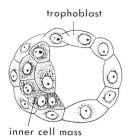

trophoblast

inner cell mass

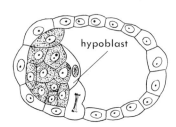

hypoblast

Figure 211.   Origin of hypoblast (endoderm) in a rat blastocyst. (After Dalcq. 1954.)

themselves in a continuous layer, as previously stated (Fig. 211). Unfortunately, the methods of local vital staining or marking cells could not as yet be applied to the study of the migration of cells which give rise to the hypoblast in mammals. Thus, the conclusion must be reached on the grounds of differences in the staining reactions of the various cells. It is important to note, therefore, that the early endodermal cells in their lack of cytoplasmic basophilia and in their negative reaction for alkaline phosphatase differ from the inner mass cells and resemble the cells of the trophoblast (Dalcq, 1954), thereby making their origin from the latter fairly plausible.

The trophoblast corresponds in position to the chorion of the later embryos of reptiles and birds, but there is the difference that the chorion, in its typical form, develops in conjunction with the amnion after all the germinal layers have been segregated, whereas the trophoblast precedes the germinal layers in its development. Obviously this is so because in mammalian development the connection with the maternal body must be established at an early stage. The hypoblast cells are at first found in the region of the inner cell mass only, but later they spread along the inner surface of the trophoblast and surround the internal cavity of the blastocyst, much as the endodermal cells in reptiles and birds surround the mass of uncleaved yolk. The inner cavity of the blastocyst thus acquires a resemblance to the yolk sac of the lower amniota, except that instead of being filled with yolk it is filled with fluid. In spite of this difference, the cavity within the enveloping endodermal layer is referred to as the yolk sac. As the hypoblast spreads out to enclose the cavity of the yolk sac, the inner cell mass also spreads out and becomes arranged into a plate, resembling the epiblast of the blastodisc of reptiles and birds. The arrangement of the formative cells in the mammalian blastocyst is at this stage similar to that in the avian blastodisc prior to the appearance of the primitive streak. Again, there is the difference that the blastodisc of a bird lies on the surface of the sphere of uncleaved yolk, whereas under the blastodisc of a mammal there is only the fluid filling the yolk sac. The situation in mammals, that after a cleavage which is not discoidal the formative cells become arranged in a disc, can only mean that the ancestors of the mammals had large yolky eggs with meroblastic discoidal cleavage.

In eutherian mammals, the blastodisc is primarily formed underneath the layer of trophoblast cells. In those mammals whose development may be considered to be more primitive (e.g., in the rabbit, the ungulates, most insectivores, and lemurs), the layer of trophoblast over the blastodisc (**Rauber's layer**) disappears, so that the epiblast becomes temporarily superficial (Fig. 218*A*). In the higher mammals, however, Rauber's layer does not disappear in the embryonic area; the formative cells are therefore never exposed to the exterior (Figs. 212 and 218*B*).

The blastodisc, consisting of an epiblast and a hypoblast, becomes quite sharply delimited from the remainder of the embryo (Fig. 212). The epiblast consists of a thick plate of columnar cells, clearly distinguishable from the flatter and more irregularly arranged cells of the trophoblast. The hypoblast cells on the underside of the blastodisc

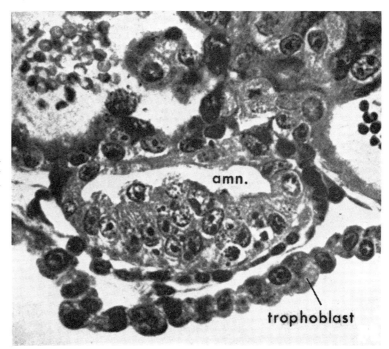

Figure 212. Monkey embryo. Formation of the amniotic cavity by cavitation. (From Heuser and Streeter, 1941.)

may become cuboidal or even columnar, and in this aspect they differ from the cells of extraembryonic endoderm lining the internal surface of the trophoblast. The hypoblast becomes especially thick, with the cells becoming high and columnar near one edge of the blastodisc. The thickening is the prechordal plate (p. 165), and it denotes what will be the anterior end of the embryo.

The embryo now enters the phase of gastrulation. The gastrulation and formation of germinal layers in mammals resemble those in birds and in reptiles. A primitive

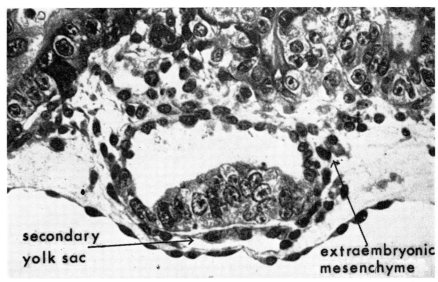

Figure 213. Monkey embryo. Development of secondary yolk sac. (From Heuser and Streeter, 1941.)

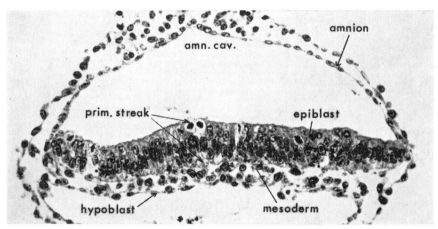

Figure 214. Monkey embryo in the primitive streak stage. Transverse section. (From Heuser and Streeter, 1941.)

streak is formed, and a Hensen's node is seen at the anterior end of the primitive streak. The primitive streak is, however, much shorter than in birds and, when fully developed, does not surpass half the length of the blastodisc, being confined to its posterior part. The cells of the primitive streak migrate downward and sideways between the epiblast and the hypoblast (Fig. 214). Some of the migrating cells appear to join the

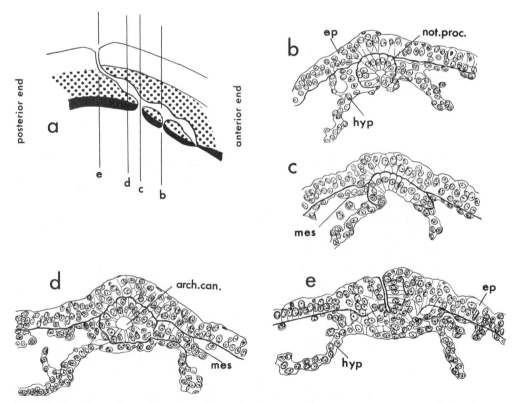

Figure 215. The archenteric canal (arch.can.) in a human embryo. *a*, Reconstructed medial section, showing the relation of the canal to epiblast (white), hypoblast (black), and mesoderm (dots). Vertical lines show the levels of sections *b* through *e*. ep, Epiblast; hyp, hypoblast; mes, mesoderm; not.proc., notochordal process. (From Grosser, 1945.)

hypoblast and thus contribute to the formation of the endoderm (Heuser and Streeter, 1941). The loose cells migrating sideways give rise to the layer of mesoderm.

The cells migrating forward from Hensen's node remain packed more closely and give rise to the "head process"—the notochordal rudiment. The notochordal rudiment in some mammals (including man) is perforated by a canal starting from Hensen's node. This is the archenteric canal which is also found in some birds, as mentioned before (p. 183), and which corresponds to the archenteron of lower vertebrates. In other mammals, either there is no archenteric canal or the canal, though present in the notochordal rudiment, does not open to the surface at Hensen's node. Where an archenteric canal is present (Fig. 215), its ventral wall fuses later with the hypoblast and is then perforated so the archenteric cavity opens into the yolk sac cavity (this also occurs in the reptiles). Subsequently, the notochord separates itself from the endoderm, and the endoderm closes underneath the notochord, forming again a continuous layer.

The primitive streak in mammals, as in birds, is a transient structure. Having given rise to the mesodermal layer and the notochordal rudiment, the primitive streak starts shrinking, its anterior end with Hensen's node receding farther and farther back, while in the anterior parts of the blastodisc the germinal layers enter the next phase of development, the formation of the primary organ rudiments. The first of these to become visible is, of course, the neural plate, followed by the somites (Figs. 216 and 217). The subsequent development of the neural tube, notochord, mesodermal segments, and so on will be treated together with the development of these parts in other vertebrates.

We must now turn to the relations between the developing embryo and the maternal body.

10–4 RELATIONS BETWEEN THE EMBRYO AND THE MATERNAL BODY IN MAMMALS

The eggs of mammals are fertilized in the funnels of the oviducts or in the uppermost part of the oviduct. Propelled by the peristaltic action of the oviduct, the fertilized eggs travel slowly down the oviduct, and the cleavage of the eggs, or at least the first stages of cleavage, take place during the sojourn of the eggs in the oviduct. During this time the eggs are still surrounded by the zona pellucida and, at first, also by the corona radiata—the follicle cells clinging to the egg membrane when the egg leaves the ovary. These cells, however, fall away rather soon. The embryo remains in the oviduct for a few (three to five) days. Toward the end of this period the blastocyst stage is attained in most mammals, and usually at this stage the embryo enters the uterus.

During cleavage and up to the morula stage the volume of the embryo remains roughly the same as the volume of the egg. With the appearance of the cavity in the blastocyst, however, the diameter of the embryo increases, in some species quite considerably. The zona pellucida becomes stretched and attenuated and eventually ruptures, though this is by no means a purely mechanical process. The blastocyst, which is by now in the uterine cavity, is set free.

The accumulation of fluid within the blastocyst implies that there is intake of water from the fluids filling the oviduct and the uterus. There is no doubt that dissolved substances are also taken in by the embryo and that some of these are essential for its development. This is an unescapable conclusion from the fact that embryos cultivated *in vitro* are very sensitive to the composition of the culture medium and that complex organic materials (provided in the form of blood serum in the culture medium) are indis-

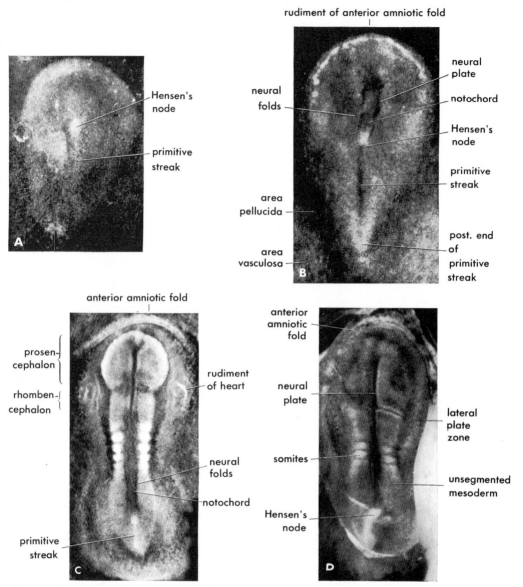

Figure 216.   Rabbit embryos in gastrulation and early organogenesis stages. *A*, Embryo with fully developed primitive streak, age 7¼ days. *B*, Embryo with neural plate anteriorly but with primitive streak in posterior half, age 8⅔ days. *C*, Embryo with three pairs of somites, age 8¾ days. *D*, Embryo with four pairs of somites and well developed neural plate, which shows the main subdivisions of the future neural tube. (Courtesy of Professor F. Seidel, 1960.)

pensable for the development of the embryo through the cleavage and blastocyst stages. (See further, p. 283.)

The nutrition of the developing embryo at the expense of the secretions of uterine glands was, of course, the transitional stage between ovoviviparity and viviparity in the ancestors of modern mammals.

The metabolism of the embryo in these early stages shows certain peculiarities. In the earliest stages of cleavage the embryo cannot use glucose as a source of energy but uses pyruvate instead. Only after the eight-cell stage can the embryo utilize glucose

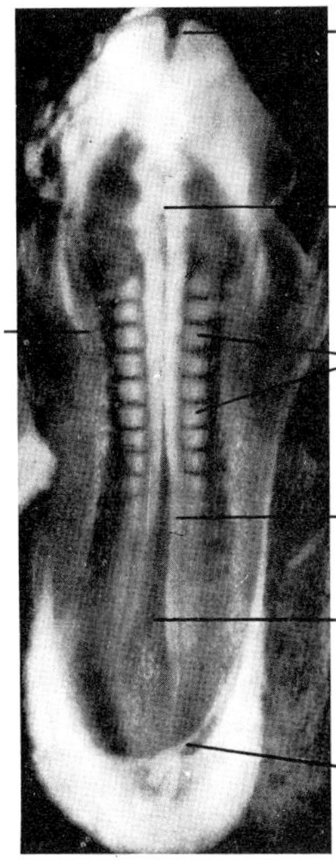

Figure 217. Rabbit embryo with neural tube closing and nine pairs of somites. (Courtesy of Professor F. Seidel.)

(Marx, 1973). The full assortment of oxidative enzymes becomes functional only after implantation of the blastocyst.

Implantation and the Development of Extraembryonic Parts. After the embryo has entered the uterus the next important step in the development of the mammalian egg must take place; the embryo becomes attached to the wall of the uterus, where it is to perform its subsequent development. This attachment of the blastocyst to the uterine wall is known as **implantation,** and it is carried out by the cells of the trophoblast. Two basically different types of attachment may be found. In most mammals, the blastocyst becomes attached to the surface of the uterine epithelium and therefore lies in the cavity of the uterus. In a smaller number of mammals—in man, but also in some of the rodents—the blastocyst penetrates deep into the uterine wall, and the development of the embryo takes place *inside* the uterine wall. The epithelium lining the cavity of the uterus at the site of implantation becomes destroyed. The destruction of the uterine epithelium is due to the activity of the trophoblast cells. The embryo is capable of imbibing the fluid filling the uterus and of deriving from it nutrient substances, just as is the case in the ovoviviparous dogfish, but this source of nutrition is insignificant as compared with the food supply that the embryo gets by diffusion from the maternal blood vessels, once the connection between the embryo and the uterine wall is established. The formation of this connection is called **placentation,** and it will be dealt with in the following section.

At the time of implantation the embryo is still in the blastocyst stage, and the processes of gastrulation take place only after implantation. The development of the embryonic membranes occurs roughly during the same period. This differs to a most amazing degree (see Mossman, 1937; Goetz, 1938) in various groups of mammals, so that only the more general features of this process can be indicated here. If the blastodisc is exposed to the exterior by the disappearance of Rauber's layer in the embryonic area, the amniotic folds may develop around the embryo proper in much the same way as in reptiles and birds. After the fusion of the amniotic folds, the body of

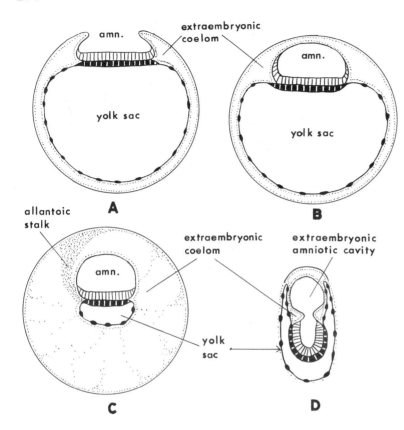

Figure 218. Diagrammatic comparison of the relation of the embryonic and extraembryonic parts (amnion, yolk sac) in a shrew (*A*), a bat (*B*), a human (*C*), and a mouse (*D*).

the embryo becomes enclosed in the amniotic cavity. A body fold later separates the embryo from the yolk sac, with which the embryo remains connected by means of the umbilical cord. The outer wall of the amniotic fold becomes the chorion. On the periphery, in the extraembryonic area, the ectoderm of the chorion is continuous with the remaining parts of the trophoblast. The trophoblast thus becomes a part of the chorion; it actually forms the epithelial layer of the chorion over most of the extraembryonic surface of the blastocyst. The mesoderm spreading from the embryonic area may supply this part of the chorion with a connective tissue layer (Fig. 218*A*).

The development of the amniotic cavity in the manner just described occurs relatively late, after gastrulation and the formation of primary organ rudiments. We find this method of amnion formation in the marsupials, ungulates (pigs), most insectivores, and lemurs. In other mammals, a tendency may be observed for the cells giving rise to the embryo proper to sink deeper into the blastocyst, so that the formative cells do not appear on the surface. Under these conditions the amniotic cavity can no longer be formed by the convergence of amniotic folds. Instead, the amniotic cavity is then formed by a crevice appearing between cells in the future blastodisc area. This mode of formation of the amniotic cavity is known as **cavitation** (Fig. 218*B* and *C*).

The split may occur either between the trophoblastic layer and the inner cell mass or inside the inner cell mass. In the first case (Fig. 212), the amniotic cavity is covered only by the trophoblast, and an epithelial lining representing the amnion proper must be developed later, either at the expense of the edges of the blastodisc (in the bats) or at the expense of cells derived from the internal surface of the trophoblast (in higher primates, according to Heuser and Streeter, 1941). The amniotic cavity appears relatively earlier; its formation precedes gastrulation, instead of following it, as in the mammals having a more primitive type of development. In some rodents, such as the mice and rats, the tendency for the formative cells to sink deeper into the blastocyst goes even further, and they become separated from the surface of the blastocyst not only by the Rauber's layer but, in addition, by a mass of cells which later become extraembryonic ectoderm. In this case, the amniotic cavity is formed (by cavitation) deep in the interior of the inner cell mass and is from the start separated from the trophoblast by cells of extraembryonic ectoderm (Fig. 218*D* and Fig. 222*A*). Further variations of amnion development have been described; references to these descriptions may be found in the papers by Mossman (1937) and Goetz (1938).

The development of the allantois also varies a great deal within the class Mammalia. With the acquisition of viviparity, the original function of the allantois as a urinary bladder becomes altogether lost. As the maternal organism supplies the embryo with all the necessary nutrition, it also takes over the removal of the waste products of metabolism from the embryo. Not only does the carbon dioxide produced by the embryo diffuse into the maternal blood, but the end products of protein metabolism also find their way into the maternal blood and are excreted by the kidneys of the pregnant mother. This, of course, is possible only if the end products of metabolism are readily soluble. Uric acid, owing to its insolubility in water, is the most suitable end product of protein metabolism in the case of the cleidoic eggs of reptiles and birds, but in the mammals we find that the end product of protein metabolism is urea which, owing to its solubility, can be passed from the embryo to the mother and can thus be disposed of. As a result, the storage of these waste products in the allantoic cavity is no longer necessary.

On the other hand, the function which was secondary in the case of birds and reptiles retains its importance for the mammalian embryo. The mammalian embryo, enclosed in the amniotic cavity and surrounded by the amnion and the chorion (trophoblast), is still in need of a pathway by which oxygen from outside (from the maternal organism) may be transported to its tissues. This pathway, in the higher mammals, is still supplied by the blood vessels of the allantois. Moreover, besides supplying the embryo with oxygen, the allantoic circulation now supplies the embryo with nutrient substances as well, since these can no longer be drawn from the yolk sac.

The result is that the endodermal part of the allantois (the epithelial lining of the allantoic vesicle) gradually becomes reduced in the higher mammals, while the mesodermal part, giving rise to the blood vessel system of the allantois, develops progressively. The endoderm of the allantoic stalk still fulfills a useful function, as it forms a bridge, growing out from the embryonic body and reaching the chorion, along which the mesoderm spreads outward and establishes a connection with the trophoblastic epithelium. The expansion of the endodermal vesicle at the outer end is, on the other hand, wholly superfluous. The development of the allantoic mesoderm, however, may proceed even in the total absence of the endodermal component (in the guinea pig).

Two special cases will be mentioned here. In the monkeys and in man, a typical blastocyst is formed with a cavity which becomes lined by hypoblast (endodermal cells). The blastocyst becomes attached to the wall of the uterus by its dorsal side, that is, by the side bearing the internal cell mass (Fig. 219). In man, the blastocyst sinks into the uterine wall, as a result of the destruction of the maternal tissues by the trophoblast, and the opening through which the blastocyst enters is closed by a clot of coagulated blood plasm and later is covered by a sheet of regenerated uterine epithelium. In the monkey, the blastocyst does not enter the uterine epithelium but, instead, becomes attached by its ventral side to the opposite uterine wall.

The amniotic cavity is formed by cavitation as described on page 276, and the epithelium surrounding it is of very unequal thickness throughout. The roof of the am-

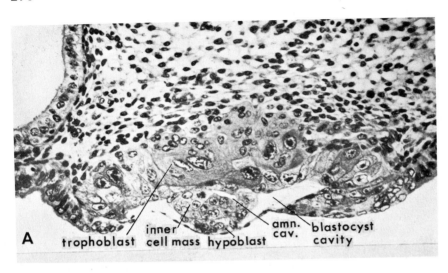

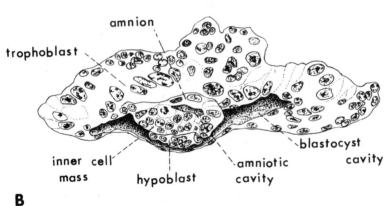

Figure 219.   Implantation of the 7-day-old human blastocyst. *A*, Photograph of section. (From Hertig and Rock, 1945.) *B*, Only the embryonic parts are shown, drawn as if the embryo were dissected in the plane passing through its main axis.

niotic cavity is formed by a fairly thin layer of epithelium; this is the amniotic ectoderm. The floor of the amniotic cavity consists of a thick plate of cells arranged in a layer of columnar epithelium. This plate, with the underlying layer of hypoblast (endoderm) which is also thickened, constitutes the bilaminar blastodisc, inside of which the primitive streak later appears and from which the body of the embryo proper later develops (Fig. 220).

In both monkeys and man, at an early stage of development, even before the primitive streak is formed, numbers of loose mesenchyme-like cells appear between the formative parts and the trophoblast. These cells, which may be compared to extraembryonic mesenchyme, are supposedly derived from the trophoblast (Heuser and Streeter, 1941), whereas in most mammals extraembryonic mesoderm spreads out from the blastodisc and has the same origin as the embryonic mesoderm (i.e., it is derived from the primitive streak). The delay in the development of the primitive streak in higher primates, in conjunction with the need for accelerating the development of the extraembryonic parts which are required to supply the embryo with food and oxygen, must have made it necessary to provide for a second unusual source of extraembryonic mesoderm.

The extraembryonic mesoderm rapidly increases in quantity and gradually fills the cavity of the blastocyst, which was recognized as the yolk sac. A secondary, smaller yolk sac is then formed by the hypoblast cells connected to the blastodisc (Fig. 213). The cavity of this secondary yolk sac is for some time much smaller than the amniotic cavity, but it catches up with the amniotic cavity later (at the time when the primitive streak appears). In the meantime, the mass of extraembryonic mesenchyme separates to form a cavity of its own, which is, of course, the extraembryonic coelom; the mesenchyme cells arrange themselves into a mesothelium lining this coelom. The extraembryonic coelom now becomes the greatest cavity in the blastocyst and takes up most of the space inside the trophoblast. In the area adjoining the posterior end of the embryo, the extraembryonic mesenchyme becomes concentrated and forms a strand leading from the blastodisc to the area of trophoblast, which enters into the formation of the placenta. The strand of mesoderm known as the **connecting stalk** or **allantoic stalk** represents the mesodermal component of the allantois, and through this strand the allantoic circulation is established later (Fig. 221). Here again, it can be noted that the necessity for early provision of the embryo with food and oxygen supplies has brought about a short cut in the established pattern of the developmental processes. The endodermal part of the allantois—the allantoic vesicle (or allantoic diverticulum)—becomes superfluous and remains rudimentary, never reaching the placenta in monkeys and man.

The second type of modified mammalian development is that found in mice, rats, and related rodents. The initial stages of cleavage and blastocyst formation in these animals do not differ essentially from those in other mammals. Implantation is interstitial but is achieved in an essentially different way from the implantation of a human embryo. The blastocysts entering the uterus from the fallopian tube, in which the earlier stages of development occur, find their way into crypts along the ventral side of the uterus, which in these animals is almost completely subdivided into two long "horns." Thereupon the uterine epithelium adjoining the blastocyst degenerates, and the em-

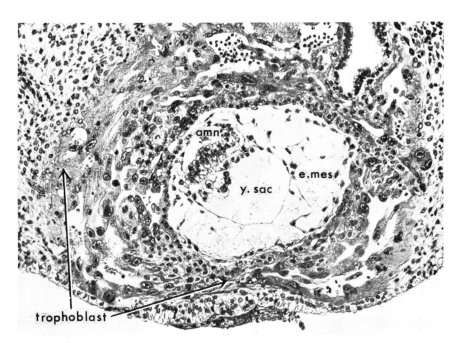

Figure 220. Implanted 11-day-old human embryo prior to primitive streak formation with amniotic cavity (amn.), secondary yolk sac cavity (y. sac), and extraembryonic mesenchyme (e. mes.). The embryo is surrounded by the trophoblast. (From Hertig and Rock, 1941.)

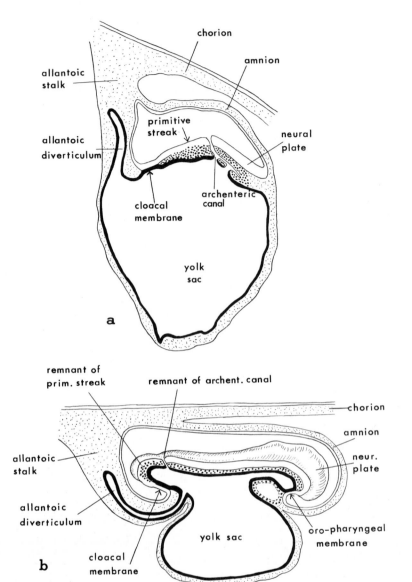

Figure 221. Semidiagrammatic median sections of the human embryo in the primitive streak stage (*a*) and the neural plate stage (*b*). The ectoderm is white, the endoderm black, the embryonic mesoderm coarsely strippled, the extraembryonic mesoderm finely stippled. (*a*, After Jones and Brewer, 1941, and Grosser, 1945; *b*, after Grosser, 1945.)

bryo comes into direct contact with the connective tissue of the uterine wall. The connective tissue then starts proliferating and completely encloses the blastocyst. The proliferation of the connective tissue is so abundant that the lumen of the uterine horn is entirely occluded in the regions in which the blastocysts come to lie, and the passage along the uterus becomes opened up again only toward the end of pregnancy.

In the meantime, the blastocyst, instead of becoming bloated by accumulation of fluid in its cavity, becomes elongated in a vertical direction (Fig. 222*A* and *B*). The embryo in this stage is known as the "egg cylinder." Both the trophoblast and the inner cell mass covered by the hypoblast (endoderm) stretch, leaving very little space between them. Eventually, the endoderm lines this cavity from the trophoblast side, converting it into the yolk sac. As the inner cell mass is attached to the trophoblast only at one end of the egg cylinder, the formative cells become almost completely surrounded by the yolk sac (Fig. 222*B*). The amniotic cavity is now formed as a rather narrow canal inside the inner cell mass. The cells of the inner cell mass form an epithelium surrounding the amniotic cavity; part of this epithelium represents the epiblast of the blastodisc. The formative cells make up about half the lining of the amniotic cavity, that half lying farther from the attachment of the egg cylinder. The other half of the egg cylinder contains only extraembryonic parts: extraembryonic ectoderm, extraembryonic endoderm, and at a later stage, also extraembryonic mesoderm.

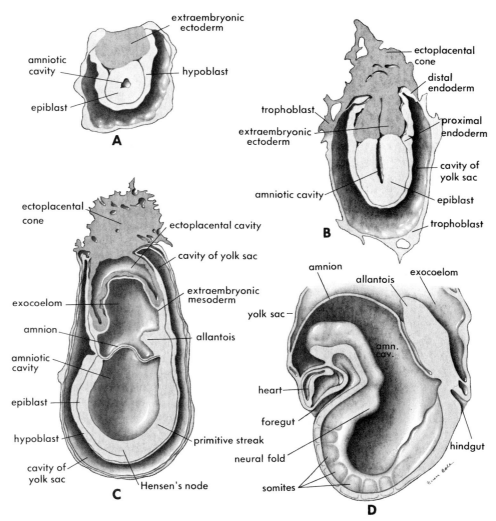

Figure 222. Four stages of development of a mouse embryo. *A*, Early egg cylinder stage; the inner cell mass has become subdivided into the rudiment of the embryo proper (epiblast) and extraembryonic ectoderm; the endodermal (hypoblast) layer has been formed. *B*, Later egg cylinder stage; the amniotic cavity has been formed by cavitation in both the embryo rudiment and the extraembryonic ectoderm. The latter has produced the ectoplacental cone, which is invading the maternal tissues; the endoderm has started spreading to cover the inner surface of the trophoblast (distal endoderm). *C*, Further development of the egg cylinder—the primitive streak stage. The cavity inside the egg cylinder has become subdivided into the amniotic cavity proper and the cavity of the ectoplacental cone, the extraembryonic coelom having developed between; the endoderm has lined the inner surface of the trophoblast, thus completely surrounding the yolk sac cavity. *D*, Embryo with open neural plate and seven pairs of somites; the embryo is curved around the amniotic cavity and is concave dorsally. The allantois is growing from the posterior end of the embryo into the extraembryonic coelom. (Modified from Snell, 1941.)

At an early stage, the extraembryonic ectoderm lying at the upper end of the egg cylinder starts proliferating and forms a mass of cells called the **ectoplacental cone.** This mass of cells invades the maternal connective tissues and establishes close contact with the maternal blood vessels. This is the beginning of the formation of the placenta. The amniotic cavity, originally stretching throughout the length of the egg cylinder (Fig. 223), later becomes constricted in the middle. One part of the cavity is now almost completely surrounded by formative cells (the cells of the epiblast), while only a small part of the lining of this cavity can be recognized as being amniotic ectoderm.

The other part of the original amniotic cavity lies immediately underneath the ectoplacental cone and is therefore referred to as the ectoplacental cone cavity. When the ectoplacental cone cavity becomes separated from the amniotic cavity proper, a further cavity is formed between them which is entirely surrounded by mesoderm and represents the **extraembryonic coelom** or "**exocoelom**" (Figs. 222C and 223). As the exocoelom enlarges, the ectoplacental cone cavity is squashed between the exocoelom and the ectoplacental cone, and eventually it disappears. The structure of the embryo at this stage may be described as resulting from an invagination of the bilaminar blastodisc into the yolk sac. The dorsal surface of the blastodisc, instead of being flat as in most other mammals, is concave, and the ventral side (lined by the hypoblast) is convex and surrounded by the yolk sac cavity (Fig. 222C).

In this abnormal position, the embryo proceeds to the formation of the primitive streak, which gives rise to the mesoderm of the embryo. The neural tube, mesodermal somites, the heart, and other parts of the embryonic body are developed in the usual way, but the body of the embryo, instead of being surrounded by the amniotic cavity, is curled around it with the dorsal side facing the cavity, while its convex ventral surface is surrounded by the yolk sac (Fig. 222D). Only at a later stage, during the eighth day of intrauterine life in the case of the mouse (total duration of pregnancy—20 days),

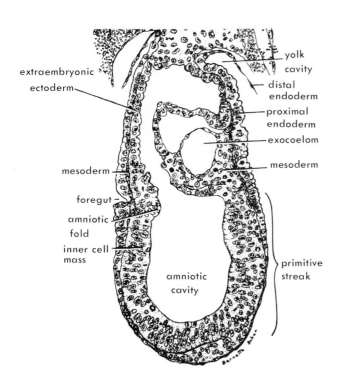

Figure 223. The primitive streak stage (section) in the development of the mouse. (Modified from Snell, 1941.)

does the embryo start twisting itself out of this unusual position, beginning with the head end, and eventually assumes the normal position in the amniotic cavity, with the body convex dorsally and concave ventrally. The amniotic wall becomes stretched so that the amniotic cavity surrounds the embryo. Before this happens, however, the allantois is formed as a proliferation of mesenchyme at the posterior end of the embryo. The allantois grows into the extraembyronic coelom and eventually establishes contact with the ectoplacental cone and enters into the formation of the placenta. The allantois in mice and rats is completely devoid of endodermal parts and consists only of connective tissue and blood vessels, which eventually serve to establish circulation between the embryo and the placenta. (For further details see Snell, 1941.)

We have seen that with modern improved methods mammalian eggs can be made to develop *in vitro* from fertilization (even from maturation) through cleavage to the formation of a blastocyst. The challenge remains of attempting to obtain further development outside the body of the female from the blastocyst onward, and as an eventual goal, to allow the fetus to develop to a stage in which it is normally born, without having to return the embryo to the uterus of a female.

The fulfillment of such a task is fraught with tremendous difficulties. The first step would be to find a means of bringing the embryo through the stage at which the implantation occurs. This has, in fact, been achieved with mouse embryos. Apart from providing a suitable chemical environment, the problem lies in finding a substitute for the uterine wall into which the embryo, or at least the trophoblast, may become embedded, as it does during implantation in normal development. It might be possible to cultivate *in vitro* a piece of uterine wall and to allow the embryo to implant into it. The best results, however, have been achieved with an artificial substrate, a layer of reconstituted rat tail collagen (Hsu, 1973). To prepare this substrate large tendons, consisting of collagen fibers, are removed from the tails of rats, and the collagen is dissolved in dilute acetic acid. The solution is made to set by exposing it to ammonia vapor, and it becomes a soft gel with a high water content. The gel ("reconstituted collagen") can be made to form a layer on the bottom of a dish, and the cultivated embryos are placed on top of this surface, which is immersed in a fluid culture medium.

Under these conditions mouse blastocysts attached themselves to the surface of the collagen, and the trophoblast invaded the collagen. Inside the blastocyst the inner cell mass started to differentiate into an "egg cylinder" and produced first the yolk sac, then proceeded through the stage of primitive streak formation, and eventually reached the organogenesis stage with a well-formed closed neural tube and several pairs of somites. The heart was formed and started pulsating, and circulation was established in a blood vessel network on the yolk sac. This condition is normally reached after nine days of gestation *in utero*. After this stage, however, the embryos degenerated and underwent autolysis.

During the period of cultivation the embryo was found to be extremely sensitive to the kind of culture medium that was used. In addition to the medium developed for cultivating mammalian tissue cells *in vitro*, the inclusion of blood serum of a specific kind was found to be indispensable. In early cleavage, calf serum was used, but at blastocyst stage this had to be replaced by fetal calf serum. The last stages, the organ formation in the embryonic disc, occurred only when human fetal blood serum was added. (This was obtained by drawing blood from the umbilical cord and placenta after human births.)

Although in the experiments just described the embryos were made to develop well beyond the implantation stage, the limitations of the method stand out very starkly: the embryo did not develop a blood vessel system invading the substrate (reconstituted collagen), which was supposed to represent the uterine wall. Thus, no step was made to form a substitute for the placenta, through which the embryo is nourished and sustained through the middle and later stages of development and growth.

10–5 PLACENTATION

The term **placenta** applies to any type of organ, composed of maternal and fetal tissues jointly, which serves for the transport of nutrient substances from the tissues of the mother into those of the embryo. Placentae are not found exclusively in mammals but appear also in animals belonging to various groups of the animal kingdom, such as *Peripatus* (Protracheata), *Salpa* (Tunicata), *Mustelus laevis* (Elasmobranchia), and certain lizards. The nature of the tissues entering into the formation of the placenta is not the same in all cases. Prerequisites for participation in the development of the placenta are the superficial position of the part in question and the possiblity of the formation of a blood vessel network, which may undertake the transport of nutritive substances from the surface of contact with the maternal tissues to the developing embryo (Fig. 224).

In mammals there exist two essentially different main types of placentae. In the first type, the connection is established between the uterine wall and that part of the chorion which is lined on the inside by the yolk sac with its network of vitelline blood vessels. This type of placenta is called the **choriovitelline placenta.** It is found in some of the marsupials *(Didelphys, Macropus).* In the second type of placenta, the connection is established by the chorion lined by the allantois, with the allantoic blood vessel system taking over the transport of substances from the mother to the embryo (Fig. 224). This type of placenta is called the **chorioallantoic placenta,** and it is found in all the Eutheria as well as in some marsupials *(Parameles, Dasyurus).* Remnants of the choriovitelline placenta may be found either temporarily or even permanently in higher mammals, playing a subsidiary part in their placentation.

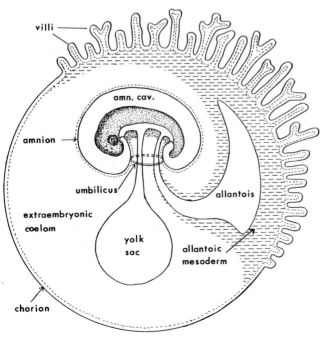

Figure 224. Embryo of placental mammal, surrounded by embryonic membranes (diagrammatic).

When the mammalian embryo enters the uterus, the zona pellucida, which previously surrounded it, becomes dissolved, and the embryo, now in the blastocyst stage, is bathed by the uterine fluid. This fluid contains organic substances produced by the tubular glands of the uterine wall and voided into the lumen of the uterus. The early embryo may absorb some of these substances through its epithelium covering so long as a closer connection with the uterine wall has not been established. For its further development, however, the embryo is completely dependent on substances supplied to it from the tissues of the uterine wall. To insure a steady and abundant flow of these substances, a close relationship between the embryonic tissues and the maternal tissue is essential. Nevertheless, it is found that the closeness of this connection between fetal and maternal tissues differs greatly within the class of mammals and even within the subclass of true placental mammals (Eutheria).

In most mammals, as mentioned on page 275, the implantation is superficial; that is, the blastocyst remains lying in the cavity of the uterus in contact with the uterine wall (Figs. 225 and 226). The contact may be made more intimate by the surface of the blastocyst forming finger-like outgrowths that penetrate into depressions in the wall of the uterus. Such outgrowths are initially formed by the trophoblast (the epithelial layer covering the blastocyst), but connective tissue and blood vessels later enter the outgrowths. The outgrowths are known as the chorionic villae, and the blood vessels are ramifications from the allantoic blood vessels (in the case of chorioallantoic placentae) (Figs. 224 and 226) or the vitelline blood vessels (in the case of the more primitive choriovitelline placentae). The chorion with its villi and the maternal wall of the uterus jointly represent the placenta in this case. At the time of parturition the chorionic villae are simply drawn out from the depressions in the wall of the uterus, and thus maternal and fetal tissues are separated without further damage to the uterine wall and without any bleeding taking place. This type of placenta, found in pigs, cattle, and some other animals, is known as a **nondeciduous placenta.**

In other mammals, however, the degree of connection between maternal and fetal tissues becomes more intimate. The wall of the uterus becomes eroded to various degrees through the action of the trophoblast, and the embryonic tissues penetrate into the uterine wall, establishing a more intimate contact and facilitating the passage of substances from the mother to the fetus and from the fetus to the mother. (See further, Section 10–6.) At the end of pregnancy the uterine wall is no longer intact, and when the fetus with its membranes, including the chorion, is removed, more or less extensive hemorrhage from the uterine wall ensues. Placentae of this type are known as **deciduous placentae,** and the uterine wall participating in the formation of such a placenta is called the **decidua.**

An extremely intimate contact between fetal and maternal tissues develops early in cases of **interstitial implantation.**

A human blastocyst, after establishing contact with the uterine wall, destroys the adjacent epithelial layer of the uterus and sinks into the underlying connective tissue. By the ninth day after fertilization this process is completed, and the uterine wall closes over the implanted blastocyst. Thereupon, the trophoblast, which is in immediate contact with the maternal tissues, starts proliferating profusely and sends outgrowths penetrating still deeper into the uterine wall. In the more external-lying part of the trophoblast, cell boundaries disappear, and the mass of trophoblast becomes a syncytium, the **syncytiotrophoblast.** The deeper-lying portion of the trophoblast (i.e., the portion nearer the embryo) retains its cell boundaries and is known as the **cytotrophoblast.** At the same time, the trophoblast ceases to be a compact mass and becomes permeated by a system of cavities—the trophoblastic **lacunae** (Fig. 220). As a result the trophoblast, especially the syncytiotrophoblast, becomes converted into a meshwork of

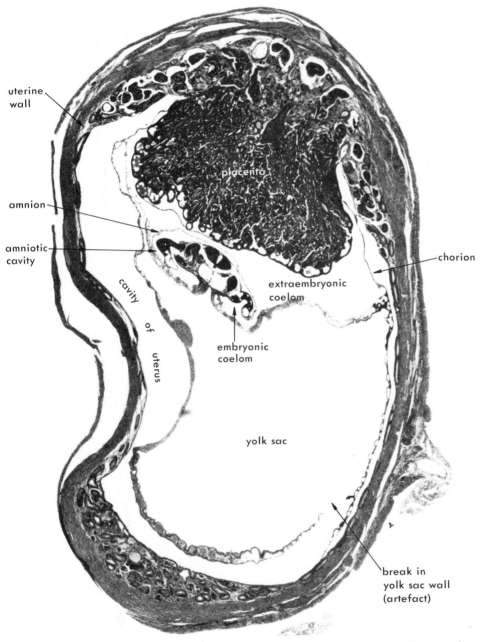

Figure 225.   Embryo of an insectivore (*Eremitalpa*) in the uterus. Section at the level between the anterior and posterior intestinal portal.

irregular strands with interstices in between. As the whole mass of trophoblast increases and thus penetrates deeper into the uterine wall, it reaches the small blood capillaries. In the same manner in which the trophoblast destroyed the uterine epithelium, it now breaks down the endothelium of the capillaries, and a communication becomes established between the lumen of the capillaries and the trophoblastic lacunae. Blood flows into the lacunae, and in this way nutritive substances contained in the blood become available to the fetal tissues over a very large surface of the reticulated trophoblast. As both venous and arterial branches of the blood vessels come into communication with the trophoblastic lacunae, blood circulation becomes established, fresh blood being pumped into the lacunae and removed from there via the veins of the uterus.

Initially the strands of trophoblast bathed by the maternal blood in the lacunae consist almost completely of syncytiotrophoblast. During the third week following fertilization, these strands become completely reorganized. First, the cytotrophoblast

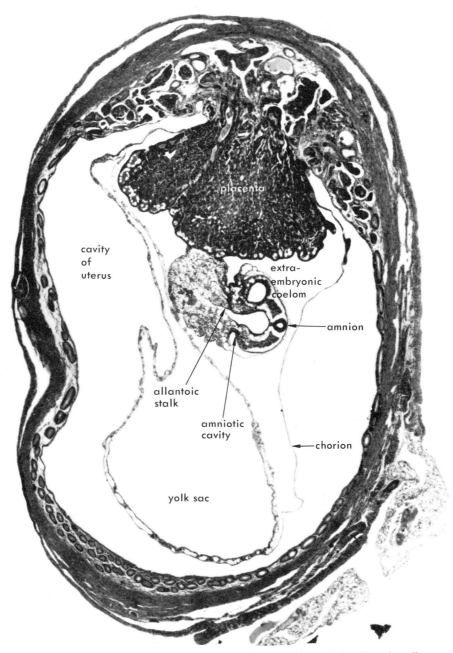

Figure 226. Same embryo as in Figure 225. Section at the level of the allantoic stalk.

penetrates into the strands, forming a cellularized core within the syncytial mass. Next, extraembryonic connective tissue underlying the trophoblast penetrates the cytotrophoblast strands and is closely followed by blood vessels, which are extensions and ramifications of the allantoic blood vessels. Lastly, the shape and arrangement of the trophoblastic strands change; instead of forming an irregular mesh, they become converted into a dendritic system, with main stems arising from the surface of the blastocyst and finger-like ramifying branches spreading out into the blood-filled trophoblastic lacunae (Fig. 227). The individual projections can now be recognized as "villi," although their mode of origin is different from the simple villi developed by lower animals. Most of the villi terminate freely in the lacunar space, but some reach the maternal tissues and serve to anchor the whole fetal system in the uterine wall. The uterine wall, in turn, becomes highly vascularized and thickened, compared with the preimplantation condition.

We will now consider the relation of the fetus and its membrane to the uterus as a whole. The normal site of implantation in the human being is in the upper part of the uterus, most often on one side of it. Since the early embryo is very minute (the blastocyst at the time of implantation is only about 0.2 mm. in diameter), it disappears within the uterine wall without causing a marked distortion of the latter. As the embryo, including the trophoblast, begins to grow, it forms a conspicuous swelling of the uterine wall, a swelling which encroaches on the uterine lumen. The blastocyst originally attaches itself to the uterine wall by the side containing the inner cell mass (Fig. 219), but after penetrating into the uterine wall it becomes surrounded by an almost uniform layer of syncytiotrophoblast and cytotrophoblast. The trophoblastic lacunae and subse-

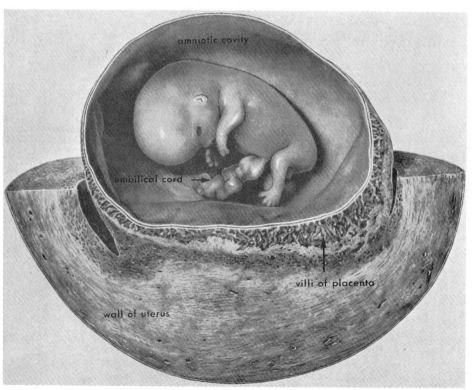

Figure 227.   Human fetus about 1½ months inside amniotic cavity. Part of the uterus cut away. (From Hamilton and Boyd, 1960.)

quently the ramified villi develop initially all over the surface of the blastocyst. The efficiency of the villi depends naturally on an adequate blood supply from the maternal side. This, however, becomes unequal as the fetus with its cavities and membranes increases in size. The side of the blastocyst facing the uterine wall is in a better position, because the maternal tissues here have a good blood supply. Both the system of embryonic villi and the uterine wall, therefore, develop progressively and form the actual functional placenta, which becomes disc-shaped (discoidal placenta). On the side of the blastocyst facing the uterine lumen, the blood supply is less efficient, and as a result the system of villi on the "outer" surface of the blastocyst becomes reduced, until eventually the villi here disappear completely.

Inside the trophoblast the actual embryo initially is relatively small, most of the space being taken up by the yolk sac, the extraembryonic mesenchyme and the extraembryonic coelom (Figs. 220 and 225). During the second month of pregnancy the amniotic cavity begins to enlarge and eventually displaces the extraembryonic coelom; the amnion is then pressed close to the inner surface of the chorion. The embryo is suspended in the amniotic fluid, attached by the umbilical cord to the disc-shaped placenta (Fig. 227). Toward the end of term the growth of the fetus catches up with the enlargement of the amniotic cavity and fills the whole cavity.

10–6 REVIEW OF PLACENTAE IN DIFFERENT GROUPS OF MAMMALS

Differences in placentae of various mammals are caused in part by the structure and arrangement of the villi and in part by the degree of connection between the maternal and fetal tissues. In the more primitive Eutherian mammals the placentae are nondeciduous. The villi may be scattered all over the surface of the chorion, in the case of the pig, and this type of placenta is known as the **diffuse placenta.** In cattle, the villi are found in groups or patches, while the rest of the chorion surface is smooth. The patches of villi are called **cotyledons,** and a placenta of this type is known as the **cotyledon placenta.** In carnivores, the villi are developed in the form of a belt around the middle of the blastocyst, which is more or less elliptical in shape. This is the **zonary placenta.** In man and the anthropoid apes, the chorion is, at first, all covered with villi, but the villi continue developing only on one side, the side turned away from the lumen of the uterus, while on other parts of the chorion the villi are reduced. The functional placenta therefore has the shape of a disc and is known as the **discoidal placenta.** A discoidal placenta has also been developed independently in the rodents (mouse, rat, rabbit, and others). In the monkeys, the placenta consists of two discs—a **bidiscoidal placenta.**

The thickness of the partition between the maternal and fetal blood may be diminished by the removal of some of the intervening layers of tissue. Depending on which layers have disappeared, several types of placentae may be distinguished. The names given to the various types indicate the two tissues—one maternal, the other fetal—which are in immediate contact.

In the more primitive cases, the following layers of tissue participate in the diffusion of substances from the mother to the embryo:

1. The blood of the mother.
2. The endothelial wall of the maternal blood vessels.
3. The connective tissue around the maternal blood vessels.
4. The uterine epithelium.

5. The epithelium of the chorion.
6. The connective tissue of the chorion.
7. The endothelial wall of the blood vessels in the chorion.
8. The blood of the embryo.

If all these tissues are present in the placenta, the chorionic epithelium is in contact with the uterine epithelium, and the placenta is designated as an **epitheliochorial placenta.** This type of placenta is found in all marsupials and in the ungulates (horses, pigs, cattle). In the case of an epitheliochorial placenta the villi, in their growth, push in the wall of the uterus and later lie in pocket-like depressions of the uterine wall.

When the blastocyst is implanted and, subsequently, when the villi begin to grow, the superficial tissues of the uterine wall may be destroyed to a greater or lesser extent. If the destruction involves the uterine epithelium and the underlying connective tissue, the epithelium of the chorion may come into direct contact with the endothelial walls of the maternal capillaries. A placenta is then formed which is called an **endotheliochorial placenta.** It is found mainly in carnivores but also in a few other mammals.

The destruction of the maternal tissues of the uterine wall may involve the endothelium of the maternal blood vessels. The cavities of the blood vessels are then opened up, and the villi of the chorion become bathed in the maternal blood, thus facilitating gas exchange and diffusion of nutrient substances from the maternal blood into the blood vessels of the chorionic villi. This type of placenta, the **hemochorial placenta,** is found in primates and in many insectivores, bats, and rodents. Actually, the chorionic villi are surrounded by spaces (sinuses) devoid of an endothelial lining, into which maternal blood enters through the arteries of the uterus and from which the blood flows into the uterine veins. The villi may be ramified dendritic structures, or they may coalesce distally and form a more or less complicated network.

In the case of epitheliochorial placentae, at parturition the villi can be pulled out of the pockets in which they have been embedded, and the fetal part of the placenta may be removed, leaving the surface of the uterine wall intact. There is therefore no bleeding at birth. This is not the case with other types of placentae; at parturition not only the fetal component of the placenta is shed but also a part of the uterine wall participating in the function of the placenta is torn away. An open wound is left on the wall of the uterus, and hemorrhage inevitably occurs. In the latter case the placenta is said to be **deciduous,** whereas in the first case the placenta is **nondeciduous.**

The hemorrhage at parturition is normally stopped by the same mechanism that serves for the expulsion of the newborn: the contraction of the muscular wall of the uterus constricts the blood vessels and thus slows down the flow of blood, until clotting of the blood stops the hemorrhage altogether.

10–7 PHYSIOLOGY OF THE PLACENTA

In the absence of yolk in mammalian eggs, the nutrition of the mammalian embryo in the uterus is wholly dependent on the flow of supplies from the maternal body via the placenta, hence the close connections between the fetal and the maternal tissues as described in the previous section. Nevertheless, the fetal and maternal tissues in the placenta do not blend together. It cannot be stressed too much that the blood of the mother and that of the embryo do not mix under normal conditons; the maternal blood does not enter the blood circulation of the embryo or vice versa. Between the maternal and the fetal blood there exists a separation—the **placental barrier.** Physically, the barrier consists of the tissue lying between the blood spaces in the embryonic and maternal parts of the placenta; this barrier may be attenuated (as in hemochorial placentae) but is not broken down. Physiologically the placental barrier is like a semi-permeable membrane, allowing some substances to pass through but keeping out others.

Small molecule substances pass through the placental barrier by simple diffusion. This applies to water, oxygen passing from the maternal into the fetal blood, carbon dioxide and urea passing from fetal to maternal blood, simple salts of sodium potassium and magnesium, and monosaccharides. Active transport of some form or another participates, however, in the penetration of more complex substances through the placental barrier. It is a well-established fact that vitamins and hormones may pass from the mother to the fetus. Passage of some very complex substances, proteins in particular, may perhaps be effected by pinocytosis at the surface of the trophoblast. Highly complex proteins are known to be able to penetrate the placental barrier. In this way antibodies, which have developed in the blood of a mother who has acquired immunity to certain diseases, such as diphtheria, scarlet fever, smallpox, and measles, are passed to the fetus, which thus becomes passively immunized and unsusceptible to these illnesses in the first period after birth. It is worth noting that in cows, which have an epitheliochorial placenta and thus a formidable placental barrier, antibodies cannot be passed from mother to offspring through the placenta but, instead, are supplied to the newborn animal in colostrum milk after birth.

Certain pathogenic organisms and viruses are able to penetrate through the placental barrier and infect the fetus if the mother is infected. Such penetration is known to happen with syphilis (causing congenital disease) and also in infections with smallpox, chickenpox, and measles. One virus infection which has been found to be very dangerous for the embryo is rubella, or German measles (p. 228).

Many drugs used medicinally may penetrate the placental barrier and may sometimes cause most adverse effects on the embryo. Thus it is believed that the drug thalidomide, which was used as a sedative, when taken by women in early pregnancy (25 to 44 days), caused extensive deficiencies in the development of limbs, the alimentary canal (nonperforation of the anus), and the heart.

Lastly, it must be pointed out that although the tissues of the mother and fetus, including the trophoblast, do not mix and the blood streams of the two are held apart, occasional penetration of individual cells across the placental barrier is not an exceptional occurrence. Small numbers of fetal blood corpuscles are sometimes found in the maternal circulation as well as maternal corpuscles in the circulation of the embryo. This may be the result of accidental breakage of the respective blood capillaries. The origin of the corpuscles can be verified since the fetal erythrocytes are nucleated and the erythrocytes of the adult female are without nuclei.

Small fragments of the trophoblast may become detached from the chorionic villi and may be carried away by the maternal blood stream; they are later found in the blood capillaries of maternal organs, such as the lungs. On the other hand, maternal cells, probably white blood corpuscles, have been found lodged in the lymphatic system (spleen, thymus, lymph nodes, bone marrow) of the fetus (Tuffrey et al., 1969).

10–8 HORMONAL CONTROL OF OVULATION AND PREGNANCY

Ripening of the egg and its release from the ovary in mammals is believed to be the result of the synergic action of two hormones produced by the pituitary gland. These are the follicle-stimulating hormone (FSH) and the luteinizing hormone (LH). The follicle-stimulating hormone acting on the ovary causes the growth of the egg follicles and the eggs. Addition of the luteinizing hormone causes the rupture of the follicle and release of the egg, that is, **ovulation.** At the same time, the follicular cells start secreting large quantities of the hormone **estrogen.** One of the functions of this hormone is to cause increased blood supply to the uterine mucosa, which becomes hypertrophic and thus better suited for the reception of the embryo.

After the ovum is released the remaining cells of the ruptured follicle, under the influence of the luteinizing hormone, become transformed into a **corpus luteum** (yellow body). This involves the deposition in the follicle cells of a carotenoid substance which gives the tissue a yellowish color. In some animals the follicle cells proliferate, so that the corpus luteum becomes larger than the follicle giving rise to it, but in other animals there is no proliferation. In animals in which more than one follicle ruptures at ovulation, the number of **corpora lutea** may be used as an estimate of the number of eggs released from the ovary.

In many rodents (mice, rats, hamsters), if the eggs are not fertilized, the corpora lutea persist for only a few days and soon degenerate and are resorbed. In most mammals, however, the corpora lutea persist much longer (up to two months in the dog).

The corpora lutea play an important role in the further fate of the eggs by secreting a hormone, progesterone. Progesterone, exerting an influence on the uterine mucosa and causing a further proliferation of its tissues, is necessary for the successful implantation of the embryo in the uterine wall. So important is this hormone that removing the ovaries or only the corpora lutea after ovulation but before the embryo is implanted in the uterus effectively prevents implantation and results in the death of the embryos. Furthermore, since the maintenance of the corpora lutea depends on the supply of the luteinizing hormone, removal of the hypophysis in ovulated animals also leads to failure of the embryos to become implanted, because corpora lutea stop producing progesterone. At the same time that progesterone favors the implantation of the developing embryo in the uterus, it inhibits the secretion of the follicle-stimulating hormone and thus prevents the ripening of new eggs.

Once the embryo has become implanted in the uterus, the proliferating trophoblast becomes a source of hormone secretion. One of the secretions produced is the hormone **luteotropin,** which acts on the corpora lutea and prevents their degeneration, thus prolonging their life throughout the period of pregnancy (in most mammals). The continuation of the existence of the corpora lutea provides for a continued secretion of progesterone. Progesterone is vitally necessary for the normal course of development of the placenta and the fetus; cessation of its supply leads to abortion. Thus in rabbits, mice and rats, pigs, goats, cows, and dogs, removal of the ovaries at any time during pregnancy causes abortion. In other mammals development of the placenta and the fetus is dependent on ovarian secretions during only the earlier part of pregnancy. In the later part of pregnancy removal of the ovaries no longer causes abortion. Thus in the cat, removal of ovaries up to the forty-sixth day of pregnancy results in abortion, but after an operation on the forty-ninth day the young are carried to term. In the monkey ovariectomy may be performed as early as the twenty-fifth day of pregnancy (term 165 days). In women removal of ovaries from the fortieth day of pregnancy (but not earlier) allows for normal development.

The explanation of this independence from the ovary is that the placenta (presumably the trophoblast) produces sufficient quantities of progesterone in the latter part of pregnancy to insure normal development. The corpora lutea become dispensable. The original corpora lutea, formed at ovulation, also may become dispensable in another way. In the horse they degenerate very rapidly, but additional corpora lutea are formed from degenerating ovarian follicles which did not have an opportunity to ovulate. In later stages of pregnancy, progesterone is produced by the placenta, and the ovarian secretions are not needed.

Lastly, the placenta also produces estrogen (the primary source of which is the ripe ovarian follicles).

As progesterone is necessary for the maintenance of pregnancy, it seems possible that parturition, in the main, results when the production of progesterone by both the corpora lutea and the placenta ceases, or when there is a sharp reduction of its quantity. This view is supported by experiments in which pregnancy (in rats) has been extended beyond the normal term by injections of progesterone. On the other hand, an increased output of estrogens, which is observed toward the end of pregnancy, may also be a factor in causing parturition. This view is supported by the fact that injections of large quantities of estrogens may cause abortion. Thus, the initiation of labor may be the result of a simultaneous change in the quantities of two hormones: the decrease of progesterone and the increase of estrogen.

Part Five

ORGANOGENESIS

The germinal layers formed as a result of the process of gastrulation are the source of material for the development of all the organ rudiments of the embryo. The germinal layers become progressively subdivided into groups of cells which have been called **primary organ rudiments,** because some of them are composite in their nature and undergo a further subdivision until all the various parts are segregated from one another in the form of **secondary organ rudiments.**

The formation of the primary organ rudiments has been described in connection with the process of gastrulation. The following chapter will be devoted to the development of the embryo as a whole during the early stages of organogenesis. This will be followed by a description of the development of various organs and organ systems in Chapters 12, 13, and 14.

Chapter 11

GENERAL INTRODUCTION
TO ORGANOGENESIS

11–1 DEVELOPMENT OF GENERAL BODY FORM

Changes in Body Shape.   While the various organs are being formed, the shape of the embryo as a whole undergoes far-reaching changes. During the period of organogenesis, in the case of the vertebrate embryo, the main changes are:

1. Elongation of the body.
2. Formation of the tail.
3. Subdivision of the body into head and trunk.
4. Development of appendages.
5. Separation of the embryo proper from the extraembryonic parts. (The latter process has already been dealt with in Sections 10–1 and 10–2.)

Some of the processes just enumerated also occur in invertebrates; in particular, the elongation of the body occurs in annelids and arthropods. The subdivision of the body into sections, such as the head, thorax, and abdomen, is typical of insects and, with modifications, of some other arthropods. The development of appendages is an essential feature of arthropod development. Other processes concerning the body as a whole may occur in invertebrates but have no counterpart in vertebrate development. For instance, in insects the body of the embryo undergoes a peculiar shifting from the surface of the egg into the interior of the yolk, from which it emerges again at a later stage.

In holoblastic vertebrates, of which the amphibians may serve as an example, the embryo retains a spherical shape (i.e., the shape possessed by the unfertilized egg) up to the end of gastrulation and the beginning of neurulation. In the neurula stage the embryo becomes slightly elongated in an anteroposterior direction, but only after the completion of neurulation does the elongation of the embryo become prominent. A tail rudiment, the **tailbud,** appears at the posterior end of the body and rapidly develops into an elongated appendage, but the rest of the body also stretches, becoming at the same time flattened laterally and, to a certain extent, lower in a dorsoventral direction (Fig. 228).

Although most of the organ rudiments of the embryo are involved in this elongation, there is experimental evidence that not all are equally active. If the notochordal rudiment of an amphibian embryo in the neurula stage is excised, the embryo remains stunted and does not stretch as usual (Fig. 229–Hörstadius, 1944; Kitchin, 1949). On the other hand, the notochordal rudiment will stretch and form an elongated rod even if it is cultivated *in vitro* and is not accompanied by other parts (Fig. 230–Holtfreter, 1939b). Isolated parts of the neural system or induced brain vesicles, when no other tissues accompany them, fail to elongate. From these facts it appears that the notochord

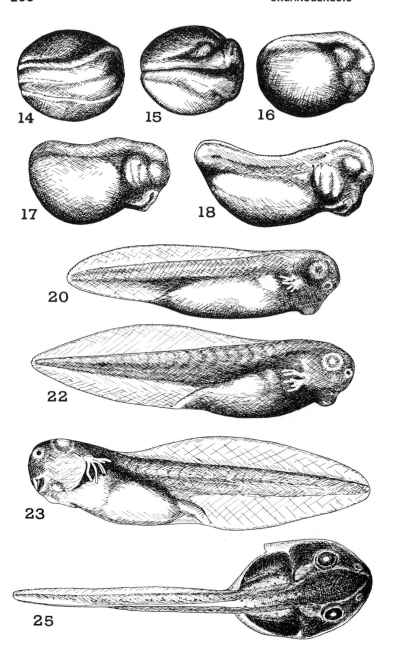

Figure 228. Stages of development of a frog (*Rana pipiens*), beginning with the neurula stage. The numbers indicate stages of development after Pollister and Moore, 1937. (Redrawn after Rugh, 1948.)

changes its shape actively, while the nervous system is pulled in length by the adjacent notochord. In amniotes the elongation starts in the primitive streak stage, so that the body of the embryo is already long and narrow by the time the main axial organs (neural tube, notochord, dorsal mesoderm giving rise to the somites) are laid down.

The formation of the tail rudiment will be dealt with in Section 12–1. In terrestrial vertebrates the subdivision of the body into the head and trunk is largely dependent on the reduction of the branchial apparatus. As will be described in Section 14–3, the system of visceral clefts and arches is fully developed in embryos of all vertebrates. In fishes and the larvae of amphibians, the visceral clefts and arches persist and take up

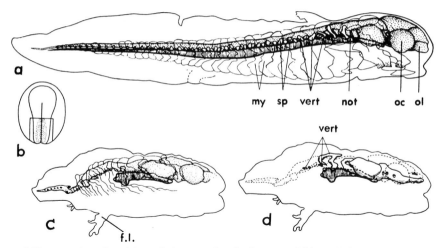

Figure 229. Results of excision of the notochord of an amphibian (*Ambystoma punctatum*) embryo. *a*, Normal larva. *b*, Diagram of operation; stippled area of the archenteron roof removed. *c*, Operated larva; drawing shows notochord (vertical lines), neural system (stippled), and contours of muscle segments. *d*, Same larva, showing notochord and cartilaginous skeleton. f.l., Forelimbs; my, muscle segments; not, notochord; oc, eye; ol, nose; sp, spinal ganglia; vert, vertebrae. (From Hörstadius, 1944.)

the area on the ventral side and posterior to the head. In terrestrial vertebrates, the branchial apparatus loses its respiratory function and becomes reduced. As a result, in later embryonic stages the area of the body posterior to the head fails to grow at the same rate as the other parts, thus producing a constricted section between the head and trunk (Figs. 231 and 232). The constriction is accentuated further by: (a) a certain amount of longitudinal stretching of the neck region, as a result of which the cervical vertebrae are, as a rule, somewhat longer than thoracic vertebrae (not true in some mammals with shortened necks, such as man or the whales); and (b) the withdrawal of the heart, originally situated in the neck region next to the branchial clefts, into the trunk (thorax) (Figs. 231 and 232). The development of appendages, particularly the paired limbs, will be described in Section 13–3.

11–2 NORMAL STAGES OF DEVELOPMENT

The changing appearance of the embryos, especially during organogenesis, invites the distinguishing of certain **stages** which can be referred to when it is desired to indicate

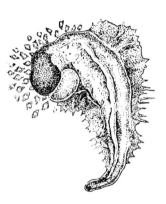

Figure 230. Explantation of the dorsal blastopore lip of an axolotl gastrula. The notochord has developed in the form of an elongated rod in the middle of the explant. Lateral parts are mesoderm. At the anterior end is a piece of ectoderm (pigmented). (From Holtfreter, 1939b.)

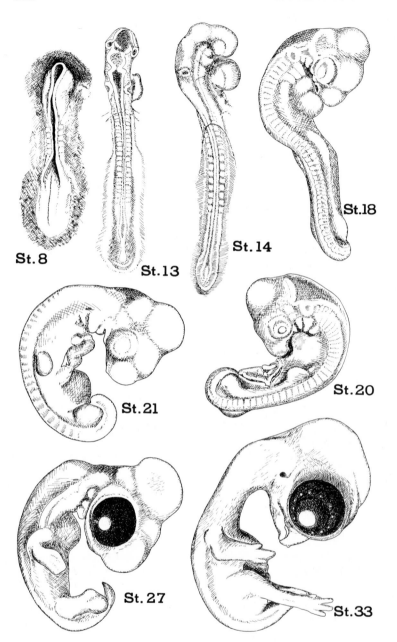

Figure 231. Stages of development of a chick. (Stages 18 and 20 after Keibel and Abraham, from Keibel in Hertwig, 1906; stages 13, 14, 21, 27, and 33 from Hamburger and Hamilton, 1951; stage 8 from Patten, 1957.)

how far an embryo has progressed in its development. Tables of "normal stages" have been worked out for a number of species of animals, especially those that are most often used for research.

In the latter half of the nineteenth century, an ambitious project on establishing series of normal stages for a large number of animals was undertaken by Keibel and his collaborators. This work was an essential contribution to the science of comparative embryology (p. 10) but went into oblivion later, when the interest of the great majority of embryologists shifted from a descriptive to an experimental approach to the development of animals. However, it soon became evident that tables of normal stages were quite as important for experimental work, as it was often necessary to indicate precisely at what stage an operation or other experiment was carried out.

The table of normal stages of *Ambystoma* (= *Amblystoma*) *punctatum*, prepared by Harrison, was the first made expressly with experimental investigations in mind. Harrison's table was published posthumously (Harrison, 1969), though earlier it was made accessible privately to many workers in the field and has been widely used. Harrison's stages were also redrawn and included in Rugh's book, *Experimental Embryology* (1948). Other tables of normal stages followed; the most widely used ones are probably the stages of *Rana pipiens* by Shumway (1940) and by Rugh (1948) and the stages of the chick by Hamburger and Hamilton (1951). There is still no up-to-date table of normal

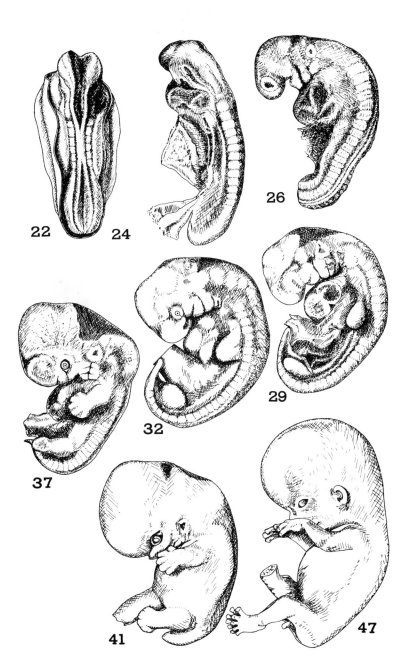

Figure 232. Stages of development of the human embryo. Figures show approximate age in days. (22-day-old embryo, after Arey: Developmental Anatomy; 37-day-old embryo redrawn from Hamilton, Boyd, and Mossman, 1947; the rest after Streeter, 1942–1951.)

stages of the development of the human embryo. (See, however, the series of papers by Streeter, 1942–1951.)

In compiling a series of normal stages the embryologist is confronted with the task of deciding what characteristics should be selected for distinguishing one stage from another. The characteristics that are often used are:

1. Age of the embryo.
2. Size of the embryo.
3. Morphological peculiarities of the embryo.

In human embryology particularly, the first two criteria are often used. It is customary for an author to refer to embryos by age (five-week embryo, two-month-old embryo) or by size (an embryo or fetus of so many millimeters crown-rump length). Both these criteria are, however, not very convenient. The age of an embryo is often not known, and in animals other than mammals, the rate of development is dependent on the temperature of the environment to such an extent that a statement about the age of the embryo is meaningless unless the temperature at which the development has proceeded is likewise indicated. The size of the embryo is no true indication of its degree of development, as the dimensions of the embryo vary to a great extent. Moreover, some variability in the size of embryos may be resolved later in the course of development, which adds to the difficulty of using size as a criterion for the definition of stages.

What remains is to base the normal stages on morphological properties of the embryo and especially on properties that can be easily ascertained by external examination of the embryo, without its fixation or dissection—that is, identification based largely on external features. In the initial stages of development (cleavage stages), the number and size of the blastomeres may conveniently be used. During gastrulation, the shape of the blastopore or its equivalent (primitive streak) may be used, and just after gastrulation, the neural plate offers easily recognizable features. During early organogenesis, the number of pairs of somites has often been used to define the stage of development of the embryo. The somites, though not strictly "external features," can be seen on external inspection, especially in the amniotes. In still later stages, the development of the appendages presents easily distinguishable and convenient characters for the definition of normal stages.

Although morphological characters appear to be the best criteria for establishing the stage of development of an embryo, there are certain limitations even to this approach. It has been found that the development of different parts (organs) of the embryo is not always strictly coordinated in time; sometimes certain ones develop more rapidly, sometimes others. So if two embryos have certain organs (e.g., the forelimbs) in exactly the same condition, they may at the same time differ in the degree of development of other organs (e.g., the nervous system or the liver). This phenomenon of **heterochrony,** or unequal rate of development of parts, must always be kept in mind when any tables of normal stages are being referred to.

11–3   THE ANATOMY OF REPRESENTATIVE STAGES OF DEVELOPMENT OF THE FROG AND CHICK EMBRYOS

A few stages of the frog and chick embryos will now be considered in greater detail. The stages chosen will serve as landmarks which, hopefully, will help the student to sort out the information concerning the development of the separate organ systems

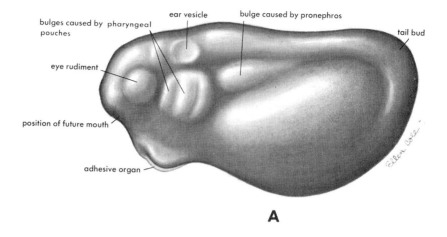

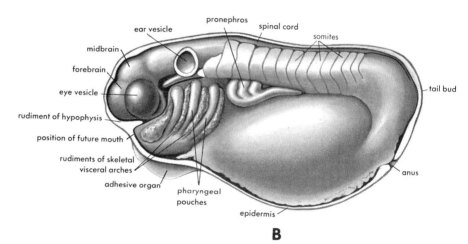

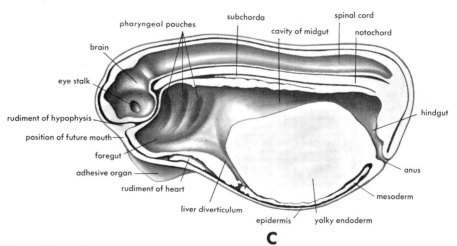

Figure 233. A frog embryo in an early tail bud stage. *A*, External view; *B*, same embryo with the skin of the left side removed; *C*, same embryo cut in the median plane.

of the embryo, as dealt with in the following chapters (Chapters 12–14). In those chapters the student will find extensive information on the various organs and organ systems of the body, some of which will be mentioned briefly for the first time in this section.

Frog Embryo After Completion of Neurulation, Stage 17 (Rugh, 1948) (Fig. 233). The neural system is in the form of a closed tube, broadened anteriorly where the brain will develop. The parts of the brain begin to be indicated by thickenings and constrictions of the neural tube, and the eye rudiments bulge laterally from the forebrain. The notochord stretches underneath the neural tube from the midbrain level to the posterior end of the body, where it fuses with other tissues which are, as yet, in an undifferentiated state. Under the notochord in the posterior part of the body there lies a strand of endodermal cells: the **subchorda,** a structure which exists for a short time and soon disappears. Lateral to the notochord the mesoderm is subdivided into segments, the somites. The endoderm surrounds the gut cavity, which is broad anteriorly and narrow in the posterior part of the body. In the anterior part of the gut, lateral outpushings represent the pharyngeal pouches. At the posterior end, an anal opening has been formed, but the mouth is not perforated, although the endoderm and ectoderm are in contact at the site of the future mouth. Above the place where the mouth will be, an ingrowth of the ectodermal layer represents the rudiment of the hypophysis. Lateral to the foregut, masses of mesectoderm can be seen which give rise to the visceral skeleton. Underneath the foregut, groups of mesodermal cells represent the rudiment of the heart. The posterior end of the body above the anus has elongated somewhat in the form of a

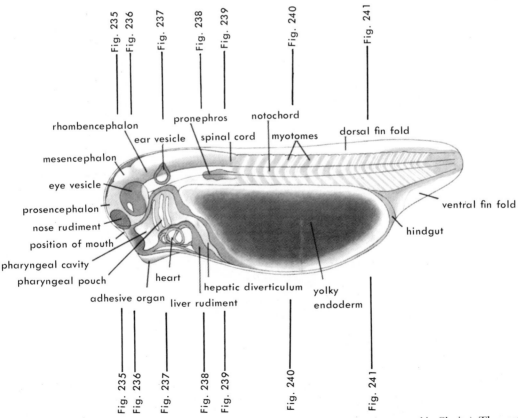

Figure 234. 4 mm. frog embryo (drawing made from a specimen cleared and mounted in Clarite). The vertical lines indicate the levels represented by photographs of transverse sections (Figs. 235 to 241). The sections used for the photographs are from different embryos; this accounts for some slight discrepancies between the drawing and the photographs.

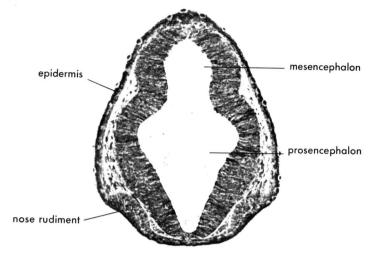

Figure 235. Transverse section of 4 mm. frog embryo at the level of the prosencephalon and nose rudiments. This and the following photographs correspond to the embryo illustrated in Figure 234 in which the levels of the sections are also shown.

tailbud, still in an early stage of development; hence this stage may be referred to as an early tailbud stage.

4 mm. Frog **(Rana pipens)** *Embryo, Stage 19 (Rugh, 1948) (Figs. 234–241).* The tailbud has elongated and has developed into a clearly recognizable tail with an axis consisting of neural tube, notochord, and segmented mesoderm, and a fin fold all around its edge. The neural system has developed further; the eye vesicles have been constricted at the base and thus subdivided from the brain. The parts of the brain are more distinctly indicated. Ear vesicles have been formed to the right and left of the hindbrain. The nose rudiments are in the form of placodes, thickenings of the epidermis. The oropharyngeal membrane separating the stomodeum from the endodermal foregut has become thin and will soon be perforated. Ventral to the mouth, the adhesive organs have become well developed. The foregut has become subdivided into an anterior part, the pharynx, and a posterior part which will give rise to the gastric and duodenal section of the alimentary canal. The pharynx has produced pharyngeal pouches laterally. Underneath the pharynx the heart rudiment is developing; in cross section (Fig. 237) one can see the thin-walled endocardial tube and the thicker parts of lateral plate mesoderm which will give rise to the myocardium and pericardium. The

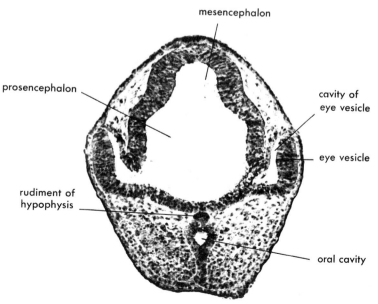

Figure 236. Transverse section of 4 mm. frog embryo at the level of the eye rudiments.

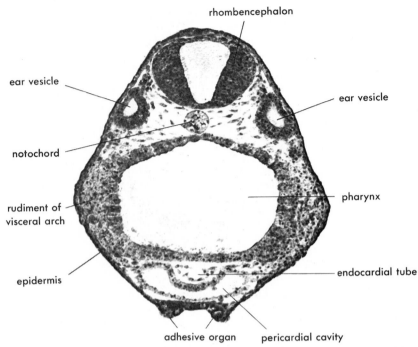

Figure 237. Transverse section of 4 mm. frog embryo at the level of the ear vesicles and the heart.

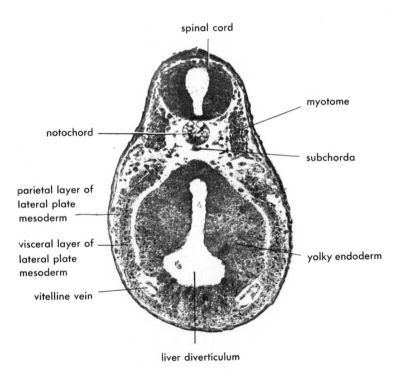

Figure 238. Transverse section of 4 mm. frog embryo at the level of the liver diverticulum.

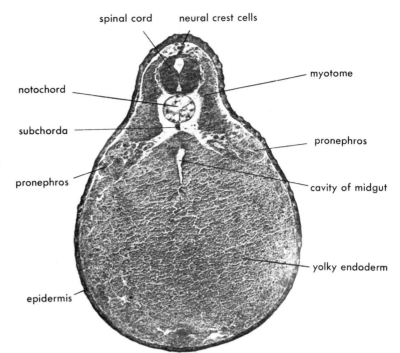

Figure 239. Transverse section of 4 mm. frog embryo at the level of the pronephros.

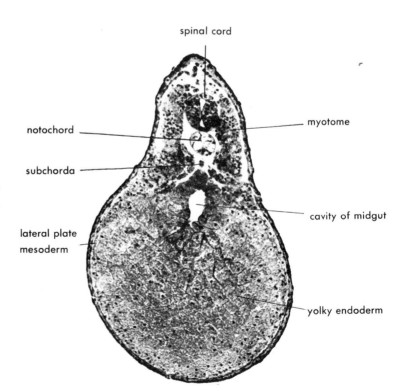

Figure 240. Transverse section of 4 mm. frog embryo at the midtrunk level.

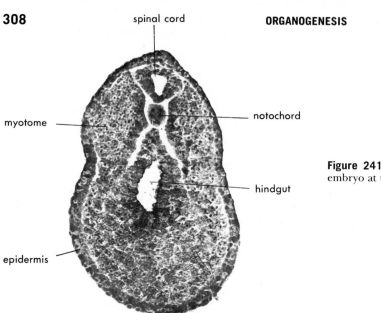

spinal cord

myotome

notochord

hindgut

epidermis

Figure 241. Transverse section of 4 mm. frog embryo at the posterior trunk level.

Fig. 243   Fig. 244   Fig. 245   Fig. 246   Fig. 247   Fig. 248   Fig. 249   Fig. 250   Fig. 251

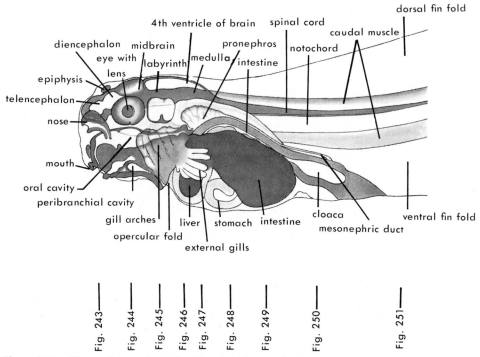

dorsal fin fold

4th ventricle of brain   spinal cord   caudal muscle

diencephalon   midbrain   pronephros   notochord

eye with   labyrinth   medulla   intestine

epiphysis   lens

telencephalon

nose

mouth

oral cavity

peribranchial cavity

gill arches   liver   stomach   intestine   cloaca   ventral fin fold

opercular fold   mesonephric duct

external gills

Fig. 243   Fig. 244   Fig. 245   Fig. 246   Fig. 247   Fig. 248   Fig. 249   Fig. 250   Fig. 251

Figure 242.   10 mm. frog embryo (drawing made from a specimen cleared and mounted in Clarite). The vertical lines indicate the levels represented by photographs of transverse sections (Figs. 243 to 251). The sections used for the photographs are from different embryos; this accounts for some slight discrepancies between the drawing and the photographs.

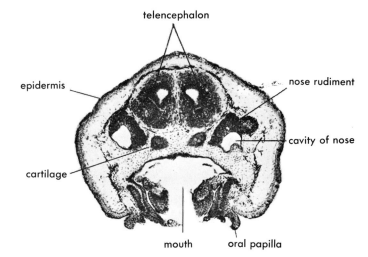

Figure 243. Transverse section of 10 mm. frog embryo at the level of the nose and mouth. This and the following photographs correspond to the embryo illustrated in Figure 242, which also shows the levels of the sections.

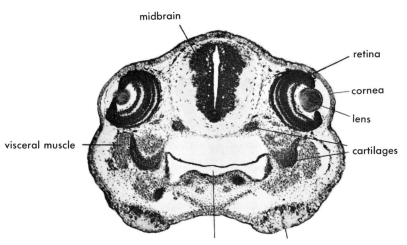

Figure 244. Transverse section of 10 mm. frog embryo at the eye level.

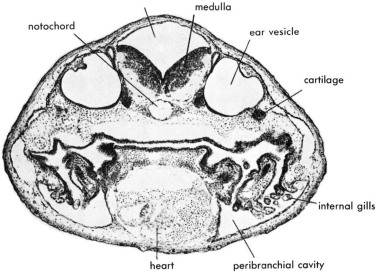

Figure 245. Transverse section of 10 mm. frog embryo at the ear level.

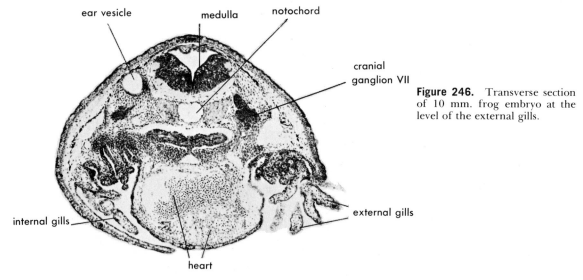

Figure 246. Transverse section of 10 mm. frog embryo at the level of the external gills.

pronephric tubules are in the process of formation (Fig. 239). The posterior part of the endodermal gut is still much in the same condition as in the preceding stages.

10 mm. Frog (**Rana pipiens**) *Embryo, Stage 23 (Rugh, 1948) (Figs. 242–251).* The embryo has changed its shape and has become a tadpole. The head and trunk are bloated and together form a more or less egg-shaped body. The tail has elongated and has become a powerful swimming organ, with segmented muscle laterally and broad fin folds dorsally and ventrally (Fig. 251). The brain is rapidly differentiating, and the forebrain has produced the two hemispheres of the telencephalon (Fig. 243). The roof of the medulla has become membranous. The nose rudiments have invaginated and are in the form of sacs connected to the exterior by nares. The eyecup is clearly differentiated into a neural retina and a pigment layer. A lens has been formed (Fig. 244). The ear vesicles are in the process of producing the labyrinths; endolymphatic ducts can be seen on the dorsomedial aspect of each labyrinth (Fig. 245). The pronephros

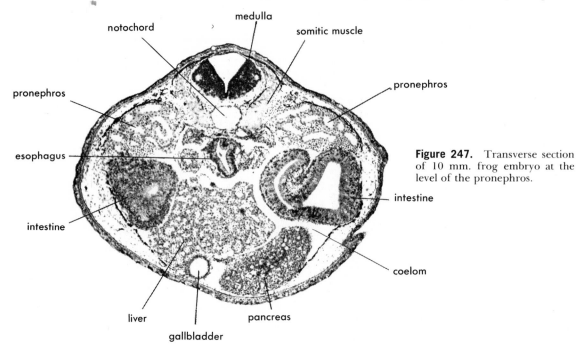

Figure 247. Transverse section of 10 mm. frog embryo at the level of the pronephros.

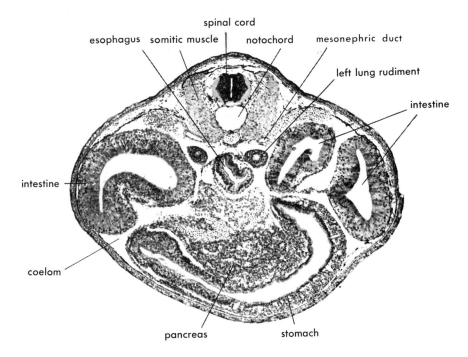

Figure 248. Transverse section of 10 mm. frog embryo at the level of the stomach.

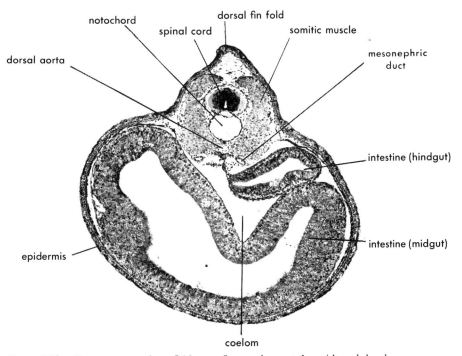

Figure 249. Transverse section of 10 mm. frog embryo at the midtrunk level.

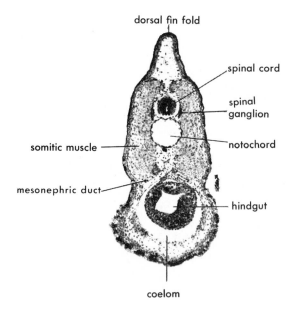

dorsal fin fold

spinal cord

spinal
ganglion

somitic muscle

notochord

mesonephric duct

hindgut

coelom

Figure 250. Transverse section of 10 mm. frog embryo at the level of the hindgut.

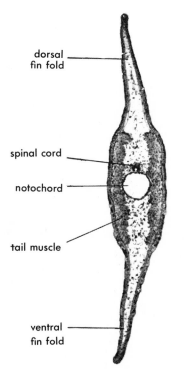

dorsal
fin fold

spinal cord

notochord

tail muscle

ventral
fin fold

Figure 251. Transverse section of the tail of a 10 mm. frog embryo.

consists of several (three) convoluted tubules which form thickenings on the sides of the body anteriorly, and the mesonephric ducts have been developed. The mouth is open and is surrounded by horny jaws, teeth, and soft circumoral papillae. The branchial clefts are perforated, and finger-like external gills project from the sides of the head in the branchial region. The endodermal gut is differentiated into its various parts. In sections one can distinguish pharynx, esophagus, lung rudiments, stomach, liver, gallbladder, pancreas, and intestine. The intestine has already become coiled spirally and largely contributes to the bloating of the body of the tadpole.

Chick Embryo, Stage 10 (Hamburger and Hamilton, 1951) (29 to 30 Hours) (Figs. 252–261). The most conspicuous parts of the chick embryo in this stage are the brain and spinal cord rudiments and somites, of which there are 9 or 10 pairs. The brain shows a subdivision into primary brain vesicles, and eye vesicles are already prominent. Nevertheless, the closing of the neural tube is not quite completed: anteriorly the cavity of the neural tube is still open to the exterior by the neuropore; at the posterior end the neural folds have not come together, and there is still an open neural plate present. Still farther back, the remnants of the primitive streak have not yet disappeared. Underneath the

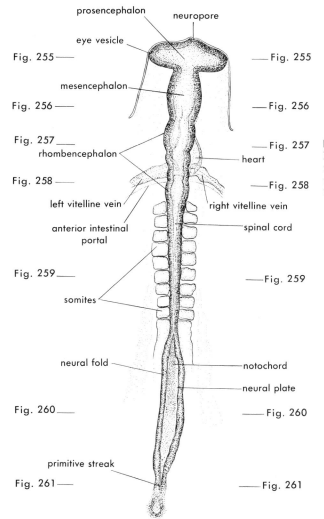

prosencephalon

neuropore

eye vesicle

Fig. 255 ——— ——— Fig. 255

mesencephalon

Fig. 256 ——— ——— Fig. 256

Fig. 257 ——— ——— Fig. 257

rhombencephalon

heart

Fig. 258 ——— ——— Fig. 258

left vitelline vein

right vitelline vein

anterior intestinal portal

spinal cord

Fig. 259 ——— ——— Fig. 259

somites

neural fold

notochord

neural plate

Fig. 260 ——— ——— Fig. 260

primitive streak

Fig. 261 ——— ——— Fig. 261

Figure 252. Chick embryo, stage 10 (age 29 to 30 hours). Drawing made from a specimen cleared and mounted in Clarite. The lines indicate the levels represented by transverse sections shown in Figures 255 to 261.

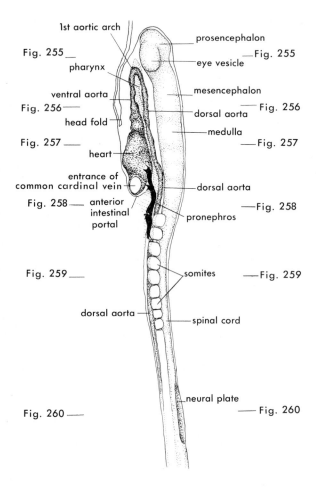

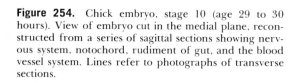

Figure 253. Chick embryo, stage 10 (age 29 to 30 hours). Lateral view reconstructed from a series of sagittal sections showing nervous system, blood vessels, somites, pronephros, and gut rudiment. Lines refer to photographs of transverse sections.

Figure 254. Chick embryo, stage 10 (age 29 to 30 hours). View of embryo cut in the medial plane, reconstructed from a series of sagittal sections showing nervous system, notochord, rudiment of gut, and the blood vessel system. Lines refer to photographs of transverse sections.

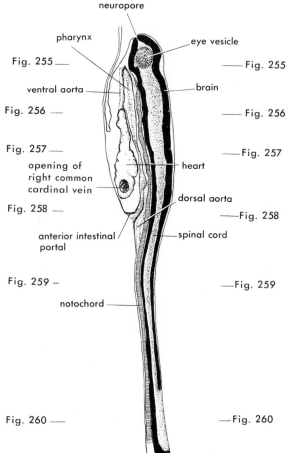

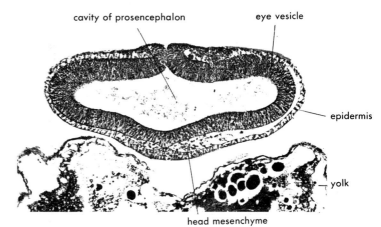

cavity of prosencephalon   eye vesicle

epidermis

yolk

head mesenchyme

Figure 255. Chick embryo, stage 10. Transverse section at eye vesicle level. This and the following photographs correspond to the embryos illustrated in Figures 252, 253, and 254, and the levels of the sections are shown in these figures.

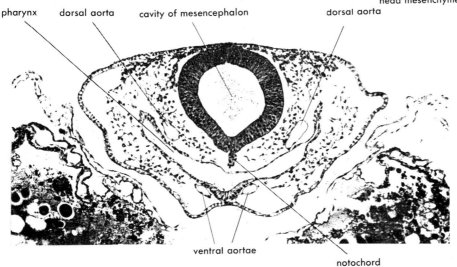

pharynx   dorsal aorta   cavity of mesencephalon   dorsal aorta

ventral aortae

notochord

Figure 256. Chick embryo, stage 10. Transverse section at midbrain level, showing the gut enclosed completely on the ventral side, and the paired dorsal and ventral aortae.

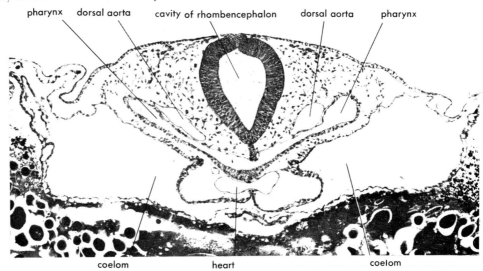

pharynx   dorsal aorta   cavity of rhombencephalon   dorsal aorta   pharynx

coelom   heart   coelom

Figure 257. Chick embryo, stage 10. Transverse section at level of rhombencephalon. The heart is single at this level, although its origin from two halves is clearly shown. The gut is completely enclosed on the ventral side.

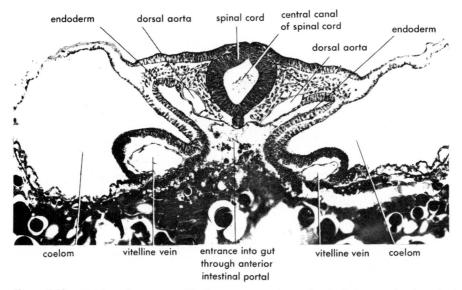

Figure 258. Chick embryo, stage 10. Transverse section at level of the anterior intestinal portal, showing also the proximal parts of the vitelline veins.

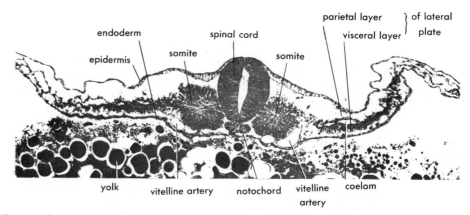

Figure 259. Chick embryo, stage 10. Transverse section in trunk region with closed neural tube and somites.

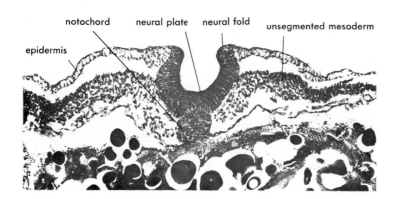

Figure 260. Chick embryo, stage 10. Transverse section, showing open neural plate and undifferentiated mesoderm.

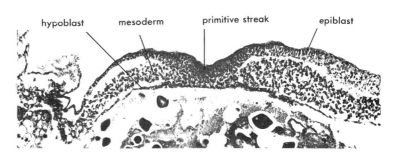

Figure 261. Chick embryo, stage 10. Transverse section at posterior end, showing primitive streak.

neural tube lies the notochord which at its posterior end is continuous with the primitive streak material. Posterior to the somites lies the still unsegmented mesoderm. The body folds have undercut the anterior end of the embryo, so that the foremost part of the head is lifted above the surface of the yolk, but farther back the separation of the body from the yolk sac has not yet begun. The anterior part of the gut is already closed and separated from the yolk sac, but from the anterior trunk region onward the gut is open ventrally. Pronephroi have started developing. The blood vessel system is represented by separate vessels on the right and left sides of the body, and only in the heart the rudiments of the right and the left sides are fusing together (Fig. 257). There is one pair of aortic arches (the first) at the anterior end of the body. The embryo as a whole is symmetrical.

Chick Embryo, Stage 15 (Hamburger and Hamilton, 1951) (50 to 55 Hours) (Figs. 262−269). The anterior end of the embryo is twisted to the right, so that the left side of this part of the embryo lies flat on the surface of the yolk, largely because the brain of the embryo has become bent at an angle at the midbrain level. The parts of the brain are clearly recognizable. The eye is already in the eyecup stage with a lens lying in the pupil. The lens is, however, not yet separated from the epidermis and is in the form of a sac with an opening to the exterior. The ear rudiments are also in the form of pockets open to the exterior. The nose rudiment is still only a thickening of the epidermis. Masses of cells lateral to the brain represent the rudiments of the large cranial ganglia. The number of pairs of somites has greatly increased, but at the posterior end of the body there is still some unsegmented mesoderm present. The neural tube is not completely closed at the posterior end, and even a remnant of the primitive streak is to be found, though immigration of cells from the streak has ceased. The first three pairs of pharyngeal pouches have been formed. The blood vessel system of the embryo has undergone a marked development. The heart is a large and conspicuous organ just underneath the head. A paired ventral aorta conveys blood into the first pair of aortic arches, which are still the only ones to carry blood. The dorsal aortae are paired anteriorly and posteriorly but are fusing in the median region. Anterior cardinal veins are well developed, but the posterior cardinal veins are very small. The yolk sac circulation is carried on by means of large vitelline veins joining the sinus venosus and the vitelline arteries, which at this stage appear to be direct continuations of the dorsal aortae.

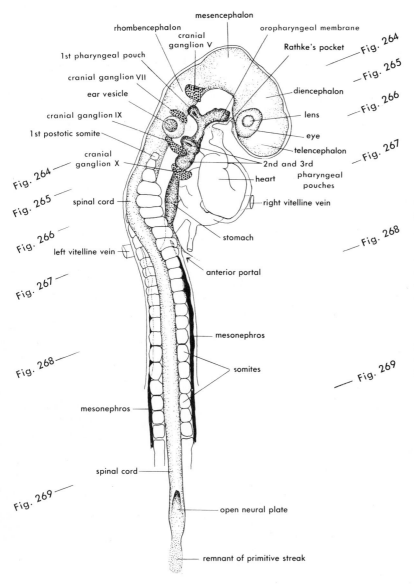

mesencephalon

rhombencephalon

cranial
ganglion V

oropharyngeal membrane

Rathke's pocket

Fig. 264

1st pharyngeal pouch

Fig. 265

cranial ganglion VII

ear vesicle

diencephalon

Fig. 266

cranial ganglion IX

lens

1st postotic somite

eye

cranial
ganglion X

telencephalon

Fig. 267

2nd and 3rd
pharyngeal
pouches

Fig. 264

heart

Fig. 265

spinal cord

right vitelline vein

Fig. 266

left vitelline vein

stomach

Fig. 267

anterior portal

Fig. 268

mesonephros

somites

Fig. 269

Fig. 268

mesonephros

spinal cord

Fig. 269

open neural plate

remnant of primitive streak

Figure 262. Chick embryo, stage 15 (age 50 to 55 hours). Reconstruction from a series of horizontal sections showing nervous system, somites, kidney rudiments, and alimentary canal with gill clefts. The lines on the sides indicate the levels represented by sections shown in Figures 264 to 269. The sections used for the photographs are from a different embryo, and this accounts for some slight discrepancies between the drawing and the photographs.

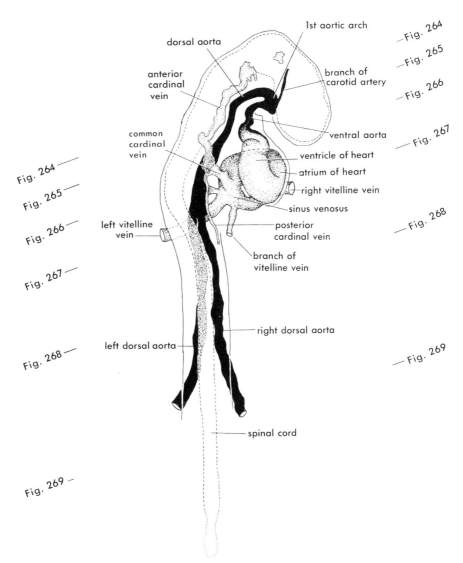

Figure 263. Chick embryo, stage 15 (age 50 to 55 hours). Reconstruction from a series of horizontal sections showing the blood vessel system. Lines on the sides indicate levels of photographs in Figures 264 to 269.

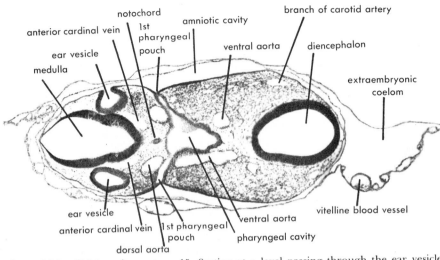

Figure 264. Chick embryo, stage 15. Section at a level passing through the ear vesicles, still connected to epidermis, and the first pharyngeal pouch. This and the following photographs correspond to the embryo illustrated in Figures 262 and 263.

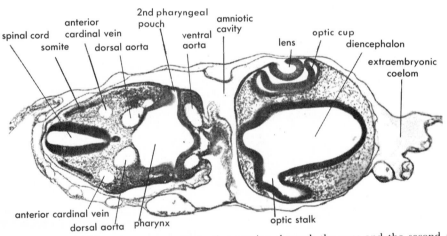

Figure 265. Chick embryo, stage 15. Section passing through the eyes and the second pair of pharyngeal pouches.

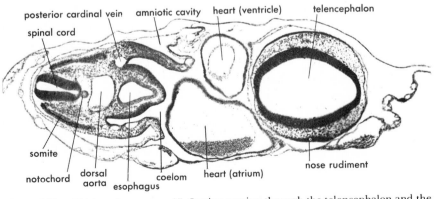

Figure 266. Chick embryo, stage 15. Section passing through the telencephalon and the heart.

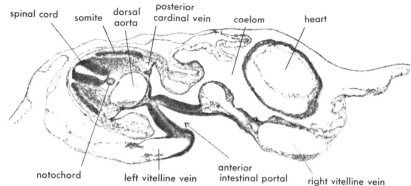

Figure 267. Chick embryo, stage 15. Section at the level of the anterior intestinal portal and the proximal parts of the vitelline veins.

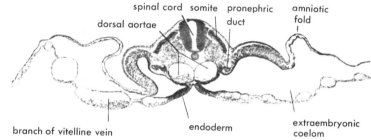

Figure 268. Chick embryo, stage 15. Section through the anterior trunk region at level of forelimb rudiments.

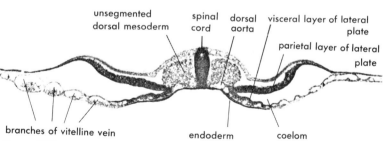

Figure 269. Chick embryo, stage 15. Section through posterior trunk region. Somites and notochord not yet differentiated at this level.

Chick Embryo, Stage 19 (Hamburger and Hamilton, 1951) (68 to 72 Hours) (Figs. 270–281). The curvature of the anterior part of the head and body has become even greater, and more of the embryo's body is lying on its left side. The formation of the neural tube and the segmentation of the mesoderm have been completed; posteriorly the body ends with the tail which is curved downward. Limb-buds—the rudiments of the fore- and hindlimbs—are conspicuous. The brain vesicles have developed further, and the hemispheres of the telencephalon are clearly discernible. The nose rudiments

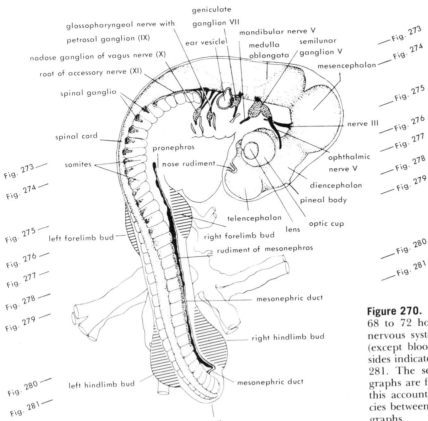

glossopharyngeal nerve with
petrosal ganglion (IX)
geniculate
ganglion VII
mandibular nerve V
medulla
oblongata
semilunar
ganglion V
ear vesicle
nodose ganglion of vagus nerve (X)
root of accessory nerve (XI)
mesencephalon
spinal ganglia
Fig. 273
Fig. 274
Fig. 275
nerve III
Fig. 276
Fig. 277
spinal cord
pronephros
nose rudiment
ophthalmic
nerve V
Fig. 278
Fig. 279
Fig. 273
somites
diencephalon
Fig. 274
pineal body
Fig. 275
left forelimb bud
telencephalon
lens
optic cup
Fig. 276
right forelimb bud
Fig. 277
rudiment of mesonephros
Fig. 280
Fig. 281
Fig. 278
Fig. 279
mesonephric duct
right hindlimb bud
Fig. 280
left hindlimb bud
mesonephric duct
Fig. 281
tail

Figure 270. Chick embryo, stage 19 (age 68 to 72 hours). Reconstruction of the nervous system and mesodermal organs (except blood vessels). The lines on the sides indicate the levels of Figures 273 to 281. The sections used for the photographs are from a different embryo, and this accounts for some slight discrepancies between the drawing and the photographs.

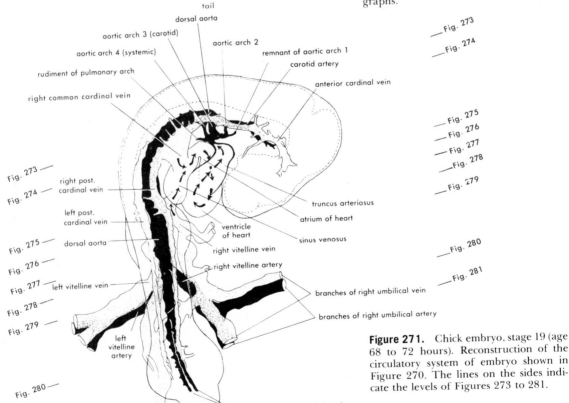

dorsal aorta
aortic arch 3 (carotid)
aortic arch 2
aortic arch 4 (systemic)
remnant of aortic arch 1
carotid artery
rudiment of pulmonary arch
anterior cardinal vein
right common cardinal vein
Fig. 273
Fig. 274
Fig. 275
Fig. 276
Fig. 277
Fig. 278
Fig. 279
Fig. 273
right post.
cardinal vein
left post.
cardinal vein
truncus arteriosus
atrium of heart
Fig. 274
ventricle
of heart
sinus venosus
Fig. 275
dorsal aorta
right vitelline vein
Fig. 276
right vitelline artery
Fig. 280
Fig. 277
left vitelline vein
Fig. 281
Fig. 278
Fig. 279
branches of right umbilical vein
branches of right umbilical artery
left
vitelline
artery

Figure 271. Chick embryo, stage 19 (age 68 to 72 hours). Reconstruction of the circulatory system of embryo shown in Figure 270. The lines on the sides indicate the levels of Figures 273 to 281.

Fig. 280
Fig. 281
paired posterior part of dorsal aorta
posterior vitelline vein

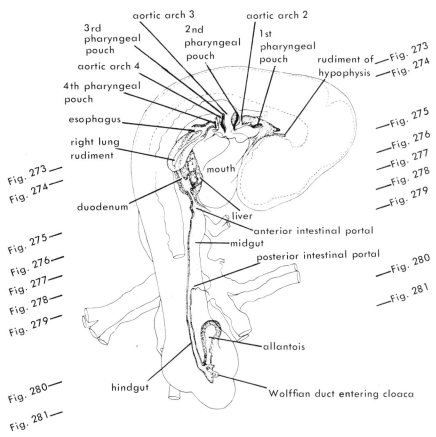

Figure 272. Chick embryo, stage 19 (age 68 to 72 hours). Reconstruction of the alimentary canal and associated structures of the embryo shown in Figures 270 and 271. The lines on the sides indicate the levels of Figures 273 to 281.

are in the form of pockets, and the lens and the ear vesicles are completely separated from the epidermis. The cranial ganglia and some of the cranial nerves can be seen clearly. The mouth is open, and there are four pairs of pharyngeal pouches present; the anterior three pouches are open to the exterior as gill clefts. The rudiments of the posterior part of the hypophysis, the epiphysis, and the thyroid gland can be found (the latter could not be shown in the figures). The alimentary canal shows subdivision into several parts: the pharynx with pharyngeal pouches, the esophagus with the lung rudiments, and the gastroduodenal part with the liver rudiment. The middle part of the gut is open ventrally, but the hindgut is separated from the yolk sac, and the rudiment of the allantois has been formed as a ventral outgrowth from the hindgut. Three pairs of aortic arches are present; they are the second, third, and fourth pairs, the first pair having been reduced by this stage. The internal carotid arteries are well developed. The dorsal aorta, unpaired along most of its length, is a very large vessel. The vitelline arteries are now seen as lateral branches of the aorta. Posteriorly, the dorsal aorta is still paired. Both anterior and posterior cardinal veins are present, the latter being in close association with the metanephric kidneys, the rudiments of which can be traced along most of the trunk region of the body. The mesonephric ducts reach the cloaca and open into it. The paired vitelline veins, before entering the sinus venosus, fuse into a single trunk — the unpaired vitelline (or omphalomesenteric) vein. The extraembryonic vitelline veins, as well as the vitelline arteries, are very conspicuous. The main branches of the veins and arteries follow the same course, the veins lying more superficially than the arteries.

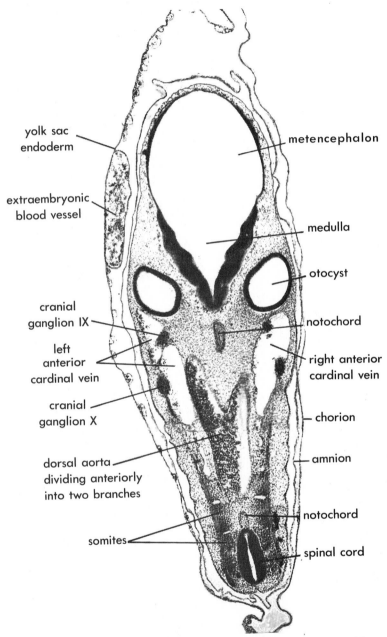

Figure 273. Chick embryo, stage 19 at level of ear vesicles.

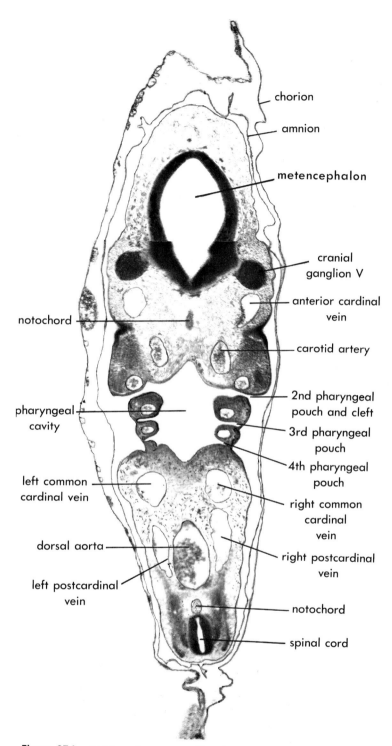

Figure 274. Chick embryo, stage 19 at level of pharyngeal pouches.

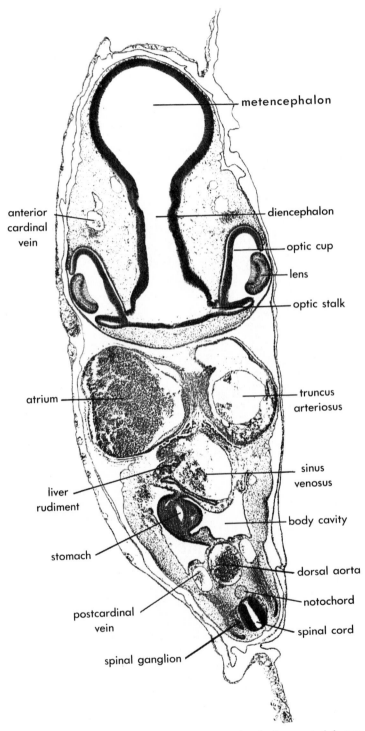

Figure 275. Chick embryo, stage 19 at level of eyes and heart.

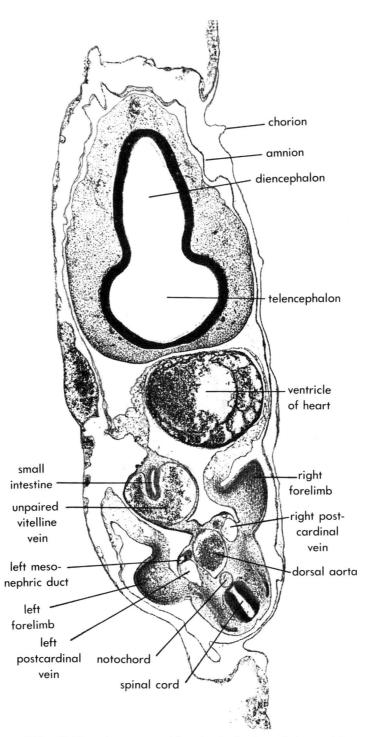

Figure 276. Chick embryo, stage 19 at level of telencephalon and heart.

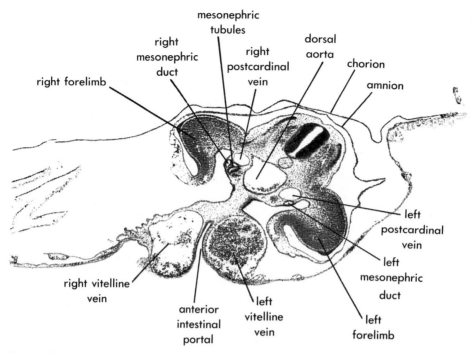

Figure 277. Chick embryo, stage 19 at level of forelimb buds and anterior intestinal portal. Note change of orientation (by 90 degrees, anticlockwise) in this and following figures, compared with Figures 273 to 276.

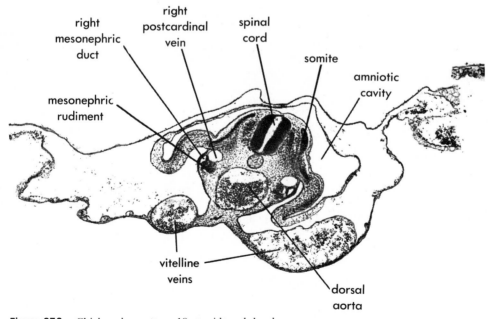

Figure 278. Chick embryo, stage 19 at midtrunk level.

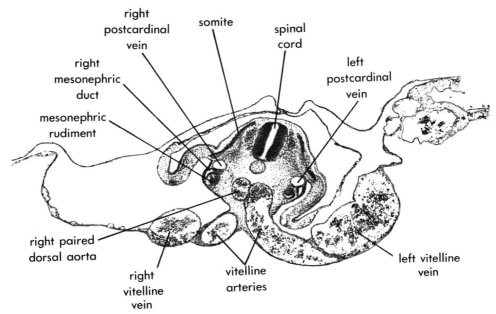

Figure 279. Chick embryo, stage 19 at level of roots of vitelline arteries.

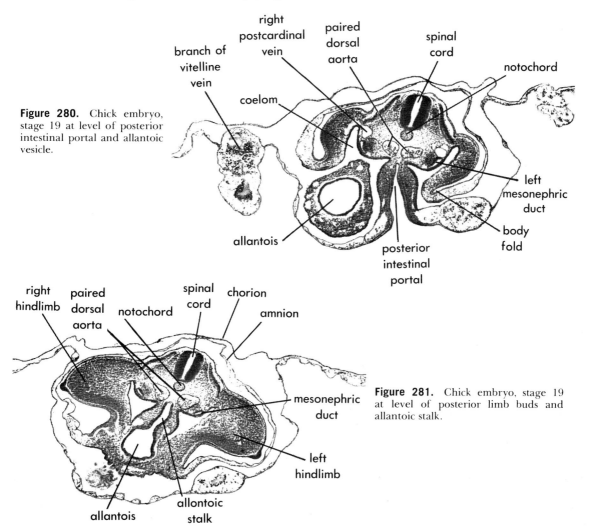

Figure 280. Chick embryo, stage 19 at level of posterior intestinal portal and allantoic vesicle.

Figure 281. Chick embryo, stage 19 at level of posterior limb buds and allantoic stalk.

329

Chapter 12

DEVELOPMENT OF THE ECTODERMAL ORGANS IN VERTEBRATES*

12–1 DEVELOPMENT OF THE CENTRAL NERVOUS SYSTEM

The central nervous system of vertebrates develops from the primary rudiment, the neural tube. The origin of the neural tube was described earlier (Section 7–5). The tube when formed is of unequal diameter throughout; its anterior end is expanded, the cavity is broader, and the walls are thicker than in the posterior part of the tube. These differences foreshadow the development of the brain from the anterior part of the neural tube and the development of the spinal cord from the posterior part.

The various parts of the brain (forebrain, midbrain, etc.) are first indicated as thickenings of the wall of the neural tube, which are followed, especially in the case of the cerebral hemispheres, by the development of pocket-like evaginations of the brain wall. Shallow constrictions develop early all around the neural tube, thus permitting the distinguishing of several "brain vesicles." Three such brain vesicles appear at the beginning. The most anterior brain vesicle, the **prosencephalon,** later gives rise to the **telencephalon** and the **diencephalon.** The second brain vesicle, the **mesencephalon,** is not subdivided further and develops into the midbrain. The third brain vesicle, the **rhombencephalon,** gives rise to the **metencephalon** (cerebellum) and the **myelencephalon** (medulla oblongata) (Fig. 282). The medulla oblongata becomes constricted by shallow furrows into a number of segments, **neuromeres.** This segmentation, which is especially clear in fish embryos, is, however, only temporary and does not leave any trace in the organization of the adult brain. It is not correlated with the metameric arrangements of the cranial nerves.

Even before the prosencephalon is clearly subdivided into the telencephalon and the diencephalon, a pair of saclike protrusions appear on its lateral walls. These protrusions are the rudiments of the eyes which are thus, basically, specially differentiated parts of the brain. In this stage, they are called optic vesicles. The optic vesicles become constricted from the remainder of the prosencephalon, and the connecting **optic stalk** later forms the basis for the development of the optic nerve. The optic stalk (and the optic nerve) join that part of the brain vesicle which becomes the diencephalon. By the method of local vital staining, as well as by observation of pecularities of pigmentation, it has been possible to trace back the cells of the optic vesicle to the open neural plate. In the neural plate the presumptive material of the optic vesicles lies far forward, close

---

*A more detailed description of organogenesis in vertebrates may be found in the books by Nelsen (1953), Witschi (1956), and Patten (1958).

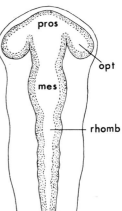

Figure 282. Primary brain vesicles and eye rudiments of a 33-hour-old chick embryo. pros, Prosencephalon; mes, mesencephalon; rhomb, rhombencephalon; opt, eye vesicles.

behind the transverse neural fold and rather near to the midline (Fig. 283). In the course of development, the eye rudiments are drawn out away from each other and into a more lateral position than they occupied initially. The subsequent development of the optic vesicles will be dealt with in a special section.

After the neural tube has been closed, it undergoes considerable stretching, together with the stretching of the embryo as a whole, especially in its posterior half. As this stretching is at the expense of the thickness of the tube, it tends to enhance the difference in bulk between the brain and the spinal cord.

A most peculiar transformation occurs in the posterior part of the neural plate and tube. Here the neural plate reaches the blastopore. The posterior part of the neural tube elongates to a greater extent than the ventral part of the embryo, and the posterior end of the neural tube is therefore carried beyond the blastopore. Because its hindmost tip is attached to the blastopore, the neural tube becomes bent on itself some distance from the blastopore. The apex of the bend then becomes the tip of the tail rudiment (Fig. 284). The major part of the neural tube, from its cranial end to the apex of the bend, differentiates as central nervous sytem (brain and spinal cord). The inflected part of the tube, however, the part lying between the apex of the tail rudiment and the blastopore, differentiates as muscle of the tail region. It loses its central canal and becomes split along the midline into two lateral masses or strips of cells which forthwith shift upward, so as to lie on both sides of the notochord and spinal cord. Cranially

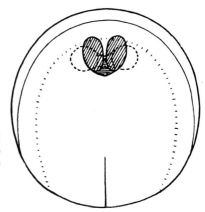

Figure 283. Position of the presumptive eye rudiments in the neural plate of an amphibian. Shadowed areas show the position of the eyes in the very early neurula. The dashed contour shows their positions in the early neurula. (From Manchot, 1929.)

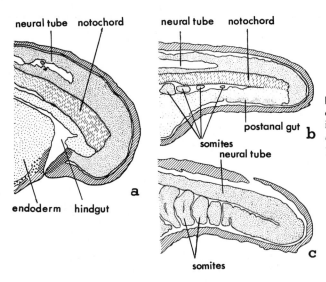

Figure 284. Transformation of the posterior end of the neural tube into the caudal somites in an amphibian embryo. *a*, Early tailbud stage; *b*, late tailbud stage, median section; *c*, same, paramedian section. (From Bijtel, 1931.)

these cell masses join up with the dorsal part of the mesodermal mantle. Together with the latter, the presumptive muscle of the tail region is subdivided into muscle segments, the somites. No stalks of somites and no lateral plate are developed in the tail region, and in this respect the segmentation of the presumptive muscle derived from the neural plate differs from the segmentation of the trunk mesoderm. By local vital staining it is possible to determine exactly which part of the neural plate differentiates as caudal muscle and which part differentiates as neural tissue. The boundary between the two parts runs straight across the neural plate at about one sixth the distance to its posterior end (Bijtel, 1931, 1936).

It has been shown previously (in Section 8–8 on gradients in amphibian development) that the roof of the archenteron is responsible for the differentiation of the various parts of the neural plate and neural tube. The anterior part of the archenteron roof—namely, the prechordal plate—induces predominantly the forebrain and eyes (archencephalic inductor), a more posterior part of the archenteron roof induces the hindbrain and associated structures (deuterencephalic inductor), and the most posterior part of the archenteron roof induces spinal cord and muscle (spinocaudal inductor). The induction of muscle by the spinocaudal inductor becomes comprehensible from what has just been stated about the fate of the posterior end of the neural plate.

A corollary to the experiments on the regional specificity of inductors is provided by the following experiments designed to test the determination of parts of the neural plate. It was desired to test whether in the neural plate, after it has already made its appearance but before it starts to differentiate into its subordinate organ rudiments, the various parts are interchangeable or whether they are already determined for their respective destination. For this purpose pieces of the young neural plate were cut out and reimplanted into the same or another embryo. The transplanted parts were placed in abnormal orientations, as for instance with reversed anterior and posterior ends, or they were placed in an altogether different region of the neural plate. The result was found to be different depending on whether the pieces of the neural plate were taken with or without the underlying archenteron roof. If the graft consisted of neural plate cells only, it differentiated in agreement with its surroundings and the graft's original polarity, or its place of origin did not manifest themselves. Rotated grafts gave rise to parts of the brain which were in complete harmony with their surroundings (Alderman, 1935). Pieces of neural plate taken from its posterior region and transplanted into the anterior region differentiated as brain parts instead of differentiating as spinal cord or muscle. In short, the development of the neural system went on as if nothing had happened (Umanski, 1935). The various parts of the neural plate were found not to be determined, or else their determination was not final and could be overridden by the influence of the surrounding tissues.

If, however, the neural plate material was taken together with the underlying archenteron roof, the graft differentiated in accord with its original prospective significance. Inverted sections of the brain developed if the graft was inverted (Spemann, 1912b). The normal location and differentiation of the eyes were disarranged if the rotation of a piece involved the eye region of the neural plate. That this result is due to the rotation of the archenteron roof together with a portion of the neural plate is clearly proved by experiments in which only a piece of archenteron roof was rotated, while the neural plate remained in its normal position. This experiment caused a derangement in the development of the eyes (Alderman, 1938). The removal of a part of the archenteron roof underlying the eye region of the neural plate causes defects in the development of the brain and eyes; parts of the forebrain are found to be missing, and the eyes are fused into one cyclopic eye. (Adelmann, 1937; see also Section 8–8.)

What has been said in respect to determination of the structure of parts developing from the neural plate (of which the eyes are the most easily recognizable) may be extended to the determination of the functional mechanism developing in the brain. As will be shown later (p. 339), the normal movements of the forelimb in salamander larvae depend on a central mechanism ("action system" of Weiss, 1955), which is located in the spinal cord at the level of the three pairs of spinal nerves III, IV, and V which supply the forelimbs. If the area of the neural plate giving rise to this segment of the spinal cord is excised and replaced by a more posterior part of the neural plate, the graft will fit into its new position and acquire the functional properties necessary for controlling the movements of the forelimbs. If a similar transplantation is carried out later, in the tailbud stage, so that the spinal cord at the forelimb level is replaced by a more posterior section of the cord, the graft can no longer fully take over the function of the more anterior section, and the movements of the forelimbs supplied by nerves from the graft are abnormal (Detwiler, 1936). At the same stage (tailbud), however, the section of the neural tube which develops into the medulla (the rhombencephalon) may be cut out and replaced, with inversion of its anteroposterior axis, and not only does the graft develop into a morphologically perfect medulla, tapering from the anterior end backward, but also it acquires a functional polarization in harmony with the rest of the central nervous system. All the nervous responses of the operated larvae, in particular the control of the swimming movements, may be perfectly normal (Detwiler, 1949).

These experiments proved that the detailed structure of the central nervous system, on which its functional properties depend, is not determined at the time when the rudiment of the nervous system is first formed, but that the peculiarities of the various parts of the brain and spinal cord are elaborated gradually throughout an extended period.

Further development of the brain and spinal cord is more or less complicated, depending on the degree of perfection that the central nervous system attains in any group of animals. Whereas in the lower vertebrates, such as elasmobranch fishes and amphibians, the adult conditions do not depart greatly, as far as the shape of the brain is concerned, from the conditions in the embryo, in higher vertebrates (and especially in mammals) the brain, which in the early embryo does not differ greatly from the brain of an amphibian or fish embryo, changes later in a most striking way as a result of progressive development of certain parts.

Later Development of the Spinal Cord.   The spinal cord in later embryonic stages is a tube with the lumen, the central canal, in the form of a vertical slit (as seen in cross section). The lumen is lined by a layer of epithelial cells, the **ependyma.** At the top and at the bottom of the slitlike central canal, the ependyma forms a roof plate and a floor plate. Laterally the ependyma participates in the proliferation of spinal cord cells which give rise to the mass of nerve cells and neuroglia cells of the cord (Fig. 285).

The walls of the spinal cord are thickened laterally, and as the proliferation of cells continues in the sides of the cord, the growth of the lateral parts soon exceeds the growth of the dorsal and ventral walls. Ventrally the tissue of the cord grows downward, leaving in the middle a narrow slit, the **ventral median fissure.** Dorsally the central canal becomes compressed; the ependymal layers of the right and left sides fuse and form a membrane-like structure, the **dorsal median septum,** which separates the masses of nervous tissue above the central canal just as the ventral median fissure separates the nervous tissue beneath the central canal.

Even before the dorsal part of the central canal becomes obliterated, owing to the fusion of the ependymal layers of both sides, a longitudinal, outwardly directed groove appears on the sides of the canal. The groove bears the name of "limiting groove," and it serves as a margin dividing each lateral wall of the spinal cord into two parts: the upper part is the dorsolateral plate, and the lower part is the ventrolateral plate (Fig. 286). When the nervous centers of the spinal cord become differentiated, the sensory centers develop in the dorsolateral plate, while the motor centers develop in the ventrolateral plate. In this way, the limiting groove indicates the subdivision of the spinal cord into a sensory region and a motor region. After the obliteration of its dorsal part, the central canal becomes approximately rounded or oval in cross section (Fig. 286).

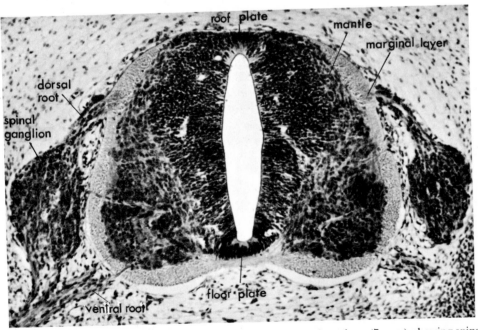

Figure 285.   Transverse section of the spinal cord of a young pig embryo (7 mm.), showing spinal ganglia and roots of the spinal nerves.

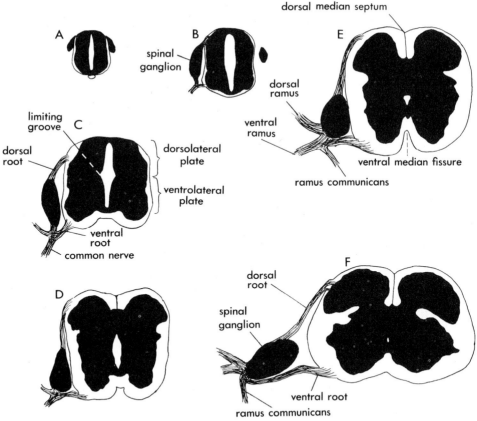

Figure 286. Changes in the gross morphology of the spinal cord during embryonic development of a mammal (the elephant shrew, *Elephantulus myurus*). Camera lucida drawings of cross sections made to the same scale. *A*, An embryo with 30 somites; *F*, fetus close to parturition. The gray matter of the cord and the ganglia are shown in black; the marginal layer and the white matter in later stages are shown in white. The nerves and spinal ganglia have been drawn only on the left side (except for first two stages).

While the gross changes in the shape of the spinal cord and its central canal are proceeding, the cells in the walls of the cord undergo a number of transformations leading eventually to the development of the spinal cord as a functioning part of the nervous system (Fig. 287). When the neural tube is first formed, its walls consist of a single layer of pseudostratified neural epithelium. All cells of this epithelium reach the inner surface of the tube (the tube lumen), but the nuclei of the cells are arranged at different distances from the lumen, giving the impression of several horizons of cells. All cells are anchored at the inner surface of the tube by intercellular connections in the form of desmosomes. All neuroepithelial cells at this stage are capable of growth and proliferation and are equivalent to one another in this respect (Fujita, 1963; Watterson, 1965). This has been proved by supplying the cells with radioactive thymidine. It was found that over a period of 8 to 10 hours all cells of the neuroepithelium take up thymidine—a clear sign that each cell is preparing for a mitotic division (Fujita, 1963; Sauer and Walker, 1959). When any cell approaches mitosis, however, it rounds off and in so doing is drawn toward the lumen of the tube, where it is connected by desmosomes to other cells at that level. Consequently, all mitoses from late prophase to telophase are seen to occur only next to the lumen of the neural tube (Fig. 287*A*). After mitosis is completed, the two daughter cells elongate again and become indistinguishable from other cells of the neuroepithelium. In this way **proliferation** of the neuro-

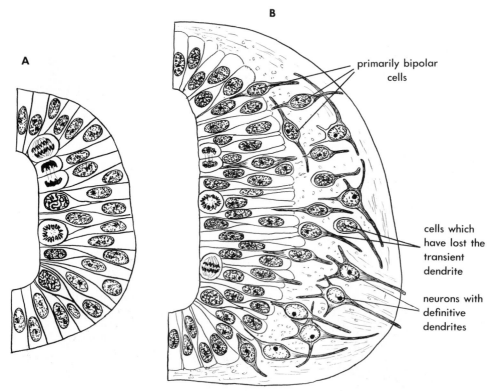

Figure 287. Stages in the development of neurons in the central nervous system (diagrammatic). *A*, Earlier stage, proliferation of neuroepithelial cells. *B*, Later stage, individual cells move out into the mantle and develop neurites and dendrites.

epithelial cells increases the number of cells in the neural tube and also increases the volume of the tube as a whole.

At a later stage some of the cells of the neuroepithelium start losing their attachment to other cells at the inner surface of the neural tube and slip outward, eventually emerging from the pseudostratified neuroepithelium. This **migration** outward is accompanied by the **differentiation** of these cells into neurons (Fig. 287*B*).

While some of the cells start migrating before showing signs of differentiation, it seems to be more common for the cells to start differentiating even before leaving the neuroepithelium (Watterson, 1965). The attenuated outer end of the columnar neuroepithelial cell becomes the neurite and develops a system of neurofibrils, which can be stained by the silver nitrate method. The inner end of the cell also develops neurofibrils and has been recognized as a "transient dendrite." The cell thus becomes distinctly bipolar.

The cells migrating out of the neuroepithelium accumulate just underneath of what will become the white matter of the spinal cord to form a loose layer of cells called the **mantle.** In addition to the primary outward migration, nerve cells may undertake more complicated migrations up or down the length of the neural tube, forming concentrations of nerve cells known to anatomists as the various nuclei of the brain and spinal cord. In the mantle the neurons attain their full functional differentiation. The "transient dendrite" becomes reduced, so that temporarily the future neurons become unipolar, but soon new, definitive dendrites are formed. After the production of the neurons a further migration of cells from the neuroepithelium gives rise to the astrocytes, the cells of the neuroglia which never acquire nervous functions. The residual neuroepithelium becomes the ependyma.

A considerable number of cells in various parts of the central nervous system degenerate during the embryonic period (during the fifth to sixth days of incubation in the chick, after Hamburger and Levi-Montalcini, 1949), thus enhancing the differences in cell numbers in the various areas. At present, there is no means of finding out in advance what will be the fate of any given cell, but the experiments previously mentioned show that the fates of individual cells are probably not firmly fixed in early stages.

The part of the spinal cord from the central canal to the outer boundary of the mantle contains numerous cells (neurons and neuroglia cells) and is the "gray matter" of the cord. Outside the mantle lies the marginal layer of the cord (Fig. 285). This layer is made up of the processes of the ependymal and neuroglia cells to which the axons of the neurons are later added in increasing numbers. When these become myelinated, the marginal layer acquires a whitish appearance in the fresh state. Hence, this part of the spinal cord is known as the white matter.

Development of Spinal Nerves and Spinal Ganglia.

The spinal cord becomes connected with the organs of the body, the limbs, and the tail by means of spinal nerves. In the development of the spinal nerves a very important part is played by the spinal ganglia. The ganglia do not form from parts of the neural tube but from cells of the neural crest, the origin of which has been described on page 173. Some cells of the neural crest migrate from their site of origin above the neural tube to the sides of the spinal cord and aggregate as small compact masses of cells segmentally arranged along the whole length of the spinal cord. The groups of cells become the rudiments of the spinal ganglia.

Most cells in the spinal ganglia differentiate as nerve cells, and long processes grow out from these cells, connecting them to the peripheral organs and to the spinal cord. The longer processes growing outward mainly to the skin are afferent and sensory and serve to bring nerve impulses to the spinal ganglia and through them to the central nervous system. The shorter processes growing out toward the spinal cord enter it as the main part of the dorsal roots of the spinal nerves and make connections with the cells in the dorsal columns of the spinal cord. The ventral roots of the spinal nerves are formed by outgrowths of the motor neurons situated in the ventral columns of the spinal cord and are efferent in nature, serving to supply nerve impulses to the somatic muscles of the body. The fibers of the motor nerves are actually the first to be formed.

When the sensory nerves start developing, the outward-growing fibers emerging from the spinal ganglia meet the nerve fibers leaving the spinal cord by way of the ventral roots, and together they form mixed sensory-motor nerves (Figs. 285 and 286). The nerves branch again to supply different areas of the body. A dorsal ramus supplies the dorsal integument and muscles of the back. A ventral ramus supplies the side of the body and the ventral surface and also gives off a connecting branch, the ramus communicans, to the ganglia of the autonomic nervous system (Fig. 286E and F). (See further, p. 352.) The limbs of vertebrates are supplied by the spinal nerves belonging to several segments in the case of both fore- and hindlimbs. Before entering the limbs, the nerves become interconnected, forming plexuses: the brachial plexus for the forelimb and the lumbar plexus for the hindlimb.

Dependence in Development of Nerves and Nerve Centers on Peripheral Organs.

It is common knowledge that the nerves and nerve centers show a certain correspondence in the degree of their development with the organs they supply. Thus, the spinal nerves supplying the limbs in terrestrial animals are stronger than the other spinal nerves, the corresponding ganglia are larger, and the spinal cord itself has swellings in the cervical and the lumbar regions, from which the fore- and hindlimbs are innervated.

It can be shown experimentally that, at least in part, this correspondence in the degree of development of the nervous system and the peripheral organs is due to a direct correlation between the two. If the forelimb rudiment of a salamander embryo is removed and the limb fails to develop, the nerves of the brachial plexus remain smaller (thinner) than they would have been if the limb were there. Also, the spinal ganglia III to V are smaller. The number of cells in each ganglion may be reduced to 50 per cent as compared with the normal or the unoperated side of the same animal. If an additional limb rudiment is transplanted, the local spinal nerves supply the nerves to the transplanted limbs, and then these nerves increase in thickness and the corresponding ganglia increase in size. The increase in the number of cells may be up to 40 per cent (Detwiler, 1926a). In similar experiments performed on the chick embryo, the sensory and motor parts of the spinal cord can be shown to be similarly reduced in the absence of a limb and increased in cases of peripheral overloading (transplantation of an additional limb-bud). (See Hamburger, 1934; Bueker, 1947.) Reduction occurs in both the gray matter (the cells) and the white matter (the fibers) of the spinal cord.

The mechanism by which the size of the nerve center (a spinal ganglion or a part of the brain or spinal cord) is altered by a change in the periphery is fairly complicated. When the periphery connected to a nerve center is increased, the increase of nerve cells in the center may be brought about by:

1. Increased proliferation—this has been found in some cases, but seems to be of minor importance.
2. Increase in the number of cells which differentiate as neurons.
3. Decrease in the number of degenerating cells.

All three of these factors may act simultaneously or may become prominent in changing proportions in various cases. The main effect of a decrease in the periphery seems to be a degeneration of large numbers of nerve cells, sometimes as many as 90 per cent in restricted areas having no other peripheral connections. One of the most remarkable facts in this dependence of the nerve centers on the periphery is that cells may apparently be affected although they have no direct connections with peripheral organs. Nerve cells may degenerate even though the organs to which they should have been related are removed before nervous connection between the central nervous system and the periphery is established. In the case of overloading, cells which do not normally send their processes into the area concerned are drawn into supplying the additional periphery. (See further, Piatt, 1948; Weiss, 1955; Hamburger, 1956.)

Not only the volume (thickness) of the nerves is dependent on the organs which they supply but also the paths which the nerves take, and thus the whole configuration of the peripheral nervous system is determined by the periphery.

The mode of development of the nerves was a subject of controversy as long as embryology relied on purely descriptive methods. The conflicting theories were those of His and Hensen. According to His, the nerves consist of processes growing out from the nerve cells and eventually reach their organs of destination or make contacts with the processes of other nerve cells in the central nervous system. According to Hensen, the nerve fibers develop from intercellular bridges connecting all the cells of the multicellular animal from the earliest stages of development. The controversy was solved by Harrison (1908), when he tried cultivating in a plasma clot pieces of neural tube taken from frog embryos. He observed directly the formation of outgrowths from the nerve cells and saw the free ends of the processes push forward through the medium. Harrison's findings were corroborated by numerous investigators, and it was also shown that the processes of nerve cells can establish new connections *in vitro*, if nervous tissue is cultivated together with a different type of tissue (e.g., muscle tissue).

The question now arises as to what directs the nerve fibers in their outgrowth from the central nervous system or from the cranial and spinal ganglia to the peripheral organ. The answer is given by the results of embryonic transplantations. If the forelimb rudiment of an amphibian embryo, prior to the outgrowth of the nerves, is cut out and transplanted to a position very near the original one, the brachial nerves will deviate from their normal paths and will be deflected in the direction of the transplanted limb. If the distance of the transplanted limb from the original position is not too great, the brachial nerves will penetrate into the limb and ramify in it just as if the limb were in its normal position. The limb in this case becomes fully functional and moves in coordination with the other limb.

The same may happen if an additional limb is transplanted into the immediate vicinity of the host limb (Fig. 288). The brachial nerves will develop branches running out to the additional limb and will supply it (Detwiler, 1926b, 1930). If the normal path of the nerves is blocked by some obstacle, the outgrowing nerves may avoid the obstacle, go around it, and still reach their normal destination. This action has been observed when a piece of mica was inserted into a frog embryo between the spinal cord and the region where the hindlimb rudiments were to develop. The nerves formed loops around the mica plate and still reached the hindlimb rudiments (Fig. 289 — Hamburger, 1929). However, if the limb rudiment is placed farther away from the normal limb site, or if the obstacle between the spinal cord and the limb rudiment is too great, the nerves fail to be attracted to the limb. If the limb rudiment is placed on the side of the embryo, too far for the normal forelimb or hindlimb nerves to reach it, it will still attract the local spinal nerves. These nerves will grow into the limb, but they cannot provide for the normal functioning of the limb: the limb cannot move.

Only the areas of the spinal cord from which the nerves of the brachial and lum-

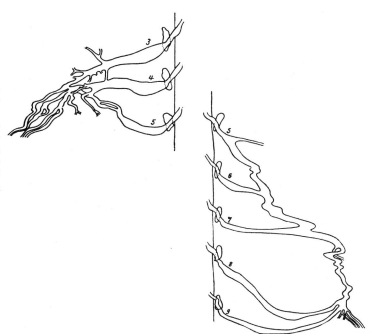

Figure 288. Nerve development after transplantation of a forelimb rudiment to a more posterior level. *Left,* nerves supplying normal (unoperated) limb. *Right,* nerves supplying transplanted limb. Only the nerves supplying the limbs are drawn. The numbers indicate the spinal ganglia. (After Detwiler, from Mangold, 1928.)

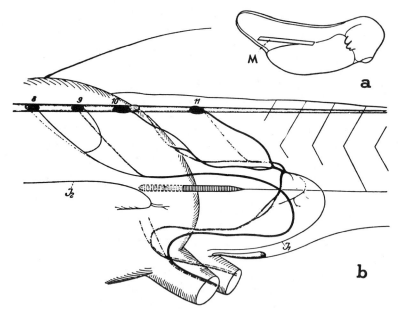

Figure 289. *a,* A mica plate (M) inserted in the path of outgrowth of the spinal nerves to the hindlimbs in a frog embryo. *b,* The nerves have grown around the obstacle and have reached the hindlimbs. (From Hamburger, 1929.)

bar plexuses originate appear to possess the properties necessary for controlling the function of limbs. These properties, as we have seen previously (p. 333), are established at some time between the neurula and the tailbud stages. The centers controlling the movements of the fore- and hindlimbs are interchangeable, however: when forelimb buds were transplanted in place of hindlimb rudiments, they acquired normal mobility. Limbs transplanted to the head may be supplied by fibers of the cranial nerves, and in that case the limbs can be seen to move synchronously with the respiratory movements of the jaws and gills. The movements of the limb are, however, rather of the nature of twitchings and differ from the coordinated movements of normal limbs (Detwiler, 1930).

The last experiment shows that the attraction of the nerves by the peripheral organs may be unspecific to a certain extent, the nerves growing out to organs other than the ones they normally supply. This conclusion is further borne out by the following experiment. An eye was transplanted into the side of a salamander embryo after the forelimb rudiment was removed. The brachial nerves were deflected from their normal path and grew out toward the transplanted eye. Having approached the eye, however, the brachial nerves failed to penetrate into the eye and to establish an actual connection with it but stopped with free ends in the tissue surrounding the eye (Detwiler and van Dyke, 1934). Two aspects of the nerve supply to organs must thus be recognized: one aspect is the outgrowth of nerves toward the organ, and the other aspect is the actual establishment of a connection between the nerve and the organ. The attraction of the outgrowing nerves to peripheral organs seems to be unspecific to a very high degree; possibly any growing mass of tissue will attract a nerve that is sufficiently near to it. The connections between the nerve and the end organ, however, can be made only if the two correspond to each other, at least in a general way.

The nature of the attraction of nerves to peripheral organs has been investigated in special experiments. It was shown that the nerves are not directed by any chemical substances diffusing from the peripheral organs; neither can their outgrowth be controlled by electrical currents or magnetic fields. The attraction is thus not a case of chemotaxis or galvanotaxis. What actually directs the outgrowing nerve fiber is the ultramicroscopic structure of the colloidal intercellular matrix through which the tip of the nerve fiber is moving. It is thus the same factor as that which directs the movements of mesenchymal cells. The nerve follows pathways consisting of bundles of oriented molecules or colloidal micellae (Fig. 290). The peripheral organs can influence the direction of nerve outgrowth by altering the submicroscopic structure of the intercellular matrix surrounding them. They probably accomplish this by withdrawing water from their surroundings and causing a shrinkage which orients the particles of the matrix along lines radiating from the intensively growing organ. This radial arrangement of particles of the matrix naturally causes the nerves to converge toward the center around which the matrix is so polarized (Weiss, 1934).

A counterpart to the dependence of the nervous system on the peripheral organs is the influence that the nerves exercise on the organs which they supply. This influence does not concern the initial stages in the development of organs but is sometimes very important for their subsequent differentiation. The muscles are originally formed before the nerves are developed, and the differentiation of muscle tissue may proceed for some time in the complete absence of nerve supply, such as when the entire neural plate is removed in an early stage of development. The histological differentiation may proceed so far that the muscles become functional; that is, they may show contractions, spontaneously or as reactions to direct stimulation. If the innervation of the muscle does not occur, however, the muscle fibers undergo a fatty degeneration and are eventually resorbed (Hamburger, 1929). Thus, some sort of **trophic** influence of the nerves is necessary for the persistence of muscle tissue.

Some sense organs depend for their persistence on the continued influence of the nerve endings supplying them. The gustatory organs in fishes and in man degenerate if the nerve which supplies them (the glossopharyngeal nerve) is interrupted. If a regeneration of the nerves takes place, the regenerating fibers upon reaching the epithelium cause the epithelial cells to differentiate as gustatory buds in place of the ones that previously degenerated (Detwiler, 1926b).

Development of the Brain. We will first trace the features manifest in the development of the brain in all vertebrates and then point out some of the peculiarities found in higher vertebrates, especially in mammals and in man.

As indicated earlier, the anterior brain vesicle, the prosencephalon, gives rise at its

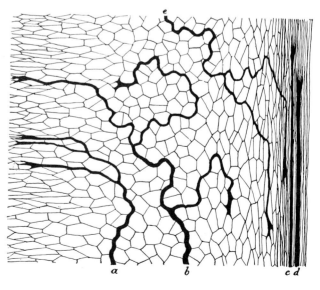

Figure 290. Outgrowth of nerve fibers in fibrous media with a different degree of ultra-structural organization (random arrangement in center turning into prevailing horizontal orientation on left and strict vertical orientation on right). (From Weiss, in Willier, Weiss, and Hamburger, 1955.)

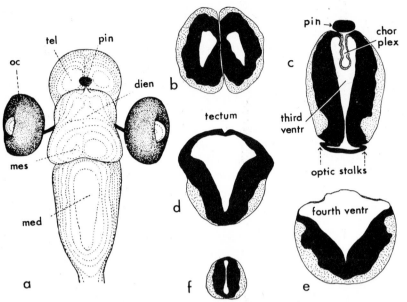

Figure 291. Reconstruction of the brain and eyes in a young frog tadpole (*a*). (After Spemann, from Huxley and de Beer, 1934.) Transverse section of (*b*) telencephalon, (*c*) diencephalon, (*d*) mesencephalon, (*e*) medulla, (*f*) spinal cord. chor plex, Choroid plexus; dien, diencephalon; med, medulla; mes, mesencephalon; oc, eye; pin, pineal body; tel, telencephalon. In the sections, masses of cells (gray matter) are shown in black, and fibers (white matter) are indicated by stippling.

anterior end to the telencephalon. The latter produces in an early stage two bulges directed anterolaterally which become the cerebral hemispheres (Figs. 291 and 292). Each bulge of the telencephalon contains a pocket-like cavity, which is an extension of the original cavity of the anterior brain vesicle (prosencephalon). The two cavities are known as the **lateral ventricles** of the brain. Originally, they are in broad communication with the rest of the cavity of the prosencephalon, but later the channels leading into the lateral ventricles may become constricted. These channels are called the interventricular foramina (also known as the foramina of Monro).

The cerebral hemispheres grow out forward beyond the original anterior end of the prosencephalon. The median strip of the anterior wall of the prosencephalon persists in a relatively unchanged state as the **lamina terminalis.**

The telencephalon in lower vertebrates is the primary center of the olfactory sense and is intimately connected with the olfactory organs which develop immediately anterior and lateral to it. The nerve fibers growing out from the sensory cells in the olfactory organs penetrate into the telencephalon near its anterior end, in a part which differentiates as the olfactory lobe. Being primarily a center for the olfactory sense, the telencephalon shows the same kind of dependence on the olfactory organ as the spinal nerves and the spinal cord show in respect to the peripheral organs to which they are connected. This has been shown by removing the olfactory rudiments and by transplanting additional ones. In the absence of an olfactory organ, the corresponding lobe of the forebrain remains underdeveloped (Burr, 1916). If an additional olfactory organ is transplanted into the region beside the normal one, the fibers of the olfactory nerve grow from the transplanted organ into the adjoining lobe of the forebrain, and this causes an increase in the size of the latter (Fig. 293 — Burr, 1930).

In the lower groups of vertebrates, the walls of the cerebral hemispheres are only moderately thickened, and the nerve cells remain accumulated on the inner surface of the walls, that is, on the surface facing the internal cavity.

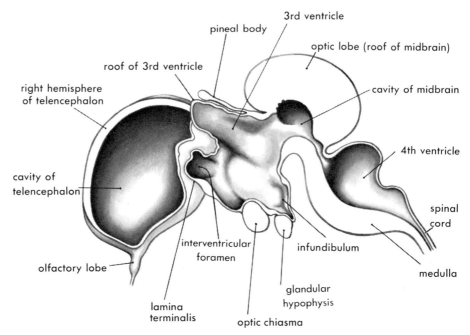

Figure 292. The brain of an 8-day chick embryo, dissected in the median plane. The right telencephalon hemisphere, however, has been dissected in a paramedian plane to show the cavity of the brain. (Redrawn from Lillie, 1919b.)

In the higher vertebrates, starting with the reptiles, the outward migration of nerve cells is not restricted to the mantle surrounding the cavity of the brain, but the migrating cells penetrate beyond the layer of nerve fibers which make up the white matter originally surrounding the gray matter and accumulate near the surface, giving rise to the **cortex** of the cerebral hemispheres.

The diencephalon in all vertebrates is remarkable in that it produces a great variety of structures with different functions, in addition to the two eye vesicles, which will be

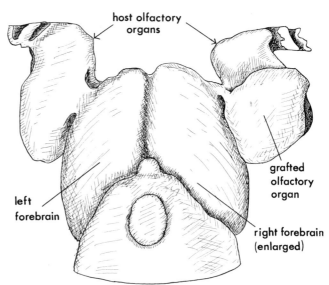

Figure 293. Hyperplasia of the right lobe of the telencephalon due to transplantation of an additional olfactory organ in *Ambystoma*. (From Burr, 1930.)

referred to later. The brain cavity in the region of the diencephalon remains fairly large and is known as the **third ventricle** of the brain. The cells in the brain wall become concentrated mainly in the sides of the diencephalon which become thickened and are known as the **optic thalamus.** The thickening is continued here until the inner surfaces of both sides meet in the middle, forming what is called the **commissura mollis.** The optic thalamus is primarily an association center, which increases in importance in higher vertebrates.

The greater part of the dorsal wall, which we may refer to as the roof of the diencephalon, becomes membranous and later does not contain any nerve cells at all. Instead it is richly supplied with blood vessels and becomes the **choroid plexus,** which later bulges down into the cavity of the third ventricle. The choroid plexus is the pathway by which nutrition and oxygen are brought into the ventricles of the brain. Processes of the choroid plexus may penetrate from the third ventricle into the lateral ventricles by way of the foramina of Monro.

Only the posterior section of the roof of the diencephalon retains the nervous character, but parts of it form dorsally directed outgrowths, of which the most important are the parietal organ and the pineal body. Both are formed as rather long, finger-like outgrowths of the brain roof, the end sections of which become transformed into more or less rounded masses of cells, while the stalks become constricted and may even be interrupted later. In lower vertebrates, either the pineal body (in some frog tadpoles—see Eakin, 1961), or the parietal organ (in reptiles), or both (in cyclostomes) become eyelike organs, but in higher vertebrates they differentiate as glandular structures. The homologies and the function of these structures have been the subject of many investigations (Stebbins and Eakin, 1958), but in neither respect has a clear answer as yet been given. Both organs develop mid-dorsally, that is, at the site where the neural folds fused at an earlier stage, thus raising the question of whether the presumptive material of each organ is contained in one of the neural folds or in both. It has been found that at least in the case of the pineal body there are originally two rudiments, one on each edge of the neural plate, and that after neurulation these two rudiments fuse into one single unpaired organ (van de Kamer, 1949).

The floor of the diencephalon produces in all vertebrates a funnel-like depression, the **infundibulum.** Part of the wall of the latter becomes segregated from the brain wall and fused with an outgrowth from the stomodeal invagination (Section 14–2), the two together forming the **hypophysis** or **pituitary gland,** the most important endocrine gland in vertebrates. The walls of the diencephalon on the sides and posterior to the infundibulum differentiate as another important nerve center, the **hypothalamus,** which is the center controlling, through the autonomic system, the vegetative functions of the body.

The midbrain remains a fairly simply organized part of the brain. The walls of the midbrain become thickened mainly ventrally, but the lateral walls and the roof are also fairly thick, and the latter gives rise to an important nerve center, the **tectum.** The cavity of the midbrain becomes narrow and is known as the **aqueduct of Sylvius.** In the lower vertebrates, fishes and amphibians, the tectum serves as the primary center of the visual organ, and the nerve fibers entering the brain from the eyes end here. Beginning with the anuran amphibians, however, the posterior part of the tectum acquires connections with nerve fibers coming in from the ear, and in the amniotes the tectum develops four thickenings, the **corpora quadrigemina,** of which the two anterior thickenings are concerned with the sense of vision, while the two posterior ones are related to the sense of hearing.

Just as the hemispheres of the telencephalon are related to the olfactory organ, the tectum of the midbrain, in lower vertebrates, is related to the eyes and is dependent in its differentiation on the nerve fibers entering the midbrain from the eyes. In amphibians and fishes, the optic tectum fails to develop normally and remains thinner than usual if the eye rudiment is removed or reduced in size in an early stage (Dürken, 1913; Harrison, 1929; White, 1948).

The ventrolateral parts of the midbrain show some resemblance to the ventrolateral plates of the spinal cord, inasmuch as they contain groups of motor cells which send out processes forming two pairs of motor nerves, the third (oculomotor) and the fourth (trochlear) cranial nerves (p. 347). In mammals, the ventrolateral parts of the mesencephalon are very enlarged, owing to the number of nerve fibers passing through this part of the mesencephalon on the way from the cerebral hemispheres to the medulla and the spinal cord. These thickened parts are known as the **cerebral peduncles.**

The neural tube may be fairly straight or only slightly curved at the time of its formation, but in later stages it becomes bent at an angle at one or more levels. These bends are known as **flexures.** The most important flexure, and the one found consistently in all vertebrates, is that at the level of the midbrain, known as the **cephalic flexure.** Here the foremost part of the brain (the telencephalon and the diencephalon) is bent downward in front of the anterior tip of the notochord.

The rhombencephalon gives rise to the metencephalon and the medulla oblongata. The cavity of the rhombencephalon expands especially anteriorly, just behind the midbrain, and becomes the **fourth ventricle.** The roof of the medulla thins out and is converted into a second choroid plexus, the **posterior choroid plexus,** which is similar to the one developed from the roof of the diencephalon. The future nerve cells are concentrated lateroventrally in the floor of the medulla but are separated into two masses by a median groove. This arrangement of nervous tissue gives the medulla a very characteristic appearance in cross section (Fig. 291e).

In spite of differences in gross configuration, the rhombencephalon in its internal organization shows an unmistakable similarity to the spinal cord. A ventral floor plate extends along the whole length of the rhombencephalon, forming a median groove, the **median sulcus.** On the sides of the rhombencephalon a lateral groove is clearly indicated which is an extension of the limiting groove of the spinal cord. This groove subdivides the mass of tissue on each side into a ventrolateral plate and a dorsolateral plate. Owing to the expansion of the membranous roof of the rhombencephalon, the dorsolateral plate lies lateral rather than dorsal to the ventrolateral plate.

The rhombencephalon, as indicated before, gives rise to the metencephalon and the medulla oblongata. The metencephalon in a younger embryo is no more than a slightly thickened section at the anterior end of the medulla (which in higher vertebrates gives rise to the **pons Varolii)** and a transverse bar in the roof of the brain, just behind the mesencephalon and anterior to the choroid plexus of the fourth ventricle.

The dorsolateral part of the metencephalon, corresponding to the dorsolateral plate of the spinal cord, gives rise to the cerebellum. The cerebellum, although a middorsal organ in its final form, develops from two swellings of the brain tissue to the right and left of the anterior end of the membranous roof of the fourth ventricle.

The development of the cerebellum starts rather late. In the human embryo, the growth of the cerebellar rudiments begins during the second month of pregnancy. At two months the swellings are very distinct, though still separate. Later the two swellings fuse, producing the **vermis** of the cerebellum in the midline, while the lateral parts become the hemispheres. As in the telencephalon, the nerve cells migrate in the cerebellum from their place of origin near the cavity of the brain toward the external surface, producing the cortex of the cerebellum.

In the medulla, the main mass of gray matter lies adjacent to the fourth ventricle. Here are situated the centers of the cranial nerves entering and leaving the medulla.

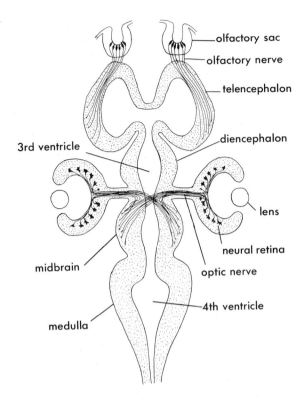

Figure 294. Diagrammatic horizontal section of the brain in a mammal. The diagram also shows the relations of the olfactory and optic nerves to the brain. The parts developing from the neural tube are shown by stippling; the parts developing from the epidermis are white.

At its posterior end, the medulla gradually merges into the spinal cord. The membranous part of the roof becomes narrower and eventually disappears, and the medioventral groove becomes deeper and is directly continued as the central canal of the spinal cord, while the dorsolateral plates converge and fuse dorsally over the canal.

Development of Cranial Nerves and Ganglia. The cranial nerves of vertebrates, when compared with the spinal nerves, show a great diversity in configuration, distribution, and function, and the same diversity can also be noted in their development in the embryo.

To start with, the two most anterior nerve pairs, nerves I (olfactory) and II (optic), are completely different in nature and cannot be in any way compared with the spinal nerves (Fig. 294).*

Nerve I (olfactory) consists of fibers which are processes of the primary nerve cells of the olfactory organ. The fibers are not joined together in a compact bundle (or two bundles), so that one could speak of many olfactory nerves rather than of only a pair of nerves being present. The nerves are very short and enter the olfactory lobe of the telencephalon. The fibers of the olfactory nerves establish connections with nerve cells in the telencephalon.

Nerve II (optic) is not really a peripheral nerve at all but must be regarded as a nerve tract in the central nervous system. The sensory retina of the eye develops at the expense of a lateral outgrowth of the brain wall and is thus essentially a part of the central nervous system. The stalk of the eye vesicle becomes converted into the optic nerve. When the cells of the nervous retina differentiate (see Section 12–2 on the development of the eye), the ganglion cells situated on the distal surface of the sensory

*The unpaired and poorly developed *nervus terminalis* is not sufficiently well studied to be considered here.

retina produce axons which grow along the stalk and enter the brain. In the optic chiasma, the fibers cross to the opposite side and end in the contralateral half of the brain, but in mammals a smaller portion do not cross and thus end in the ipsilateral side.

The following nerve pairs (III to X in fishes and amphibians and III to XII in amniotes) may be compared to a certain extent with spinal nerves, even though the comparison cannot be complete in any case (Figs. 295 and 296).

The three nerves supplying the eye muscles develop much like the ventral (motor) roots of the spinal nerves; they are mainly constituted of axons growing out from groups of motor neurons located in the ventrolateral plates of the brain.

The mixed and sensory cranial nerves—V, VII, VIII, IX, and X—possess ganglia, and in this respect they resemble the dorsal roots of the spinal nerves. The development of the cranial ganglia differs, however, from that of the spinal ganglia. It has been proved experimentally, and is generally recognized, that the spinal ganglia develop from cells

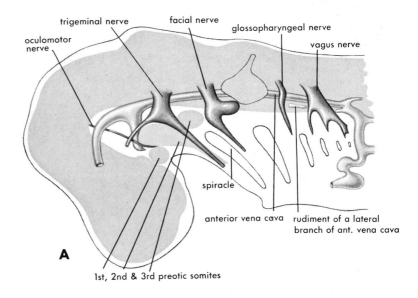

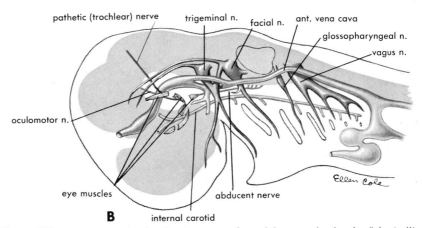

Figure 295. Two stages in the development of cranial nerves in the dogfish. *Scyllium. A,* Earlier stage, with preotic head somites intact and not all cranial nerves present yet. *B,* Later stage with preotic somites differentiating into eye muscles and all cranial nerves present. (After de Beer, 1924.)

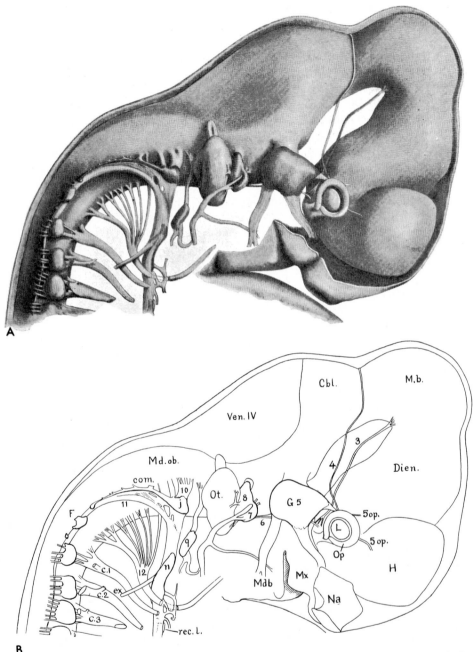

B

Figure 296. The brain and cranial nerves of a 12 mm. pig embryo. c.1, c.2, c.3, Spinal (cervical) nerves 1–3; Cbl., cerebellum; com., commissure joining accessory ganglia of vagus nerve; Dien., diencephalon; ex, external branch of the spinal accessory nerve; F, "Froriep's ganglion," ganglion of a vestigial spinal nerve preceding the "first"; G, gasserian ganglion (V); H, cerebral hemisphere; j, jugular ganglion (X); L, lens; M.b., midbrain; Mdb, mandibular process (rudiment of lower jaw) with mandibular nerve (V); Md.ob., medulla oblongata; Mx, maxillary process (rudiment of upper jaw) with maxillary nerve (V); n, ganglion nodosum (X); Na, nasal pit; Op., optic cup; Ot., otocyst (ear vesicle); rec. 1., recurrent laryngeal nerve (branch of X); Ven. IV, roof of fourth vesicle; 3, oculomotor nerve (III); 4, trochlear nerve (IV); 5 op., branches of ophthalmic division of trigeminal nerve (V); 6, abducent nerve (VI); 7, geniculate ganglion of facial nerve (VII); 8, vestibular ganglion (VIII); 9, petrosal ganglion (IX); 10, vagus nerve (X); 11, spinal accessory nerve (XI), partially fused with the vagus nerve; 12, hypoglossal nerve. (From Lewis, 1903.)

of the neural crest. The cranial ganglia, on the other hand, derive their material from two sources: from the neural crest and from thickenings of the epidermis on the sides of the head, known as the **placodes.** Two rows of such thickenings can be distinguished in the lower vertebrates (fishes and amphibians), an upper row of **dorsolateral placodes** and a lower row of **epibranchial placodes** (Fig. 297).

The dorsolateral placodes play a part in the development of the lateral line sense organs (p. 365) and also the ear, which may be said to develop from one of the placodes (p. 367). The epibranchial placodes are located just above the gill slits. In amniotes the placodes are not as distinct as in lower vertebrates, but nevertheless, migration of cells from slightly thickened portions of the epidermis and their contribution to the formation of nerve ganglia have been definitely established (Fig. 298).

Simple observation proved to be inadequate to establish the contribution of the two sources to individual ganglia, with the result that widely divergent views were held on the subject by different authors. In recent years, however, experimental methods have been applied, and the question of the origin of the various cranial ganglia has been solved satisfactorily.

The experimental approach consists in cutting out parts of the neural tube with the neural crest in early embryos and observing whether particular ganglia are formed without the neural crest (Levi-Montalcini and Amprino, 1947; Hamburger, 1961; Hammond and Yntema, 1958). A counterpart of such an experiment is the removal of a strip of epidermis on the side of the head containing the presumptive material of the placodes (Hamburger, 1961). Cranial ganglia developing after operations of the first type must be derived from the placodes; those which are found after operations of the second type can be presumed to be derived from the neural crest. Much of the work on the origin of cranial ganglia was done on chicken embryos.

Another possible approach to the problem is the transplantation of epidermis between amphibian embryos differing in the amount of pigment in cells. If epidermis of a darkly pigmented species is transplanted to a host having light pigmentation, the gan-

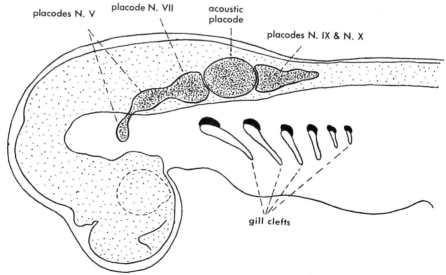

Figure 297. Diagram showing position of the dorsolateral placodes (close stippling) and epibranchial placodes (black) in a fish embryo. The brain and spinal cord are shown by sparse stippling. (Combined from Ihle, 1947 [after Froriep] and from Goodrich, 1958.)

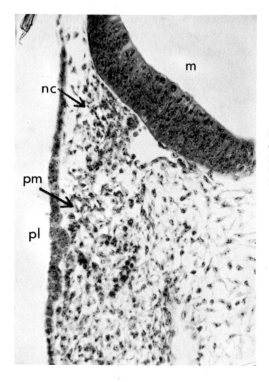

Figure 298. Origin of the trigeminal ganglion in the chick embryo from neural crest and from a dorsolateral placode: m, medulla oblongata; nc, neural crest mesenchyme; pl, trigeminal epidermal placode; pm, mesenchyme derived from placode. (From Hamburger, 1962.)

glia formed from the transplanted epidermis can be recognized at once by their dark pigment (Balinsky, unpublished experiments).

By applying experimental methods, it has been established (Hamburger, 1961) that the ganglion of the trigeminal nerve (V), the **ganglion semilunare,** is composed of both neural crest cells and cells from a placode (presumably a dorsolateral placode). The cells of placodal origin are mainly responsible for the outgrowth of the purely somatic sensory ophthalmic and maxillary nerves and for the somatic sensory fibers of the mandibular nerve.

The facial nerve (VII) in aquatic vertebrates supplies innervation to the lateral line canals of the head and the muscles in the region of the hyoid arch. In terrestrial vertebrates the lateral line sense organs disappear, and the muscles of the hyoid arch change their function, becoming the facial muscles in mammals and man. Accordingly, the facial nerve becomes mainly a visceral motor nerve in higher vertebrates. The ganglion of the facial nerve in the higher vertebrates, the **geniculate ganglion,** is derived almost exclusively from an epibranchial placode, while in fishes it receives a contribution from a dorsolateral placode forming a special upper section of the ganglion. This section disappears with the disappearance of the lateral line sense organs.

The acoustic nerve (VIII) is a special nerve and may be considered as a specially differentiated section of the facial nerve. The roots of the two nerves lie very close to each other in fishes and amphibians, though they are clearly separated in amniotes. The **acoustic ganglion** is derived from cells of the ear rudiment, the ear vesicle (van Campenhout, 1935; Batten, 1958). As the ear vesicle is formed from a thickening of the epidermis belonging to the dorsolateral placodal system, the acoustic ganglion may be said to be of placodal origin.

The glossopharyngeal nerve and the vagus have much in common in their composition and development. In both, the cells of the ganglia are derived from two sources, the neural crest and the placodes, but contrary to what is seen in the semilunar ganglion of the trigeminal nerve, cells from different sources do not join together in one ganglion but, instead, form separate ganglia. The glossopharyngeal nerve (IX) has an upper **"root" ganglion,** consisting of cells of neural crest origin, and a lower **petrosal ganglion,** derived from an epibranchial placode.

The vagus nerve (X) also has two main ganglia, the upper **ganglion jugulare,** corresponding to the root ganglion of the glossopharyngeal nerve and consisting of cells derived from the neural crest, and a lower **ganglion nodosum,** derived from epibranchial placodes. In addition to these a number of **accessory ganglia** adjoin the jugular ganglion posteriorly. These ganglia are connected to the medulla by independent rootlets, but the fibers of the accessory ganglia connecting them with the periphery are joined to the fibers of the main vagus nerve. The presence of several ganglia connected to the vagus is additional proof of the polysegmental nature of this nerve. The accessory ganglia of the vagus represent a transition from the cranial to the spinal nerve system.

The **spinal accessory nerve (XI)** and the **hypoglossal nerve (XII)** of amniotes are formed of fibers growing out from motor cells lying in the floor of the medulla.

The Human Brain. The development of the brain in higher vertebrates can best be illustrated by a brief description of the changes which the brain rudiment undergoes in the human embryo.

The brain of the human embryo toward the end of the first month after conception is not very different from the brain of an amphibian embryo, except that it is distinctly more elongated (Fig. 299a). The eye rudiments are separated from the prosencephalon, and the cephalic flexure is indicated, but there is, as yet, no trace of the progressive development of the hemispheres of the forebrain or the cerebellum.

Soon after the beginning of the second month after conception, the telencephalon forms a conspicuous bulge dorsally in front of the eye rudiments. The bulge is slightly bilobed, the first indication of the future hemispheres of the brain. The **cephalic flexure** is increased to such an extent that the brain appears to be bent on itself (Fig. 299b). In addition to the cephalic flexure, the brain now shows two more flexures. At the level of the anterior part of the rhombencephalon, the brain is bent with the convexity facing downward, forming the **pontine flexure.** The metencephalon with the pons Varolii (after which the flexure is named) lies in front of the flexure, and most of the medulla remains posterior to the flexure. A third flexure, with the convexity facing dorsally (the same as the cephalic flexure), appears at the junction between the medulla and the spinal cord. It is called the **cervical flexure.**

About the middle of the second month after conception, the flexures of the brain become much more distinct, especially the pontine flexure. The development of the cerebellum has not progressed much, but the midbrain enlarges considerably and attains its largest relative size. The main advance, however, is shown by the telencephalon. The two lobes (indicated previously) enlarge greatly and spread out forward, upward, and backward, partially covering the laterodorsal surfaces of the diencephalon (Fig. 299c).

By the beginning of the third month after conception, the hemispheres of the telencephalon constitute, by far, the greatest part of the brain (Fig. 299d). They have expanded backward to such an extent that they almost completely cover the diencephalon. A broad shallow groove on the outer surface of each hemisphere (the future lateral fissure) indicates the separation of the temporal lobe of the brain. The mesencephalon is greatly expanded dorsally and forms a large mass posterior to the cerebral hemispheres.

The cerebellum is the last part of the brain to become conspicuous on inspection from the outside, since the rudiments of the cerebellum are formed originally as masses

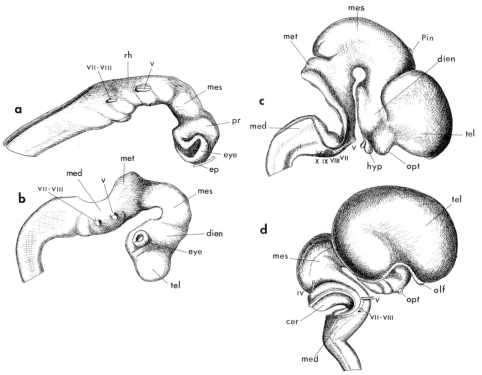

Figure 299. Four stages in the development of the brain in a human embryo, cer, Cerebellum; dien, diencephalon; ep, epidermis; hyp, hypophysis; med, medulla; mes, mesencephalon; met, metencephalon; olf, olfactory bulb; opt, optic stalk (or nerve); pin, pineal body; pr, prosencephalon; rh, rhombencephalon; tel, telencephalon. Roots of the cranial nerves are indicated by means of Roman numerals. (After Hochstetter, from Grosser, 1945.)

of brain tissue bulging into the fourth ventricle from the sides of the metencephalon. These masses increase and later fuse together above the cavity of the brain, and only after this does the rudiment of the cerebellum swell to the exterior of the fourth ventricle. This occurs toward the third month after conception.

At four months after conception, the cerebral hemispheres have grown so large that they cover the midbrain from the sides and touch the cerebellar hemispheres, which by this time have become clearly discernible. Even at this time the surface of the cerebral hemisphere, apart from the lateral fissure, is quite smooth, but during the second half of the period of pregnancy the surface becomes wrinkled and folded, giving rise to the characteristic gyri of the human brain.

Development of the Autonomic Nervous System. The autonomic nervous system consists exclusively of efferent neurons conveying nervous impulses to the cardiac and smooth muscles, to digestive and sweat glands, and to some endocrine glands.

The bodies of the neurons of the somatic motor nerves and visceral motor nerves are located in the ventrolateral plates of the spinal cord and brain, and the axons of these neurons lead directly to the effector muscles. In the sympathetic system, the pathway from the spinal cord to the effector organ is made up of at least two neurons. The body of the first neuron lies in the spinal cord, in the lateral part of the ventrolateral plate. The axons leave the spinal cord in the ventral motor roots and then, by way of the rami communicantes (p. 337), reach a system of sympathetic ganglia lying in two rows to the right and left of the dorsal aorta, where these fibers terminate. Because the neurons whose bodies are located in the central nervous system reach only as far as the ganglia, they are known as **preganglionic neurons.** Nerve cells located in sympathetic ganglia, receiving impulses from the preganglionic neurons, forward them to the effector organs directly or by means of further neurons situated in the secondary and even tertiary sympathetic ganglia, the latter being located in the immediate vicinity of the effector organs (on the gut, the bladder, and so forth).

The origin of the preganglionic neurons of the sympathetic system, as well as of the cranial components of the parasympathetic system (visceral efferent components of nerves VII, IX, and X), presents no special problems. The origin of the ganglia of the autonomic system, on the other hand, in particular the origin of the chain of sympathetic ganglia, has been a subject of much controversy. At least three possible sources of these nerve cells have been considered—namely, the neural crest, the neural tube, and the local mesoderm.

After much inconclusive work in which descriptive methods were used, a satisfactory solution to the problem has been reached by the use of experimental methods (Yntema and Hammond, 1947). When parts of the neural crest are removed in young embryos (of the chick), there is an immediate effect on the development of the sympathetic ganglia, which either are reduced or in some cases may be completely lacking (Hammond and Yntema, 1947; Nawar, 1956). The other two sources under consideration (the neural tube and the local mesoderm) were not tampered with in these experiments, so that the defects are clearly an indication that the sympathetic ganglia and the neurons contained in them are of neural crest origin. Even the tertiary sympathetic ganglia in the walls of the gut are of neural crest origin and are absent when access of the neural crest cells to the gut rudiment is precluded by operation (Andrew, 1963). In a separate investigation with the same methods, the ciliary ganglion (the parasympathetic ganglion connected to the oculomotor nerve) also was found to be formed from cells of the neural crest originating in the mesencephalon area of the neural tube (Hammond and Yntema, 1958).

The actual formation of the sympathetic ganglia takes place in rather early stages of the development of the embryo, before the spinal nerves start growing out. When the neural crest cells migrate downward from their site of origin, dorsal to the neural tube, groups of the migrating cells accumulate along the dorsal aorta. The cells first aggregate into paired ganglia, and these become linked with one another lengthwise to form a longitudinal nerve cord on each side of the dorsal aorta. The cord is secondarily extended forward into the head region. When the spinal nerves start growing out, they eventually establish connection with the sympathetic trunks by means of the rami communicantes, as already stated.

12–2 DEVELOPMENT OF THE EYES

The Optic Cup. The origin of the optic vesicles has already been described. The optic vesicles push outward until they reach the epidermis, displacing the intervening mesenchyme, so that they come into direct contact with the inner surface of the epidermis. Next, the external surface of the optic vesicle flattens out and invaginates inwards, so that the vesicle is transformed into a double-walled, cuplike structure—the **optic cup** (Fig. 300). The invaginated wall of the optic cup is much thicker than the remaining external wall. The first develops into the **sensory retina** of the eye, and the second develops into the **pigment coat** of the retina **(tapetum nigrum).** The rim of the eye cup later becomes the edge of the pupil. The cavity of the optic cup is the future posterior chamber of the eye, which is filled by the vitreous body. The opening of the eye cup is very large at first, but later the rims of the cup bend inward and converge, so

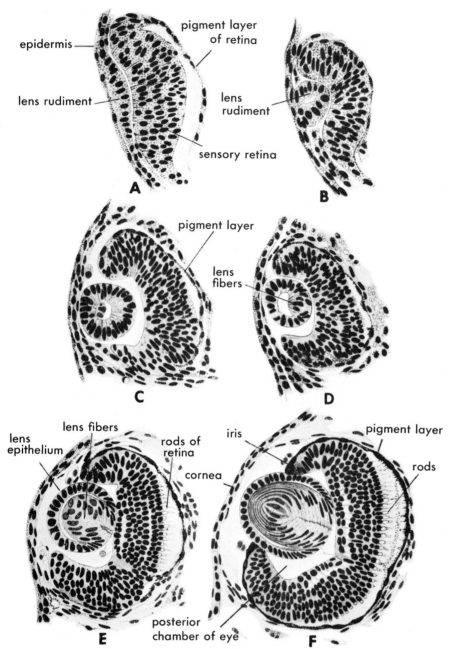

Figure 300.   Stages of the development of the eye in the axolotl. (After Rabl, from Spemann, 1938.)

that the opening of the pupil is constricted and reduced to its final relative dimensions. The rim of the optic cup surrounding the pupil becomes the **iris.** The constriction of the pupil does not take place equally all around the circumference of the eye, but is delayed on the ventral edge of the eye cup. A groove, the **choroid fissure,** remains here, cutting through the otherwise approximately circular edge of the eye cup and reaching inward as far as the optic stalk (Fig. 301). This fissure serves for the entry into the posterior chamber of the eye of a blood vessel and of mesenchymal cells which are found later

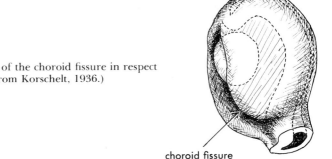

Figure 301. Optic cup, showing the position of the choroid fissure in respect to the lens and the eye stalk. (After Froriep, from Korschelt, 1936.)

choroid fissure

in the vitreous body. The fissure normally closes during embryonic life; however, it may persist, and as a result a gap is left on the ventral edge of the pupil. This deformity, which in uncomplicated cases does not substantially affect vision, is known as the coloboma of the iris. In humans it is caused by hereditary factors.

The size of the optic vesicles relative to the rest of the prosencephalon may vary considerably in different vertebrates. Even in one order among the vertebrates, such as the frogs, it was found by direct measurement that the mass of cells used for the formation of the eye vesicles may, in different species, range from 10 per cent to 50 per cent of the volume of the prosencephalon (Balinsky, 1958). As a general rule, the eye rudiments are large in bony fishes, reptiles, and birds, smaller in amphibians, and relatively very small in mammalian embryos. The determination of the optic cup is due to the action of the underlying roof of the archenteron on a part of the neural plate. In the late neural plate stage the determination appears to be irrevocable, and parts of the optic vesicle cannot differentiate in any other way than by developing into retina, iris, or pigmented epithelium of the eye. The eye rudiment may be excised and transplanted into any region of the embryo, and its development will continue more or less normally, producing a heterotopic eye.

A transplanted eye rudiment, although no longer capable of being transformed into other tissues, cannot develop normally unless it is surrounded by mesenchyme. Without mesenchyme in its environment, the differentiation of the optic rudiment remains extremely poor (Holtfreter, 1939b; Lopashov, 1956).

The determination of the eye as a whole does not mean that all the parts of the eye rudiment are also determined. The determination of the parts of the eye occurs much later. The eye rudiment in the neural plate stage and in the optic vesicle stage may be split in two, and each half will then develop into a complete eye. This can best be demonstrated if a part of the eye rudiment is transplanted. It is found then that the remaining and the transplanted parts each develop into a small eye. Experiments have also been performed on the eye rudiment at the stage when the optic vesicle is being transformed into an optic cup and the two parts of the cup, the future sensory retina and the future pigment coat, become morphologically distinguishable. If a piece of the presumptive pigmented epithelium is excised and transplanted into the vicinity of a normal eye of another embryo, it will develop into a complete eye, consisting of both a pigment coat and a sensory retina (Fig. 302 – Dragomirow, 1933). However, a prerequisite for this is that the piece not be too small. Very small fragments of the eye rudiment usually develop into pigmented epithelium only.

The ability of a part of the organ rudiment to develop as a whole is analogous to a similar ability of parts of the egg in the early stages of cleavage in some animals. The pro-

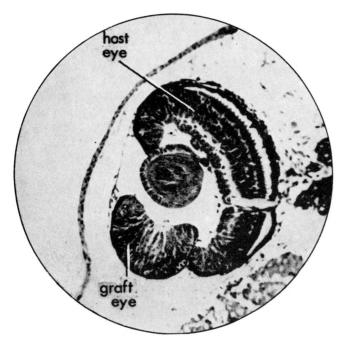

Figure 302. A whole eye, developed from a transplanted piece of presumptive pigment coat. (From Dragomirow, 1933.)

cess observed in both cases is known as **self-regulation.** It is found in many organ rudiments; in fact, it can be considered as a common property of organ rudiments. The ability to self-regulate presupposes that the parts of the rudiment (or of the egg in the early cleavage stages) are not determined.

In the case of the eye cup, the absence of determination of parts can also be demonstrated in another way. If a suitable inductor is applied to the outer surface of the optic cup—that is, to the surface which normally differentiates into the pigmented epithelium—it may be induced to develop as sensory retina, so that the eye has two retinas, a normal one and an additional one. The latter is never as large as the normal one. A suitable inductor for this purpose is the sensory epithelium of an ear (Fig. 303). The experiment therefore consists in the transplantation of an optic vesicle into the immediate vicinity of the ear vesicle to insure the close contact of the epithelia of the two organ rudiments (Dragomirow, 1936).

With advancing development it is apparent that the rim of the optic cup becomes increasingly different from the deeper-lying parts. The constriction of the pupil (already referred to) takes place at the expense of a considerable thinning out of the wall of the rim of the optic cup. The thinned-out portion becomes the iris of the eye, while the remaining part, which stays considerably thicker, gives rise to the retina proper.

In the iris, large amounts of pigment are deposited in the outer epithelial layer. This layer is actually a part of the pigment coat of the eye cup. In addition to cells carrying pigment, this layer also gives rise to the smooth muscle fibers of the sphincter and dilator muscles of the iris. In the sensory retina, the cells start differentiating into the sensory and nerve (or ganglion) cells. The first trace of this differentiation is seen in the arrangement of the nuclei of the cells in several layers. The nuclei situated in the layer nearest to the pigment coat belong to the future rod and cone cells. The rudiments of the rods and cones appear as cytoplasmic processes on the inner ends of these cells. The remaining nuclei, arranged in two or more layers nearer to the cavity of the eye cup, give rise to the various types of intermediate and ganglion cells of the

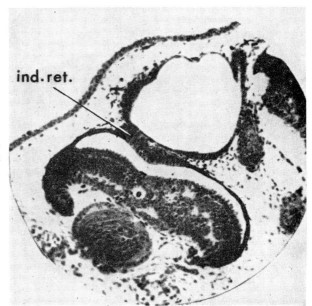

Figure 303. Induction of an additional sensory retina (ind. ret.) in an eye by means of contact with an ear vesicle. (From Dragomirow, 1936.)

retina. Nerve processes arising from the ganglion cells of the retina grow out toward the brain, and the path which they take is along the stalk of the eye cup. In this way the stalk of the optic cup becomes transformed into the optic nerve. On reaching the floor of the diencephalon, in all vertebrates except mammals, the nerve fibers do not enter the same side of the brain but cross over to the opposite side and there penetrate into the wall of the diencephalon and the mesencephalon. Where the nerve fibers of the two eyes cross and bypass each other on their way to the contralateral parts of the brain arises the **optic chiasma.** In mammals, the crossing of the fibers of the optic nerve in the chiasma is not complete; a small proportion of the fibers enter the brain on the same side as the eye in which they originated.

The optic cup, even when fully differentiated, is not yet a complete eye. Certain accessory structures have to be added to it to make the eye fully functional. The most important of these structures is the **lens,** which serves for the refraction of light rays entering the eye. The lens is not developed from the optic cup, but from the epidermal epithelium with which the optic vesicle comes in contact, as mentioned before. As the outer wall of the optic vesicle begins to invaginate to become the retinal layer of the optic cup, a thickening appears in the epithelium which is in contact with the invaginating part of the optic cup (Fig. 300a). This thickening is the rudiment of the lens.

The Lens.   The way in which the lens rudiment is separated from the remainder of the epidermis varies in different classes of vertebrates. In birds and mammals, the epidermal thickening folds in to produce a pocket which is for a short time open to the outside and later becomes a vesicle lying in the opening of the iris (in the pupil). In amphibians and bony fishes, the thickening in the formation of which only the inner layer of epidermis takes part is nipped off from the epidermis as a solid mass, but later the cells of this mass rearrange themselves into a vesicle. In both cases the vesicle must undergo further differentiation before the lens can function as a refracting body. This happens in such a way that the cells on the inner side of the lens vesicle elongate, become columnar at first, and later are transformed into long fibers. During this transformation, the nuclei of the cells degenerate, and the cytoplasm becomes hard and transparent. The fibers are arranged in the lens in a very orderly way, forming the spherical or ellipsoid refracting body of the lens. Part of the lens epithelium remains unchanged and covers the sphere of fibers distally. The junction between the unchanged lens epithelium and the mass of fibers is the growth point of the lens. Here the epithelial cells are continuously transformed into fibers, so that the refracting body grows by apposition of new fibers.

Between the development of the optic cup and the development of the lens there exists, in most of the vertebrates studied experimentally, a direct causal relationship; the development of the lens is dependent on an induction from the optic vesicle. As the optic cup touches the epidermis, it gives off a stimulus of some kind, which causes the epidermal cells to develop into the lens rudiments (Lewis, 1904). Any epidermal cells are able to react to the induction of the optic vesicle, and without this induction the lens does not develop at all, or at least the development is defective. The dependence of lens development on the action of the optic cup can be shown by several types of experiments. One type of experiment is to remove (excise) the eye rudiment before it can reach the epidermis. Such an operation usually leads to the absence of the lens. Another pertinent experiment is to remove the epidermis which normally would have formed the lens and to replace it with a piece of epidermis taken from another part of the body, from the head or even from the belly. In this experiment it was observed that the epidermis, if in contact with the optic vesicle, develops into a lens. A third type of experiment is to transplant the optic vesicle, without the epidermis normally covering it, under the epidermis in an abnormal position. In this case the local epidermis may be caused to develop a lens. (See Mangold, 1931b.)

For a long time it was believed that an intimate contact between the optic vesicle and the epidermis is indispensable for lens induction. A thin layer of cellophane inserted between the optic vesicle and the epidermis in a chick embryo completely stopped the inducing action of the optic vesicle (McKeehan, 1951). Insertion of a porous membrane, which presumably allowed for the passage of macromolecules, between the eye vesicle and the epidermis in a frog embryo also precluded lens induction (de Vincentiis, 1954). However, experiments have been reported (McKeehan, 1958) in which a partial screening of the epidermis from the eye vesicle in a chick embryo by a thin slice of agar did not prevent the complete development of the lens. The inducing agent could thus either get through or get around the agar. This suggests that the inducing agent is a chemical substance.

Apart from this, not much is known as to the nature of the stimulus responsible for the induction of the lens. It has not been possible to extract a lens-inducing substance from the optic vesicle. On the other hand, some of the "abnormal" inductors of neural plates are known to induce "free" lenses, that is, lenses without an eye. The thymus of the guinea pig seems to be especially suitable for this purpose (Toivonen, 1940, 1945). That the stimulus in this case is exactly the same as that exercised by the optic cup remains to be proved.

There is one peculiar similarity between induction of the neural plate and induction of the lens, concerning the distribution of the cytoplasmic ribonucleic acid in the components participating in the process. The eye vesicle at the time when it comes in contact with the presumptive lens epidermis contains a large amount of ribonucleic acid. At the same time, the presumptive lens epidermis has little ribonucleic acid, and in this respect it is no different from the rest of the epidermis. After the contact is established, the ribonucleic acid in the cells of the eye vesicle is found to be concentrated near the outer margin of the cells, that is, where the cells touch the epidermis. Large amounts of ribonucleic acid now appear also in the presumptive lens cells, at first only at their proximal ends, where they are in contact with the eye vesicle, but later in the outer parts of the epidermal cells. In subsequent stages, the ribonucleic acid content in the retinal cells decreases, while it continues to increase in the cells of the lens rudiment (McKeehan, 1956). There is no proof, however, that the ribonucleic acid actually passes from the eye cup cells into the lens rudiment cells.

In another experiment, performed on *Xenopus* embryos (Sirlin and Brahma, 1959), the proteins of the eye cup cells were marked by supplying them with radioactively tagged phenylalanine (containing C^{14}). Subsequently, radioactive material was found to have passed into lens rudiments induced by the marked eye cups which were transplanted into normal embryos. Although this experiment is open to the same criticisms as the experiments with radioactively marked inductors of the neural plate (p. 000), the results are compatible with the assumption that a chemical substance passes from the eye cup into the reacting epidermis during lens induction and that this substance is a protein. There would thus be a similarity in this respect between the mechanism of neural plate induction and that of lens induction.

The relation between lens development and optic cup development is complicated considerably by the fact that in a few amphibians—*Rana esculenta, Xenopus laevis, Rhacophorus schlegelii,* and others to a lesser extent (Spemann, 1912a; Balinsky, 1951; Tahara, 1962)—the lens shows a certain degree of independent development ("self-differentiation") even in the absence of the optic cup, that is, when the optic cup has been previously removed. The degree of independent development varies from a tiny nodule of epidermal cells to a rather typical lens with fiber differentiation. The development is never completely normal, as lenses without eyes undergo a far-reaching degeneration once the initial stages of development have occurred. Nevertheless, the independent development of lenses shows that the eye cup is not the only part which may be involved in lens development. Experiments in which free lenses have been induced by abnormal inductors (thymus) suggest that the "self-differentiation" of lenses, when it occurs, is due to some other influence on the differentiation of the epidermal cells and that this influence is responsible, in some species, for "independent" lens development (i.e., lens development which is independent of the eye cup).

What these influences might be may be concluded from experiments in which belly ectoderm was not capable of developing a lens when transplanted just before the formation of the optic vesicle but was able to react if transplanted in the neurula stage. It follows that to be able to react the epidermis must be in position some time before contact with the optic vesicle is established. During this time, the head mesoderm lies immediately underneath the presumptive lens epidermis, and the influence of the head mesoderm was concluded to be the factor which prepares the epidermis for the subsequent induction emanating from the optic vesicle (Liedke, 1951, 1955). Presumably, in some species, the "preparation" goes so far that the development of the lens may start even if the eye vesicle is not present.

Our conclusion may perhaps be stated in another way: we may say that the development of the lens is a result of induction emanating from two sources, the head mesoderm and the eye vesicle. Normally, both are necessary for successful lens development. There is some evidence, however, that the relative importance of the two inductors may be changed by environmental factors. Keeping the developing embryos at a low temperature seems to favor induction by the head mesoderm, so that induction by the eye vesicle becomes unnecessary (Ten Cate, 1953; Jacobson, 1955).

The differentiation of the lens cells into lens fibers in normal development is caused by the same induction as that involved in the development of the lens as a whole. It has been proved that contact with the presumptive retinal layer of the eye can induce lens fiber differentiation. The epithelial cells of the lens rudiment are all capable of fiber differentiation if they are exposed to a suitable stimulus. Contact with retinal tissue or with the sensory epithelium of an ear vesicle (Fig. 304) (compare what has been said on the induction of the retina) may cause the formation of an additional mass of lens fibers, so that under experimental conditions lenses may develop having two independent masses of fibers (Dragomirow, 1929).

Accessory Structures. The other accessory structures of the eye which are present in all vertebrates having functional eyes are the **choroid coat,** the **sclera,** and the **cornea.** The choroid coat and the sclera develop from mesenchyme accumulating around the eyeball, in the way mesenchymal cells accumulate around many organs giving rise to their connective tissue capsules. (See Section 9–5.) In the case of the eye, the interior layer of mesenchymal cells gives rise to a network of blood vessels surrounding the pigment epithelium. The outer layer of mesenchyme forms a fibrous capsule around the eye, which serves for its protection and for the insertion of the eye muscles. The capsule may either remain fibrous or develop cartilage or even bone (in reptiles and birds).

The cornea originates, in part, from mesenchyme, but the epidermal epithelium also has an essential role in its formation. The connective tissue part of the cornea is continuous with the sclera, while the corneal epithelium is continuous with the skin epidermis or with the epithelium of the eyelids, where such are present. Both the epithelium and the connective tissue of the cornea become transparent, so that the light rays may enter the eye. The development of the cornea can easily be traced in living amphibian embryos. Initially the epidermis covering the eye is pigmented, for the epidermal cells contain granules of pigment derived from the egg. In the cells of the presumptive cornea these pigment granules become dissolved, and later, when the chromatophores develop in the connective tissue of the skin, the cornea remains free of them.

Transformation of the skin into the cornea is caused by an induction, the source of which is the eyeball. This can be proved by transplanting the eyeball heterotopically or by replacing the normal cornea by skin from another part of the embryo. The stimulus can be given off by both the eye cup and the lens. If the lens alone is transplanted, the

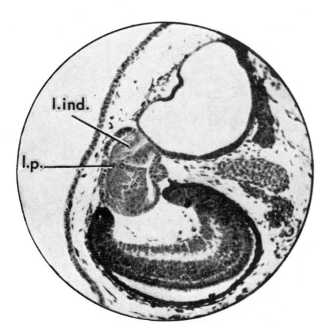

Figure 304. Induction of an additional mass of lens fibers (l. ind.) by means of contact with an ear vesicle. l.p., Primary lens fibers. (From Dragomirow, 1929.)

epidermis over it loses its pigment and differentiates as cornea. If the eye is removed, the cornea does not develop at all (Spemann, 1901; Fischel, 1919; Mangold, 1931b).

Induction of the cornea presents an interesting peculiarity as compared with neural plate and lens induction in amphibians. The competence to differentiate as cornea is found in the skin not only during a short period of embryonic development but for a long time, long after the normal differentiation of the cornea has taken place. Also, the eyeball retains its inductive ability for a long time, probably permanently. Moreover, the persistence of the cornea is dependent on the continuous presence and influence of the eyeball. If, in a late amphibian larva or an adult, the eye is removed, the cornea soon loses its transparency, is invaded by chromatophores, and becomes more or less normal skin. On the other hand, a fully differentiated piece of skin will lose its chromatophores and become transparent cornea if it is transplanted over the eye.

In the development of the eye, induction takes place repeatedly, and some parts after having been induced themselves become a source of inducing stimuli. A whole chain of inductors can thus be noted:

1. The roof of the archenteron induces the neural plate and therefore also the eye cup rudiment which is part of the neural plate.
2. The eye cup rudiment, becoming the optic vesicle, induces the lens (acting together with head mesoderm).
3. The lens induces the cornea (acting together with the optic cup).

Parts developing as a result of induction, and inducing in their turn, may be called secondary, tertiary, etc., inductors, or organizers of the second grade, third grade, and so forth.

12–3 THE FATE OF THE NEURAL CREST CELLS

The neural crest at the time of its formation is represented by a mass of loose cells lying dorsal to the neural tube. Almost at once after the formation of the crest, the cells of which it consists start migrating in a lateral and ventral direction from the place of their origin (Fig. 305). As the neural crest cells move, they form streams, bypassing, as they go, certain organs (viz., the eye, the gill pouches). These streams of neural crest cells are especially conspicuous in the head and neck region, while in the trunk region the neural crest cells are more scattered right from the start (Stone, 1926; Raven, 1931). Some of the neural crest cells move into the space between the epidermis and the layer of mesoderm; others penetrate into the interstices between the neural tube and the inner surface of the somites and down to the dorsal aorta and beyond. In their movements the neural crest cells follow mesenchyme-filled spaces between organ rudiments (Weston, 1963). Sooner or later, the most advanced neural crest cells reach the midventral line of the body. Not all of them, however, travel as far as this: rather they become spread all along the path, some of the cells even retaining their original position dorsal to the neural tube.

The migration of the neural crest cells is a complicated process, dependent on many factors. If the neural crest cells are cultivated *in vitro*, the cells spread evenly in the available medium (Twitty and Niu, 1948). In the organism, however, the even distribution of neural crest cells is disturbed by the surrounding tissues and organ rudiments which lie in the path of the migrating cells and are bypassed. Other rudiments seem to attract the neural crest cells, in some way, or to keep them fixed once they have reached a certain position. Thus, the neural crest cells are held along the upper edges of the somites and also along the upper edge of the lateral plate. Later, when part of

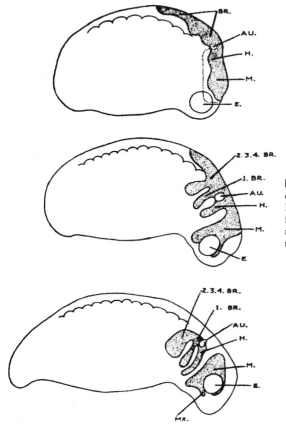

Figure 305. Three stages in the migration of neural crest cells in the salamander, *Ambystoma punctatum*. AU., Ear vesicle; BR., branchial neural crest cells (numbers refer to particular branchial arches); E., eye; H., hyoid neural crest cells; M., mandibular neural crest cells; MX., maxillary neural crest cells. (From Stone, 1926.)

the neural crest cells differentiate into chromatophores, these accumulations of neural crest cells become conspicuous as longitudinal strips of pigment (Fig. 306; Twitty, 1949).

The migration of neural crest cells can be observed in several different ways. The neural crest cells, being ectodermal cells, are, in amphibians, distinguishable from the mesodermal cells by their smaller content of yolk granules and by a greater amount of pigment derived from the egg. Another way of tracing the neural crest cells in amphibians is by means of local vital staining. In birds, migration of neural crest cells has also been traced by using grafts labeled with radioactive substances (Weston, 1963). If the stain is applied to the neural folds, the neural crest cells derived from these can readily be seen against the background of unstained ectoderm and mesoderm. Another method is based on the ability of the neural crest cells to differentiate into melanophores. By isolating parts of the embryo in sufficiently early stages and cultivating them in suitable surroundings, it is possible to prove that without the neural crest cells no pigmentation can develop in the skin or in other organs and tissues. If, however, part of the embryo is isolated and transplanted after the migrating neural crest cells have reached this part, the pigment cells later differentiate in the graft. The actual differentiation of pigment cells occurs after the migration has been completed. While on the move the neural crest cells do not differ in their pigment content from other cells, and this of course makes it difficult to observe their migration directly. This last method of investigating the migrations of neural crest cells has been applied in mammals and birds (Rawles, 1948).

Figure 306. Larva of the newt, *Triturus cristatus,* with longitudinal stripes of pigment. (From Balinsky, 1925.)

In fishes and amphibians, the pigment cells are found predominantly in the connective tissue—in that of the skin, but also in the peritoneum, in the walls of blood vessels, and elsewhere. In birds and mammals, the pigment is found predominantly in the epidermal derivatives: the hairs and the feathers. Nevertheless, the production of the pigment is also due to the activity of the neural crest cells. They penetrate into the hair and feather follicles and deposit the pigment granules in the hairs and the feathers as they grow out of the follicles. If the access of the neural crest cells to the hair and feather follicles is precluded, the hairs and feathers may develop normally, but they are completely devoid of pigment (Rawles, 1947). (For further information on the role of neural crest cells in the pigmentation of feathers in birds, see also Willier, 1952.)

Besides the pigment cells, other types of cells are also differentiated from the neural crest cells. The visceral skeleton is almost completely developed from neural crest cells. The visceral arches occupy approximately the same position as the streams of neural crest cells in the early embryo. The mass of neural crest cells behind the eye, that is, between the eye and the first branchial pouch, becomes the mandibular arch, the upper part of which becomes differentiated as the **quadrate,** and the lower part of which gives rise to the **Meckel's cartilage** or the mandible proper (Fig. 307). The mass of neural crest cells between the first and second branchial pouches becomes the hyoid arch, the next mass becomes the first branchial arch, and so forth. The mass of neural crest cells moving downward in front of the eye contributes to the formation of the anterior half of the trabeculae of the skull (Stone, 1926; Raven, 1931; Hörstadius and Sellman, 1946). The neural crest cells of the trunk region, on the other hand, do not participate in the development of skeletal tissues (Raven, 1936), both the axial skeleton (vertebral column) and the limb skeleton being derived from the mesoderm. Rather peculiarly, one element of the visceral skeleton in the amphibians, the second basibranchial, is also of mesodermal origin.

Figure 307. Skull of a salamander larva, indicating parts developing from neural crest material (stippled) and from mesoderm (white). (From Stone, 1926.)

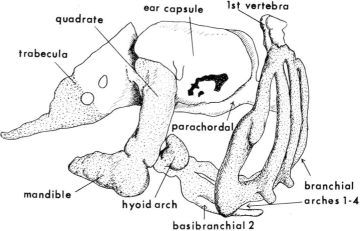

The papillae of the teeth in urodele amphibians have been shown to be derived from neural crest cells (de Beer, 1947). It is highly probable that the papillae of the teeth in all other vertebrates are of the same origin.

The role of the neural crest cells in the development of the peripheral nervous system has been a matter of controversy. At present, it is accepted that the spinal ganglia and the ganglia of the autonomic nervous system are derived from the neural crest cells (pp. 337–353) but that the ganglia of the cranial nerves (V, VII, IX, X) are developed only in part from neural crest, the other part being contributed by the dorsolateral or epibranchial placodes (pp. 349–350). In addition to the ganglion cells, the neural crest participates in the development of the nervous system by contributing material for the sheaths of the nerves (Schwann cells) and the meninges (at least the pia mater and arachnoidea) (Piatt, 1951). Lastly, the neural crest cells are found to differentiate as subcutaneous connective tissue, although in this case they are joined by mesenchymal cells derived from the mesoderm.

As the neural crest cells may take such divergent paths of differentiation, the question arises as to whether the fate of individual crest cells is determined by environmental influences or whether these cells already differ at the time when they leave the neural folds. Apparently, both the alternatives are partially true. If different parts of the neural fold are explanted in a culture medium (Niu, 1947) or transplanted to the side of an embryo (Hörstadius and Sellman, 1946; Hörstadius, 1950), different results are obtained, depending on the area from which the neural fold has been taken. Pieces of cranial neural folds under these conditions produce only small numbers of melanophores, but they give rise to cells which may develop into cartilage. Neural folds of the trunk region, when explanted or transplanted, give rise to numerous melanophores, but no procartilage cells are formed. On the other hand, it has been shown that cartilages develop from neural crest cells only when they are induced to do so by adjoining tissues. Pieces of cranial neural crest were cultivated in an epithelial vesicle either alone or together with other tissues, such as neural plate, notochord, foregut endoderm, midgut endoderm, and lateral mesoderm. Under these conditions, cartilage developed from neural crest cells only when they were cultivated together with foregut endoderm. Trunk neural crest under the same condition produced only melanophores and mesenchyme and no cartilage (E. W. Okada, 1955). It follows that:

1. Only cranial neural crest is competent to produce cartilage.
2. It can produce cartilage only under the influence of foregut endoderm. (See also Section 14–3.)

It has been noticed further that spinal ganglia are formed only in the immediate vicinity of the neural tube. It would appear that some influence of the tube is necessary for the crest cells to become differentiated in this way (Weston, 1963). In the sympathetic ganglia, however, crest cells acquire neuron differentiation at a distance from the neural tube.

There is some evidence (Stevens, 1954) that among the pigment cells in amphibians the two types, melanophores and guanophores, are already distinct while the cells are migrating from the site of their origin, the neural folds.

12–4 THE FATE OF THE EPIDERMIS AND THE STRUCTURES DERIVED FROM IT

When the epidermis is first segregated from the other parts of the ectoderm (neural plate, neural crest) during the process of neurulation, it is still a very complex rudiment. Most of it becomes the epidermis of the skin, but in addition, a number of other structures are derived from it. Some of these have been mentioned already: the lens, the cornea, and the cranial ganglia.

The epidermis itself gives rise to quite a large number of special differentiations, such as various unicellular and multicellular skin glands, including the sweat glands and the sebaceous glands, the hairs, feathers and scales, and various other special structures derived from these. The development of some of these parts involves histogenesis rather than organogenesis and will not be dealt with here.

In the early embryo the epidermis is a sheath of epithelium. In frogs the epithelium consists of two layers of cells: the outer **covering layer** or **periderm** and the inner so-called **sensory layer.** The latter name is used not because the layer as such has nervous functions but because some sensory organs are derived from parts of this layer. In birds and mammals, the epidermis of early embryos also consists of two layers of cells, but the cells of the outer layer, the peridermis, are flattened and are shed eventually. In all vertebrates the inner layer proliferates, producing the stratified epithelium, the epidermis of the skin. The innermost cells of the epidermis, the ones adjoining the basal membrane, become the **generative** (or **Malpighian**) **layer** of the epidermis. The skin is then composed of the epidermis and the layer of mesenchyme, partly derived from the neural crest and partly from the dermatomes (p. 375), which gives rise to the **dermis.**

The Placodes. Many structures derived from the epidermis make their first appearance in the form of plate-shaped thickenings of the epidermal epithelium. Such thickenings are called **placodes.** That the ganglia of the cranial nerves are derived in part from placodes has been stated. When the lens rudiment first appears as a thickening of the epidermis, it bears a great similarity to the other placodes. A pair of placodes, appearing in front of the anterior end of the neural plate and probably deriving their material from the neural fold itself, develop into the olfactory sacs. A placode appearing against the side of the hindbrain invaginates and produces a vesicle which is eventually separated from the epidermis. This structure is the ear vesicle, the rudiment of the internal ear (the ear labyrinth), and the placode from which the ear vesicle is developed is the **auditory placode.** Parts of the epidermis adjoining the auditory placode also become thickened, and from these placodes develop the **lateral line sense organs.** In the aquatic vertebrates (fishes, aquatic larvae of amphibians), the lateral line organs are distributed over the head (in several rows), and a row of the same organs stretches backward along the side of the entire body and tail. Wherever the lateral line organs are found, the cells of which they consist come from the placodes of the ear region.

In amphibians, the backward migration of the lateral line organ cells was demonstrated in a grafting experiment. Two embryos belonging to different species of frogs were cut transversely in halves, and the anterior half of a darkly pigmented species (*Rana sylvatica*) was grafted onto the posterior half of a lightly pigmented species (*Rana palustris*). The darkly pigmented cells of the lateral line rudiment of the anterior half could then be observed to migrate into the posterior half and along the trunk and tail (Fig. 308; Harrison, 1903).

In the case of lateral line sense organs, the pathway of the migrating cells is dependent on the surroundings through which they migrate. If the anterior half of a frog embryo is transplanted onto another embryo whose own lateral line organ rudiment was removed previously, the lateral line cells of the anterior half grow out into the second embryo, and once they have reached the path of migration normally taken by the lateral line cells, they start moving along this path, even though their new direction is at an angle to the one they had been following before (Fig. 309). This experiment shows

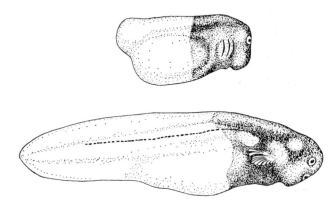

Figure 308. Migration of cells of the lateral line rudiment, demonstrated by uniting the anterior half of an embryo of *Rana sylvatica* and the posterior half of an embryo of *Rana palustris*. (After Harrison, from Weiss, 1939.)

that the path of migration is determined by a factor lying outside the migrating cells themselves (Harrison, 1903). It has been shown that the path of migration in this case is determined by the mesoderm.

The development of the placodes is probably always dependent on a stimulus from the tissues situated under the epidermis. This has been proved in some instances, as in the case of the lens of the eye. The ear is also dependent in its differentiation, a fact which can be deduced from experiments in which ear vesicles have been induced heterotopically as the result of the transplantation of various inductors. The ear vesicle is among the structures that are often induced when the primary organizer is transplanted, but it can also be induced in experiments involving transplantation of adult tissues and of parts of the neural plate and neural tube (Guareschi, 1935; Gorbunova, 1939; Kogan, 1939). The latter experiments suggest that in normal development the ear vesicle is induced by the medulla oblongata. This, however, cannot be the sole inductor, as the ear vesicles may develop in their normal position after the medulla oblongata is removed at an early stage, either alone or together with most of the central nervous system rudiment. It is concluded therefore that the development of the ear vesicle is dependent on multiple induction, and that similar stimuli are emitted both from the medulla oblongata and from the mesoderm developing from the roof of the archenteron (Harrison, 1935); Albaum and Nestler, 1937).

The induction of the ear vesicle probably proceeds in two stages: first, the presumptive ear ectoderm is acted upon by the underlying mesodermal mantle in the late gastrula and early neurula stages, and later, the determination is finally stabilized by the influence of the medulla which, as a result of the closure of neural folds, comes into

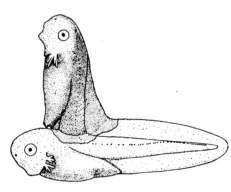

Figure 309. Migration of cells of the lateral line rudiment in an experiment in which the anterior part of a *Rana sylvatica* embryo was grafted on to the back of a *Rana palustris* embryo. (After Harrison, from Huxley and de Beer, 1934.)

close contact with the epidermis in the ear region (Yntema, 1950). We have seen a similar case in the "independent" development of the lens in some species of frogs. (p. 359).

A similar duplication in the sources of induction has been postulated for the nose rudiment—namely, an earlier induction by mesoderm and a later induction by the forebrain. It has been claimed, for instance, that the anterior portion of the neural plate if transplanted under the epidermis on the flank may induce a nose rudiment locally. In more careful experiments it was found, however (Zwilling, 1940; Schmalhausen, 1950), that nose rudiments may develop in the absence of brain tissues, and thus the alleged inductions must be due to the nose rudiment material being grafted together with the presumptive forebrain. The two rudiments are thus induced simultaneously by the underlying roof of the archenteron and initially lie very close to each other.

The Olfactory Organ. Although the nose rudiments seem to be determined at a very early stage (late gastrula), they first become discernible morphologically after the closure of the neural tube, in the form of two thickenings of the epidermis, the **olfactory placodes,** just anterolateral to the hemispheres of the telencephalon (Figs. 235 and 266). The central part of each placode becomes invaginated, and the olfactory placode thus becomes converted into an olfactory sac which is open to the exterior by the external naris. Parts of the wall of the olfactory sac, especially the dorsal and lateral wall, are differentiated as olfactory epithelium. The primary sensory cells of the olfactory epithelium develop on their proximal ends nerve processes (axons) which converge to form the olfactory nerve. The olfactory nerve grows into the adjacent wall of the telencephalon, bridging the narrow gap between the olfactory organ and the brain.

In most fishes the olfactory organ retains essentially the same structure in the adult state, but in the group of Choanichthyes among the fishes and in the terrestrial vertebrates, the structure of the olfactory organ is further complicated by the internal nares, and in the mammals also by the nasolacrimal duct.

The internal nares or primitive **choanae** arise by a perforation of the nose sac cavity into the oral cavity. The actual perforation is preceded by the formation, from part of the ventral wall of the nose sac, of an elongated tube stretching backward and downward toward the oral cavity. This tube is the nasal canal, and the epithelium lining it becomes thin and is thus fairly sharply segregated from the thicker epithelium giving rise to the sensory part (Fig. 243). With the elongation of the nasal canal the sensory part appears to be a dorsolateral growth of this canal, although it is actually the older portion of the olfactory organ. The internal (posterior) end of the nasal canal fuses eventually with the epithelium lining the oral cavity, and the intervening membrane becomes perforated as internal nares. **Jacobson's organ,** where it is present, is another section of the olfactory sac retaining sensory function. It develops from a medioventral part of the sac.

The development of the nasolacrimal duct will be discussed together with the development of the face in Section 14–2.

The Ear. The **auditory placode,** from which the internal ear is developed, initially shows considerable similarity to the olfactory placode and is also converted by invagination into a saclike structure (Fig. 310), but in the early stages there already are important differences. In the amniotes the whole epidermal layer is involved in the formation of the auditory placode, and it later invaginates to form a sac which is, at least temporarily, open to the exterior. In the frogs, however, the auditory placode is formed by the thickening of the interior "sensory" layer of the epidermis, while the external covering layer is not involved at all (Fig. 237). As a result, when the placode invaginates, there is no opening or pit on the surface of the skin. In both cases, however, the opening of the sac becomes constricted and closed, so that eventually the rudiment of the ear takes the form of a completely closed vesicle, the **ear vesicle.** In the bony fishes

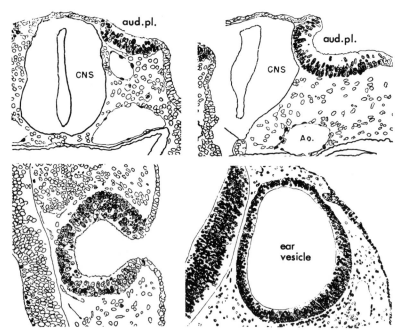

Figure 310. Four stages in the development of the inner ear from auditory placode—aud. pl.—to completed ear vesicle in human embryo. CNS, Central nervous system; Ao., dorsal aorta. (From Streeter, 1942, 1945.)

the auditory organ is formed not by invagination but as a solid mass of cells on the inner surface of the epidermis and is hollowed out secondarily.

The ear vesicle is the rudiment of the most essential part of the internal ear, the **labyrinth.** When first formed, it is somewhat pear-shaped, the pointed end directed upward. This pointed end later gives rise to the endolymphatic duct. Soon the ear vesicle starts expanding, pushing away the surrounding loose mesenchyme. Parts of the wall of the vesicle become very thin, and the epithelial cells become flat. These parts will be the membranous areas of the labyrinth. Other parts, particularly those of the medioventral wall of the vesicle, remain thick or even become thicker; the cells in these areas become columnar and give rise to patches of sensory epithelium which form the maculae of the internal ear. Even before it is subdivided into membranous and sensory parts, the ear vesicle gives off on its median surface a group of cells which become the acoustic ganglion (ganglion of nerve VIII) (van Campenhout, 1935).

The expansion of the ear vesicle is unequal, and thus it becomes constricted in some places and bulges out in others. As a result, the shape of the organ becomes increasingly complicated, so that it eventually deserves its name—the labyrinth (Fig. 311). The sacculus is subdivided by a constriction from the utriculus. The utriculus becomes drawn into three mutually perpendicular folds, the rudiments of the semicircular canals. The sides of the folds eventually stick together and become perforated, while parts of the original cavity along the edges of the folds remain open and become the semicircular canals, opening at both ends into the cavity of the utriculus. A hollow outgrowth of the sacculus forms the rudiment of the lagena in lower vertebrates, and in higher vertebrates this outgrowth becomes very elongated and coiled to give rise to the cochlea. As the ear vesicle changes its shape and produces the various parts of the labyrinth, the sensory areas become subdivided and further differentiated until each of the maculae have taken up their final positions in the fully developed labyrinth.

As the ear vesicle expands to produce the labyrinth, it becomes surrounded by mesenchymal cells which later give rise to cartilage and produce the cartilaginous ear

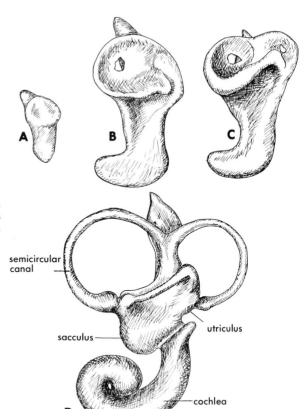

Figure 311. Development of the labyrinth from the ear vesicle in a human embryo. The drawings show lateral views of the left ear. (Modified from Streeter, 1906.)

semicircular canal

sacculus

utriculus

cochlea

D

capsule which surrounds and protects the inner ear. There is a direct causal relationship between the ear vesicle and the development of the ear capsule; if the ear vesicle is removed, the ear capsule does not develop, and if a foreign ear vesicle is transplanted in the tailbud stage, the local mesenchymal cells may aggregate around it and produce an additional cartilaginous capsule (Lewis, 1907). The mesenchyme that is used for the ear capsule is of mesodermal origin and is derived from the sclerotomes (Section 13–1). Mesenchyme of neural crest origin, as well as subcutaneous mesenchyme, is apparently not capable of reacting to induction by the ear vesicle. As a result, an ear vesicle transplanted heterotopically does not always cause a good capsule to be developed around it. The most complete capsules develop around ear vesicles transplanted in the immediate vicinity of the normal ear, between the ear and the eye, where the grafted vesicle can draw on the same supply of mesenchyme as the normal ear vesicle. However, the sclerotome mesenchyme of the trunk, that is, the mesenchyme giving rise to the cartilages of the vertebral column and the ribs, reacts to the ear vesicle by forming large masses of cartilage, which may partially surround the grafted ear vesicle (Balinsky, 1925; Syngajewskaja, 1937).

The concentration of mesenchyme on the surface of the ear vesicle may be partly the result of its expansion, which would lead to the mesenchyme being compressed against its surface. However, there is no doubt that mesenchymal cells may travel considerable distances to reach the ear vesicle and to invest it with cartilage. In experiments on transplantation of the ear vesicle to the trunk region it can be seen that thick bars of cartilage grow out from the vertebral column to the ear vesicle, presumably indicating the pathway which had been followed by the mesenchyme. In the same experiments it can also be seen that the total amount of cartilage in the area is greatly increased, so that the ear vesicle must either stimulate the proliferation of skeletogenic mesenchyme or increase the proportion of mesenchymal cells which become chondroblasts. The reverse occurs in the case of the removal of the ear vesicle. Not only does the ear capsule not develop, but there is no superfluous cartilage in the area. In the absence of the ear vesicle the proliferation of the procartilage cells falls short of the normal, or else the cells that should have become cartilage cells differentiate along other paths.

The development of the ear provides another example of a chain of inductions:

1. The primary inductor—the roof of the archenteron, consisting of presumptive chordomesoderm—causes the development of the hindbrain.
2. The hindbrain, as a secondary inductor, stimulates the development of the ear vesicle (in conjunction with the direct action of the mesoderm on the presumptive ear ectoderm).
3. The ear vesicle, as a tertiary inductor, causes the formation of the cartilaginous capsule.

One might have expected that the development of the middle ear would be related to the development of the inner ear, but this is not the case. The middle ear, consisting of the eustachian tube, the ear ossicles (columella in the frog) in the cavity, and the tympanic membrane, develops normally after the removal of the ear vesicle. This may be because the middle ear is derived from the branchial apparatus, which is an essential part of the vertebrate organization, deeply rooted in the basic mechanism of vertebrate development and thus not in need of stimulation from the inner ear. In this case the functional apparatus, the organ of hearing, is made up of two parts not causally connected in development but linked together only through the medium of their definitive functioning. (See Yntema in Willier, Weiss, and Hamburger, 1955.)

Besides the organs developing from placodes, a number of structures are developed from the epidermis which in their early stages can be classified as "outgrowths" or, more correctly, as protrusions. These are the unpaired fin fold, the external gills (in aquatic vertebrates), the "balancer" (in the larvae of salamanders), and the paired limbs.

The Fin Fold. The unpaired fin fold is a structure found in all fishes and in the larvae of amphibians. It is a vertical fold of skin which starts in the posterior head region or the anterior trunk region, extends backward all along the back and dorsal side of the tail, bends over to the ventral side of the tail, and can be traced forward, along the ventral side of the tail and the belly to the middle of the trunk. The fold consists of epithelium and connective tissue. In fishes, parts of the fold are later invaded by skeletogenous tissue which produces fin rays, thus transforming these parts into the unpaired fins of the adult (Fig. 312). Parts of the larval fin fold in between the unpaired fins of the adult disappear. In amphibians which metamorphose into terrestrial adults (all Anura and some of the Urodela), the fin fold disappears at metamorphosis, but it may persist in neotenic species or in some purely aquatic salamanders (*Cryptobranchus*).

The fin fold first appears as a longitudinal thickening of the epidermis, which is seen as a ridge on external inspection. The thickening increases by shifting upward and toward the midline of the adjacent strips of the epidermis. The cells moving in from the right and left flanks remain separated as two layers of epithelium, except at the crest of the ridge. There is, however, no hollow in between the two layers. The fold is hollowed out shortly thereafter by the two epithelial layers separating in the middle, and then connective tissue cells of neural crest origin penetrate into the fold.

Although the neural crest cells enter into the formation of the fin fold in a later

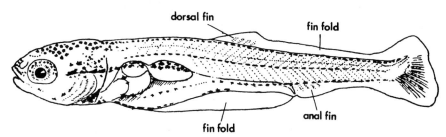

Figure 312. Development of the unpaired fins from parts of the fin fold in a fish larva. (From Balinsky: Proc. Zool. Soc. London, *118*, 1948.)

stage, they are actually responsible for the determination of the whole structure. If the neural crest cells are removed shortly after their formation, or if the neural folds, from which they arise, are cut away, the fin fold is not developed in the region of the defect (Fig. 313*A* and *B*). If a piece of the neural fold or a mass of the neural crest cells is transplanted under the epidermis in any part of the body, a fin fold develops at the site of the transplantation (Fig. 313*C* and *D*) (Terni, 1934; Du Shane, 1935). The ability to induce the development of the fin fold is found in the trunk neural crest cells only. The neural crest cells of the head cannot induce a fin fold, although the epidermis of the head region is fully competent to react by developing a fin fold, if it is exposed to the action of the necessary inductor. The extent to which the fin fold extends anteriorly is thus dependent on the extent to which the neural crest possesses the ability to induce the fold (Terentiev, 1941).

The External Gills. The external gills are protrusions of epidermis, with connective tissue, blood vessels, and muscle inside. They develop as outwardly directed pockets above the gill clefts and are found in some fishes (*Polypterus*, the lungfishes, *Misgurnus*) and in amphibian larvae. The original outpushing forms the shaft of the external gill.

Figure 313. Experiments showing dependence of the fin fold on the neural crest in salamander embryos. *A*, The neural folds removed on both sides in the neurula stage; *B*, result: no fin fold and no melanophores in the region of the operation. *C*, Transplantation of the neural fold onto the side of another embryo in the neurula stage (graft shaded); *D*, result: fin fold developed on the side. (*A* and *B* from Du Shane, 1935.)

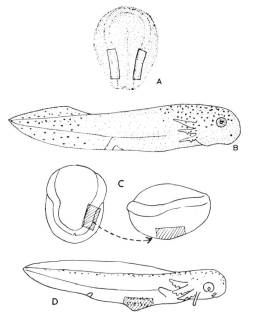

On this shaft secondary branches develop, which are formed at first as solid, outwardly directed thickenings of the epidermis of the shaft and are subsequently hollowed out (Fig. 314). Sometimes the shaft is so short that the gill appears to be a bunch of filaments which in their turn may develop branches.

The pattern manifested in the development of the gills has been found in the Urodela to be dependent not on the epidermis but on the inner layers of the embryo. A piece of epidermis taken from the flank may be transplanted over the gill region to replace the local epidermis, and if the transplantation is carried out early enough (soon after the closure of the neural folds), the normal development of the external gills is not impeded. The epidermis of the gill region may be lifted and replaced again after it has been rotated 90 or 180 degrees. If such an operation is carried out before gill development begins, the external gills will appear in their normal positions, as if nothing had happened (Harrison, 1921a). If the mesoderm of the gill region is included in the graft which is implanted in inverse orientation, the gills will still develop in their normal posi-

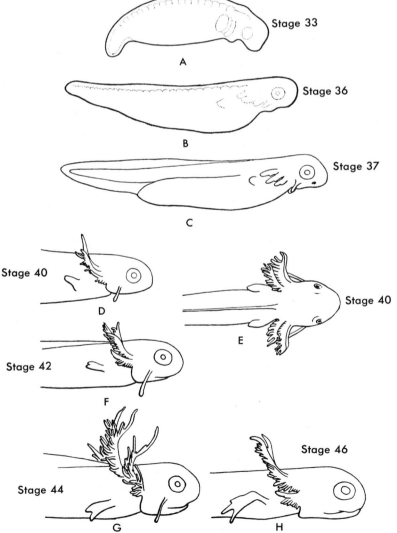

Figure 314. Development of the forelimbs, gills, and balancer in *Ambystoma punctatum.* The stages are shown after Harrison's normal table. (After Harrison, from Mangold, 1929.)

tions although the development may not go as smoothly as when only the epidermis is involved. If, however, the endoderm of the gill region is rotated together with the other germ layers, the developing gills are dependent on the new orientation of the graft, even if it is disharmonious with the other parts of the embryo (Fig. 315). It is therefore the endoderm that determines the position of the developing external gills (Severinghaus, 1930).

The part played by the endoderm in the development of the external gills may also be tested by completely removing the endoderm of the gill region, while leaving the other two layers intact. The result is that the external gills do not develop at all (Mangold, 1936).

The Balancers and Adhesive Organs. The **balancers** are tentacle-like organs present in the larvae of many species of urodele amphibians (newts and salamanders). The organ is situated, one on each side, just behind the angle of the mouth underneath the eye. It is a slightly curved cylindrical process, consisting of epithelium and a connective tissue core. The connective tissue is especially dense just underneath the epithelium, where it forms a cylindrical supporting membrane. Proximally this membrane is attached to the quadrate. The epithelium on the tip of the balancer produces a mucous secretion and is therefore slightly adhesive. A newly hatched larva of a newt or salamander uses the balancers for support, when resting on the ground, and to prevent the body from falling on one side in the stages when the forelimbs are not yet developed (Fig. 314F and G). When the forelimbs become functional, they take over the support of the body, and the balancers gradually degenerate (Fig. 314H; Harrison, 1925a).

The balancers would not deserve our attention if it were not that these simple organs lend themselves to some experiments of considerable interest. The epidermis of the balancer is normally derived from a part of the ectoderm lying just outside the neural fold in the vicinity of the eye rudiment. Other parts of the ectoderm, however, possess the ability to develop into a balancer when stimulated by an inductor. Balancers have often been induced in experiments on the "primary organizer" and on the inducing ability of the archenteron roof (Fig. 151, p. 204). The development of balancers is an indication that the inductor possesses the regional specificity of an archencephalic inductor. (See Section 8–4.)

The part of the organizer actually responsible for the induction of the balancer in normal development is probably the archenteron roof, but the adjacent portion of the neural plate, once it has been determined, also possesses the ability to induce a balancer (Mangold, 1931a). Further experiments on the development of balancers are described in Section 16–6.

The larvae of anurans do not possess balancers, but tadpoles of frogs and toads develop a different organ with somewhat similar biological functions. It is the **adhesive organ,** also known as the **oral sucker.** Actually, the adhesion which is effected by the organ is not due to a sucking action but to the secretion of a sticky slime. The adhesive

Figure 315. Transplantation of the gill rudiment in the salamander. *a*, Stage of operation and area which was excised for transplantation. *b, c,* Gills in abnormal positions developed from grafts rotated 180 degrees during operation. (From Severinghaus, 1930.)

organ is a glandular structure consisting of very elongated columnar cells which produce a copious secretion of mucoprotein (Eakin, 1963). The swimming powers of a newly hatched tadpole are quite limited, and after swimming for a few seconds the tadpoles tend to become attached either to some submerged objects (to the glass in an aquarium) or even to the surface film of the water. The sticky secretion of the adhesive organ keeps them suspended; otherwise, they would be lying most of the time on the bottom, where the oxygen supply may be considerably depleted.

When fully developed, the adhesive organs vary from one species to another. (See Figures 237 and 244.) Most often they are in the form of a V-shaped furrow with thickened edges (especially in toad tadpoles), but in some species of frogs they have the shape of two conical projections or even the form of one unpaired conical outgrowth (in *Xenopus*).

It is interesting to note that the adhesive organs are the first part of the body of a frog embryo to become functionally differentiated. (See Section 15–1.) The differentiation of these organs begins in some toad species immediately after neurulation, and they start secreting before the tadpole can swim. This is advantageous for the animal because, after hatching, the tadpoles for some time remain hanging on the outer surface of their egg membranes. At this stage the tadpoles possess a limited power of locomotion owing to the ciliary action of their epidermis.

As the tadpole develops and its swimming ability improves, the adhesive organs cease secreting and degenerate.

The Paired Limbs.   The early rudiments of paired limbs are similar to the gill rudiments in that they are outpushings of the epidermis which are filled with a mass of mesenchymal cells. In the case of the limbs, however, it is the differentiation of the mesenchyme into parts of the skeleton and muscle of the limb that deserves the greatest attention. The limbs will therefore be dealt with in conjunction with the organs derived from the mesoderm.

There is still one more structure of great importance which is derived from the epidermis, namely, the mouth invagination (the stomodeum). Its formation and further development can be most conveniently treated together with the other parts of the alimentary canal, which are derived from the endoderm.

Chapter 13

DEVELOPMENT OF THE
MESODERMAL ORGANS IN
VERTEBRATES

13–1 THE FATE OF THE SOMITES AND THE ORIGIN OF THE SOMATIC MUSCLES

The somites, when first formed, are masses of mesodermal cells with a small cavity in the middle. The cells are arranged radially around the central cavity. With further development, the shape of the somites changes; they become extended in the dorsoventral direction and flattened mediolaterally. In the vertebrates with discoidal cleavage, this change in shape leads to the elevation of the dorsal parts of the embryo above the general level of the blastodisc.

The flattening of the somite is accompanied by a change in the shape of its central cavity (the myocoele). Instead of being spherical, the cavity becomes a narrow vertical slit. An inner wall and an outer wall become clearly distinguishable, corresponding to the parietal and visceral layers of the lateral plates. The inner wall of the somites becomes very much thicker than the outer wall.

The fate of the inner and outer walls of the somites is completely different. The outer wall contributes to the formation of the connective tissue layer of the skin and is therefore called the **dermatome.** The inner wall produces skeletogenous tissue and the voluntary striated muscles of the body. The skeletogenous tissue develops from the lower edge of the inner wall, and this part of the somite is therefore called the **sclerotome** (Fig. 316). The sclerotome breaks up into a mass of mesenchymal cells. The cells migrate into the spaces surrounding the notochord and the spinal cord, envelop these organs, and later differentiate into cartilage, thus forming the bodies and the neural arches of the vertebrae. The hemal arches in the tail region and the ribs are of the same origin.

The dorsal part of the inner wall of the somite is the source of somatic muscle in the vertebrate's body and is therefore called the **myotome.** The cells of the myotome rearrange themselves so that they become elongated in a longitudinal direction. These longitudinally elongated cells differentiate subsequently into the striated muscle fibers. Originally each myotome becomes a muscle segment, separated from the one anterior and the one posterior to it by a connective tissue layer, the vertical myocomma. At one time in its development, the whole somatic musculature of the vertebrate consists of such segments arranged in linear order.

In the lower vertebrates, the myotome is the largest part of the somite, the sclerotomes being small and rather inconspicuous. In the Amniota, however, the sclerotomes are much larger. Only the upper edge of the inner wall of the somite adjoining the

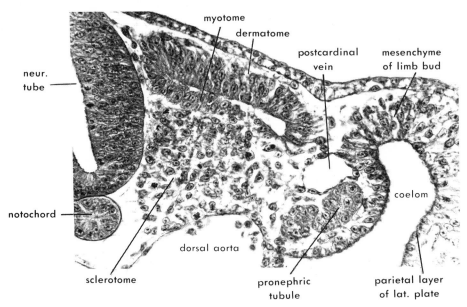

neur. tube

notochord

sclerotome

myotome

dermatome

postcardinal vein

mesenchyme of limb bud

coelom

dorsal aorta

pronephric tubule

parietal layer of lat. plate

Figure 316. Differentiation of the somite in a chick embryo.

dermatome becomes the myotome. The size of this part increases rapidly; the myotomes grow downward to assume the same position, lateral to the neural tube and the notochord, that they occupy in the amphibians right from the start.

Since they develop from the somites, the myotomes are originally dorsal in position. In the course of subsequent development, the muscle segments spread downward in the space between the skin on the outside and the somatic layer of the lateral plate on the inside, until the muscle segments on the right and the left sides meet ventrally. With minor alterations, the somatic muscles persist in this condition in the fishes and in the aquatic larvae of amphibians. The segmentation of the lateral and ventral muscles, as well as of the dorsal muscles, is directly derived from the segmentation of the mesodermal mantle in somites.

In terrestrial vertebrates, the primitive segmentation of the somatic muscles is more or less obliterated in connection with a change in locomotion. The segmented lateral bands of muscle are adapted for locomotion by lateral inflections of the body and the tail. The locomotion by means of two pairs of legs requires a completely different organization of the muscle system. Consequently, only traces of the original muscle segmentation can be discovered in the terrestrial vertebrates.

In *Amphioxus*, muscle segments continue anteriorly almost to the tip of the snout. With the development of the brain and the skull in vertebrates, the head region is exempted from the process of propelling the body by lateral inflections. The somatic muscles become superfluous in the head region. Nevertheless, mesodermal somites are formed in the head region of the embryo. The muscle segments derived from the somites lying posterior to the ear (the postotic somites) may persist and be linked with the muscle segments of the neck region. The somites lying anterior to the ear vesicle are always very transitory structures. Part of the cells of these somites, however, differentiates as somatic muscle. The muscles derived from this source do not serve for locomotion but are the six pairs of oculomotor muscles: the four rectus muscles and the two oblique muscles on each side. (See Figure 295*A* and *B*.)

13–2 THE AXIAL SKELETON: VERTEBRAL COLUMN AND SKULL

In most vertebrates, the axial skeleton passes through three phases in its development. In the first phase, the supporting system of the body is represented by the notochord. In the second phase, cartilage develops partly in direct connection with the notochord and partly independently of it. This condition is preserved in the adult state of contemporary cyclostomes and elasmobranch fishes. (We are not concerned with the question of whether this condition is primitive or secondary.) In the remainder of the fishes and in tetrapods, the cartilaginous skeleton is later replaced or supplemented by the bony skeleton. The bony skeleton and its relationship with the cartilaginous skeleton are amply covered in courses on comparative anatomy of vertebrates and may, therefore, be entirely left out in this text. We will consider here only the following aspects in the development of the skeleton:

1. The origin of cells giving rise to the cartilaginous skeleton.
2. The arrangement of the parts of the early cartilaginous skeleton in relation to the other organ rudiments of the embryo.
3. The dependence of the formation of the skeletal parts on the adjoining structures.

The material for the cartilaginous axial skeleton in the body and tail of vertebrates is derived, as mentioned on page 375, from the sclerotomes. The sclerotomes, being parts of the somites, are segmental in origin, but once they become transformed into mesenchyme the segmental arrangement is largely lost, and the mesenchyme spreads out as a continuous sheath along the notochord, enveloping the spinal cord above and the caudal artery and vein below the notochord in the caudal region. In this continuous mesenchymal sheath, nodules of cartilage, known as the **arcualia,** appear later in close apposition to the external surface of the notochord. Typically the arcualia appear in double pairs; one pair is formed dorsolaterally and another pair ventrolaterally to the notochord. The dorsolateral arcualia grow out dorsally beside the spinal cord and unite above the spinal cord to form the **neural arch.** The ventrolateral cartilages in the caudal region grow downward and unite underneath the caudal vein to form the **hemal arch.** In the cervical and thoracic region they give rise to lateral outgrowths, the rudiments of the ribs. At the same time, the proximal parts of all four cartilages spread out around the surface of the notochord, contributing in varying degrees to the formation of the body of the vertebra (Fig. 317).

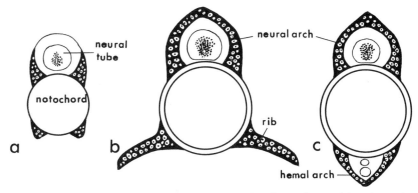

Figure 317. Diagram of the dorsal and ventral pairs of arcualia (a) giving rise to the neural arch and the ribs in the trunk (b) and to the neural arch and the hemal arch in the tail (c).

The cartilaginous neural and hemal arches do not bear a simple relationship to the somites (or sclerotomes). In many groups of vertebrates there are two series of cartilages (two pairs of dorsal elements and two pairs of ventral elements) developed in each mesodermal segment (cf. in fishes, in part also in tetrapods). In other cases, although there is only one set of cartilages formed, they are situated intersegmentally at the junction of two myotomes. (The myotomes retain the original segmentation of the somites.) The way in which the original cartilage rudiments cooperate in the formation of the definitive vertebrae will not concern us here.

The fact that the cartilaginous (and later bony) vertebral column replaces the notochord functionally suggests that the location and arrangement of the vertebral cartilages should be dependent on the notochord. In the absence of the notochord, the cartilaginous axial skeleton is very irregular. (See Figure 229.) Cartilage, however, is not completely absent in those sections of the body which do not have the notochord, and thus the notochord is not indispensable for the formation of axial cartilages. The irregularity of the latter could be a secondary effect of the extirpation of the notochord, since other systems, such as the spinal cord and segmented muscles, are greatly distorted as a result of the failure of the embryo to stretch normally.

In some experiments (Holtzer and Detwiler, 1953) in which the spinal cord, instead of the notochord, was removed from salamander embryos, it was found that the axial cartilages were either completely absent or reduced to insignificant vestiges. This result could not be caused by damage to the sclerotomes, as in other experiments extensive defects of the somites, including the sclerotomal region, were completely restored at the expense of the remaining fragments of the somites. The results of extirpation experiments were corroborated by the transplantation of pieces of spinal cord into an incision on the lateral surface of the somites of another embryo (Fig. 318a). It was found that complete and well-developed neural arches were formed in association with the grafted spinal cord (Fig. 318b).

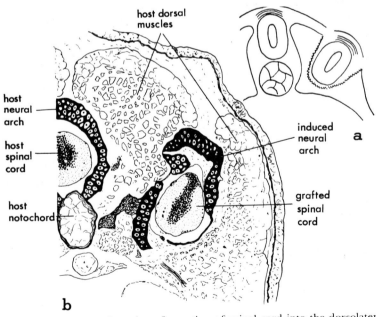

Figure 318. Transplantation of a section of spinal cord into the dorsolateral mesoderm of a salamander embryo. *a*, Diagram of operation; *b*, result: the graft surrounded by an induced vertebra. (From Holtzer and Detwiler, 1953.)

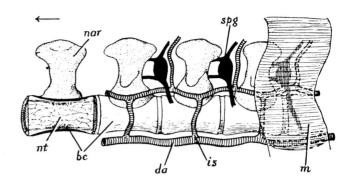

Figure 319. The relation between the muscle segments (m), the spinal ganglia (spg), the intersegmental arteries (is), and the vertebrae in a larva of *Ambystoma*. bc, Bony cylinder of centrum of vertebra; da, dorsal aorta; nar, neural arch; nt, notochord. (From Goodrich, 1930.)

The segmentation of the neural arches has not as yet been considered. In normal development there is a definite relationship between the segmentation of the longitudinal dorsolateral muscles, the segmentation of the spinal nerves and ganglia, and the segmentation of the vertebral skeletal elements, in particular that of the neural arches. There is a pair of spinal nerves with spinal ganglia corresponding to each muscle segment, the ganglia being situated opposite the median surface of the muscle segment. The neural arches alternate with the spinal ganglia and are thus situated intersegmentally with respect to the myotomes, so that each neural arch (and subsequently each vertebra) is connected to two consecutive muscle segments. The blood vessels also bear a relationship to this common pattern, an intersegmental artery arising from the dorsal aorta along each myocomma, close to the vertebral column (Fig. 319).

Experimental evidence is available to show that this whole system of metamerically arranged parts is originally dependent on the segmentation of the muscle rudiments (Detwiler, 1934). In a salamander embryo at the tailbud stage, the anterior somites are somewhat broader than the posterior ones, so that if a block of several somites in the brachial region is removed and replaced by a block of somites from the posterior trunk region, more somites may be fitted into the wound than had been cut out. Thus, somites 7 to 12 could be substituted for somites 3 to 5 (Fig. 320). As a result, the number of muscle segments on the operated side was increased in comparison with the normal number. It was found that the number of spinal nerves and spinal ganglia also increased, but not necessarily in strict correspondence with the number of muscle segments, one nerve sometimes supplying more than one muscle segment. On the other hand, the number of neural arches was found to be strictly in accord with the number of spinal ganglia, a bar of cartilage always appearing between two adjacent ganglia (Fig. 321). Since the ganglia are formed earlier than the cartilages, there can be no doubt that the neural arches are dependent on the ganglia and not the other way around.

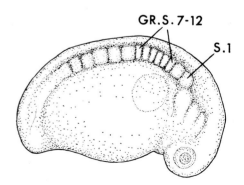

Figure 320. *Ambystoma punctatum* embryo with somites 7–12 (G.R.S. 7–12) transplanted in place of somites 3–5. (After Detwiler, 1934.)

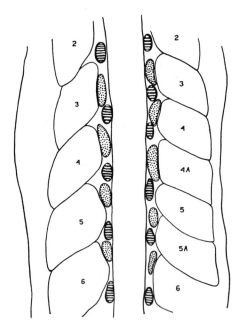

Figure 321. Result of increasing the number of somites (on right side). The number of spinal ganglia (shaded) and neural arches (stippled) increased, as compared with control side (left). (From Detwiler, 1934.)

The whole chain of reactions thus appears to be as follows. The neural crest cells which are produced along the whole length of the neural tube become aggregated opposite the median surfaces of the somites (or myotomes), and these aggregations become the rudiments of the spinal ganglia. Next, the cells of the skeletogenic mesenchyme, produced by the sclerotomes, spread out over the notochord and neural tube, but they are apparently repulsed by the spinal ganglia. Consequently, instead of forming a continuous sheet of cartilage enclosing the spinal cord, they give rise to a series of disconnected elements (the dorsal arcualia) alternating in position with the spinal ganglia. This is obviously not the complete picture, as it does not account for the cases in which two pairs of cartilages appear between each consecutive pair of spinal ganglia. It is, therefore, likely that some structures other than the spinal ganglia, the neural tube, and the notochord take part in determining the position of early cartilaginous rudiments.

The neural portion of the cranium in vertebrates in part bears the same relationship to the notochord and to the neural tube as does the axial skeleton in the posterior parts of the body. However, there is little, if any, trace of segmentation in the development of the cranium, and important parts of the cranium are quite peculiar in this respect.

In the lower vertebrates at the earliest stages of development, the cartilaginous cranium consists of several independent rudiments. They are: (1) the trabeculae, (2) the parachordals, and (3) the capsules of the sense organs—the nose, the eye, and the ear (Fig. 322).

The **trabeculae** (or **trabeculae cranii**) are a pair of elongated cartilages which appear in the most anterior part of the head, in front of the hypophysis. The trabeculae lie ventrally and ventrolaterally to the diencephalon and telencephalon, and their upper edges are wedged in between the brain on the inside and the rudiments of the nose and the eye on the outside. It has been established both by observation and experiment that the mesenchyme from which the trabeculae are developed comes from two different sources (cf. p. 363). The anterior part of the trabecula is formed of neural crest cells migrating forward and downward anterior to the eye cup. The posterior part of the trabecula is of mesodermal origin and is derived from the prechordal plate mesenchyme.

The **parachordals,** or parachordal cartilages, are derived from the mesenchyme produced by the sclerotomes of the somites in the head region. This mesenchyme spreads out on both sides of the notochord and eventually chondrifies in the form of two longitudinal rods, situated alongside the notochord and ending anteriorly at the same level as the notochord, that is, just posterior to the infundibulum and the rudiment of the hypophysis. The parachordals are similar in origin and position to the rudiments of the cartilaginous vertebral column, but they lack the segmentation of the latter (possibly because the cranial ganglia, owing to the greater breadth of the neural tube in the head region, lie much farther laterally and away from the region in which development of the parachordals takes place).

The cartilaginous **capsules of the sense organs** (the nose, the eye, and the ear) develop from skeletogenic mesenchyme accumulating around the surface of the epithelial parts of these organs. The source of the mesenchyme may not be the same in all three cases. It is fairly certain that the ear capsule is formed by mesenchyme derived from the same sclerotomes as those which give rise to the parachordals. In fact, the ear capsules are formed in close proximity and even in continuity with the parachordals, and strands of mesenchyme have been observed to lead from the rudiments of the parachordals to the site in which the cartilaginous ear capsules start to develop (Filatoff, 1916).

It has been claimed (O. Schmalhausen, 1939) that cartilage of the nose capsules is derived from the epithelial nose rudiment itself. This statement seems to need further corroboration. On the other hand we have seen that the nasal placodes develop in very close proximity to the anterior transverse neural fold. As the neural fold is the source of neural crest cells, it would not be very astonishing if the neural folds could produce the nasal cartilages as well as the epithelial parts of the olfactory organ.

In most vertebrates, the eyeball is surrounded first by a connective tissue capsule, the sclera, which becomes cartilaginous in later stages. The source of the cells forming the sclera has not yet been established.

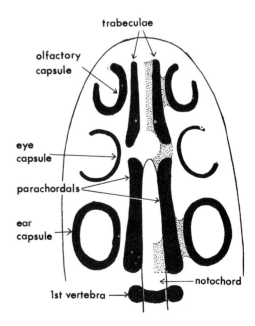

Figure 322. Diagram of the component elements of the vertebrate chondrocranium. Black areas represent cartilages in the initial stage. The fusions of the components are indicated by the stippling on the right side of the diagram.

The capsules of all three sense organs are dependent in their development on the epithelial parts of the organs. This dependence has been clearly shown in the case of the ear capsule (cf. Section 12–4, The Ear) and is probably true in the case of the olfactory capsule and the sclera of the eye, as these capsules fail to be formed if the epithelial parts of these organs are removed.

The parachordals, though spatially intimately associated with the notochord, show a high degree of independence from the latter. In experiments in which the notochordal rudiment is removed (p. 297), the development of the base of the skull does not seem to be affected to any extent. In particular, there is no foreshortening of the posterior part of the head, in sharp contrast to the shortening and stunting of the trunk and tail regions in the absence of the notochord.

The further development of the cartilaginous skull (chondrocranium) is characterized by enlargement and fusion of the initially formed cartilages. The right and left trabeculae fuse across the midline underneath the forebrain, and their posterior ends fuse with the tips of the parachordals. The parachordals envelop the notochord, particularly its dorsal side, and so give rise to the **basal plate** of the skull. At the point where the infundibulum and the hypophysis are situated, the cartilages leave a ventral opening (the hypophyseal fenestra) which persists for a long time and is closed only much later by cartilage or bone on the ventral side. The posterior ends of the parachordals grow upward and eventually fuse above the medulla, thus enclosing the foramen magnum of the skull. The nose capsule and the otic (auditory) capsule become firmly joined to the trabeculae and the parachordals respectively, thus contributing to the formation of the lateral walls of the chondrocranium. Gradually the lateral edges of the cartilaginous skull grow upward, and in the more primitive vertebrates (cyclostomes, many fishes and amphibians) form a roof over the dorsal surface of the brain. In teleost fishes and in all amniotes, however, the cartilaginous skull remains incomplete on the dorsal surface, and the cranial roof is formed by bone at a later stage. In higher vertebrates, mammals in particular, the initial stages in the development of the cartilaginous skull may be speeded up in such a way that the trabeculae and parachordals are fused right from the start—a condition which is achieved in lower vetebrates secondarily.

13–3 DEVELOPMENT OF THE PAIRED LIMBS

The paired limbs of vertebrates are very complex organs, built up of components derived from several different sources—from the lateral plate mesoderm, the epidermis, and the somites, to name only the main components. Nerves and blood vessels are, of course, also indispensable components of differentiated limbs.

The first trace of limb development may be found in the lateral plate mesoderm. The somatic layer of the lateral plate becomes thickened just underneath its upper edge. The cells of this thickening soon lose their epithelial connections and are transformed into a mass of mesenchyme without the somatic layer having lost its continuity. It is therefore a case of migration of mesenchymal cells from an epithelial layer, rather than that of the breaking up of epithelium into mesenchyme. The mesenchyme accumulates between the remaining lateral plate epithelium and the epidermis and soon becomes firmly attached to the inner surface of the epithelium (Fig. 323). The thickening of the lateral plate mesoderm and the subsequent formation of a mass of mesenchyme under the epithelium may coincide rather closely with the positions of the two pairs of limbs; that is, they may appear in two disconnected regions—just behind the branchial region and just in front of the anus. This is the case in amphibians. In other vertebrates, however, the thickening and the mesenchyme accumulation may

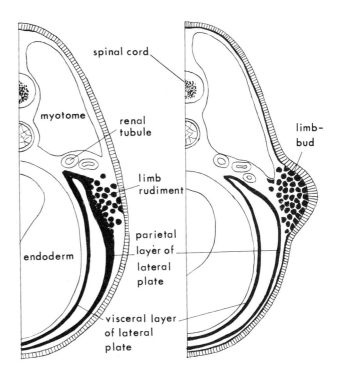

Figure 323. Diagram of the origin of limb mesoderm in an amphibian embryo.

spread far beyond the actual region of limb development. In fishes, early limb rudiments are more elongated anteroposteriorly in the earlier stages than in the later stages
of development. In the amniotes the thickenings and the mesenchyme gatherings are
continuous throughout the whole length of the body in the form of horizontal ridges,
the **Wolffian ridges.** (See Figure 231, Stage 18.) However, the most anterior and most
posterior parts of the ridge are thicker than the intermediate part, and it is only these
anterior and posterior parts that develop progressively, giving rise to the forelimbs and
hindlimbs. The intermediate part of the Wolffian ridge later disappears.

The epidermis over the mesenchyme mass becomes slightly thickened and bulges
outward. This occurs over the Wolffian ridge as well, but in the intermediate parts of
the ridge the epithelial thickening disappears together with the gathering of mesenchyme. In the regions where the fore- and hindlimbs are to develop, the protrusion,
consisting of a thickened epithelial covering and of an internal mass of densely packed
mesenchyme, increases and becomes the **limb-bud.**

Of the two components contributing to the formation of the limb-bud, the mesoderm is determined as such at an early stage shortly after the closure of the neural
tube. Pieces of lateral plate may be cut out in this stage and transplanted under the
epidermis on the flank or on the head. The local epidermis will then become the epithelial component of the limb-bud, and a limb will develop heterotopically. The presumptive epidermis of the limb in the same stages, that is, before a limb-bud has been
formed, does not possess any special properties and, if transplanted alone, will not give
rise to a new limb. Epidermis from any part of the body is able to cooperate with the
presumptive limb mesoderm in forming a limb-bud. This can be shown by removing
the epidermis in the limb region and then covering the wound with a flap of epidermis
taken from any part of the body (Harrison, 1918; Balinsky, 1931).

The epidermis is, however, by no means a passive component in limb development. This is especially clearly shown by some peculiarities of limb development in

higher vertebrates. In the amniotes, the limb-bud becomes slightly flattened at an early stage, and an epidermal thickening develops along the edge of the flattened bud. The thickening is in the form of a sharply defined ridge and sometimes (in reptiles) even takes the form of a solid fold of the epidermis. In cross section the ridge looks like a nipple (Fig. 324). It is referred to as the ectodermal apical ridge.

The cells of the ridge differ from the ordinary epidermal cells not only in their arrangement but also in their physiological properties. It was found that they contain more ribonucleic acid and more glycogen, and they differ conspicuously from surrounding epidermal cells in their high content of the enzyme alkaline phosphatase (Fig. 325 – Milaire, 1956). All these biochemical properties may be taken as indications of active metabolism.

The ectodermal apical ridge is indispensable for the normal outgrowth of the limb rudiment. If the apical ridge ectoderm of a wingbud of a three-day chick embryo is removed without causing damage to the underlying mesoderm (Fig. 326), the distal parts of the wing fail to be formed, although the proximal parts develop quite normally (Saunders, 1948). If the ectoderm covering the limb-bud in a chick embryo is removed and replaced by epidermis from another part of the body, the latter is found to be incapable of developing an apical ridge. As a result, the development of the entire distal part of the limb is suppressed. The girdle may, however, develop normally, and a short piece of cartilage representing the proximal part of the humerus or femur may be formed. This is in some contradiction to the conditions found in amphibian embryos, where flank or head ectoderm, as was indicated, may participate in the development of a limb, but then there are no apical ridges on amphibian limb-buds.

The investigation of the role of the epidermal apical ridge in the development of limbs in birds was extended further after it had been discovered that the mesoderm and ectoderm of a limb-bud may be separated very neatly by chemical instead of mechanical means (Zwilling, 1955). Treating a limb-bud with a trypsin solution causes the epidermis to separate from the mesodermal core of the bud. The mesoderm after this

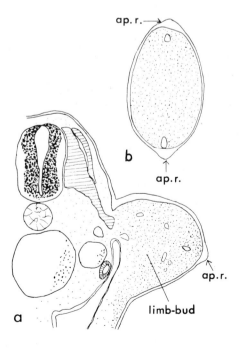

Figure 324. Advanced limb-bud with apical ridge (ap.r.), in a chick embryo *(a)* (from Saunders, 1948) and cross-section of the forelimb-bud in a rat *(b)* (from Milaire, 1956).

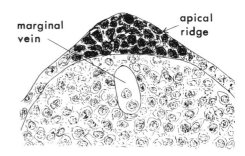

Figure 325. Part of cross-section of the limb-bud in a rat stained for alkaline phosphatase. The apical ridge shows intense positive reaction. (From Milaire, 1956.)

treatment is not fully viable, but the epidermis is quite healthy and may be used to cover a mesodermal part from which the epidermis is removed by immersion in a solution of Versene (the latter treatment destroys the epidermis, which comes off in flakes, but leaves the mesoderm in a very good condition).

After the mesoderm of a limb-bud is covered by epidermis of a different origin, the two stick firmly together, and such a composite limb-bud may be implanted onto the flank of a third embryo and there allowed to grow into a limb (Fig. 327). In this way it is possible to combine the mesoderm of a legbud with the epidermis of a wing-bud and vice versa. The structure of the developing limb has in every case been determined by the origin of the mesodermal component. A wing developed if the mesoderm was taken from a wingbud, and a leg if the mesoderm was that of a posterior limb rudiment. The origin of the ectoderm did not affect the nature of the developing limb. This experiment stresses the major role of the mesoderm in limb development.

The next experiment emphasizes the importance of the epidermis. In a "wingless" mutation in fowls, the forelimb-bud appears approximately at the same time as the normal limb-bud but fails to grow and produce a limb, though parts of the limb girdle

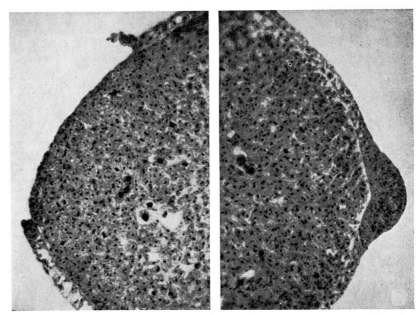

Figure 326. Apex of wingbud from which the epidermis with apical ridge has been removed (left), and control normal wingbud with apical ridge (right). (From Saunders, 1948.)

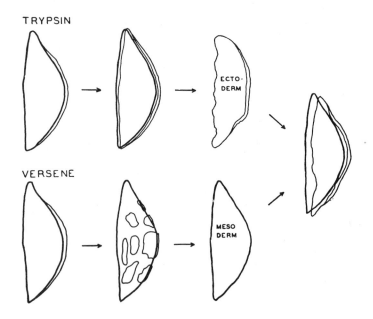

TRYPSIN

ECTO-DERM

VERSENE

MESO DERM

Figure 327. Separation and recombination of the limb-bud mesoderm and epidermis in the chick embryo. (From Zwilling, 1956.)

may be present. It was noticed that the epidermis covering the wingbud in the affected embryos did not have an apical ridge. An experiment was therefore undertaken in which the mesoderm of a normal wingbud was combined with the epidermis of an embryo of the wingless strain. As had been expected, in the absence of an apical ridge the bud did not grow, and no distal parts of the wing were formed (Zwilling, 1956).

The competence for limb development can be shown to be present all along the flank of the embryo between the forelimb and hindlimb regions, even if it does not manifest itself in normal development. In urodele amphibians supernumerary limbs have been induced in the area between the forelimb and the hindlimb by transplanting an ear vesicle or a nose rudiment (Fig. 328) (Balinsky, 1925, 1933). The graft serves as a kind of "abnormal inductor" (p. 210) which activates latent potencies in the flank tissues. Supernumerary limb induction is also possible in toad embryos, but with these the complex of organ rudiments of the anterior part of the head must be used as the inductor (Perri, 1951; Balinsky, 1974) (Fig. 329). The active components of the grafts in this case appear to be the nose rudiment, as in urodeles, but also the rudiments of the mouthparts (Balinsky, 1974). Under the influence of the induction, the local mesodermal cells accumulate as a compact mass under the epidermis, the epidermis is also made to react, and an additional limb develops. Depending on whether the induced limb lies nearer to the normal forelimb or hindlimb region, it may resemble a forelimb or a hindlimb in its structure (Balinsky, 1933; Perri, 1951).

Experiments on limb induction suggest that also in normal development there must be some factor determining which part of the mesoderm component for limb development actually produces a limb rudiment. Limb-buds are often induced together with other structures when "spinocaudal" inductors are introduced into embryos in the gastrula stage (p. 220). By transplanting the presumptive somite mesoderm together with the notochord into the lateral plate region of the embryo in the neurula stage, it is possible to induce the lateral plate mesoderm to develop into kidney (normally produced by the stalks of the somite). In these experiments it has often been observed that additional limb-buds develop together with the kidney tubules (Yamada, 1937). It follows that the determination of the limb mesoderm occurs in conjunction with the

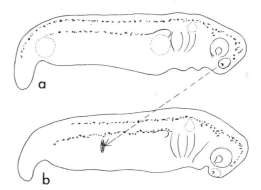

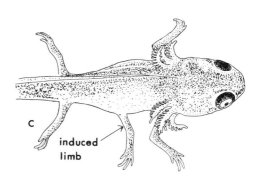

Figure 328. Induction of a supernumerary limb in *Triturus taeniatus* by means of a grafted nose rudiment. *a, b,* Diagrams of the operation. The position of the normal rudiments of the fore- and hindlimb shown in *a. c,* Larva with induced limb.

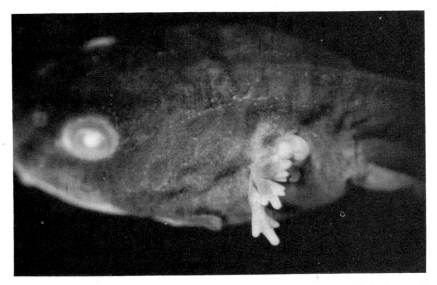

Figure 329. Induction of supernumerary limbs in a toad tadpole by transplanted anterior neural fold and adjacent ectoderm and endoderm. (From Balinsky, 1974.)

determination of other parts of the mesodermal mantle. The kidney in itself cannot, however, induce a limb.

Differentiation in the Limbs.   After the limb-bud has grown so far that its length exceeds its breadth, the differentiation of the subordinate parts of the limb sets in. We have already noted the slight flattening of the limb-bud. Now the distal portion of the bud becomes flattened even more, and at the same time it becomes distinctly broader than the proximal part of the limb rudiment. The flattened and broadened distal part is the hand (or foot) plate. The edge of the plate is initially circular, but soon it becomes pentagonal, the projecting points indicating the rudiments of the digits. While the tips of the digit rudiments continue to grow out farther, extensive necrosis occurs in the intervening sections. Mesodermal and ectodermal cells (including those of the apical ridge) in these sections die and are consumed by macrophages (Saunders and Fallon, 1967). As a result, the digits become separated by distinct incisions (Fig. 330). The five digits appear simultaneously in all amniotes with pentadactyl limbs, but in amphibians (especially urodeles) the first two digits appear earlier and digits 3, 4, and 5 are formed one after another on the posterior edge of the limb. When less than five digits are present in the adult limb, or more than five in cases of hyperdactyly, this condition is reflected in the structure of the hand (foot) plate.

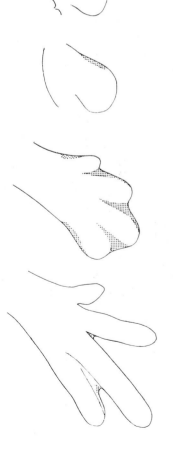

Figure 330. Development of digits in the hindlimb of a chicken embryo. The areas undergoing necrosis are shown by stippling. (From Saunders and Fallon, 1967.)

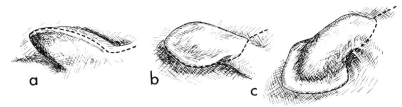

Figure 331. Rotation of the developing forelimb in the lizard. (After Braus, from Hertwig, 1906.)

In the early limb rudiments, the future flexor surface is ventral and the future extensor surface is dorsal, but as the limb elongates a rotation takes place so that the flexor surface is turned posteriorly, and eventually it may even face in a posterodorsal direction. The preaxial edge of the limb, which is originally anterior, is then turned downward (Fig. 331). With the elongation of the limb, it becomes bent at the elbow joint (or knee joint). A less pronounced flexion develops at the base of the carpus (or tarsus). The three main sections of the limbs thus become recognizable externally.

Concurrently with changes in the external appearance of the limbs, differentiation occurs in the interior of the limb rudiments. The mesenchymal cells which are closely packed in a young limb-bud become segregated into areas in which they lie more loosely and into other areas in which they are crowded. The latter are the rudiments of the skeletal parts of the limb. The concentrated masses of mesenchyme eventually become converted into procartilage, and then, by further deposition of intercellular matrix, into cartilage. Whereas in the initial stage of mesenchyme concentration large sections of the limb skeleton are represented by a common mass of mesenchyme, in the procartilage stage individual elements of the skeleton are laid down as separate units, which may fuse together later.

The differentiation of the limb skeleton generally proceeds in a proximodistal direction, though some deviations from this order are a fairly general occurrence. In amphibians, the first skeletal part to become recognizable is the **stylopodium** (humerus or femur). Parts of the **zeugopodium** (radius and ulna in the forelimb, tibia and fibula in the hindlimb) are laid down next, and the **autopodium** differentiates considerably later. The girlde rudiments appear after the stylopodium, but earlier than the autopodium. In higher vertebrates the girdle tends to be developed simultaneously with the proximal elements of the limb. In the autopodium, the proximodistal sequence is upset by the larger skeletal elements, the metacarpals and metatarsals differentiating more rapidly than the smaller elements, namely, the carpals and the tarsals. In the digits, however, the proximal phalanges are laid down earlier than the distal ones (Sewertzoff, 1931).

The blood vessels appear in the limb-bud at an early stage, and although the pattern of the arteries, veins, and capillaries is too variable to deserve much attention here, one blood vessel may be mentioned. It is situated along the edge of the hand (or foot) plate, just underneath the ectodermal apical ridge, and it is possibly responsible for supplying nutriment to this rapidly growing area of the limb (Fig. 332).

The pattern according to which the various components (bones, muscles, blood vessels, nerves) are arranged in each limb is asymmetrical—proximal and distal ends, dorsal and ventral surfaces, anterior and posterior sides of the limb being different from one another. The first indication of this asymmetrical pattern can be noted when the tip of the limb-bud grows out in a slanting posterior direction instead of growing straight out away from the side of the body (Fig. 314).

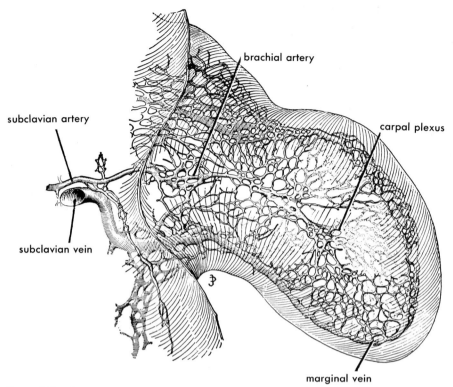

subclavian artery

brachial artery

carpal plexus

subclavian vein

marginal vein

Figure 332. Reconstruction of the blood vessels in a limb rudiment of a pig embryo. (From Woollard: Carnegie Contrib. Embryol., *22*, 1922.)

Special experiments have been performed to find out how the asymmetry of the limb, and thus the basic pattern of its differentiation, is determined. Experiments were originally carried out on the embryos of the salamander *Ambystoma punctatum*. Forelimb rudiments were transplanted at different stages after the end of neurulation in such a way that either one limb rudiment axis or two or all three were inverted (disharmonious) with respect to the host's body (Harrison, 1921b). Inversion of the proximodistal axis could be carried out only with the mesodermal part of the limb rudiment, but this is not of importance, since the mesoderm is the carrier of the limb determination (Harrison, 1925b; Swett, 1937). The transplantations were done either **orthotopically,** that is, in place of a normal limb rudiment, or **heterotopically** on the flank. The experiments showed that the three axes of the limb rudiment were not determined simultaneously. In the earliest rudiments immediately after neurulation, the anteroposterior axis is already fixed. Limb rudiments, transplanted disharmoniously, with this axis inverted, had the tip of the limb-bud growing forward instead of backward, and the limbs later had their postaxial (ulnar) side placed anteriorly (Fig. 333). At the same time, the inversion of the dorsoventral axis of the transplanted limb rudiment was still able to be corrected; that is, the original upper part of the rudiment developed into the ventral (palmar) surface of the limb. This shows that the pattern of the limb differentiation was not yet fixed with respect to its dorsoventral axis and that this pattern could be imposed upon the limb rudiment by the host, that is, by the parts which were in connection with the limb rudiment in its new position. In a later stage of development when the tail rudiment begins to elongate, the dorsoventral axis of the limb was also found to be

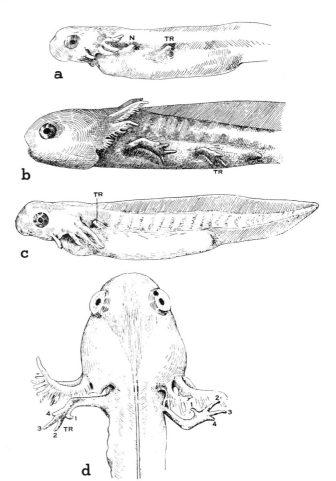

Figure 333. Experiments demonstrating the properties of the forelimb rudiment of the salamander. *Ambystoma punctatum*, in the tail-bud stage. N, Normal limb; TR, transplanted limb. *a*, Transplantation of a right limb rudiment to the left side, with inversion of the dorsoventral axis. *b*, Same animal later. *c*, Left limb transplanted orthotopically with inverted anteroposterior and dorsoventral axes. *d*, Same animal later; a.p. axis determined and retained in new position. (From Harrison, 1921b.)

determined. If the limbs were transplanted in an inverted position, with the dorsoventral axis of the limb rudiment being the inverse of the dorsoventral axis of the host, the limb developed with an abnormal orientation, its plantar surface facing upward. The surroundings could no longer change the pattern of limb differentiation with respect to the dorsoventral axis, as well as with respect to the anteroposterior axis (Fig. 334).

At the same time that the dorsoventral axis of the limb rudiment is determined, the proximodistal axis may still be inverted without impairing the normal development of the limb. Only in a later stage, when the limb-bud begins to be visible from the outside, does the proximodistal axis show signs of being determined, and if the limb mesoderm is transplanted with the axis inverted the limb shows abnormalities in its develop-

Figure 334. Heterotopic transplantation of the limb rudiment in the late tailbud stage, with inverted dorsoventral axis. The palmar surface of the transplanted limb (partially reduplicated) facing upward. (From Swett, 1927.)

ment. (A limb cannot actually grow inward into the body because it would then no longer be in contact with the epidermis, and this contact is indispensable for the differentiation of parts of the limb.)

In cases in which the transplanted limb rudiment grows in a disharmonious orientation to the host, it is often observed that a kind of regulation occurs by means of the formation of a second limb-bud, whose orientation and differentiation are harmonious with the host's body. The appearance of the second limb-bud is the result of a sort of splitting of the original rudiment and is due to an influence of the host on the graft. This influence is not strong enough to invert the axial structure of the transplanted rudiment as a whole but is sufficiently strong to divert part of the cells of the transplanted rudiment and to cause them to take on the axial structure of the host. If the original disharmonious limb were then to degenerate (as sometimes happens), the host would be in possession of a normal set of limbs.

The splitting of a limb rudiment (or any other organ rudiment) to produce two similar rudiments is known as **reduplication.** That reduplication is possible shows that the cells of the rudiment are not each determined to fulfill a definite part in the developing organ, even though the axial pattern of the rudiment as a whole is determined. From this fact we may further infer that the determination of the axial pattern is not a matter of determining what each cell or group of cells must do in the process of development of the organ, but it is rather a matter of polarity, of the heteropolar structure of the rudiment as a whole. This property immediately recalls the polarity of the egg in the early stages of development, and some similarity between the rudiment of an organ and the early egg as a whole is indeed shown. Both possess the ability to develop a number of subordinate parts, each within its own scope of action (the whole animal in the case of the egg, one organ in the case of an organ rudiment), without these subordinate parts being represented by discrete particles in the initial system.

Just as the egg can be split mechanically into parts and each part will develop into a miniature whole, an organ rudiment, in a suitable stage of development, may also be split mechanically into halves, and each half will develop into a complete organ. We have already mentioned similar results in the case of the eye rudiment, and the same applies to the limb rudiment. The limb rudiment up to the stage of limb-bud formation may be cut in two, and each half may be transplanted separately, or the two halves may be left in place and kept apart by inserting a piece of extraneous tissue between them. Each will develop into a whole limb (Fig. 335—Swett, 1926). The two limbs resulting from such splitting may later grow to the normal size. Splitting of limbs may sometimes occur accidentally in young amphibian larvae developing in nature, or it may possibly be caused by the pressure of folds of the amnion in higher vertebrates, and abnormalities will be the result, i.e., limbs that are completely or partially reduplicated. If the splitting of the rudiment is due to a mechanical cause rather than being spontaneous, the two halves retain the same axial structure (polarity), thus being replicas of each other.

The limb girdles normally develop in intimate connection with the limbs themselves. The mesodermal material for the girdle is derived from the peripheral parts of the mesenchyme mass, the central part of which becomes the limb-bud. However, in their determination the girdles are partially independent of the limbs. They may develop independently of the limbs, if the development of the limb itself is suppressed in some way, such as by the removal of the limb-bud (Detwiler, 1918). In limb induction experiments, limb girdles are sometimes found where the limb itself fails to develop. Development of the girdle is in no way dependent on the existence of an interaction with the epidermis of the skin, as is that of the limb proper. In birds, the limb girdle may also develop fairly well even if the distal part of the limb is absent as a result of

Figure 335. Splitting of the forelimb rudiment with a strip of extraneous tissue (T) inserted across it (*a*) to produce two limbs (*b, c*). (From Swett, 1926.)

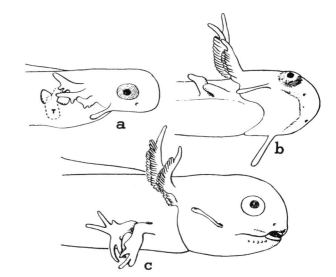

failing cooperation between mesoderm and ectoderm, such as in cases in which the epidermis over the limb-bud lacks an apical ridge.

If the girdle (shoulder girdle or pelvic girdle) develops in the absence of the limb, it may be normal in its peripheral parts, but the fossa for the articulation with the humerus or femur does not develop in the absence of at least the proximal part of the stylopodium (humerus or femur) (Balinsky, 1931). The thickening of the girdle in the region of the articulation also does not occur in the absence of a limb. As far as these parts are concerned, the amphibian girdles are dependent on the limbs. In bony fishes, in which the cartilaginous part of the shoulder girdle is reduced to a small skeletal element in the region of articulation, the bony **cleithrum** can develop independently of the pectoral fin (Balinsky, unpublished).

The mesenchyme of the limb-bud does not contain all the mesodermal cells used for the development of a limb. After the limb-bud is formed it receives an additional supply of mesodermal cells from the lower edges of the myotomes. These cells produce the muscles of the limb or at least a part of them. In fishes it is possible to observe **muscle buds** being formed at the lower edges of the myotomes. These muscle buds are protrusions of the cell masses of the myotomes. They push downward and outward until they enter the limb-buds. In the limb-bud itself, the muscle buds derived from different myotomes fuse into a common mass of cells (myoblasts) from which the muscles of the limb subsequently develop. In the higher vertebrates, beginning with amphibians, muscle buds are not found, although the migration of individual cells from the myotomes into the limb rudiments may take place.

In the fishes, if the lateral plate mesoderm of the limb region is transplanted to near the midventral line on the abdomen, it gives rise to fins which are devoid of muscles (Lopashov, 1950), thus proving that the lateral plate mesoderm is not capable of producing the limb muscles. Muscles are developed, however, if part of the somite is included in the graft. In amphibians, on the other hand, when the limb rudiment, consisting of lateral plate mesoderm with or without epidermis, is transplanted heterotopically, it will develop into a complete limb with muscles. This development, however, does not definitely exclude the participation of the myotome material, since under experimental conditions some regulation could have taken place, just as half the rudiment may produce a whole limb in cases of reduplication.

The nerve supply to the developing limbs has been dealt with previously in Section 12–1 on the differentiation of the nervous system.

13–4 DEVELOPMENT OF THE URINARY SYSTEM

The excretory organs in vertebrates are essentially aggregates of uriniferous tubules, connected originally at their proximal ends with the coelomic cavity by ciliated funnels (the **nephrostomes**) and communicating to the exterior by a system of ducts which, in lower vertebrates, open into the cloaca. Both the tubules and the ducts are of mesodermal origin, and they develop from the stalks of the somites (the **nephrotomes**). In the most primitive vertebrates, such as the cyclostomes, and also in the Gymnophiona, one uriniferous tubule develops from the nephrotome in each mesodermal segment.

The nephrotome, prior to the development of a uriniferous tubule, is a strand of cells connecting the somites to the lateral plate mesoderm. The cells become separated into the parietal and visceral layers, and the cavity, which we may call the **nephrocoele,** is for a while continuous with both the myocoele and the definitive coelom between the two sheets of the lateral plate mesoderm. The connection of the nephrocoele with the cavity of the somite is soon obliterated, but the connection with the cavity of the lateral plate persists in the more primitive type of vertebrate excretory organs and becomes the nephrostome. The dorsolateral wall of the nephrotome becomes drawn out into a hollow tube, the cavity of which is an extension of the nephrocoele. The tube, which is in open connection with the coelomic cavity, becomes the uriniferous or renal tubule. The distal (outward) ends of the most anterior tubules (that is, the tubules formed in the anterior part of the trunk) soon turn backward and then fuse with one another, thus giving rise to the common excretory duct, known as the **pronephric duct** (Fig. 336).

An essential feature in vertebrates is that the renal tubules are associated with bunches of fine blood vessels (the **glomeruli**) through the endothelial walls of which the blood plasma containing excretory products is filtered into the uriniferous tubules or into the coelom in the immediate vicinity of the nephrostomes, so that the nitrogenous waste products may be carried through the tubes and the excretory ducts and removed from the body. Either the mass of blood vessels is invaginated into the wall of the renal tubule, which enlarges to contain the glomerulus and becomes the **Bowman's capsule,** or else the blood vessels form a bulge on the wall of the coelom. In this case

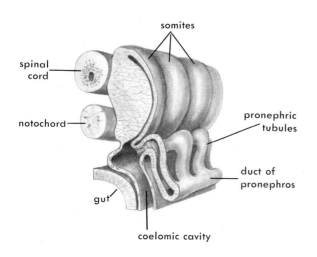

Figure 336. Diagram showing relation of pronephric uriniferous tubules to other parts of the mesoderm and to the pronephric duct.

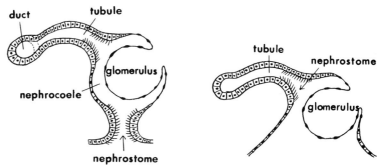

Figure 337. Diagrams of two main types of an excretory unit, with internal glomerulus (left) and external glomerulus (right). (From Fraser, 1950.)

the structure is referred to as the **external glomerulus,** or if glomeruli of several segments are joined together, as the **glomus.** It is believed that the segment of the coelomic cavity into which the glomus projects is itself derived from an expansion of the nephrocoele, or several nephrocoeles (Fig. 337).

The basic pattern of development of the excretory organs as just described becomes modified in various degrees. As with many other organ systems, in vertebrates the development of the excretory tubules progresses in a craniocaudal direction, the tubules differentiating earlier in anterior segments than in posterior ones. The tubules which develop and begin functioning earlier, that is, those of the anterior part of the trunk, tend to show more primitive conditions in their organization, while those developing later are modified to a greater degree. In the most primitive vertebrates, the cyclostomes and some fishes, the differences between the anterior and posterior parts of the excretory system are only gradual, but in the higher vertebrates we may distinguish more or less clearly three sections of this system: the **pronephros,** the **mesonephros,** and the **metanephros.**

The most anterior nephrotomes give rise to the pronephros, the mesonephros develops from those of the midtrunk region, and the metanephros is derived from the nephrotomes of the posterior part of the trunk. The resulting three types of kidneys can thus be distinguished as anterior, middle, and posterior with respect to their place of origin. The pronephros is the first to develop, the mesonephros second, and the metanephros, if present, appears latest of all.

The three types of kidneys differ also in the way in which they are produced by the stalks of the somites, in the presence or absence of the nephrostomes, in the position of the glomerulus, and in the origin and connections of the excretory ducts. In the pronephros, the nephrostomes are well developed, and the glomeruli tend to be replaced by a common glomus (see previous description), though in some primitive forms, such as the Gymnophiona, each tubule of the pronephros may be supplied with a separate glomerulus. In the mesonephros, a separate glomerulus is, as a rule, intercalated in the course of each uriniferous tubule. Nephrostomes are present at first but may disappear later. In the metanephros, the nephrostomes are not formed at all, and the uriniferous tubules begin with the Bowman's capsules.

Further differences in the development of the tubules and ducts can best be described separately in relation to each section of the excretory system. Since all three types of kidneys are nothing but local differentiations of one system stretching throughout the body of a vertebrate, morphological features do not necessarily change abruptly from one section to another, and transitional conditions may sometimes be found. (See Fraser, 1950.)

The Pronephros. The description of the origin of a uriniferous tubule given on page 394 follows the observations made on the embryos of Gymnophiona (Brauer,

1902) and may be taken as representing the typical development of the pronephric tubules in the most archaic groups of vertebrates. The strictly segmental origin of the tubules is an essential feature of this development. In amphibians, the pronephric tubules are formed from a common mesodermal thickening appearing beneath the third and fourth somites (in salamanders) or second, third, and fourth somites (in frogs). Nevertheless, the number of the pronephric tubules corresponds to the number of segments participating in the development of the pronephros.

The rudiment of the pronephros can be traced back in amphibians to the neurula stage. By means of vital staining, the presumptive material of the pronephros was found to lie in the mesodermal mantle just outside of the edge of the neural plate, posterior to the middle of the embryo (Fig. 338 – Yamada, 1937; Muchmore, 1951). In the neurula stage, mesoderm of this region is capable of self-differentiation when transplanted heterotopically (Fales, 1935). At the same time, however, other parts of the mesodermal mantle also possess the ability to develop into renal tubules, as for instance when an inductor (the notochord) is transplanted into the lateral plate region of the embryo. The presumptive somites, when isolated from the notochord and cultivated *in vitro* (surrounded by a coat of skin epidermis for protection), may develop into renal tubules even though their normal destiny would be a different one (the development of muscle). After the end of neurulation (closure of the neural tube), the competence for the development of pronephric tubules is restricted to the presumptive material of the pronephros. From this stage onward the pronephric rudiment cannot be replaced, and removal of the rudiment leads to the absence of the pronephros.

The sequence of events during the transformation of the nephrotomes into the pronephric tubules in frogs, salamanders, and higher tetrapods is not as clear as in the lower vertebrates. It is carried out by means of a rearrangement of cells, that is, by morphogenetic movements. The end result is, however, the same: several pairs of pronephric tubules are formed. They open by means of the nephrostomes into the coelomic cavity (Fig. 339) and fuse distally to form the pronephric duct (Fig. 340).

The formation of the pronephric duct, by fusion of the distal ends of the pronephric tubules, is a very important phase in the development of the excretory system, as this duct not only serves the pronephros but also is instrumental in providing pathways for the outflow of urine from the mesonephros and metanephros and, in the males, for the passage of spermatozoa.

In the salamanders, it has been shown that the pronephric duct develops right from the start from a more caudal part of the mesoderm than the pronephric tubules: while the two pronephric canals develop from the mesoderm lying under the third and fourth somites, the rudiment of the pronephric duct lies under the fifth, sixth, and seventh somites. If the rudiments of the pronephric tubules are removed, or if the embryo is bisected transversely between the levels of the fourth and fifth somites, the pronephric

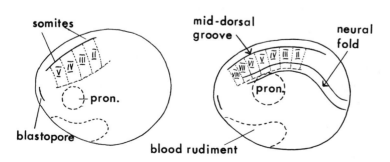

Figure 338. Position of the presumptive pronephros in the very early (left) and middle neurula (right) of a urodele amphibian. (After Yamada, 1937.)

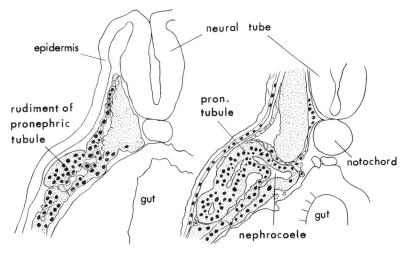

Figure 339. Early development of a pronephric tubule in the frog. (After Field, from Brachet, 1935.)

duct can still develop, thus showing that the pronephric tubules and the pronephric duct are determined independently of each other (Holtfreter, 1943c).

Once they have been formed, both the pronephric tubules and the pronephric duct elongate very considerably. The pronephric tubules, as a result of their elongation, develop numerous loops and eventually form a more or less spherical body, consisting of tangled and intertwined tubules. The pronephric duct, on the other hand, remains straight, and while it elongates, its posterior free end pushes itself backward along the lower ends of the somites. This backward movement of the tip of the pronephric duct ends when the duct reaches the cloaca and fuses with its wall, while the lumen of the duct opens into the cavity of the cloaca.

The backward elongation of the pronephric duct may be interrupted in various ways, such as by making a deep incision across its path. If the wound remains gaping, the pronephric duct cannot spread beyond the wound, and it is not continued into the posterior part of the body (Waddington, 1938; Holtfreter, 1943c). If, however, the wound is too small or is covered to a sufficient extent, the tip of the elongating duct may find its way around the wound and may penetrate into the posterior part of the body. Behind the wound, the duct returns to its level under the lower edges of the somites and eventually reaches the cloaca (Fig. 341a, b). This experiment shows that the elongation of the pronephric duct is largely independent of the surrounding parts, although the latter do seem to direct the duct by furnishing a suitable path for its move-

Figure 340. Reconstruction of the pronephric canals in a newt embryo (median view). nst. 1, nst. 2, First and second nephrostomes; t. 1, t. 2, first and second pronephric tubules; 1., loop formed by joint part of tubule; pr. d., pronephric duct. (After Mangold, from Spemann, 1938.)

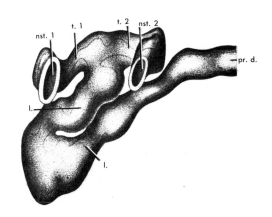

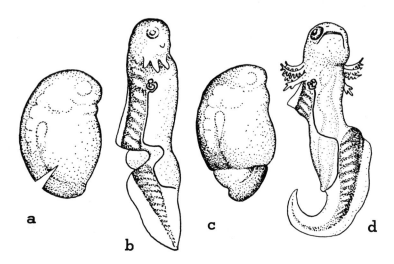

Figure 341. Pronephric duct reaching the cloaca by an abnormal path after transection of the embryo *(a, b)* and after rotation of the posterior part of the embryo through 180 degrees *(c, d)*. (From Holtfreter, 1943c.)

ment. The same is borne out with the utmost clarity by further experiments (Holtfreter, 1943c), in which embryos in the closed neural tube stage were bisected transversely. One half was reversed, and the two parts were caused to heal together with opposite dorsoventral orientation (Fig. 341 *c, d*). When the pronephric duct reached the level of the operation in its backward elongation, the tip of the duct changed its direction and struck a new path across the lateral body wall, until it reached the somites of the inverted posterior part of the embryo. It then elongated along the edge of the somites and reached the cloaca as usual.

The development of the pronephric duct shows a great similarity to the development of the lateral line (Section 12–4). In both cases the migration of the posterior end of the organ rudiment along the length of the body is due to intrinsic tendencies of the cells of the rudiment, but the organs and tissues with which the rudiments come in contact influence the path taken by the migrating cells. (See also Section 9–2.)

The kidneys of the most primitive living vertebrates, the cyclostomes and elasmobranch fishes, in their development and organization in the adult, show features characteristic of the pronephros, although they may perhaps be more correctly described as representing a stage in which the differentiation into pronephros and metanephros has not yet taken place. The typical pronephros is a functional kidney of the larval stages of bony fishes and amphibians. In frog tadpoles, the pronephric canals form a bulky convoluted mass on both sides in the anterior part of the trunk at the level of the forelimbs, on the median aspect of the pectoral girdle (Fig. 247). In frog tadpoles aproaching metamorphosis, the pronephros gradually degenerates; the glomus shrinks, the nephric tubules are resorbed, and the anterior portions of the pronephric ducts are also resorbed. The function of excretion is taken over by the mesonephros.

In the embryos of amniotes, the pronephros develops in the anterior trunk region, but it is not functional at any stage. In the human embryo, about seven pairs of rudimentary pronephric tubules are formed. They soon degenerate, but not before giving rise to the pronephric ducts, which remain after the pronephric tubules have disappeared.

The Mesonephros. The mesonephric tubules are derived from the nephrotomes, as in the case of the pronephros, but their interrelation is by no means so simple. Only in the more archaic groups of vertebrates (selachians, Gymnophiona, ganoids) are the renal tubules formed directly from the nephrotomes (Fig. 342). In most amphibians and in all higher vertebrates, the mesodermal cells of the nephrotomes dissolve into a mass of mesenchyme extending on each side of the body along the dorsal edge of the lateral plates from the level where the pronephros ends to the pelvic region. This elongated mass of mesenchyme is known as the **nephrogenic cord** or nephrogenic tissue (Figs. 262 and 270). The mesonephric tubules are developed from the nephrogenic tissue by a secondary aggregation of the mesodermal cells.

The aggregating cells first form epithelial vesicles which stretch and elongate to become tubes. One end of such a tube becomes connected to the pronephric duct, while the other end becomes invaginated to form a Bowman's capsule, or in rarer cases, opens into the coelomic cavity by a nephrostome, while a Bowman's capsule is formed higher along the course of the tubule. Bowman's capsules become supplied by small branches from the dorsal aorta. The number of mesonephric tubules does not correspond to the number of segments, several tubules being developed in the region of each segment. Even if only one tube is initially formed in each segment, it soon gives rise by budding to secondary and tertiary tubes, and so on. If present at first, the nephrostomes may disappear later. In the frogs, the nephrostomes lose their connection with the rest of the nephric tubule but open secondarily into the veins (connecting the veins to the coelom).

The nephrogenic tissue is not at once determined for differentiation as mesonephros. If the presumptive material of the mesonephros of a newt embryo in the neurula stage is transplanted into the pronephric area of a tailbud stage embryo, the grafted tissue gives rise to excretory tubules of the pronephric type. On the other hand, when the same material is taken from a tailbud stage embryo and grafted, it will produce mesonephric tubules even if it comes to lie at the site where a pronephros should have developed (Machemer, 1929).

The mesonephric tubules do not produce a duct of their own. As the tubules are formed, their distal, free ends join up with the pronephric duct, which thus becomes the duct of the mesonephros as well and is then called the **mesonephric duct** (or **Wolffian duct**). As the mesonephric tubules reach the mesonephric duct, the walls of the latter bulge out, forming collecting ducts. The mesonephric tubules, especially the ones formed later, open into these collecting ducts rather than into the mesonephric duct itself.

The development of the mesonephros has also been found to be dependent on the pronephric duct in another way. The nephrogenous tissue develops into the mesonephric tubules only if it is stimulated by the pronephric duct. In the preceding section,

Figure 342. Relation of the mesonephric tubules to other mesodermal parts in the embryo of a sturgeon. The rudiment of the mesonephric tubule (MT) is shown as still connected to the somite (S) on one side, and just separated on the other. C, Coelom; N, nephrostome; MD, mesonephric duct. (After Maschkowzeff, from Fraser, 1950.)

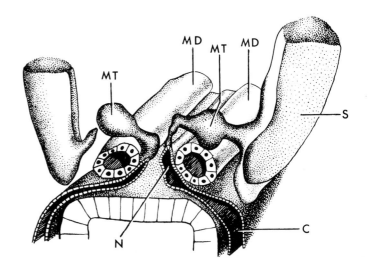

experiments were mentioned in which the penetration of the pronephric duct into the posterior half of the body was prevented by placing an obstacle (in the form of a gaping wound) in its path. When this operation is successful—that is, if the duct does not reach the region where the mesonephros normally develops—the nephrogenous tissue fails to form the renal tubules or forms only poorly developed tubules (Waddington, 1938; O'Connor, 1939). (See Figure 375, p. 437.) Apparently some stimulus (induction) from the pronephric duct is necessary for the normal development of the mesonephric tubules from the nephrogenous tissue.

The most anterior mesonephric tubules degenerate later in many animals, so that the kidney becomes shortened at its anterior end and somewhat more compact. In many fishes and in amphibians, the posterior part of the mesonephros shows some traits found in the metanephros: complete absence of nephrostomes and joining of the nephric tubules to large collecting ducts—outgrowths of the mesonephric duct. For this reason, some authors prefer to regard the kidney in these animals as a joint mesonephros and metanephros, giving it the special name of **opistonephros.** In fishes and amphibians, the mesonephric (or opistonephric) kidney is the excretory organ of the adult animal.

In reptiles and birds, on the other hand, the mesonephros functions only during the embryonic period of development and loses its excretory function at the time of hatching. In mammals, the placenta takes on the task of removing excretory products from the blood of the embryo and passing them into the blood of the mother, from where they are removed by the maternal kidneys. In mammals such as the pig, in which the connection between the fetal and maternal tissues is not very close, the mesonephros is still active as an excretory organ, and a certain quantity of urine eventually reaches the cavity of the allantois and the amniotic cavity. In mammals having a very close connection between the fetal and maternal tissues (e.g., in rodents with a hemochorial placenta) there is no sign of excretory activity in the mesonephros, which is thus a rudimentary organ like the pronephros.

The Metanephros. The metanephros develops from the posterior part of the nephrogenic cord, the part adjacent to the cloaca. There is no trace of a relation of the tubules to the nephrotomes, although the nephrogenic tissue is primarily derived from the latter. As in the case of the mesonephros, the metanephros does not develop a duct of its own but uses the mesonephric duct as a means of removing the urine that it excretes. However, the connection between the metanephros and the mesonephric duct is established not directly but by means of a special outgrowth or branch of the duct. Before the metanephros starts differentiating, a bud is formed on the mesonephric duct a small distance in front of the point where the duct joins the cloaca. The bud elongates in the direction of the posterior part of the nephrogenic tissue (Fig. 343). The duct formed in this way becomes the **ureter,** and the bud from which it develops is the **ureteric bud.** Having reached the nephrogenic tissue, the end of the ureter begins to branch, the branches later becoming the collecting tubules of the kidney (Fig. 344). At the point where the branching begins, the ureter expands to form the renal pelvis. The nephrogenic mesenchyme accumulates around the tips of the collecting tubules and eventually differentiates into the renal tubules with their glomeruli. No trace of nephrostomes can be discovered in the metanephros at any stage.

As in the mesonephros, the conversion of the metanephrogenic tissue into a system of renal tubules is dependent on a stimulus from the excretory duct—in this case from the ureter and its branches, which develop from the ureteric bud. In experiments on chick embryos, in which the backward growth of the pronephric duct is prevented by destroying its growing tip, no mesonephric duct develops in the posterior part of the body, and naturally no ureteric bud is formed (Gruenwald, 1952). The result is that

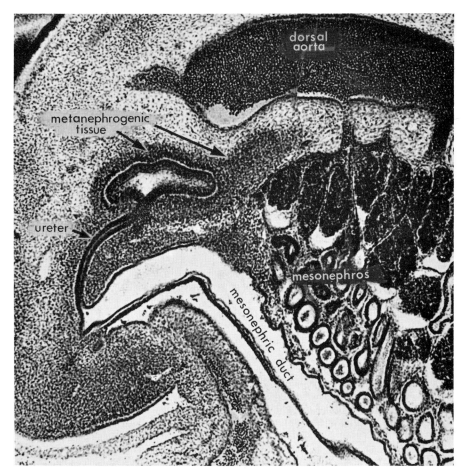

Figure 343.   Development of the metanephros in a 10 mm. pig embryo. Photograph showing ureteric duct as an outgrowth of the mesonephric duct, and the expansion of the duct into the mass of metanephrogenic mesenchyme, as well as the functionally differentiated mesonephros.

no metanephric kindey is formed on the operated side of the embryo. A confirmation of results gained by operation may be found in certain defects caused by mutations (p. 517).

The metanephros is the functional kidney in the postembryonic life of the reptiles, birds, and mammals. In lizards, the fully differentiated metanephros largely retains the position of the metanephrogenic part of the nephrogenic cord; the functional kidney lies at the level of the hindlimb and even projects into the base of the tail. In birds and mammals, however, at an early stage the metanephric kidneys become displaced in an anterior direction. The metanephros shifts past the posterior end of the mesonephros and eventually comes to lie at about the same level as the latter, in the lumbar region in mammals, and filling the space underneath the synsacrum and the pelvic girdle in birds. Concurrently with the displacement of the kidneys, the ureters elongate to a very great extent as compared with their length at the time when the ureteric bud first contacts the metanephrogenic tissue (Fig. 344).

The excretory ducts of the kidney in most vertebrates acquire close relations to the reproductive organs. These relations and the modifications of the excretory ducts in this connection will be dealt with together with the development of the sexual organs.

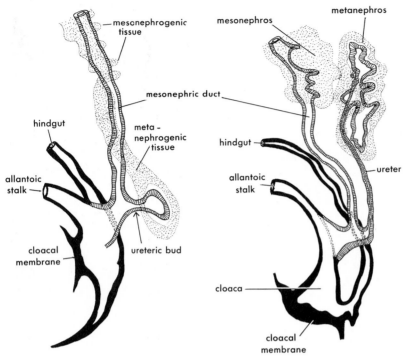

Figure 344. Development of the ureter and the metanephros in a rabbit embryo. (After Schreiner, from Brachet, 1935.)

13–5 DEVELOPMENT OF THE HEART

The heart in vertebrates develops from the mesoderm forming the ventral edges of the lateral plates in the pharyngeal region of the body. It will be most convenient to describe first the development of the heart in animals such as the amphibians and to consider later the development of the heart in higher vertebrates.

The Heart in Lower Vertebrates. During gastrulation in the amphibians, the sheet of mesoderm advances forward from the blastopore, penetrating between the ectoderm and the endoderm. The rate of movement of the mesoderm is greatest in the dorsal region of the embryo, is intermediate laterally, and is least ventrally. In the neurula stage, the dorsal and dorsolateral parts of the mesodermal mantle have reached the head region of the embryo; ventrally an approximately triangular area remains in which there is no mesoderm intervening between ectoderm and endoderm. This triangle, roughly corresponding to the oral and pharyngeal region of the embryo, has its apex posteriorly and its broad base anteriorly. The posterior part of the area is the site of development of the heart, and it is later filled in by the mesoderm participating in the formation of the heart. The most anterior part of the area remains free of mesoderm, and here the mouth breaks through after a fusion between the endoderm and the ectoderm has taken place.

The presumptive material of the heart is found in the edges of the mesodermal mantle bordering the mesoderm-free area on the right and on the left. By vital staining in the neurula stage, it has been shown that the presumptive material for the heart is located in the edge of the mesodermal mantle, rather high on the flank on each side, where it adjoins the parts of the neural folds which become the hindbrain portion of the neural plate (Fig. 345).

If the mesoderm of this area is stained with nile blue sulfate or neutral red (by lifting a flap of ectoderm and mesoderm and applying a piece of cellophane soaked in stain to the *inner* surface of the mesodermal layer), the mesoderm may be observed to shift in later stages in a ventral direction and eventually to come to lie midventrally in the area where the heart can be seen to differentiate. The particles of vital stain remain in the cells long enough to make it possible to ascertain that the stained material on each side of the embryo gives rise to exactly one half of the heart rudiment (Fig. 345*D*), the other half being derived from a similar area of mesoderm on the other side of the embryo.

Parts of the free edge of the mesodermal mantle may be excised in the neurula stage and cultivated in a saline solution (Goerttler, 1928; Bacon, 1945: Jacobson, 1960).

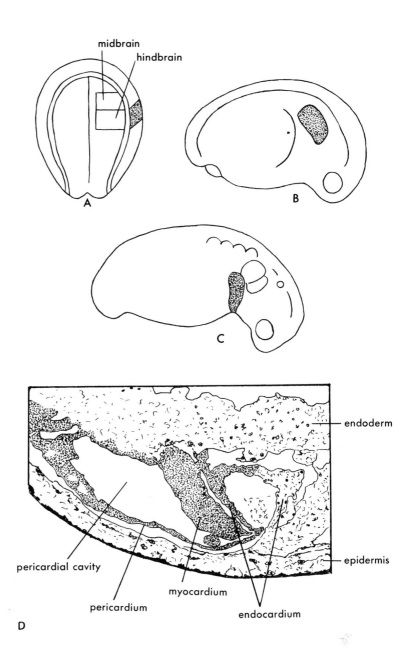

Figure 345. Origin of the heart mesoderm after experiments of vital staining in *Ambystoma punctatum. A*, Diagram showing the position of stained mesoderm (stippled) just outside the neural fold, in the hindbrain region of a neurula. *B*, Embryo in early tailbud stage with stained mesoderm in the ear region. *C*, Embryo in late tailbud stage with the heart mesoderm shifted to the ventral position, where the heart rudiment is actually formed. *D*, Cross-section through the heart region of an embryo in which the heart mesoderm was stained on the right side in the neurula stage (as in *A*). The stained material (stippled) on the right half of the heart includes all layers: endocardium, myocardium, and pericardium. (After Wilens, 1955.)

To prevent the mesoderm from disintegrating, it is sometimes isolated together with a flap of ectoderm. The ectoderm then closes into a vesicle with the mesoderm inside. The mesodermal cells under these conditions differentiate into muscle tissue, which begins to pulsate rhythmically—an unequivocal indication that the developed muscle tissue is cardiac muscle, as only this type of muscle is capable of autonomous rhythmical contraction. It has been claimed (Bacon, 1945) that in some experiments the heart rudiments explanted in neurula and even late gastrula stages can produce pulsating tubes even having an obvious resemblance to normal hearts.

It is rather remarkable that there is a definite discrepancy between experiments using the vital staining technique and explantation experiments. While vital staining indicates that the material forming the heart lies in the upper part of the mesodermal mantle, heart differentiation has been observed in explanted pieces taken from different parts of the free edge of the mesodermal mantle: from the upper part (Jacobson, 1960), from the middle part (Bacon, 1945), and from the ventral part (Goerttler 1928; Fig. 346). It would seem that mesoderm in a wide region of the early embryo has a potentiality for heart development. From this wider area the potentiality for heart development is later restricted to the actual material participating in the formation of the heart in normal development.

In these early stages, development of the heart is in some way dependent on the endoderm. By cutting through the ectoderm and the mesodermal mantle and removing the entire endoderm, it is possible to produce newt embryos which consist of ectoderm and mesoderm only (Mangold, 1936; Balinsky, 1939). The embryos survive the operation quite satisfactorily, but their term of life is limited because they do not possess the food supply normally contained in the yolky cells of the endoderm. The endodermless embryos show various defects in the ectodermal and mesodermal organs owing to the lack of inductive influences emanating from the endoderm. One such defect is the complete absence of the heart, although the mesodermal layer, from which the heart rudiment is derived, may remain intact. (See p. 410.)

The dependence of heart development on the endoderm and also on other adjacent tissues has been tested in hanging drop cultures (Jacobson and Duncan, 1968). The techniques used were similar to those employed in studying the induction of nervous tissues (p. 210). Pieces of mesoderm of newt embryos containing the presumptive heart mesoderm were explanted alone or with other embryonic tissues in "Niu and Twitty solution." If explanted alone, the presumptive heart mesoderm showed signs of progressively increasing determination, depending on the stage at which the material was removed from the embryo. There was no heart differentiation in material taken

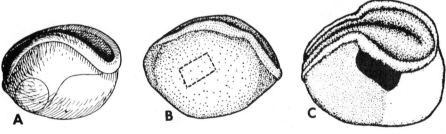

Figure 346. Diagrams showing origin in the neurula stage of mesoderm that differentiated into cardiac muscle tissue in different experiments. *A,* Goerttler's experiments (Goerttler, 1928), *B,* Bacon's experiments (Bacon, 1945); *C,* Jacobson's experiments (Jacobson, 1960). Position of mesoderm differentiating as heart in explant is shown by circle in *A,* by quadrangle in *B,* and in black in *C.* Only in the last case did the explant consist of material normally giving rise to the heart according to Wilens. (Compare with Figure 334.)

from gastrulae, but during neurulation the percentage of positive cases, as shown by spontaneous contractions, increased rapidly and became maximal (though not quite 100 per cent) when late neurulae with closing neural folds were used. Addition of bits of anterior endoderm, the suspected heart inductor, increased the percentage of positive cases and accelerated the beginning of contractions.

Instead of bits of intact endodermal tissue, an extract could be added to the hanging drop. The extract was prepared by homogenizing embryonic tissues and passing them through a cephadex column with large particle size, so that small bits of cytoplasmic organoids (up to the size of pigment granules) could get through. The extract was found to be able to promote heart differentiation, thus suggesting that in this case, as in others, the induction is mediated by soluble substances, although the action of fragments of cytoplasmic membranes in this process has not been ruled out. Anterior epidermis was also able to stimulate heart development, but its effect can be considered as only a supporting factor, since hearts fail to develop in endoderm-free embryos when the epidermis is present. Posterior endoderm, on the other hand, had no effect, and the transverse neural fold seems to exercise even a depressing effect on heart development.

After the end of neurulation, the free edges of the mesodermal mantle gradually converge toward the middle of the mesoderm-free area and become thickened in the heart region, foreshadowing the formation of the heart rudiment. At this stage a number of loose cells, similar to mesenchymal cells in their structure, can be noticed to lie between the free edges of the mesodermal mantle converging from the right and the left. These cells are derived from the ventral edge of the mesodermal mantle. They are the rudiment of the **endocardium,** that is, the endothelial lining of the cavity of the heart. The endocardial cells soon accumulate in the midline as a longitudinal strand and eventually become arranged in the form of a thin-walled tube. The lumen of the tube is the cavity of the heart (Figs. 237 and 347). The endocardial tube bifurcates at both ends. At the anterior end its two prolongations are the ventral aortae, and at the posterior end it receives the two vitelline veins, the first venous blood vessels to reach the heart. All these vessels are at first similar to the endocardial tube in that they consist of only a thin layer of endothelium, produced by the mesenchymal cells joined together.

While the endocardial tube is being formed, the edges of the mesodermal mantle, which at this stage may also be called the edges of the lateral plate mesoderm, close in along the midline and fuse with each other. The fusion first occurs under the endocardial tube, between it and the ectoderm. Soon, however, the visceral layer of mesoderm envelops the endocardial tube on the dorsal side as well. By fusion of the mesodermal layers of the right and left sides, epithelial partitions are formed above and below the endocardial tube. In analogy to the dorsal and ventral mesentery, these are called the **dorsal** and **ventral mesocardium.** The ventral mesocardium has an ephemeral existence. It becomes perforated very soon, and the coelomic cavities of the right and left sides become continuous underneath the endocardial tube. The dorsal mesocardium persists longer but is also dissolved at a later stage. The coelomic cavities expand in the heart region to form the pericardial cavity. The pericardial cavity is initially only a part of the general body cavity, the coelom. It becomes completely separated from the remainder of the coelom, largely owing to the failure of the coelom to develop in the region of the branchial pouches, which lie immediately dorsal to the heart rudiment. Posteriorly the connection of the pericardial cavity with the rest of the coelom becomes occluded by the developing liver, to which the posterior end of the endocardial tube becomes closely connected. A connective tissue wall developing in this position (at the anterior boundary of the liver) is the **septum transversum.**

From the preceding description it will be clear that the pericardial cavity is lined by

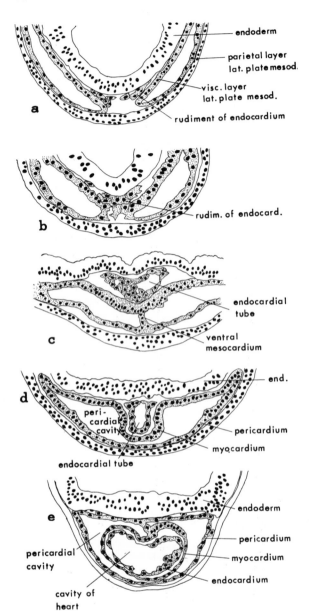

Figure 347. Development of the heart in amphibian embryos. *a, d, e, Triturus; b, Salamandra; c, Rana.* (From Mollier, in Hertwig, 1906.)

lateral plate mesoderm. The parietal layer of this mesoderm persists as the epithelial wall of the pericardial cavity, or the **pericardium** proper. The visceral layer adheres to the endocardial tube. This layer differentiates as muscle tissue and thus gives rise to the **myocardium** of the heart.

The heart is at first an almost straight tube and does not show a subdivision into its various chambers. Later, the tube becomes inflected in a very characteristic way. Starting from behind, the tube first runs forward, then bends downward and to the right, and eventually bends again to the left, upward and forward. The heart thus becomes coiled in the shape of an S. The degree of the twisting in higher vertebrates is greater than in lower ones, so that in the former the tip of the second inflection comes to lie well posterior to the tip of the first inflection (Fig. 348). The tubular heart rudiment

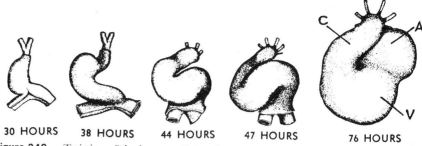

Figure 348. Twisting of the heart rudiment in a chick embryo (ventral view). A, Atrium; V, ventricle; C, conus arteriosus. (Modified from Patten, 1958.)

becomes constricted in some places and dilated in others and is thus subdivided into its four main parts. The sinus venosus lies posteriorly; the atrium develops at the tip of the first inflection of the heart; the descending part, from the first to the second inflection, becomes the ventricle; and the part going forward from the second inflection becomes the conus arteriosus.

Before this subdivision occurs, however, the functioning of the heart starts; it begins to pulsate at a regular rhythm. The pulsations of the heart start very early in the development of the embryo, even before the peripheral blood vessels are ready to receive the blood stream.

As the presumptive heart rudiment is being formed, its capacity for performing the further stages of development independently of the normal surroundings increases perceptibly. If the heart rudiment is excised after the end of neurulation and transplanted into an abnormal position or allowed to develop *in vitro*, enclosed in a vesicle of skin, the differentiation goes much further than in the experiments previously referred to. Not only pulsating muscle tissue develops, but a cardiac tube is formed, and the tube becomes inflected in the same manner as the heart in normal development (Stöhr, 1924). The rudiment of the heart in this stage, however, is by no means strictly determined in all its parts. The left or right half of the heart rudiment may be excised, and the remaining half then develops into a complete whole (Ekman, 1925; Copenhaver, 1926). Moreover, half of the heart rudiment may be explanted or transplanted, and it still develops into a complete heart. The ability of half of the heart rudiment to form a whole heart may be used to produce two hearts in the same embryo. To achieve this an incision should be made lengthwise before the two halves of the heart rudiment unite in the middle. Inserting a piece of extraneous tissue, a somite for instance, or leaving the wound to gape may prevent the halves of the heart rudiment from coming together. Two complete pulsating hearts develop under these conditions (Ekman, 1925). The reverse can also be done: a complete heart rudiment may be superimposed on the intact presumptive heart mesoderm of a host embryo. The two rudiments then fuse into one whole and produce a normal heart (Ekman, 1925; Copenhaver, 1926).

The main points emerging from the foregoing will now be summarized:

1. The heart develops in a very anterior position, in the pharyngeal region. The position of the heart in the thorax (as in an adult tetrapod) is thus secondary and is due to a displacement of the organ in later development.
2. The heart develops from a paired rudiment, uniting in the midline secondarily.
3. Each half of the heart rudiment is able to differentiate even before the two halves fuse.
4. Two heart rudiments may fuse into one organ of normal structure.
5. The functioning of the heart begins at an early stage of development.

The Heart in Higher Vertebrates.   With these facts in mind, it will be easy to understand the development of the heart in the vertebrates having yolky eggs and partial cleavage.

Because of the physiological requirements of the developing embryo (the necessity of establishing circulation), the heart in meroblastic vertebrates develops precociously, before the body of the embryo becomes separated from the yolk sac. The embryo in this stage is still lying flat on the surface of the yolk, and its lateral plate mesoderm is found toward the outer parts of the blastodisc. The lateral plate mesoderm is prevented from uniting on the ventral side of the embryo by the intervening yolk. As a result, the two halves of the heart rudiment begin differentiating independently of each other. Two endocardial tubes are actually formed; each becomes invested by the myocardium and surrounded laterally by the pericardial cavity. When the body folds undercut the anterior end of the embryo and the foregut becomes separated anteriorly from the yolk sac, the right and left heart rudiments are able to meet in the midline under the pharynx. The two endocardial tubes come to lie alongside each other and soon fuse into one tube in the cardiac region, while in front of and behind the heart the endothelial tubes remain separate, thus leading to a state described earlier for the lower vertebrates (Fig. 349). The visceral walls of the right and left pericardial cavities also meet and fuse above and below the endocardial tubes, forming a single pericardial cavity. The single endocardial tube is now completely surrounded by the myocardium. In view of the experimental results previously mentioned, the fusion of two heart rudiments to form one single organ is not at all surprising.

Of all the organs of a vertebrate, the heart is the one which starts its definitive function earliest. It is also greatly dependent in its development on function. It is essential

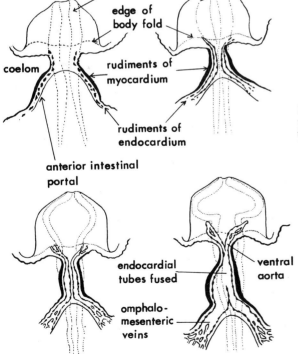

Figure 349. Development of the heart from a paired rudiment in the chick embryo, viewed from the ventral side (semidiagrammatic). (After Patten, 1958.)

for the development of the heart that a blood stream actually flow through it. It is true that a heart of a typical shape with recognizable parts will develop even in complete isolation, as in the explantation experiments (Stöhr, 1924), but such an isolated heart soon stops developing further. Also, a heart in its normal position is arrested in its development if in some way or other (interruption of the afferent blood vessels) it is deprived of circulation. The degree of the heart's development and growth appears to be dependent on the volume of blood passing through it or on the size of the animal which the heart supplies with blood. Heart rudiments have been transplanted reciprocally between large and small species of salamanders (Copenhaver, 1930, 1933). It was found that the hearts of small species grew beyond their usual size in large hosts and that the hearts of large species were undersized in small hosts. In every combination the transplanted heart grew to approximately the same size as the host heart would have grown.

In the vertebrates developing a pulmonary circulation, the heart becomes separated to a greater or lesser degree into a right half, carrying blood to the lungs, and a left half, receiving blood from the lungs by way of pulmonary veins and sending it to the rest of the body. The first indication of this separation is found in the lungfishes (Dipnoi). In the amphibians, the atrium becomes subdivided by a partition wall arising between the points of entry into the atrium of the sinus venosus and pulmonary vein. After the separation is completed, the right atrium receives blood from the sinus venosus and the left atrium from the lungs by way of the pulmonary vein. Both atria, however, pour out their blood into the ventricle, which remains undivided.

13−6 DEVELOPMENT OF THE BLOOD VESSELS

In the adult vertebrate, blood vessels, large and small, permeate almost all parts of the body. Supernumerary parts of the body, whether they are grafts or results of induction, become vascularized, and the blood vessels, which in this case are additional to the ones produced in normal development, link up with host arteries and veins and are included in the host's blood circulation. It would appear that blood vessels are attracted to penetrate any parts which are in need of a blood vessel supply. Furthermore, it appears that a suitable situation for blood vessels is duly reserved in the pattern of organization of any organ. It must be noted, however, that in vertebrates the blood vessels are invariably situated in spaces occupied by connective tissue or its derivatives. Where capillaries seemingly penetrate into epithelium, this is actually tantamount to the invasion of channels or other spaces between epithelial cells by connective tissue from which the endothelial cells of the blood vessels are derived.

The rudiments of the blood vessels are laid down as aggregations of mesenchymal cells. The cells participating in the formation of blood vessels are called **angioblasts.** They are probably always of mesodermal origin. We have seen that in the vascular area of amniotes the blood vessel walls are developed from the blood islands in conjunction with the development of the first blood cells (Section 10−1). Similar, though more concentrated, blood islands occur in vertebrates with holoblastic cleavage (in particular in amphibians) on the ventral side of the abdomen. In the amniote body proper and in all areas except for the ventral body wall in the anamniotes, the blood vessels develop independently of blood corpuscles. The aggregations of angioblasts become arranged in the form of a flat epithelium surrounding a cavity. The epithelium is the endothelium of the blood vessel; the outer layers of the blood vessel walls are differentiated much later. A student of adult vertebrate anatomy is accustomed to finding the blood vessels in the form of tubes of a constant diameter over relatively long stretches. The first blood vessels laid down in the embryo are only rarely in the form of straight tubes. Over large areas the blood vessels are initially laid down in the form of a network. (See Figures 197 and 332.) The further development of individual canals in the network depends on the amount and the direction of blood flow. Those channels which happen to come in the line of the greatest blood flow become increased in diameter, develop the connective tissue and the muscular layers, and become arteries or veins. The channels that receive less blood flow remain in the form of capillaries or degenerate and disappear completely. Another form in which early blood vessels appear is as sinuses—extensive irregular spaces surrounded by endothelium. Large sections of some of the larger veins appear in this form (Fig. 350). Some of these sinuses later acquire a more regular tubular form; others remain as such in the adult animal. For instance, the postcardinal "vein" becomes a tubular blood vessel in urodeles but remains essentially in the form of a sinus in the dogfish.

Once the network of endothelial tubes, the rudiments of the blood vessels, has been established, new blood vessels continue to be formed by sprouting and outgrowth of those already present. Lateral branches may form on already existing capillaries, or a capillary may become interrupted and each free end may grow out in a new direction. Two outgrowths from different blood vessels may contact each other and fuse, thus establishing a new channel for circulation (Fig. 351).

As a result of the extreme plasticity of the blood vessel system, the eventual arrangement of arteries, veins, and capillaries in any part of the body is largely dependent on the amount and direction of blood flow in the part in question. In their earliest formation the embryonic blood vessels also appear in conjunction with other rudiments, suggesting a dependent mode of differentiation, though the dependence is not of a functional nature since the blood vessels develop prior to the establishment of circulation. If the heart rudiment in an amphibian embryo is removed before circulation starts, the main blood vessels continue to develop for some time, until the embryo dies. The exact position of the blood vessel rudiments is probably determined by a process similar to an induction, though this has not been shown conclusively.

Certain locations in the embryo offer preferential conditions for the development of blood vessels. These locations are: (1) between the visceral mesoderm and the endoderm (Fig. 199); and around the kidneys, especially the pronephros and the mesonephros. Close networks of capillaries are always found in these two locations, and they are the sites of development of some of the major blood vessels.

The heart obviously belongs to the first group. We have seen that the determination of the heart occurs at an early stage of development and that as a consequence it shows a high degree of autonomy in its further differentiation. It is noteworthy, nevertheless, that the heart does not develop in embryos from which the endoderm has been removed (Mangold, 1936; Balinsky, 1939; p. 404). The blood vessels which form the (paired) continuation of the heart tube anteriorly and posteriorly also develop in conjunction with the endodermal parts. The two posterior vessels are the vitelline veins, which collect the blood from the network on the surface of the gut and, in amniotes, from the network on the yolk sac. The anterior prolongations of the heart tube are the ventral aortae, which become connected along the partition walls between the endodermal pharyngeal pouches (Section 14–3) with the dorsal aortae, a pair of blood vessels developing on the dorsal side of the endodermal gut (Figs. 352, 353, 354). The vessels conveying blood from the ventral aortae to the dorsal aortae are the **aortic arches.**

The aortic arches develop in a craniocaudal sequence, and the first to appear is the mandibular arch which ascends between the edge of the mouth and the first (spiracular) pharyngeal pouch. The second aortic arch develops between the first spiracular and the second (first true branchial) pouches and subsequent aortic arches de-

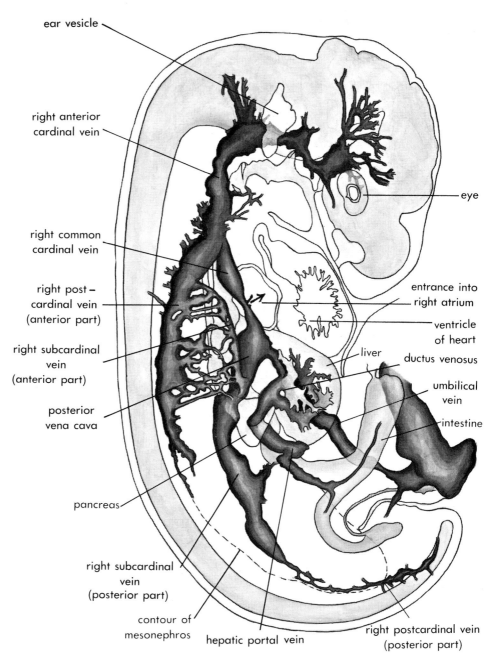

ear vesicle

right anterior
cardinal vein

right common
cardinal vein

right post –
cardinal vein
(anterior part)

right subcardinal
vein
(anterior part)

posterior
vena cava

pancreas

right subcardinal
vein
(posterior part)

contour of
mesonephros

hepatic portal vein

eye

entrance into
right atrium

ventricle
of heart

ductus venosus

umbilical
vein

intestine

liver

right postcardinal vein
(posterior part)

Figure 350.   Venous system of 10 mm. pig embryo. Lateral view. The arrow indicates the direction
of entry of the blood from the common cardinal vein into the atrium of the heart.

Figure 351. Outgrowth of new capillaries, establishing new channels of circulation. (After Clark, from Arey, 1947.)

velop between the more posterior branchial pouches. Six aortic arches are formed in the embryos of all vertebrates except for some very archaic forms (cyclostomes and some selachians), whose number of gill clefts is larger than six (Figs. 355 and 356). The dorsal aortae extend forward into the head and backward throughout the length of the trunk and to the base of the tail. In the trunk region the two dorsal aortae later fuse into one unpaired dorsal aorta, while in the branchial and head region the two aortae

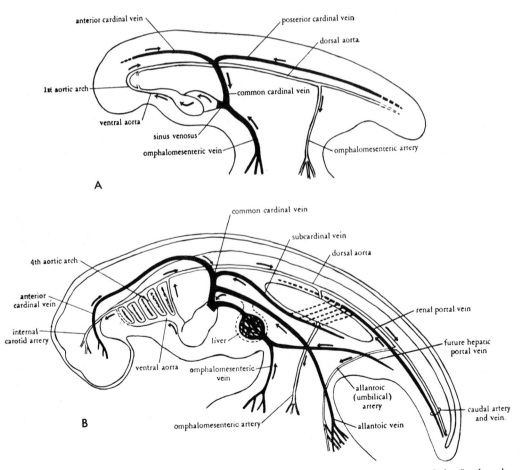

Figure 352. General arrangement of blood vessels in an early amniote embryo *(A)* and the further development of the circulatory system *(B)*. The arteries are shown in white, the veins in black. (From Moog, 1949.)

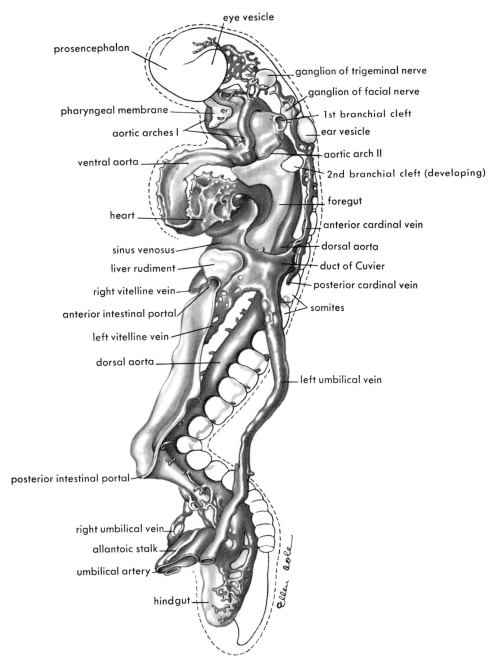

Figure 353. The circulatory system of a 20 somite human embryo. This reconstruction of the actual blood vessels of an early embryo should be compared with the diagrammatic presentation of the circulatory system in Figure 352. The kink in the posterior part of the body is an artefact caused by fixation. The kink has brought the left umbilical vein further out than it should have been in life. (After Davis, 1923.)

remain separate. In the trunk, the dorsal aorta gives off an intersegmental artery at the level of each myocomma. Larger arteries convey blood from the dorsal aorta to the viscera, to the limbs, and in embryonic stages, to the yolk sac (vitelline arteries) and to the allantois (umbilical arteries).

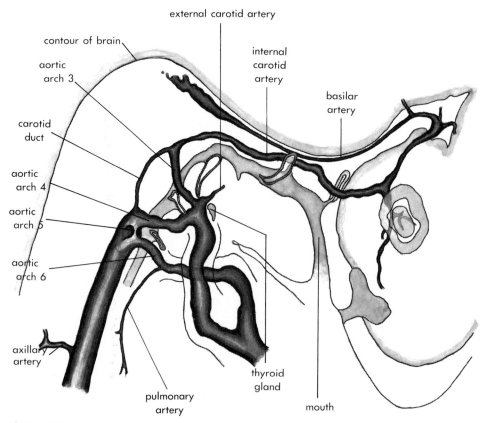

Figure 354. Arterial system of a 10 mm. pig embryo. Lateral view.

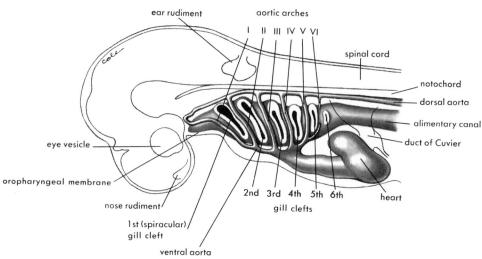

Figure 355. Relation between the aortic arches and the branchial clefts in a dogfish embryo. (After Balfour, 1881, and Goodrich, 1958.)

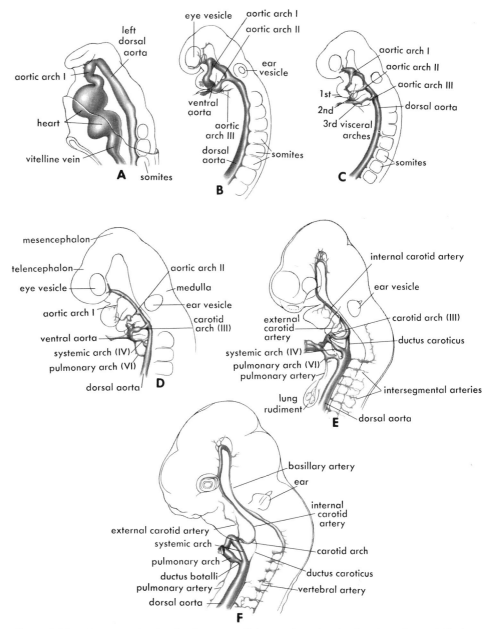

Figure 356. Five stages in the development of the aortic arches in the pig embryo. *A*, Embryo with 10 pairs of somites. *B*, Embryo with 19 pairs of somites. *C*, Embryo with 28 pairs of somites. *D*, Embryo with 36 pairs of somites, 6 mm. *E*, Embryo 12 mm. *F*, Embryo 14 mm. (*A*, Enlarged 30 times; *B*, enlarged 14 times; *C*, enlarged 11 times; *D*, enlarged 8 times; *E* and *F*, enlarged 6 times.) (After C. H. Heuser, 1923.)

The blood vessels developing in conjunction with the excretory organs are the **cardinal veins.** The first rudiments of these appear, in Amphibia, in the form of a venous sinus around the pronephros. Soon prolongations of this sinus are found anteriorly and posteriorly. The posterior prolongation is the postcardinal vein, which develops along the groove between the somites and lateral plates dorsolateral to the nephrotomes and later becomes closely associated with the mesonephric kidney. The anterior prolongation is the anterior cardinal vein, which runs forward at the same level as the postcardinal vein, just above the pharyngeal pouches into the head. The anterior cardinal vein and postcardinal vein join at the level of the anterior edge of the pronephros to form the common cardinal vein **(duct of Cuvier),** which runs inward to join the vitelline veins where they enter the heart (Figs 352 and 353).

At a later stage, a second pair of veins develops in tetrapods on the median side of the mesonephric kidney—the **subcardinal veins.**

This basic pattern of the main blood vessels is established with great regularity in embryos of all classes of vertebrates, but it becomes greatly modified as development proceeds, especially in the higher vertebrates.

In the fishes and amphibian larvae, a number of aortic arches from the third onward supply blood to the gills. The blood flow from the ventral portion of each aortic arch involved is directed into a network of capillaries which develop in the gill lamellae attached to the gill septum or in the filaments of the external gills. From these, the blood then returns by way of collecting vessels into the upper part of the arch and through it to the dorsal aorta. The middle part of the aortic arch becomes greatly reduced in diameter and is lost in the capillary network in the branchial septum, but it may not be completely interrupted, though the amount of blood which passes through it is small compared with the volume of blood going through the gills.

In amphibians which lose their gills on metamorphosing into the adult stage, the connection between the ventral and dorsal portions of the aortic arches is reinstated by the enlargement of the narrow blood vessels connecting them during the larval stages. The capillary system in the gills is reduced and disappears, and the aortic arches again assume the shape of simple tubes leading from the ventral aorta to the dorsal aorta.

Of the six pairs of aortic arches, the anterior two pairs degenerate to a greater or lesser extent, but even in the higher vertebrates remnants of the ventral portions of these arches take part in the formation of the branches of the external carotid in the ventral region of the head. The anterior portions of the ventral aortae, which gave rise to the first and second aortic arches, become the **external carotids** (Fig. 356*E, F*).

In the terrestrial vertebrates (including amphibians after metamorphosis), the third pair of aortic arches become the carotid arches (Figs. 354 and 357). The anterior parts of the dorsal aortae now serve to forward the blood from the carotid arch to the dorsal part of the head and to the brain. This blood vessel becomes the **internal carotid artery** (Figs. 356 and 357). The sections of the ventral aortae between the points of origin of the fourth and the third (carotid) aortic arches carry blood both to the carotid arches (now the internal carotids) and to the external carotids. They thus become the **common carotids.**

The fourth pair of aortic arches become the main channel for the blood flow from the heart to the dorsal aorta and hence to the body. They are the systemic pair of arches (in amphibians and reptiles).

The fifth pair of aortic arches degenerate in all tetrapods except in urodeles, in which they carry part of the blood to the dorsal aorta. The general direction of blood flow in the dorsal aorta is from the anterior to the posterior end. This direction of blood flow is patent from the level at which the fourth (systemic) arches join the dorsal aortae. However, the anterior prolongations of the paired dorsal aortae serve for supplying blood to the head. This blood enters the dorsal aortae mainly through the carotid arches; the anteriorly directed flow of blood starts, therefore, from the point of entry of these arches. Between the level where the carotid arches join the dorsal aorta and the level where it is joined by the systemic arches, there is an area of blood stagnation. As in other cases, in which a blood vessel is devoid of blood flow, the sections of the dorsal aortae between the carotid and the systemic arches tend to become atrophied; the ves-

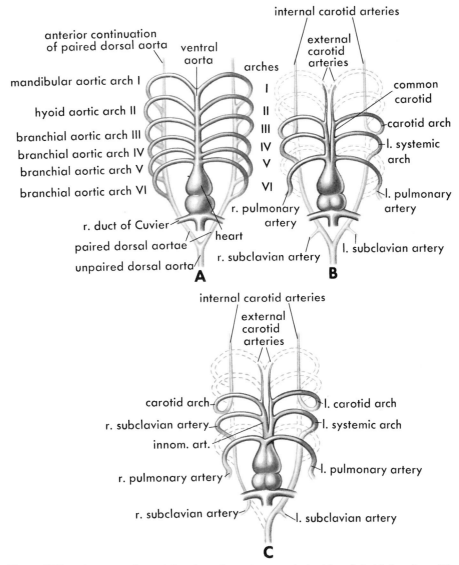

Figure 357. Diagrams of arterial arches (shown in ventral view) in a fish *(A)*, in a frog *(B)*, and in a mammal *(C)*.

sels become attenuated or even completely interrupted. In some urodeles and reptiles this section of the dorsal aorta, known as the **carotid duct,** persists as a narrow vessel in the adult stage. In most terrestrial vertebrates, however, the duct, though existing in the embryo (Figs. 354 and 356), becomes completely reduced, and the anterior sections of the dorsal aortae, now becoming a part of the internal carotid artery, are completely separated from the remainder of the dorsal aorta (Fig. 357*B*).

The sixth pair of aortic arches gives off blood vessels to the lungs (and to the skin in amphibians). It is thus the **pulmonary arch.** The upper part of the pulmonary arch, connecting the base of the pulmonary artery to the dorsal aorta, is known as the **arterial duct** (ductus arteriosus or ductus Botalli) (Fig. 354). In the larvae of amphibians and in the embryonic stages of amniotes, the entire pulmonary arch is a fairly large vessel carrying blood from the ventral aorta to the dorsal aorta. This condition changes drastically, however, when the lungs start functioning as respiratory organs. The pulmonary arteries, which had been insignificant blood vessels until this stage, greatly increase in size and take over the blood flow entering the pulmonary arch, while the arterial ducts become reduced. They remain as narrow blood vessels in adult urodeles and some reptiles, but in adult frogs and the majority of amniotes the ducts are closed and reduced to a fibrous strand connecting the pulmonary and systemic arteries.

In amniotes, and in mammals in particular, the pattern of aortic arches is modified still further (Fig. 357C). The fifth aortic arch is already greatly reduced at the time of its formation. It is still recognizable as such in embryos of reptiles and birds but is completely lacking in mammals.

A characteristic feature of amniotes is the asymmetry in the development of their systemic arches. Although the aortic arches are laid down as symmetrical pairs of blood vessels, the left systemic arch degenerates in birds, and the right systemic arch degenerates in mammals. In mammals, originally both systemic arches reach the paired dorsal aortae, and the latter fuse together posteriorly into the single dorsal aorta. Before their fusion, the paired aortae give off the subclavian arteries laterally. When the right aorta becomes reduced, it is interrupted at its posterior end after the origin of the subclavian artery and just before its connection to the unpaired aorta. In this way, the right systemic arch becomes the channel carrying blood to the right subclavian artery. In later stages, the distinction between the aortic arch and the right subclavian artery is lost, and the whole vessel is then known as the right subclavian artery. It is, therefore, not entirely homologous to the left subclavian artery, which starts, as in lower vertebrates, from the left paired dorsal aorta.

The section of the paired ventral aorta between the branching of the common aorta and the origin of the right common carotid artery becomes what is known as the **innominate artery.**

The reduction of the aortic arches is obviously dependent on the blood flow through these vessels, in the same way that the blood flow has been found to mold the development of vessels from the original capillary networks. The following experiments prove this point. In frog tadpoles, all four (third to sixth) aortic arches persist up to metamorphosis, and only then do the transformations occur which lead to the adult condition. In the tadpole of the frog, *Xenopus laevis*, the systemic arch on one side was destroyed shortly before metamorphosis. As a result, the carotid arch of the same side took over the work of the systemic arch. The connection of the carotid arch to the dorsal aorta, which is interrupted in normal development, persisted, and the blood from the heart flowed to the dorsal aorta by way of the carotid arch (Fig. 358 — Millard, 1945). Even more interesting, perhaps, is an experiment performed on the chick embryo, in which the normally persisting right systemic arch was ligatured. The effect was that the left systemic arch took over the blood flow from the heart to the dorsal aorta and became permanent instead of degenerating (Fig. 359).

At the time of its origin, the ventral aorta is a paired blood vessel (p. 405) diverging anteriorly from the heart. The two ventral aortae may be quite independent, each giving off a number of aortic arches. Subsequently they fuse in various degrees, forming a single unpaired ventral aorta, especially in fishes. In terrestrial vertebrates, however, the two aortae usually remain separate, except for the most posterior section, where they are connected to the heart.

In diagrams, it is convenient to show the ventral aorta as being a long tube extending forward from the heart (Fig. 357). Actually this representation is not very accurate.

In adult fishes, the ventral aorta is a fairly long blood vessel extending in an anteroposterior direction, but in embryos of all vertebrates the heart is situated underneath (ventral to) the pharynx. The ventral aorta lies more or less in a vertical position

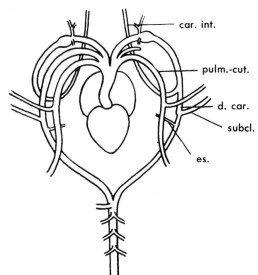

Figure 358. Arrangement of the arterial arches in a frog, *Xenopus laevis*, which had the fourth left aortic arch destroyed in the tadpole stage (ventral view). The blood flows to the dorsal aorta through the third (carotid) arch and the persisting carotid duct (d. car.); car. int., internal carotid artery; es., esophageal artery; pulm.-cut., pulmo-cutaneous artery; subcl., subclavian artery. (From Millard, 1945.)

and soon splits to give rise to the roots of the aortic arches, of which only the first two or three pairs lie anterior to the heart, while the rest are developed on a posterior extension of the aorta (Fig. 355). The pulmonary arteries, when these develop, may be described as being on the extreme backward prolongation of the ventral aortae (Fig. 271). When the last aortic arches are formed, the pulmonary arteries shift upward and can then be seen as originating from these last (sixth) aortic arches.

A further important change in the arterial system in higher vertebrates accompanies the separation of the heart into right and left halves. The pulmonary arteries acquire a separate connection to the right ventricle, while the systemic and carotid arteries are linked to the left ventricle. The separation occurs from the branching points of these arteries backward toward the heart. The ventral aorta becomes split lengthwise by a partition cutting in at the point where the pulmonary arches branch off from the systemic arches, the latter carrying the carotid arteries with them.

In fishes, the heart remains in the pharyngeal region, but in terrestrial vertebrates the heart is shifted into a more posterior position in the thoracic cavity of the animal. The shift is, in part, due to the emancipation of the heart from its connection with the gills and, in part, is a result of the development of the neck as a distinct region of the body (Dalcq, 1960). The posterior position of the heart causes a considerable elonga-

Figure 359. Ligating the aortic arches in a chick embryo. *a,* Normal arrangement of aortic arches in later stages of development (ventral view); *b,* thread in position to ligate the aortic arches on the right side; *c,* suppression of the fourth right aortic arch and retention of the fourth left arch. (After Stephan, from Waddington, 1952.)

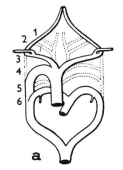
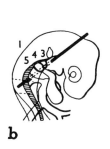
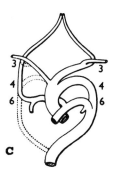

tion of some of the blood vessels conveying blood to the head. The carotid arteries are the ones affected most. There is, however, some variety in the way that the elongation is achieved. In the amphibians, mammals, birds, and most reptiles, the carotid arch is carried backward together with the heart (and the other aortic arches). It is then the external and internal carotids which become elongated. In snakes and some lizards, on the other hand, the carotid arches remain in the original position, just posterior to the head, and it is the common carotids which elongate. The systemic and pulmonary arches in all terrestrial vertebrates move into the thoracic cavity together with the heart.

The disappearance of the branchial respiratory apparatus and the development of the neck separate parts of a previously closely knit system and bring them into widely different positions in the body. The ganglia of the cranial nerves originally connected to the branchial apparatus and developing, in part, from epibranchial placodes (ganglia VII, IX, and X) lie in the head, inside the brain case. The remnants of the visceral skeleton (hyoid and branchial arches) are to be found in the neck as the hyoid bone and the cartilages of the larynx and trachea. The aortic arches lie in the thoracic cavity, and close to them also lies the thymus, another derivative of the branchial pouches. (See p. 455.)

The vitelline veins on their way to the heart pass through the region where the liver is developed at a later stage. When the rudiment of the liver is formed (Section 14–4), it envelops the vitelline veins. These break up into a system of hepatic sinusoids permeating the liver lobules and thus give rise to the hepatic portal system. The veins in front of the liver and leading to the heart become hepatic veins, and the parts of the veins caudal to the liver give rise to the **hepatic portal vein.**

In amniotes, the development of the vitelline veins is asymmetric, the right vein tending to degenerate posteriorly and the left vein disappearing, in its most anterior part, in front of the liver. Through a system of anastomoses, however, the continuity of the blood flow is preserved, and a single blood vessel may eventually be formed from parts of the two vitelline veins. The parts of the vitelline veins collecting blood from the yolk sac of the embryo disappear at the time of birth, but a branch from the vitelline veins, the **mesenteric vein,** takes over the drainage of blood from the intestinal tract and becomes the main component of the hepatic portal vein of the adult animal.

The blood vessels of the cardinal system undergo extremely complicated transformations in the course of ontogenetic development and of evolution in vertebrates. The anterior cardinal veins are least changed; they become the internal jugular veins. The common cardinal veins become the anterior venae cavae. In some mammals, including man, a transverse anastomosis is formed between the anterior cardinal veins of the right and left sides, whereupon the connection of the left anterior cardinal to the heart disappears, and all the blood from the head, neck, and forelimbs is directed into the right common cardinal vein which becomes the single anterior vena cava.

The transformations of the veins in the posterior part of the body are much more drastic. The postcardinal veins, which are the main veins of the posterior trunk region in fishes and in early embryos of all vertebrates, are gradually reduced in the embryos of tetrapods.

The first stage in this reduction, as it occurs during the embryonic development in amniotes, is the formation of a venous network on the ventromesial side of the mesonephric kidneys. This plexus gives rise to veins lying in the same position—the **subcardinal veins.** The subcardinal veins become linked to one another by anastomoses in the midline (Fig. 360B). Initially, the subcardinal veins join the postcardinal veins anteriorly and use the latter as a channel for blood outflow. At a later stage, however, a connection becomes established between the network of the subcardinal veins, with their associated plexuses, and the system of veins in the liver. This opens a more direct route to

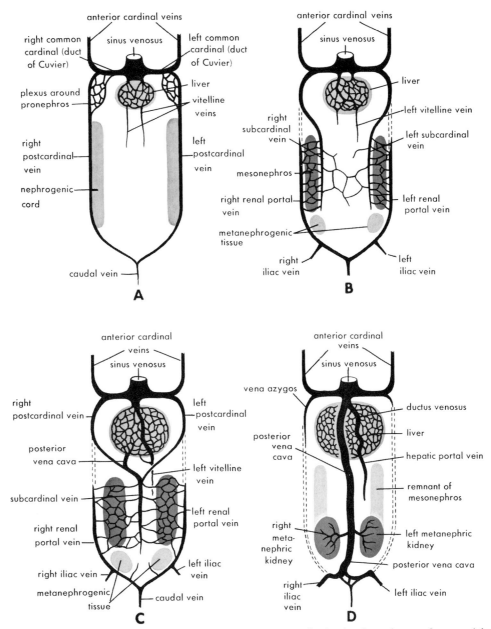

Figure 360. Stages in the transformations of the postcardinal veins in embryos of terrestrial vertebrates (diagrammatic). *A*, Early stage, slightly later than that shown in Figure 352*A*; the vitelline veins have already given rise to the hepatic sinusoids. *B*, Stage corresponding to that in Figure 352*B;* allantoic veins not shown. *C*, First appearance of posterior vena cava formed from a connection between the right subcardinal vein and the venous system of the liver. *D*, Posterior vena cava fully developed.

the heart for the blood collected from the kidneys into the subcardinal veins (Fig. 361).

At first, the blood from the subcardinal veins passes through the hepatic sinusoids in the tissue of the liver, but as more and more blood goes through, the sinusoids enlarge and develop into one straight channel, passing through the liver directly to the

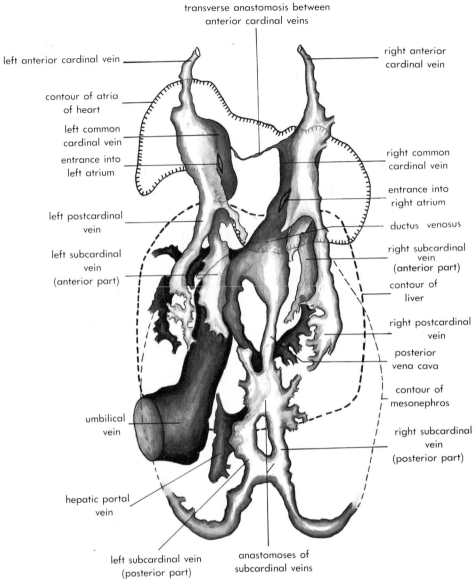

Figure 361.   Venous system of a 10 mm. pig embryo. Dorsal view.

sinus venosus. This channel is the anterior portion of the **posterior vena cava** (Figs. 360*D* and 361). (As a result of the reduction of part of the hepatic tissue, the posterior vena cava later comes to lie on the edge of the liver.) The middle portion of the posterior vena cava is formed by the joining together of smaller vessels belonging to the sub-cardinal vein system. Eventually, the posterior vena cava, through a system of interme-diate smaller vessels, establishes a connection with the iliac veins which collect the blood from the hindlimbs. The posterior vena cava is then established in its final form as it is found in adult amniotes.

   The postcardinal veins, in the meantime, lose their significance as channels convey-ing blood from the posterior part of the body, and their anterior halves leading to the ducts of Cuvier degenerate, leaving behind only insignificant remnants. The posterior parts of the postcardinal veins remain intact as renal portal veins, in amphibians throughout life, and in embryos of reptiles and birds as long as the mesonephric kidneys remain functional. By means of these veins, the mesonephric kidneys are supplied with blood returning from the hindlimbs through the iliac veins and from the caudal vein (where it is present). With the replacement of the mesonephric kidney by the metanephros, the renal portal veins degenerate and disappear, and the iliac veins gain direct connection to the posterior vena cava, as had been explained.

The development of the posterior vena cava is a good illustration of the formative influence of the blood flow on the differentiation of blood vessels. We can observe how the blood flow is not restricted by the arrangement of existing blood vessels but causes the blood vessels to be fashioned in correspondence with the direction and amount of blood flow. Thus, the largest vein of the vertebrate body develops from blood vessels of capillary dimensions where they lie in the line of blood flow.

In addition to the hepatic portal vein and the inferior vena cava, a third system of veins becomes associated with the liver in amniotes. The allantoic (or umbilical) veins go toward the heart in the body wall and enter the common cardinal veins. Before joining the latter, the allantoic veins pass very close to the liver. As the liver grows, it makes contact with the allantoic veins, whereupon a connection is established between them and the system of hepatic sinusoids in the liver. More and more blood from the allantoic veins enters the liver, until the main channel into the common cardinal veins becomes reduced and obliterated, so that all the blood from the allantois (from the placenta in mammals) goes to the heart by way of the liver. The right allantoic vein later degenerates in all amniotes, the left remaining as the only functional one.

The large amount of blood entering the liver from the left allantoic vein cannot be accommodated in the hepatic sinusoids and eventually makes for itself a broad direct channel which joins the posterior vena cava near the point where it leaves the liver at its anterior end. This direct channel is referred to as the **ductus venosus.** After birth and the cessation of placental circulation, the ductus venosus degenerates and becomes transformed into a strand of connective tissue (the **ligamentum venosum**), as often happens with large blood vessels when they cease to be functional. (Compare the fate of the arterial duct, p. 418).

13–7 DEVELOPMENT OF THE REPRODUCTIVE ORGANS

The gonads in vertebrates develop from the upper edge of the visceral layer of lateral plate mesoderm in the posterior half of the body. The first rudiment of the gonad appears as a thickened longitudinal strip of mesodermal epithelium lining the body cavity immediately lateral to the dorsal mesentery. This thickening is called the **germinal ridge.** Initially the ridge is produced by the mesodermal cells, which assume a high columnar shape, but soon the cells become arranged in a compact mass several cells thick, and the ridge protrudes into the coelomic cavity (Fig. 362). Still later, the connection between this cell mass and the peritoneal wall becomes constricted laterally, and the gonad remains suspended from the peritoneal wall by only a double layer of peritoneum, the **mesorchium** or **mesovarium.**

The germinal ridges are situated in the same region of the body as the rudiment of the mesonephros and in close spatial association with the latter, the germinal ridge being medial to the mesonephros. Although in the later development of vertebrates the genital apparatus becomes associated with the excretory system, the initial stages of the two organ systems are completely independent of each other.

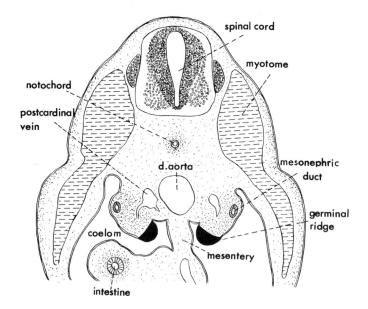

Figure 362. Transverse section of a mouse embryo to show the position of the germinal ridges and their relation to the mesonephric rudiment and the dorsal mesentery.

What has been said so far about the formation of the gonad rudiment refers only to its gross morphology. A more detailed microscopic examination of the germinal ridge epithelium discloses that it consists of two types of cells. The majority of cells are similar to the other cells of the peritoneal epithelium, even though they may become columnar in the initial stages of the development of the ridge. The second type of cells has a very different appearance. They are much larger than the ordinary mesodermal cells and contain large, more or less vesicular nuclei. Their cytoplasm differs from that of the surrounding cells in its staining properties (amphophil instead of acidophil in birds) and, in amphibians, in its higher yolk content. The shape of the cells is nearly spherical, and they do not participate in the epithelial arrangement of the typical mesodermal cells but appear to be interspersed between the cells of the mesodermal epithelium. Cells of this peculiar type are known as the **primordial germ cells** (Fig. 363).

The Origin of the Primordial Germ Cells. According to views held by many embryologists, only the primordial germ cells are destined to give rise to the gametes (eggs and spermatozoa), while the ordinary mesodermal cells differentiate into the somatic cells of the adult gonad: the Sertoli cells in the testes, the superficial epithelium, and the follicle cells in the ovaries. Whatever their eventual fate, it appears certain that the primordial germ cells have a different origin from the rest of the cells in the germinal ridges, that they first appear in parts of the embryo other than those in which the germinal ridges lie, and that the primordial germ cells reach the germinal ridges after a more or less extensive migration.

Although in this text the germ cells are dealt with in the chapter on mesodermal organs, it should not be concluded that they are actually cells of mesodermal origin. It is possible that in phylogenesis the differentiation of the animal body into somatic and reproductive cells preceded the development of the three germinal layers. In this case, the generative cells would not really belong to any of the germinal layers. This interpretation is supported by observations that in some animals the future generative cells become distinguishable from future somatic cells already in early cleavage stages. The difference concerns such an important feature as the structure of the chromosomes. In *Ascaris* the long V-shaped chromosomes, appearing immediately after fertilization and in the first division of the egg, are retained only in the blastomeres giving rise to the germ cells (blastomeres P_1, P_2, P_3, and so forth), while in all other blastomeres, the blastomeres of the somatic line, the long chromosomes become fragmented, the middle part of each chromosome is broken up into a number of small elements, and the ends are discarded. (See Figures 73 and 74.) A similar phenomenon occurs in some insects. Also, primordial germ cells of some insects have been found to be formed at a very early stage of cleavage at the posterior end of the egg. When the blastoderm becomes established (after superficial cleavage, p. 114), the primordial germ cells may be left for some time *outside* of the blastoderm (Fig. 364). Only later do they penetrate through the blastoderm into the interior of the embryo and eventually take up their position inside the gonads.

In amphibians, the primordial germ cells may be traced back in microscopic sections to a position just above the dorsal mesentery. Here the primordial germ cells lie surrounded by mesenchyme before they migrate into the mesodermal epithelium of the germinal ridges. The median position of the primordial germ cells is probably already a secondary one. In the birds and reptiles, cells identical in their cytological properties to the primordial germ cells have been found in early developmental stages — before the beginning of segmentation of the mesoderm — in the endodermal layer of the extraembryonic part of the blastoderm (Fig. 365). Similarity in appearance cannot, however, be considered as proof that the cells found in the extraembryonic endoderm are the same cells as those which appear later in the germinal ridges.

To test whether or not the primordial germ cells are derived from the cells in the extraembryonic endoderm, the latter may be destroyed prior to their migration. In the chick, the cells in question occupy a crescentic area in front of the head end of the

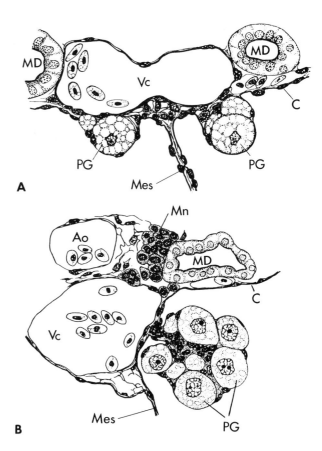

Figure 363. Early stages in the development of gonads in the frog. *A,* Primordial germ cells embedded in the genital ridges. *B,* Indifferent gonad stage. Ao, Dorsal aorta; Mn, mesonephrogenic tissue; Pg, primordial germ cells; Mes, dorsal mesentery; Vc, posterior cardinal veins; MD, mesonephric duct; C, coelomic epithelium. (From Witschi, 1929.)

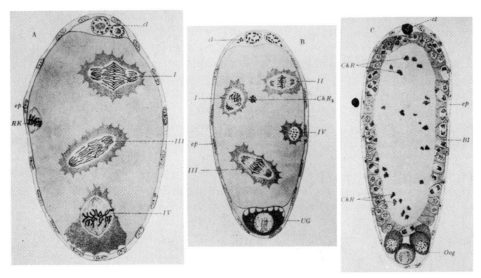

Figure 364. Origin of germ cells in the midge, *Miastor. A,* Third cleavage stage. The mitotic figure at posterior edge (bottom) will give rise to primordial germ cell. *B,* Primordial germ cell (UG) completely separated from the rest of the egg. *C,* Primordial germ cell has given rise to several oogonia, which lie outside the blastoderm now being formed in the rest of the embryo. I, II, III, IV, First four cleavage nuclei; ep, follicle epithelium; cl, nurse cells; RK, polar body; ChR, chromatin discarded from somatic cells; Bl, blastoderm; Oog, oogonia. (From Korschelt, 1936, after Kahle.)

embryo (Fig. 366). This area can be destroyed in various ways, such as by cauterizing it with a hot needle or irradiating it with radium emanation (Dantschakoff, 1941). As the crescentic area lies well beyond the embryo proper, the development of the latter is unimpaired. At the time when the germinal ridges are formed, no primordial germ cells are to be found in and around the germinal ridges, although the area of the germinal ridges had not been tampered with. The experiment shows that the peculiar cells of the extraembryonic endoderm are actually the primordial germ cells, which are later incorporated into the gonad.

This conclusion is further supported by the following experiment: If the primordial germ cells are exposed to very weak irradiation with x-rays, while still in the extraembryonic endoderm, they are damaged to a certain extent but not killed at once. Some of the damaged cells succeed in reaching the site of gonad formation and may be found there for a short time. Then all the primordial germ cells perish, and a gonad remains consisting of mesodermal cells only. The gonad may grow and develop further without showing any signs of producing generative cells (Dantschakoff, 1941). Such a gonad, devoid of generative elements. may be called a sterile gonad.

As to the means by which the primordial germ cells travel from the extraembryonic endoderm to their later position in the germinal ridges, it is supposed that in birds they do so largely through the blood vessel system. The cells have been observed to leave the endodermal epithelium, to move slowly into the space between endoderm and mesoderm, and then to penetrate the blood vessels of the area vasculosa. The primordial germ cells may actually be seen in the blood stream (Fig. 367). Although most of them eventually reach the germinal ridges, some go astray and are found in various parts of the body, where they eventually degenerate.

Further experimental work on the origin of primordial germ cells has been carried out in amphibians. According to the view held by the majority of embryologists, the

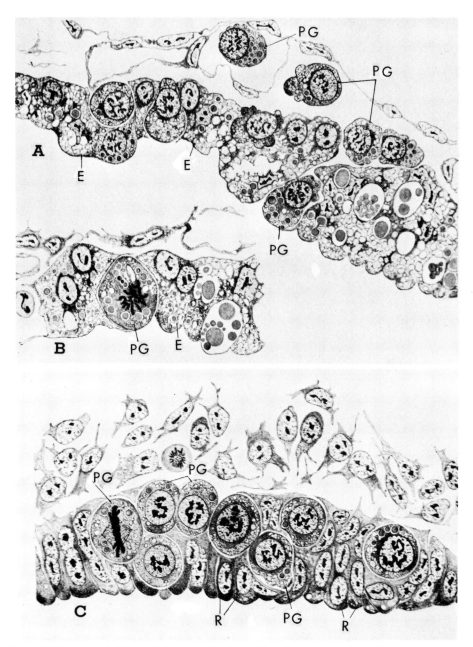

Figure 365.   *A* and *B*, Primordial germ cells of a chick embryo; two sections of extraembryonic endodermal layer of a chick embryo containing primordial germ cells. Some of these escape into the space between the endoderm and mesoderm and penetrate into the lumen of blood vessels. PG, Primordial germ cells; E, endodermal cells. *C*, Germinal ridge of a 3-day-old chick embryo containing primordial germ cells, one of them in mitosis. R, Cells of a germinal ridge. (From Dantschakoff, 1941.)

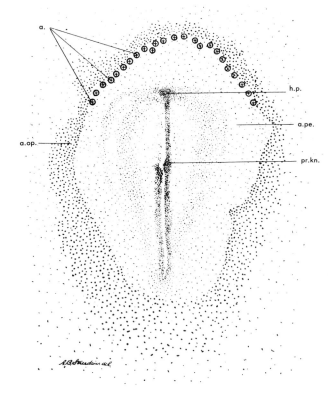

Figure 366. Surface view of a chick embryo in the head process stage, showing the original position of the primordial germ cells on the margin between the area pellucida and the area opaca: a, primordial germ cells (diagrammatic); a.op., area opaca; a.pe., area pellucida; h.p., head process; pr.kn., Hensen's node (primitive knot). (From Swift, 1914.)

cytoplasmic material included in the primordial germ cells in amphibians can be traced back as far as the uncleaved egg. Before cleavage begins, cytoplasmic inclusions of a particular kind can be found in the subcortical layer at the vegetal pole of the egg (Fig. 368A). The inclusions are very rich in RNA. During cleavage, these inclusions move upward along the cleavage furrows and surround the nuclei of a group of endodermal cells lying right in the middle of the mass of yolky endoderm (Fig. 368B, C, and D). The cells receiving this particular cytoplasmic substance are the primordial germ cells, and the substance of the cytoplasmic inclusions has been called the **"germinal plasm"** (Bounoure, 1934; Blackler and Fischberg, 1961).

By irradiating the vegetal pole of newly fertilized frog's eggs with ultraviolet light, it is possible to destroy the "germinal plasm" and in this way to reduce greatly the number of primordial germ cells formed in later stages (Bounoure, 1939). By removing the endoderm in the neurula stage, the embryos are completely sterilized (Monroy, 1939; Nieuwkoop, 1946). After the completion of neurulation, the primordial germ cells migrate first to the dorsal region of endoderm, to the dorsal crest, the part of the endodermal wall lying above the gut cavity. From here the primordial germ cells move into the mesenchyme of the dorsal mesentery and eventually into the germinal ridges.

The area containing primordial germ cells can be transplanted in the late neurula stage from one embryo to another (Fig. 369), and if the tissues of the donor embryo are distinguishable from the tissues of the host, it can be shown that the eggs and spermatozoa are actually derived from the grafted material. In one experiment, the transplantation was made between two subspecies of the frog *Xenopus laevis*, differing, among other characteristics, in the properties (size, color) of their eggs. Operated female embryos were reared to the adult stage and laid eggs which in color and size clearly showed the properties of the donor species. Thus, the primordial germ cells transplanted with a section of endoderm at the end of neurulation not only found their way into the gonads but also preserved intact their genetic characters, in spite of the fact that during their development and maturation in the gonads they were surrounded by somatic cells of a different subspecies (Blackler, 1962).

Surprisingly, the primordial germ cells in urodele amphibians appear to have an entirely different origin from those in frogs and also from those in amniotes. They develop not from the endodermal cells but from cells originally located in the posteroventral parts of the lateral plate mesoderm.

As stated earlier (Section 13–5), in salamander embryos it is possible to remove the entire endoderm in the neurula stage. The endoderm from another embryo may then be inserted into the empty ectomesodermal shell. If the operation is carried out heteroplastically, and if the two species are chosen so that the cells of each are distinguishable by some peculiarity, then the derivation of the primordial germ cells or any other cells from the endoderm or the mesoderm respectively can be determined with certainty. The experiments were carried out with the purpose of investigating the origin of the primordial germ cells using *Triturus cristatus* as one of the components and *Triturus alpestris* or *Ambystoma mexicanum* as the other. The eggs of the latter two species

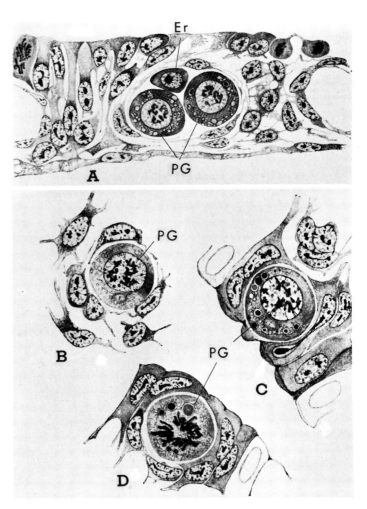

Figure 367. *A,* Migration of primordial germ cells in the chick embryo. Two primordial germ cells, together with a young erythroblast (Er), in a small blood vessel near the coelomic lining in the gonad area of a chick embryo. *B, C, D,* Three primordial germ cells in a chick embryo, one (in *B*) surrounded by mesenchyme, the other two (*C* and *D*) already embedded into the coelomic epithelium. One cell (in *D*) in mitosis: PG, primordial germ cell. (From Dantschakoff, 1941.)

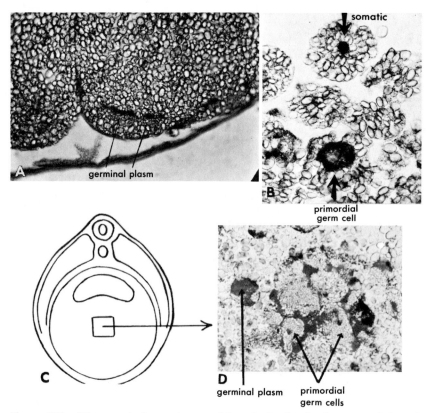

Figure 368. The germinal cytoplasm and its role in the development of the primordial germ cells in frog embryos (*Xenopus*). *A,* Germinal cytoplasm (dark areas) near the vegetal pole in the two-cell stage. *B,* Primordial germ cell, near floor of blastocoele, in a midgastrula stage. The germinal cytoplasm (black) surrounds the nucleus, which is light colored. Above is an ordinary endodermal cell with small dark nucleus. *C,* Position of primordial germ cells (square) in a completed neurula (diagrammatic transverse section). *D,* Primordial germ cells in the interior of the endoderm of a completed neurula. Photograph corresponds to the diagram in *C*. (Courtesy of Blackler and Fischberg.)

contain in their cytoplasm considerable amounts of pigment granules, while in the eggs of *Triturus cristatus* pigment granules are either completely absent or very few in number. The operation is a rather difficult one, but a sufficient number of operated embryos survived (Nieuwkoop, 1946). The result was a very clear-cut one: All the primordial germ cells found in the operated embryos had the specific properties (pigmentation) of the species to which the ectoderm and mesoderm belonged and not of the species to which the endoderm belonged. The primordial germ cells were, in this experiment, embedded in germinal ridges consisting of cells of the same species.

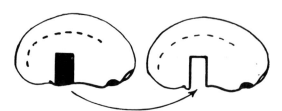

Figure 369. Diagram of germ cell transfer operation in a *Xenopus* embryo. The black section of the left embryo, containing the primordial germ cells, is being transplanted into the same position in the embryo on the right. (From Blackler and Fischberg, 1961.)

In another series of experiments, the lateral plate mesoderm was transplanted in the same heteroplastic combination (Nieuwkoop, 1946), and after such operations numbers of primordial germ cells were found to develop from the graft. The primordial germ cells, although mesodermal, do not arise from cells in the germinal ridges, but as in the case of endodermal primordial germ cells, they migrate from their site of origin (ventrolaterally in the posterior part of the body) to the area above the dorsal mesentery and then into the germinal ridges.

The conclusion seems inevitable that the primordial germ cells are derived from different germinal layers in the urodele amphibians and in the frogs. In many invertebrates, however (*Ascaris, Sagitta*, some insects), the primordial germ cells are known to be distinguishable in very early cleavage stages, even before the germinal layers become segregated. This is perhaps another confirmation of the view that the primordial germ cells do not belong to any of the germ layers, in contrast to the somatic parts of the gonads and the auxiliary parts of the sexual organs (ducts, etc.) which are mesodermal.

As to the mechanism by which the primordial germ cells in amphibians reach their destination, it is ameboid movement of the cells themselves. There is no reason to suppose that blood vessels play any part in their transport.

No experimental investigations have been carried out on the origin of the primordial germ cells in human embryos, but careful observations show that they are first found in the endodermal epithelium of the yolk sac in the vicinity of the allantoic stalk (Fig. 370), and that from there the germ cells migrate into the adjoining mesenchyme and eventually take up their position in the germinal ridges (Witschi, 1948).

In mice, on the other hand, investigations comparable to those done on birds and amphibians have been carried out. As in man, the primary germ cells first occur in the endoderm and later migrate into the germ ridges. These cells contain large quantities of alkaline phosphatase in early embryonic stages when this enzyme is absent from other cells of the embryo. (Compare p. 498.) By using a special histochemical stain for alkaline phosphatase on sections, therefore, it is possible to distinguish the primary germ cells and to follow their migration from the gut into the dorsal mesentery and then into the germinal ridges (Mintz, 1960). Originally there are less than a hundred primary germ cells in a mouse embryo, but during their migration the cells increase in number by repeated mitosis, so that eventually 5000 or more cells are present. The

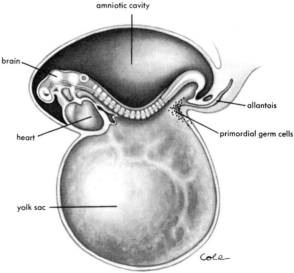

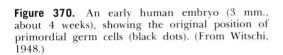

Figure 370. An early human embryo (3 mm., about 4 weeks), showing the original position of primordial germ cells (black dots). (From Witschi, 1948.)

primary germ cells are extremely sensitive to x-rays, and if the female mice are irradiated between the eighth and twelfth days of gestation with a dose of 400r, all or almost all the germ cells are killed before they reach the germinal ridges. The germinal ridges develop, nevertheless, and differentiate into gonadal tissue which is, however, completely sterile—devoid of sex cells or their predecessors (Mintz, 1960).

Differentiation of the Gonads. After arriving in the germinal ridges, the primordial germ cells become embedded in the epithelium of the germinal ridge. At the same time, the germinal ridge epithelium becomes convex toward the coelomic cavity and, on its dorsal side, forms a hollow which is filled initially with loose undifferentiated mesenchyme. Subsequently, this loose mesenchyme is partly replaced by compact strands of cells migrating into the gonad from the mesonephrogenic cord. These are called the **primitive sex cords** (Witschi, 1956; Nieuwkoop and Faber, 1956). (Students of human embryology, however, usually assume that the compact strands of cells are derived from the germinal ridge epithelium.) The epithelium of the germinal ridge (originally a thickening of the coelomic epithelium) which forms the surface of the gonad is known as the **cortex** of the gonad. The primitive sex cords (the strands of compactly arranged cells in the interior of the gonad) constitute its **medulla** (Fig. 371).

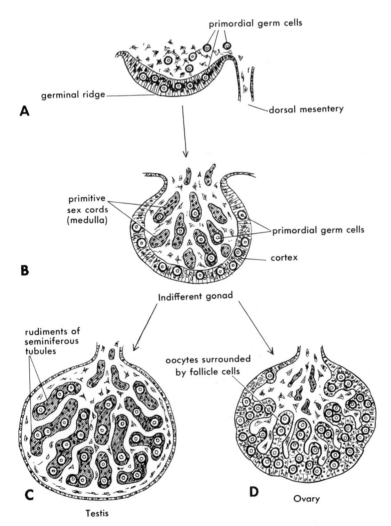

Figure 371. Diagram showing development of gonads in higher vertebrates. *A*, Genital ridge stage; primordial germ cells partly embedded in epithelium of the ridge and located partly in the adjacent mesenchyme. *B*, Indifferent gonad; germ cells in the cortex and in primary sex cords. *C*, Gonad differentiating as testis; cortex reduced; germ cells in sex cords (future seminiferous tubules). *D*, Gonad differentiating as ovary; primary sex cords reduced; proliferating cortex contains the germ cells.

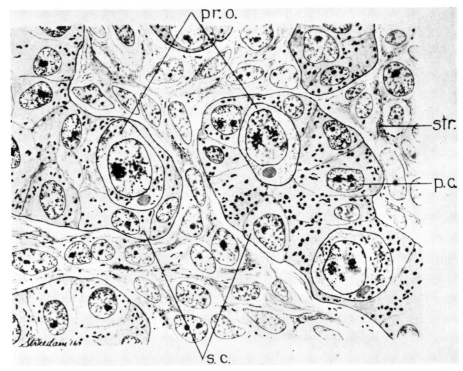

Figure 372. Primordial germ cells embedded in the primitive sex cords (future seminiferous tubules) in a male, 11-day chick embryo. pr.o., Primordial germ cells; str., mesenchyme; s.c., sex cords; p.c., cell in sex cord. (From Swift, 1916.)

The remainder of the loose mesenchyme serves as the pathway for the blood vessels supplying the gonad.

In the foregoing statements, no distinction was made between the development of the ovaries and the testes. In the early stages of organogenesis with which we are concerned here, there is practically no difference in the development of the gonads of the two sexes, and the gonad in these stages may be called the **indifferent gonad.** Only with the onset of histological differentiation do the testes and the ovaries become increasingly different from each other.

In male embryos, the primordial germ cells migrate from the cortex of the gonad, in which they were originally embedded, into the primitive sex cords of the medulla (Fig. 372), which forthwith become hollowed out and are thus converted more or less directly into the seminiferous tubules. The primordial germ cells give rise to the spermatogonia and subsequently to the spermatozoa, while the Sertoli cells are derived from the sex cords. The seminiferous tubules become connected to the **rete testis,** a system of very thin tubules developing in the dorsal part of the gonad, probably from the same material from which the seminiferous tubules are formed. The canals of the rete testis, in their turn, form connections to the adjoining tubules of the mesonephros. Thus, a pathway is established from the male gonad, via the mesonephros, to the mesonephric duct, which in vertebrates serves as the outlet for the sperm. While the medulla of the testis becomes its functional part, the cortex undergoes reduction and is converted into a thin epithelial layer covering the coelomic surface of the testis.

In female embryos, the medulla of the gonad becomes reduced, the primary sex cords are resorbed, and the interior of the mature gonad is filled with loose mesenchyme permeated by blood vessels. The primordial germ cells remain embedded in the cortex, which greatly increases in thickness. During the growth of the gonad, the masses of cortical cells on the inner surface of the cortex split up into groups and strands of cells surrounding one or several primordial germ cells. These then become the primary oocytes which enter in the initial stages of oogenesis (p. 32). The primordial germ cells lying nearer to the surface of the cortex retain their undifferentiated state longer and serve for the production of eggs at a later stage in the female's life.

Special features in the development of gonads in some groups of animals will now be noted.

In the ovaries of amphibians, the degenerating medulla becomes cavitated, so that the mature gonad acquires the form of a sac with a hollow space (or rather with several hollow spaces) in the interior (Fig. 373). The mature eggs bulge into these internal cavities. When ovulation occurs, however, the external surface of the gonad becomes ruptured, and the eggs are shed not into the internal cavity but into the coelom.

In birds, with some rare exceptions, only the left ovary becomes fully developed. From the start, more primordial germ cells enter the left germinal ridge, and only here does the ovarian cortex acquire a typical structure. The gonad on the right side remains in the state of an indifferent gonad, with the primary sex cords not reduced. The primordial germ cells in the right gonad eventually degenerate. However, if the left gonad is removed by operation, the right gonad enters into a stage of compensatory growth and differentiates progressively. The right gonad then acquires the general structure not of an ovary but of a testis and even starts to secrete male hormones. Under the influence of the male hormones, the bird shows some external features of the male sex (especially in color and shape of the feathers), even though it was and remains a genetic female.

Something similar has been observed in male toads in which the anterior ends of the germinal ridges do not differentiate fully (as testes) but retain the state of an indifferent gonad. This part of the gonad is known as **Bidder's organ.** If the testes are

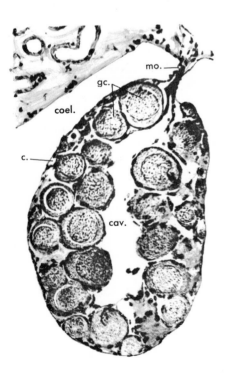

Figure 373.   Female gonad of a frog tadpole in early stages of differentiation: c., cortex of gonad; gc., germ cells; mo., mesovarium; cav., cavity of ovary; coel., coelom. (From Witschi, 1929; photograph retouched.)

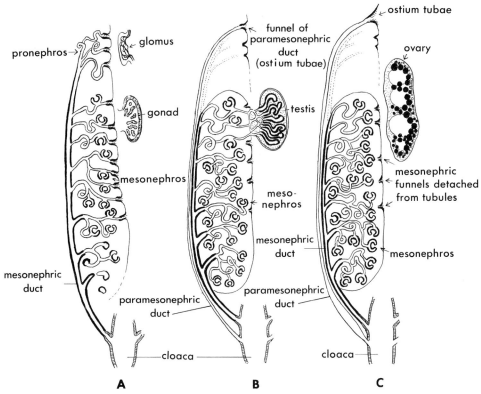

Figure 374. Diagrams showing the transition from the indifferent stage of the urogenital system (*A*) into the male condition (*B*) and the female condition (*C*) in frogs. Note the presence of the paramesonephric duct in both sexes.

removed, the Bidder's organs start growing and differentiate into fully functional ovaries. Sex reversal occurs here in the opposite direction to that in birds.

The sex conversions show that the primordial germ cells are not determined to become gametes of any sex in particular. Their ultimate fate depends on whether they remain in the cortex of the gonad or whether they are transferred to the medulla. Where the medulla develops progressively, the germ cells give rise to spermatozoa. Where the medulla is reduced and the cortex is retained and enlarged, the primordial sex cells develop into eggs.

Development of the Genital Ducts. Essential parts of the genital organs are the **genital ducts** by means of which the sex cells (ova and spermatozoa) are passed to the exterior or to another location where fertilization can take place.

The channels by which the spermatozoa are conveyed from the testes are the mesonephric ducts (Fig. 374). The seminiferous tubules are connected to the mesonephric ducts through the rete testis and a number of mesonephric tubules which become the **epididymis.** In fishes (except for the bony fishes, whose gonaducts are not comparable with those of other vertebrates) and in amphibians, the mesonephric ducts serve simultaneously for the passage of urine from the mesonephros and of the spermatozoa from the testes. There may be, however, a certain degree of segregation between the genital and the excretory functions, the anterior part of the kidney and the corresponding part of the mesonephric duct serving mainly for reproductive functions, while the posterior part of the mesonephros carries out the main part of excretion (in selachians and in urodeles). Nevertheless, both the spermatozoa and the urine pass eventually into the cloaca through the same duct.

In amniotes, the mesonephros ceases to be an excretory organ, and the only part remaining functional is that which in males develops as the epididymis. The mesonephric duct becomes the vas deferens, serving only for reproduction.

In embryos of the female sex, the development of the mesonephric duct starts in the same way as in male embryos. In anamniotes, it functions throughout life as the pathway for urine. In amniotes, the mesonephric duct is formed in connection with the development of the pronephros and mesonephros. When the mesonephros ceases to function as an excretory organ, the mesonephric duct degenerates.

In addition to the mesonephric ducts, a second pair of genital ducts, the **paramesonephric** (or Müllerian) **ducts** (Fig. 374), are formed in all vertebrates (again with the exception of bony fishes). At their anterior ends, the paramesonephric ducts open by a funnel-shaped **ostium tubae** into the coelomic cavity, and in adult females the paramesonephric ducts serve as channels for the eggs, that is, as **oviducts.** Although functional only in females, the paramesonephric ducts develop initially in embryos of both sexes. Their development starts at the anterior end with the ostium tubae. In lower vertebrates (selachians, amphibians) the ostium tubae is the surviving funnel (nephrostome) of one of the pronephric tubules, which remains intact after the rest of the pronephros degenerates and is resorbed. (There are indications that two or more pronephric nephrostomes may fuse to form the ostium.) The deep end of the ostium starts growing in a backward direction retroperitoneally and reaches the mesonephric duct. It then continues its growth alongside the mesonephric duct until it reaches the cloaca and fuses with the cloacal wall.

In higher vertebrates, no nephrostomes remain after the degeneration of the pronephros, but the coelomic epithelium becomes thickened in the area where the pronephros was connected to the coelom. This thickening, which is elongated in a craniocaudal direction, sinks in, forming a longitudinal groove, and later it rolls into a tube. The tube remains open and in communication with the coelom anteriorly but is completely closed posteriorly, where its free end is embedded in the retroperitoneal connective tissue. At this stage, the rudiment of the paramesonephric duct becomes essentially similar to the initial stage of the duct in anamniotes, in which it starts growing from a remnant of the pronephric nephrostome. The opening of the tube anteriorly, of course, becomes the newly formed ostium tubae. There follows backward growth of the paramesonephric duct alongside the mesonephric duct and its connection to the cloaca.

The details of the growth of the paramesonephric ducts are not very well known, and they appear to vary to some extent in different animals. It is questionable to what extent the mesonephric duct contributes to the development of the paramesonephric duct, if at all. It has been found, however, that if the mesonephric duct is incomplete because its outgrowth was interrupted in early stages (p. 400), the paramesonephric duct, developing at a much later stage, grows backward only as far as the posterior end of the mesonephric duct (Gruenwald, 1952; Fig. 375). It appears that the mesonephric duct in some way provides a path for the subsequent growth of the paramesonephric duct, even if it does not itself contribute material for the elongation of the latter.

The paramesonephric duct initially develops in both male and female embryos, although it becomes functional in females only. In this way a most peculiar situation arises—namely, that embryos of both sexes possess a double set of genital ducts, the mesonephric ducts, which in the male sex function as the vasa deferentia, and the paramesonephric ducts, which become the oviducts in the female sex. This condition of the genital ducts can be called the indifferent stage with respect to sex. In chick embryos this condition prevails up to the ninth day of incubation, and in a human embryo until the age of about two and one-half months. After this age, the ducts of the opposite sex regress (Fig. 376). The paramesonephric ducts in male embryos of amniotes disappear

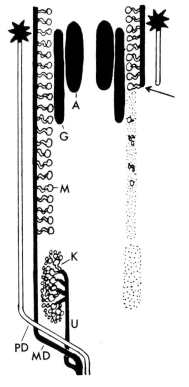

Figure 375. Diagram showing the relations between different parts of the urogenital system, based on experiments with bird embryos. The left side shows the normal urogenital system in the indifferent stage. The right side shows the effects of cessation of the growth of the mesonephric duct at the point indicated by arrow. The undifferentiated nephrogenic tissues are stippled. A, Adrenal gland; G, gonad; K, metanephric kidney; M, mesonephros; PD, paramesonephric duct; U, ureter; MD, mesonephric duct. (From Gruenwald, 1952.)

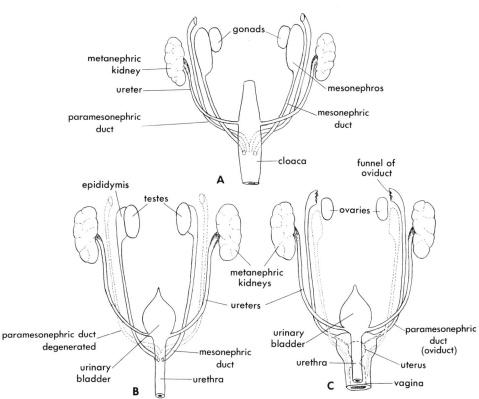

Figure 376. Diagram showing transformations of the genital ducts in mammalian embryos in transition from an indifferent stage *(A)* to the male *(B)* and female *(C)* conditions.

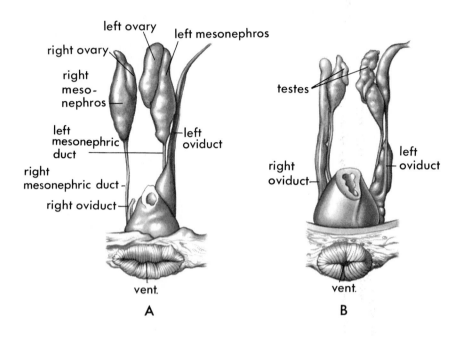

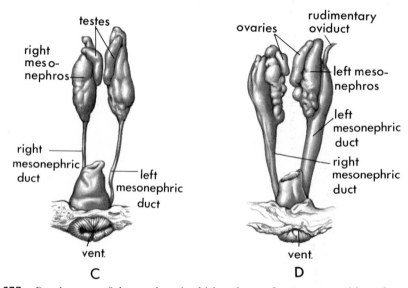

Figure 377. Development of the sex ducts in chick embryos after treatment with sex hormones. *A*, Normal female embryo incubated for 18 days. *B*, Male embryo treated with female hormone; both oviducts are present and greatly hypertrophied. *C*, Normal male embryo of 17 days, showing complete absence of oviducts. *D*, Female embryo treated with male hormone; the oviducts are absent except for small fragments anteriorly, and the mesonephric ducts are greatly hypertrophied. (From Willier, Weiss, and Hamburger, 1955.)

practically completely. The mesonephric ducts in female embryos are reduced to a lesser degree. In birds, the ducts persist as rudimentary structures even in the adult state, and in mammals and man, small remnants of mesonephric ducts may also remain. In birds, the right paramesonephric duct becomes reduced in correspondence with the reduction of the right gonad, while the left duct becomes the functional oviduct. In female mammals the distal portions of the paramesonephric ducts fuse to a greater or lesser extent to form the uterus and the vagina (Fig. 376C), which acquires a separate opening to the exterior. In male mammals both the products of excretion (via the ureters) and the spermatozoa (via the mesonephric ducts) enter a common channel, the urethra (Fig. 376B). (For the origin of the urethra, see p. 448.)

The progressive development of the genital ducts, corresponding to the sex of the embryo, and the reduction of the genital ducts of the opposite sex are the result of the action of sex hormones which start being secreted by the gonads when they emerge from the indifferent state and differentiate into either testes or ovaries. The hormones are secreted not by the germ cells (future eggs and spermatozoa) but by the somatic cells of the gonads. The female hormones are believed to be secreted by cells of the cortex; the male hormones are produced by some of the cells of the medulla (the future interstitial cells).

Proof that the differentiation of the genital ducts is under hormonal control has been provided by experiments in which sex hormones were injected into embryos of birds or mammals, or in which sex hormones were added to water in which the larvae of amphibians were being reared. Figure 377 shows the result of one such experiment performed which chick embryos. The injection of a female hormone into a male embryo caused the retention and hypertrophy of the paramesonephric ducts (compare with normal male of same age; Fig. 377B and C). Injection of a male hormone into a female embryo caused the paramesonephric ducts to disappear almost completely and the mesonephric ducts to increase greatly in size. In these experiments the gonads were not affected, so that the original sex of the treated embryos could be easily ascertained. Essentially similar results were produced in numerous experiments on other animals. (For a review see Willier, 1955.)

The results of sex hormone injections may be checked by experiments on castration of early embryos. In birds, castrated embryos, female as well as male, retain both paramesonephric and mesonephric ducts and remain thus in an indifferent condition with regard to the genital duct system. Castrated mammalian embryos of both sexes tend to develop according to the female sex pattern; their paramesonephric ducts develop progressively, but the mesonephric ducts degenerate. This is interpreted as showing that in mammals it is the male hormone which mainly causes differentiation of the secondary sex characters, while the female characters do not need special stimulation for their development. (See Willier, 1955.)

The dependence of the development of sexual characters on hormones has been the subject of very numerous investigations, which could not be reviewed here. For further information on this subject the student is referred to Burns, 1961.

Chapter 14

DEVELOPMENT OF THE ENDODERMAL ORGANS IN VERTEBRATES

14–1 THE RELATION BETWEEN THE ARCHENTERON AND THE DEFINITIVE ALIMENTARY CANAL

The relation between the archenteron and the definitive alimentary canal is very different in vertebrates with complete cleavage from that in vertebrates with incomplete (meroblastic) cleavage.

The Relation in Lower Vertebrates. The archenteron in *Amphioxus* is originally lined with presumptive endoderm, presumptive mesoderm, and presumptive notochordal cells. When the notochord and the mesoderm are segregated from the endoderm, the latter closes the gap on its dorsal side (Section 7–3), and the resulting cavity becomes the cavity of the alimentary canal.

In holoblastic vertebrates—cyclostomes, ganoid fishes, lungfishes, and amphibians—the presumptive notochord and mesoderm also participate in the lining of the archenteron, forming its roof. This roof is segregated from the endoderm and the endodermal cavity which is closed by the fusion of the free edges of the endodermal layer.

The resulting endodermal cavity consists of three unequal portions. The anterior portion is dilated and lined by a relatively thin endodermal epithelium. This part is usually referred to as the **foregut.** The following portion is called the **midgut.** The cavity of the midgut is narrower. Dorsally it is lined by a rather thin epithelium, but the ventral wall consists of a mass of large cells containing abundant yolk, so that the wall here is very thick. The most posterior part of the cavity which adjoins the blastopore may be distinguished as the **hindgut.**

The fate of the various parts of the endodermal lining of the fore- and midgut in amphibians has been elucidated by the method of local vital staining (in newts by Balinsky, 1947; in frogs by Nakamura and Tahara, 1953; Tahara and Nakamura, 1961—Fig. 378). To stain the inner surface of the gut, cuts were made through the body wall (in some experiments through the neural plate), and pieces of agar soaked in vital stain were applied to the endoderm from the inside. After the stain was taken up by the endodermal cells, the agar was removed. The wound healed easily, and the embryos continued to develop apparently quite normally.

In the neurula stage, the ventral wall of the foregut becomes flattened out and later even folded slightly upward. The cavity is thus subdivided into two pocket-like recesses. The larger anterior one, lying immediately underneath the brain, gives rise to the cavities of the mouth and of the branchial region. The posterior pocket, bordering on the mass of yolk-laden cells of the midgut, is known as the **liver diverticulum** (or hepatic diverticulum), although in addition to giving rise to the rudiments of the liver

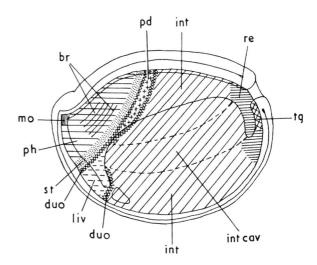

Figure 378. Fate map of the endodermal organs in the neurula of a frog. The embryo is represented as being cut in half in the median plane and viewed from the side of the cut surface. br, Branchial pouches; duo, duodenum; int, small intestine; int cav, region in which the cavity of the intestine will later develop; liv, liver; mo, mouth; pd, dorsal pancreas; ph, pharynx; re, rectum; st, stomach; tg, postanal gut. (From Tahara and Nakamura, 1961.)

and the pancreas it also participates in the formation of the stomach and the duodenum (Balinsky, 1947). It is rather remarkable that in frogs the liver diverticulum is extended downward and backward until the endoderm is perforated, and the gut cavity opens into the space between the endoderm and mesoderm on the ventral side of the embryo. The significance of this perforation is not known.

The liver diverticulum becomes extended as a funnel-shaped invagination farther and farther in a posteroventral direction (compare Fig. 379 c, d, and f), drawing into it the rudiments which were originally located in the walls of the broadened cavity of the foregut (such as the rudiments of the anterior part of the duodenum, stomach, and esophagus). These rudiments become reshaped in the form of a narrow tube. This transformation can best be shown by considering the development of the stomach in the frog embryo. In the neurula stage, the presumptive material of the stomach is arranged in the form of a narrow ring (Fig. 380a), the ventral part of which lies just in front of the liver diverticulum, the dorsal part occupies the roof of the gut just at the mouth of the midgut, and the right and left parts obliquely cross the lateral walls of the foregut. Figure 380b shows how this ring contracts and at the same time broadens until eventually the rudiment of the stomach becomes more or less barrel-shaped, with a fairly narrow cavity. The tube is constricted to an extreme degree between the pharynx and the stomach and may even be temporarily occluded in this region, which becomes the rudiment of the esophagus, a very short section of the gut in the embryo.

The part of the gut in front of the stomach rudiment also elongates, but instead of becoming tubular it is flattened dorsoventrally and expanded sideways. Most of this cavity becomes the pharynx. The floor of the pharynx is raised, partially as a result of the increase of the heart rudiment, which develops just underneath. The lateral edges of the pharynx are drawn outward even more and form the pharyngeal pouches, which will be dealt with in Section 14–3. The epithelium lining the pharynx becomes rather thin, in contrast to the lining of the rest of the alimentary canal.

The part of the liver diverticulum posterior to the stomach retains a fairly broad cavity. At the same time, the cavity of the midgut (in urodeles and in most frogs) becomes occluded by the yolky endodermal cells. For a time there is no cavity in the midgut, and the gut cavity ends with the liver diverticulum. Subsequently, the centrally lying yolky cells of the midgut break down and become resorbed, and the remaining cells are rearranged around a new cavity which extends in a posterior direction from

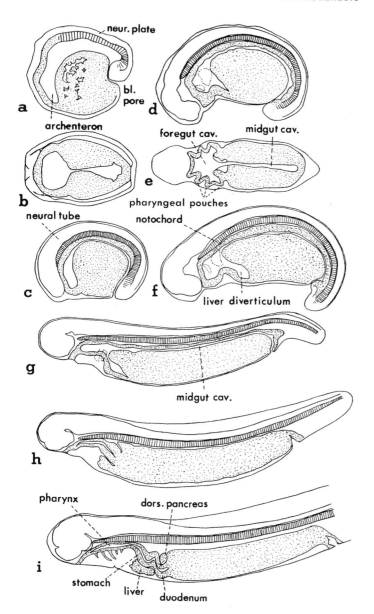

Figure 379. Development of the endodermal organs in the newt *Triturus taeniatus*, from the neurula stage to swimming larva. *b, e,* Frontal sections; all others are median sections. (From Balinsky, 1947.)

the liver diverticulum. The liver diverticulum thus becomes incorporated into the main channel of the alimentary canal.

This mode of development of the alimentary canal of the amphibians was first described for urodeles (Balinsky, 1947), and the findings were confirmed in all essential points by work done on frogs (Tahara and Nakamura, 1961). In some frogs, however, such as *Xenopus laevis*, the midgut cavity is not completely closed at any stage and later expands to form the intestinal cavity (Balinsky, in Nieuwkoop and Faber, 1956).

At the time when the midgut becomes occluded, the hindgut does not lose its cavity, and the latter persists as the cavity of the cloaca. The dorsal wall of the hindgut becomes extended into the tail rudiment as a postanal gut (Fig. 132*A*). The postanal gut has only an ephemeral existence and is soon broken up and disappears. The hindgut also gives rise in amphibians to the urinary bladder, which develops as a ventral evagination of the gut in late stages of larval life.

As the main portions of the alimentary canal begin to take shape, the canal as a whole becomes twisted in a characteristic way. At an early stage, the stomach has already assumed a slanting position (Fig. 380). Subsequently, the posterior end of the stomach is shifted to the left, while the adjoining part of the duodenum comes to lie transversely, going from left to right. The distal part of the duodenum is then bent in such a way that it leads to the anterior end of the intestine, which is more or less in a dorsal position. The alimentary canal thus performs a complete spiral revolution, which may be referred to as the gastrointestinal loop and which occurs with greater or lesser modifications in all vertebrates. Where the loop is in its lowest position, it leaves a space or saddle on the dorsal side, and this space is taken up by the rudiment of the pancreas. (See Section 14−4.) The posterior part of the alimentary canal also becomes twisted into folds and loops as a result of the elongation of the alimentary canal which exceeds the elongation of the body. The folding and twisting of the duodenum and intestine are especially prominent in the tadpoles of frogs which, in connection with their herbivorous diet, have a very long intestine (Fig. 381). The pattern of twisting varies somewhat in different frogs and need not be considered here. In urodeles and also in fishes, the intestine does not elongate to the same degree, and its folding may be very limited, though the gastroduodenal loop is always present.

The Relation in Higher Vertebrates.   In vertebrates having a meroblastic type of cleavage, the development of the alimentary canal presents very different problems, and the processes leading to the formation of the definitive cavity of the alimentary canal are quite peculiar. In the following paragraphs we will describe this process as it occurs in birds and mammals.

It must be noted in the first instance that in birds the archenteric cavity is often lacking altogether, and if it is present, as a canal leading forward from Hensen's node, it is very small and its walls are not endodermal. The starting point for the development of the alimentary canal proper is a sheet of endoderm lying flat under the ectodermal and mesodermal parts of the embryonic region of the blastodisc. The sheet of endoderm lies flat on the yolk of the yolk sac or forms the roof of the mammalian yolk sac, which is a space filled with fluid. In both cases, the cavity of the alimentary canal is separated from the cavity of the yolk sac by a process of infolding.

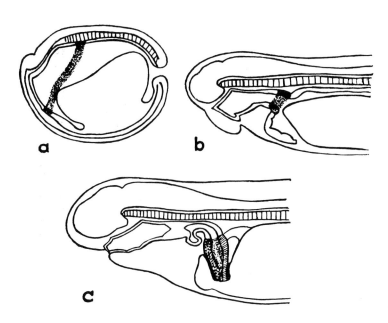

Figure 380.   Development of the stomach in a frog embryo (after experiments with vital staining). *a*, Position of the presumptive gastric endoderm (stippled) in the neurula stage; *b*, constriction of the alimentary canal in the stomach region; *c*, elongation of the stomach rudiment. (From Nakamura and Tahara, 1953.)

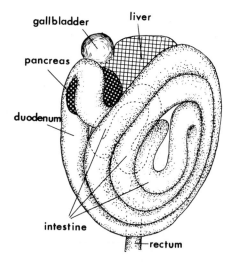

Figure 381. Coiling of the intestine in a tadpole of the frog *Xenopus laevis* (ventral view).

During this infolding, the median strip of the endoderm lying immediately under the notochord and somites remains in this position, while the immediately adjoining strips on the right and left become inflected downward, and the crests of the folds converge toward the middle and eventually fuse (Fig. 382). The inner surfaces of the folds contribute to the formation of the floor of the gut, while the outer surfaces of the folds are continuous with the endodermal lining of the yolk sac. Concurrently with the movement of the endodermal layer in a transverse plane, a complicated shifting has been found to occur longitudinally. The median strip which forms the dorsal wall of the gut slides forward, while in the folds closing the gut laterally and ventrally, endodermal material moves obliquely backward. During their movement downward, the lateral strips of endoderm are accompanied by the visceral layer of the lateral plate mesoderm, which closely adheres to the endoderm throughout this whole series of formative movements. It is probable that the dynamic force of the movement is due to the endoderm and mesoderm jointly (Bellairs, 1953).

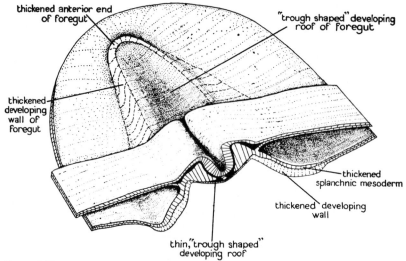

Figure 382. Stage in the development of the foregut in a chick embryo. Ectoderm with neural folds and the parietal mesoderm are shown as being cut away at the anterior end of the embryo. (From Bellairs, 1953.)

As a result of the downward movement of the visceral layer of the lateral mesoderm, the coelomic cavity becomes considerably expanded locally. The coelom is reduced again, however, when the body folds undercut the embryo, as explained on page 256. The alimentary canal becomes separated from the yolk sac cavity initially at the anterior end of the embryo. This part corresponds to the foregut of the amphibian embryo. Somewhat later, the posterior part of the endodermal groove closes into a canal that becomes the posterior part of the alimentary system, which in the amniotes is called the **hindgut**, although most of it corresponds to the midgut of amphibian embryos. Between the foregut and the hindgut a gap remains where the endoderm does not close to form a canal, or even a groove, and where it is in open communication with the yolk sac (Fig 383). This part, in amniotes, is referred to as the **midgut.**

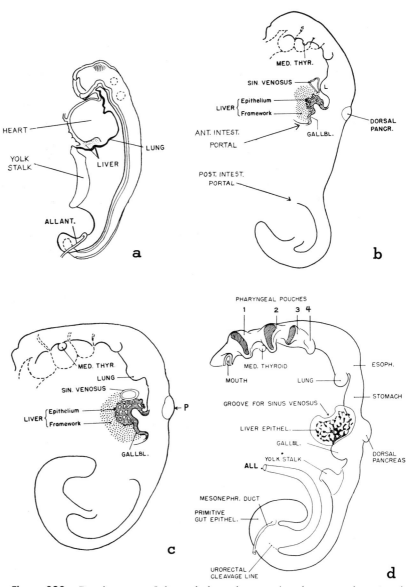

Figure 383. Development of the endodermal organs in a human embryo (semi-diagrammatic). (From Streeter, 1942.)

Where the edges of the folds separating the foregut from the yolk sac meet in the middle they form a ridge, known as the **anterior intestinal portal.** A similar edge at the anterior end of the hindgut is the **posterior intestinal portal.** The gap between the two is very large at first, but its relative size diminishes with the growth of the embryo, and eventually it is reduced to the opening of the **yolk stalk,** connecting the gut cavity to the cavity of the yolk sac.

From the beginning the foregut is much broader than the hindgut and is flattened in cross section. As in amphibians, the foregut gives rise to the endodermal lining of all anterior parts of the alimentary canal, including most of the duodenum, the liver, and pancreas developing just in front of the anterior intestinal portal. The oral and especially the pharyngeal parts of the foregut remain expanded in a transverse direction, and the pharynx becomes drawn out to form the **pharyngeal pouches,** but the posterior part of the foregut, corresponding to the esophageal, gastric, and duodenal parts of the alimentary canal, eventually becomes round in cross section. The esophagus in higher vertebrates is soon greatly elongated, in connection with the development of the neck. The hindgut is, from the start, narrower than the foregut and soon also becomes round in cross section (Fig. 384).

Although the foregut in higher vertebrates is quite different in shape from the foregut of amphibians, it develops the gastroduodenal loop in much the same way. The anterior (cardiac) end of the stomach is displaced to the left side, and the posterior (pyloric) end is turned downward and toward the middle, while the anterior part of the duodenum assumes a transverse direction, thus bringing about the familiar position of the stomach in adult mammals and birds. The intestine becomes convoluted in a pattern that varies greatly not only between different classes but even within one class, such as in various mammals. A peculiarity occurring in mammals is that sections of the intestine adjoining the yolk stalk, both anteriorly and posteriorly, sink down into the umbilical cord (Fig. 384) and lie for a time practically outside the body of the embryo proper as a sort of umbilical hernia. The convolutions of the intestine begin forming inside the umbilical cord, but well before birth the definitive intestine is withdrawn into the body, and only the yolk stalk remains in the cord. (This occurs in the human embryo during the third month of pregnancy.)

The posterior end of the hindgut gives rise to the cloaca and also to a postanal gut, which disappears later. The ventral wall of the cloaca produces the allantoic diverticulum (the endodermal part of the allantois), as has been indicated in Section 10–1. In higher mammals (in particular in man), however, the allantoic diverticulum is formed very early as an outgrowth of the yolk sac at the posterior end of the embryo, even before the embryo becomes subdivided into the embryonic body and the extraembryonic parts. Later, the allantoic diverticulum is incorporated into the ventral floor of the midgut and assumes the same position that it has in lower amniotes. (See Fig. 221.)

At the time of their formation, both the foregut and the hindgut are blind diverticula, without openings to the exterior at the front and hind ends of the embryo. The formation of the mouth opening at the anterior end of the foregut occurs in much the same way in both holoblastic and meroblastic vertebrates. It will be dealt with summarily in the next Section (14–2). The development of the anal or cloacal opening, however, is different in the two types of vertebrates. In some vertebrates with holoblastic cleavage, the blastopore (or part of it) persists as the anal (cloacal) opening. In others the anus is formed early in embryonic life near the spot where the blastopore opening had been (p. 172). As, in higher vertebrates, there is no patent blastopore leading into an endodermal archenteron, the cloacal opening has to develop by a perforation of the body wall at the posterior end of the hindgut. The point at which this perforation

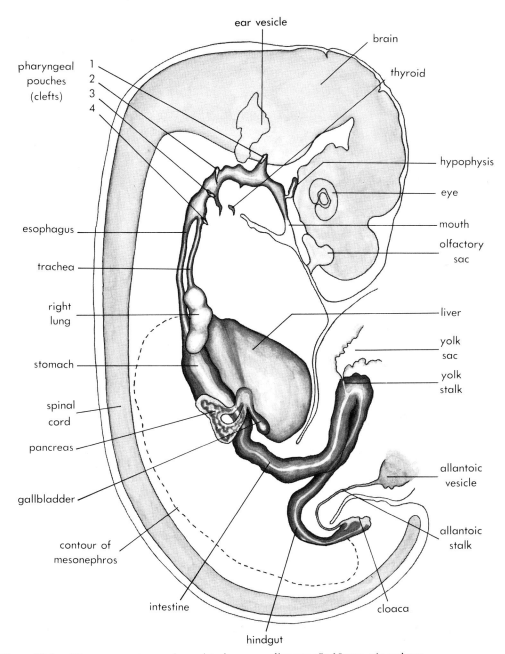

Figure 384. Alimentary system and associated organ rudiments of a 10 mm. pig embryo.

occurs is discernible as early as the primitive streak stage and lies at the posterior end of the streak (Fig. 221a).

When the primitive streak shrinks in the late gastrulation stages, it leaves in front the three germinal layers: the ectoderm, the mesoderm, and the endoderm, lying one above the other. At the posterior end, however, the separation of germinal layers does not occur; the ectodermal and endodermal layers do not become separated by the intervening mesoderm and remain in close contact with each other. The resulting double-layered plate is the **cloacal membrane.** When the hindgut becomes separated from the yolk sac by folds, the cloacal membrane is incorporated into the wall of the gut. The ectodermal side of the cloacal membrane is originally dorsal, but this position is inverted by the development of the tailbud, occurring just anterior to the cloacal membrane (Fig. 221b). The tail rudiment protrudes backward, and the hindgut above the cloacal membrane develops a diverticulum entering the tail rudiment—the postanal gut. As a result, the cloacal membrane comes to lie at the root of the tail with the ectodermal side facing downward. The part of the hindgut adjoining the cloacal membrane is somewhat dilated and becomes the rudiment of the **cloaca.** The ectoderm is slightly depressed in the region of the cloacal membrane, forming the external cloaca or **proctodeum.** The cloacal membrane separates the cavity of the cloaca from the cavity of the proctodeum till late in embryonic development, but it is eventually ruptured and thus a free passage from the alimentary canal to the exterior is allowed.

In amphibians, reptiles, and birds, the cloaca receives the ducts carrying excretory products (the mesonephric ducts and ureters) as well as the ducts carrying eggs and sperm (the oviducts and the vasa deferentia). The cloacal opening thus serves, in these animals, for the exit of feces, urine, and sex cells. In mammals the cloaca becomes subdivided in the course of embryonic development, so that the channel for feces is separated from the pathway serving for the conveyance of urine and sex products. This subdivision is achieved by the backward growth of a septum arising at the angle formed by the ventral surface of the gut and the allantoic stalk (Fig. 383d: marked "urorectal cleavage line"). Extending backward, the septum reaches the cloacal membrane and fuses with it, thus subdividing the cloaca into a dorsal and a ventral compartment. The dorsal compartment is in continuity with the gut, and the ventral compartment receives the openings of the excretory and genital ducts, as well as the opening of the allantoic stalk. Consequently, this compartment becomes the **urogenital sinus.** The inner part of the urogenital sinus with the openings of the ureters and the adjacent part of the allantoic stalk expands to form the **urinary bladder.** The canal leading from the bladder to the umbilicus later closes and degenerates. The portion of the urogenital sinus immediately following the bladder becomes narrow and develops as the **urethra.**

14—2 DEVELOPMENT OF THE MOUTH

In all vertebrates, the mouth opening appears rather late in embryonic life after all the primary organ rudiments have already been formed. The mouth opening breaks through where the anterior end of the endodermal part of the alimentary canal touches the ectodermal epidermis beneath the front end of the neural tube. Here the ectodermal epidermis sinks in to form a pocket-like depression, the ectodermal mouth invagination or the **stomodeum.**

The ectodermal epithelium and the endodermal epithelium fuse in this area and become an **oropharyngeal membrane.** (See Figure 221b.) The membrane becomes very thin and eventually disappears. Forthwith, the boundary between the ectodermal and endodermal epithelia becomes indistinguishable, and it is not an easy matter to determine later which part of the oral cavity is derived from the stomodeum and which from the endodermal gut. There are, however, indications that the stomodeal epithelium extends (in mammals) to about the middle of the tongue ventrally and to the beginning of the pharynx on the dorsal side.

The determination of the stomodeum begins at the end of gastrulation, and the development of the ectodermal mouth invagination is induced by the endoderm

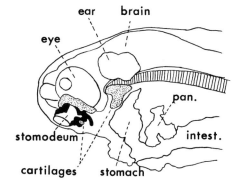

Figure 385. Development of the mouth and stomodeal invagination in a newt embryo after removal of the anterior endoderm in the neurula stage. (From Balinsky, 1939.)

when the anterior end of the archenteron comes in contact with the ectoderm of the presumptive mouth region (Ströer, 1933). No mesoderm ever penetrates here between the ectoderm and endoderm, and the endoderm remains in contact with the ectoderm throughout the subsequent stages until the rupture of the pharyngeal membrane. After the endoderm and ectoderm come in contact in the oral region, the oral ectoderm acquires, to some extent, the ability to differentiate as stomodeum. The anterior part of the archenteron may now be removed, and the remaining ectoderm produces a mouth invagination, although there are no endodermal oral and branchial parts behind. Mouth invaginations thus developed are small, and their shape is not normal. If, however, the operation is performed slightly later, in the neurula stage, the stomodeum may develop in an almost completely normal way, with the sole reservation that it does not lead into the endodermal alimentary canal (Fig. 385). As the oral ectoderm acquires the ability for self differentiation, all the rest of the epidermis loses the competence for development into stomodeum. If the region where the mouth is to develop is covered with a flap of epidermis taken from a different part of the body, the development of the stomodeum is suppressed, in spite of the presence of oral endoderm.

Presumptive epidermis of the stomodeum may develop a mouth invagination in an abnormal position, but it can do so only under certain conditions. In newts, stomodeal epidermis of a neurula, when transplanted alone, will not develop a mouth, but when it is transplanted together with some adjoining endoderm, and if the site of transplantation is not too far from the oral region of the host, an additional stomodeal invagination will be formed by the graft, and this invagination may break through into the endodermal cavity of the host (Fig. 386) (Balinsky, 1948).

Figure 386. Transplantation of the oral ectoderm alone (a) and with the underlying endoderm (b); diagram of operation. c, Position of the graft on the host embryo several days after operation. The position indicated is favorable for development of a stomodeum from grafted tissue.

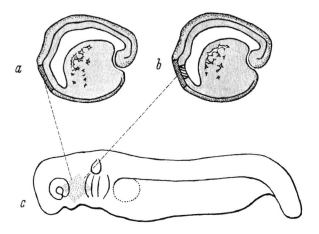

In frogs, the determination of the mouth ectoderm proceeds more rapidly than in urodeles, and the capacity for differentiation in the determined direction is higher. Already in the early neurula, the removal of oral endoderm does not interfere with mouth development, and transplants of presumptive mouth ectoderm may develop in a typical way, producing the characteristic horny jaws and horny teeth not only in the vicinity of the normal mouth but also in other locations, such as on the belly. For the mouth to be developed in abnormal regions, however, it is necessary to transplant a piece of the transverse neural fold together with the presumptive mouth ectoderm. Only a narrow median piece of the neural fold can support mouth development; lateral neural fold does not have this ability (Fig. 387). In addition to the mouth, the adhesive organ also develops (Cusimano, Fagone, and Reverberi, 1962).

The transplantation experiments show that the oral endoderm is not the only part which induces the development of the ectodermal mouth invagination. Another factor operating in the immediate environment of the mouth, namely, the transverse neural fold, is also involved. In this case, there seems to be some analogy to the dual dependence in the development of the lens: the dependence on the head mesoderm as well as the dependence on the eye cup (p. 359). In the case of mouth differentiation, as with the lens inductor, a certain variation of modes of development in different animals may be attributed to the relative strength of one or the other inducing factor.

Besides giving rise to part of the oral epithelium, the stomodeal invagination furnishes the cells which become the rudiment of the anterior lobe of the **hypophysis.** The rudiment is formed as a solid bud, or a small pocket **(Rathke's pocket),** on the dorsal side of the stomodeal invagination, just in front of the oropharyngeal membrane. (See Figures 233, 262, and 391.) The rudiment pushes backward through the connective tissue and eventually comes to rest underneath the diencephalon. (See Figure 272.) The original connection with the stomodeum becomes interrupted, while the floor of the diencephalon furnishes the posterior lobe of the hypophysis.

One of the most characteristic differentiations of the oral cavity is the **teeth.** The rudiments of the teeth consist of an epithelial cap (the **enamel organ** which secretes the enamel) and a connective tissue **papilla** which produces the dentine (Fig. 388). The enamel organs of the teeth may develop from both the ectodermal and the endodermal epithelia. This can be proved by experiments similar to the ones already described, in which either the stomodeum or the endodermal oral epithelium is prevented from developing. The mouth invaginations, developing in the absence of the anterior end of the archenteron and the parts derived from it, often have well-differentiated teeth. When the endodermal oral cavity develops without a stomodeum being formed, or when the two are not in contact, teeth can be found developing in the endodermal epithelium. In cyprinid fishes, one or two rows of teeth develop on the inner surface of the last branchial arch (the pharyngeal teeth). There can be no doubt that the enamel organs in these teeth are of endodermal origin.

In lower vertebrates, the fishes and amphibians, the connective tissue papillae of the teeth project from the inside into the stratified epithelium lining the mouth, and the enamel organs of the teeth develop from the malpighian layer of this epithelium. In mammals, however, the epithelium sinks down at the edges of the jaws into the connective tissue in the form of ridges. The rudiments of the individual teeth are formed at the edge of the dental ridges, at the expense of a connective tissue papilla and of the layer of innermost cells of the dental ridge which adjoin the papilla and form the enamel organ. The teeth are thus formed and begin to grow deep in the tissue of the jaw. They erupt to the surface only when they have almost reached their full development.

While the cells of the enamel organ may be ectodermal or endodermal, the

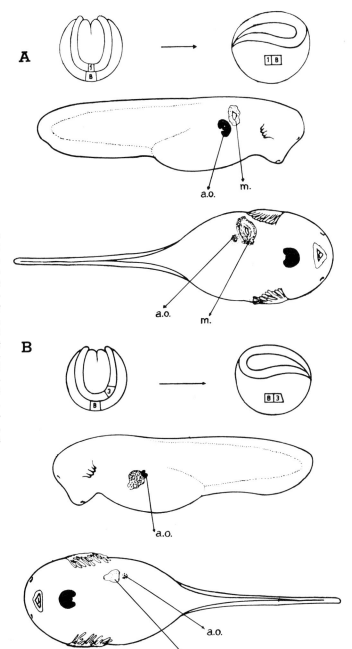

Figure 387. Experiments proving that in frogs the transverse neural fold plays a part in causing the development of the mouth. In *A*, the oral ecto- and endoderm were transplanted together with the medial part of the transverse neural fold. In *B*, the oral ecto- and endoderm were transplanted together with a part of the lateral neural fold. In *A*, the graft developed a perfectly formed mouth and adhesive organ. In *B*, only the adhesive organ developed, as well as some skin epidermis; no mouth was formed. In both *A* and *B*, drawings above show diagrams of the operation, and below are the host embryos after 4 and 6 days of development respectively. a.o., Adhesive organ; B, presumptive buccal ectoderm; e.p., transplanted epidermis; m., mouth; 1,3, parts of the neural fold. (From Cusimano, Fagone, and Reverberi, 1962.)

connective tissue cells of the dental papillae are derived from the neural crest. The neural crest is also the source of skeletogenous cells for the development of the mandibular arch: the quadrate and Meckel's cartilage (Section 12–3). The development of the mouth as a whole depends therefore on the harmonious cooperation of cells coming from different primary rudiments—the foregut, the epidermis, and the neural crest.

The development of the mandibular arch is, at least in part, directly dependent on the influence exercised by the ectodermal mouth invagination. If the ectodermal

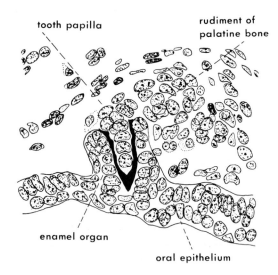

tooth papilla

rudiment of
palatine bone

Figure 388. Rudiment of tooth in a salamander embryo. (After de Beer, 1947.)

enamel organ

oral epithelium

mouth invagination fails to be formed, the ventral part of the mandibular arch is not formed. An additional Meckel's cartilage can develop in connection with a mouth invagination resulting from the transplantation of the presumptive stomodeal ectoderm. The quadrate, on the other hand, may develop even in the complete absence of any sort of oral cavity; it is thus not dependent on the latter (Balinsky, 1948).

The development of structures surrounding the edges of the mouth contributes very much to the formation of the face in man and of corresponding parts in higher mammals. The structures in question are a number of swellings, consisting of actively growing mesenchyme and covered by ectoderm, which are formed around the stomodeal invagination (Fig. 389). At the dorsal edge of the stomodeum medially, there is a slight swelling, the **medial frontal process.** Lateral to the frontal process on each side there develops a U-shaped swelling encircling the nasal pit. The free ends of the U are directed downward. The inner branch lies just alongside the frontal process and impinges on the edge of the mouth as the **medial nasal process.** The outer branch, lying lateral to the nasal pit, is the **lateral nasal process,** and it does not quite reach the edge of the mouth. Around the angle of the mouth on each side, another U-shaped swelling develops, with the upper branch ending on the edge of the mouth as the **maxillary process.** The lower branch extends along the lower edge of the mouth as the **mandibular process.**

The two medial nasal processes grow downward and toward the midline until they fuse and exclude the frontal process from participating in the formation of the edge of the mouth. The maxillary processes grow forward and eventually fuse with the lateral edges of the medial nasal processes, thus completing the upper edge of the mouth. The opening of the nasal pit remains just above the line of fusion of the maxillary and medial nasal processes. The upper edge of the maxillary process also fuses with the lateral nasal process. The furrow lying between the maxillary process and the lateral nasal process, as can be seen in Figure 389, leads from the angle of the eye to the nasal pit. The infolded epidermis lining this furrow gives rise to a ridge of epithelial cells, which later becomes hollowed out and establishes a communication between the space underneath the eyelids and the nasal cavity. This is the **nasolacrimal duct.**

The lower edge of the mouth acquires its final shape after the median fusion of the two mandibular processes.

The nasal, maxillary, and mandibular processes, as previously stated, are essentially proliferating masses of mesenchyme covered externally by ectodermal epithelium. The mesenchyme later ossifies and gives rise to some of the most important parts of the facial skeleton. The mesenchyme in the lower portions of the medial nasal processes ossifies to form the pair of premaxillary bones. The maxillary bones are produced by the ossification of the mesenchyme of the maxillary processes, and the mesenchyme of the mandibular processes gives rise to the mandibular bones. The upper parts of the medial nasal processes together with the medial frontal process develop into the back of the nose.

14–3 DEVELOPMENT OF THE BRANCHIAL REGION

The development of the branchial region is no less dependent on the cooperation of parts of different origin than is the mouth region. The leading part in the development of this area belongs to the endoderm. The endodermal cavity in this region, from its beginning, is distended in a transverse direction. In amphibians it is derived from the inflated part of the foregut. In the stage immediately following the closure of the neural

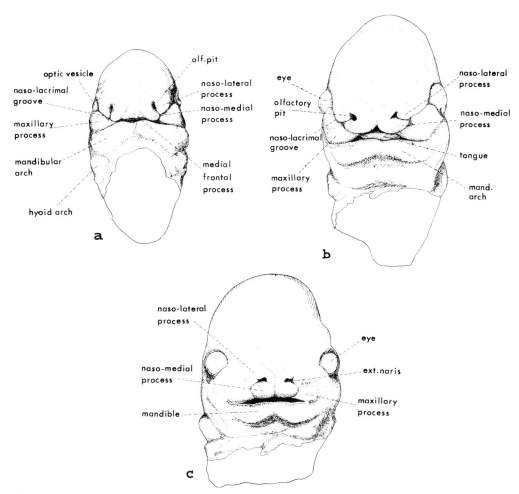

Figure 389. Development of the face of a pig embryo. (After Patten, 1944.)

tube, the lateral walls of the pharyngeal cavity bulge out and produce a series of outwardly directed pockets on each side. These pockets are the **pharyngeal pouches** or **branchial pouches.** The pharyngeal pouches are developed one after another, beginning with the first pair, that is, the one lying just posterior to the mandibular arch. As the endodermal pharyngeal pouches reach the epidermis, having pushed aside the intervening mesoderm, the epidermis folds inward to meet the pharyngeal pouches. A series of branchial grooves is thus developed on the surface of the embryo, each groove corresponding to an endodermal pouch (Fig. 390). The outer wall of the endodermal pouch and the inner wall of the epidermal groove fuse into a **branchial membrane,** similar to the oropharyngeal membrane. A **gill cleft** is formed when the branchial membrane becomes perforated, so that an open communication is established between the pharyngeal cavity and the outer medium. In aquatic vertebrates gill filaments are then developed on the walls of the gill clefts. These filaments are the **internal gills.**

The development does not always reach the final stage (the formation of the gill filaments serving for respiration). Even in aquatic vertebrates, some of the pharyngeal pouches may not develop into gill clefts. Thus, in bony fishes and amphibians, the first pair of pharyngeal pouches (the one lying behind the mandibular arch) does not reach the epidermis and remains a blind diverticulum of the pharyngeal wall. In the amniotes, the gills never function as respiratory organs, but four pairs of pharyngeal pouches nevertheless develop (Fig. 391). Branchial grooves and branchial membranes are formed, and in the first three pairs of pharyngeal pouches the branchial membranes are broken so that gill clefts are formed (Fig. 383*d*). The fourth pouch does not open to the outside. The existence of open gill clefts is, however, of very short duration, and the clefts are again closed by membranes. The new closing membranes develop farther out than the original branchial membranes and are produced as folds of ectoderm only, whereas the original branchial membranes consisted of a layer of ectoderm on the outside and endoderm on the inside.

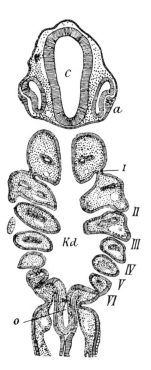

Figure 390. Frontal section of an embryo of a skate, *Raja,* showing pharyngeal pouches and gill clefts, which are indicated by Roman numerals. a, Eye; c, cavity of brain; Kd, cavity of pharynx; o, esophagus. (After Maurer, in Hertwig, 1906.)

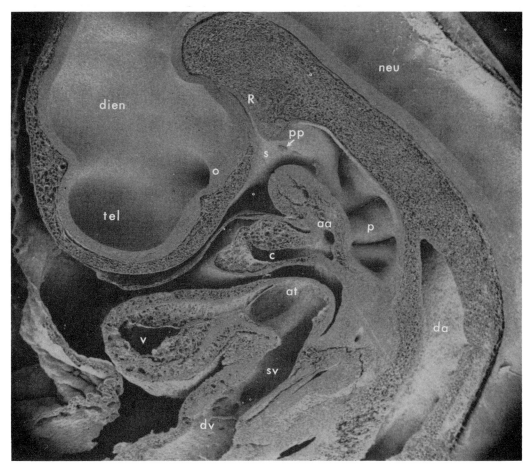

Figure 391. Scanning electron microscope photograph of a 72-hour chick embryo dissected in the median plane. aa, Base of aortic arches 3 and 4; at, atrium; c, conus arteriosus; da, dorsal aorta; dien, diencephalon; dv, ductus venosus; neu, neuromeres; o, optic vesicle; p, pharynx; pp, remnants of the oropharyngeal membrane; R, Rathke's pouch; s, stomodeum; sv, sinus venosus; tel, telencephalon; v, ventricle of heart. (From Armstrong and Parenti, 1973.)

The whole system is later reduced to a greater or lesser extent, and its remnants are used up for the development of parts having nothing to do with the respiratory function. Thus, the first pair of pharyngeal pouches becomes the Eustachian tube in the terrestrial vertebrates. The third and fourth pharyngeal pouches give rise to a series of glands of internal secretion, the thymus and the **parathyroids.** These organ rudiments develop from masses of cells that become detached from the walls of the pharyngeal pouches and are then shifted downward and backward until they come to lie in the neck or in the anterior trunk region. The main part of the thymus is derived from the third pair of pharyngeal pouches. The pouches actually become extended lateroventrally, and this extension is later used in the formation of the thymus rudiment. Its development is of great importance, as it is now known that the lymphocytes which in higher vertebrates and in man perform the cell-mediated immune responses are differentiated exclusively in the thymus. (See Cooper and Lawton, 1974.)

It has been generally believed that the cells giving rise to the parathyroid glands and the thymus are derived from the endodermal part of the pharyngeal pouches. There are indications, however, that at least part of the thymus rudiment is derived from the ectodermal cells of the branchial grooves. (For a survey see Ruth, Allen, and Wolfe, 1964.) The importance of these glands for the well-being of the animal may account for the persistence of the branchial pouches in the embryos of terrestrial animals, although they have long since lost their original functional significance.

Another important gland of internal secretion associated with the pharynx in its development is the **thyroid gland.** The thyroid gland develops in vertebrates as a ventral pocket in the floor of the pharynx. Subsequently the pocket becomes closed and separated from the pharyngeal wall. The thyroid rudiment is then displaced in a caudal direction, until in terrestrial vertebrates, it comes to lie ventral to the trachea.

It has already been stated that the endodermal pharyngeal pouches are the initiators of all the developments in the branchial region. The epidermal branchial grooves are induced by the endodermal pouches when they touch the epidermis. Without the endodermal pharyngeal pouches, the epidermal grooves do not develop, nor do they develop if the endodermal pouch, though present, does not reach the epidermis (as in the case of the first pouch in amphibians and bony fishes).

The relationship between the endoderm and the external gills has previously been stated (Section 12–4). For the development of the external gills in the urodeles, it is also necessary that the endodermal pharyngeal pouch reach the epidermis; otherwise, the gills fail to appear. The external gills in the urodeles are later supplemented by internal gills: gill filaments developing on the branchial arches. Both external and internal gills function simultaneously, until both are reduced during metamorphosis. In the anurans, the external gills function only temporarily, during a short period after the hatching of the larvae. Soon after the small tadpoles begin to swim, a fold of skin, the **opercular fold,** appears anterior to the external gills. The opercular fold spreads backward over the gill region, covering both the external gills and the gill slits. Both are thus included in a branchial cavity. The posterior edge of the opercular fold becomes attached to the skin behind the branchial region, so that only a narrow opening, the **branchial aperture,** leads from the branchial cavity to the exterior. At the same time, the external gills are reduced in size, and the internal gills develop on the branchial arches beneath the external gills and function throughout the whole period of larval development.

The visceral skeleton is an important integral part of the branchial region. As has been stated, the visceral skeleton develops from cells of the neural crest. In their downward migration, these cells are split by the pharyngeal pouches into several streams, moving in between the adjacent pharyngeal pouches. Later, the branchial arches are formed by the chondrification of the neural crest mesenchyme in about the same position as the masses of migrating cells were to be found. The dependence of visceral arch development on the gill clefts is, however, a more intimate one than would follow from this description. In the absence of the pharyngeal pouches, the neural crest mesenchyme does not chondrify, and no branchial arches are formed (Balinsky, 1948). If the number of pharyngeal pouches is reduced (after an operation in which part of the endoderm of the branchial region has been removed), the number of visceral arches is similarly reduced; one skeletal arch is developed on each side of the remaining pharyngeal pouches. Thus, the number of arches is one more than the number of pouches present.

The hyoid arch shows the same kind of dependence on the pharyngeal pouches as the branchial arches proper.

For the development of the visceral arches, it is not necessary for the endodermal pharyngeal pouches to establish a connection with the epidermis and for the gill cleft to break through. The presence of the endodermal pouches, even if they are represented by blind pockets, is sufficient to induce the development of the skeletal arches.

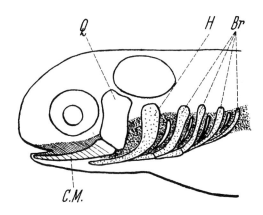

Figure 392. Diagram of the head of a urodele amphibian, showing the dependence of parts of the visceral skeleton on different sections of the alimentary canal. The ectodermal stomodeum and that part of Meckel's cartilage (C.M.) that is dependent on it are hatched; the endodermal foregut and the visceral arches dependent on the endoderm are stippled. Q, Quadrate; H, hyoid arch; Br, branchial arches.

We can now review the dependence of the development of the various parts of the visceral skeleton on the adjoining parts of the alimentary canal (Fig. 392).

1. The upper part of the mandibular arch (the quadrate and the region of articulation with the lower jaw) develops independently of parts of the alimentary canal.
2. The lower part of the mandibular arch is dependent on the ectodermal mouth invagination.
3. The hyoid arch and the branchial arches are dependent on the endodermal pharyngeal pouches.

A very peculiar feature in the development of the amphibian visceral skeleton is presented by the second basibranchial. The skeletal element is developed from mesodermal mesenchyme, not from neural crest mesenchyme, and it is also independent of the endodermal pharyngeal pouches. It is formed even if the whole of the endoderm of the branchial region has been removed.

The visceral arches, which are dependent in their development on the epithelial parts of the alimentary canal, appear to exercise some influence on the development of the teeth. The tooth rudiments, consisting of the ectodermal or endodermal enamel organ and the papilla, derived from neural crest mesenchyme, are formed in connection with certain skeletal elements: Meckel's cartilage, the rudiments of the vomer and palatine bone, and later the rudiments of the premaxilla, maxilla, and dental bone.

With the reduction of the pharyngeal pouches in adult terrestrial vertebrates, the visceral skeleton becomes modified. The lower end of the hyoid arch persists as the body of the hyoid bone, and the lower part of the first branchial arch is utilized in the formation of the horns of the hyoid bone. Parts of the subsequent branchial cartilages contribute to the formation of the thyroid cartilages and the cartilages surrounding the trachea.

14–4 DEVELOPMENT OF THE ACCESSORY ORGANS OF THE ALIMENTARY CANAL: LUNGS, LIVER, PANCREAS, BURSA FABRICII

The Lungs.   The lungs develop from a rudiment which is a pocket-like evagination of the endodermal epithelium on the ventral side of the alimentary canal, just posterior to the branchial region. The pocket at first projects straight downward. At its tip it bifurcates, and the two branches grow out to the sides and backward. The unpaired medial part of the rudiment becomes the trachea, and the two branches give rise to the two bronchi and to the lungs themselves.

In the lower vertebrates, the lungs are developed as saclike expansions at the ends of the bronchi, the walls of which become folded to various degrees. In warm-blooded vertebrates, birds, and mammals, in which lung respiration attains the highest efficiency, the greater degree of differentiation of the air spaces in the lungs already becomes manifest in the earlier stages of lung development. In mammals, the distal ends of the bronchi, as they grow out, become branched in a more or less dichotomous fashion, the branches representing the secondary, tertiary, etc., bronchi and the bronchioles. The alveoli are eventually developed on the terminal branches of this system (Fig. 393).

The unpaired part, the rudiment of the trachea, may elongate greatly by subsequent growth (Fig. 384). Although the first visible rudiment of the lungs is ventral and unpaired, there is good reason to believe that the lungs are derived from originally paired and lateral rudiments. By local vital staining in amphibians, the presumptive epithelium of the lungs has been found to lie in the neurula stage, in the lateral walls of the foregut, just posterior to the presumptive endoderm of the pharyngeal pouches. The presumptive lung endoderm later shifts downward toward the midline (Balinsky, 1947).

In the frog *Xenopus* (and possibly also in other frogs), the lung rudiments first become noticeable as two separate lateral pockets at a stage when the parts of the alimentary canal are not yet clearly separated from one another. When the gastric part of the gut becomes inflected downward, the lung rudiments are found just behind and ventral to the crest of the transverse fold separating the pharyngeal section of the gut from the esophagus. Following this, the part of the gut cavity connected to the lateral lung rudiments protrudes forward and becomes a distinct pocket-like evagination (Fig. 394). This cavity is the rudiment of the trachea (which is very short in frogs). The cavity of the trachea, which is continuous with the cavities of the lung rudiments, becomes temporarily separated from the esophagus and later opens into the pharyngeal cavity. This new opening is the glottis. It does not coincide with the mouth of the original invagination which gave rise to the trachea (Nieuwkoop and Faber, 1956).

The lateral and independent origin of the lung rudiments in amphibians makes it

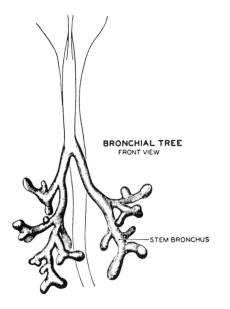

BRONCHIAL TREE
FRONT VIEW

STEM BRONCHUS

Figure 393. Branching of the lung rudiment in a 35-day-old human embryo. (From Streeter, 1948.)

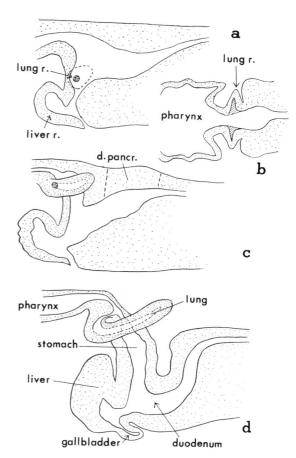

Figure 394. Development of the lungs in a frog, *Xenopus laevis* (semidiagrammatic). *a, c, d,* Projection on the median plane; *b,* frontal section.

probable that in the early history of the terrestrial vertebrates the lungs developed from the last pair of pharyngeal pouches, which failed to break through to the exterior and became adapted to the retention of air gulped in through the mouth. Thus, they became organs in which the oxygen could diffuse into the blood vessels supplying the organ.

The swim bladder of fishes is similarly a pocket, growing out from the endodermal wall of the alimentary canal posterior to the branchial region. In many fishes the cavity of the swim bladder remains permanently in communication with the esophagus, and air can be taken into the swim bladder through this canal. It is fairly obvious that the unpaired lung of the lungfish *Neoceratodus* is the same organ as a swim bladder, except that it is adapted to respiration. Whether the rudiment of the swim bladder may be compared to a pharyngeal pouch has not been investigated.

The Liver. The liver in all vertebrates develops from the endodermal epithelium on the ventral side of the duodenum.

In amphibians, the site of liver development is the anterior wall of the liver diverticulum, referred to on page 440. At the stage when the main parts of the alimentary canal begin to take shape, the anterior wall of the liver deverticulum bulges forward, so that the slitlike cavity of the diverticulum enlarges locally. This pocket-like enlargement of the gut cavity is the **primary hepatic cavity** (Fig. 395). At the time when this occurs, in frogs, the liver diverticulum is still in communication with the space between the endoderm and the mesoderm. Next, the front wall of the hepatic rudiment becomes thrown

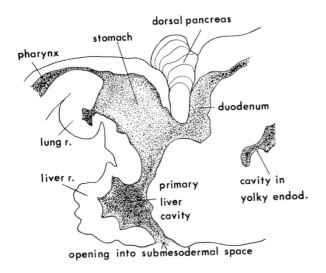

Figure 395. Reconstruction of the lung, liver, and pancreas rudiments of the frog *Xenopus laevis* (median section).

into folds, which occlude most of the primary hepatic cavity, leaving only the most posterior part open. At the same time, the original communication between the primary hepatic cavity and the duodenum is constricted and is gradually transformed into the **bile duct.** Simultaneously, the opening into the submesodermal space becomes closed by a sheet of cuboidal endodermal epithelium, and the adjoining posterior remnant of the primary hepatic cavity becomes the cavity of the gallbladder. The epithelial folds of the anterior wall of the liver rudiment soon break up into strands of cells, which for a short time may appear as tubules, with their lumen opening into the remainder of the primary hepatic cavity and thus also communicating with the rudiment of the bile duct and the gallbladder. The strands of liver cells become interwoven with blood vessels and sinuses produced by ramifications of the vitelline veins (p. 420).

In the embryos of amniotes, the structure of the gut in early stages is very different from that of amphibians, owing to the absence of yolk in the endoderm. The liver rudiment accordingly also has a different appearance, but it develops in a corresponding position, namely, in the ventral wall of the gut, posterior to the section which gives rise to the stomach. The first visible rudiment of the liver can be found as a pocket-like evagination on the anterior intestinal portal (p. 446; Fig. 383a), at a stage when the opening from the yolk sac into the definitive gut is still quite wide. As the floor of the gut continues to close in an anteroposterior direction, the liver rudiment is later found some distance in front of the anterior intestinal portal, well within the foregut. The endodermal cells then begin migrating forward from the original pocket-like evagination in the form of solid strands or cords of cells (Fig. 383b and c). The strands of liver cells form a meshwork and enclose the blood vessels, the vitelline (omphalomesenteric) veins and their ramifications, lying in the region posterior to the heart (Figs. 352 and 353). From an interaction between the cords of liver cells and the blood vessels, the complicated structure of the adult liver eventually emerges.

The gallbladder is formed as a secondary hollow outgrowth at the posterior edge of the original hepatic rudiment.

The liver increases in size very rapidly and soon becomes a large and massive organ, although the part of the duodenal wall from which it develops is relatively a very small one and even of that a large proportion is used up for the formation of the gallbladder and cystic duct.

The Pancreas.   The pancreas develops from two rudiments, a ventral rudiment (or two ventral rudiments) and a dorsal one. The ventral rudiment develops from part of the ventral wall of the duodenum just posterior to and in close association with the rudiment of the liver. It may appear as a pocket-like evagination and soon becomes subdivided distally into a system of proliferating epithelial tubules. The original evagination becomes the duct of the ventral pancreas.

The dorsal pancreatic rudiment in amniotes is also a pocket-like evagination of the duodenum, but it appears on its dorsal side slightly in front of the liver rudiment.

In frogs, the dorsal pancreas develops from a section of the roof of the midgut, posterior to the rudiment of the stomach. Part of the roof becomes cut out by transverse crevices reaching from the cavity of the gut right into the submesodermal space above the gut. At the same time, that part of the gut wall which is destined to become the pancreas makes an abortive attempt to form a pocket-like evagination, but eventually the dorsal pancreatic rudiment in amphibians, as well as the ventral rudiment, is a solid mass which becomes transformed into a system of alveoli and ducts secondarily by rearrangement of cells.

The dorsal and ventral pancreatic rudiments may remain completely independent throughout life, as in the dogfish, but in amphibians and amniotes the two rudiments approach each other and fuse completely. The system of pancreatic ducts becomes reorganized in later life and is very variable in different vertebrates.

Bursa Fabricii. In birds a dorsal diverticulum of the cloaca becomes the bursa fabricii, which is the source of antibody-producing plasma cells. (See Burnet, 1962; Cooper and Lawton, 1974.)

14–5 DETERMINATION OF THE ENDODERMAL ORGANS

Experimental investigation of the endoderm in early stages of development shows that the endodermal organ rudiments, like those derived from other germinal layers, are not initially determined, but that the endodermal cells destined to participate in the formation of the various organ rudiments are no more determined for their respective fates than are the cells of the other germinal layers. In the earlier stages of development, their fate is a function of the position that each cell or group of cells occupies in the embryo as a whole. This can be proved by isolating parts of the presumptive endoderm and cultivating them apart from the rest of the embryo or by transplanting them into an abnormal position.

In an extensive series of experiments, pieces of presumptive endoderm of young gastrulae were cultivated in the "Holtfreter solution." Various tissues were observed to differentiate from such isolated pieces; some conformed to the normal destiny of the isolated parts, and some did not. The range of differentiations included not only orobranchial epithelium, stomach epithelium, liver, pancreas, and intestine, but also notochord and muscle, which should not have developed from the presumptive endoderm if it had kept its prospective significance (Holtfreter, 1938a, 1938b). Thus, the fate of the endoderm is not established finally in the early gastrula stage. Much greater deviations from the prospective significance of the various endodermal parts could be observed when these parts were placed in surroundings which, unlike the saline solution, could actively influence the differentiation of these parts.

In the early neurula stage, it is possible to separate the entire endoderm of a newt embryo from the ectoderm and mesoderm. The endoderm is removed as a whole through a slit on the ventral side of the embryo, leaving the ectoderm and mesoderm as an empty shell. The isolated endoderm can then be inserted again into the ectomesodermal shell of the same embryo or of another embryo of the same species, or even into the ectomesodermal shell of an embryo of another species. The endoderm of the small *Triturus taeniatus* has been successfully implanted into the ectomesodermal shell of the larger *Triturus alpestris* (Mangold, 1949).

The implantation may be carried out so that the orientation of the endoderm is in harmony with the orientation of the ectomesoderm, or the endoderm may be implanted in an inverted position. In the first case, a completely normal larva has been observed to develop. A normal embryo also developed if the endoderm was implanted with its dorsoventral orientation reversed. This result shows that the determination of the dorsal and ventral parts in the endoderm is not fixed in the endoderm itself but is imposed on the endoderm by the surrounding ectomesoderm. Here we may recall that in the experiments involving transplantation of the dorsal lip of the blastopore (primary organizer) the endoderm was often observed to develop a secondary lumen of the midgut, just underneath the notochord developed from the transplanted organizer. This secondary lumen was, of course, part of the dorsal differentiation of the endoderm. However, if the anteroposterior axis of the endoderm was inverted with respect to the axis of the ectomesodermal shell, the development was highly abnormal, thus showing that the differentiation of the endoderm along the anteroposterior axis cannot be dominated by the ectomesoderm.

In another set of experiments, small pieces of endoderm taken from a late gastrula or early neurula stage were implanted in various positions into another embryo. When the pieces of endoderm were taken from embryos in the gastrula stage, the grafts were often smoothly incorporated into the endoderm of the host. The use of heteroplastic transplantation made it possible to distinguish the grafted cells from the host cells (by differences in cell size in grafts between *Triturus taeniatus* and *Ambystoma mexicanum*) and thus to make sure that the graft was not destroyed but had fitted into the construction of local tissues. Thus, presumptive orobranchial endoderm was found to be able to develop into intestinal epithelium and vice versa. Stomach epithelium was developed from endoderm having a different prospective significance. Occasionally, however, grafts differentiated out of harmony with their surroundings, and the later the stage of the embryo from which the graft was taken, the oftener this occurred. After the end of neurulation the grafts differentiated, in the main, according to their prospective significance (Balinsky, 1948).

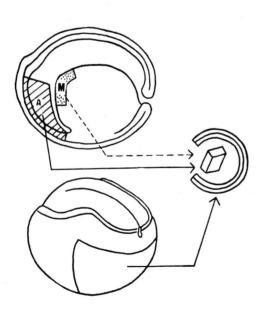

Figure 396. Testing the determination of parts of the neurula endoderm in the newt *Triturus pyrrhogaster* (diagram of operation). A, Anterior endoderm; M, middle endoderm. (From Okada, 1955.)

Similar results were obtained when different parts of the neurula endoderm were transplanted into parts of the ectomesodermal shell, either from the anterior half of the neurula or from the posterior half (Fig. 396; Okada, 1955a, 1955b). The endoderm taken for this experiment was either part of the foregut endoderm, mainly destined to become pharynx, or endoderm from the midgut, normally differentiating as stomach and intestine. It was found that midgut endoderm grafted into the anterior ectomesoderm produced pharynx (in addition to other parts). Foregut endoderm surrounded by posterior ectomesoderm was in part differentiated as intestine. In both cases, endoderm produced parts which were not in accord with the prospective significance of the endodermal cells, and it seems plausible that these differentiations were induced by the adjoining mesoderm. Again we find that the endoderm, as well as the ectoderm, is dependent on the mesoderm in its differentiation.

There is, as yet, very little information concerning the earliest determination of endodermal organs in higher vertebrates. In birds, some information has been derived from experiments in which parts of the chick blastoderm were grafted to the chorioallantoic membrane of another chick embryo. Various endodermal organs were observed to differentiate from the grafts, namely, pharyngeal epithelium, thyroid, lung, liver, and large and small intestine. However, these tissues developed without a very definite relationship to the origin of the grafts.

It has been concluded (Rudnick and Rawles, 1937; Rudnick, 1952) that in itself the endoderm has a very low power of differentiation. Liver and thyroid usually differentiate in explants which also show the presence of the heart (Willier and Rawles, 1931), and intestine is accompanied by mesoderm forming coelomic spaces (Rudnick and Rawles, 1937). This broadly corresponds with what has been found in the amphibian embryo. The epiblast alone, without the hypoblast, when cultivated on the chorioallantois, produces various endodermal tissues, such as thyroid, liver, pancreas, and intestine, almost in the same way as a whole blastoderm consisting of both epiblast and hypoblast. This is in agreement with the origin of definitive endoderm from the epiblast (p. 186).

Only when the endodermal gut becomes separated from the yolk sac during the second day of incubation does the differentiation of endodermal explants correspond to their prospective significance.

Earlier work on mammalian embryos, employing the method of explantation of parts of the blastoderm, did not yield very clear results. To date, there is no overall picture of the early determination of the endoderm as a whole in mammalian embryos. Some more recent experiments are concerned with the development of particular parts of the alimentary tract. In mouse and rat embryos the determination of the pancreas was studied by the method of explantation in a culture medium (Wessells, 1968; Wessells and Rutter, 1969). Normally the (dorsal) pancreas appears as an evagination on the ninth day of gestation when the gut is already closed into a tube and has become segregated from the yolk sac. A piece of gut, including the presumptive material of the pancreas, if explanted on the eighth day, will differentiate, producing pancreating tissue, provided that the adjoining mesodermal tissue is explanted together with the endodermal epithelium. Pieces of gut from younger embryos, in their seventh day of gestation (embryos with five pairs of somites—compare Fig. 216D), cannot develop pancreatic tissue, although they produce liver, lung, stomach, and intestinal tissues (Wessells, 1973). Presumptive pancreatic endoderm explanted without mesoderm fails to differentiate, however, even if taken from 11-day embryos in which the pancreas had already attained the stage of a pouchlike evagination. The mesoderm is thus necessary to promote the differentiation of pancreatic tissue. (Compare with results obtained in amphibians and birds!)

Part Six

DIFFERENTIATION AND GROWTH

Chapter 15

GENERAL CONSIDERATIONS ON GROWTH AND DIFFERENTIATION

15—1 DEFINITIONS

With the formation of all or most organ rudiments of the embryo, the main features of the organization of the animal are already indicated. What may be called the **morphological plan** of the animal is established. By morphological plan we mean the kind and number of organs, their relative positions, and the general features of each organ's structure. However, organ rudiments at this stage are not capable of performing their specific functions, on which depends the ability of the animal to lead an independent existence. The cells of the organ rudiments lack the peculiar structures that are necessary for specific functions; the organ rudiments are usually too small, and the animal as a whole is likewise far from the adult size. All the developmental processes dealt with so far may be grouped together as the **prefunctional** stages of development. Next, a new phase of development sets in, which brings the animal to its **functional** state. The main processes involved are **growth** and **differentiation.** Some new organs may appear in late stages of development, especially in animals passing through a larval stage, and minor morphological adjustments may occur in the organs formed earlier, but the processes of growth and differentiation are predominant.

Growth is the increase in size of an organism or of its parts due to synthesis of **protoplasm** or of **apoplasmatic substances.** Protoplasm in this definition includes both the cytoplasm and the nucleus of cells. Apoplasmatic substances are the substances which are produced by cells and which form a constituent part of the tissues of the organism, such as the fibers of connective tissue or the matrix of bone and cartilage, as opposed to substances produced by the cells and subsequently removed from the organism, such as the secretions of digestive and skin glands, or substances stored as food, such as fat droplets in cells of the adipose tissue. Imbibition of water or taking food into the alimentary canal before the food is digested and incorporated into the tissues of the animal, although they may increase the weight of the animal, do not constitute growth.

Growth is the result of a preponderance of the anabolic (synthetic) over the catabolic (destructive) processes in the organism. If synthesis and decomposition go on at the same rate, there is no increase in the bulk of the organism—no growth. Under certain conditions, decomposition dominates over synthesis, as for instance in prolonged inanition, when synthetic processes are impossible because of a lack of food supply, while catabolic processes (oxidations, etc.) continue to satisfy the current requirements for energy. After the internal food reserves (fat in the adipose tissue) have been exhausted, energy is produced at the expense of the proteins of the protoplasm, and the result is a decrease in the mass of living matter which may be called **degrowth** (Needham, 1942).

Differentiation is a term that is somewhat ambiguous, as it is used at least in two senses: a broader, more general sense and a narrower sense. In the broad sense, differentiation is the process in which the cells or other parts of an organism become different from one another and also different from their previous condition. For example, while the neural plate as a whole is being induced by the roof of the archenteron, its cells become different from both the presumptive epidermis and the parts of blastoderm which gave rise to the neural plate. In this broad sense, almost the whole of development may be said to be essentially a process of differentiation, and if an understanding is gained of how cells of the embryo become different from one another and from their original condition, then such an understanding would be equivalent to understanding development, or at least a very essential part of it.

At the same time, the term differentiation is also used in a narrower sense, in the sense of **histological differentiation.** Histological differentiation is the process as a result of which the parts of the organism acquire the ability to perform their special functions. In the case of multicellular animals, the parts in question are the cells and groups of cells.

The special functions of cells in this definition are distinguished from the basic functions of life which are common to all living cells. Every cell is capable of performing the processes of metabolism (respiration, synthesis, and so on), possesses to a certain degree the ability for ameboid movement, shows irritability, and is able to react to external stimuli. These functions are found in both undifferentiated and differentiated cells. Differentiated cells, however, are able to perform special functions or to perform them in such a way that other cells cannot. Thus, the nerve cells are capable of conducting nervous impulses to great distances and at a high speed. The liver cells secrete bile (besides their other functions). The melanophores produce granules of pigment in their cytoplasm. These are the special functions of nerve cells, hepatic cells, and melanophores.

The neural plate of the early embryo, although differentiated in a general way, having become different from other parts of the embryo, is not differentiated histologically, since its cells are not yet capable of functioning as nerve cells.

The ability to perform special functions is dependent on the existence of specific mechanisms in the differentiated cells. These mechanisms are sometimes visible in the form of organoids of the cell, such as the myofibrils of muscle cells, the cilia of epithelial cells of the trachea, and the long processes of nerve cells. These tangible morphological properties of the cells are also called **differentiations,** a practice that is legitimate as they are actually the visible expression of the process of differentiation. In other tissues, the differentiation becomes visible not so much as a change in the structure of the cell itself but as the result of the production by the cells of intercellular structures, such as fibers in the connective tissues, matrix of cartilage and bone, and cuticle on the outer surface of the skin in invertebrate animals. These extracellular parts are called differentiations in the same sense as the organoids of the cells. The special function of a cell may be the secretion of some substance that does not remain in the tissue as a permanent part but is removed or dissolved in the surrounding medium. In this case, granules of secretion may sometimes be seen in histological preparations as a morphological expression of the function of the cell.

The functional mechanisms of histologically differentiated cells are cytoplasmic. The building up of these mechanisms therefore causes a shift in the relative volume of the nucleus and the cytoplasm. As previously stated, the nuclei of the cells are their conservative part; they do not change essentially with the onset of development (Section 6–5). It is believed that even in the case of fully differentiated cells, the nuclei in all the various tissues of the animal's body have the same chromosomes and genes. In differentiated cells, the mass of cytoplasm increases while the nuclei do not increase or do not increase in the same proportion. The ratio—mass of cytoplasm to mass of nucleus—increases with differentiation. This ratio gives a rough quantitative estimate of the degree of differentiation. The estimate may be made even more exact if chemical substances and not morphological parts are taken into consideration. The basic substances of the nucleus are the chromosomes, with deoxyribonucleic acid as their essential component. The cytoplasmic structures are composed of proteins. Furthermore, the enzymes, which are the essential part of the functional mechanism of a cell, are of a protein nature. It is thus possible to substitute the amount of deoxyribonucleic acid for the basic structure of the cell and the amount of protein for its changing functional mechanisms. Direct measurements show that the relation, protein/deoxyribonucleic acid, changes with differentiation in agreement with expectation (Davidson and Leslie, 1950).

15–2   MECHANISMS OF CELL REPRODUCTION*

To clarify the relationship between growth and differentiation it will be useful to consider the simplest case of cells growing in a constant medium, such as in a medium used for tissue cultures. Under such conditions animal cells increase in size and divide at regular intervals. Cell divisions have been observed to occur in a culture at intervals of 10 to 30 hours. With a regular supply of fresh culture medium a strain of cells may be kept growing and proliferating indefinitely. In the famous experiment performed by Carrel and his collaborators, a strain of fibroblasts was kept growing for 34 years. Under these conditions it may be assumed that between two mitotic divisions the cells grow and attain a state, both qualitatively and quantitatively, which the cells had at the time immediately preceding the previous mitosis. We can distinguish several aspects of this process.

1. The chromosomes must be brought to the same state that they were in in the mother cell.
2. Proteins in the nucleus and the cytoplasm must be doubled, so that their amount would equal that in the mother cell.
3. Other cell components must be replenished accordingly.

The three aspects are satisfied by entirely different mechanisms which have to be considered separately.

The necessity of (1) follows from the postulate that all cells during development are supplied with a full complement of hereditary information; therefore, they must possess at least one set of double-stranded DNA molecules. We may thus expect that the amount of chromosomal DNA per cell will remain constant in spite of repeated mitotic divisions. This can actually be proved by direct measurement. The amount of this acid per cell may be estimated either chemically or spectroscopically (by ultraviolet light absorption due to the presence of nucleic acid). It was found that in any species of animal all somatic cells having a diploid set of chromosomes possess the same amount of deoxyribonucleic acid. Table 11 shows the amounts of deoxyribonucleic acid per cell

*For a review see Mazia, 1974.

found in different vertebrates. The average quantities of deoxyribonucleic acid per cell found in different somatic tissues of the same animal are the same. The figures apply to the nuclei in the resting stage.

Between two mitotic divisions the amount of deoxyribonucleic acid is doubled, as the daughter cells show the same amount of the substance generation after generation (Swift, 1950). By using radioactively labeled DNA precursors, the part of the cell's life cycle during which the DNA synthesis occurs has been ascertained with a high degree of exactness.

It is believed that molecules of DNA isolated *in vitro* and provided with precursors and the enzyme DNA polymerase can replicate very rapidly, perhaps in a matter of a few minutes or less. The replication of DNA in intact chromosomes, however, takes quite a considerable time. This is because different parts of the chromosomes replicate asynchronously, groups of genes or possibly even individual genes replicating at different times (Taylor, 1960a).

As a result, the duplication of chromosomes takes a considerable part of the life cycle of a cell. This part of the cell's life is called phase S, and it is found to occur toward the second part of the interphase. Immediately after mitosis, which may be called phase M, the nucleus enters a phase in which it does not incorporate DNA precursors and thus does not make new DNA. This is phase G_1. The duration of this phase depends on the length of the whole cycle, but it continues in any case for several hours. Next follows phase S, during which all the DNA in the chromosomes is replicated. The duration of this phase is again variable; times of from six to eight hours have been recorded in different kinds of cells. After the completion of DNA replication, the cells enter a second period during which no new DNA is made and which is designated as phase G_2. This phase continues for another two to six hours and is terminated by the onset of the next mitosis (Taylor, 1960a; Kuyper, Smets, and Pieck, 1962; Prescott and Bender, 1963).

Figure 397 shows the approximate length of the phases during a cell's life cycle in several kinds of cells grown in tissue culture.

The way in which the chromosomes become duplicated has been directly deduced

TABLE 11   Amounts of Deoxyribonucleic Acid in Single Nuclei in Various Animals (in mg $\times 10^{-9}$)

Bull	6.4
Pig	5.1
Guinea pig	5.9
Dog	5.3
Man	6.0
Rabbit	5.3
Horse	5.8
Sheep	5.7
Mouse	5.0
Duck	2.2
Fowl	2.4
Turtle	5.1
Toad	7.3
Frog	15.7
Carp	3.3
Trout	5.8

Data from Vendrely and Vendrely, 1949; Mirsky and Ris, 1949.

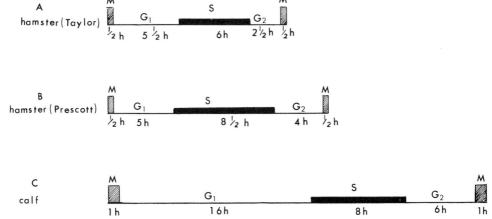

Figure 397. The length of different phases during the life cycle of cells grown in tissue culture. *A*, Embryonic Chinese hamster cells; *B*, embryonic Chinese hamster cells; *C*, fetal calf liver cells. M, Mitosis; G_1 and G_2, phases in which there is no synthesis of DNA; S, phase of DNA synthesis (duplication of DNA strands). (*A*, From Taylor, 1960a, 1960b. *B*, From Prescott and Bender, 1963. *C*, From Kuyper, Smets, and Pieck, 1962.)

from the double helix structure of DNA. It is assumed that during the S phase the two component chains of a double DNA molecule separate and that each then serves as a template for the construction of a new polynucleotide chain. The new polynucleotide chains are built from a pool of free mononucleotides available in the cell in such a way that each of the mononucleotides in the old chain picks out a complementary mononucleotide from the pool. Where an adenylic acid mononucleotide is present in the old chain, a thymidylic acid mononucleotide becomes attached to it. A thymidylic acid mononucleotide in the old chain attracts an adenylic acid mononucleotide, and the guanylic acid and cytidylic acid nucleotides of the old chain become associated respectively with cytidylic acid and guanylic acid mononucleotides from the pool. It is obvious that in this way each half of the original DNA molecule can supplement itself with a new half which is in every respect identical to the alternate half of the original molecule (Fig. 398).

The result is the formation of two double-intertwined DNA molecules, each similar to the original one. It should be assumed that at least one replication of this kind occurs between each two mitotic divisions of a cell, with the result that at the beginning of a mitotic division each chromosome is composed of two chromatids, each chromatid consisting of a double-stranded molecule of DNA. In anaphase the two chromatids separate, and each becomes a daughter chromosome which is then drawn into one or the other daughter cell. In this way every peculiarity of each DNA molecule available in the mother cell becomes transmitted to both daughter cells. There is experimental evidence to show that replication of DNA molecules does, in fact, proceed as just described.

In one experiment, cells derived originally from hamster embryos were grown as a tissue culture *in vitro* (Prescott and Bender, 1963). A nucleic acid precursor, thymidine, labeled with tritium (radioactive hydrogen), was added to the culture. The radioactive thymidine was taken up by the cells and was used in the replication of the chromosomal DNA. After 30 minutes, the cultures were washed with fresh medium containing only unlabeled thymidine. They were then left in the "cold" medium for further growth.

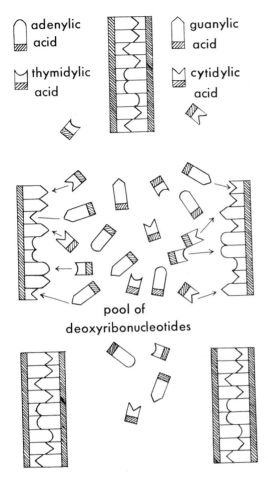

adenylic acid

guanylic acid

thymidylic acid

cytidylic acid

pool of
deoxyribonucleotides

Figure 398. Diagram showing the mode of duplication in a section of a DNA molecule. The two strands of the molecule separate, and each picks up a set of complementary mononucleotides from a deoxyribonucleotide pool. The cross-hatched stripes represent the phosphate-deoxyribose "backbones" of the molecules; the differently shaped figures represent the bases—adenine, thymine, guanine, and cytosine. (System of representation of DNA molecule components based on the paper of M. B. Hoagland, 1959.)

The fate of the radioactive thymidine could be traced when the treated cells entered the metaphase stage after varying lengths of time. When the treated cells were placed in contact with photographic film, the dots in the emulsion appeared along the whole length of the treated chromosomes, if the test was performed within the period of one cell generation after the original treatment of the cells with labeled DNA precursor (Fig. 399).

If, however, the cells, after being transferred into a "cold" medium, were allowed to remain there for over one cell generation, the results of a test for incorporated tritium were essentially different. In this case, black dots in the emulsion appeared over only one chromatid in each chromosome, while the other chromatid did not cause the appearance of dots in the emulsion, thus showing that it did not contain any radioactive hydrogen. After another cell generation (at the third mitosis), about half of the chromosomes were clear of radioactive material, while the other half of the chromosomes had one chromatid labeled and the other unlabeled.

These results can easily be understood. In the first replication, the new halves of the DNA double molecules were built at the expense of a pool of mononucleotides containing radioactive thymidine. The old half in each case (the one serving as template) did not include any radioactive material, but this could not be seen as it was closely intertwined with the new half (Fig. 400). Both chromatids of each chromosome showed the same labeling. In the next generation of cells the two halves of each double molecule separated again, and each served as a template for the formation of another polynucleotide chain. As the pool of nucleotides by this time did not contain labeled molecules, the new chains were free of radioactive materials. In this way the original unlabeled DNA chain, supplemented by a new nonradioactive complementary chain, showed no radioactivity. On the other hand, the chain built at the time when the mononucleotide pool contained radioactively labeled precursors still showed radioactivity, although its complementary partner, built at a later stage, did not contain radioactive hydrogen.

In the next generation the nonradioactive chromatid became a separate chromosome, consisting in its turn of two chromatids, which were both free of radioactivity. The chromatid containing a readioactive strand in its DNA passed it on to one of the daughter chromatids. Similar results were obtained with a culture of human leukocytes (Prescott and Bender, 1963) and also with plant cells (Taylor, Woods, and Hughes, 1957).

To the question of why the DNA double-stranded molecules separate and duplicate themselves at a particular moment during phase S in the life cycle of a cell, there is practically no answer, although many factors are known which stimulate growth and division of cells in general. None of the conditions favoring growth and division can be related directly to the duplication of DNA as the causative agent of such a duplication.

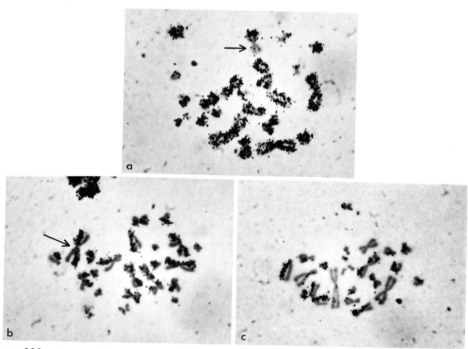

Figure 399. Results of experiment showing the mode of duplication of chromosomes; autoradiographs of chromosome sets freed from cytoplasm. *a*, A complete set of chromosomes at the first metaphase following DNA labeling. All chromatids are labeled equally except one arm of the X chromosome (arrow) which synthesizes DNA out of step with the other chromosomes. *b*, Metaphase chromosomes of second division after labeling; in each pair only one of the chromatids is labeled. At the arrow the two chromatids of a long chromosome are twisted across one another. *c*, Metaphase chromosomes of third division after labeling. Roughly half of the chromosomes are completely free of labeling; the other half have one chromosome labeled and one unlabeled. (Courtesy of Dr. D. M. Prescott, from Prescott and Bender, 1963.)

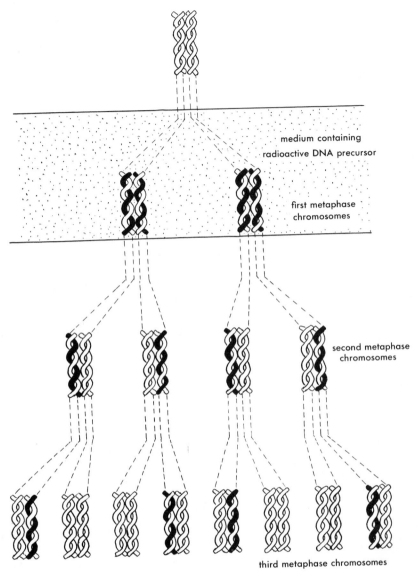

Figure 400. Diagram showing mode of duplication of chromosome DNA strands as revealed by experiments incorporating radioactive precursors into DNA. Chromosomes shown as consisting of two chromatids, each constituted of a double-stranded DNA helix. The stippled zone represents the sojourn of the cells in a medium containing a radioactive DNA precursor. As the chromosomes duplicate in this medium, the newly synthesized strands become labeled (black). During the subsequent two replications, the labeled strands are transmitted intact to some of the chromatids. A chromatid in which at least one strand is labeled registers as radioactive in an autoradiograph.

The formation of new, complementary strands of DNA requires the presence of an enzyme, DNA polymerase. The latter, being a protein, falls under the next section.

With regard to the second aspect (2), the synthesis of proteins in the growing cell, including all the proteins with enzymic action, requires the following conditions in order to proceed: messenger RNA, transfer RNA, ribosomes, and precursors in the form of a full complement of amino acids. It is thus evident that rapid growth requires considerable amounts of RNA to be present in the cytoplasm. The amount of ribonucleic acid per cell is by no means constant; the ratio of ribonucleic acid to deoxyribonucleic acid varies in different kinds of cells from 0.2 in the thymus to as much as 8.1 in the pancreas (Davidson, 1947). The observation has been made (Brachet, 1941) that the ribonucleic acid content is highest in those cells which are especially active in synthesizing proteins. Such are some of the secretory cells (pancreas), and also the actively growing cells. Cells that cease growing lose their ribonucleic acid. This has been directly demonstrated in the development of some tissues, in which growth and reproduction of mother cells ("stem cells") precedes the specific histological differentiation of the cells produced in this way. The differentiated cells lose the ability to grow and divide, as in the case of red blood corpuscles or of the permanently growing teeth of rodents, in which the dentine is perpetually produced at the expense of a group of reproducing cells at the root of the tooth. In the stem cells, from which the red blood corpuscles are eventually produced, the content of ribonucleic acid is 5 per cent. As the differentiation progresses, the percentage of ribonucleic acid drops almost to zero (Thorell, 1947).

On the other hand, if cells are stimulated to more intensified growth and reproduction, the percentage of ribonucleic acid increases. This happens when differentiated cells are explanted *in vitro* and allowed to grow as a tissue culture. Measurements of the ribonucleic acid have been made on explants of chick heart tissue, and the amount of ribonucleic acid was compared with the amount of deoxyribonucleic acid, which as we know remains constant per cell. The measurements showed that at explantation the amount of ribonucleic acid was about double that of deoxyribonucleic acid, but it rose to about five times the amount of deoxyribonucleic acid after four days of cultivation *in vitro* (Davidson and Leslie, 1950).

All these earlier data refer mainly to the ribosomal RNA. There are very few data on the amounts of messenger RNA and transfer RNA in different types of cells. By using radioactive tracers, it has been shown that there is no synthesis of RNA in the chromosomes while they are in the strongly spiralized and contracted form, that is, from the later prophase to middle telophase. RNA synthesis is resumed in late telophase, when the chromosomes become despiralized, continues through the interphase, and becomes gradually diminished and eventually ceases during the prophase (Taylor, 1069b; Prescott and Bender, 1962).

Table 12 shows the incorporation of H^3-labeled uridine into RNA and also the incorporation of H^3-labeled histidine into cellular proteins during interphase and mitosis of cells grown in a tissue culture.

TABLE 12   Incorporation of Tritiated Uridine into RNA after a 5-Minute Exposure and Incorporation of Tritiated Histidine into Proteins after a 10-Minute Exposure of Chinese Hamster Cells Grown *in Vitro*. (Figures Show Average Counts of Silver Grains in Photographic Emulsion over the Cells.)

Stage, Site	*H^3 Uridine*	*H^3 Histidine*
Interphase nucleus	39 ± 2	45 ± 2
Interphase cytoplasm	0	—
Early prophase, nucleus and cytoplasm	29 ± 2	40 ± 2
Late prophase, nucleus and cytoplasm	0	25 ± 1
Metaphase, nucleus and cytoplasm	0	15 ± 1
Anaphase, nucleus and cytoplasm	0	11 ± 1
Early telophase, nucleus and cytoplasm	0	10 ± 1
Late telophase, nucleus and cytoplasm	4 ± 1	17 ± 1

From Prescott and Bender, 1962.

RNA precursors are initially incorporated into the nucleus only, showing that synthesis of RNA is carried out in the nucleus. After some time, the newly synthesized RNA passes out into the cytoplasm. The passage of the RNA into the cytoplasm is fairly slow and takes 20 to 60 minutes. However, toward the end of prophase there is a more rapid release of RNA from the nucleus, and most of the remaining nuclear RNA passes into the cytoplasm in a short time.

RNA synthesis does not stop during the S phase when chromosomal DNA is being replicated, though there are indications that it may be retarded and that, at the points where DNA is replicating, no RNA synthesis takes place (Prescott and Kimball, 1961). The synthesis of RNA continues probably because at any time during the S phase only part of the genes in the chromosomes are actually in the process of replication (Mazia, 1974); the rest are available for use as templates for the RNA. RNA synthesis is, however, completely blocked when the chromosomes are in a condensed state. Although synthesis of RNA is blocked during mitosis, synthesis of protein goes on though at a diminished rate, as shown by the continued incorporation of labeled histidine. This protein synthesis is due to the survival in the cytoplasm of messenger RNA produced during the preceding interphase.

With regard to (3), the synthesis of nonprotein components in the cell is presumably not directly dependent on the nuclear DNA but is regulated by the enzymes which have been produced on the messenger RNA templates in cooperation with the ribosomes. The proteins do not have a very rapid turnover; therefore, a cell emerging from a mitosis would have supplies of enzymes inherited from the mother cell and capable of carrying out synthetic processes in a continuous fashion. Among the nonprotein cell constituents which are synthesized in the cytoplasm are the nucleic acid precursors—the phosphoribonucleotides and also the nonessential amino acids. (The essential amino acids, that is, those which cannot be synthesized by animal cells, have to be provided in ready-made form from the environment.)

15–3 RELATION OF CELL PROLIFERATION TO DIFFERENTIATION

With certain restrictions arising from sexual processes and environmental factors, the reproduction of unicellular organisms proceeds similarly to the growth and reproduction of clones of metazoan cells in tissue culture; that is, the daughter cells eventually become similar to the parent cell. In whole multicellular animals, however, the cells do not remain the same for prolonged periods of cell division and growth but eventually become different, both from the initial state and from each other (Holtzer, 1963). This is essentially the process of differentiation, as defined on the previous pages. In a general way, the whole of ontogenetic development may be considered in this light, starting, as it does, with cell divisions (cleavage) and going on to the formation of germinal layers and organ rudiments.

In organ rudiments it very often happens that the initial group of cells laid down for an organ or part of the embryo undergoes a period of proliferation by mitotic division. After a while the divisions cease, and the cells of the organ rudiment begin to differentiate. This happens, for instance, in the development of the nervous system of vertebrates (as described on p. 335), in the development of the pancreas (Wessells and Rutter, 1969), and in the development of many other organ rudiments. In many cases only part of the cells of an organ rudiment become differentiated, while a residue of undifferentiated cells retains the ability to grow and reproduce by mitosis, thus providing a supply of new cells that can join in the process of differentiation. In the vertebrate epithelium the generative layer presents such a residue of undifferentiated cells. In the case of bone, the differentiated cells (osteocytes) do not grow and divide, but the cells of the periosteum continue to divide and provide for additional layers of bone. In the intestine, reserves of undifferentiated dividing cells are present at the ends of the intestinal crypts. Lastly, the blood corpuscles, in particular the erythrocytes, are differentiated nondividing cells, and their supply is continuously replenished at the expense of the proliferating nondifferentiated cells in the erythropoietic (and granulopoietic) tissues.

The transition from the undifferentiated and proliferating cells to nonproliferating and differentiating cells may be quite abrupt, occurring in a matter of hours, but there are certain complications of this transition which must be kept in mind. The complications may be classed under three headings:

1. In some tissues, cells, after obviously having entered the phase of differentiation, continue to divide by mitosis, at least for some time. An example of this is the cells of hyaline cartilage that may divide several times after being enclosed in cartilage matrix, which they produced as a visible sign of differentiation. Similarly, fibroblasts of the connective tissue are supposed to possess the ability to proliferate, although they are already surrounded by the connective tissue fibers that are the product of their secretory activity. (See Holtzer et al., 1972.)

2. Some differentiated cells, although having lost the ability to divide by mitosis, retain the capacity for growth (increase in size). An excellent example of this is nerve cells, which never divide but, after divisions stop, increase in size to a very considerable degree, becoming the largest cells in the vertebrate body, except for the oocytes. In some invertebrates, notably nematodes, cell divisions, apart from those in the gonads, are restricted to early stages of development, while a very considerable growth occurs later by an increase in size of practically every cell of the body.

3. A special case of growth, after mitosis has ceased, is presented by cells becoming polyploid in certain tissues; the genome becomes duplicated once or several times without the cells dividing. This is a fairly common occurrence in insects and leads to the formation of giant cells, which sometimes contain giant chromosomes, produced by multiple duplication of the normal double DNA strand (as in salivary gland cells of the fruitfly and other Diptera). In vertebrates some liver cells contain a double amount of DNA and thus are probably tetraploid, and the megakaryocytes of the bone marrow, producing blood platelets, are believed to be polyploid. Whether these cells possess multiple sets of chromosomes or whether the chromosomes become many-stranded is not known, as the cells do not divide and their chromosomes are not visible in the interphase state.

Chapter 16

DIFFERENTIATION

16-1 THE CHEMICAL BASIS OF DIFFERENTIATION

Whatever the type of differentiation of the cell, it is doubtless based on the chemical constitution of the cell. In every case in which the function of the cells consists in the elaboration of some substances, whether in the form of structural elements or in the form of secretions, there must be a specific enzyme or specific combination of enzymes responsible for the reaction. As the substances produced by cells are very diverse, a correspondingly varied assortment of enzymes may be postulated as being present in different tissue cells. It may therefore be said that "differentiation is the production of unique enzymatic patterns" (Spiegelman, 1948; see also Boell, 1948, 1955).

All enzymes are known to be proteins, so that the last sentence may be paraphrased to read: "differentiation is the production of unique protein patterns." This wording is a more general one, as it includes not only proteins having enzymatic properties but also the proteins which are not enzymes. An important group of the latter kind are the structural proteins, which are particularly prominent in some differentiations of cells and tissues. Some of the structural proteins are permanently intracellular, such as the keratin produced in the epidermis cells of terrestrial vertebrates. Other structural proteins, although produced by the cells, are eventually extruded from the cells and accumulate in the intercellular spaces. Collagen is the best known of such proteins.

Other organic substances, such as lipids and carbohydrates, alone or in conjunction with proteins, play a very important role in the life of cells, but between these substances and proteins there is a significant difference. The synthesis of carbohydrates, lipids, and also smaller molecules, such as organic acids (including amino acids), occurs under the control of enzymes and enzyme systems, which are of a protein nature. Given a certain pattern of proteins in a cell, the other components are then determined by the enzymatic activities of the proteins (provided, of course, that the necessary precursors are present and the environmental conditions are of a specified nature).

The proteins of the cell, however, are not produced by the action of other proteins, but are synthesized according to a code contained in the DNA of the chromosomes, which is carried to the sites of protein synthesis by messenger RNA. The molecules of messenger RNA are molded on the DNA molecules much in the same way that DNA replicates itself. The two strands of DNA become separated from each other, and then instead of picking up deoxyribonucleic acid mononucleotides, one of the DNA strands (Roth, 1964) serves as a template for the fixation of ribonucleic acid mononucleotides. The assembly of these mononucleotides into a continuous chain molecule is assisted by an enzyme, RNA polymerase. In the synthesis of the ribonucleic acid molecule the complementarity of the bases is made use of, just as in the replication of DNA, except that uracil, not thymine, pairs with adenine. In this way the molecule of RNA, being complementary to a DNA molecule, comes to contain the complete genetic information of the latter (Fig. 401).

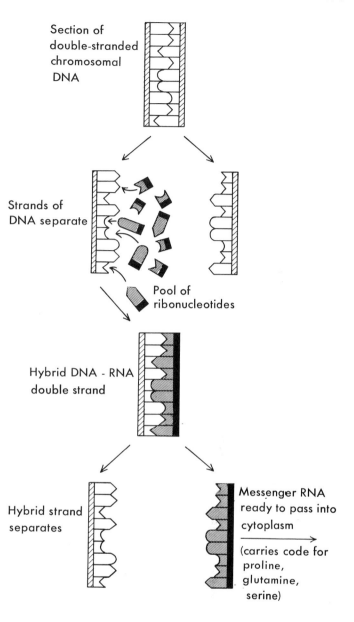

Section of
double-stranded
chromosomal
DNA

Strands of
DNA separate

Pool of
ribonucleotides

Hybrid DNA - RNA
double strand

Hybrid strand
separates

Messenger RNA
ready to pass into
cytoplasm

(carries code for
proline,
glutamine,
serine)

Figure 401. Diagram showing the formation of messenger RNA on a section of a DNA molecule. The two strands of the DNA molecule separate, and then complementary RNA mononucleotides are picked up from the RNA mononucleotide pool and assembled on a single-stranded DNA molecule. After the assembly is complete, the RNA strand separates from the DNA strand. The RNA components have been darkened. Nucleotides represented by the same symbols as in Figure 398.

The RNA molecule then splits off from the DNA molecule and travels from the nucleus into the cytoplasm (probably making use of the pores in the nuclear membrane!), and in the cytoplasm it becomes associated with a ribosome or rather with a group of ribosomes. The messenger RNA molecule is then ready to direct the arrangement of amino acids in the formation of a particular protein molecule. The base triplets of the messenger RNA correspond in arrangement to the base triplets of the chromosomal DNA. The corresponding amino acids are placed opposite the base triplets of the messenger RNA and are joined together to produce a protein of a specific kind. From what has been said, it is evident that the "genetic code" in the DNA determines the sequence in which the amino acids are joined together to form a polypeptide chain. The functional properties of the proteins depend largely on the way in which the polypeptide chain is folded to form a three-dimensional structure, the parts of which are held together by cross-linkages. The folding of the polypeptide chain to produce the final structure of the protein molecule is thought to occur spontaneously, given the particular sequence of the amino acid residues in the chain.

Genes Determining Protein Structure: The Hemoglobins. The relationship between gene structure and protein structure has so far been best studied in the case of hemoglobins contained in the red corpuscles of human blood.

By applying methods of chemical analysis (stepwise fractionation and analysis of fractions) on the one hand, and of methods of x-ray diffraction (analysis of x-ray scattering by hemoglobin in crystalline form) on the other, it has been possible to get a complete picture of the organization of the molecules of human hemoglobin. The hemoglobin—or, to be more exact, hemoglobin A—which makes up about 95 per cent of the hemoglobin of normal human blood, has a molecular weight of about 65,000. The molecule is complex and consists of four subunits: there are two identical α subunits and two identical β subunits, lying in a tetrahedral arrangement. The whole molecule is roughly spherical (Fig. 402). Each of the subunits is a polypeptide chain, folded in a complicated three-dimensional figure, to which is attached a disc-shaped "heme" group carrying in its center an atom of iron, which serves for the binding of oxygen in oxyhemoglobin. The polypeptide chain consists of a number of amino acids linked by typical peptide bonds. The exact composition, that is, the sequence of amino acids, is now known for both kinds of subunits. The α subunit, apart from the heme group, consists of 141 amino acids; the β subunit consists of 146 amino acids. (For a list of the amino acids in each chain, see Perutz, 1961.)

In addition to the normal hemoglobin A, other kinds of hemoglobins also occur in man. Of particular interest in connection with gene action are the hemoglobins called hemoglobin S and hemoglobin C. Hemoglobin S has been found to occur in persons suffering from an inheritable condition known as sickle-cell anemia, a disease which

Figure 402. View of the complete hemoglobin molecule, reconstructed on the base of electron density distribution. The molecule consists of four subunits; the two on top are shown white and the two below, black. Two hem groups are shown as discs (on one of them is the mark O_2, indicating the oxygen combining site). (After Cullis et al., from Perutz, 1961.)

in some cases may be fatal. The red blood corpuscles of persons suffering from sickle-cell anemia have a shrunken, irregular shape often resembling a sickle—hence the name. The condition is inherited in a typical Mendelian fashion and is dependent on a recessive gene, which is allelic to one of the genes involved in determining the normal condition of the red blood corpuscles.

It was found by Pauling and his collaborators (1949) that the hemoglobin S contained in the blood of persons suffering from sickle-cell anemia differs from normal hemoglobin A. The difference is, however, only in one kind of subunit in the hemoglobin molecule, namely, in the β subunit which is changed from β^A to β^S, while the α subunit is identical with the subunit of normal individuals. Furthermore, it was found that persons heterozygous for the gene of sickle-cell anemia do not show the symptoms of this abnormality either in their physical state or in the shape of their red blood corpuscles, but their blood contains both hemoglobins A and S in roughly equal proportions. The presence of subunits β^A and β^S in the two kinds of hemoglobin indicates that in the heterozygous state the sickle-cell gene and its normal allelomorph produce polypeptide units independently of each other, and these then combine with the α polypeptide units which are not affected by the sickle-cell gene. This situation indicates a direct relationship between the gene and the protein (or polypeptide) chain which is synthesized under the gene's control, presumably through the separate emission of a specific messenger RNA by each gene, which then serves as a template for the assembly of the protein chain.

Another abnormal hemoglobin, hemoglobin C, does not cause any deficiencies in persons carrying the gene and can only be discovered by a physicochemical investigation of the blood. The action of the gene causes an alteration of the β component of hemoglobin, while the α component remains unchanged, as in sickle-cell disease. The inheritance of the condition closely resembles the inheritance of sickle-cell anemia; the gene is allelic to the gene responsible for the formation of component β of normal hemoglobin. In persons homozygous for gene Hb-C, most of the hemoglobin molecules have the composition $\alpha_2^A + \beta_2^C$; in heterozygous persons, roughly half* the molecules have this abnormal composition, and the other half have the normal structure $\alpha_2^A + \beta_2^A$. Again, the presence of both kinds of β chains in heterozygous persons shows that each gene acts independently in producing the specific kind of polypeptide molecule in the cell containing such a gene.

A further understanding of the close relation between the gene structure and the structure of the corresponding protein emerges from a study of the amino acid composition of the abnormal hemoglobins S and C. By comparing the amino acid sequence in the β chains of normal hemoglobin A and of the abnormal hemoglobins S and C, it was found that the latter two differ from normal hemoglobin only in one amino acid (Gerald, 1961).

In sickle-cell hemoglobin, the glutamic acid residue which is the sixth from the beginning of the β chain is replaced by a valine residue. In hemoglobin C the same glutamic acid residue is replaced by lysine. The section of the β chain in which the replacement occurs can be shown as follows:

Normal hemoglobin A, β chain: Val-His-Leu-Thr-Pro-*Glu*-Glu-Lys
hemoglobin S, β chain: Val-His-Leu-Thr-Pro-*Val*-Glu-Lys
hemoglobin C, β chain: Val-His-Leu-Thr-Pro-*Lys*-Glu-Lys

---

*Actually, the amount is somewhat less than half in the case of both hemoglobin C and hemoglobin S, thus suggesting that the rate of production of hemoglobin by different genes is not exactly the same.

According to the "genetic code" (p. 14):

The (RNA) code for glutamic acid is   GAA

The (RNA) code for valine   is GUA

The (RNA) code for lysine   is   AAA

It is evident that a change in one base only could account for the transformation of the normal hemoglobin into one or another of the alternative forms. In the case of sickle-cell hemoglobins, a uracil nucleotide could have been substituted for an adenine nucleotide; in the case of hemoglobin C, an adenine nucleotide could have been substituted for a guanine nucleotide.

The abnormal hemoglobins S and C most probably owe their origin to mutations. It thus becomes likely that "point" mutations (mutations of a single gene) may, at least in some cases, result from a change in only one base pair of DNA, which changes a code triplet in such a way that it determines a different amino acid from the one coded by the previously existing sequence of base pairs in the chromosomal DNA. As a result of a change in one code triplet, the resulting protein would have one amino acid residue in the chain replaced by a different one. Sometimes such a replacement might not seriously affect the functional properties of the protein concerned, but occasionally the change may be very considerable. We have seen that the substitution of only one amino acid residue in the β chain of human hemoglobin not only leads to a distinct morphological expression, in the shape of the red blood corpuscles, but also changes the metabolism of the whole organism. The change in one base pair in the chromosomes may become a matter of life or death for the carrier of the abnormal gene.

16–2 THE MESSENGER RNA IN METAZOA

The mechanism of protein synthesis and the roles of chromosomal DNA and messenger RNA in this process have been unravelled in experiments performed on bacteria and phages. It is therefore of some importance to know that the same mechanism is in operation in eukaryotes (organisms possessing nuclei in their cells) and multicellular animals in particular. The study of messenger RNA in higher organisms has proved to be a formidable task, perhaps in part because of the much greater complexity of these organisms, which involves tremendous numbers of different proteins and, as each protein has to be synthesized on its own special messenger RNA, of tremendous numbers of *different* messenger RNA's. Because larger protein molecules are modeled on a longer mRNA, the lengths of different mRNA's, and thus also their molecular weights, differ to a great extent. For this reason messenger RNA's fall into a class referred to often as *heterogeneous* RNA, as compared with the ribosomal RNA and transfer RNA whose molecular weights are more nearly constant.

The best chances of isolating and studying a specific mRNA are to be sought in cells and tissues that synthesize only a limited number of proteins or possibly only one kind of protein. Fortunately, some progress in this direction has already been achieved.

Messenger RNA has been isolated and purified which directs the synthesis of the silk fibroin, produced by silk glands of pupating caterpillars of the silkworm, *Bombyx mori* (Suzuki and Brown, 1972; Suzuki et al., 1972). Advantage was taken of two favorable circumstances. Firstly, in the posterior part of the salivary gland of the pupating caterpillar, the silk fibroin is practically the only protein produced. Secondly, silk fibroin is a very peculiar protein; most of it consists of repeating units made up predominantly of only a few amino acids. Glycine comprises 45 per cent of all residues, and it alternates predominantly with alanine and serine. If the RNA codons for these three amino acids are considered (GGX for glycine, GCY for alanine, and UCZ or AG^U_C for serine, where X, Y, and Z are third bases in the triplets which do not change the meaning of the code; see Table 1, p. 14), it can be calculated that the messenger RNA for a protein of this composition should have an unusually high proportion of guanine and cytosine nucleotides. In fact, 40 per cent of them should be guanine and 17 per cent cytosine. This is very different from the usual proportions of nucleotides in the same animal; in the DNA of the animal as a whole the proportion of G + C is 39 per cent, and in the ribosomal RNA it is 50 per cent.

After chemical purification of total RNA, a rapidly sedimenting fraction of RNA was isolated by ultracentrifugation which corresponded to expectation, having 40 per cent guanine nucleotides and 19 per cent cytosine nucleotides. Furthermore, this particular fraction of RNA is present only in the posterior part of the salivary gland of the silkworm and not in the middle silk gland, which does not produce silk fibroin, nor in the rest of the caterpillar's body. From this and from the high proportion of G + C, as explained previously, it could be concluded that the isolated RNA was, in fact, messenger RNA for silk fibroin.

Another type of cell which produces only a limited number of proteins at a time is the reticulocyte, which becomes a red blood corpuscle and in the process, synthesizes the two types of polypeptide chains (α and β chains) composing the hemoglobin molecule. This case differs from the previous one in that the ability of the mRNA to direct protein synthesis was actually tested in living cells. The fraction of RNA presumed to be hemoglobin messenger RNA was extracted from rabbit reticulocytes and separated from other kinds of RNA by centrifugation on a sucrose gradient. The preparation, designated as 9s RNA, was injected into frog *(Xenopus laevis)* oocytes. Oocytes in their growth synthesize large amounts of protein and thus can be expected to have available all the chemical apparatus for protein synthesis, including mRNA of their own. By supplying the cells with what was expected to be hemoglobin mRNA, it was possible to test whether this foreign mRNA, in combination with other necessary components supplied by the host cells (ribosomes, transfer RNA's, pool of amino acids, etc.), could cause the synthesis of hemoglobin. Any quantities of hemoglobin discovered in the injected oocytes would be due to the injected mRNA, as oocytes do not normally produce hemoglobin. In order for the oocytes to synthesize molecules of hemoglobin, and not only blood globin polypeptides, the system had to be supplied with ready-made heme groups, which, not being part of a polypeptide chain, are not directly coded for by mRNA. This requirement was taken care of by injecting some heme molecules in solution simultaneously with the reticulocyte 9s RNA. The results of the experiment fully justified expectations (Lane, Marbaix, and Gurdon, 1971; Moar et al., 1971). A substance could be extracted from the injected oocytes after three or more hours of incubation, which, tested in a variety of ways, short of determining the amino acid sequence, was found to be identical with rabbit hemoglobin. There appears to be very little doubt that the 9s RNA fraction which was tested was indeed, or contained, the messenger RNA for rabbit hemoglobin. Consequently, it is not possible to interpret the results of the experiment in the sense that the production of hemoglobin was due to some kind of activation of the host cell's hemoglobin genes. It is noteworthy that the efficiency of the synthesis in the frog oocytes was several hundred times greater than can be achieved *in vitro*, in a cell-free system (judged by the amount of protein per amount of RNA), and only several times less than the efficiency of the hemoglobin synthesis in the reticulocytes.

The frog oocytes were thus shown to be an excellent system for testing mRNA's, whether fully purified or in a crude form (Gurdon et al., 1971). Following the work of Gurdon's group, nuclear RNA from frog *(Xenopus)* neurulae and RNA extracted from ribosomes of young larvae of the same animal were injected into oocytes and caused synthesis of collagen in the oocytes. This was detected by supplying the oocytes with radioactive hydroxyproline, an amino acid which could only be incorporated into collagen. RNA from gastrulae did not cause collagen to be synthesized. Even though the mRNA for collagen had not been obtained in anything like a pure form, the conclusion may be drawn that mRNA for collagen starts being produced in the neurula stage and not earlier (Rollins and Flickinger, 1972).

It has been convincingly shown by a variety of methods that by no means all the chromosomal DNA is or can be transcribed to produce messenger RNA. In the first place, in addition to codes for mRNA, the chromosomal DNA contains the codes for the other kinds of RNA, ribosomal RNA and transfer RNA. These, however, do not make up a very large proportion of the chromosomal DNA. The situation in the oocyte, as outlined in Section 3–1 (p. 35), is a special case. In ordinary somatic cells, the DNA concerned with the production of the ribosomal RNA makes up no more than 0.15 to 0.2 per cent of the total chromosomal DNA. (See Brown and Dawid, 1968.) The proportion of DNA coding for the transfer RNA's would probably be of the same order of magnitude.

A surprising fact revealed by the detailed study of chromosomal DNA in eukaryotic organisms is that a very substantial part of it consists of repetitive sequences of nucleotides. Some sequences of nucleotides occur in different parts of the chromosomes and are repeated thousands of times. These sequences do not code for proteins, and the function of the RNA's transcribed from these sequences is unknown. The repetitive sequences constitute from 15 to 80 per cent of the total DNA (Britten and Davidson, 1969; Britten and Kohne, 1970).

A counterpart to these data is the fact that up to 80 per cent of the RNA manufactured in the nucleus never passes into the cytoplasm and is broken down (Shearer and McCarthy, 1967). At least part of this "rapid turnover nuclear RNA," however, is directly connected with the production of the RNA's which later pass into the cytoplasm.

The ribosomal RNA, as it is found in the cytoplasm of eukaryotes, consists of three kinds of molecules: 18s RNA, 28s RNA, and 5s RNA. The codes for the first two RNA's are contained in two genes lying close together in one of the repetitive units of the nuclear DNA. About 450 such units are contained in one of the chromosomes of the frog, *Xenopus laevis*. The two genes on the chromosome are separated by a short "spacer," and between each pair of genes there is a longer "spacer" sequence. The genes for the 5s RNA are completely independent of the 18s and 28s RNA genes, and they lie on different chromosomes (Brown, 1973). The genes for the 18s and 28s RNA's are transcribed together, resulting in a large molecule, which in addition to the ribosomal RNA's proper includes a transcript of the small "spacer" sequence. After separating from the DNA, this large (40s) molecule is broken up. About 20 per cent of the constitutent ribonucleotides are discarded, and the rest, that is, the 18s and the 28s RNA molecules, pass into the cytoplasm. The larger "spacer" sequence between pairs of genes is not transcribed at all.

In the case of transfer RNA, the section of DNA involved in its production is also larger than the eventually formed RNA. The transfer RNA for tyrosine consists of 85 nucleotides, but it is synthesized originally as a precursor consisting of 126 nucleotides. Forty-one nucleotides are split off from the precursor before the final functional molecule is ready. The gene which produces tyrosine transfer RNA, in addition to the sequence coding for the whole precursor molecule, also consists of an "initiator" sequence of unknown length and a "terminator" sequence of at least 24 nucleotides (work by Korana and associates, reported by Maugh, 1973).

Finally, the messenger RNA is likewise synthesized on the DNA template in the form of a precursor, which may be considerably longer than the section actually carrying the code for a specific protein. Exactly how much longer the precursor is is not known for certain at present and is probably not the same in different mRNA's. These precursors constitute an important part of the intranuclear heterogeneous RNA — HnRNA for short. It is not known at this stage whether each precursor molecule contains only one sequence coding for a protein, or whether it may include two or more such sequences (as the ribosomal RNA precursor contains the sequences of both the 18s and the 28s RNA's). It seems probable, although it has not been proved for certain, that the protein coding sequence is situated at the end of the precursor molecule which is assembled *last*, before the molecule is separated from the DNA template. After separation from the DNA, the sequence lying at the end of the precursor molecule is extended by the addition of a number (about 200) of adenylic acid nucleotides. This additional poly(A) region is not coded for in the chromosomal DNA but is added nucleotide by nucleotide through the activity of a special enzyme.

After the poly(A) region is added on to the mRNA precursor molecule, the sequence of nucleotides preceding the protein-encoding sequence is split off and disintegrates, while the messenger RNA with the attached poly(A) region is ready to be passed into the cytoplasm. The poly(A) region remains a part of the mRNA molecule while it attaches to the ribosome and directs protein synthesis, but the poly(A) region is not translated — it has no equivalent in the resulting protein molecule (Darnell, Jelinek, and Molloy, 1973).

16–3 SELECTIVE ACTION OF GENES IN DIFFERENTIATION

We now have to consider a very serious contradiction between the properties of cleavage nuclei and the fates of cells differentiating in various directions. In Section 6–5 it has been shown that during cleavage and also in subsequent stages the nuclei of all the cells of the embryo retain the ability to support the development of a whole embryo with all its various differentiations. When nuclei of cells are taken from swimming frog larvae or even from adult animals and transplanted into enucleated eggs, the eggs can develop into perfectly normal frogs which do not lack any kind of tissues or cells. This is a clear indication that the cells of animals which have advanced far in development still contain in their nuclei the full set of information necessary for any differentiation occurring normally in that particular kind of animal. The same conclusion may be reached by considering the processes of regeneration and asexual reproduction, which will be dealt with in Chapters 19 and 20. It will be shown that in the repair of damage to the body of adult animals and in the production of new individuals from somatic cells, already differentiated cells may participate in the production of tissues of a different type.

If we accept — as it appears that we must — that the whole genotype with a complete set of information covering all possible differentiations of a certain kind of animal is contained in the nucleus of every cell, or at least in most cells, even when the adult stage is reached, then how is it that in any given cell only part of its potentialities is revealed, while others do not manifest themselves? In other words, how is differentiation of cells possible?

The obvious solution of this contradiction is that while the nucleus of every normal cell contains the full complement of genes, not all of them are in an active state. In terms of the mechanism of gene action, this means that not all genes are producing messenger RNA in any given tissue at any given time. The transition of a gene from an inactive to an active state may well be called "activation."

We know that at the beginning of development, immediately after fertilization and during cleavage, there is no indication that the nuclear genes are directly involved in the control of the properties of cells. No new tissue specific kinds of proteins are formed during this period, indicating that no corresponding messenger RNA is being produced. (See Chapter 6.) The genes are almost completely inactive (repressed), and they become derepressed or activated in the course of development. A very important step is the beginning of gastrulation, at which time there occurs a tremendous increase in the synthesis of new proteins, showing that a number of genes have become active—that they have been activated or derepressed. It is not the same genes that become active in different parts of the embryo, with the result that different kinds of enzymes and other proteins are synthesized in different tissues.

On the basis of the preceding rather general statements, two main questions must be answered, namely:

1. What is the actual mechanism of the transition of a gene from an inactive or repressed state to an active state?
2. What is the cause or causes of particular genes becoming active at specific times and in certain groups of cells?

In connection with the first question, it may perhaps be relevant to point out that DNA, although the most important, is not the sole constituent of the chromosomes. The chromosomes as they are seen in ordinary microscopic preparations consist of nucleoproteins, the nucleic acid component being mainly DNA; the protein components are mostly a particular class of proteins known as histones. (In some cells, such as spermatozoa, protamines—a simpler kind of polypeptide—are associated with the DNA instead of histones.)

The histones are proteins with an unusually high proportion of basic amino acids (arginine, lysine, and histidine). The double-stranded molecules of DNA have free acidic groups of phosphoric acid on their outer surface, and these can establish firm bonds with the NH_2^+ groups of the basic amino acids of histone chains. Hence the association of the histones with deoxyribonucleic acids may be a very close one. It is likely that the molecule of histone may be wrapped around the double-stranded DNA molecule, following the shallow groove on its surface (Fig. 403). It has been suggested (Stedman and Stedman, 1950; Bloch, 1962) that by associating themselves with the DNA molecules (either by wrapping themselves around them, as shown in Figure 403, or in some other way), the histones prevent the DNA from interacting with other substances in the cell and thus from serving as templates for the production of messenger RNA. There are some experimental results which are relevant to this hypothesis.

Cell-free preparations containing DNA (from calf thymus) and RNA polymerase from *Bacillus megagaterium,* in the presence of RNA precursors, will synthesize RNA. If histones (also derived from calf thymus) are added to this mixture, the synthesis of RNA is inhibited. Different fractions of histones prepared by chromatography and various precipitation and extraction procedures inhibit RNA synthesis in an unequal degree, one particularly active fraction inhibiting the synthesis completely.

A different type of experiment consists in treating isolated calf thymus nuclei with a proteolytic enzyme, trypsin. The treatment removes most of the histone associated with the nucleic acid in the chromosomes. The trypsin-treated chromosomes are then used as primers in preparations containing RNA precursors. It was found that the

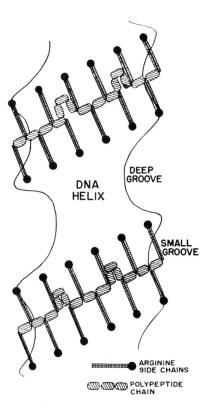

Figure 403. Diagram showing how a protamine molecule becomes wound around a double DNA molecule, following the small groove between the two strands of DNA. The phosphate groups of DNA are at the black circles and coincide with the basic ends of the arginine side chains. Nonbasic residues are shown as folds in the polypeptide chain. (From Wilkins, 1956.)

amount of RNA synthesized was increased roughly threefold when 70 per cent of the histones were removed (Allfrey, Littau, and Mirsky, 1963). In an experiment on plant material, removing all the histone from chromosomal DNA resulted in a fivefold increase in the synthesis of RNA (Huang and Bonner, 1962). The conclusion seems justified that histones associated with the DNA in the chromosomes repress their ability to serve as RNA templates.

Gene regulation has been studied in bacteria, which have the ability to adjust the production of various enzymes, depending on the nutrients available to them (Jacob and Monod, 1961). However, differentiation of cells (i.e., production of cells specializing in different activities) is a feature which is specific for the multicellular organisms, and it is only these that require some of their genes to be permanently or semipermanently repressed in some kinds of cells and activated in others. It is interesting to note, therefore, that histones appear to be present in the chromosomes of multicellular animals and plants only. No true histones have been found with certainty in a number of bacteria (de Reuck and Knight, 1966), in protozoa (trypanosomes — Beck and Walker, 1964), and in blue-green algae (Makino and Isuzuki, 1971). We can thus make the statement: no histones — no differentiation.

It has not been possible, however, to prove that histones show sufficient specificity to discriminate which genes are to be repressed and which genes are allowed to be in an active state. If the histones are, in fact, bonded to the acidic groups of the phosphate radicals of the DNA, they would be facing the "backbone" of the DNA double strand which is uniform throughout. Other mechanisms must be sought to account for the specificity of the attachment. The exact role of histones in gene regulation is not clear. If they prevent the transcription of the DNA in genes, thus making them inactive, then the activation of a gene must be due to the arrival of a signal, which by some means dislodges the histones. In any case, the histones and other chromosomal proteins do not appear to be the deciding agents, but only tools in manipulating the chromosomal DNA, themselves dependent on the truly determining agents, which have the ability of sequence-specific recognition of parts of the DNA.

Signals to Genes in Differentiation. The answer to the second question in a general form can be derived from facts dealt with in the earlier chapters of this book. One fact is that cytoplasmic differences in the egg determine which path differentiation will take in the various parts of the embryo. As the differentiation of cells in various parts of the embryo involves activation of genes responsible for the manufacture of corresponding kinds of messenger RNA, it follows that the kinds of cytoplasm present in the cells determine in the last instance which genes are to be activated.

Nuclear transplantations present further proof that the cytoplasm is responsible for the kind of activity performed by the chromosomes and thus by the genes. To what was stated in Section 6–5 on transplantation of nuclei into an egg, it may be added that nuclei can also be transplanted into amphibian ovarian oocytes, which present them with an essentially different environment. An oocyte nucleus does not replicate and therefore does not synthesize DNA. On the other hand, it synthesizes large quantities of RNA, partly ribosomal RNA and partly messenger RNA. Nuclei transplanted into oocytes, like those transplanted into an egg, conform in their behavior to their surrounding cytoplasm. This is shown to best advantage by transplanting a cleavage stage nucleus into an oocyte. Cleavage nuclei synthesize much DNA and little or no RNA. Transplanted into oocytes, such nuclei stop DNA production and start producing RNA instead. Furthermore, at the time of maturation divisions, transplanted nuclei enter into mitosis synchronously with the host nucleus of the oocyte (Gurdon, 1968c).

Control of the nucleus as a whole by the cytoplasm makes it necessary to assume that some substances pass from the cytoplasm into the nucleus which cause the chromosomes and the genes to change their activity, repress some genes and activate others. In fact, passage of proteins from the cytoplasm into the nucleus has actually been shown to take place (Gurdon, 1969).

Instead of transplanting nuclei into another cell, cytoplasm of a different nature may be added to a cell to test the influence of cytoplasm on the nucleus. If cells of two different kinds are cultivated *in vitro* together, some of them fuse, producing composite binucleated cells. Fusion may occur of cells differentiated in different ways, and also of cells of different species. Hybrid cells are produced in the latter case. Mouse liver (hepatoma) cells have been caused to fuse with human leukocytes. Liver cells synthesize blood serum albumin; leukocytes do not do so. The hybrid mouse liver + human leukocyte cells continued producing mouse blood serum albumin, but also produced some human blood serum albumin (which could be detected by immunological methods). The result shows that the cytoplasm of mouse liver cells activated the genes in human chromosomes to produce mRNA for blood albumin (Ruddle and Kucherlapati, 1974).

The messages regulating the behavior of nuclear genes, however, do not all originate in the cytoplasm of the cell containing the nucleus in question. Differentiation of cells is to a very large extent dependent on influences reaching the cells from outside. Of primary importance in this respect are the phenomena of embryonic induction. It has been proved that embryonic induction is mediated by the transmission of a substance (a protein or a nucleoprotein) from the inductor to the reacting tissues, as for instance in the induction of the neural plate by the roof of the archenteron in vertebrates. (See Section 8–5.) It is an open question, whether the inducing substance emitted by the archenteron reaches the nuclear DNA as such, or whether it only starts a chain reaction in the recipient cells, the end link of which is the signal received by the genes.

There is no doubt, however, that gene behavior is changed in the process of induction; this is evident because induction is suppressed by the action of actinomycin D, thus new mRNA must be synthesized as part of the cell's reaction (p. 220). It is also evident that the signal generated by the inducing substance must have entered the nuclei of the reacting cells via their cytoplasm!

In later stages of development and in the adult organism the behavior of the cells, including their genes, may be under the influence of hormones or hormone-like substances, originating sometimes in distant parts of the body. While facts of this kind transgress into the field of adult physiology, one or two cases must be mentioned here.

Erythropoiesis is a continuous process in healthy animals, but it can also be stimulated to a higher level by a hormone-like substance, erythropoietin, which is produced in the kidneys in response to insufficiency of oxygen supply. Treatment of hemopoietic tissue *in vivo* or *in vitro* with erythropoietin stimulates the transformation of stem cells into erythrocytes. The initiation of hemoglobin mRNA production is prevented by actinomycin D, which proves that DNA-dependent mRNA synthesis is part of the cell's reaction to the external stimulus. Again, the message could reach the nuclear genes only by first entering the cytoplasm of the reacting cells.

The reaction of various tissues to stimulation by sex hormones is known to be sensitive to actinomycin D and thus must involve activation of the genes and DNA-dependent synthesis of mRNA. A system favorable for study has been found in the oviducts of birds (hens) which produce massive quantities of ovalbumin (the main constituent of egg white) in response to estrogen. The estrogen causes the oviducal cells to manufacture ovalbumin mRNA, followed by ovalbumin synthesis. The reaction can be suppressed by chemicals which block DNA transcription (O'Malley and Means, 1974).

The general conclusion from the preceding facts is that the genes become active as a result of signals of some kind which reach them from the cytoplasm of the cells. These signals differ depending on the condition of the cells and depending on the environment—the presence of various substances in the cells' surroundings.

It is possible now to pass on to some more speculative concepts concerning the ways by which the genes can be brought into an active state.

The signals activating the genes need not always come directly from outside the nucleus. Britten and Davidson (1969) pointed out that in the process of differentiation not one gene at a time becomes activated, thus causing the synthesis of only one protein, but rather whole batteries of genes come into action, thus accounting for a whole array of enzymes which are characteristic of a particular tissue. Thus the function of the liver in vertebrates involves the production of at least 147 different enzymes, not even counting the enzymes present in each cell, such as those taking part in oxidation, cell growth, and replication.

A single external factor, such as an inducing substance in early embryogenesis or a hormone in later development, must be able to activate more than just one gene. To explain this, Britten and Davidson postulated the existence of a system of regulatory genes. They suggested that the initial signal goes to a "sensor gene," which through a system of "regulator genes," attached to it in sequence in the chromosome, produces molecules of "activator RNA." These molecules have no other function than to seek out the initiator sequences (or "receptor genes," in Britten and Davidson's terminology) of the "producer genes." The latter are then transcribed into mRNA. A particular producer gene could be called into action starting from several sensor genes, and this would account for the same proteins being synthesized in different types of tissues.

One of the arguments which Britten and Davidson have adduced in support of their theory is that in the cells of multicellular organisms there is far more DNA than is necessary to serve as templates for the mRNA actually used in the synthesis of cytoplasmic proteins. The rest of the DNA could serve as part of a vast and complicated regulatory system which controls the activity of the mRNA-producing genes. The intranuclear fast turnover RNA could in part be made up of the "activator RNA" carrying messages from the sensor system of genes to the producer genes.

We have seen, however, that some of the fast turnover RNA may be part of the precursor sequences of the different RNA's (ribosomal, transfer, and messenger). This remark, of course, does not eliminate the logical necessity of the existence of a regulatory system, which would allow batteries of genes to be activated together and in accord with one another in the different tissues of a multicellular organism.

There is also a completely different way of interpreting the nature of the fast turnover nuclear RNA. It has been suggested (Kijima and Wilt, 1969; Aronson and Wilt, 1969) that all genes (the whole length of the DNA strands) are transcribed into RNA at all times, but that the RNA produced by some genes is rapidly destroyed and does not leave the nucleus, while the RNA's modeled on other genes are stabilized (by some substances entering the nucleus from the cytoplasm?), and only these RNA's are passed into the cytoplasm. If this interpretation were correct, the regulation of gene action would be performed not at the transcription stage but subsequently between transcription and translation. The end result, as far as cytoplasmic differentiation is concerned, would be exactly the same: the repressed genes would be those the RNA modeled on which is destroyed; the derepressed genes would be those whose RNA is stabilized and passed into the cytoplasm.

The latter concept is perhaps supported by the fact that none of the RNA's leave the DNA template in a ready-to-use form, but rather in the form of precursors, which have to be processed before they pass into the cytoplasm. Parts of the precursor molecules have to be split off, and poly(A) must be attached to the future mRNA, as explained on page 485. These processes could be used for controlling the kind of RNA's that pass into the cytoplasm. It is not yet known whether the processing of the various precursors is automatic or not (Darnell, Jelinek, and Molloy, 1973).

In terms of molecular processes, bringing a gene into an active state means allowing a molecule of an enzyme RNA polymerase to attach itself to a DNA molecule at a specific point, from where it starts moving along the length of a DNA molecule, placing the RNA nucleotides in their correct positions and binding them together (Miller, 1973). From what has been said about the formation of RNA precursors, it appears that the point of attachment of the RNA polymerase molecule is not within the section of the DNA coding for the meaningful (functioning) RNA sequence, but within a region specially adapted for this purpose—the "spacer section," the "initiator sequence," or the sector coding for the later discarded part of the mRNA precursor molecule. Once attached and functioning, the RNA polymerase molecule works its way along the whole length of the sequence, until it is stopped by a "terminator sequence" or its equivalent.

What is required of the signal coming in from the cytoplasm into the nucleus is (1) to recognize a certain sequence on the DNA molecule which *precedes* the meaningful sequence to be transcribed as RNA, and (2) to enable the RNA polymerase molecule to be attached to the DNA and to start the synthesis of RNA. The sequence on the DNA molecule which is recognized by the signal may be the same as the "initiator sequence" or the "receptor gene" (Britten and Davidson, 1969).

Obviously, the recognition of the "receptor gene" or the "initiator sequence" is the central point in understanding the mechanism of gene regulation. As the specificity of any particular section of the DNA lies in the order in which the four deoxyribonucleotides are arranged along its length, the most plausible way of recognizing this order is by means of another nucleic acid molecule (perhaps an RNA molecule) with a complementary sequence to that of the DNA. This is what happens in DNA-RNA hybridization (p. 192). The idea is used in Britten and Davidson's concept of the intranuclear "activator RNA." We must not exclude, however, the possibility of a protein molecule recognizing a specific DNA sequence. In fact, it has been proved that in the regulation of gene action in bacteria a protein molecule, the "lac repressor," finds a specific locus on the chromosome and, by attaching itself to it, represses a particular gene. (See Ptashne and Gilbert, 1970.)

Studies on the structure and mode of action of actinomycin D, often used to suppress DNA-dependent RNA synthesis, furnish further proof that proteins are able to recognize a specific DNA sequence. Molecules of actinomycin D consist of a "phenoxazone" ring (a triple ring of carbon, nitrogen, and oxygen atoms) to which a loop of 5 amino acids is attached on each side. The loops are identical, but rotated in respect of each other by 180°. The exact shape of the molecule and the way it attaches itself to the DNA helix were determined by the x-ray diffraction method. It was found that the phenoxazone ring becomes inserted in between two adjacent deoxynucleotide pairs, and each of the pentapeptide (amino acid) loops establishes firm hydrogen bonds with a guanine molecule above and below the phenoxazone ring. Because of the shape of the actinomycin D molecule, it can only "fit in" between two guanosine-cytosine groups, provided that these groups are in the "trans" position, that is, that the guanosines (and cytosines) are in opposite DNA chains:

guanosine – cytosine

cytosine – guanosine

It is then this arrangement of four bases that is recognized by the actinomycin D molecule (Sobell, 1974). The attachment of the actinomycin D to the DNA double helix distorts its shape, and makes it impossible for the DNA to serve as template for RNA synthesis. It is noteworthy that the actinomycin molecule attaches itself to an intact double helix, without the two strands of DNA becoming separated. It does so through the narrow groove of the DNA (compare p. 486 and Fig. 403). Evidently, enough of the steric structure of the nucleotides is accessible in between the "backbones" (sugar-phosphate chains) of the DNA double helix to make it possible for the guanosine-cytosine nucleotides to be recognized. If a small molecule, with partly protein properties (in its pentapeptide loops), can recognize a specific sequence of two nucleotide pairs, there is no reason to doubt that a larger protein molecule could recognize a long nucleotide sequence by the same mechanism — that is, by conformation of the arrangement of atoms in the molecules.

It is doubtful that substances in the environment of cells, such as inducing substances, hormones, etc. possess the properties necessary for the recognition of specific sequences on the chromosomes. Rather, it is likely that the signal molecules entering the cells from without become associated with some sort of molecule within the cell which leads the signal to the correct spot on the DNA. J. Bonner (1971) suggested that such a molecule plays the part of a "seeing eye dog" (the analogy is to guide dogs of blind people).

The "seeing eye dog" could be an RNA molecule, a view to which Bonner was inclined, or it could be a purely protein molecule, like the lac repressor. Some recent work appears to be in favor of the latter view. In studies on the activation of the ovalbumin gene in cells of the hen oviduct (O'Malley and Means, 1974), it was found that the steroid hormone, estrogen, enters the cells of the oviduct and in the cytoplasm of these cells forms a complex with a "receptor" protein. The complex of the steroid hormone and the receptor protein then enters the nucleus, and there it binds with the chromosomes. The result is the activation of the ovalbumin gene, transcription of ovalbumin mRNA, and synthesis of ovalbumin.

This field of study is very actively pursued at present; thus it may be expected that in the foreseeable future, theories and hypotheses may be replaced by firmly established facts.

Relation of Differentiation to Mitosis. In a double helix DNA molecule the nitrogenous bases, the arrangement of which provides the specificity of particular parts of the chromosome (genes), are joined to one another pairwise in the interior of the molecule. Many biologists believe that the recognition of the nucleotide sequences would be greatly facilitated if the two strands of DNA were to separate, thus exposing the inner surfaces of the genes. Such a separation occurs normally in the S phase of the mitotic cycle, when the two strands of the double helix in the chromosome replicate before mitosis. It is therefore very suggestive that the start of differentiation seems to be dependent on a previous mitotic division of the cell.

As mentioned previously (p. 489), actinomycin D prevents the start of hemoglobin synthesis in erythroblasts treated with erythropoietin. Hemoglobin synthesis can, however, also be stopped at an earlier stage by supplying substances which upset the DNA replication, such as fluorodeoxyuridine (Paul and Hunter, 1968, 1969). It seems therefore that the DNA replication, which normally occurs in the mitotic reproduction of the stem cells, is somehow necessary for the subsequent DNA-dependent mRNA synthesis. Accordingly, it has also been observed that when erythropoietic tissues are treated by erythropoietin, the first reaction is rapid DNA synthesis which is detectable within the first hour after treatment, whereas hemoglobin synthesis gets underway after two hours and attains its maximum even later (Fig. 404).

Very similar observations have been made on mouse pancreatic tissue differentiating *in vitro* (Wessells and Rutter, 1969). The production of pancreatic enzymes is prevented if the tissue is treated with actinomycin D in the early stages of differentiation, but once differentiation has got under way, actinomycin D has no effect, showing that mRNA has already been produced in the cells, and protein synthesis can proceed unimpeded. However, if the pancreatic tissue, at the time when the cells are dividing, is treated with fluorodeoxyuridine, which prevents DNA replication and thus stops cell

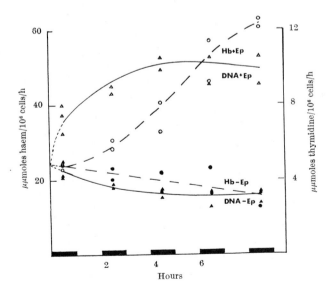

Figure 404. Rate of synthesis of DNA and of hemoglobin in erythropoietic cells after treatment with erythropoietin (+ Ep) in tissue cultures, compared with control untreated cells (− Ep). (From Paul and Hunter, 1968.)

divisions that normally precede differentiation, the tissue will be unable to differentiate and start its secretory activity.

In numerous other cases differentiation seems to follow a series of mitotic divisions of cells. We may refer to the differentiation of neurons in the central nervous system (Fujita, 1963; Watterson, 1965 – p. 335). There may be the same kind of link between the period of cleavage and the subsequent period of gastrulation, during which extensive gene activation takes place. It is also suggestive that a series of lymphoid cell divisions is somehow involved in the mechanism of production of antibodies in the immunization process (Speir, 1964).

Lastly, it is possible to quote an observation on a type of cancer, a liver hepatoma. Cells of this tissue grown *in vitro* are able to synthesize an enzyme, tyrosine aminotransferase, in response to a treatment with corticosteroids. The reaction involves the production of a messenger RNA for the enzyme. The important point, however, is that the induction of tyrosine aminotransferase is possible only during the latter half of the G_1 phase and during the S phase of the tumor cells (Darnell, Jelinek, and Molloy, 1973). This observation, in conjunction with all the facts previously stated, makes it plausible that the "seeing eye dog" with the signal molecule attached makes use of the separation of the two strands of the double helix DNA in the chromosomes, replicating in preparation for the next mitosis, to slip in, recognize the gene, and become attached to it (Gurdon, 1968a).

Now it is believed that of the two strands of DNA of which the double helix is composed, only one strand is used in transcription. Which of the two strands is recognized by the incoming signal? It has been suggested that the signal molecule recognizes and becomes attached to the "passive" strand of the helix. By blocking one strand, the signal molecule (or the complex containing the signal molecule) would then leave the other strand permanently or semipermanently open. This strand would then be free to serve as a template for transcription into ribonucleic acid. Thus, the gene would become active (Frenster, 1965). The majority of researchers, however, visualize the attachment of the signal complex to the active strand of the DNA. It is also increasingly likely that the signal molecules are not nucleic acids but proteins (p. 491).

Numbers of Genes in Differentiation. One further point has to be discussed in connection with gene regulation in development. In the course of differentiation of certain tissues, some proteins are manufactured in very large quantities in proportion to the overall size of the cell. This is true, for instance, in muscle differentiation, in which up to 10 per cent of the mass of the cell consists of one protein, myosin. Other cases are hemoglobin in erythrocytes and silk fibroin in the silk gland cells of moths. If the genomes of all cells of one animal are the same, how is it that one pair of genes, constituting only an infinitesimally small proportion of the chromosomal DNA, manage to give rise to a sufficient number of mRNA molecules to account for the synthesis of such prodigious amounts of a particular protein?

We have seen that at least in one case, the oocytes, which have to manufacture very large quantities of ribosomal RNA, this is achieved by means of an amplification of the rRNA-producing genes. Is it possible that in cells synthesizing very large quantities of a few proteins the corresponding genes become amplified, so that not just two, but large numbers of the same kind of gene can be transcribed in order to furnish the necessary quantities of mRNA?

Such a mechanism is feasible in principle, in view of the possibility of gene amplification in the oocytes. The evidence is, however, against the amplification of producer genes. The messenger RNA for silk fibroin (p. 482) has been hybridized with DNA from silk glands on the one hand and with DNA from the rest of the body of the caterpillar on the other. It was found that the levels of saturation were the same in both cases. This result shows that the number of producer genes for silk fibroin in the silk glands is no greater than the number of these genes in parts of the body which do not produce silk (Suzuki, Gage, and Brown, 1972). The results of transplantation of nuclei of differentiated cells into eggs point in the same direction. If larval gut cells had some of their genes amplified in a way appropriate for gut differentiation, the amplification would have shown itself in some way after implantation into the egg; this was not the case (Gurdon, 1960; Gurdon and Uehlinger, 1966). It appears, therefore, that the cell has some other mechanism for increasing production of proteins starting from only two copies of a gene present in the diploid set of chromosomes.

The Time Factor in Progressive Differentiation.   Once the mechanism of gene activation is set in motion, it is not difficult to imagine that when the messenger RNA reaches the cytoplasm it would cause new proteins to be synthesized there. These proteins either could be instrumental themselves in activating further genes, or could, by enzymatic action, produce substances which could act on the genes or their products, thus causing yet new kinds of messenger RNA to be released into the cytoplasm. In this way more and more genes would become active as development proceeded, and of course the kinds of genes put into an active state would depend upon what direction the differentiation had already taken in the previous stages. More variety in the results of this nucleocytoplasmic interaction would be introduced by the influences of already distinct parts of the embryo on one another, by the activity of "organizers of higher degrees" (p. 361), by tissue interactions, which will be dealt with on pages 508 to 512, by the actions of hormones, etc.

The state of a cell is obviously not defined solely by the condition of the genes at any given time (whether a particular set of genes is active or repressed). Rather the previous activities of the genes, as well as the conditions of the environment to which a cell has been exposed, contribute to the state in which the cell is found. These previously existing conditions may in certain cases be decisive and may override subsequent events. One very important contributory factor to this state of affairs is that mRNA produced by the genes and passed into the cytoplasm may sometimes fail to be immediately involved in protein synthesis but begins to be active sometime later. Messenger RNA was first discovered in bacteriophages and was characterized as "fast turnover RNA." The mRNA of animals is by no means as short-lived and may persist in the cytoplasm for days and possibly even for months. We have seen that messenger RNA produced in the oocytes remains inactive ("masked") until fertilization occurs, when the oocyte mRNA starts directing protein synthesis throughout the period of cleavage (p. 90). There are indications that also in later development mRNA may be produced a considerable time before it becomes active in the synthesis of proteins, which give the visible expression to the differentiation of the respective embryonic parts. This may be the explanation of determination, which as we have seen often precedes visible differentiation (Tyler, 1967). For instance, in the development of the amphibian embryo the neural plate becomes determined by the middle of gastrulation, when it is underlaid by the chordomesoderm, but the neural plate does not appear until several hours later, in the early neurula stage (p. 202).

The use of actinomycin to suppress transcription of DNA into RNA and of puromycin to suppress the translation of mRNA into protein structure helps to distinguish between the two processes of determination and differentiation. A particularly clear experiment of this kind has been performed on chicken embryos (Ranzi, 1968). Treatment of chicken embryos with puromycin at the primitive streak stage causes a retardation of development of all systems at the beginning of organogenesis. The nervous system is thinner than usual, the brain vesicles are reduced in size, and the neural tube is seldom closed. The heart is reduced in size, and often there is no hemoglobin produc-

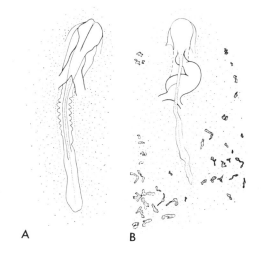

Figure 405. Effects on the development of organs of the chick embryo when the embryo is treated at the primitive streak stage with puromycin (in *A*) and actinomycin D (in *B*). (From Ranzi. 1968.)

A   B

tion in the blood islands (Fig. 405*A*). Treatment of the embryo in the same primitive streak stage with actinomycin D produces very different results. The neural system is reduced to an extreme degree and in some cases is completely suppressed. The somites are also suppressed, while the heart develops very well and sometimes is the only organ to be differentiated. Blood islands produce hemoglobin as normal (Fig. 405*B*). The fairly obvious interpretation of this experiment is that at the primitive streak stage the mRNA necessary for heart and blood island development has already been produced, but the mRNA necessary for nervous system and somite development has not yet formed, and its production is eliminated by the actinomycin. The differentiation of the heart, however, takes place later roughly simultaneously with the differentiation of the nervous system and somites, and as differentiation involves protein synthesis, all organs are retarded by puromycin treatment in about the same degree.

16–4 CHANGING PATTERN OF PROTEIN SYNTHESIS

In this and the following sections, the changing pattern of processes of differentiation will be shown in examples taken from the development of vertebrates.

Hemoglobins.   How specific genes change their activity during development is illustrated by the composition of hemoglobins in man. As previously explained (p. 480), the hemoglobin in adult humans is in the form of composite molecules consisting of four units (polypeptides), two of which are of the α type and two of the β type. Each type of polypeptide is produced by a separate gene. The human fetus, however, has a different hemoglobin, which has been shown to consist of four polypeptides, two of which are identical with the α chains of adult hemoglobin; the other two have a different polypeptide sequence (not yet known in every detail) which has been designated as chain γ. Having a different polypeptide sequence, the γ chain is produced by a separate gene. The amounts of fetal hemoglobin in blood diminish sharply after birth and are reduced to about 1 per cent for the rest of life. Instead, there appears the adult hemoglobin composed of the α and β chains. In view of the very direct relationship existing between the genes and the hemoglobin components, it is evident that after birth the gene producing the γ chain is "switched off" and the gene producing the β chain is "switched on," while the gene for the α chain continues to be active both before and after birth (Gerald, 1961).

Lactate Dehydrogenases.   A somewhat similar situation to that of the hemoglobins is presented by the production, in mammals, of the enzyme lactate dehydrogenase (Markert, 1961, 1963). The enzyme has a molecular weight of 135,000, and the molecule can be split into four component polypeptides of about equal size. The isolated subunits do not have any enzymatic action. The enzyme is commonly found in five forms, differing in electrophoretic behavior but having practically an identical enzymatic capacity. The five enzymes have therefore been called **isozymes.** Furthermore, it has been found that the subunits of the five isozymes are of two kinds only and that the subunits, named subunit A and subunit B, combine at random to produce the five complete isozymes. The composition of the isozymes can be represented in the following way:

Isozyme	I	II	III	IV	V
Subunits	A^0	A^1	A^2	A^3	A^4
	B^4	B^3	B^2	B^1	B^0

The isozymes I and V are composed of four subunits of one kind only; the isozymes II, III, and IV contain both kinds of subunits in different proportions.

Unfortunately, the genetic background of the formation of the two subunits of lactate dehydrogenase is not yet known, but it is a reasonable assumption that each of the two kinds of subunits is coded by a separate gene. The subunits are then produced on ribosomes independently of each other and combine to form the five active isozymes.

If the combination of the subunits were random, and if both subunits were available in equal quantities, the five isozymes should be produced in the proportion 1:4:6:4:1. This has actually been confirmed experimentally by mixing together equal quantities of subunits A and B. In the tissues, however, the proportions of the enzymes are not random, but depending on the particular tissue, the enzymes contain more either of component B or of component A. The first case is found in the skeletal muscles, while the kidney has a more balanced proportion of B and A. Furthermore, there is a general increase in isozymes with a high proportion of component B during life.

It is thus seen that although both of the genes responsible for the production of dehydrogenase isozyme components are active practically in all cells of a mammal, their degree of activity changes from one tissue to another and in the course of an animal's life.

Myosin.   The manufacture of specific substances may be correlated with visible changes in the organization of tissues in the embryo. This can be illustrated by considering the build-up of the protein **myosin,** which is the specific substance of muscle tissue. In the adult muscle, myosin amounts to from 10 to 12 per cent of the fresh weight or nearly 50 per cent of the dry weight.

The amount of myosin in the protoplasm of the cells undergoing differentiation into striated muscle fibers has been estimated and compared with the morphological differentiation of the muscle (Nicholas, 1950). In the rat embryo, the differentiation of striated muscle proceeds in the following way:

12th to 13th days after fertilization: The myotomes expand into the lateral and ventral body wall.

14th day: The syncytial muscle fibers have been formed; on the same day the muscles become functional: they start contracting.

17th day: Transverse striation of the myofibrils becomes visible.

TABLE 13   Changes in Chemical Composition of Muscle Tissue during Development in the Rat

Stage	Dry Weight in % of Fresh Weight	Deoxyribo- nucleic Acid in % of Fresh Weight	Myosin in % of Fresh Weight
13th day after fertilization	5.5	0.57	0.1
At term (21 days after fertilization)	9.0	0.34	1.0
At weaning	14.0	0.22	4.5
Adult rat	20.0	0.08	10.0

The figures have been recalculated from the data of Nicholas and his collaborators. See Nicholas, 1950; Herrmann and Nicholas, 1948, 1949.

The changes in the chemical constitution of the muscle tissue during embryonic and postembryonic development are shown in Table 13.

Several conclusions can be drawn from the preceding data:

1. In the prefunctional stage, the amount of myosin in the presumptive muscle cells is very small, and this amount increases greatly in the period when the functioning of the muscle begins (between the thirteenth and twenty-first days after fertilization).
2. The differentiation of the muscle tissue is by no means accomplished when the tissue begins to function. This is shown by the tenfold increase in myosin in the postembryonic stages of development. This increase is due mainly to the elaboration of additional quantities of myofibrils in the muscle fiber. As a result, the strength of the fiber increases greatly. As measured by the breaking load of the fibers, their strength increases more than 400-fold between the seventeenth day after fertilization and the adult stage.

The specific physical organization of the myofibrils can be studied by testing the differentiating muscle for birefringence. Positive birefringence was found to be present in muscle tissue on the fourteenth day after fertilization. The birefringence is definite proof that the myosin molecules are arranged in long chains. This arrangement thus precedes the appearance of contractility, while the transverse striation of the fibrils appears after the muscles have started to contract.

The data on the development of striated muscle again stress the change in proportions between the nuclear apparatus and the functional mechanism of differentiating cells. While the amount of myosin (the functional substance) increases, the amount of the deoxyribonucleic acid decreases as compared with the other substances of the cells (fibers).

There is another lesson to be learned from the development of muscle tissue. It concerns the relationship between organ rudiment formation and histological differentiation. The organ rudiments—the myotomes, in this case—are formed in an early stage of embryonic development, about the seventh or eighth day after fertilization. The cellular materials for the development of the muscles take up their final position in the lateral and ventral body wall during the twelfth to thirteenth days of development. The elaboration of myosin and its arrangement in long chains (as shown by positive birefringence) follows only two days later.

Digestive Enzymes.   The development of enzymatic mechanisms may be illustrated by the appearance of the proteolytic enzymes, pepsin and trypsin, in the digestive tract of the salamander, *Ambystoma punctatum* (Dorris, 1935). The gastric glands which se-

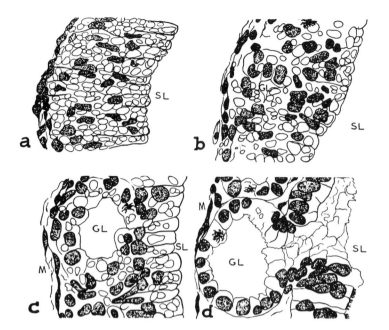

Figure 406. Development of the gastric glands in the stomach of *Ambystoma punctatum* embryos, stages 40(*a*), 41(*b*), 42(*c*), and 43(*d*). GL, Cavity of gastric gland; M, visceral mesoderm lining alimentary canal; SL, stomach lumen. (From Dorris, 1935.)

crete pepsin are developed in the walls of the stomach after all the parts of the alimentary canal are already clearly recognizable (Section 14–1). In stage 40 the walls of the stomach consist of columnar epithelium, with a large number of yolk platelets in the cells (Fig. 406*a*). At stage 41 the deeper-lying cells of the epithelium become clumped together, each clump representing the rudiment of one gastric gland (Fig. 406*b*). In stage 42 the cells of the gland rudiments are clearly arranged in a spherical layer, and a cavity appears in the middle—the lumen of the gland (Fig. 406*c*). In stage 43 the lumen of the gland is lined by a smooth cuboidal epithelium and communicates by means of a narrow duct with the cavity of the stomach (Fig. 406*d*). The yolk granules in the cells have disappeared by this time.

Pepsin first appears in the gastric glands between stages 42 and 43. Prior to and in stage 42 no trace of pepsin can be found in the stomach, and at no stage either earlier or later is pepsin present in other parts of the body. We see that the glands are already clearly distinguishable morphologically (in stage 42) before they can perform their specific function, the production of pepsin. The morphological differentiation of the glands is, however, not quite complete by that stage, as there is some further progress till stage 43, when the glands attain their final structure. The presence of trypsin in the pancreas can first be discovered in stage 43, slightly later than pepsin. The pancreatic acini first become distinguishable in parts of the organ (especially in the dorsal pancreas) at stage 41; they are more distinct in stage 42 and well differentiated in the dorsal pancreas in stage 43. As in the gastric glands, the structure of the secretory parts is laid down first, and the specific function (production of trypsin) sets in after that. It may be added that the mouth in salamander larvae breaks through at about stage 42, and feeding begins normally in stage 44.

Alkaline Phosphatase. We shall now consider an enzyme which is more widely distributed in the body of the animal, namely, alkaline phosphatase, the enzyme which causes hydrolytic splitting of monoesters of the phosphoric acid in an alkaline medium (Moog, 1946) and which may also function as a phosphotransferase; that is, it transfers the phosphate radical from one molecule to another. In early mammalian embryos, the enzyme is present in only small quantities and in a diffuse state, except that it seems to be always present in the nuclei of the cells (Danielli, 1953).

At the time of the onset of differentiation, it appears in large quantities but in only a few tissues. It is found to be concentrated in the subcutaneous tissue of the embryo in cells concerned with the development of the subcutaneous connective tissue layer. A little later, alkaline phosphatase is found in cartilages and in the hair papillae (Hardy, 1952). In all three sites the enzyme is supposedly connected in some way with the elaboration of fibrous proteins—collagen fibers in the connective tissue and cartilage matrix and fibers of keratin in developing hairs.

In later stages of development, the enzyme is very abundant in the periosteum of bone and in the matrix of bone. In the latter position (where the enzyme is extracellular), the alkaline phosphatase splits off the phosphoric acid from glucose phosphates in the form of calcium phosphate, which impregnates the bone matrix. Other sites of alkaline phosphatase concentrations are in the brush border of proximal convoluted tubes of the kidney and in the cells of the intestinal mucosa. In both of these sites, the alkaline phosphatase is concerned with the transfer of glucose from the lumen (of the renal tubule and the intestinal lumen respectively) into the internal medium of the body. In the adult, the quantities of alkaline phosphatase in the kidney, the intestinal mucosa, and the bone surpass by far the quantities found in other tissues, as may be seen in Table 14.

It is worthwhile to trace the timing of the appearance of the alkaline phosphatase in the hair papillae. The rudiments of hairs in the mouse embryo first appear 14 days after fertilization in the form of epithelial thickenings. There is no trace of the connective tissue papilla in this stage. In 15-day-old mouse embryos, mesenchymal cells accumulate under the epithelial thickenings, thus forming the rudiments of the future papillae. The differentiated papilla, as previously indicated, contains large amounts of alkaline phosphatase, which is connected with the function of the papilla as the organ supplying the materials for the formation of the hair itself, which largely consists of fibrous keratin (Birbeck and Mercer, 1957).

TABLE 14 Distribution of Alkaline Phosphatase in Different Tissues

Tissue	Enzymatic Activity of Alkaline Phosphatase (In Arbitrary Units)
Liver	4
Hyperplastic breast	9
Lymph nodes	8
Bone marrow	23
Spleen	17
Kidney	1072
Skeletal muscle	2
Cardiac muscle	12
Skin	5
Lung	36
Intestinal mucosa	2789
Gastric mucosa	17
Thymus	3
Pancreas	1
Brain	12
Bone	420

After Greenberg, from Spiegelman, 1948.

In the rudiment of the papilla, when it is first detectable, there is no alkaline phosphatase. Only in the more advanced hair rudiments of a 15-day-old embryo can small amounts of alkaline phosphatase be demonstrated. In the hair papillae of 16-day-old embryos, large quantities of the enzyme may already be found (Fig. 407; Balinsky, 1950). The formation of the rudiment of the hair papilla thus precedes the appearance of the specific substance (enzyme) which is part of the functional mechanism of the differentiated organ.

Detection of New Proteins by Immunological Methods.   In cases in which the specific substances of differentiated tissues cannot be determined chemically, they can still be traced by the use of immunological methods. A suitable experimental animal, usually a guinea pig or rabbit, is immunized against the tissue that is being studied. For this purpose the tissue, crushed into a brei or in the form of an extract, is injected into the animal that is to be immunized. The injected animal develops **antibodies** against the protein of the tested tissue in its blood plasma. The proteins which are used for immunization are called **antigens.** If the antibodies are again brought in contact with the same antigens, a reaction of a high degree of specificity will take place — the antibodies

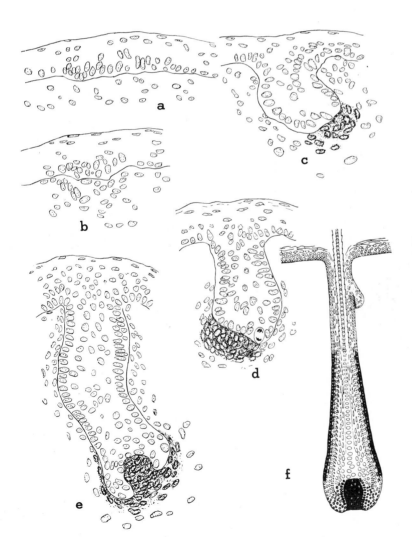

Figure 407. Distribution of alkaline phosphatase (shown by dark stippling) in developing hair follicles (a, b, c, d, e) and in a fully differentiated hair follicle of a mouse (f). (f, From Hardy, 1952.)

reacting only with the same antigen or with very closely related substances (and then to a weaker degree).

If a tissue extract containing antigens is mixed with the blood plasma of the immunized animal, a precipitation reaction takes place; the antigen is agglutinated by the antibodies and forms a precipitate. (For review see Cooper, 1948; Tyler, in Willier, Weiss, and Hamburger, 1955; Holtzer, 1961.)

The following experiment may serve as an example of the results obtained by the precipitation method (Burke, Sullivan, Petersen, and Weed, 1944). Experimental animals were immunized by injecting them with tissues of the adult chick. In this way antiorgan sera were prepared against brain, lens, kidney, bone marrow, erythrocytes, ovaries, and testes. The antisera were then tested against tissue extracts from organs of the chick embryo at different stages of development. A reaction between the two proves that the specific tissue substances (antigens) of the adult animal are already present in the organs of the embryo. The following list shows the earliest stage (expressed in days of incubation) at which the specific adult tissue antigens could be discovered:

Brain	11 days
Lens	7 days
Kidney	9 days
Erythrocytes	4 days
Ovaries and testes	11 days

In every case the antigen was found to be present when the tissues were well on their way to histological differentiation. Some "cross reactions" were also found, that is, reactions of an antiserum with tissue other than the one used for immunizing the donor of the serum. This means that the substances of the various organs could not be completely told apart.

The technique used in this experiment appears not to have been up to the highest standard of perfection which is possible by means of the precipitation method, as was shown by subsequent work (Ten Cate and van Doorenmaalen, 1950). The latter work, however, was restricted to the development of the specific substance of the lens. Rabbits were immunized by injecting them with extract of chick and frog lens respectively. Tables 15 and 16 show the results obtained. In the chick, the earliest unambiguously positive reaction was with lens tissue of a 58-hour-old chick, and in the frog, with the lens of an embryo in stage 19.

The first positive reactions were obtained in these experiments with lenses in a very early stage of differentiation. The lens of a 58- to 60-hour-old chick embryo is in the form of a vesicle which is still connected with the epidermis, with the inner wall not yet thickened by the initial stages of fiber differentiation. The lens rudiment of a frog

TABLE 15   Adult Lens Antigen in Chick Embryos

Age of Embryo (Hours)	Number of Experiments	Number of Lenses	Result of Reaction
60	3	40	+ + ++
58	1	60	+
54	3	50, 30, 30	± ± −
51	1	50	−
48	1	50	−

TABLE 16   Adult Lens Antigen in Frog Embryos

Stage of Development*	Antigen	Result of Reaction
25	lens + eye cup	+
23	lens + eye cup	+
23	rest of embryo	−
21–22	lens + eye cup	+
21–22	ventral epidermis	−
19–20	lens + eye cup	+
19	lens + eye cup	+
19	lens + eye cup	+
19	lens + eye cup	−
17–18	presumptive lens ectoderm	−

*Shumway, 1940.

embryo in stage 19 is still less developed; it has not yet even formed a vesicle but only a thickening lying in the pupil of the eye cup. Thus, the experiments prove that the organ-specific substance may be present as soon as the organ rudiment is formed and before it shows any visible traces of histological differentiation.

16–5   CONTROL OF DIFFERENTIATION BY THE INTRAORGANISMIC ENVIRONMENT

The Problem of Reversibility of Differentiation.   Differentiation may be reversible to a certain extent. The morphological and physiological peculiarities of tissues require for their maintenance the environment which surrounds them in the normal organism. Except for manifestly nonliving parts, such as the chitin cuticle in insects or the hairs in mammals, all other animal structures may become changed or may even dissolve and disintegrate if the normal conditions in the organism are changed. A disintegration of the animal's morphological organization is even more to be expected if the integrity of its body is interfered with, such as in the case of wounding or in experiments with explantation of parts of the animal's organs and tissues.

The cultivation of small portions of tissues in a clot of blood plasma or any other suitable medium (tissue culture) is an especially effective method of investigating the extent to which the histological differentiation of tissues may be reversed. In tissue cultures, intercellular structures (fibers of connective tissue, matrix of the bone and cartilage) become destroyed, the normal arrangement of cells in the tissues becomes dissolved, and the specific organoids of cells may also disappear (e.g., myofibrils or cilia). Cells derived from different tissues may acquire a very similar appearance, an appearance not unlike that of cells which have not yet undergone differentiation. These phenomena may be conveniently called **dedifferentiation.**

We have seen that in differentiation there is an increase of proteins in the cells in relation to nucleic acids. The reverse is true in the case of dedifferentiation, which involves a breakdown of functional mechanisms of cells (composed of protein). Thus, it can be expected that the amount of protein in the cells would diminish as compared with the amount of nuclear material. This is actually the case. Determinations of protein nitrogen and deoxyribonucleic phosphorus were carried out on a culture of fibroblasts taken from a chicken's heart and grown *in vitro*. After six days of cultivation, the amount of protein nitrogen per unit of deoxyribonucleic acid phosphorus was diminished by one half (Davidson and Leslie, 1950).

The dedifferentiation of cells under conditions of tissue culture or the retention of a lower state of differentiation, if embryonic cells are being cultivated, is the result of the change in environment in which the cells are being kept. The separation of a part of the body from other parts and the damage to the tissues caused by cutting them may contribute toward these changed conditions. (The latter factor will be dealt with in the discussion on regeneration, Chapter 19.) However, the main factor which causes the cells to lose their differentiation and to start growing and proliferating instead is the medium which surrounds them in tissue culture, and which is different from the medium (the fluid bathing the cells) in the intact organism.

The standard medium for tissue cultures consists of blood plasma, embryo extract, and some modification of the Ringer saline solution. The salts are necessary for the upkeep of ionic balance between the cells and the surroundings. The blood plasma contains fibrin, which clots and forms the solid substrate on which the cells can spread out and which also becomes slowly dissolved and supplies some of the necessary nutrients for the cells. The embryo extract is added as a growth-stimulating agent. Without embryo extract, on blood plasma and saline alone, the cells grow only very slowly, if at all. With embryo extract added, the cells start growing and proliferating rapidly, and dedifferentiation occurs, as previously described.

The embryo extract naturally contains some of the substances surrounding the cells in the early embryo. The behavior of cells in tissue cultures may thus be attributed to the fact that the medium bathing the cells tends to keep them in a condition characteristic of the embryo in the stage from which the extract had been taken. In a seven-day-old embryo, often used for the preparation of embryo extract, growth is rapid (Section 16–5), and there is as yet not much histological differentiation. If this interpretation were essentially correct, it should be possible to produce progressive differentiation of cells by exposing them to media containing extracts from consecutively older embryos.

This has been done in the following experiment (Gaillard, 1942). Osteoblasts derived from a 16-day-old chick embryo were kept in cultures in two series. The tissues were grown in flasks (Carrel flasks) on the surface of a blood plasma clot and suffused with embryo extract, which was changed every two days. In one series, the embryo extract was always the same, prepared from seven-day-old chick embryos. In the other series, older embryos were used to prepare the extract at each subsequent change, namely: 10-day, 12-day, 15-day, and 18-day embryos, then the extract from the heart of a newly hatched chick, and lastly blood serum of an adult hen. The changing extracts were to imitate the changes in the tissue fluids with progressing development. The experiment yielded results according to expectation; the first culture grew and proliferated without any differentiation, while bone developed in the second culture.

In spite of the apparent simplification of the cells which have undergone dedifferentiation, they do not revert to the state of embryonic cells. Evidence acquired from numerous experiments, which cannot be considered here in detail, proves that dedifferentiated cells retain their histological specificity and do not acquire new competences. If cultivated alone, renal tissue may become disintegrated, and its component cells grow out in the form of a disorganized sheet or layer, but the cells remain kidney cells, and given suitable conditions, they again arrange themselves into the shape of tubules. (See also p. 508.)

Similarly, cartilage may be dedifferentiated, and the cartilage cells then grow as a disorderly mass, scarcely distinguishable from a mass of connective tissue cells. But if the culture is kept under conditions which do not favor rapid growth, such as when the culture medium is poor in growth-promoting substances (small amounts of embryo extract) or if it is not changed often enough to a fresh medium, a new differentiation becomes possible. When this happens, the former cartilage cells again secrete cartilage matrix, thus showing that they retained their functional specificity in spite of morphological simplification.

In tissue cultures, cells are dissociated from one another as a result of the breakdown of the bonds which bind them together in a normal tissue. The dissociation is not complete, especially in the case of epithelial tissues, which remain joined in a sheet even when cultivated *in vitro*, nor is it quite under the control of the experimenter. Therefore, the attempt was made to separate tissue cells by more direct methods in order to test to what extent individual cells can retain the peculiarities that they had acquired previously in conjunction with other cells. By grinding tissues, especially embryonic tissues, in a specially prepared small glass mortar, they may be disaggregated, and a sufficient number of individual cells remain alive and may be used for further study (Weiss and Andres, 1952). A more delicate method is to treat tissues with a weak solution of trypsin in a calcium- and magnesium-free saline. This treatment causes the cells to separate from one another. A suspension can be obtained in this way consisting almost exclusively of completely separated individual cells which appear to be quite healthy (Fig. 408 — Moscona, 1952). The suspension may then be put into a medium in which the cells can reaggregate and, under favorable conditions, resume differentiation. In some experiments, the medium was that used for tissue culture with reduced embryo extract to facilitate differentiation (see preceding discussion). In other experiments, the cell suspension was injected into the veins of chicken embryos, and the cells became disseminated through the vascular route. Individual cells or small clusters of cells then settled at various sites in the body of the embryo or on the chorioallantois and either were incorporated into the tissues of the host or gave rise to small local growths, **teratomas** (Andres, 1953).

The main result of these experiments was that the cells, which had passed through a condition of complete disaggregation, were able to resume specific tissue differentiation, whether on a plasma clot or on the chorioallantois of a living embryo. Masses of brain tissue, muscle, cartilage, bone, nephric tubules, glandular tissue, or epidermis,

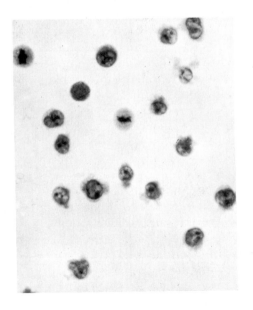

Figure 408. Suspension of completely dissociated cells, produced by treating chick embryo tissues with trypsin. (Courtesy of Dr. A. A. Moscona.)

usually in the form of cysts with clearly differentiated feather germs, were observed in various experimental series. The assortment of tissues appearing in any experiment depended on the origin of the cell suspension, the stage of the embryo from which the cells were derived, and the part of the embryo taken. The results available so far are compatible with the assumption that every type of cell differentiates after disaggregation in conformity with its previous differentiation. Thus, if cell suspensions were prepared from whole embryos, the teratomas contained a wide variety of tissues including nervous tissue, musculature, and glands. If only limb-buds were used for preparing the suspension, the teratomas contained epidermis, cartilage, bone, and mesenchyme, but no nervous tissue, muscle, or glands (Andres, 1953).

It is remarkable that structures developing from cell suspensions do not present a chaotic assemblage of different types of differentiation. Rather, they produce parts resembling organ rudiments of a normal embryo, with the various tissues segregated from one another and each tissue arranged in a recognizable morphological unit: nerve cells form brain vesicles with a central lumen; cartilage cells are sometimes arranged in elongated rods with perichondrial ossification; epidermis cells are arranged in layers with a clear distinction between proximal and distal surfaces; and feather germs may show a very high degree of internal organization (Fig. 409). As it is unlikely that each unit of tissue is always derived from one single cell, it follows that the cells sort themselves out in some way—that cells of any one kind join together and group themselves anew in an order similar to what they had before they had been separated. This is the same process that we have found in relation to cells at the time of gastrulation and neurulation (Section 9–2) and is further evidence of the "affinities" between particular kinds of cells.

Chemical Substances as Means of Controlling Differentiation. Since the actual differentiation of tissue cells is dependent on conditions in their environment, it is possible, as we have seen (p. 503), to direct their development in definite pathways by exposing them to appropriate treatments. Further experiments along these lines will now be described, first of all in relation to the control of differentiation by such chemical substances as vitamins and hormones.

Stratified epithelium in vertebrates appears to be a suitable tissue for experiments of this kind. Under normal conditions, stratified epithelium takes on various forms, both in different groups of vertebrates and in different parts of the same animal. In terrestrial vertebrates, stratified epithelium is squamous and cornified on its surface, but in fishes the epidermis, though stratified, is not squamous and contains mucus-secreting cells. The degree of cornification in terrestrial vertebrates varies, being very strong on the surface of the body but weak in the lining of the oral cavity, the pharynx and the esophagus. In the vagina, the stratified epithelium undergoes cycles of cornification accompanying the menstrual cycle. The epidermis gives rise to a number of glands, among them the mammary glands, in which the epithelium becomes simple columnar, but under pathological conditions it may revert to the cornified type, such as in some kinds of cancer (Pullinger, 1949). Lastly, in the endodermal part of the alimentary canal of mammals the esophagus is lined with stratified epithelium, while the posterior parts, starting with the stomach, are lined with columnar mucus-secreting epithelium.

The transformation of stratified epithelium from the cornified to the noncornified type may be achieved by purely chemical methods under conditions of explantation *in vitro*. If small pieces of vaginal wall of juvenile female mice are cultivated in the standard media for tissue cultures, the epithelium remains without any traces of cornification. If, however, the female sex hormone, $3,17\beta$-estradiol, is added to the culture medium, the epithelium becomes squamous and cornified on its surface. Other preparations of estrogens have a similar effect (Hardy, 1953).

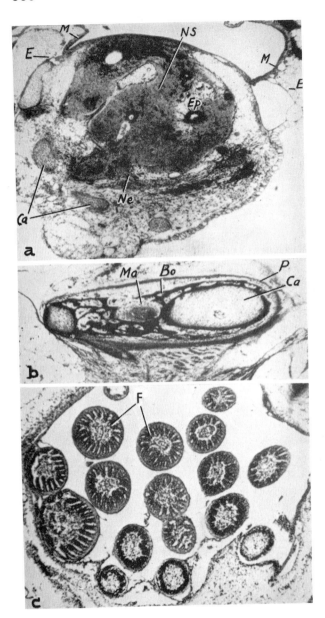

Figure 409. Teratomas developed from cells of a chick embryo completely dissociated and injected into the blood vessels of an intact chick embryo. *a*, Teratoma consists mainly of a large mass of brain tissue (NS) surrounded by some muscle tissue and cartilages (Ca); M, visceral mesoderm of the host; E, endodermal epithelium of the host; Ne, nerves; Ep, ependyma. *b*, Teratoma consists of an elongated piece of skeleton with cartilage (Ca) and bone (Bo), surrounded by a layer of perichondrium (P) and also of masses of muscle and adipose tissue; Ma, bone marrow. *c*, Teratoma is a large epidermal cyst with feather germs (F) seen in cross-section. (From Andres, 1953.)

The opposite transformation may be achieved by treatment of the cells with vitamin A. If the skin from young seven- to eight-day incubated chicken embryos is grown in the ordinary tissue culture medium (blood plasma + embryo extract), it develops a squamous cornified layer on the surface of the epithelium. An addition of vitamin A to the medium causes a complete transformation of the epithelium, which now becomes a cuboidal or columnar mucus-secreting one. It is remarkable that the cells do not need to be continuously in a vitamin A–enriched medium to be converted into the nonkeratinized type; a short treatment with the vitamin suffices to switch the development from the one channel into the other. To enable the vitamin to reach each epithelial cell in a short time, the skin of a chick embryo was dissociated into single cells by trypsin treatment (as described on p. 504), and these were then immersed in a 0.06 per cent solution of vitamin A for 15, 30 or 60 minutes. After this, the cells were put onto a plasma clot and cultivated for several days. The cells reaggregated and in the best cases formed cysts or vesicles, with the distal surface of the epithelium turned inward and the outer surface surrounded by connective tissue. Typical stratified squamous epithelium developed in controls not exposed to vitamin A (Fig. 410*a*). As to the vitamin A–treated preparations, it was found that even a 15-minute sojourn in the vitamin solution is sufficient to transform the cells into the nonkeratinizing type, while additional periodic

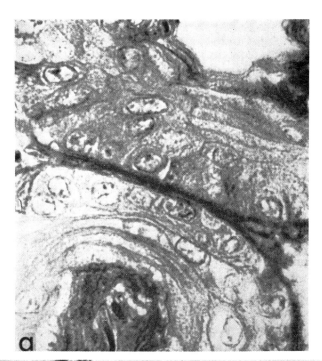

Figure 410. Influence of a short treatment with vitamin A on the differentiation of dissociated and reaggregated epidermal cells. *a,* Control culture, differentiating as typical stratified squamous epithelium. *b,* Treated culture, developing into columnar epithelium with goblet cells. (From Weiss and James, 1955.)

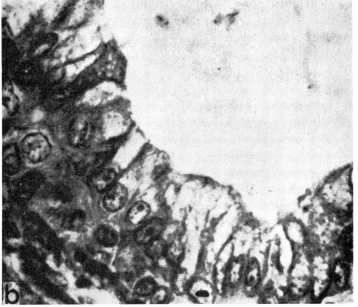

washings (30 minutes every two days) furthered the development of the epithelium into a typical columnar epithelium with goblet cells (Fig. 410 – Weiss and James, 1955).

A special type of condition for the differentiation of some tissues is the interaction between epithelium and mesenchyme. The organized growth of many epithelial structures and the very preservation of the epithelial arrangement of cells are dependent on the presence of connective tissues on one (the proximal) surface of the epithelium. An example of this is presented by the embryonic epidermis of the early amphibian embryo (p. 202). When isolated without mesoderm or mesenchyme of any kind, embryonic epidermis soon loses its epithelial arrangement; the cells acquire a reticulate arrangement and eventually degenerate and die off. In the presence of mesodermal mesenchyme, the epithelial arrangement is preserved; the ectoderm remains healthy and differentiates as normal skin epidermis.

When epithelial tissues are grown in tissue culture they tend to grow as sheets of cells which, though preserving contact with one another, do not follow the arrangement that was present in the original tissue or that should have arisen in the course of development, if the part taken for cultivation was the early rudiment of some epithelial structure. For example, if epithelial cells of renal tubules are cultivated alone, they form a disorganized sheet or layer spreading out on the surface of the plasma clot. If, however, some connective tissue cells are added to the culture, the spreading out of the sheet of renal cells is arrested, and they become reconstituted into tubules, which bear a similarity to the normal tubules of the kidney (Drew, 1923).

The dependence of epithelia on the connective tissue can be demonstrated very clearly when rudiments of glands are cultivated *in vitro* with or without connective tissue. The submandibular glands of a mouse embryo appear on the thirteenth day of gestation in the form of a pair of solid, club-shaped buds, growing from the buccal epithelium down into the connective tissue layer, one on each side of the tongue. Soon the epithelial bud becomes surrounded with dense mesenchyme forming a "capsule." On the fourteenth day, the tip of the epithelial bud becomes indented. This is the beginning of the branching of the gland rudiment, and the two primary branches grow out, each bearing a knoblike thickening at its end. The end knobs divide repeatedly and eventually give rise to the secreting acini of the gland, while the more proximal parts likewise become split and form the ramified system of ducts. While this outgrowth and branching of the epithelial parts goes on, the capsule mesenchyme surrounds the branches and penetrates between them, forming the connective tissue of the gland. The ducts and acini become hollowed out at a later stage.

This development can be observed on whole gland rudiments cultivated *in vitro* in a plasma clot (Fig. 411 – Borghese, 1950). By treatment of the early thirteenth day rudiment of the submandibular gland with a 3 per cent solution of trypsin for three to five minutes in the absence of calcium and magnesium ions, the cohesion of the mesenchyme with the epithelium can be destroyed, and the two components of the rudiment may be separated from each other. On placing them in a culture medium, both components are found to be fully viable, but their differentiation is no longer normal: the capsule connective tissue produces a typical culture of mesenchyme, with individual cells spreading out radially from the initial piece of tissue; the epithelial bud loses shape and becomes transformed into a sheet of growing cells. This behavior, however, does not show that the cells of either the capsule or the epithelial part have changed in their essential nature. If the two components are set in the culture medium near to each other, the mesenchyme comes to surround the epithelial rudiment, whereupon the latter starts sprouting and branching in a very nearly normal fashion. The interaction between the epithelium and mesenchyme restores the system to its normal state (Fig. 412 – Grobstein, 1953a, 1953b).

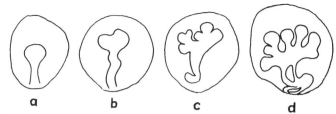

Figure 411. Camera lucida drawings of the capsule and the epithelial part of a mouse's submandibular gland growing *in vitro*. (From Borghese, 1950.)

The mutual influence of the epithelium and mesenchyme of the submandibular gland rudiment may be considered to be a special case of induction, and as with inductions occurring in earlier stages, it may be questioned whether the influence of one component on the other is specific or not. To answer this question, isolated epithelium of the submandibular gland was cultivated with mesenchyme of different origins: mesenchyme from the rudiment of the maxilla, from the somites, from the lateral plate, or from the lung rudiment. With all foreign mesenchyme, the epithelial rudiment of the submandibular gland failed to sprout and ramify. The sheetlike spreading out of the epithelial cells was arrested, however, and the rudiment developed eventually into an epithelial cyst. When the epithelial rudiment was implanted into a culture of capsular mesenchyme, which had been previously killed by heat, the rudiment behaved much the same as when it was surrounded by foreign mesenchyme; the flattening out and spreading of the epithelium was suppressed, but no growth or ramification took place. Apparently the stimulus necessary for the normal development of the duct and acini system of the submandibular gland can only be given off by the capsular mesenchyme of this organ and only in the living state (Grobstein, 1953b).

In another case of tissue interaction, the inducing stimulus has been found to be less specific. We have seen (p. 400) that the development of renal tubules of the meta-

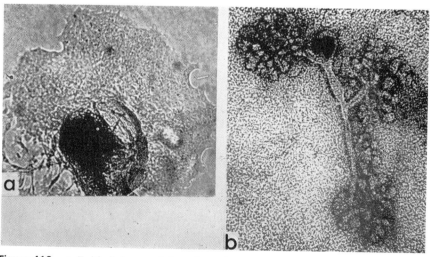

Figure 412. *a,* Epithelial part of the submandibular gland separated from the capsular mesenchyme and grown *in vitro*. *b,* Epithelial part of the submandibular gland separated from the capsular mesenchyme as in *a* but then reunited with the capsular mesenchyme in the culture. (*a,* From Grobstein, 1953a; *b,* from Grobstein, 1953b.)

nephros is dependent on the presence of the growing and ramifying ureter bud. Under the influence of the latter, the loose mesenchyme of the metanephric rudiment (the metanephrogenic mesenchyme) becomes partially converted into epithelium and forms convoluted tubes, the renal tubules, which in normal development link up with the terminal ramifications of the ureteric ducts. The interaction of the ureteric bud and the metanephrogenic mesenchyme can be observed *in vitro* in the same way as the interaction of epithelium and mesenchyme of the submandibular gland, that is, after separating the epithelial and the mesenchymal components by trypsin treatment and reuniting them in the culture medium, though the results of the reaction here are, of course, essentially different. The experiment previously referred to (p. 508) differs from the ones which will now be described in the stage at which the renal tissue was taken for cultivation. In the previous experiment the epithelium was derived from already differentiated renal tubules; in the present experiment it is the initial formation of the renal tubules that is under consideration.

By combining metanephrogenic mesenchyme taken from an 11-day-old mouse with other tissues, it was found that the ureteric duct is not the only part that causes the mesenchyme to be converted into renal tubules. The epithelial part of the submandibular gland cultivated together with metanephrogenic mesenchyme produces the same effect, although there was no reciprocal action; the gland epithelium remained unbranched. The epithelium of the submandibular gland can thus serve as an "abnormal inductor" of renal tubules, just as adult liver can serve as an abnormal inductor of a neural plate. Furthermore, the spinal cord, especially its dorsal half, proved to be a very efficient inductor of renal tubules, when placed in a culture of metanephrogenic mesenchyme (Grobstein, 1955).

The renal tubules in the preceding experiments always developed in the immediate vicinity of the inducing tissue. An attempt was made therefore to test whether immediate contact is necessary for the induction to take place. For this purpose, the inducing and reacting tissues were separated by thin membranes of various degrees of porosity. Cellulose ester membrane filters were used, varying in thickness from 20 to 150 μ, and with pores approximately 0.8 μ, 0.4 μ, and 0.1 μ in diameter. The experiment consisted essentially in arranging for two tissue cultures to grow, one on each side of the membrane filter. The inducing culture (spinal cord) was grown on one side of the membrane and the reacting tissue (the metanephrogenic mesenchyme) on the other side. It was soon discovered that the inductive influence could easily pass through the coarser filters (with pores 0.8 μ and 0.4 μ in diameter) of up to 60 μ in thickness but could not pass if the filter was 80 μ thick (or consisted of four or more layers each 20 μ thick) (Fig. 413). With finer filters (pores approximately 0.1 μ in diameter) the induction became weaker, and the influence could cross only a thin membrane not exceeding 30 μ in thickness. A still finer filter, such as a cellophane membrane 20 μ thick, effectively stopped the inducing influence.

Positive results have also been obtained when the submandibular gland components were cultivated on the opposite sides of a membrane filter, the epithelial part on one side and the capsule mesenchyme on the other. The epithelial bud produced numerous ramified outgrowths, thus showing that the inducing principle could penetrate through the filter (Grobstein, 1953c).

Similar methods were used in a third and very interesting case, the differentiation of the pancreatic rudiment in the mouse embryo (Wessells, 1968; Wessells and Rutter, 1969). It has already been mentioned (p. 463) that the epithelial part of the pancreatic rudiment does not differentiate if it is separated from the adjoining mesoderm. This case is of particular interest, as the signs of differentiation were not confined to morphological changes (formation of tubules and the like) as in the kidney or salivary gland tis-

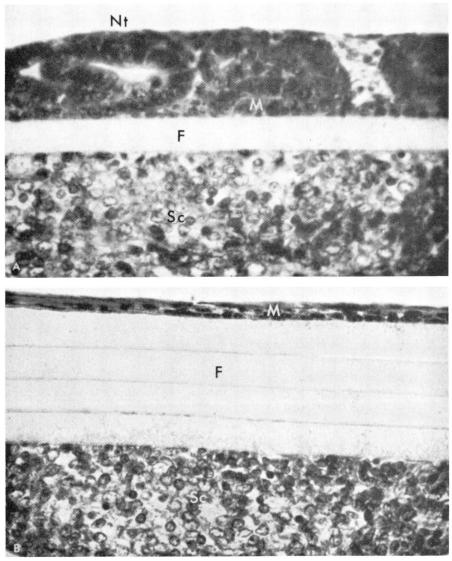

Figure 413.   Induction of kidney tubule differentiation through cellulose ester membrane filters. Below the filters is the inductor, spinal cord tissue; above the filters is the reacting nephrogenic mesenchyme. *A*, Successful induction through one layer of filter (20 μ thick). *B*, Four layers of filter preclude induction. F, Layers of filter; M, nephrogenic mesenchyme; Nt, induced nephric tubule; Sc, spinal cord tissue serving as inductor. (From Grobstein, 1957.)

sue, but the actual process of specific pancreatic secretion, as seen in the formation of secretory granules, could also be detected. Thus, the mechanism of specific protein synthesis was obviously involved.

The tissues were put to the same test as the metanephrogenic tissues; that is, the epithelium of the pancreatic rudiment and the mesodermal tissue were allowed to grow in culture on opposite sides of a porous membrane. The pancreatic epithelium of 11-day embryos, which when cultivated alone did not produce pancreatic enzymes, differentiated fully with the formation of secretory granules when exposed to the influence of mesodermal tissue acting through the porous membrane. The action of the mesoderm turned out to be even less specific than in the case of kidney tissue, since not only was pancreatic mesoderm able to stimulate pancreatic cells to secretion, but so were other kinds of mesoderm (from the kidney, from the salivary gland, and from the lung).

Another interesting finding was that if the pancreatic epithelium was allowed to develop in the presence of mesoderm to the fourteenth day, it proceeded to produce pancreatic enzymes even if separated from the mesoderm. Apparently the mechanism for producing the messenger RNA necessary for the synthesis of the pancreatic enzymes was already set in motion and did not need further extraneous support.

In experiments with nephrogenic mesenchyme the electron microscope was applied to see what was going on in the pores of the filter membrane separating the inductor and the reacting tissue. In ultrathin sections made perpendicular to the separating membrane, it could be seen that in the case of coarser membranes, the pores contained cytoplasmic outgrowths of cells, coming both from the neural tissue and from the mesenchyme. The possibility was therefore not excluded that the processes of the two kinds of cells met somewhere inside the membrane and thus established a direct contact between the inducing and reacting cells. These processes were, however, more scarce when filters with 0.4μ pores were used, and in the filters with 0.1μ pores there was practically no penetration of the filter by cytoplasmic processes of the cells, except for a few small inpocketings which did not go more than 1 to 2μ into the substance of the filter (Grobstein and Dalton, 1957).

The results of these experiments are very important: they show that the interaction of cells responsible for the normal differentiation of tissues can cross a narrow gap between the cells. The inducing principle is therefore a diffusible substance. Since this substance does not penetrate through cellophane, it must be a macromolecular substance, most probably a protein or a nucleoprotein. The similarity of these results to those obtained with the neural inductor in early amphibian development (p. 208) is obvious and very suggestive. Much remains to be done; we still have to learn why the inducing substance does not spread over greater distances—whether this is the result of extremely small quantities of the substance produced, of its instability, or of some other unknown factor. The chemical nature of the inducing substance has to be elucidated. The way for further research is now open, and new discoveries will probably not be long awaited.

16–6 CONTROL OF THE REACTIVE ABILITY OF TISSUES BY THE GENOTYPE

While the initial step in the differentiation of cells is made by the cytoplasm, it can be shown that the final result cannot be achieved without the involvement of hereditary factors, the nuclear genes. This is true in particular in the case of differentiations caused by induction. It has been shown (p. 211) that inducing substances are not highly specific, and tissues of one species of animal may induce structures in another species even if the two are, systematically, very far apart. The structure induced, on the other hand, is as a general rule one that is peculiar to the reacting species.

In Section 12–4 it was indicated that the balancer of salamander larvae develops as a result of an induction, the stimulus being given off probably by the roof of the archenteron, and the reacting system being the epidermis in the neurula stage. Balancers are found in most newts and salamanders, but a few species do not possess them. Experiments have been performed to determine the cause of this difference. If, in the neurula stage, the epidermis of a species possessing a balancer (viz., *Triturus taeniatus*) is taken from any part of the body and transplanted to the side of the head of a species which does not have balancers (viz., *Ambystoma mexicanum*), the transplanted epidermis will develop a balancer in exactly the same position in which it is usually found, that is, underneath the eye near the angle of the mouth. A reverse transplantation, that is, transplantation of epidermis from *Ambystoma mexicanum* to the site of balancer development in *Triturus taeniatus*, results in the absence of a balancer on the operated side of the head (Mangold, 1931a; Rotmann, 1935).

These results show that the absence of balancers in *Ambystoma mexicanum* is due to failure on the part of the epidermis to react to the stimulus of the inductor, or in other words, that the epidermis lacks an appropriate competence. The inducing stimulus, on the other hand, is present both in species possessing balancers and in species not having these organs. As the inability of *A. mexicanum* to develop balancers is genetically fixed, we may conclude that the hereditary factors (genes or their combinations) responsible for this particular peculiarity of *A. mexicanum* directly affect the competence of the ectoderm, while the inducing systems remain unchanged.

Another example illustrating the same principle is found in the development of the mouth in anurans and urodeles. The larvae of urodeles (salamanders and newts) have typical teeth consisting of the pulp and the layers of dentine and enamel. The teeth are situated inside the mouth and are attached to the jaws and the bones of the palate. The tadpoles of frogs and toads have no true teeth; instead, the edges of the jaws are covered by horny sheaths, and rows of small horny teeth and epidermal papillae are developed on an oral disc surrounding the mouth. It was shown in Section 14–2 that the ectodermal mouth parts develop under the influence of an induction from the oral endoderm. The inductor, however, determines only the position of the ectodermal oral invagination, not its specific peculiarities. It is possible to transplant ectoderm from an early frog embryo to a salamander embryo in such a way that the grafted ectoderm covers the mouth region. As a result, the graft is induced to develop a mouth. This mouth is, however, in every respect the mouth of a frog, with horny jaws and rows of horny teeth and oral papillae (Fig. 414–Spemann and Schotté, 1932).

It is thus evident that the influence of the mouth inductor is similar in salamanders and frogs, or at least sufficiently similar for frog ectoderm to be able to react to a salamander's inductor. What is different is the competence of the ectoderm; to the same stimulus, the ectoderm of different animals reacts in its peculiar way. It is the competence or nature of reaction of the ectoderm that is affected by the hereditary factors responsible for the differences in the development of the mouth in urodeles and anurans. The power of the inductor is limited; it can evoke only such differentiations as are provided by the hereditary constitution of the reacting cells. Since this constitution is embodied in the structure of the nuclear DNA, it may be concluded that induction eventually is consummated by changing the activity of the genes in the reacting cells.

It is also possible, however, for the hereditary factors to modify an inducing system without changing the competence of the reacting tissues. It has been proved that the dorsal fin fold in amphibians is induced by the neural crest (Section 12–4, The Fin Fold). In most salamander larvae, the fin fold reaches anteriorly almost to the occipital region, but in *Eurycea bislineata* it is present only on the tail, the trunk being devoid of a fin fold. This peculiarity is not a result of the inability of the trunk epidermis to develop a fin fold. If the trunk epidermis of *Eurycea bislineata* is transplanted to the back of an embryo of *Ambystoma maculatum* (a species having a fin fold in the trunk region), the transplanted epidermis will participate in the formation of a fin fold (Bytinski-Salz, 1936). The peculiarity of the fin fold of *Eurycea* is thus due to failure of the trunk neural crest to act as an inductor.

We have seen that the position of the external gills found in amphibian larvae, and

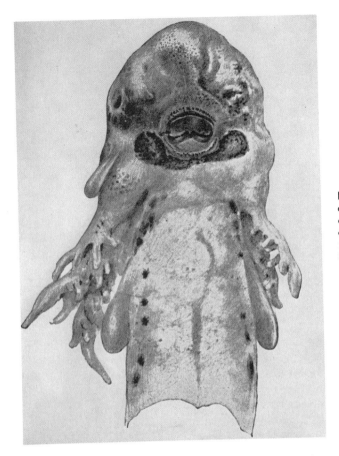

Figure 414. Larva of a newt in which the ectoderm in the oral region was replaced by ectoderm
of a frog embryo. The grafted ectoderm developed horny jaws and teeth and the adhesive organ
found in frog tadpoles. (After Schotte, from
Spemann, 1938.)

the very fact of their development, is dependent on the endoderm, while both the epidermis which covers the gills and the mesenchyme which forms the connective tissue
and the blood vessels of the gills react to the inducing influence of the endoderm. The
number of pairs of external gills in tadpoles of frogs and toads varies in different species from one to three. In the South African toad, *Bufo carens,* there are three pairs of
external gills, while in another species, *Bufo regularis,* only two pairs are formed. If the
presumptive epidermis of *Bufo regularis* is transplanted to an embryo of *Bufo carens,*
and if it lies in the branchial region, it will form gills in response to an induction from
the host endoderm. It was found that three gills were formed on the operated side, the
number typical for the host and not for the reacting epidermis (Fig. 415). In the reciprocal transplantation, that is, transplantation of the presumptive epidermis from *Bufo
carens* to *Bufo regularis,* two external gills develop, again the number typical for the host.
It is evident that the epidermis can form any number of external gills, and the number
that actually develops is determined not by the reacting system (grafted epidermis) but
by the inductor, the host endoderm (Balinsky, 1956). The ability of a part of the embryo to act as inductor may thus be affected by the specific genetic constitution of an
animal.

    There is no contradiction between these results and the experiments showing that
an induction can produce only such differentiations as are provided for by the genotype
of the reacting cells. Although in the last two experiments the inductor was responsible
for some features of the induced structure (extent of fin fold, number of gills), the result

of the induction did not go beyond changing the spatial distribution of certain structures. The heteroplastic inductor did not produce any qualitatively new differentiations that were foreign to the species supplying the reacting tissue.

16–7 SEQUENCE OF GENE ACTION IN DEVELOPMENT

It was suggested at the beginning of this chapter that when development starts the genes are in a repressed stage. During cleavage the genes are inactive, and whatever hereditary traits can be observed are due to gene action during the growth and maturation of the oocyte.

From gastrulation onward, some genes become derepressed, and the action of individual genes becomes evident. By observing normal development, it is usually not possible to state whether a particular advancement in the organization of the embryo is due to the action of a gene or genes. (Some exceptions to this statement have been dealt with on pages 196 and 495.) In a mutant, however, one or more genes become different from the "normal" allele, or the arrangement of the genes in the chromosomes is changed (in translocations), and this change is made apparent by producing a deviation from the normal development of the organism. Quite often the deviation is

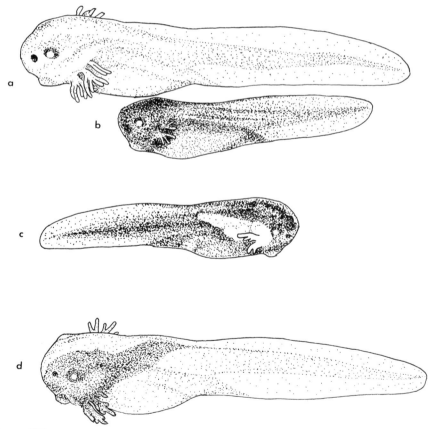

Figure 415. Results of exchanging ectoderm between two species of toads differing in the number of external gills. *a*, Normal tadpole of *Bufo carens*; *b*, normal tadpole of *Bufo regularis*; *c*, tadpole of *B. regularis* with grafted ecotderm of *B. carens*; *d*, tadpole of *B. carens* with grafted ectoderm of *B. regularis*.

of the nature of a developmental arrest, and from this it may be inferred that the gene or genes in their normal, unmutated state are somehow involved in producing the normal course of development. If, therefore, we find that in a mutant some process leading to formation of organ rudiments is disturbed, we may conclude that this process is under the control of the normal allele of the mutated gene or is dependent on the normal arrangement of genes in the chromosomes.

Genes Affecting the Earliest Stages of Organogenesis.   There are several mutations in mice which produce a partial or complete duplication of the whole body. The duplication may be posterior, involving the tail, sacral, and hindlimb region (Danforth's posterior reduplication, Danforth, 1930), or more generalized ("kinky" homozygotes, symbol *Ki*, Gluecksohn-Schoenheimer, 1949). The duplications could not have been produced later than the time of primary organ formation, possibly even during the gastrulation stage. The mechanism by which duplication is achieved is not known, but in this connection it is sufficient to know that the genetic constitution of the embryo may influence the morphogenetic processes involved in the formation of primary organ rudiments.

A mutant line is known in guinea pigs which shows various degrees of abnormalities of the head. The abnormalities are of the nature of cyclopic defects (Section 8–7); paired organs of the head tend to approximate each other on the ventral side and fuse into unpaired organs. In animals defective to a greater degree, the more anterior parts of the head disappear altogether, and even the entire head may be absent, while the organs of the trunk region are more or less normal (Fig. 416; Wright and Wagner, 1934). These defects are so similar to cyclopic defects which can be produced in amphibians by removing the anterior part of the archenteron that it is hardly possible to doubt that in the mutant guinea pigs the origin of the abnormalities is a similar one.

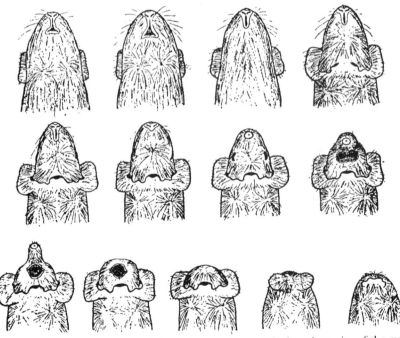

Figure 416.   Series of cyclopic defects of increasing severity in guinea pigs of the mutant strain studied by Wright. (From Needham, J., 1942.)

The abnormal genetic constitution in some way inhibits the action of the primary organizer (the chordomesoderm and the endoderm of the head region). The external ear is also involved in abnormal guinea pigs, and this is a direct indication that the branchial region of the foregut developed abnormally, as the external ear is connected in its development with the first pharyngeal pouch, and its abnormal position could arise only if the arrangement of the pharyngeal pouches were defective right from the start. A similar mutation is known in mice (Little and Bagg, 1924), and it probably occurs in other animals as well.

Genes Affecting Organ Rudiments in Later Organogenesis. An example of genetic control of organogenesis in a later stage is presented by Danforth's short tail mutant in the mouse (symbol *Sd*). Externally, the mice carrying this gene differ from the normals by having a shortened tail or by the complete absence of the tail. Another feature of interest is the reduction in size or complete absence of one or both kidneys found in these mice. As explained previously (Section 13–4), the metanephros in mammals develops from two separate rudiments, the metanephrogenic tissue and the ureter, which sprouts from the mesonephric duct. In Danforth's short-tailed mice, the ureter buds off from the mesonephric duct as usual, but it tends to remain shorter and sometimes does not reach the metanephrogenic tissue at all. If the ureter does not reach the metanephrogenic tissue, the kidney does not develop, as induction from the ureter is necessary for the kidney tubules to be differentiated. If the ureter reaches the metanephrogenic tissue, at least a small kidney develops. The size of the developing kidney depends on the degree of branching taking place at the end of the ureter, because only those parts of the metanephrogenic tissue which lie in the immediate vicinity of the tip of the ureter or its branches (the latter becoming the collecting tubules of the differentiated kidney) become differentiated into renal tubules. Thus, the cause of the kidney defect is the arrest in development of the ureter as a result of the changed genetic composition of the affected animals. The arrest in development of the ureter prevents the establishment of the spatial relationship between inductor and reacting system (the ureter and the metanephrogenic tissue), which is necessary for the induction of the kidney to take place (Gluecksohn-Schoenheimer, 1943, 1945).

The genes responsible for excessive development may be represented by the genes causing an increase in the number of digits on the forelimb or hindlimb. Quite a number of such genes are found in different animals. The development of this condition has been studied in a mutation of the mouse called "luxate" (symbol *lx* — Carter, 1954). In these mice additional toes (one or two) appear on the preaxial side of the foot, that is, on the inner side of the hallux. The anomaly can be traced back to the limb-bud stage, when the hindlimb-bud is excessively broad on its anterior edge. In the next stage, when mesenchymal condensations appear, indicating the rudiments of the digits, the number of these condensations is greater than normal. As the size of each digit rudiment corresponds to the size in normal limbs, it seems plausible that the excessive number of digit rudiments is the result of the excessive amount of material provided for digit development in the abnormally broad limb-bud.

Effects of Genes on Growth. Differences in size of animals are hereditary, but the control of growth and size is mostly determined by a multiplicity of genes. Occasionally, however, a single gene may have a marked effect on growth. The dwarf mutation in mice (pituitary dwarf, symbol *dw*) presents an example of this kind. The character is dependent on a single recessive gene. The homozygotic mice, which alone show the character, are born indistinguishable from normal individuals. At the end of the first week of postembryonic life, however, the dwarf individuals begin to show a slightly retarded growth as compared with the controls. In the third week, the dwarfs become conspicuously smaller than normals, and after weaning they increase in weight only slightly. The adult dwarfs are only one third to one fourth of the weight of normal mice. Retardation of growth in this case has been traced to an abnormality of the hypophysis, which is reduced in size and fails to produce the **growth hormone.** This has been proved experimentally by transplanting pieces of fresh rat hypophysis subcutaneously into dwarf mice. The result was that the treated mice resumed growth and reached the size of normal individuals (Smith and MacDowell, 1930). In this case, therefore, the gene does not directly affect the intrinsic growth rate or the cells' ability to proliferate, but the primary effect of the gene is to modify in a certain way the differentiation of one of the organs (the anterior lobe of the hypophysis). The arrest of growth is then the visible expression of the deficiency of a growth-stimulating substance normally produced by the hypophysis.

Sequence of Gene Action and the "Biogenetic Law."   The sequence of gene action in the course of development throws a new light on some generalizations which played a considerable role in embryology in the nineteenth century, namely, the "laws" of Baer and of Müller-Haeckel (pp. 7, 8). Essentially, both "laws" stress the greater conservatism of the earlier stages of ontogenetic development rather than that of the later stages, as a result of which the earlier stages show features general to large groups of animals, while the more specialized features distinguishing lower taxonomical units become apparent during later stages. These features distinguishing closely related animals may be considered as later acquisitions in the course of evolution.

We have seen that mutations can affect all stages of ontogenetic development and that even the earliest stages may thus become changed. In spite of this, the cumulative effect of mutations must necessarily show a greater influence on the later than on the earlier stages. If a mutation occurs which changes any developmental process in an earlier stage, and if the embryo nevertheless remains viable, then a number of later processes might also be modified. Mutations of genes which normally become derepressed during later stages of ontogenesis obviously cannot affect the earlier stages. In this way, the earlier stages of development would be modified by only a minority of mutant genes, while the great majority of the genes, both those acting in earlier stages and those acting later, would leave a mark on the later stages. The greater conservatism of the earlier stages can thus be given a rational genetic explanation.

It is obvious, however, that exceptions to the general rule are always possible. One obvious exception is presented by the structure of the egg. Lying, as it does, at the end of a life cycle (as well as at the beginning of a new one), the structure of the egg may be under the influence of genes becoming active throughout a lengthy period of life. It is, of course, well known that the structures of eggs, including egg membranes, show a great amount of variation. It is sufficient to point to the endless variety of sculpture found on the surface of insect eggs.

Chapter 17

GROWTH

17-1 MEASUREMENT OF GROWTH AND ITS GRAPHIC REPRESENTATION

In ordinary life, growth is often identified with increase in height. From a biological viewpoint, however, growth is an increase of the mass of living substance; therefore, only weight can be considered as an index of growth. The increase in linear dimensions (height, length) may accompany increase of mass, but the connection is by no means a simple one, and this must always be kept in mind, if for any reason the height or length of a growing animal is measured instead of its weight.

By weighing a growing animal at regular intervals and plotting the weight against time on a diagram, we get a **growth curve,** showing the increase of the mass of the animal with time. The shape of the growth curve very often resembles the letter S: There is an initial part, where the curve rises very gradually, then a middle part, where the curve rises steeply, and the last part, where the rise of the curve is again slowed down and the curve asymptotically approaches a horizontal line signifying the limit of growth in each particular case. A curve of this shape is known as the **sigmoid curve** (Fig. 417).

If the increments of growth for equal time intervals are measured, the increase during different periods of life of the same individual may be estimated. The increase, taken as a difference between the final and initial sizes (weights) of an animal for any period of time, irrespective of other factors, is called the **absolute increase.** The sigmoid curve is actually a graphic representation of the absolute increase, and it shows that the absolute increase per unit of time is, in this case, greatest in the middle part of the growth cycle. The absolute increase is, however, not a correct indication of the rate of growth and cannot be used for comparing the growth at different periods of life or for comparing the growth of different organisms. It is obvious that if a small animal and a big animal show the same absolute increase in a given time, their rates of growth are not the same: the small animal has to grow at a greater rate to have the same absolute increase as the big one. The rate of growth is measured as the relative increase—the increase related to the initial mass of growing substance. Therefore, the exact definition of the rate of growth is

$$v = \frac{dW}{dt} \cdot \frac{1}{W} \qquad (1)$$

where v is the rate of growth and W is the size (weight) of the animal at any given time, t.

Growth thus defined is in its essence an increase in geometrical progression, the increase being proportional to the initial quantity of growing substance.

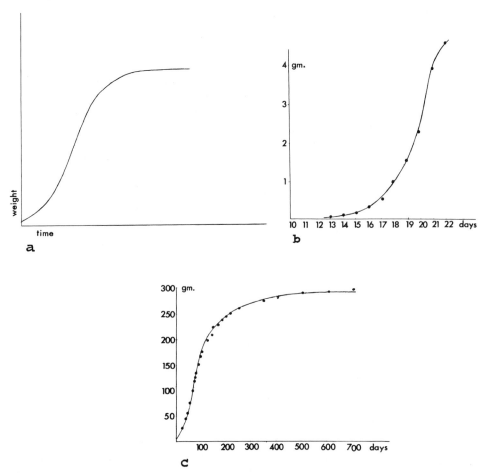

Figure 417. Sigmoid growth curves, *a,* Ideal sigmoid curve (after Robertson, 1908). *b,* Intrauterine growth of the white rat. *c,* Postnatal growth of the white rat. The points show the actual measurements. (After Donaldson, Dunn, and Watson, from Fauré-Fremiet, 1925.)

Increase in geometrical progression is called exponential; growth is thus an **exponential process.** The general formula for exponential growth, showing the increase of the growing mass with time, is

$$W = e^{vt} \qquad (2)$$

where W, as before, is the weight of the animal at any given time t, v is the observed rate of growth, and e is the base of natural logarithms (equal to 2.71828 . . . , accurate to the fifth decimal place). This formula is correct if the initial quantity of growing matter is infinitely small, and very nearly accurate if the initial size is negligible. If the initial size of the growing animal cannot be ignored, the formula must be slightly changed by the introduction of a constant, as follows:

$$W = be^{vt} \qquad (3)$$

where b is a constant and is equal to the initial size of the growing organism.

The rate of growth may be calculated from the same data as those on which the sigmoid curve of growth is based.

17–2 GROWTH ON THE CELLULAR AND ORGANISMIC LEVELS

Growth of individual cells is the most essential component of the growth of multicellular bodies. It is therefore of some importance to know the quantitative characteristic of cell growth. Unfortunately, because of the size of cells, measuring the growth of individual tissue cells is very difficult, although the rhythm of cell multiplication, especially *in vitro* in tissue cultures, can be observed quite easily.

Actual measurements of growth of single cells between two mitoses have been made on unicellular organisms. The growth of infusorians with elongated bodies was studied by measuring their length at regular intervals between two divisions (Schmalhausen and Syngajewskaja, 1925), and in a heliozoan, *Actinophrys*, which has a spherical shape, the growth was estimated by measuring the diameter of the cell (Syngajewskaja, 1935). Increase in weight of individual cells is even more difficult to measure. However, by means of a very fine technique, the weighing of individual cells of *Amoeba* throughout its life cycle has been accomplished (Prescott, 1957). The results of all these measurements conform well with one another and show that in these unicellular organisms growth is most rapid after a cell division and slows down later (Fig. 418).

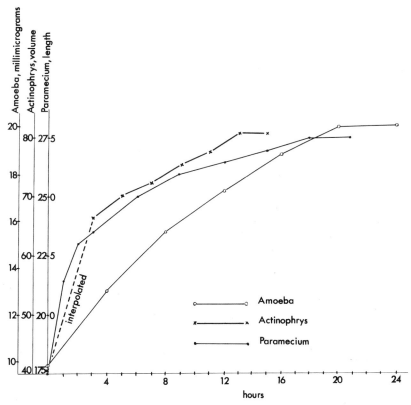

Figure 418. Growth curves of three protozoans. *Paramecium*—growth in length, in arbitrary units (divisions of ocular micrometer). (After Schmalhausen and Syngajewskaja, 1925.) *Actinophrys*—growth in volume, recalculated from original measurements of diameter. (After Syngajewskaja, 1935.) *Amoeba*—growth in weight, in millimicrograms. (After Prescott, 1957.) In *Actinophrys*, growth could not be measured during division, which takes about three hours to complete, and is interpolated for that period.

It is reasonable to assume that the course of growth of cells in multicellular organisms follows the same general course; this, however, is an extrapolation, which needs further experimental confirmation.

Although the growth of multicellular organisms is based on the growth of its cells, the relation between cell growth and organism growth is by no means a simple one. If the conditions for the growth and reproduction of cells are favorable and constant, cells tend to grow and divide by mitosis after regular intervals. If after the completion of a growth cycle each cell produces two cells, the whole population of cells would be doubled each time after a period equal to the average length of one cycle. The increase of the cell population would follow the principle of **exponential growth** defined by the expression

$$P = 2^n \qquad\qquad (4)$$

where P is the size of the population and n is the number of cycles of cell reproduction. The living mass would be proportional to the size of the population if the size of individual units (cells) does not change.

Exponential growth is observed in the reproduction of unicellular organisms (bacteria, protozoa, unicellular plants) living in constant and favorable conditions. The principle of exponential growth applies in the same way to the increase of populations of multicellular animals reproducing in optimal conditions, provided that the food supply is infinite and that the animals can disperse themselves in space to avoid overpopulation. Theoretically, the growth of living matter (in this case proportionate to the number of animals) may continue indefinitely, always having an exponential nature. In reality, of course, such an indefinite increase can never occur, since the conditions for it cannot be provided.

Tissue cells growing *in vitro* in a suitable nutritive medium, with sufficiently frequent transfers to new media, may grow exponentially, but entirely different conditions prevail in the case of cells that are components of a multicellular body (a multicellular animal). We may distinguish three basic ways in which the growth of a multicellular animal may be related to the growth of its cells.

Type 1. The volume of the animal's body may increase owing to the growth of the individual cells, without an increase in their number (**auxetic growth:** Needham, 1942). This rather rare case is found in nematodes, rotifers, and Larvacea among the tunicates. In the development of nematodes, cell divisions stop in the early stages of organogenesis. The number of cells in the fully grown nematode is thus the same as in a young one which has just emerged from the egg. (See Hyman, 1951.) The number of cells in each rudiment may be definitely fixed in this type of development. Thus, the whole excretory system of a nematode consists of only three cells. In the rotifer *Hydatina senta*, the total number of somatic cells of the body could be estimated and was found to be 959 (Martini, 1912). The gonads, however, are an exception among the other organ rudiments. The mitoses in the sex glands are not restricted to the early embryonic period, and the gametocytes continue to proliferate in the adult animals.

While the number of the cells is restricted, owing to the cessation of mitoses, the growth of the individual cells continues, and the growth of the animal is proportional to the increase in size of its constituent cells.

Type 2. The animal's growth may be the result of an increase in the number of its constituent cells (**multiplicative growth;** Huxley, 1932). The increase in number is brought about by mitotic division of all the cells, while the average size of the cell remains the same or nearly so. This type of growth is found quite commonly in embryos, and it is especially characteristic of the prenatal growth of the higher vertebrates. If the size of the cells remained strictly constant and no other processes were involved, growth of the animal's body would be directly proportional to the number of its constituent cells. Actually the size of the cells does not remain constant. As the embryo develops and its tissues become differentiated, the cells, or many of them, increase in size. This increase in size of the individual cells is, however, usually very limited and can account for only a very small proportion of the overall increase of the body. As the animal develops, the mechanism of its growth becomes modified by the differentiation of its tissues, so that a further type of growth should be recognized.

Type 3. In this type, the growth of the animal is based to a greater or lesser degree on the activity of special cells, retaining their ability to divide mitotically, while other cells have lost this ability more or less completely and thus cannot proliferate any more (**accretionary growth;** Huxley, 1932). These latter cells are the differentiated cells of the body, performing various physiological functions necessary for the maintenance of the animal's life. The former cells may be called the reserve cells, for they provide a supply of new cells, capable of reinforcing and replacing, in case of necessity, the functioning differentiated cells. The multiplying cells can also be called undifferentiated cells, inasmuch as they lack the specific morphological and physiological properties of functioning cells. However, the competence of the proliferating cells may already be curtailed to a large extent, so that they can differentiate into only one type of functioning cell or into only a limited number of types.

The epidermis of terrestrial vertebrates may serve as an example of this type of growth. In the outer layers of the epidermis, the cells do not divide and do not grow. Their cytoplasm becomes keratinized, thus forming a protective layer on the surface of the skin. The fully keratinized cells are no longer vital and are perpetually being peeled off. The malpighian or generative layer of the epidermis consists of cells which are not keratinized and which possess the capacity to proliferate. Later, the outermost cells of the layer replace the keratinized cells, undergoing keratinization in their own turn.

The cells of the bone tissue (the osteocytes) do not grow and do not proliferate. The growth of the bones is dependent on the activity of the cells of the periosteum, which are capable of proliferation and can become osteocytes, while they secrete additional quantities of intercellular bone matrix.

In a number of organs centers are found as regions or layers which alone contain proliferating cells. In the vertebrate eye, the proliferating cells are found in a ring in the region of the ciliary body. From there new cells are added on to both the retina and the iris. In the intestine of vertebrates, proliferating cells are found at the bottom of the intestinal glands, and the same is true in other organs.

17–3 INTERPRETATION OF GROWTH CURVES

If the growth of an animal were a strictly exponential process, the growth curve would have the shape of a hyperbola, the curve produced by an exponential increase of a quantity. The curve would begin at a point near zero of the coordinate system and would rise with ever-increasing steepness.

A prerequisite for growth to be truly exponential is that the rate of growth be constant. This, in turn, is possible only if neither the animal itself nor the environmental conditions change as growth proceeds. Such conditions apply to the growth of populations in perfectly favorable conditions, as mentioned previously. Purely exponential growth has been found in rod-shaped bacteria (Schmalhausen and Bordzilowskaja, 1930). In round bacteria, cocci, growth becomes retarded as the organism increases in size, presumably because the relative decrease of the surface in relation to the mass of protoplasm places the coccus in a worse position with regard to exchange of substances with the surrounding medium.

In multicellular organisms, exponential growth is never found in a pure form. The nearest approach to it is the growth of larvae of insects having a complete metamorphosis. In caterpillars of moths, growth starts at about the same rate after each molt; the absolute increase, therefore, is most rapid toward the end of larval development, as would be expected with exponential growth. However, the growth curve in this case is not a smooth hyperbola, as growth is very much retarded before and during molting. The curve is thus broken up into a series of spurts and level portions (Levitt, 1932).

In most animals, the rate of growth does not remain the same but diminishes quite regularly from the beginning to the end of the cycle of growth. As already mentioned, the resulting curve in many cases is in the shape of a letter S, the sigmoid curve. The absolute increase is small at the beginning, becomes greatest in the middle of the period of growth, and diminishes again later, until growth ceases altogether. If, however, the rate of growth is calculated for the various parts of the growth cycle, it is found that the rate of growth is highest at the beginning and declines throughout the period of growth. The low absolute increase at the beginning is due to the fact that only a small mass of living substance is growing. In the middle part of the cycle, the mass of growing substance has increased, and the absolute increase is, therefore, greater even though the rate of growth is not as high as at the beginning. In the last part of the growth period, the growing mass is even greater than at the middle of the cycle, but the rate of growth has sunk so low that the absolute increase slows down and comes to a standstill. The growth curve approaches asymptotically a horizontal line, which is the limit of growth or, in other words, the maximal size that can be attained by the animal whose growth is being considered.

The limit to growth is very distinct in such animals as birds and mammals but not in some of the lower vertebrates, such as reptiles and fishes. In these animals growth can go on indefinitely, although with an ever-retarded rate. As a result, some individual fishes and reptiles may attain an extraordinarily large size, if they escape accidental death which puts an end to the growth of other members of the same species. The growth curve in these animals approaches the shape of a parabola; it starts a little above the zero point (the initial mass of growing substance can never be 0) and gradually rises. The concavity of the curve is greatest near the starting point, and the curve flattens out as growth proceeds, approaching asymptotically a straight line inclined at an angle to the axes of the system of coordinates. This type of growth has been called **parabolic growth** (Fig. 419; Schmalhausen, 1927, 1930a).

It is interesting to note that the sigmoid curve is also found in the growth of populations. Animal populations under ideally favorable environmental conditions should increase exponentially, but under actual conditions a shortage of food and space soon causes a decline in the rate of reproduction, with the result that the growth curve of the population becomes sigmoid. In the end, the population may stop increasing further, having achieved the limit which is permitted by the environment in each case.

The progressive decline in the rate of growth has attracted much attention, and several theories have been proposed to account for this decline. The theories fall into two main groups, depending on whether the sigmoid curve or the parabolic curve has been considered as better representing the nature of the growth in metazoa.

The first interpretation, mainly propounded by Robertson (1908) and later followed by L. von Bertalanffy (1948, 1949), starts with the assumption that the organisms possess an intrinsic potential growth rate. This potential growth rate is curtailed by obstacles arising in the process of growth and setting a limit to it. As the organism

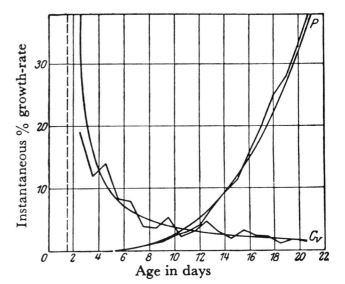

Figure 419. Increase in weight (P) and true rate of growth (Gv) in parabolic growth (growth of the chick embryo). The curves based on actual measurements and the smoothed theoretical curves are shown side by side. (After Schmalhausen, from Needham, 1931.)

approaches the limit of growth, the growth rate diminishes, and eventually growth stops altogether. These relationships have been given the following mathematical form:

$$\frac{dW}{dt} = bW\frac{L - W}{L} \qquad (5)$$

where W is the size of the organism at any given time t, b is the potential growth rate, and L is the maximum size (limit of growth or asymptote of the growth curve). The equation (5) is known as the **logistic equation.** The sigmoid curve is sometimes also referred to as the logistic curve.

The logistic equation represents growth as an essentially exponential process, the increase $\left(\frac{dW}{dt}\right)$ being proportional to the growing mass (W). On the other hand, the potential growth rate can actually be observed only in the initial stages of growth, while W is very small in relation to L (the actual size being very small as compared with the final size). At this stage the factor $\frac{L - W}{L}$ is near unity. As W increases, this factor becomes less than unity, and when W = L, it becomes 0. The increase then also becomes 0.

The logistic equation has been applied both to the growth of individuals and to the growth of populations. It appears, however, that the equation is much more suited to the latter case. Several objections may be made against the application of the equation to the growth of individuals:

1. The nature of the limiting factor has not been made clear. Bertalanffy suggests that the decline in the rate of growth is due to a discrepancy between the increase of surfaces and volumes. (The surfaces increase as the square of linear dimensions, the volumes as the cube of linear dimensions.) The ultimate growing unit, however, is the cell, and the sizes of cells as a rule do not change so as to account for the retardation of growth. The change in the relative size of any other surfaces in the animal's body would not explain the universal nature of the slowing down of growth.

2. Animals having no upper limit of size (having a parabolic growth curve) still show a retardation of growth. The logistic equation is not applicable in this case.

3. Holometabolic insects do not show a retardation of growth, although their growth is limited. The existence of a growth limit thus does not, in itself, cause a slowing down of growth during the earlier periods of development.

The second way of interpreting the growth curve makes use of the fact that the rate of growth (specific rate of growth) is inversely proportional to the time that elapses since the beginning of growth:

$$v = \frac{k}{t} \qquad (6)$$

$$\text{or } v \cdot t = k \qquad (7)$$

where v is the growth rate, t is time, and k is a constant which is different for each organism.

The time since the beginning of growth is an indicator of the age of the animal. Minot (1891, 1908) was the first to make this observation, and he accordingly stated that the decrease in the rate of growth is due to increasing age of the animal. The decline of growth is thus one of the many expressions of **aging**. Minot also showed the connection between aging and the differentiation of cells in the developing animals. The more the tissues of the animal become differentiated, the less they are able to grow. This aspect of Minot's theory has been further elaborated by Schmalhausen.

According to Schmalhausen (1930b), the differentiation of parts of the developing animal, and not age as such, is the cause of retardation of growth. Inasmuch as the degree of differentiation of some cells of the animal's body may be preserved at a certain level, their rate of growth may remain constant, and exponential growth would take place in such groups of cells or tissues. This would account for the type of growth found in caterpillars: The caterpillar of a later instar is not more differentiated than one of the first instar, as far as its larval organs are concerned. The organ rudiments of the adult moth (the imaginal discs) do not change very much until the pupal stage, and in any case, they form only an insignificant part of the bulk of a caterpillar and do not essentially influence the growth curve. Given equal levels of differentiation at the start of every instar, the growth rate remains constant, and this has actually been observed.

Growth may also be exponential in the growth centers of a vertebrate's body (e.g., malpighian layer of the epidermis, the ciliary body of the eye), but the proportion of proliferating cells to those which have partially or completely lost their ability to proliferate may diminish with age, and this becomes noticeable in the decrease of the overall rate of growth. Besides, the cells of the growth centers are already partially differentiated and do not proliferate as rapidly as the cells of an early embryo. Lastly, growth centers may disappear altogether. Growth of the long bones of the limbs in vertebrates is dependent on the presence of an actively proliferating layer of cells between the diaphysis of the bone and its epiphyses. At about the time of sexual maturity the proliferating layer disappears, and the bony cores of the epiphyses are firmly joined to the diaphysis. The result is that the growth in length stops.

The residual growth centers are, however, an exception rather than the rule among the tissues of a higher animal. Most of the cells become differentiated, which is, of course, essential for the functioning of the various organs, and this leads to retardation of growth.

There appears to be an inverse proportion between the degree of differentiation and the rate of growth; the more highly an organ or tissue is differentiated, the slower it grows. The most highly differentiated tissue of the vertebrate's body is the tissue of the brain and spinal cord. The nerve cells in the adult state completely lose the ability to proliferate, and in the brain and spinal cord there is no residue of undifferentiated cells functioning as a growth center. The result is that the brain and spinal cord, throughout life, grow more slowly than other parts of the body. The sense organs have a degree of differentiation that is scarcely lower than that of the brain tissue, and they also grow quite slowly. Organs like the heart and the kidneys grow at a higher rate than the brain and sense organs. The heart and the kidney, however, are rather highly differentiated organs, with no growth centers consisting of undifferentiated cells to provide for proliferation, and their growth is consequently not very fast. Higher rates of growth are found in the muscle and skeletal system. Although bone tissue itself does not grow and striated muscle possesses a highly specialized differentiation, in both cases there are residues of undifferentiated cells providing for rapid growth. There is the periosteum in the case of bone (perichondrium for cartilage, in addition to a certain degree of interstitial growth). New elements of muscle tissue, new cells or muscle fibers, are not normally produced after the earlier part of the embryonic period has passed, but the individual muscle fibers, or muscle cells, in the case of the smooth muscle tissue, have the ability to increase in size to a very great extent and thus to produce a rapid growth of the muscular tissue.

The highest rates of growth are found in parts of the intestinal tract and in the skin of vertebrates. The skin possesses a permanent growth center in its malpighian layer. There are groups of growing and proliferating cells in the gut (at the bottom of the intestinal glands and crypts); furthermore, the epithelial cells of the intestinal tract do not have a complicated morphologic differentiation. The connective tissue is also capable of rapid growth. While the products of differentiation in this tissue are the various intercellular fibers, the cells themselves (fibroblasts and histiocytes) do not become highly differentiated, and even in the adult animal they can at any time be mobilized to proliferate and produce new connective tissue fibers. Such a mobilization of the connective tissue cells takes place in wound healing.

In any given organ, the rate of growth is highest before histological differentiation has set in, or at least when it has not progressed too far. Even in the case of very highly differentiated organs, such as the brain, the early rudiment of the organ may and actually does possess a very high growth rate, as shown in Table 17.

The inverse relationship between growth and differentiation can also be shown under experimental conditions. In the intact organism, the tissue cells differentiate,

TABLE 17   Growth of Organs of the Chick Embryo from the Fourth
Day of Incubation to Hatching

	Initial Weight (In Milligrams)	Final Weight (In Milligrams)	Weight Increased by a Factor of	Growth Constant $k = v \cdot t$
Brain	10	1020	×102	2.10
Lens	0.08	8.8	×110	2.10
Forelimb	0.75	540	×720	2.93
Metanephros	1.4 (on ninth day)	130	× 93	3.58
Whole embryo	53	41,000	×773	

After Schmalhausen, 1927.

and their growth becomes restricted. When bits of tissue are cultivated *in vitro*, the tissues become dedifferentiated, and at the same time they regain the ability to proliferate at a rapid rate. It is possible, however, to retard the growth of a tissue culture by supplying it with a medium poor in growth-promoting substances or by keeping it longer on the same plasma clot. Under these circumstances, the cells may show signs of renewed differentiation; intercellular fibers (or cartilage matrix) are produced in the culture, depending on the nature of the cultivated cells. Differentiation goes hand in hand with reduced proliferation.

Dedifferentiation and a concomitant increase of the proliferation rate of cells also occurs during regeneration of lost parts, as will be described in Chapter 19.

The contention that progressive differentiation is the actual cause of the decline in the growth rate of animals with time is thus well substantiated. On the other hand, this does not quite explain the existence of a definite growth limit found in many animals, such as birds, mammals, and also insects. The nature of the limitation of growth may perhaps be different in the arthropods and in the vertebrates. In the arthropods (insects) the limit to growth is quite obviously connected with the mechanism of metamorphosis and will be discussed in Chapter 18. In the higher vertebrates, growth is regulated by means of hormones. The hypophysis is known to produce a growth-stimulating hormone. Animals (dogs) treated with this hormone may be caused to grow in excess of the usual size (Evans and collaborators, 1933). In humans, dwarfism is attributed to an insufficient production of the growth hormone by the hypophysis. The thyroid hormone is also necessary for normal growth, and removal of the thyroid in a young animal causes growth to be below normal. The cessation of growth in mammals (and probably also in birds) seems to be the result of an interplay of perhaps numerous hormones controlling growth. The details of this mechanism are not yet completely understood.

A new aspect of growth control, which might contribute to the progressive slowing down and eventual cessation of growth, may possibly emerge from studies on substances called chalones (pronounced kay-lones), which are claimed to inhibit cell proliferation in tissues. (See Bullough, 1967, 1973.) The chalones were originally discovered in studies on cell reproduction in epidermis, where differentiated (and keratinizing) cells are continuously replaced due to cell proliferation in the deep generative layer (Bullough and Laurence, 1960). Extracts from epidermis, administered to epidermal cells in tissue culture, or even better, to cultivated *in vitro* pieces of skin (bits of mouse ears), cause within hours a reduction in the number of mitoses (Laurence, 1973). Similar chalones have been found to influence mitosis in other tissues, such as granulocyte-producing tissue, liver, and kidney. The active substance appears to be a protein or mucoprotein of comparatively low molecular weight (30,000 to 40,000 in epidermal chalones, but in other tissues possibly as low as 5000 to 2000). The peculiar feature of the chalones, it is claimed, is that they are tissue-specific (though not species-specific). Thus, an increase in the mass of any particular tissue would reduce the proliferation of that tissue by producing greater amounts of the particular tissue chalone.

The relevance of this concept to the interpretation of growth curves is fairly obvious. In any animal, organ, or tissue which increases by accretionary growth (p. 523), the proportion of differentiated cells to undifferentiated, dividing cells continuously increases. If a tissue-specific chalone is produced by differentiated cells, as has been implied, its concentration would become higher, and it would depress the mitotic activity of the undifferentiated cell reserve. Although this seems to be an attractive idea, much more has to be learned about chalones before definite conclusions can be reached. There are indications that different agents may be lumped together under the name of chalones; the substances have not yet been purified and characterized sufficiently, and their mode of action is not clear (whether they suppress the replication of chromosomes in the S phase or prevent cells already in the G_2 phase from actually entering mitosis). (For a discussion see Marks, 1972, Laurence, 1973.)

17–4 PROPORTIONAL AND DISPROPORTIONAL GROWTH OF ORGANS

The growth of different organs and of parts of the same animal (embryo) very seldom goes on at the same rate. As a rule, some grow faster and some grow slower. The result is that the proportions of the animal change with growth. In the case of vertebrates, the central nervous system and the sense organs are distinguished by their particularly low rate of growth, and the size of these organs diminishes relatively throughout the whole embryonic and postembryonic development. In the early embryo (of a chick, for instance), the head is at first quite as large as the rest of the body, and a large portion of the head is made up by the relatively enormous eyes. At the time of hatching, the head is already much smaller than the body, and in the adult fowl the head is relatively small.

To compare the growth of different organs, their rates of growth may be estimated. However, the rate of growth changes with age, so that it is not a very convenient quantity for purposes of comparison. The growth constant k (see preceding section) may be calculated for each organ as well as for the whole animal, and the constants of different organs may be conveniently compared.

A different method of analyzing the unequal growth of organs has been proposed by J. Huxley (1932). Huxley found that if two parts of an animal grow at different rates, their sizes at any given moment are in a simple relationship to each other. The relationship is expressed by the following formula, known as the formula of **allometric growth:**

$$y = bx^k \qquad (8)$$

where y is the size of one of the organs, x is the size of the other organ, b is a constant, and k is known as the **growth ratio,** because it shows the relation of the rates of growth of the two parts which are being compared. If the growth ratio equals 1, the two organs grow proportionally or **isometrically.** The constant b in this case shows the proportion of one organ to the other: $b = \frac{y}{x}$. If the growth ratio does not equal 1, the growth is disproportional or **allometric,** the relative size of the two organs changing as growth proceeds. If the growth ratio is greater than 1, organ y grows at a quicker pace than organ x. In this case, organ y is called **positively allometric** to organ x. If the growth ratio is less than 1, organ y grows at a slower pace than organ x; it is **negatively allometric** to x.

The formula for allometric growth can be given in a logarithmic form:

$$\log y = \log b + k \log x \qquad (9)$$

In other words, the logarithms of the sizes of two organs growing at different rates are proportional to each other. This form of the equation is especially instructive, as it

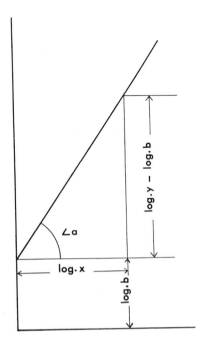

Figure 420. Graphic representation of allometric growth (in logarithmic transformation).

allows for a simple graphic representation of allometric growth (Fig. 420). If the logarithms of the sizes of one organ are plotted against the logarithms of the sizes of the other organ, the points lie on a straight line crossing the ordinate axis at a distance from the zero point equal to log b and ascending with an inclination a, determined by the equation:

$$\tan \angle a = \frac{\log y - \log b}{\log x} = k \quad (10)$$

The latter formula discloses the true significance of the exponent k and also shows how this exponent can be determined in practical work.

A great advantage of the formula for allometric growth is that it eliminates time, which enters into all the other growth equations as the independent variable. As a result, the formula can be applied for the analysis of growth of parts in groups of specimens whose timing (age) is not known, such as when a series of animals of different sizes is collected in nature and not bred and reared under permanent observation. The formula of allometric growth is also applicable to linear dimensions, as well as to volume or weight, and is in this respect more pliable than the other equations used for analysis of growth.

The organs to be compared, x and y, may be chosen at will. A special case of the application of the formula is when the size of an organ is compared with the size of the body less the size of the organ, the growth of which is being investigated. It can then be learned whether or not an organ grows in excess of the other parts (the remainder of the body).

It has been found that organs attaining especially large proportions in certain animals often owe their increased size to positive allometric growth. In the fiddler crab, *Uca pugnax,* one of the chelae in the males attains an extraordinarily large size, up to 38 per cent of the weight of the body as a whole. The large chela is positively allometric with respect to the rest of the body, with a growth ratio of 1.62. This means, by the

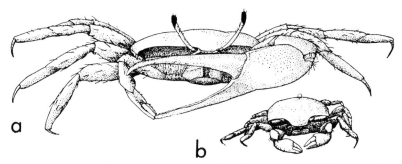

Figure 421. Positive allometric growth of the left chela in a fiddler crab, *Uca*. *a*, Adult male with disproportionately large left chela. *b*, Young male crab with both chelae of equal size. (From Morgan, 1927.)

way, that in a young male crab the chelae are very nearly equal in size and that the relative size of the larger chela increases as the crab grows (Fig. 421).

In most mammals the facial region of the head, including the jaws, grows at a greater rate than the cranial part of the head. The length of each of these parts of the head has been measured and the growth compared during the postembryonic life of sheep dogs. In a puppy, the length of the cranium is almost double the length of the facial region (42 mm. and 22 mm.). In the adult dog, the cranial region is only slightly longer than the facial region (120 mm. and 112 mm.). The facial region is thus positively allometric with respect to the cranial region. The growth ratio is, in this case, 1.49. The lower rate of growth of the cranial region is mainly due to the rate of growth of the brain, which is enclosed by the skull and which is noted for its low rate of growth. It is negatively allometric with respect to the body as a whole.

The formula of allometric growth can be used not only for the comparison of the growth of morphologically definable parts of the animal's body but also for comparing the increase of various chemical components of the body. In differentiating tissues, the amount of protein increases with respect to the amount of nuclear chromatin, as indicated in Section 15–1. This can be represented as a case of allometric growth. In Table 18, the quantities that have been compared are the deoxyribonucleic acid phosphorus and the protein nitrogen in several organ rudiments of the chick embryo.

TABLE 18   Comparison of Increase of Protein Nitrogen and Deoxyribonucleic Phosphorus in Different Organs of the Chick Embryo

Organ	Incubation Period in Days	Growth Ratio of Protein Nitrogen to Deoxyribonucleic Phosphorus
Brain	8–13	2.55
Brain	15–19	1.05
Heart	10–19	1.05
Liver	8–19	1.10
Muscle (leg)	12–19	1.40

After Davidson and Leslie, 1950.

The high rate of increase of the protein nitrogen (large growth ratio) corresponds, according to what was previously said, to a rapid rate of differentiation. The high figures for the brain and the leg muscle can therefore be easily understood. The heart shows little increase in differentiation because in the 10-day chick the heart is already fully functional, that is, already differentiated. The liver is an example of an organ which becomes differentiated late, and even then its level of morphological differentiation is rather low.

Part Seven

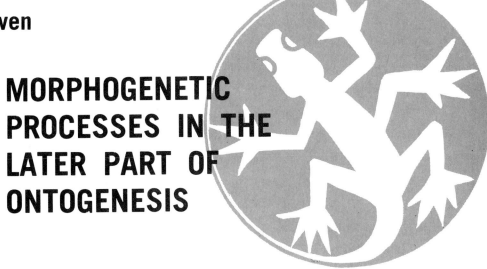

MORPHOGENETIC PROCESSES IN THE LATER PART OF ONTOGENESIS

Taken as a whole, ontogenetic development appears to slow down with the increasing age of the individual animal. In the later periods of life, spectacular changes such as those occurring during cleavage, gastrulation, and organ formation do not take place. When histological differentiation starts, it gives rise to various tissues, but further differentiation serves mainly to support and only sometimes to increase the already established initial differences. Even growth, as previously stated, diminishes in rate with age. There are, however, special cases in which morphogenetic processes may be aroused again at a late stage of ontogenesis, so that the changes produced by these processes may equal in volume those observed in the earlier stages of the individual's development. These special cases will now be considered.

Chapter 18

METAMORPHOSIS

The first case in which morphogenetic processes may be reactivated after development has almost reached a standstill is observed in animals in which the embryo develops into a larva, and the larva is transformed into the adult by way of **metamorphosis.** Larval forms and an accompanying metamorphosis are found in most groups of the animal kingdom, though by no means in all representatives of each group. The larvae usually have special names distinguishing them from the adult forms. The following list is a survey of the occurrence of larval forms, with their names:

1. Porifera —— amphiblastula
2. Coelenterata —— planula (Fig. 422*A*)
3. Platyhelminthes
   Turbellaria polycladida —— Müller's larva
   Trematodes —— miracidium, cercaria, redia, sporocyst
   Cestodes —— onchosphaera
4. Nemertinea —— pilidium
5. Annelida —— trochophore (Fig. 422*B*)
6. Mollusca —— trochophore and veliger
7. Crustacea
   Entomostraca —— nauplius (Fig. 422*D*)
   Malacostraca —— zoea

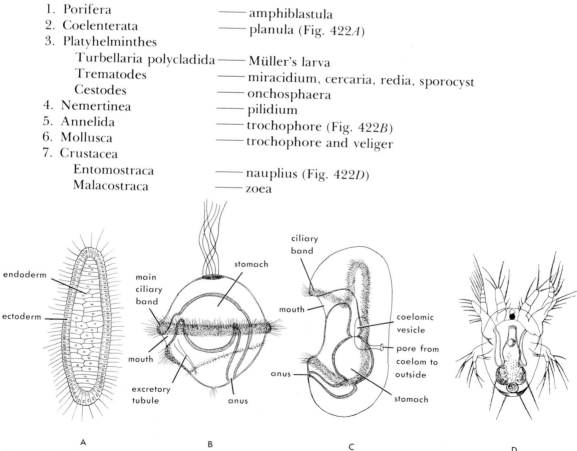

Figure 422. Some types of larvae in different animals. *A*, Planula of a coelenterate; *B*, trochophore of an annelid; *C*, larva of an echinoderm (starfish); *D*, nauplius larva of a crustacean.

8. Insecta	——nymphs of insects with incomplete metamorphosis (Fig. 430); caterpillars, grubs, etc., of insects with complete metamorphosis (Fig. 434)
9. Bryozoa ectoprocta	——cyphonautes larva
10. Echinodermata	——pluteus, bipinnaria (Fig. 422C), auricularia, etc.
11. Enteropneusta	——tornaria larva
12. Ascidiacea	——tadpole larva
13. Cyclostomata	——ammocoete
14. Amphibia	——tadpoles of frogs; aquatic larvae of salamanders

The morphogenetic processes in all these groups differ both in the nature of transformation and in the mode of causation of the whole sequence; it is thus impossible to describe them in common terms. Two cases are chosen here for closer study: the metamorphosis in amphibians and the metamorphosis in insects.

18–1   CHANGES OF ORGANIZATION DURING METAMORPHOSIS IN AMPHIBIANS

In amphibians, metamorphosis is associated in typical cases with a transition from an aquatic to a terrestrial mode of life. Superimposed on this change in environment, there is in the anurans (frogs and toads) a change in feeding. The tadpoles of most frogs and toads feed on vegetable matter—particles of plants, living and decaying—which they scrape off submerged objects with the aid of the horny teeth surrounding their mouths. Some are detritus feeders, passing through their guts the mud and detritus collected from the bottom, and others, such as the tadpole of *Xenopus*, are plankton feeders. Adult frogs are carnivorous, living on insects, worms, and the like, but sometimes also on larger prey, such as smaller frogs and even little birds and rodents, which they catch, overpower, and swallow. In the case of urodeles, there is no substantial change in diet, the larvae being as carnivorous as the adults, though naturally feeding on smaller animals (mainly crustaceans and worms). The changes in the organization of the animals during metamorphosis are in part progressive and in part regressive and may be grouped into three categories:

1. The organs or structures necessary during larval life but redundant in the adults are reduced and may disappear completely.
2. Some organs develop and become functional only during and after metamorphosis.
3. A third group of structures, while present and functional both before and after metamorphosis, become changed in order to meet the requirement of the adult mode of life.

In **anurans,** the differences between the modes of life of the larva and the adult are much more profound, and accordingly the changes at metamorphosis are more extensive than in the urodeles. We will consider them first (Fig. 423).

The regressive processes occurring during the metamorphosis in frogs are the following: The long tail of the tadpole with the fin folds is resorbed and disappears without a trace. The gills are resorbed, the gill clefts are closed, and the peribranchial cavities disappear. The horny teeth of the perioral disc are shed as well as the horny lining of the jaws, and the shape of the mouth changes. The cloacal tube becomes shortened and reduced. Some blood vessels are reduced, including parts of the aortic arches, as indicated in Section 13–6.

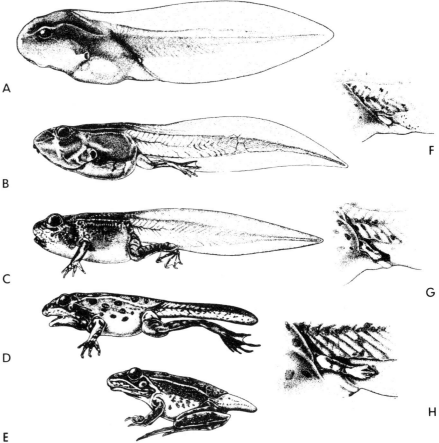

Figure 423. Metamorphosis in the frog. *A*, Tadpole with rudiment of hindlimb in the form of a small bud. *B*, Tadpole with fully developed hindlimbs; *C,D*, stages of metamorphosis; *E*, metamorphosed froglet with remains of the tail; *F,G,H*, three stages of hindlimb development (these fall between the stage of A and the stage of B). (From Witschi, 1956.)

The constructive processes involve, first, the progressive development of the limbs, which increase in size and differentiation. The forelimbs, which in the frogs develop under cover of the opercular membrane, break through to the exterior. The middle ear develops in connection with the first pharyngeal pouch (the pouch situated between the mandibular and the hyoid arches). The tympanic membrane develops, supported by the circular tympanic cartilage. The eyes protrude on the dorsal surface of the head and develop eyelids. The tongue is developed from the floor of the mouth. The organs which function both in the larva and the adult, but change their differentiation during metamorphosis, are primarily the skin and the intestine. The skin of the tadpole is covered with a double-layered epidermis. During metamorphosis, the number of layers of cells in the epidermis increases, and the surface layers become cornified. Multicellular mucous and serous glands develop as pockets sinking from the surface into the subcutaneous connective tissue layer. The lateral line sense organs, present in the skin of tadpoles, disappear during metamorphosis in most frogs. The pigmentation of the skin is changed; new patterns and colors appear. The intestine, which is very long in tadpoles, as in most herbivorous animals, becomes greatly foreshortened, and most of the coils which it forms in the tadpole become straightened out. The metamorphosis is very rapid and takes only a few days.

In **urodeles,** the changes at metamorphosis are far less striking. The tail is retained; only the fin fold disappears. The branchial apparatus is reduced, the external gills become resorbed, and the gill clefts close. The visceral skeleton becomes greatly reduced. The head changes its shape, becoming more oval. The progressive changes are less conspicuous than in the metamorphosing frog tadpoles. They are restricted mainly to changes in the structure of the skin and the eyes. The eyes bulge more on the dorsal surface of the head and develop lids. The skin becomes cornified, and multicellular skin glands become differentiated. The pigmentation of the skin changes. The legs, contrary to those of tadpoles, undergo hardly any change at all, and the same may be said of the alimentary canal. The metamorphosis is, on the whole, more gradual and may take up to several weeks (Fig. 424).

It is worth noting that the lungs do not undergo drastic changes during metamorphosis both in anurans and urodeles. They develop very gradually and become fully functional in the larval state. Long before metamorphosis, the larvae of frogs as well as of salamanders start coming up to the surface and gulping air into their lungs and thus supplementing their aquatic respiration. This may be of considerable importance where the larvae develop in stagnant and polluted waters, as is often the case.

Hand in hand with morphological changes during metamorphosis go physiological

Figure 424. Metamorphosis in the axolotl. *A*, Animal in the larval condition; *B*, fully metamorphosed animal. (Courtesy of V. V. Brunst, Roswell Park Memorial Institute.)

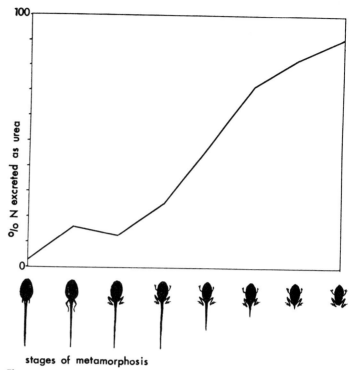

Figure 425. Changes in the amounts of ammonia and urea excreted by tadpoles of the frog *Pyxicephalus delalandii* during metamorphosis. (After unpublished data by J. Balinsky.)

changes, of which we shall point out the following. In frog tadpoles, the endocrine function of the pancreas starts at metamorphosis and this is connected with the increased role of the liver in the turnover of carbohydrates (glycogen). (See Abeloos, 1956.) A profound change takes place in the excretory mechanism. In the tadpole, the end product of nitrogen metabolism is ammonia, which is easily disposed of (by diffusion) in an aquatic medium but which in a terrestrial animal might accumulate and become dangerous because of its high toxicity. Metamorphosed frogs, however, excrete most of their nitrogen in the form of urea and only small amounts as ammonia. The change-over occurs in the late stages of metamorphosis and is, of course, due to a changed function of the liver, which performs the synthesis of urea (Munro, 1939: Fig. 425).

The reduction of the gills and the tail in tadpoles is effected by autolysis of the component tissues of these organs, with active participation of ameboid macrophages which phagocytose the debris of the disintegrating cells. The same mechanism, though of a limited scope, is in action when the external gills and fin folds are reduced in urodele amphibians.

Since destructive processes play such a considerable role in metamorphosis and since, in addition, food intake may be interrupted during the crucial part of the transformation, especially in tadpoles, the mass of the body at the end of metamorphosis is smaller than at the beginning ("degrowth," p. 468). The reduction of the body mass is due not only to loss of some parts (gills, tail), but the remaining parts, excepting the actively growing organs, also appear to shrink during metamorphosis: the head and trunk of metamorphosed amphibians are smaller than in larvae just before the commencement of metamorphosis.

18–2 CAUSATION OF METAMORPHOSIS IN AMPHIBIANS

The concurrent changes in so many parts of the animal's body during metamorphosis suggest the existence of some common cause for all the transformations. It has been found that this common cause is a hormone released in large quantities from the thyroid gland of the animals entering the stage of metamorphosis. The first indication of this was obtained when Gudernatsch fed frog tadpoles on dried and powdered sheep thyroid gland and observed that they metamorphosed precociously (1912). Feeding tadpoles with preparations of other glands did not have the same effect. This experiment made it seem very probable that tadpoles are capable of reacting to the thyroid hormone by metamorphosing.

That the thyroid hormone is actually the cause of metamorphosis in normal development was further proved by the following two experiments. The rudiment of the thyroid gland was removed in frog embryos in the tailbud stage (through an incision on the ventral side, Fig. 426A). The operated tadpoles were fully viable and showed normal growth but failed to metamorphose (Allen, 1918), although they were kept alive almost a year after the control animals had become little froglets. The thyroidless tadpoles continued to grow and attained a much greater size than normal, having a total length of up to 123 mm. instead of about 60 mm., as at the beginning of metamorphosis in normal tadpoles. It was thus proved that metamorphosis cannot set in without a stimulus emanating from the thyroid gland. The final experiment consisted in supplying the thyroidless tadpoles with thyroid hormone from without, either by feeding them on dried thyroid gland (Allen, 1918) or by immersing them in water containing soluble extracts from thyroid glands. The tadpoles treated in this way immediately proceeded to metamorphose, thus showing that their own thyroid glands were not necessary as long as they were supplied in some way or other with thyroid hormone (Allen, 1938).

Similar experiments were carried out on urodeles. A very suitable animal for the

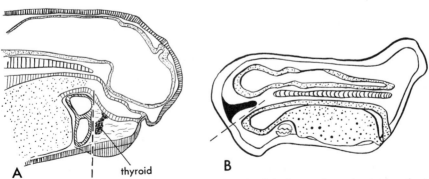

Figure 426. Removal of rudiments of endocrine glands in frog embryos (sagittal section). *A,* Late tailbud stage, showing how the rudiment of the thyroid was removed through a ventral incision (broken line); *B,* Early tailbud stage showing how the rudiment of the hypophysis (black) was removed through an incision indicated by a broken line. (From Allen, in Willier, Weiss, and Hamburger, 1955.)

experiments is the axolotl *Ambystoma mexicanum,* which under ordinary conditions does not metamorphose at all but may be induced to metamorphose by thyroid treatment (Marx, 1935).

As already indicated, the active principle of the thyroid gland may be introduced into the animal's body in several different ways. Normally it emanates from the animal's own thyroid gland. The same effect may be produced by implanting bits of live thyroid gland, feeding the animals on thyroid gland, injecting them with preparations of thyroid gland, or keeping them in water containing soluble extracts of thyroid glands. The last fact clearly shows that the active principle of the thyroid gland is a chemical substance—a hormone.

A saline extract of fresh thyroid tissue contains protein, **thyroglobulin,** which retains the activity of the thyroid gland. An important characteristic of thyroglobulin is that it contains iodine, which, as we will see, is of great importance for the working of the thyroid hormone. Thyroglobulin has a molecular weight of about 675,000. Its molecules are thus very large, and it is unlikely that thyroglobulin can penetrate as such through cellular membranes, which would be necessary if it were to leave the thyroid gland and reach the cells eventually reacting to the thyroid treatment.

To become active, the iodine-containing compounds are released from their connection to the thyroglobulin. Several compounds of smaller molecular weight are liberated in this way. In chemical structure, all these compounds are combinations of molecules of the amino acid tyrosine with one or more atoms of iodine. The most important of them are tri-iodothyronine and thyroxine. In both of these compounds two residues of tyrosine are joined together, and to these are attached three (in tri-iodothyronine) or four (in thyroxine) atoms of iodine. The structural formulas of these substances follow:

tri-iodothyronine:

$$\text{HO}-\text{C}\underset{\underset{\text{H}}{\overset{|}{\text{C}}-\underset{\text{H}}{\overset{|}{\text{C}}}}{\overset{\overset{\text{I}}{\overset{|}{\text{C}}}=\overset{\text{H}}{\overset{|}{\text{C}}}}{}}\text{C}-\text{O}-\text{C}\underset{\underset{\text{I}}{\overset{|}{\text{C}}-\underset{\text{H}}{\overset{|}{\text{C}}}}{\overset{\overset{\text{I}}{\overset{|}{\text{C}}}=\overset{\text{H}}{\overset{|}{\text{C}}}}{}}\text{C}-\overset{\overset{\text{H}}{|}}{\underset{\underset{\text{H}}{|}}{\text{C}}}-\overset{\overset{\text{H}}{|}}{\underset{\underset{\text{NH}_2}{|}}{\text{C}}}-\overset{\overset{}{}}{\underset{\underset{\text{O}}{||}}{\text{C}}}-\text{OH}$$

thyroxine:

$$\text{HO}-\text{C}\underset{\underset{\text{I}}{\overset{|}{\text{C}}-\underset{\text{H}}{\overset{|}{\text{C}}}}{\overset{\overset{\text{I}}{\overset{|}{\text{C}}}=\overset{\text{H}}{\overset{|}{\text{C}}}}{}}\text{C}-\text{O}-\text{C}\underset{\underset{\text{I}}{\overset{|}{\text{C}}-\underset{\text{H}}{\overset{|}{\text{C}}}}{\overset{\overset{\text{I}}{\overset{|}{\text{C}}}=\overset{\text{H}}{\overset{|}{\text{C}}}}{}}\text{C}-\overset{\overset{\text{H}}{|}}{\underset{\underset{\text{H}}{|}}{\text{C}}}-\overset{\overset{\text{H}}{|}}{\underset{\underset{\text{NH}_2}{|}}{\text{C}}}-\overset{\overset{}{}}{\underset{\underset{\text{O}}{||}}{\text{C}}}-\text{OH}$$

Of the two compounds, thyroxine is produced in very much greater quantities and is thus the main active substance released by the thyroid gland, although tri-iodothyronine is more highly active on tissues.

It has been tested whether iodine alone can cause metamorphosis in amphibians, and positive results have been obtained either by keeping frog tadpoles or newt larvae in water containing the element iodine (in solutions as weak as 0.0000003833 per cent!), by injecting animals with iodine, or by implanting iodine crystals into the body cavity. The element iodine caused metamorphosis even in thyroidectomized axolotls; thus, it must have been acting directly and not through the increased production of hormone by the animal's own thyroid gland. On the other hand, it appears that the degree of activity of the iodine atoms may be greatly influenced by the type of amino acid with which the iodine is bound. This can be clearly shown by comparing the activity of the two amino acids di-iodotyrosine and thyroxine. By placing tadpoles in solutions of each amino acid separately, it was found that the same amount of iodine was 300 times more active when forming a part of the thyroxine molecule than when incorporated into the di-iodotyrosine molecule. The activity of tri-iodothyronine is three to five times as high as that of thyroxine (L. Wilkins, 1960).

The thyroid gland is not the only gland which is involved in the causation of metamorphosis in amphibians. It has been discovered that the hypophysis plays an important part as well. If, in frog tadpoles, the hypophysis is destroyed (Adler, 1914) or if the rudiment of the hypophysis in a late frog embryo is excised (Fig. 426B — Allen, 1929), the tadpoles do not metamorphose, just as if their thyroid glands had been removed. The removal of the hypophysis rudiment may later be compensated for by implanting pieces of hypophysis from metamorphosed or adult frogs, but only if the thyroid gland of the animal is intact. If the thyroid gland has been removed, no amount of implanted hypophyseal tissue can induce metamorphosis. From this it may be concluded that the hypophysis does not act upon the tissues directly but acts only by way of stimulating the thyroid gland. In fact, the thyroid gland of hypophysectomized animals remains underdeveloped and does not accumulate the thyroid hormone (in the form of "colloid") in its follicles. Implantations of active hypophyses can also be used to stimulate metamorphosis in the axolotl.

The agent necessary for activating the thyroid gland is produced in the anterior (epidermal) lobe of the hypophysis and has been isolated in the form of a **thyrotropic hormone.**

In the case of larval amphibians, the hypophysis does not produce the thyrotropic hormone until the time when metamorphosis normally occurs. This has been proved by taking hypophyses from tadpoles of various ages and transplanting them into tadpoles whose own hypophyses had been removed previously. Whereas hypophyses taken from tadpoles in stages of metamorphosis or from metamorphosed frogs compensated for the removal of the animal's own hypophysis and restored the tadpole's ability to metamorphose, hypophyses taken from younger stages were not effective.

The hypophysis of tadpoles also secretes another hormone which acts as an antagonist of thyroxine during the larval life of the tadpole. The hormone stimulates growth and retards metamorphosis. The chemical nature of the hormone is very similar to or possibly even identical with prolactin (Bern, Nicol, and Strohman, 1967). In mammals, prolactin causes growth of milk glands and milk secretion. It is quite remarkable that the same or a very similar chemical substance is present in such different animals as mammals and amphibians but is used to control entirely different processes.

The following pattern of amphibian metamorphosis emerges from preceding studies. The initial signal for metamorphosis is given by the anterior lobe of the hypophysis when it reaches a certain degree of differentiation and becomes capable of producing the thyrotropic hormone. The thyrotropic hormone activates the thyroid gland, which synthesizes and releases the thyroid hormone. The thyroid hormone (of which thyroxine is the most important component) overcomes the action of the prolactin-like hormone and affects the tissues directly, causing the degeneration and necrosis of some cells and stimulating the growth and differentiation of others.

18-3 TISSUE REACTIVITY IN AMPHIBIAN METAMORPHOSIS

One of the most intriguing aspects of amphibian metamorphosis is the diverse nature of reactions in different tissues to one and the same condition—the presence of the thyroid hormone. Whereas some of the tissues (the tail, the gills) become necrotic owing to the action of the hormone, others (the limbs) react to the hormone by increased growth and progressive differentiation. It can easily be shown that the character of the reaction is not due to the position of the parts in question, nor to an uneven distribution of the active principle, but solely to the nature of the reacting part. Parts of a tadpole's tail transplanted to the trunk undergo metamorphosis together with the host's tail and become absorbed. On the other hand, an eye transplanted to the tail of a tadpole before metamorphosis remains healthy, while all the surrounding and underlying tissues undergo necrosis. As the tail shrinks, the eye is brought nearer to the trunk and eventually becomes fused to the body in the sacral region after the entire tail has disappeared (Fig. 427—Schwind, 1933). The same experiments, incidentally, prove that the stimulus of the thyroid gland is carried through the blood vascular system, as it is only by this route that it can reach any part independently of its actual position. The secretion of the thyroid gland is thus a hormone in the narrow sense of the term. What happens to a tissue under the influence of the thyroid hormone is determined by the reactive properties of the tissue itself, or by what we have earlier called its **competence** (p. 202). The competence of tissues to react to the thyroid hormone is not directly dependent on their histological differentiation. In tadpoles, while the myotomes of the tail become resorbed during differentiation, the myotomes of the trunk are not so affected.

Furthermore it has been noticed that different parts of the body reacting to the thyroid hormone (whether by degeneration or progressive development) are not equally responsive to the dosage of the hormone. Very weak dosages applied to frog tadpoles

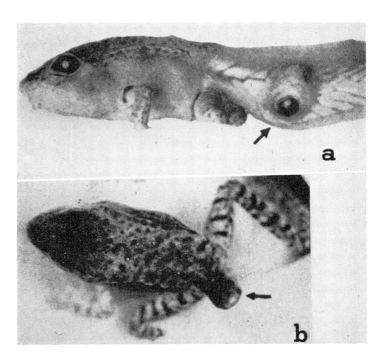

Figure 427. *a*, Metamorphosing frog (*Rana sylvatica*) tadpole with a grafted eye in the tail. *b*, Metamorphosed froglet: the tail has disappeared, but the eye is retained in the sacral region. Eye indicated by arrow. (From Schwind, 1933.)

cause an acceleration of the growth and differentiation of hindlimbs and a shortening of the intestine. Further processes may not be set into motion at all or follow only
after a lengthy treatment. A higher dose of the thyroid hormone causes the breakthrough of the forelegs. An even greater dose of the hormone is necessary to cause the
resorption of the tail. There is evidently some kind of threshold value for each part
which has to be attained before the reaction sets in. Different parts of the tail have different threshold values, the tip of the tail reacting more readily than the proximal
parts. In general, the degree of sensitivity to the thyroid hormone is reflected in the
order in which the metamorphic changes proceed in normal development: the parts
which have a low threshold (legs, reacting by growth) respond earlier than the parts
having a high threshold value (tail, reacting by reduction). (See Etkin in Willier, Weiss,
and Hamburger, 1955; Abeloos, 1956.)

When heavy doses of thyroid hormone are supplied to young tadpoles, all processes start at once, and the normal sequence of events becomes upset, the destructive
processes being capable of proceeding faster than the constructive processes. The forelimbs break through before becoming differentiated; the tail becomes reduced before
the legs are sufficiently developed to take over locomotion. The result is, of course, the
death of the animal.

In urodeles, the bulging of the eyes seems to be the reaction which may be elicited
by the weakest doses of the thyroid hormone. Next follow, more or less concurrently,
the reduction of the fin fold and the shortening and disappearance of the external gills.
The closure of the gill clefts and the transformation of the skin are the results of maximal stimulation and accompany complete metamorphosis in the course of normal development. The earlier stages of metamorphosis, including the shortening of the external gills, are partly reversible: if the thyroid treatment is stopped, the gills may
elongate again to a certain extent.

18–4 PROCESSES OF INDUCTION DURING AMPHIBIAN METAMORPHOSIS

Although, as a general rule, the processes of metamorphosis are a direct reaction
to the thyroid hormone which reaches each tissue, there are some notable exceptions to
this rule. The skin covering the tail of a tadpole, although subject to necrosis in normal
metamorphosis, remains healthy if transplanted to the body, provided that it is transplanted without the underlying muscles. Skin transplanted with underlying tail muscle
will necrotize in any position on the body. It is thus evident that the direct action of the
thyroid hormone is on the muscle tissue, and the overlying skin becomes involved in the
process of resorption secondarily.

A more complicated case is presented by the development of the tympanic membrane in frogs. The middle ear, with its cavity connected by the eustachian tube to the
pharynx, is one of the structures that develop progressively during metamorphosis.
The tympanic membrane first becomes differentiated toward the end of metamorphosis. It is supported in frogs by a cartilaginous ring, **the tympanic cartilage,** which develops as an outgrowth from the posterior edge of the quadrate cartilage. The skin
which later participates in the formation of the tympanic membrane is originally no
different from the skin covering the rest of the body. During metamorphosis, the
connective tissue layer of the skin in the area of the tympanic membrane becomes reorganized, the original layer of fibers, the stratum compactum, is broken up with the participation of phagocytes, and a completely new, somewhat thinner fibrous layer is developed in its place. In the fully differentiated tympanic membrane, the skin is less than
half the thickness of ordinary skin but much more compact, and it also differs in its
pigmentation.

It has been found that the differentiation of the tympanic membrane is not due to a direct action of the thyroid hormone but that it is induced by the tympanic cartilage (Helff, 1928). If the tympanic cartilage is removed before metamorphosis, the tympanic membrane does not develop. If the area in the otic region is covered by skin from the flank or from the back, the skin will develop a tympanic membrane. Lastly, if the tympanic cartilage is inserted under the skin on the flank or the back of a tadpole approaching metamorphosis, the local skin becomes differentiated as tympanic membrane (Fig. 428).

We can draw up a complicated chain of interactions that must take place before a tympanic membrane becomes differentiated. The first step is the formation of the rudiment of the hypophysis. The latter is developed in conjunction with the stomodeal invagination (p. 450) and presumably is induced by the oral endoderm (p. 448). The hypophysis in due time secretes the thyrotropic hormone and activates the thyroid gland. The thyroid gland releases the thyroid hormone, which causes the posterior edge of the quadrate to become differentiated as the tympanic cartilage. Lastly, the tympanic cartilage induces the skin to differentiate as tympanic membrane.

18–5 MOLTING AND ITS RELATION TO METAMORPHOSIS IN INSECTS

In any consideration of metamorphosis in insects, it must be taken into account that metamorphosis in these animals is a special form of molting (the periodic shedding of the cuticle of the skin which necessarily accompanies growth, because the strongly sclerotized parts of the cuticle cannot stretch or cannot stretch sufficiently to accommodate the growing mass of the body). A large proportion of the external features of an insect is embodied in the sclerotized parts of the cuticle, such as the details of shape of parts of the body, the hairs and spines on the surface of the skin, the sculp-

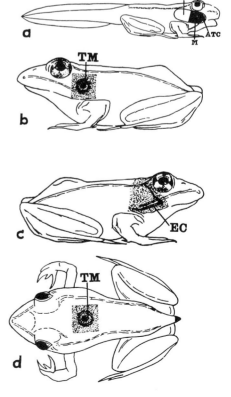

Figure 428. Dependence in the development of the tympanic membrane on the tympanic cartilage. *a*, Stage of operation: skin (S) turned back to reveal tympanic cartilage (ATC). *b*, Normal tympanic membrane (TM) developed on control side. *c*, No tympanic membrane developed after tympanic cartilage had been removed (EC, scar from operation). *d*, Tympanic cartilage transplanted subcutaneously on the back has caused development of tympanic membrane from local skin. (From Helff, 1928.)

ture of the surface of the cuticle, and to a certain extent, the pigmentation. In the process of molting, these features are lost with the discarded cuticle. The external characters of the insect, as far as they find their expression in cuticular structures have to be produced anew at each molt, though on a larger scale. The new cuticle is secreted by the epidermis of the skin, and therefore it is this layer of cells which is directly responsible for the external features of the insect emerging from a molt, whether the molted insect is an enlarged copy of the previous stage or whether it shows some new characters.

The molt in every case is quite a complicated process. In between two molts the cells of the epidermis are quiescent, they are more or less flat, and the epithelial layer may be rather thin. The epidermal cells adhere closely to the inner surface of the cuticle. Before each molt, however, they become activated, detach themselves from the cuticle, and enter a phase of rapid growth and proliferation. Numerous mitoses are observed. (Proliferation of epidermal cells is not observed, however, during the larval molt of cyclorraphe dipteran larvae, in which the cells of the larval epidermis do not divide between the stages of the egg and the pupa and are eventually discarded and replaced by imaginal epidermis.)

The number of epidermal cells produced by mitosis may be in excess of what is necessary, and some of the cells at this stage undergo degeneration by pyknosis. In spite of the degeneration of some of the cells, the layer of epidermis becomes thicker, and the remaining cells become arranged in a regular columnar epithelium. The surface of this epithelium foreshadows the shape of the animal emerging from the molt. In those parts of the body which are to be increased as a result of the molt, the epidermis is thrown into folds which expand and straighten out after the insect has emerged from its old skin. The folding is especially great where new or greatly increased appendages (such as the wings) are to be developed.

The epidermal cells then produce on their surface a thin layer of hardening secretion which becomes the outermost layer of the new cuticle, the **epicuticle,** consisting of a substance of lipoprotein nature, **cuticulin.** A fluid produced mainly by special molting glands is now poured into the space between the surface of the new epicuticle and the innermost surface of the old cuticle (Fig. 429). The fluid contains enzymes which digest the inner layers of the old cuticle until little more than the old epicuticle is left. The fluid with the substances digested from the old cuticle later becomes reabsorbed into the body of the insect. At the same time that the old cuticle is being digested, the epidermis produces further layers underneath the new epicuticle: the **exocuticle,** containing large quantities of cuticulin and also phenolic substances which are later oxidized to produce the dark pigment in the cuticle, and eventually the **endocuticle,** consisting of protein and chitin, which is a nitrogenous polysaccharide.

When the old cuticle is reduced to a thin shell, it is ruptured at the back of the head and thorax, and the insect crawls out of its old coat. The new cuticle is by no means complete at this stage: after molting the cuticle hardens, and visible pigment is produced in it from colorless precursors (phenolic substances). Further layers of endocuticle are deposited by the epidermal cells on the inner surface of the cuticle for days and even weeks after the molt has taken place.

It will be seen that some elements of amphibian metamorphosis, namely, destructive processes (resorption of the old cuticle, necrosis of part of the epidermal cells) as well as constructive processes (rearrangement of epidermal cells, formation of new cuticle), are present in an ordinary molt in insects. It depends on the condition of the epidermal layer whether the new structures produced during the molt are similar to or different from the old ones. In the first case the molt contributes to the growth of the animal; in the second, it becomes a mechanism for progressive development. If the changes achieved after a molt are considerable, the result is metamorphosis. In the primarily wingless archaic insects, the Apterygota, the young insect emerging from the

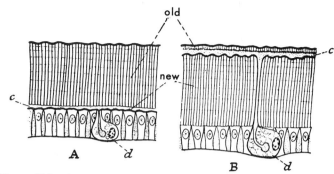

Figure 429. Two stages of molting in an insect. *A*, Old cuticle separated from epidermis which has produced a new epicuticle (thick black line). *B*, Deep layers of old cuticle dissolved, and layers of new endocuticle produced beneath the new epicuticle. d, Molting glands; c, molting fluid. (From Wigglesworth, 1939.)

egg is essentially similar to the adult, differing only in size and in the immature state of the sexual organs. Molting in these insects leads only to growth, and the advent of sexual maturity is not related in any way to molting, in fact, molting and growth continue even after the attainment of reproductive ability.

In all other insects, the Pterygota (winged or secondarily wingless forms), there is a distinct imaginal stage, which is attained after a specific imaginal molt, after which the insect does not molt any more. Except for secondarily wingless insects, the imaginal stage differs from the larval stages by the presence of wings. The imago also differs from the larval stages by the full development of external genital organs. (The gonads, on the other hand, may become fully functional only some time after metamorphosis.) In the more primitive winged insects, the wings appear gradually, the rudiments of the wings in the form of flat outgrowths of the second and third thoracic segments being visible in the later larval, or as they are often called, **nymphal** stages. These rudiments increase with every subsequent molt, but at the last imaginal molt there is an abrupt and very marked increase in the size of the wings, and after this molt the wings become functional. (Only in the mayflies, the first winged stage, the **subimago,** molts again before it turns into an imago.) The insects in which the rudiments of the wings develop on the surface of the body are called Exopterygota; these comprise the locusts, cockroaches, dragonflies, mayflies, bugs, and other related groups (Fig. 430). In the most advanced

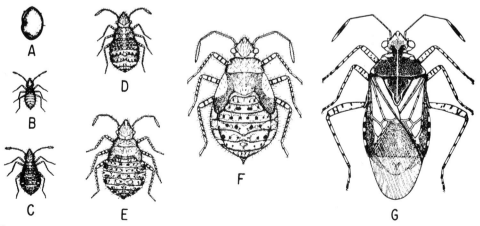

Figure 430. Development of an insect (bug) with incomplete metamorphosis. *A*, Egg; *B–F*, nymphal stages; *G*, adult. (From Borror and De Long, 1964.)

orders of insects, however, the wings develop internally, as folded appendages concealed during the larval stages in deep pockets (infoldings) of the epidermis (Fig. 431). The epidermis covering these wing rudiments retains an embryonic character throughout larval life, and although the rudiments continue growing slowly, their epidermis does not participate in the formation of the external cuticle of the larva and comes into action only when the larval stage is drawing to an end. Such rudiments, concealed under the surface of the body in the larval stage and reaching full differentiation in the imago, are called **imaginal discs.** The insects in which the wings develop internally as imaginal discs are called Endopterygota; here belong caddis flies, beetles, butterflies, bees and wasps, mosquitoes and flies.

Although the development of the wings in the adult insect attracts the greatest attention, the other parts of the body also change at the time of metamorphosis from a larva or nymph to the adult (imago). Even where the larva or nymph leads the same mode of life as the adult and has a fairly similar general appearance, as in the locusts or bugs, many finer features of structure change. In the bug, *Rhodnius prolixus,* for instance, which has been studied by Wigglesworth (1954), the fine structure of the cuticle and the pigmentation of the adult bug are very different from those in the last larval stage, so that even small areas of skin of the larva and the adult can be easily distinguished.

In the endopterygote insects, the difference between the larvae and adults is much greater. Not only wings but also mouth parts, antennae, and legs may be developed from imaginal discs, while the larval appendages become discarded, as in the case of butterflies (Fig. 432). In some parasitic wasps and in flies such as *Drosophila* and *Musca,* the whole larval epidermis is discarded and replaced by the imaginal epidermis derived from a series of imaginal discs (Fig. 433). Concurrently with the formation of appendages and other external parts, the internal organs are also reorganized. As the locomotion of the winged adult is so completely different from the locomotion of a crawling larva, the muscle system may have to be radically changed. During metamorphosis of higher insects, the larval muscles become broken down, and their remnants are consumed by phagocytes. The adult muscles, in particular the muscles operating the wings (the flight muscles), are then developed.

The eyes of adult insects in the more advanced orders are quite different from those of the larvae and are developed from special imaginal discs. Most of the cells of the alimentary canal of the larva may undergo resorption, and the alimentary canal of the imago is lined by a new epithelium produced at the expense of pockets of small reserve cells, which are found between the functioning cells of the larval intestine.

While the moderate amount of transformations that occur in exopterygote insects can be performed in one molt, the reorganization that is needed to produce the imago in most endopterygote insects is so profound that a resting stage, the pupa, is intercalated between the larval and the adult condition. In the pupa, the pockets containing

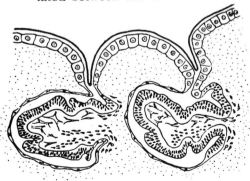

Figure 431. Longitudinal section through the imaginal discs of the wings in full-grown larva of the ant *Formica.* (After Perez, from Wigglesworth, 1954.)

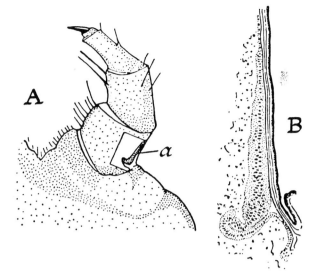

Figure 432. *A*, Position of the imaginal disc (a) in the leg of caterpillar of the butterfly *Vanessa*. *B*, Detail of the imaginal disc. (After Bodenstein, from Wigglesworth, 1939.)

the imaginal discs—wings, limbs, antennae, etc.—are brought to the surface. Internally, the formation of adult parts is, however, not yet completed, and while the reorganization takes place in the pupal stage, the insects do not take food and are very restricted in their movements, if they move at all.

When the reorganization is completed, another molt takes place, and the imago emerges from the pupa (Fig. 434). Metamorphosis which includes a pupal stage is called complete metamorphosis, and the insects having this type of metamorphosis are called Holometabola. Those not possessing a pupal stage and thus having an incomplete metamorphosis are Hemimetabola. The holometabolous insects are the Endopterygota; the terms, as far as systematics are concerned, are synonyms, though they stress different properties of the same group of insects.

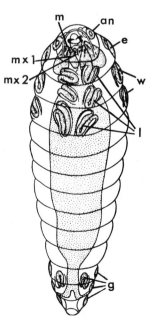

Figure 433. Imaginal discs in the last stage larva of the parasitic wasp *Encyrtus.* an, Antennae; e, compound eyes; m, mandibles; g, genital organs; mx1, mx2, first and second pairs of maxillae; l, legs; w, wings. (After Bugniol, from Kühn, 1955.)

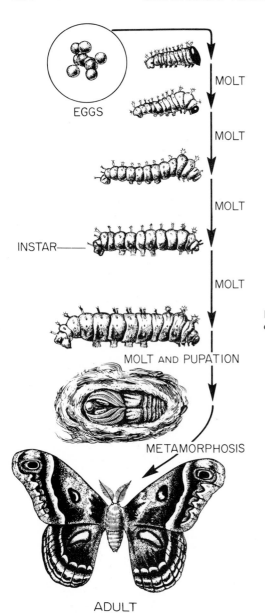

EGGS

MOLT

MOLT

MOLT

INSTAR——

MOLT

MOLT AND PUPATION

METAMORPHOSIS

ADULT

Figure 434. Development of an insect (moth *Cecropia*) with complete metamorphosis. (From Turner and Bagnara, 1971.)

18—6 CAUSATION OF MOLTING AND METAMORPHOSIS IN INSECTS

In an ordinary molt (**larval molt**) all parts of the body must participate in the process and carry it out at the same time, if the molting is to be successful. This suggests a common cause to which all parts of the insect are subjected. The existence of a common cause is even more obvious in the case of metamorphosis in which the involvement of both external and internal organs may be more far-reaching and radical. This common cause may be expected to be either external or internal. Cases are actually known in which, under natural conditions, an external factor is necessary to start a molt. In the bloodsucking bug, *Rhodnius*, such a factor is the intake of food. The bugs of this species feed only once in the interval between two molts, taking up so much blood that their body weight may increase many fold. Molting occurs regularly, 12 to

15 days after a feed in the case of the first four larval stages. The same dependence of molting on food intake holds true for the last, the fifth, larval stage, only the interval is somewhat longer, about 28 days, and the result is different: the molt transforms the larva into a winged imago. Another example in which an external factor is necessary to initiate a molt is the case of the pupa of the moth *Platysamia cecropia*. After pupation, the insect falls into a quiescent state with a reduced rate of metabolism—the **diapause**—which continues throughout winter. It is essential that during this time the pupa be exposed to cold, otherwise the diapause is prolonged indefinitely. However, the diapause may be broken precociously if the pupa is treated with cold (3° to 5° C.) for at least two weeks. The temporary cooling activates the vital processes in the pupa, and on return to a warmer environment the pupa molts, and in this way the development is completed with the emergence of the imago.

In the overwhelming majority of insects, however, no external cause of any molt can be found, and the molts follow one another at intervals which appear to be determined entirely by internal processes in the animal. In many insects the body weight increases in a fixed proportion between two molts, often by a factor of two, and it would appear that a certain amount of synthesis has to be performed after each molt before the stimulus for a new molt is generated in the organism. However, even in cases in which an external factor triggers off the mechanism of molting, it can be shown that the factor in question does not affect all parts of the body directly but that it is mediated by the brain of the insect. If a larva of *Rhodnius* is decapitated within a day or two after feeding it does not molt, although it may remain alive for over a year. If, however, it is decapitated five or more days after a meal, molting takes place. By that time a stimulus generated by the brain reaches beyond the level of decapitation and is able to spread throughout the body and cause the molt to proceed (Wigglesworth, 1954). A corresponding experiment in the moth *Platysamia* consists in activating one pupa by exposure to cold and then transplanting parts of the body of the activated pupa into an untreated pupa. The transplantation of the brain but not of other organs will cause the second pupa to molt and the adult moth to emerge, thus showing that the cold directly affects only the brain but that the rest of the body is stimulated to molting through the mediation of the latter (Williams, 1946).

The question naturally arises as to how the brain affects the rest of the body. It is now established that, as with amphibian metamorphosis, molting and metamorphosis in insects are controlled by hormones and that at least three organs of internal secretion are involved: the brain **(protocerebrum)**, the **corpora allata**, and the **prothoracic gland.** In the brain, a hormone is produced by **neurosecretory cells**, which are arranged in four groups: two groups near the midline and one group on each side (Fig. 435). Behind the protocerebrum, alongside the dorsal aorta, there are in most insects two pairs of bodies connected by nerve strands to the protocerebrum: first the **corpora cardiaca**, which are of the nature of nerve ganglia, and more posteriorly, the **corpora allata**, consisting of secretory cells. The corpora allata may be fused into one body in some insects. The third endocrine gland, the **prothoracic gland,** is an irregular branching mass of glandular cells located in the thorax, in close association with the tracheal tubes (Fig. 436). The glandular cells of all three centers show regular secretory cycles preceding each molt, and the three types of secretion are necessary for the normal course of larval molts. The molt is initiated by the neurosecretory cells of the protocerebrum, but all that the hormone of the protocerebrum does is to activate the prothoracic gland. The prothoracic gland then produces a hormone which sets in motion the mechanism of molting in the epidermis: the growth and proliferation of epidermal cells, the shedding of the old cuticle, and the production of the new one. The hormone produced by the prothoracic gland is therefore called the growth and molting hormone or **ecdysone.**

We have described some of the evidence proving that the initial stimulus for molt-

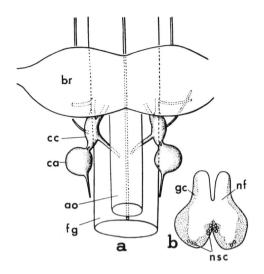

Figure 435. The endocrine glands associated with the brain in moths. *a*, Dorsal view of the brain in a hawk moth pupa. *b*, Transverse section of the protecerebrum in a caterpillar of the meal moth. ao, Aorta; br, brain; ca, corpora allata; cc, corpora cardiaca; fg, foregut; gc, ganglionic cells; nf, nerve fibers; nsc, neurosecretory cells. (After Cazal and Rehm, from Kühn, 1955.)

ing is given off by the brain (or the neurosecretory cells of the protocerebrum). One of the experiments consisted in transplanting the brain of an activated pupa of *Platysamia* into an untreated pupa, whereupon the latter molted and produced the moth. A variation of the same experiment has been used to prove that the secretion of the brain cannot act directly but only through the activation of the prothoracic gland. Instead of transplanting the activated brain into a whole pupa, it was implanted into the posterior half of a pupa which had been cut in two (the cut surface was sealed with paraffin wax). Under these conditions the graft was powerless; no metamorphosis took place. The reason for this is that the prothoracic gland is absent in the posterior half of the pupa. If in addition to the brain the prothoracic gland was also grafted, metamorphosis occurred (Williams, 1947; Wigglesworth, 1954). An analogous experiment has been performed on the bug *Rhodnius* (Fig. 437). After the neurosecretory cells of the brain had been activated by the bug's having a meal of blood, the brain was transplanted into the abdomen of a decapitated specimen. A decapitated larva is still in possession of the prothoracic gland, which could react to the implanted brain and cause the molt to proceed. If, however, the activated brain was implanted into an isolated abdomen (Fig.

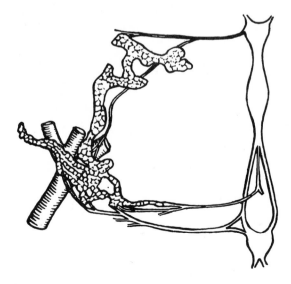

Figure 436. Prothoracic gland in the moth *Saturnia*, and its associations with the ventral nerve cord and the tracheal system. (After Lee, from Wigglesworth, 1954.)

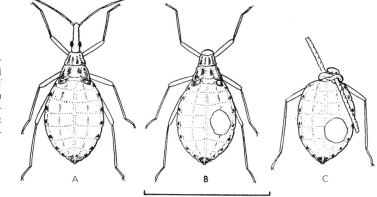

Figure 437. Experiment showing necessity of thoracic gland for molting and metamorphosis. *A*, Normal larva of *Rhodnius*. *B*, Decapitated larva with an implant in the abdomen. *C*, Larva ligated through metathorax, with implant in isolated abdomen. (From Wigglesworth, 1954.)

10 mm.

437*C*), no molting occurred. On the other hand, molting could be induced in the isolated abdomen by the implantation of a prothoracic gland (Wigglesworth, 1954).

The roles of the brain and the prothoracic gland as causative agents of molting can also be demonstrated in insects in which the time of molting is not dependent on any particular external factor. If the brain is removed from caterpillars sufficiently early before the next expected molt, the caterpillars may remain alive for over two months but do not molt and do not pupate. The implantation of a brain from another caterpillar restores the ability of a brainless caterpillar to complete its development (Kühn and Piepho, 1936). Once the prothoracic gland has become activated, the brain is no longer necessary for initiating the molt. Only those parts molt (or pupate), however, to which the hormone of the prothoracic gland has been able to gain access. If a caterpillar in the last larval stage is constricted behind the thorax, the anterior part of the body will pupate, but the posterior part, which the molting hormone could not reach, remains in the larval state (Fig. 438). A little later, when the hormone has already spread throughout the body, a transverse constriction does not prevent pupation of the posterior end of the caterpillar.

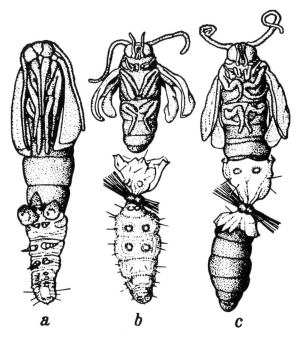

Figure 438. Experiment showing necessity of thoracic gland for pupation. *a*, Normal pupa of the meal moth *Ephestia kühniella*. *b*, Result of ligating the caterpillar before the hormone of the thoracic gland has been released: the part posterior to the ligature remains in the larval stage. *c*, Result of ligating the caterpillar after the molting hormone has been released: both parts pupate. (After Kühn and Piepho, from Kühn, 1955.)

a   *b*   *c*

The hormones emitted by the protocerebral neurosecretory cells and the prothoracic gland induce an insect to molt, but they do not determine whether it will be a larval molt, producing the next larval stage, the pupal molt, converting the larva to pupa, or the imaginal molt, leading to the eclosion of the imago. The third endocrine gland, the corpora allata, controls the nature of the change that takes place at the time of molting. Curiously enough, the first two glands, the protocerebral neurosecretory cells and the prothoracic gland, when acting alone cause immediate metamorphosis — the development of the imago in hemimetabolous insects or of the pupa in holometabolous insects.

It is possible to remove the corpora allata from caterpillars of moths. Independently of the stage in which the operation is performed, the caterpillars proceed to pupate at the next molt. Eventually, the moth emerges from the pupa, although it may have reached only a fraction of the normal size (Fig. 439; Bounhiol, 1937). Apparently the presence of the corpora allata is necessary to **prevent** metamorphosis, to keep the insect in the larval state. Accordingly, the secretion of the corpora allata has been called the **juvenile hormone.** The cells of the corpora allata show signs of secretory activity (swelling of cells, appearance and discharge of vacuoles, etc.) at every larval molt, but no such activity is present during the pupal or imaginal molt. Accordingly, it would seem that at the time metamorphosis takes place the corpora allata do not produce their secretion or at least are less active. That it is actually the absence of the juvenile hormone that is the necessary condition for metamorphosis can be proved by implanting corpora allata from a young larva into the last stage larva which should be metamorphosing with the next molt. The larva may molt in due course, but under the influence of the juvenile hormone secreted by the graft, it is not transformed into an imago (in the case of a hemimetabolous insect) but instead produces an abnormally large larva (Fig. 440).

In the case of holometabolous insects, the conditions are more complicated inasmuch as there are two molts accompanied by profound morphological changes — the pupal molt and the imaginal molt. The removal of the corpora allata from a caterpillar causes the caterpillar to be transformed into a pupa. Some experiments, the details of which cannot be related here, indicate that the subsequent transformation of the pupa into the imago is probably connected with a further decrease of the juvenile hormone in the blood of the insect. After the ablation of the gland, small amounts of the juvenile hormone could have still been present in circulation but would have been used up by the time of the second molt. (See Wigglesworth, 1954.)

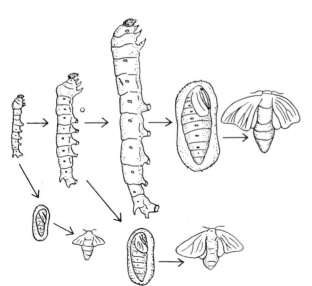

Figure 439. Precocious metamorphosis in the silkworm, caused by the extirpation of the corpora allata in the third or fourth stage, respectively. The larva pupates normally after the fifth stage. (After Bounhiol, from Abeloos, 1956.)

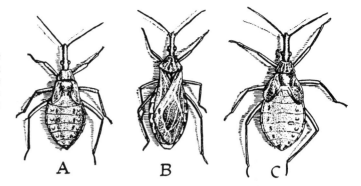

Figure 440. *A*, Normal last (5th) stage nymph. *B*, Normal adult *Rhodnius*. *C*, Giant 6th stage nymph produced by implanting the corpus allatum from a 4th stage nymph into the abdomen of a 5th stage nymph. (From Wigglesworth, 1939.)

18–7 NATURE OF THE FACTORS CONTROLLING MOLTING AND METAMORPHOSIS IN INSECTS

In the previous section, it was assumed that the agents produced by the prothoracic gland and the corpora allata are hormones, that is, chemical substances emitted by the cells and circulating in the body fluids. This could be deduced from the fact that the effect of the gland does not depend on whether it is in its normal position, with all its connections to neighboring organs and to the nervous system intact, or on whether it has been transplanted to an abnormal site. Further evidence in favor of a diffusible substance being the means of action of the gland will now be presented.

An equivalent of the glands' independence of their positions is the independence of the reaction from the position of the reacting organs. When molting or metamorphosis occurs, not only do all parts of the body of the intact animal react together, but also transplanted parts do the same. Imaginal discs and other parts of the body may be transplanted between animals in different developmental stages, and they always molt and metamorphose together with the organs of the host and independently of their own age. A very elegant experiment of this kind, carried out on the developing moth *Ephestia kühniella*, consists in transplanting pieces of skin into the body cavity of another individual (Kühn, 1939; Piepho and Meyer, 1951; Kühn, 1955). The edges of the implanted piece of skin curl so as to form a cyst, with the original distal surface of the skin turned inward. The proximal surface of the epidermis is bathed by the body fluids of the host and by the host's hormones if any are present in the body fluids. The necessary conditions are thus provided for the graft epidermis to react to any hormones circulating in the body of the host. It was found that with every molt of the host the cyst epidermis molted also, discarding the old cuticle into the cavity of the cyst. Not only was the molting of the graft simultaneous with that of the host, but the nature of the new cuticle was always the same as that of the host. When a larval molt occurred, the cyst epidermis produced a thin cuticle like the one that covered the body of the caterpillar. When the host pupated, the cyst produced a thick pupal cuticle. When the host metamorphosed into the adult moth, the epidermis of the cyst developed an imaginal cuticle with scales! All the successive cuticles could be seen later one inside the other, on sectioning the cyst (Fig. 441).

Even after reaching the stage of producing the cuticle of the adult moth, the epidermis does not lose its capacity for molting, provided that ecdysone (the molting hormone) or both ecdysone and the juvenile hormone are present in the surrounding fluid. A cyst which had gone through the pupal and imaginal molts may be excised from the first host and transplanted into a second one. If the second host is a caterpillar, the cyst will undergo a new molt simultaneously with the pupation of the host, will shed the imaginal cuticle with scales, and will again produce a thick pupal cuticle. This may be

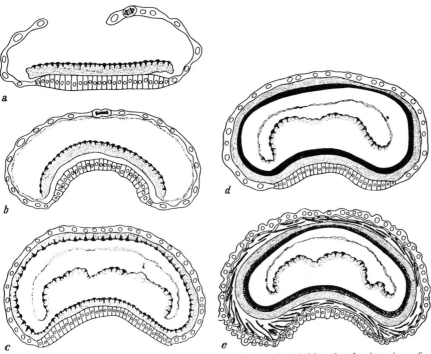

Figure 441. Hormones in the development of a moth. Molting in the interior of an implanted epidermal cyst under the influence of hormones circulating in the host's body. *a*, Cyst in the process of formation. *b, c,* Shedding of a larval cuticle. *d,* Production of a pupal cuticle. *e,* Pupal cuticle shed, an imaginal cuticle with scales produced underneath. (From Kühn, 1955.)

followed, at the time of metamorphosis of the host, by a second imaginal cuticle with a new set of scales (Fig. 442).

Apparently metamorphosis is fully reversible, at least with regard to the skin epidermis, and the nature of differentiations produced by the latter is solely dependent on the balance of hormones present in the blood. A reversal of metamorphosis, even a partial one, can occur, however, only under experimental conditions. In the normal life of

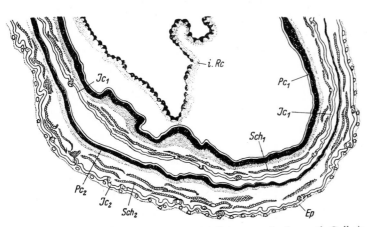

Figure 442. Control of metamorphosis by hormones in the moth *Galleria melonella.* Reversal of metamorphosis in an epidermal cyst transplanted into a second, younger host. The consecutive cuticles are: larval (i. Rc); pupal (Pc_1); imaginal (Ic_1) with scales (Sch_1); then again pupal (Pc_2); and lastly, imaginal for a second time (Ic_2), with a second generation of scales (Sch_2). Ep, Epidermis. (After Piepho and Meyer, from Kühn, 1955.)

an insect, metamorphosis marks the end of morphogenesis and of growth (except for the growth of the gonads, which may continue in the adult). The reason for the cessation of further development is that the prothoracic gland degenerates and breaks up after causing the last (imaginal) molt. With the prothoracic gland gone, no other factors can reawaken the morphogenetic activity of the epidermis, and no further molting can occur. This is also an explanation for the existence of a growth limit in insects, which was referred to at the end of Section 17–3.

The action of the juvenile hormone is not restricted to the control of qualitative changes in the body of the insect, but it appears to have also a direct influence on growth. It is found that in the imago, after metamorphosis, the corpora allata resume their secretory activity and that their secretion is necessary for the growth of the ovaries and the oocytes.

In vertebrates, the chemical nature of the agents emitted by the endocrine glands has been proved by preparing active extracts from the glands, containing chemically definable substances. The extraction of hormones from insects is much more difficult, owing to the small mass of the endocrine glands. However, some success can be noted in this field also. The molting hormone, ecdysone (the hormone of the prothoracic gland), has been prepared in a chemically pure form by Butenandt and Karlson (1954). Five hundred kilograms of silkworm pupae were used to isolate 25 mg. of the pure substance, and 0.0075 mg. of this material was sufficient to cause the pupation of a fly larva. The chemical analysis of the substance gave the empirical formula $C_{27}H_{44}O_6$ (Karlson and Hoffmeister, 1963). The substance belongs to the steroid group, and its approximate structural formula is:

The chemical analysis of the other two hormones, the brain hormone and the juvenile hormone, has not yet been as successful. The results obtained by different research workers are somewhat contradictory. It appears, however, that the juvenile hormone is chemically related to ecdysone and is also a steroid or a substance of the terpene group which is a possible precursor of steroids (Schneiderman and Gilbert, 1964).

As in the case of inducing substances in early amphibian embryogenesis, it has been found that the action of the juvenile hormone may be imitated by extracts of tissues of various animals (Williams, Moorhead, and Pulis, 1959) and by certain substances of known chemical composition, such as the terpene farnesol. The latter is also able to stimulate the prothoracic gland, an effect normally produced by the brain hormone. It has been suggested, therefore, that all three hormones regulating growth and development in insects have something in common in their chemical structure and activities (Schneiderman and Gilbert, 1964).

18–8 MECHANISM OF ACTION OF INSECT HORMONES

The study of insect molting and metamorphosis has contributed some very interesting information on the way in which chemical substances act on cell differentiation.

It has been known for many years that in the salivary glands of some insects in the order Diptera certain cells grow to a relatively large size, and that in such cells the chromosomes become visible, although the cells do not undergo mitosis. The "giant chromosomes" in such cells are the result of repeated duplication of the DNA molecules, so that many hundreds of DNA molecules lie side by side. Cells of malpighian tubules and some other tissues also grow excessively and produce giant chromosomes. The chromosomes show characteristic transverse bands which may be related to specific genes as known from breeding experiments. Even before the relation of the bands on the giant chromosomes to genes was known, conspicuous thickenings were found on some giant chromosomes. These thickenings were known for many years as the "Balbiani rings," so named after the scientist who discovered them first in salivary chromosomes of the midge *Chironomus*. Later, it was found that the thickenings occur in all cases where giant chromosomes are found, though not all are as large as the Balbiani rings. The thickenings, large or small, are now called **puffs** (Figs. 443 and 444).

A puff is actually a section of the chromosome in which the numerous strands of DNA, of which the giant chromosome consists, become separated from one another and form loops, extending outward from the main axis of the chromosome. This loosening of the chromosomal structure is favorable for the chemical interaction of the DNA strands with their environment inside the nucleus. In fact, it has been proved (by using radioactive precursors) that there is a very rapid synthesis of RNA on the puffs. The RNA synthesized on the puffs is different in base composition from the cytoplasmic ribosomal RNA and is believed to be messenger RNA (Edström and Beermann, 1962). In other words, the puffs represent sections of chromosomes in which the genes are in a derepressed state and are particularly active in producing messenger RNA. (Compare p. 486.)

A further important discovery was made (Clever, 1961; Becker, 1962; Beermann and Clever, 1964) that the pattern of puffing changes with the stage of development of the insect; puffs appear at different points on the chromosomes and may disappear

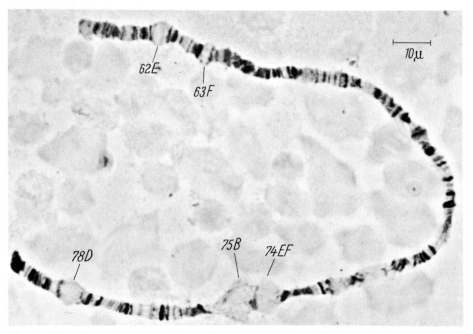

Figure 443. Giant chromosome from salivary gland of *Drosophila* larva, showing several puffs as they appear in a period just preceding pupation. (Courtesy of Dr. H. J. Becker.)

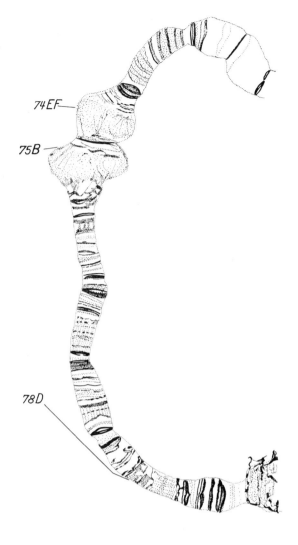

Figure 444. Drawing of part of the same giant salivary gland chromosome of *Drosophila* larva shown in Figure 443. (Courtesy of Dr. H. J. Becker, 1962.)

again at a later stage, while new puffs develop at other points. The stage of metamorphosis is characterized by a specific pattern of puffs which can be recognized if the structure of the giant chromosomes of a species under investigation is known in detail.

The question now arises as to whether there is a connection between the structural changes in the chromosomes and the morphogenetic changes (molting, metamorphosis) of the whole animal. *A priori* one could expect that changes in the activity of genes must bear some relation to the transformations of the animal's organization. On the other hand, we have seen that molting and metamorphosis in insects is under the control of hormones (ecdysone and juvenile hormone).

A breakthrough in this field was achieved when it was shown that injection of pure ecdysone into *Chironomus* larvae caused the giant chromosomes to acquire a puffing pattern identical to the one occurring during pupation (Fig. 445 – Clever and Karlson, 1960). The changes in the chromosomes occurred very rapidly; the first reactions were visible within 15 to 30 minutes after the injection, whereas the changes in the cuticle of the larvae leading to pupation occurred in a matter of a few days (Clever, 1961). It seems reasonable to conclude that the hormone acts initially on the genes and that the changed activity of the genes then causes a changed behavior of the cells and tissues.

Evidence has been obtained to show that different loci on the chromosomes do not

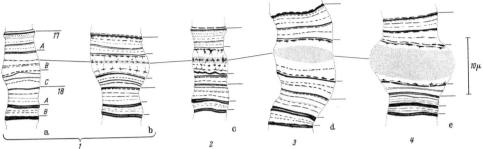

Figure 445. Induction of a puff on the giant salivary gland chromosome of *Chironomus tentans* by injecting the larva with ecdysone; *a, b, c,* chromosomes of untreated (control) larvae; *d, e,* same region of the chromosome in a larva after ecdysone injection. (From Clever, 1961.)

react to ecdysone in the same way. There are one or two loci which produce puffs very early after ecdysone injections and appear to be the immediate reacting sites to the hormone. Other loci, in which puffs may arise or increase in size later, are claimed to be dependent on the action of the genes activated in the first place. In this way it would appear that the action of ecdysone may consist in activating only two or possibly even only one gene, and that this starts a chain reaction, involving activities of other genes, which eventually ends in molting or metamorphosis (Clever, 1961, 1965).

Doubts have been expressed, however, whether ecdysone as such acts on the gene or whether it changes the intranuclear milieu, which, in turn, then affects the gene. It was found that the appearance of puffs on giant chromosomes can be invoked by means other than specific hormone action. Changes in ion concentration, certain narcotics, and even the simple fact of explanting the salivary gland in hemolymph may influence the puffing pattern of the chromosomes. The changes may be similar to those occurring in the chromosomes under natural conditions at certain periods of development. It has been claimed, in particular, that explantation of salivary glands in hemolymph causes the chromosomes to revert to a puffing pattern which is similar to that found in earlier stages of development ("rejuvenation"). Some narcotics, and $ZnCl_2$ in particular, when administered at a sensitive early pupal stage may produce a pattern of puffing characteristic of a later stage of pupation (Kroeger, 1963, 1964). The puffs caused by $ZnCl_2$ show an increased incorporation of RNA precursors (Kroeger, 1963), but it has not been shown that the unspecific agents, such as the ones mentioned, can cause metamorphosis or molting, nor indeed any morphological effect except on the chromosomes.

It is interesting to note the close similarity between the results obtained in experiments on insect metamorphosis and the conclusions which emerge from work on the primary organizer (neural inductor) in amphibian embryos. In both cases, the obvious reaction involves the activities of groups of cells, which change their behavior and acquire new differentiations. (In insects, cells change from the larval condition to that of a pupa or an imago; in an amphibian embryo, gastrula ectoderm becomes differentiated as neural tissue.) In both cases we have been led to the conclusion that the change is produced by a modification of the activity of nuclear genes. This modification is not a spontaneous change in the genes themselves but is evoked by an external factor: ecdysone in the metamorphosing insect, neuralizing substance in the amphibian embryo. Lastly, the action of the natural inducing substances may in both cases by imitated by "abnormal inductors," though in the case of insect metamorphosis and molting the "abnormal inductors" have not been able to copy the effect of the natural agents in their entirety.

18–9 FINAL REMARKS ON METAMORPHOSIS

The metamorphosis of amphibians and insects is an excellent example of the control of morphogenetic processes by hormones. The dependence of differentiation on diffusible chemical substances, as revealed in studies on metamorphosis, should be compared with the results of experiments on the influence of diffusible substances on the differentiation of cells in tissue cultures (Section 16–5).

When comparing the interrelation of different hormones in insect and amphibian metamorphosis, one is astonished to find that these two groups of animals have developed causative mechanisms having a distinct general similarity. In both cases, the transformation is initiated by a secreting organ closely associated with the brain: the hypophysis in amphibians, the neurosecretory cells of the protocerebrum in insects. In both cases, the secretion of this primary center does not act on the tissues directly but stimulates the activity of a second endocrine gland: the thyroid gland in amphibians, the prothoracic gland in insects. Lastly, the juvenile hormone of insects which has the function of checking and preventing precocious metamorphosis has an equivalent in the prolactin-like hormone of frog tadpoles.

There is very little information as to the causative agents of metamorphosis in animals other than amphibians and insects, and we do not know whether their transformations are controlled by hormones. One would expect that at least in the cyclostomes the transformation of the larva (ammocoete) into the adult should be caused by hormones, but this has not been found to be the case; thyroxine does not accelerate transformation of the ammocoete into a lamprey (Gorbman and Bern, 1962).

In the ascidian tadpole, the absorption of the tail is in some way dependent on the anterior end of the body, since cutting off the anterior tip with the adhesive papillae will prevent the necrotization of the tail (Oka, 1943). Treating ascidian tadpoles with thyroid hormone accelerates the metamorphosis (Weiss, 1928), but this action is hardly specific, as the tadpole does not have a thyroid gland of its own and treatment of the tadpoles with narcotics or even with distilled water has the same effect (Oka, 1958).

Numerous experiments have been carried out on the metamorphosis of larvae of various invertebrates (*Tubularia,* Bryozoa, sea urchins, molluscs). Treatment by various chemicals (e.g., salts of copper) has had a positive effect. (For a review, see Lynch, 1961.) However, it has not been possible to find what factors cause the metamorphosis under normal conditions, and there is so far no indication that in any of these organisms a hormone plays a decisive role in the process. An exception to the later statement is apparently presented by the transformation of the asexual form of the annelid worm, *Platynereis dumerilii,* into the sexual form. This transformation is dependent on a neurosecretory hormone released in the prostomium of the asexual individual. The hormone is an inhibitor of the transformation, and the sexual form is developed when the prostomium hormone is eliminated (Hauenschild, 1964).

Chapter 19

REGENERATION*

The second mode of reawakening the morphogenetic processes at an advanced stage of the ontogenetic cycle is by means of partial destruction of the system which has evolved as a result of the previous development. The animal organism possesses the ability to repair more or less extensive damage incurred by the body either accidentally in natural conditions or willfully imposed by the experimenter. The damage repaired may be a wound which severs or partially destroys the tissues of the animal's body, or the damage may involve the loss of an organ or larger part of the body. These can sometimes be renewed, and in this case the process of repair is known as **regeneration.**

19—1 TYPICAL CASE OF REGENERATION: THE RENEWAL OF A LIMB IN A SALAMANDER

The limbs of newts and salamanders, both in the adult stage and in larvae, are capable of regeneration to a very high degree and have often been used for various experiments on the subject. Limbs, especially those of larvae, are sometimes bitten off by other members of the same species, or the limbs may be cut off at different levels for research purposes. Whatever the cause of the loss of a limb, the first stage of repair is that the epidermis from the edges of the wound starts spreading over the wound and soon covers the open surface. The closing of the wound is relatively a very rapid process, and it is accomplished in the course of one or two days, depending on the size of the animal. During the next few days, the epidermis covering the wound begins to bulge outward, becoming more or less conical in shape (Fig. 446). A mass of cells accumulates under the epidermis. These cells are in a state of active proliferation, and together with the epidermal covering they form what is known as the **regeneration blastema** or **regeneration bud.** The blastema grows rapidly, at first retaining its conical shape, but later it begins to be flattened dorsoventrally at the end. The flattened part is the rudiment of the carpus or tarsus, called the hand- or footplate, depending on whether a forelimb or a hindlimb is regenerating. Soon the rudiments of the digits appear, separated by slight indentations at the edge of the plate (Fig. 446). In the meantime, the mass of cells in the interior of the limb-bud becomes segregated to form the rudiments of the internal parts of the limb—the various bones of the limb skeleton and the arm (or leg) muscles. These then undergo histological differentiation. The rudiments of the digits elongate, and the whole regenerating limb continues to grow until it attains the size of a normal limb. As the tissues of the regenerating limb differentiate, the limb resumes normal function (movement). Eventually the regenerated limb becomes completely indistinguishable from a normal limb. The time necessary for the completion

*Much of the work on this subject is reviewed in *Regeneration*, edited by D. Rudnick, 20th Symposium, Society for the Study of Development and Growth, New York, 1962.

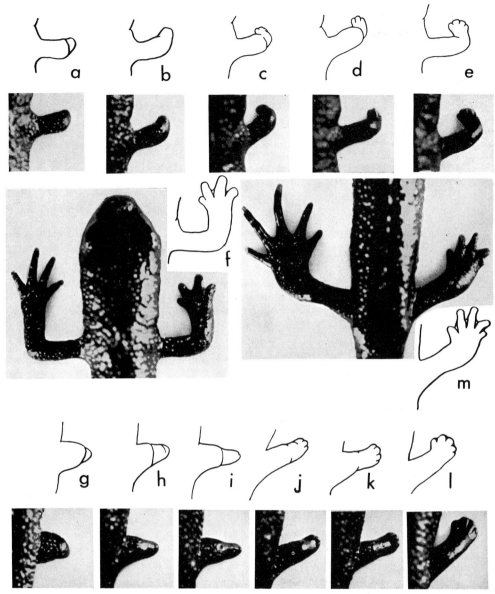

Figure 446 Regeneration in the limbs of the newt *Triturus cristatus. a, b, c, d, e, f,* Consecutive stages of regeneration of a forelimb amputated above the elbow. *g, h, i, j, k, l, m,* Stages of regeneration of a hindlimb amputated above the knee. (From Schwidefsky, 1934.)

of the process depends on the size and stage of development of the animal. Regeneration is most rapid in small larvae, and in this case the regenerated limb may be complete after three weeks (in the case of an axolotl larva). In older and larger larvae the process may take longer, and in the adult salamander the completion of regeneration takes several months.

Thus, the new limb is produced from a rudiment, the regeneration blastema, which, when formed, acquires the potentiality for development otherwise found only in the organ rudiments of the early embryo. There is the obvious difference, however, that the cells of the regeneration blastema develop into parts of one organ only (the limb), whereas the cells of the early embryo produce the whole animal. Reversion to a state somewhat similar to that of early embryonic cells obviously takes place in the formation of a regeneration blastema. How far this "rejuvenation" goes will be discussed later.

What is developed by regeneration supplements the residue of the animal's body (in our case by far the major part). The regenerated organ is thus an addition to the other parts. A type of regeneration exemplified by the renewal of limbs in salamanders is therefore called **epimorphosis** or **epimorphic regeneration** (Morgan, 1901). It is the common type found in higher animals.

19–2 REGENERATIVE ABILITY IN VARIOUS ANIMALS

Although present throughout the entire animal kingdom, the ability to regenerate lost parts differs both in scope and in its course in the various groups of animals.

In the coelenterates, the regenerative ability is exceedingly high. It was the regeneration of the fresh water polyp, *Hydra*, first discovered by Trembley in 1740, that attracted the scientists' attention to this phenomenon. A hydra may be cut in two or more parts, and each part will reconstitute itself into a new and complete individual of diminished size. The posterior end of a cut hydra regenerates the mouth and tentacles; the anterior part of the body regenerates the posterior end with the foot and adhesive disc. Even small sections of the body, comprising as little as 1/200 part of the original individual, can regenerate a complete animal. In the latter cases, it is especially clear that the new individual is produced not by addition of parts to the remaining piece (**epimorphosis**) but by remodeling the whole available mass of cells into a new whole. This process thus differs from the epimorphic regeneration of an amphibian limb. A type of repair involving a reorganization of the remaining part of the body of an animal is known as **morphallaxis** or **morphallactic regeneration.** If only smaller parts are removed from the hydra, the remodeling of the remaining body is not so extensive, and with very small defects the regeneration approaches the epimorphic type, even though a typical regeneration bud cannot be clearly distinguished.

Other coelenterates regenerate to varying extents, the regenerative power being highest in the polypoid forms and much reduced in the medusoid forms.

The next group remarkable for their high regenerative ability are the planarians. Planarians may be cut across or lengthwise, and each part of the body will regenerate the missing half. Any part of the body may be replaced in this way: the head, the tail, or the middle part with the phyarynx. When the cut is made, a regeneration blastema is formed at the cut surface, and the missing part is developed from the blastema. The remaining part is, however, reorganized on a diminished scale, so that the individual resulting from regeneration is smaller than the original one. The regeneration is thus carried out in a way that combines epimorphosis and morphallaxis. Other platyhelminths do not regenerate to any great extent.

The nemerteans have a high regenerative ability, and a complete worm may be formed starting even from very small fragments.

Many experiments have been carried out on regeneration in the annelids. Both polychaetes and oligochaetes regenerate anterior and posterior ends after an amputation. If an earthworm or other oligochaete is cut in two halves, the posterior half regenerates the anterior end with the mouth, and the anterior end regenerates a new posterior end. Two new individuals may thus be produced from the original one. In the majority of the annelids regeneration is, however, somewhat restricted; at the anterior cut surface only a limited number of segments are formed, the number being typical for every species. (See Berrill, 1952.) In the earthworm *Allolobophora foetida* this number is four or five. If five segments or less are cut off from the anterior end of the worm, the regeneration is complete; if more than five segments are removed, only four or five segments will be regenerated, and the worm will thus not attain the same overall number of segments as it had before the operation. If the cut is carried out behind the genital segments (tenth to fourteenth), only four or five anterior segments are regenerated, and the genital organs are thus never renewed. On the other hand, there is no restriction in the regeneration of posterior segments, and about as many are formed as have been removed. The process of regeneration is an epimorphosis, a regeneration bud being formed and the new parts developing at the expense of this bud. Hirudinea do not regenerate at all (Weiss, 1939).

In molluscs, the regeneration is relatively poor. In gastropods, eye stalks with eyes may be regenerated as well as parts of the head or of the foot. The whole head does not regenerate, and if the cerebral ganglia are removed together with a part of the head, these will not regenerate. The arms of cephalopods may regenerate but not other parts of their body.

In nematodes, the regenerative ability is very low. This may be connected with the high degree of differentiation of the cells of their bodies and the fixed limit to the total number of cells (cf. Section 17–2). Only the closure of superficial wounds is still possible.

In arthropods, regeneration is limited to the renewal of lost appendages, but this form of regeneration is fairly widespread. In most crustaceans, the limbs may regenerate at any stage of development, including the adult. In insects, limb regeneration occurs only in the larval stages, and the regenerated limb often does not reach the size of a normal limb. The legs of crabs and some spiders are readily shed if seized by an enemy (or the experimenter). The legs break off at a preformed breaking point, across the second leg joint. At this point there is a constriction which is a modified joint. The leg is broken off by the violent contraction of a muscle (the extensor muscle of the leg). This self-mutilation is known as **autotomy.** Autotomy is probably a special adaptation which helps the animal to escape being caught by a predator. If the predator gets hold of one limb, he succeeds in capturing only the limb but not the animal, the latter escaping at the expense of the loss of a limb.

After the amputation or loss of an appendage in an arthropod, the wound is covered by a chitinous plug. Underneath this, a regeneration bud is formed which later reproduces the limb by way of epimorphic regeneration. The new limb does not, however, become apparent until the next molt. The regenerated limb is small at first and attains normal proportion as a result of accelerated growth in the course of several molts.

Among the echinoderms, the starfishes, brittle stars, and sea lilies can regenerate arms and parts of the disc. The arms appear to be lost rather often in natural conditions, as individuals regenerating one or more arms are found quite frequently. The Holothuroidea are capable of ejecting through the anus parts of their internal organs—the respiratory tree and the alimentary canal. These can be regenerated later.

In vertebrates, the regenerative power is most spectacular in the urodele amphibians. In newts and salamanders, and especially in their larvae, not only limbs can regenerate (as has been described) but also tails and external gills and, furthermore, the upper and lower jaws. Parts of the eye can likewise regenerate, such as the lens and the retina. If most of the eye is removed, it can be regenerated as long as a small part is left, and the new eye develops at the expense of the remaining fragment. In the anuran amphibians, the faculty for regeneration is restricted to the larval stage of their development. Frog and toad tadpoles are able to regenerate their limbs and their tails. The legs of adult frogs and toads do not normally regenerate at all.

In fishes, regeneration is very restricted. The fins can regenerate if cut off or damaged, but the tail (apart from the tailfin) does not regenerate.

The lizards are known to regenerate their tails. Such regeneration follows autotomy. The tails are broken off at a preformed level near the base of the tail. To release the mechanism of autotomy, the distal part of the tail must be injured or grabbed with such force as to cause the animal discomfort. After the tail is shed, a regeneration bud is formed on the wound surface, and this gives rise to a new tail. The latter, however, differs from the original tail; the vertebral column is of a simplified structure, and the scales covering the regenerated tail differ from the normal ones. The legs and even the digits in these animals cannot regenerate completely, though very rudimentary structures are sometimes formed on the site of an amputation.

In birds, parts of the beak can be regenerated, but otherwise their regenerative ability is rather poor.

In the mammals, limbs do not regenerate spontaneously even if amputated in a fetal stage (Nicholas, 1926). However, in infant opossums, which, like other marsupials, are rather incompletely differentiated at birth, amputated hindlimbs were found to possess a considerable ability for regeneration when stimulated by the implantation of a fragment of brain tissue into the limb. (See also p. 570.) Otherwise, regenerative ability in mammals is reduced to tissue regeneration, that is, the restoration of defects and lesions in various tissues but not the restoration of lost organs. Tissue regeneration is often equivalent to wound healing. Thus skin wounds may be covered by newly formed epidermis and connective tissue. In the case of large skin wounds, however, the newly formed connective tissue differs from the normal dermis and can be distinguished as scar tissue. The skeletal tissue has a high regenerative ability. Large defects in the bones, especially those of the limbs, can be made good. If parts of individual muscles are removed, the defects can be repaired by a proliferation taking place at the expense of the remaining part of the muscle. Lesions in the tendons can be replaced by connective tissue in the first place; the connective tissue later acquires the structure and mechanical properties of the tendon. The tissues of some internal organs are able to proliferate to a great extent and thus to compensate for the loss of large parts of the organ. This is found to a striking degree in the case of the liver. The greater part of this organ may be removed, and the remaining part then proliferates and restores the normal mass of liver parenchyme. The newly proliferated tissue does not, however, assume exactly the same shape that the liver had before the operation.

From this survey it will be seen that, on the whole, representatives of the lower forms of animal life regenerate better—may restore their normal structure from smaller parts of the original individual—than is possible for more highly organized animals. The rule is, however, not without numerous exceptions. Some exceptions may be explained by the high degree of histological differentiation in animals otherwise standing low in the scale of the animal kingdom (viz., the nematodes), a high degree of histological differentiation being antagonistic to the ability of cells to proliferate. In other cases this explanation does not hold; it does not tell us, for instance, why the fishes regenerate worse than the amphibians, although the latter are supposed to be derived from the former in the course of evolution.

With one and the same species of animals, the ability to regenerate may be greater in the earlier stages of the ontogenetic cycle. Legs may be regenerated in tadpoles but not in adult frogs. Adult insects cannot regenerate legs, although the larvae or nymphs are capable of regeneration. It may be recalled in this connection that parts of early embryos are sometimes able to develop into whole animals, whereas corresponding parts at a later stage are not even viable, such as when the cleavage stage of an amphibian egg is compared with an adult amphibian, or even with an embryo after the end of gastrulation. (See Chapter 7.) It may be concluded that a renewal of the morphogenetic processes in regeneration occurs more easily if normal morphogenesis is still under way. If the morphogenetic processes of normal ontogenesis have come to a complete standstill, it is more difficult or even impossible to renew them again.

19–3 STIMULATION AND SUPPRESSION OF REGENERATION

Although the general level of regenerative ability of an animal is determined by its constitution, by the degree of differentiation of its tissues or the stages of ontogenetic development, and by other factors, this ability may be increased or diminished by the environment or by special treatment to which the animal is exposed.

The rate of regeneration is naturally dependent on temperature, as most biological processes are. Increase in temperature, up to a certain point, accelerates regeneration. In *Planaria torva*, regeneration is scarcely possible at a temperature of 3°C. Of six individuals kept at this temperature, only one regenerated a head, and it was defective; the eyes and brain were not fully differentiated after six months. Regeneration was most rapid at 29.7°C.; at this temperature new heads developed in 4.6 days. A temperature of 31.5°C. was too high, and the heads regenerated after 8.5 days. A temperature of 32°C. proved to be lethal for the animals (after Lillie and Knowlton, from Morgan, 1901).

Food, on the other hand, does not affect regeneration very much. Even a fasting animal will regenerate at the expense of its own internal resources. In such diverse cases as rats regenerating parts of the liver, salamanders regenerating limbs, or hydras and planarians regenerating parts of their bodies, depriving the animals of food does not prevent regeneration and may even accelerate it to a certain extent. If planarians are deprived of food for a long time, they can live by metabolizing constituents of their own body. The animal, of course, diminishes in size as a consequence (degrowth; Needham, 1942). In this state a planarian can still regenerate. Although the overall size decreases, the missing parts are gradually rebuilt, so that a complete, even if very small, worm is eventually developed. Figure 447 shows the result of an experiment (Morgan, 1901) in which *Planaria lugubris* was cut lengthwise, and while the left half was fed, the right half was left entirely without food. Both regenerated. It appears, therefore, that regeneration is given top priority in the utilization of resources available to the organism. Although restriction of feeding seems to be favorable for regeneration, if anything, extreme degrees of emaciation by starving prevent regeneration, except in organisms such as the planarian, which is able to utilize its own body as a source of energy without deleterious results.

The nervous system appears to have a special influence on regeneration. In the amphibians, the early stages of regeneration cannot proceed normally in the absence of an adequate nerve supply to the region of the wound. If the nerves supplying the leg or the arm of a newt are destroyed simultaneously with the amputation of the limb or during the early stages of regneration, the development of the regenerating limb is arrested, and the blastema ceases to grow or may even be resorbed (Fig. 448; Schotté, 1923, 1926). If the nerves are cut before amputating the limb, processes of dedifferentiation set in, but instead of leading later to the formation of a limb blastema, they continue unchecked until most of the limb is destroyed. The constructive part of the regeneration does not start at all (Schotté and Butler, 1941; Schotté and Harland, 1943). The nervous system exerts its influence mainly on the earlier stages of regeneration. Once the regenerating limb has reached the stage when differentiation commences, it

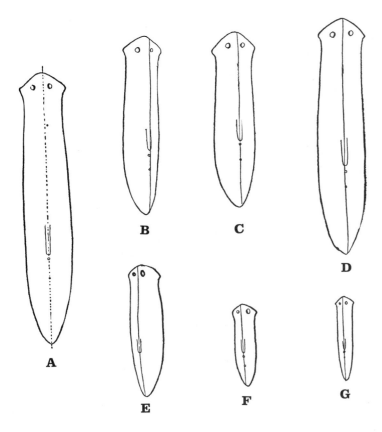

Figure 447. Regeneration of a planarian which was cut into two, lengthwise (*A*). The left half was fed (*B, C, D*); the right half was kept without food (*E, F, G*). (From Morgan, 1901.)

can proceed with its development even in the absence of nerve supply (if the nerve is cut at that stage).

The action of nerves, however, can go even further than supporting the process of regeneration. In a newt, it is possible to deviate a nerve from its normal position. In the case of limb nerves, the experiment is done so that the nerve is transected at a distal level, then separated along part of its course, from the distal end to the shoulder or pelvic level. A cut is then made through the skin of the trunk, starting from the pelvis or the shoulder, and the nerve is placed in the cut, so that when the wound heals the free end of the nerve is in a position under the skin and away from the limb it normally supplies. It was found that if the end of the nerve was not too far from the base of the limb, a limb rudiment, similar to a regeneration blastema of the limb, was formed over the end of the nerve. In successful cases this grew out into a complete new limb (Locatelli, 1924). Thus, the presence of a limb nerve may cause the formation of a limb at a spot where otherwise no limb development could have been expected.

The influence of the nervous system on regeneration is by no means restricted to

Figure 448. Arrest of regeneration in limbs with transected nerves (left limb in each pair). Limbs with intact nerves regenerate normally. (After Schotté, 1926.)

the regeneration of limbs in amphibians. It is very obvious in annelids. After section of the body, when the neoblasts migrate to the surface of the wound to form a regeneration blastema, they follow the ventral nerve cord in their migration. If the nerve cord is excised some distance from the level at which an earthworm has been transected, no regeneration will occur at that level. A regeneration blastema may be formed, however, at the spot where the nerve cord ends, and a new anterior end of the worm will be formed some distance from the anterior cut surface (Morgan, 1901).

The anterior end of the nerve cord can also be deflected similarly to the deviation of a limb nerve in a newt. The free end of the deflected nerve cord then causes the formation of a regeneration blastema and a new head (cf. Berrill, 1952). Even a simple section of the ventral nerve cord, without the removal of either end of the body, is sufficient to release regeneration in a polychaete (Okada, cited after Berrill, 1952).

Much the same holds true for planarians; regeneration blastemas are formed in conjunction with cut nerve cords. The cerebral ganglion of a planarian excised with a piece of surrounding tissue and transplanted to a different body level causes the formation of a complete new head at the site of transplantation (Santos, 1931).

The experiments described so far prove beyond doubt that the influence of the nervous system is necessary for regeneration. They do not show, however, in what way the nervous system acts—whether it acts in the same way as when transmitting stimuli that cause movement or release secretion or in some other way. An indication in this respect has been obtained in an experiment in which a limb was confronted with nerve fibers of various origins (Weiss, 1950). A piece of nervous tissue taken from the brain or the spinal cord was cultivated in the loose parenchyme filling the dorsal fin of urodele amphibians. After an initial partial degeneration the tissue started growing, and nerve fibers were produced by the surviving nerve cells. If a limb-bud was simultaneously transplanted into the dorsal fin, connections were established between the piece of nervous tissue and the limb; the outgrowing nerve fibers supplied the skin and the muscles of the limb. The fibers in question have been found to be neither the normal motor nor the sensory fibers but rather are equivalent to the "association neurons" which connect different parts of the central nervous system. If the piece of nervous tissue supplying the nerves was derived from the spinal cord or the medulla, the limb muscles could perform contractions, either spontaneously or in answer to an irritation of the limb. The contractions were, however, entirely uncoordinated, in the nature of epileptic seizures rather than normal movements. No contractions occurred, however, if the nervous tissue was taken from the forebrain, diencephalon, or midbrain. The transplanted limbs could then be amputated, and they regenerated no matter what part of the brain or spinal cord supplied the nerves to the limbs. Nerves which cannot release muscle contractions are thus adequate for supporting regeneration. The conclusion is thus reached that probably any kind of nerve supply can be the source of influence necessary for regeneration and that the nature of the influence is different from the ordinary transmission of stimuli by nerves.

It may be further suggested that it is hardly possible for the nerve fiber to be in contact with all the cells of a regeneration blastema while they are in motion taking up their positions under the wound surface. Under these conditions the normal mechanism of transmission of impulses could hardly operate, and it is thus more likely that the nerve endings act through the release of some substance. What this substance may be remains a problem for future research.

From both theoretical and practical viewpoints, it would be very important to know whether it is possible to excite regeneration in organisms or their parts which do not normally regenerate. The legs of tailless amphibians are a very suitable object for experiments of this type, since they regenerate in tadpoles but not in adult frogs. The ability to regenerate disappears in the legs of tadpoles some time before metamorphosis sets in, at a stage when all the digits have been formed and the skeleton of the limb is in a state of chondrification (Polezhayev, 1946). At a later stage, when the cartilaginous skeleton of the limb is fully differentiated and the limb becomes bent at the knee joint, an amputated leg does not regenerate any more, but the wound is covered with skin, and no regeneration bud develops. A regeneration bud may, however, be caused to form if after amputation the stump is traumatized by sticking a needle into it several times. This additional stimulus is sufficient to initiate the process of regeneration, and the regeneration bud subsequently develops into a limb in the ·usual manner (Polezhayev, 1946). In the adult frog, traumatization with a needle does not appear to be sufficient to incite regeneration, but treatment of the wound after amputation of the leg with a hypertonic salt solution may cause regeneration (Fig. 449; Rose, 1942). In all cases, the essence of the treatment used for inciting regeneration is to increase the destruction of tissue beyond that caused by the amputation. The resulting dedifferentiation of tissues favors the formation of a regeneration bud (cf. Section 19–4).

The influence of the nervous system on regeneration may be used to cause regeneration of otherwise nonregenerating parts. For this purpose, the sciatic nerve of a young metamorphosed frog was dissected (as was done in the aforementioned experiment on the newt) and deviated into the forelimb, which was then amputated. The presence of the sciatic nerve exerted the same action as traumatization of the limb stump: a blastema was formed and this developed into an incomplete limb "with a tendency to hand or finger formation" (Singer, 1950).

Encouraged by results obtained in frogs, researchers have attempted to incite regeneration of limbs in mammals, in which such regeneration does not normally occur. Use was made of the influence of the nervous system, and partial success was obtained in work done in the newly born opossum (Mizell, 1968). Marsupials are born in an early stage of differentiation, and in the opossum the hindlimbs at the time of birth possess a skeleton which is in a state of chondrification—this is the condition of the hindlimbs of frog tadpoles at the time when they lose the ability to regenerate. A simple amputation of the limbs in the newly born opossum does not lead to regeneration, but regeneration of a hindlimb may be stimulated by implanting a piece of brain tissue into the limb prior to amputation. Following this procedure distal parts of the limb start regenerating but do not obtain the perfection of a normal limb.

The method applied to the legs of older tadpoles, that is, traumatization by pricking with a needle, has been used to stimulate regeneration in the hydra (Tokin and Gorbunowa, 1934). Although a hydra is a paragon of high regenerative ability, the aboral tip of its body—that is, the foot with the adhesive disc—if severed from the rest of the

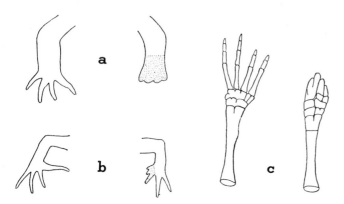

Figure 449. Regeneration of limbs in the frog after chemical stimulation. *a,* In a one-year-old frog. *b,* In a frog one month after metamorphosis. Regenerated parts stippled. *c,* Skeletons of a normal and a regenerated forelimb. (From Rose, 1942.)

body does not normally regenerate a new hydra but remains unchanged until it dies. Traumatization of an isolated foot in the manner described succeeded in causing it to reorganize itself into a complete new individual.

If the attempts to provoke regeneration where it does not naturally occur have been successful only in a few special cases, we possess some universally effective methods of preventing or inhibiting regeneration. One of the methods is by irradiating tissues with x-rays. If a leg or an arm of an adult newt is amputated, it readily regenerates. If, however, a newly formed regeneration bud of a limb is irradiated with x-rays, the development of the limb is either retarded or stopped altogether, depending on the amount of irradiation administered. There is some individual variation, but a dose of 5000 to 7000 r (r = roentgen, the unit of dosage of x-rays) suppresses regeneration in every case. An irradiated regeneration bud does not grow, and instead of developing, the skeleton and muscle of the limb become filled with connective tissue. The bud may actually decrease in size and become partially or wholly resorbed (Brunst and Scheremetjewa, 1933; Butler, 1933). The mechanism of the action of x-rays on the regeneration bud is obviously connected with the action of the rays on mitosis. X-rays are known to inhibit or suppress the mitotic activity of cells. If the cells of the regeneration bud cannot divide by mitosis, the growth of the rudiment becomes impossible; the number of cells available remains too small for the development and differentiation of the organ.

It is still more remarkable that the irradiation of a limb with x-rays may precede the amputation of the limb by months and even years and nevertheless can make the regeneration impossible. A dose of 7000 r applied to a normal limb of an adult newt does not usually cause any visible effect. Neither in appearance nor in its function does an irradiated limb differ from a nonirradiated one. The treated animals may be kept for months without showing any deleterious effect of the irradiation. If, however, an irradiated limb is amputated, regeneration is completely lacking, and the wound is instead covered by skin, leaving a permanent stump (Brunst, 1950).

It has been found that x-rays have the same effect on regeneration in all animals in which this effect has been tested. The effect is thus on some very fundamental property of living cells.

19–4 HISTOLOGICAL PROCESSES CONCERNED IN REGENERATION

Regeneration involves a complicated sequence of histological transformations in the stump of the amputated organ and later in the regeneration blastema. The immediate effect of an amputation is that tissues and cells lying normally in the interior of the body emerge on the surface. Some of the cells are squashed, torn, or otherwise destroyed; others become damaged by exposure to an unfavorable environment. The surface of the wound is thus covered with the debris of dead cells. In animals with a developed blood vessel system, the blood from the damaged vessels flows onto the surface of the wound and there coagulates, thus stopping further loss of blood.

The next stage is the covering of the wound surface with epithelium. Skin epithelium spreads over the wound surface, penetrating underneath the blood clot, between it and the intact living connective tissue. The spreading of the epithelium is due to ameboid movement of the cells and does not involve growth at the edges of the wound. No mitoses are found in the epithelium at this time. The time needed for the epithelium to cover the wound surface depends on the size of the regenerating animal and the size of the wound, besides such external factors as temperature. In the salamander larva, after the amputation of a leg the wound becomes closed by the epithelium in about one or two days. In invertebrates, the closure of the wound may be assisted by the contraction of the subepidermal muscle layer, so that the surface which has to be covered by the epidermis is diminished.

After the closure of the wound, a very important step in the process of regeneration sets in: dedifferentiation of the tissues adjoining the cut surface. This dedifferentiation has been best studied in vertebrates, where it is perhaps most pronounced. Dedifferentiation proceeds in conformity with what was said on this subject in Section 16–5. The intercellular matrix of bone and cartilage becomes dissolved, and the cells come to lie freely under the epithelium which has covered the wound. The connective tissue fibers likewise disintegrate, and the connective tissue cells become indistinguishable morphologically from cells derived from the disintegration of the skeletal tissues. The muscles also undergo dedifferentiation, the myofibrils disappear, and the nucleocytoplasmic ratio greatly increases.

Profound changes in the metabolism of the tissues in the amputation stump accompany this destructive phase of regeneration. The most prominent feature is the increase in the activity of the proteolytic enzymes, mainly cathepsin, but also the dipeptidases (Orechowitsch, Bromley, and Kozmina, 1935). As a result, there is a great increase in the amount of free amino acids (from 16.8 to 35.1 per cent of total nitrogen — Orechowitsch and Bromley, 1934). Another important change is that anaerobic glycolysis partially replaces oxidation in dedifferentiating tissues. Consequently, lactic acid accumulates in the tissues (Okuneff, 1933), thus lowering the pH from the normal 7.2 to a minimum of 6.6 at the time when a limb blastema is being formed (Okuneff, 1928).

The cells of dedifferentiated tissues become similar in appearance to embryonic cells. Whether this similarity is a superficial one or whether the cells acquire the properties of embryonic cells in every respect is a different matter and will be dealt with in Section 19–6.

The formation of the blastema or regeneration bud is the next step. Undifferentiated cells accumulate under the epidermis covering the wound, and together with it they form the regeneration bud, mentioned in Section 19–1. There has been much disagreement as to the origin of the cells of the blastema. Two main theories on this subject have been proposed. According to one theory, the cells of the blastema are all of local origin, that is, derived from the tissues immediately adjoining the wound surface. Except for the epithelium covering the wound, the rest of the cells, according to this theory, are set free from the dedifferentiating connective tissue, skeleton, and so on.

According to a second theory, the differentiated cells at the wound level have nothing to do with the formation of the blastema (except for the epithelium, which is obviously derived from the adjoining intact parts of the skin). The blastema is supposedly derived from cells migrating to the regeneration site from more or less distant parts of the body by ameboid movement or brought there with the blood stream. These cells are supposed to be special "reserve" cells, which still possess a capacity for development that has been lost by the differentiated tissue cells.

It is now known that neither of these two theories has universal application but that both sources of cells of the blastema may occur in different animals. The problem thus resolves itself into finding what the local tissues contribute to the formation of the regeneration blastema and what is contributed by migratory cells in any given animal.

Hydroid polyps are normally in a continuous state of "physiological regeneration." The cells of which the tentacles and the hypostome are composed continuously become worn out and discarded and are replaced by the cells shifting from a region lying below the hypostome (Fig. 450). Most of the mitoses in the body of the hydra, in all tissues,

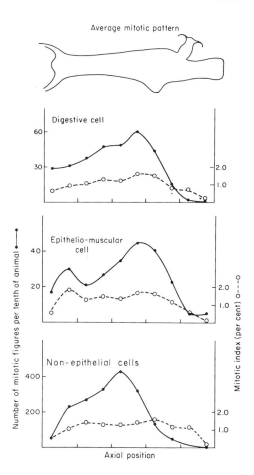

Figure 450. Distribution of mitoses in the different parts and the different tissues of a *Hydra*. Digestive cells are cells of the endoderm, epitheliomuscular cells are cells of the ectoderm, and nonepithelial cells are mainly the "interstitial" cells. Full dots show actual numbers of mitoses per section of the body; hollow dots show the percentage of cells of a given group which were found to be in mitosis. (From Campbell, 1967.)

occur in the area between the base of the tentacles and the region, toward the proximal end of the animal, in which buds are formed in asexual reproduction (Campbell, 1967). The base of the polyp (peduncle and foot) also depends on the supply of fresh cells from the growth zone, so that from the middle of the body a steady migration of cells goes on in both directions: to the hypostome and tentacles and to the peduncle and foot. On the hypostome and tentacles, the cells that need frequent replacement in the highest degree are the **cnidoblasts,** the cells producing the nematocysts which serve for the immobilization and capture of the small animals on which the hydroid polyps feed. A nematocyst can be used only once, so that the cnidoblasts degenerate after once having thrown out their thread and must be continuously replaced by new ones. The replacement of the cnidoblasts is performed by a special kind of cell, known as the **interstitial cells,** so called because they are found lying in the intercellular spaces at the base of the ectodermal epithelium and also in smaller numbers between the cells of the gut epithelium (Fig. 451). The cells are small with relatively very large nuclei (as in most undifferentiated cells) and basophilic cytoplasm.

In addition to their role in the replacement of cnidoblasts, the interstitial cells serve generally as a pool of undifferentiated cells which can be used for various morphogenetic processes in the organism of the hydroid. They are supposed to give rise to the sex cells, are the leading elements in asexual reproduction, and take part in processes of regeneration of lost parts. It appears, however, that the role of the interstitial cells in replacement of lost parts has been somewhat overrated.

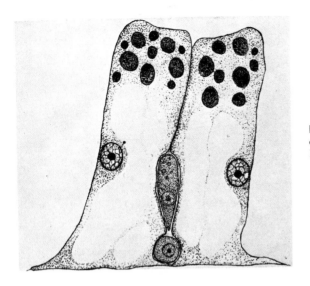

Figure 451.   Interstitial cells between two differentiated epitheliomuscular endodermal cells in *Hydra*. (From McConnell, 1936.)

When a hydroid polyp is cut transversely, the open wound is initially closed by a contraction of the epitheliomuscular cells of the ectoderm. The ectoderm next covers the damaged surface with a sheet of epithelial cells. In the next few hours mitotic activity is increased in the growth zone, and more than the usual number of cells are given off toward the damaged area, be that at the oral or the aboral end of the body. The cells thus mobilized at the site of the wound become molded to form the missing parts. Some of the interstitial cells moving toward the cut edge invade the endodermal layer and become converted into endodermal cells, thus supplementing the local endoderm. Others serve for the production of cnidoblasts for the regenerating parts (in the case of regeneration of the oral end). The epithelium of the newly built part is derived from similar epithelium adjoining the wound, and the endodermal epithelium nearest the wound contributes in part to the development of the regenerating structure (Burnett, 1962).

It has been repeatedly observed by many investigators that agents such as x-rays or mustard gas, which has a somewhat similar action, kill off interstitial cells of hydroids without causing immediate damage to differentiated types of cells. The animals continue living for a few weeks, but their ability to regenerate (and also to reproduce by budding) is reduced. This result has been interpreted as proving that regeneration is impossible without the participation of interstitial cells. In view of the moderate role of interstitial cells in the restoration of lost parts, as shown by direct observation, the preceding conclusion seems doubtful. Perhaps a more cautious interpretation would be that the x-ray or mustard gas treatment precludes the growth and mitosis in all cells of the body, and of course this would make the formation of new parts impossible.

There is, at present, sufficient evidence to show that the interstitial cells are not the only ones that can be transformed into other cell types. Attempts have been made to test whether the ectoderm or the endoderm of a hydra alone can reconstruct a complete organism. In this connection some experiments with isolation of endoderm are of particular interest. In *Pelmatohydra oligactis* and *Hydra viridis*, the endoderm was isolated from ectoderm by treatment with trypsin, which digests the mesoglea binding the two layers together. Isolated ectoderm degenerated after such treatment, but isolated endoderm underwent a progressive differentiation and produced a new complete animal. It is important that, in the species used, interstitial cells are present only in the ectoderm. Thus the new individual was produced at the expense of endodermal cells only. Some of these lost their differentiation and turned into ectodermal cells. Other dedifferentiating cells became transformed into interstitial cells, and the latter proceeded to produce cnidoblasts in the reconstituted polyp (Haynes and Burnett, 1963). It has also been claimed that isolated ectoderm of another hydroid, *Cordylophora*, is capable of producing a complete animal (Zwilling, 1963), but in this case the participation of interstitial cells has not been excluded.

It appears that in the hydroids the cells of various types show a very broad potentiality, an evidence that, in spite of being functionally differentiated, they retain a complete genotype and that all parts of their genetic information can be put to use (activated — p. 486). Nevertheless, the existence of a cell type such as the interstitial cells, which under normal conditions lend themselves especially readily to morphological transformations, is worth noting and leads us to the next example of the same kind.

In the planarians, the processes of regeneration take a somewhat similar course. The wound is initially covered by the epidermis of the skin. The cells move tangentially over the wound surface without proliferating. This occurs in the first 24 hours after infliction of the wound. The replacement of the lost parts occurs later at the expense of undifferentiated cells derived from the parenchyme (mesenchyme) of the worm. Whether these cells are "reserve" cells, as in the hydra, or are constituent cells of the normal connective tissue, which undergo dedifferentiation and become capable of participating in new morphogenetic processes, is not known for certain. They form the blastema under the sheet of epithelium, proliferate, and later differentiate into the various organs (Fig. 452). The blastema, and thus in the last instance the migrating parenchymal cells, give rise to most of the regenerating organs: the connective tissue, the pharynx, the nervous system, as well as the muscle of the body. The skin epithelium is derived from the skin at the edge of the wound, and also, it is claimed, the intestine in the regenerated part is derived from the cut edge of the old intestine (Bandier, 1937).

That the blastema is formed by cells migrating from remote parts of the body is definitely proved by the following experiment. If a planarian is irradiated with x-rays prior to wounding, no regeneration occurs, since, as we know, irradiation with x-rays makes the cells incapable of participating in the process of regeneration (Section 19–3). If, however, a small part of the worm is shielded from x-rays and then a part of its body is amputated, regeneration will proceed even if the wound is far away from the

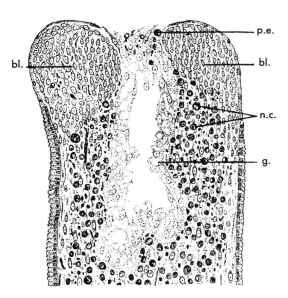

Figure 452. Regeneration in the land planarian *Rhynchodemus bilineatus*. bl., Regeneration blastema; g., gut; n.c., necrotic cells; p.e., proliferating endoderm cells. (From Bandier, 1937.)

p.e.

bl.

n.c.

g.

region which had not been irradiated. This can only mean that cells which had not been exposed to x-rays are capable of migrating long distances and then forming a healthy blastema on the cut surface (Fig. 453 – Dubois and Wolff, 1947).

In the annelids, there is a special type of cells called **neoblasts** which serve for the formation of regenerating parts. The neoblasts are large cells found normally in the peritoneum of the posterior surface of the intersegmental septa. When the body of the worm is cut, the neoblasts become activated and migrate toward the cut surface, following the course of the ventral nerve chain. The neoblasts accumulate under the epidermis which closes the wound and thus form the regeneration blastema. Not all the organs of the regenerated part are developed from the neoblasts, however, but only the mesodermal organs. The nervous system of the regenerated part is developed from the epidermis (from a thickening on its inner surface). Also the gut regenerates from the cut surface of the old gut. The competence of the neoblasts is thus more restricted than that of the mesenchymal cells in a planarian.

After treatment with x-rays, the neoblasts are the first to suffer damage and degenerate, and they may be destroyed with weaker doses of the rays, which leave the other tissues intact. Irradiated worms will not regenerate (Zhinkin, 1934). Epidermis and endoderm of the gut, which normally contribute to the formation of the regenerate, are not visibly damaged by x-rays, but they do not produce new parts in the absence of neoblasts. This may be either a result of the failure of some stimulating action normally exercised by the neoblasts or a result of direct damage by the x-rays, making these tissues incapable of proliferation even if this does not interfere with their normal activities.

The origin of the cells forming the regeneration bud in vertebrates is naturally of special interest to us. This has been most exhaustively investigated in the case of regenerating legs of newts and salamanders. A study of microscopic sections of regenerating stumps of salamander legs does not allow for an unequivocal solution to the problem. Whereas a dedifferentiation of tissues adjoining the wound can easily be observed, it is difficult to prove that the cells from dedifferentiating tissues actually form the regenerating blastema. This leaves the way open to those who claim that the regeneration blastema has a different origin and that the cells of the blastema are brought to the site of regeneration with the blood stream. The question has been settled, however, by the application of local irradiation with x-rays (Butler, 1935; Brunst, 1950). We have seen earlier (Section 19 – 3) that irradiation with a sufficiently high dosage of x-rays suppresses

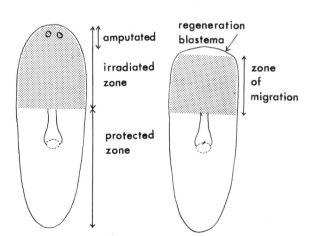

Figure 453. Regeneration in a planarian after partial x-ray irradiation. (From Dubois and Wolff, 1947.)

the ability of salamanders to regenerate. A modification of this basic experiment could be used for the solution of the problem under discussion.

Two types of experiments have been applied. In the first experiment, the whole animal was shielded from the x-rays (with a sheet of lead), and only one leg was exposed to the rays. The leg was then amputated inside the irradiated area. There was no regeneration (Brunst and Chérémétieva, 1936). In another experiment, the whole animal was irradiated, and an untreated leg was transplanted onto the irradiated body. Subsequently, the leg was amputated (inside the untreated area), and it regenerated normally (Butler, 1935). It is thus obvious that the irradiation does not interfere with the powers of regeneration of the animal as a whole but only affects the cells which are directly exposed to the rays. If the regeneration cells had been brought to the site of regeneration from other parts of the body, the local irradiation of only one leg would not have prevented its regeneration. By careful application of the lead screen, it could be proved that even a very thin layer of normal (unirradiated) tissue is sufficient to ensure a normal regeneration of the leg (Fig. 454 — Scheremetjewa and Brunst, 1938).

The irradiation of hydras or annelids in the first instance causes the degeneration of the interstitial cells or the neoblasts, and the ability to regenerate disappears as a consequence. It may be questioned in the case of newts and salamanders whether there is a special type of cells which is destroyed when the animals are made incapable of regeneration by the action of x-rays. The answer to this question should probably be a

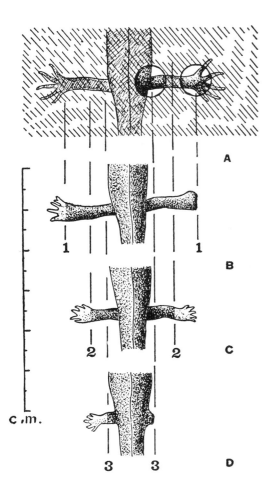

Figure 454. Local irradiation of parts of the limb in a newt, and results of amputation at different levels in the irradiated limb (right) and in the control limb (left). Drawing above (*A*) shows shielding plate covering the whole animal except for two circular openings through which the rays could pass. The limb fails to regenerate when amputated inside irradiated areas. (From Brunst, 1950.)

negative one: it is impossible to find by histological methods any cells in the tissues of the newt or salamander which could be identified as "reserve cells" or "undifferentiated cells."

By the use of radioactive tracers, it has been possible to give positive proof that all the dedifferentiating tissues are actually involved in the formation of the regeneration blastema and not one particular cell type. Tissues of amputated limbs of newts were supplied with tritiated leucine, a protein precursor, in one experiment (Anton, 1961) and with tritiated thymidine, a DNA precursor, in another experiment (Hay and Fischman, 1961). In both experiments it was observed that the radioactively tagged precursors were taken up by the cells adjoining the cut surface (to a depth of about 1 mm.). The intake started while the tissues had not yet lost their histological characteristics, and it was possible to ascertain that all kinds of tissues participated in the upsurge of synthetic activity. In particular this was found to be the case with muscle, fibroblasts, periosteum, endosteum, Schwann cells of the nerves, and the epidermis. The blastema was later found to consist of labeled cells. It follows that none of the tissues of the amputation stump are excluded from participation in regeneration, but that all respond to the wounding by growth (synthesis) and later by proliferation.

While the processes of dedifferentiation of tissues underneath the covering epidermis are still going on, the next step in regeneration is inaugurated: Cells in the region adjoining the wound start proliferating and dividing by mitosis. Mitotic activity starts as soon as tissues begin to dedifferentiate, and the rate of proliferation, as expressed by the proportion of cells in mitosis, rapidly increases (Table 19; Chalkley, 1954; Hay, 1962). This process reaches its peak at the time of the formation of the regeneration blastema in which, as Hay (1962, p. 182) put it, "muscle and other cell types have given up their structural and functional complexities to engage wholeheartedly in the pleasures of proliferation." For a time the divisions of the cells outstrip growth, and this leads to a slight diminution of the size of the cells in the blastema. Later, growth catches up with mitosis and both go hand in hand, so that a very distinct period of growth follows. The growth rate is at its highest immediately after the blastema has been formed, and subsequently the rate of growth diminishes inversely proportionally to the time that has passed since the beginning of growth (Syngajewskaja, 1936). As growth diminishes, differentiation sets in; thus, the inverse relationship between growth and differentiation holds good for regeneration as well as for normal development.

The reservation must be made that a local increase in mitotic activity is not observed where regeneration occurs by way of morphallaxis. In a hydra, as previously

TABLE 19   The Mitotic Activity in the Mesodermal Blastema and the Adjoining Part of the Stump During Regeneration of the Forelimb of the Newt, Triturus viridescens

Days After Amputation	Number of Cells Counted	Number of Mitoses	"Mitotic Index" (% of Cells in Mitosis)
1	52,487	4	0.0
7	78,533	415	0.53
13	111,705	748	0.67
19	139,322	900	0.65
25	180,043	953	0.53
31	215,035	625	0.29
37	213,432	365	0.17

After Chalkley, 1954.

stated, an increase of mitoses does occur during regeneration, but these take place away from the wound surface, mainly in the growth center (Kanajew, 1926; Burnett, 1962).

The constructive phase of regeneration is marked by a reversal of the physiological changes which accompany the initial destructive phase. The amount of proteolytic enzymes returns to the normal level. The oxidation becomes complete, and lactic acid disappears. The pH reverts to neutral.

The processes of histological differentiation taking place in the regenerating parts are, as a rule, similar to those occurring in normal ontogenetic development and need not be described here. In special cases, however, the processes may differ in two respects:

1. Certain tissues in regeneration may be developed from a source different from that in normal ontogenesis. An example of this has been mentioned previously: The regeneration cells of a planarian, which are derived from the parenchyme (mesenchyme) and are thus of a mesodermal nature, give rise in regeneration not only to the parenchyme and the muscles but also to the nervous system and the pharynx. In normal development from the egg, the nervous system and the pharynx are produced by the ectoderm. Similar cases are known to occur in some other animals. A very peculiar case, and one that has attracted much interest among embryologists, is that of lens regeneration. If the lens of the eye is removed in adult or larval urodele amphibians, a new lens is produced from the edge of the iris. The tissues of the iris lose their pigmentation, proliferate, and form a rounded body. This body separates from the rest of the iris and differentiates as a new lens, taking up the position and the function of the removed part. The remarkable feature in this case is that in embryonic development the lens forms from the epidermis, while the eye cup serves as inductor. In regeneration the parts derived from the epidermis (the cornea) do not participate in replacing the lost lens, and instead the lens is derived from a part of the eye cup itself. (See Reyer, 1962.)

2. Sometimes the regenerated structure is formed in a way different from the normal one. The notochord is the predecessor of the vertebral column during the normal development of a salamander larva. In an adult salamander, the notochord is reduced and replaced by the bony vertebrae, and it does not reappear in a regenerating tail. During regeneration, the cartilages representing the rudiments of the vertebrae are formed around the spinal cord instead. The normal segmentation of the vertebral column is not restored in the regenerated tail of a lizard.

19–5 RELEASE OF REGENERATION

According to definition, regeneration is the replacement of lost parts. One could have expected, therefore, that the loss of some part of the body would be the adequate stimulus to set in motion the mechanism which restores the part and thus the normal structure of the animal. This is by no means always the case. If a deep incision is made on the side of a salamander's limb or on the side of the body of an earthworm or a planarian, a regeneration blastema may be formed on the cut surface. The blastema then proceeds to grow and develop into a new part, as in ordinary regeneration. In the case of a limb, the new part thus developed will be the distal part of the limb, from the wound level outward. The development of the regenerating part proceeds just as if the entire distal part of the limb were cut off.

In the case of a planarian, a lateral incision may cause the development from the wound surface of either a new head, a new tail, or both. If both a head and a tail are regenerated, the head forms from that part of the wound surface which faces anteriorly, and the tail develops from the wound surface facing posteriorly. This results, of course, in the regenerated head lying more anteriorly than the regenerating tail. A somewhat similar reaction is produced by lateral incisions in the earthworm, with a restriction that lateral incisions near the head end of the worm give rise to additional heads, incisions in the middle part of the animal cause the development of both heads and tails, while incisions in the posterior part of the animal's body cause the formation of tails only. Another peculiarity in the case of an earthworm is that the incision must be deep enough to sever the ventral nerve chain if any regeneration at all is to take place. (Compare Section 19–6.)

In each of these cases, the original parts of the animal (heads, tails, limbs) had not been removed, so that the regenerated parts were additional and therefore superfluous to the animal. The experiments allow us to conclude that not the absence of an organ but the presence of a wound is the stimulus for regeneration.

The development of a superfluous number of organs or parts of the body, as a result of regeneration, is called **super-regeneration.** The clue given by super-regeneration has been followed up to analyze still further the stimulus leading to regeneration. It has been found that regeneration can be started even without inflicting an open wound. This has been demonstrated by ligaturing a limb in a salamander (axolotl). (See Nassonov, 1930.) A tight ligature causes considerable destruction of the tissues immediately affected by the pressure. The muscles and portions of the skeleton of the limb disintegrate, so that the part of the limb distal to the ligature becomes bent at an angle and is dragged about by the animal without being capable of movement of its own. The skin, however, turns out to be more resistant and preserves its integrity. After some weeks, the region of the limb just proximal to the ligature begins to swell, and it soon becomes evident that a regeneration blastema has been formed. The regeneration blastema then develops into the distal part of a new limb, although the old distal part of the limb is still present. With the processes of differentiation setting in, the skeleton of the old distal part of the limb may be joined again to the proximal skeleton.

The experiment on ligaturing the limb teaches us that the presence of an open wound is not essential for regeneration. What is really necessary is damage to the tissues of an organ that is capable of regeneration. Usually tissues are damaged by wounding, but if extensive damage to the tissues can be caused without an open wound, this suffices to start the sequence of processes leading eventually to regeneration.

Having reached thus far we may suggest that damaging the tissues is necessary so that some substance or substances be released from the damaged and disintegrating tissues, which are the immediate cause of the processes that follow. Various types of experiments have been adduced in support of this concept.

It has been found that if a regeneration blastema of an axolotl limb is dried at low temperatures, so that all the cells of the blastema are killed, and the blastema is then transplanted under the skin of an axolotl limb, it causes an outgrowth on the surface of the host limb. This outgrowth, which is covered by skin and has a cartilaginous axis in the middle, is comparable to a very rudimentary limb or at least to a digit. A similar outgrowth may be caused by an implanted piece of cartilage and also by introducing under the skin the products of alkaline hydrolysis of cartilage (Nassonov, 1936). These experiments immediately remind us of the experiments on the primary organizer in early amphibian development: there, as here, it was found that the stimulus for causing certain morphogenetic processes was not necessarily dependent on the integrity and vital activity of the cells of the inducing part and that nonliving substances could exert a similar action.

Another line of research consists in the treatment of the amputation surface with a solution of beryllium salt (beryllium nitrate). Beryllium nitrate applied to the amputation surface of a tadpole tail or of the limb of an *Ambystoma* larva completely suppresses regeneration (A. E. Needham, 1941, 1952). It has been suggested that beryllium in some way binds the substances released from the damaged cells at the wound surface that would have normally initiated the whole sequence of the processes of regeneration—the dedifferentiation in the first place and subsequently the formation of the regeneration blastema. Such an interpretation is supported by the following details of the experiments with beryllium.

First, the beryllium treatment must be carried out immediately after the amputation; an hour later the treatment is without any effect. This might mean that the substances released from the damaged cells have already started the next step of the reaction (the processes of dedifferentiation) or at least created the conditions in which this next step inevitably follows.

Second, if after treatment of the wound surface the stump is again amputated 0.5 mm. proximal to the original cut, regeneration proceeds normally. Thus, the action of the beryllium is only very local. If new cells are damaged they may in their turn release the same substances and "trigger off" regeneration. The treatment of normal tissues prior to amputation does not have any effect either; it is only after the cells are damaged that beryllium can in some way affect the results which could have been produced by the damaged cells.

In this respect there is a profound difference between the inhibition of regeneration by x-rays (Section 19–3) and the inhibition of regeneration by beryllium nitrate. X-rays make the cells incapable of growth and reproduction, and this makes regeneration impossible. The destructive processes are not checked by the rays. On the other hand, beryllium does not impair the ability of the cells to grow and differentiate, but it checks all the stages of regeneration by preventing the initial steps of regeneration from occurring. It is thus very probable that it interferes with the specific factor releasing regeneration and that this factor is a substance or substances given off by the damaged cells at the wound surface.

19–6 RELATION OF THE REGENERATING PARTS TO THE REMAINDER OF THE ORGAN AND TO THE ORGANISM AS A WHOLE

The factors on which regeneration is dependent can be subdivided into two groups:

1. The factors which are responsible for the regeneration taking place.
2. The factors which determine that the right sort of organ regenerates.

So far, we have been dealing mainly with the first group of factors. It has been shown, however, that sometimes a regenerating organ does not fit into the organization of the whole animal, such as when a new limb regenerates without the original limb having been removed (Section 19–5). Still, as a general rule, what regenerates corresponds to what has been lost. This can only mean that the position of the wound in some way determines the nature of the regenerating part. If the cut is through a limb, a limb will regenerate (if any regeneration occurs at all); if the cut is through a tail, a tail will regenerate. If the limb is cut at the lower arm level, parts of the lower arm, wrist, and digits will develop; if the cut is at the upper arm level, the upper arm will regenerate as well. A part of the shoulder girdle may also be removed, and the entire arm with the shoulder girdle will be renewed. If, however, the entire shoulder girdle together with the muscles of the shoulder is completely removed, the arm can no longer regenerate. This shows that a remnant of the original organ must remain in order to enable the organ to regenerate. The arm and the shoulder girdle with its muscles form one unit with respect to regeneration. Such a unit has been termed a "regeneration territory" (Guyénot and Ponse, 1939) or "regeneration field" (Weiss, 1926a). A similar regeneration territory is necessary for the regeneration of the tail in adult newts; as long as a small piece of the tail remains, regeneration is possible. If, however, the tail is amputated at the level of the last sacral vertebra, the whole "tail territory" is removed, and no regeneration takes place. The wound heals without any regeneration blastema being formed.

It is not inevitable that the regeneration be completely repressed if the whole of the regeneration territory is destroyed. Regeneration may occur, but in a different way. In the shrimp *Palinurus*, the eye may regenerate after being removed. The cut, however, must be made through the eye stalk, just proximal to the eye and distal to the nerve ganglion lying at the base of the eye (inside the stalk). If the cut is made at the base of the stalk, so that the ganglion is removed with the eye, the eye will not regenerate. A regeneration blastema is formed, however, but instead of the lost eye, it develops into an antenna-like organ (Fig. 455). This phenomenon—a different organ developing from the one that has been removed—has been called **heteromorphosis.** The most plausible explanation of heteromorphosis in this case is that the ganglion and the eye together constitute one regeneration territory. If a part of the territory (the ganglion) remains intact, the complete system may be restored. If the whole territory is lost, it cannot be renewed at the expense of other parts of the body.

A different result, yet illustrating the same principle, has been observed in the regeneration from anterior cut surfaces in the earthworm. As shown previously (Section 19-2), a new "head" is regenerated at the anterior end of the earthworm if this part of the earthworm is cut off. This is possible, however, only if the cut is not too far from the anterior end of the animal. If the cut is beyond the middle of the worm, the posterior part can no longer regenerate a new "head." Regeneration takes place, but what is regenerated is a second tail. As this tail is at the anterior end of the sectioned worm, it is a case of heteromorphosis (Morgan, 1901). Again, we see that a new anterior end of the animal can be restored only if not too much of the anterior body part has been removed. With more than half of the animal removed, the regeneration territory or

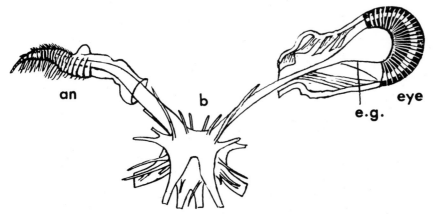

Figure 455. Heteromorphic regeneration of an antenna in place of an amputated eye in *Palinurus*. an, Antenna, regenerated; b, brain; e.g., eye ganglion. (After Herbst, from Hartmann: Allg. Biol., 1947.)

regeneration field of the head is completely gone, and what regenerates can only be a tail end. There is an interesting peculiarity in this case which should be noted. By making deep lateral incisions in the body of the earthworm, it has been observed that the ability to regenerate a head diminishes gradually in an anteroposterior direction. The heads which are formed at the site of the wound become smaller and smaller as the wound is made further away from the anterior end, and beyond the middle of the animal's body no head is regenerated, although a tail may be formed.

The same is the case in the regeneration of planarians. If the body is cut transversely at different levels, the tail piece will regenerate a head only if the cut is not too far posterior. Conversely, a very short piece cut from the anterior end of the body of a planarian does not regenerate a tail, at its posterior surface, but a second head. At the anterior end there is thus no competence for tail development present, or else the tendency to develop a head is so strong that the morphogenetic processes cannot proceed in any other way.

If the cut surface determines the nature of the organ or part of the body to be produced in regeneration, it may be questioned whether there are any special cells or tissues at the cut surface which are responsible for the course of regeneration. The problem may be approached experimentally by removing parts of the organ at the level of amputation one by one and observing the result. A classical experiment of this kind was performed by Weiss (1925). After making a slit through the skin and muscle of the arm in the newt, Weiss removed the humerus. After such an operation, the humerus was not restored. The wound healed, and then the limb was amputated through the upper arm. Due to the first operation, there was no bone or cartilage at the level of amputation. Nevertheless, the regeneration proceeded in the normal way. A regeneration bud was formed, and this developed all the parts of the limb distal to the level of amputation. The skeleton of the regenerated part was complete, including the distal part of the humerus. The proximal part of the humerus was lacking, as before (Fig. 456). The first conclusion that can be drawn from this experiment is that the tissues of the regenerated limb are not derived from corresponding tissues at the level of amputation; rather the cells of the regeneration blastema are to be considered as undifferentiated, insofar as the various parts of the limb are concerned, and capable of fitting into any part of the limb.

This has also been confirmed in a very different way. The distal part of a newt's limb may be split lengthwise by a longitudinal cut and then each half amputated and allowed to regenerate. Two regeneration buds are then formed, one at the end of each portion of the limb. Each develops subsequently into a complete distal part, with a complete carpus or tarsus, as the case may be, and not into half of one. Thus the cells derived from one half of a transverse section of the limb are capable of producing a

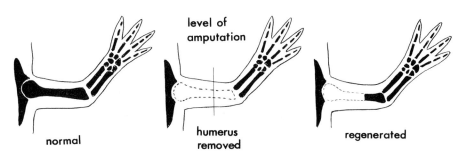

Figure 456. Regeneration of a forelimb in a newt from which the humerus has been previously removed (diagrammatic). (After Weiss, 1925.)

complete organ (Weiss, 1926b). It is interesting to compare this result with the splitting of the early limb-bud of an embryo (Section 13–3). With regard to the animal as a whole the development of two limbs, where only one was amputated, is excessive; it is a case of super-regeneration.

Returning to the experiment on regeneration of a complete distal part of a limb in the absence of skeletal parts at the amputation level, we may note a second conclusion: the skeleton at the amputation level does not appear to be necessary for determining the nature of the regenerating organ.

With the exclusion of the skeleton as carrier of the factors determining the course of regeneration, our attention turns naturally to the muscles and connective tissue of the limb. Excluding these tissues from the level of amputation cannot easily be done, but the action of muscles has been tested in a different way. Muscles from one regeneration territory (e.g., the tail) were transplanted in place of muscles of a different regeneration territory (e.g., the limb) and vice versa. If the organ was then amputated, the regeneration was no longer normal; the transplanted muscles exerted a specific influence on the nature of the regenerating organ. Thus, if the arm muscles were removed as completely as possible and pieces of tail muscle were stuffed into the space between the arm skeleton and the skin, and then the limb was amputated, the distal part of the regenerating organ was no longer an arm but a structure resembling a tail (Liosner and Woronzowa, 1936).

If the cells of the regeneration blastema are capable of producing any part (any tissue) of the regenerating organ, the question further arises as to whether different kinds of organs may be produced from a regeneration blastema, other than the organ from the stump of which the blastema had been derived. Two opposing views have been held on this subject by different investigators.

According to one view, the cells of a regeneration blastema are completely undifferentiated and capable of developing into any part of the animal's body, except perhaps that mesodermal cells do not give rise to skin epithelium, though the transformation of epithelial cells into connective tissue cells has been considered as possible by a number of zoologists. (See Rose, 1970.) In support of this view it is said that a very young blastema, corresponding in stage of development to those depicted in Figure 446*a*, may be transplanted to the amputation stump of a different organ and then may develop in correspondence with its new position. Transplantations have been carried out to exchange the regeneration blastema of the fore- and hindlimbs in the newt. An early forelimb blastema transplanted to the stump of an amputated hindlimb is said to be capable of developing into a hindlimb, and vice versa. Furthermore, the regeneration blastema of the tail has been transplanted to the shoulder region, and it was reported that it developed into a forelimb (Weiss, 1927). As regeneration does not proceed in the absence of nerve supply, a limb nerve had to be diverted to the site of transplantation to make the transplanted blastema grow and differentiate. The latter procedure, however, introduces a serious source of error in the experiment. We have seen that the deviation of a nerve into the area around the basis of a limb alone, without the transplantation of a regeneration blastema, may induce the development of a supernumerary limb (experiments by Locatelli, Section 19–3).

It may well have happened that the grafted regeneration blastema of the tail was gradually destroyed and replaced by local cells, which were activated for limb formation by the diverted nerve. The results of transplantation of regeneration buds between the fore- and hindlimbs have also been proved to be inconclusive: the mobilization of regeneration cells in the early stages to which the preceding experiments refer is not yet completed, and therefore the grafted cells could be replaced by local cells, with the result that the nature of the regenerated organ conforms to the position in which it develops.

Careful experiments have proved that no change can be produced in the specific type of development of a regeneration blastema, no matter in what way it is transplanted, at least in the case of amphibian limb regeneration.

In planarians, on the other hand, the tail regeneration blastema has been grafted onto the anterior amputation surface, with the result that it developed into a new head (Gebhardt, 1926). This result has not yet been challenged, and it may be possible that the regeneration cells in lower animals are more plastic, retaining broader potentiality for differentiation. This is perhaps to be expected, as the regeneration territories in these animals are not strictly delimited; the competence to produce a tail or a head fades away gradually, starting from the anterior or posterior end of the animals.

It is thus fairly safe to conclude that the capacities of regeneration cells have very definite limits and that these limits are more narrow in the highly organized animals than in the lower forms of life. There is no question of the regeneration cells acquiring the same abilities for development as the egg or the early cleavage cells. A further difference between the early embryonic cells and the regeneration cells is that the first may develop in complete isolation; the second can develop only in conjunction with the remainder of the organism, which provides the regeneration blastema with nourishment, supplies it with nerves, and exerts a certain degree of influence on the processes of differentiation of the regeneration blastema.

19–7 POLARITY AND GRADIENTS IN REGENERATION

The role of gradients in controlling morphogenetic processes has been invoked in two previous sections of this book: to explain the organization of the egg cytoplasm (Section 6–8), and to account for the regional differentiation of the primary organ rudiments in the embryo at the start of organogenesis (Section 8–8). The theory of physiological gradients, as proposed initially by C. M. Child (1929, 1941), leaned heavily on phenomena observed in the course of regeneration, mainly in invertebrates, the planarians and the hydroids. In these animals the ability to regenerate lost parts shows a regular decrease along the main axis of their bodies. The head end in flatworms and the oral (distal) end in coelenterates regenerate best if the section which removes part of the body is made closer to the anterior (distal) end. The further away from this end, the weaker is the regeneration process. In planarians the regeneration ability increases again toward the posterior end of the body. The weaker regeneration ability expresses itself in a slower tempo of regeneration (in hydroids—Barth, 1938), in a smaller percentage of successful regenerations, and in a less complete structure of the regenerated part (in planarians—Child and Watanabe, 1935).

Child put forward the idea that the regeneration capacity of parts of the body is closely linked with what he called the "physiological activity" of the parts in question, so that the parts of the body with higher physiological activity have a higher morphogenetic potential, that they are able to produce more elaborate structures and produce them easier and faster. The nature of the physiological activity remains ill-defined; Child was inclined to equate it with the rate of respiration, but in other tests the measure of physiological activity was the sensitivity to various noxious influences (poisons, excessively high temperatures, ultraviolet radiation, deprivation of oxygen). In more lowly organized animals (worms, coelenterates), but also in embryos of more highly organized animals, the physiological activity, as defined, is actually found to be distributed in the form of gradients along the length of the body. In particular, it was found that the oxygen uptake in the hydroid *Tubularia* (used in many experiments on regeneration) is highest at its distal end and decreases gradually along the body of the animal (Tardent,

1964). In the planaria *Dugesia dorotocephala* the uptake of radioactive precursors ($C^{14}O_2$ and C^{14} glycine) into proteins shows a gradient, with the high pole at the anterior end of the animal (Flickinger and Coward, 1962). This would appear to demonstrate a gradient of protein synthesis.

On the other hand both in planarians and hydroids the anterior or distal end of the body is the one with the greatest morphological complexity. In the planarians the anterior end contains the brain and sense organs (eyes and, in some species, tentacles or auricles, bearing chemoreceptors) but no mouth, the latter being situated toward the middle of the body. In the hydroids the distal end bears the mouth, at the tip of the hypostome, and the tentacles. In the hydroid *Tubularia* there are two rows of tentacles. The sequence of the organs starting from the distal end is then: the hypostome, the first row of tentacles, and the seond row of tentacles, arising from a widened gastric part. These organs constitute the "hydranth," demarcated by a constriction from the hydrocaulus or stem, which is of considerable length compared with that of the hydranth (Fig. 457). Thus, the higher level of morphogenetic activity coincides with the high pole of the physiological gradient.

The second proposition of Child's theory of physiological gradients is that the morphogenetic activity associated with the high pole of the gradient exercises domination over the lower levels of the gradient. While the morphogenetic processes associated with the high poles proceed by self-differentiation, they establish adjacent to themselves a regular gradient of diminishing physiological activity, and through this they prevent high levels of differentiations from occurring further down the gradient. At the same time, structures corresponding to the lower levels of the gradient are allowed to develop, thus leading to the formation of a complete whole, with all its parts.

The hydrocaulus (stem) of the hydroid *Tubularia* is capable, under certain circumstances, of producing a new hydranth when transected at any level. Normally, however, if a piece of stem is cut out, it regenerates a hydranth only at one end, the one facing the original distal end. Similarly, when a section from the middle of the body of a planaria is cut out (with the head and the tail end removed), a head will regenerate at the anterior end and a tail at the posterior end. Thus, the **polarity** of the body is preserved. In a hydra, the middle part of the body will regenerate a hypostome and tentacles at the distal end and a "foot" at the proximal end. (*Tubularia* does not have a foot but has a hydrorhiza, which does not regenerate in the short term experiments done on this genus.)

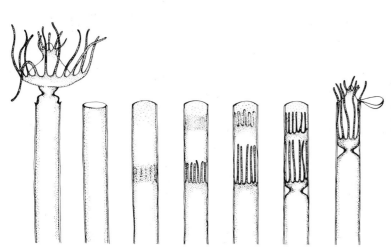

Figure 457. Stages of regeneration of *Tubularia* following amputation of the hydranth. (After Tardent and Tardent, 1956; from Goss, Principles of Regeneration, Academic Press, London and New York, 1969.)

It is an essential part of Child's theory that both the polarity and the dominance of the high pole of the gradient are causally linked with the level of physiological activity. Therefore, if the physiological activity of the high pole of the gradient is reduced in some way, the morphogenetic activity corresponding to the top level disappears, and so does the dominance of this pole over the originally lower levels. The higher physiological activity of the distal end of a piece of stem of *Tubularia* can be reduced by sticking it into the sand or by ligaturing it (in both cases the oxygen access is interrupted), and then the regeneration of the hydranth takes place at the original proximal end. The polarity is in this way reversed. A more elaborate experiment consisted in constructing a double chamber and sticking pieces of *Tubularia* stem through holes in the partition wall. Oxygenated sea water was allowed to flow through one compartment and nonoxygenated water through the other. Hydranths regenerated from the ends bathed with oxygenated water. Alternately, water in one of the compartments was warm and in the other cool. Hydranths regenerated from ends in the warm water, independently of whether the end was originally distal or proximal (Miller, 1937).

If lowering the physiological activity can depress a morphogenetic process characteristic of the high pole of the gradient, then increasing physiological activity may induce a high level of morphogenesis at a position where otherwise it could not occur. In Child's experiments this was achieved by making a jagged wound on the side of a stem of *Tubularia* or by grafting into the stem at a lower level of a piece of stem taken from nearer to the hydranth. In both cases a new hydranth was developed on the side of the stem (Fig. 458). The locally raised level of the gradient released the tissues from domination by the intact terminal hydranth (Child, 1929).

The gradient of protein synthesis in a planaria (previously mentioned) can be "flattened out" by treating the animal with metabolic poisons (mercaptoethanol and

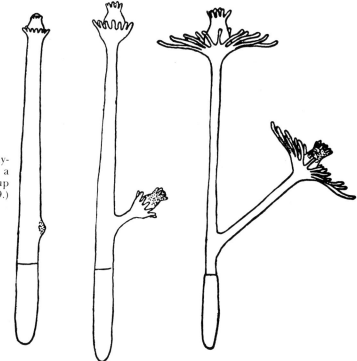

Figure 458. Induction of a lateral hydranth in *Tubularia* by implantation of a small piece of stem taken from higher up the gradient (stippled). (After Child, 1929.)

colcemide). As a result, the dominance of the anterior levels of the body is reduced, and worms which had both the head and the tail cut off regenerated, in a substantial percentage of cases, heads at both ends. A head can be made to regenerate at the posterior end even if the original head is left intact. To accomplish this the worms were pretreated with colcemide to reduce the dominance of the original head. Then their tails were amputated, and the worms were embedded in agar to immobilize them. The anterior end was then immersed in a solution of colcemide. In just over 3 per cent of the cases a head regenerated at the posterior end, thus producing two-headed worms (Fig. 459, Flickinger and Coward, 1962).

The dominance of the anterior (or distal) part of the body requires that there be a substantial difference in the level of the gradient between the two surfaces of a cut piece. If a very short transverse piece is cut out from a planaria, heads regenerate at both ends, producing a misshapen animal with two heads and practically no body. Similarly, a very short piece of stem of *Tubularia* regenerates hydranths at both ends. On the other hand, the dominance of the high pole of the gradient does not extend beyond a certain distance. When long pieces of stem are cut out from the hydrocaulus of *Tubularia*, hydranths regenerate at both the distal and the originally proximal ends.

Naturally, the nature of the control by the dominant high point of the gradient which maintains the polarity of the animal is of great interest. There is evidence that this control is exercised by means of diffusible chemicals, some activating morphogenesis and others acting as repressors.

From crude extracts of oral ends of hydra by repeated column chromatography, an activating substance has been isolated which has a potency 500,000 times that of the crude extract, and which, when administered to regenerating hydras, accelerates head regeneration and increases the number of tentacles by a maximum of 15 per cent. It also increases budding. The substance is distributed in a gradient, from hypostome (maximum) to base. It has a molecular weight of about 900, is destroyed by proteolytic enzymes, and is thus a polypeptide, possibly with some other radicals added. In the

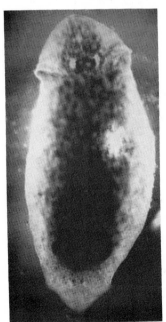

Figure 459.   Planaria regenerating a head at the posterior end after the anterior end of the body has been repressed by immersion in colcemide. (From Flickinger and Coward, 1962.)

living tissue of the hydra it is contained in nerve cells, in vesicles 1200 Å in diameter. The substance may thus be considered as being neurosecretory! (Schaller, 1973; Schaller and Gierer, 1973.)

Perhaps more important are the repressor substances which are instrumental in maintaining the dominance of the high pole of the gradient in morphogenesis. Such substances have been obtained from the hydranths of hydroid polyps and from different levels of the bodies of worms. The common property of these substances is that they repress the development of those structures and body parts which are identical to those from which they have been obtained. Extracts from hydranth homogenates of *Tubularia* inhibit regeneration of new hydranths in the same animal (Tardent, 1960). The repressor substances appear to have a high degree of specificity, as preparations from the distal part of the hydranth, from its proximal part, and from the stem have different properties, each exercising its effect on the homologous organ (Rose, 1970). It should be noted that the inhibitors act on the parts in the initial stages of their differentiation (during regeneration). Once the differentiation has proceeded to a certain stage, the structure becomes insensitive and instead starts emitting the corresponding inhibitory substance. This explains why the inhibitor produced by a structure, such as a hydranth, does not inhibit the producing part, the hydranth itself, but only prevents a similar structure from arising within its field of dominance. Similar inhibiting substances have been found to exist in worms. (See Rose, 1970.)

The inhibiting substances have not yet been completely purified, but some of their properties are known. They are destroyed by trypsin and therefore are protein-like substances. In electrophoresis they move to the positive pole, are thus negatively charged, and most probably are of the nature of histones (Rose, 1970). If the substances are in fact histones, the question may be raised as to whether they have anything to do with the nucleohistones, which have been proved to play a role in repressing nuclear genes. In other words, do the repressors of regeneration act by directly repressing genes or groups of genes?

Here the student should be reminded of what has been said earlier about the lack of specificity which makes it unlikely that histones are the primary factor in regulating gene action (p. 487). On the other hand, it seems plausible that the activating substance and the inhibiting substances, as shown in the above-mentioned experiments, are the means of maintaining the gradients responsible for the control of polarity in regeneration. Part of such a concept must be that the substances move along the length of the animal's body, so that a dominant part may exercise its influence at a distance. In a flatworm, such as a planaria, one could conceive of the regulatory substances diffusing intercellularly through the body parenchyme. In hydroids, on the other hand, there is practically no parenchymatous tissue. The gastric cavity is hardly a suitable channel for conveying finely dosed minute quantities of regulatory substances, and moreover, it was found that the distal end can continue to dominate morphogenesis while the gastric cavity is continuously flushed with a stream of water (Miller, 1959).

It would be interesting, if it could be proved, that the gradients in regenerating worms and hydroids are maintained by substances passing directly from cell to cell. In this connection it is pertinent to refer to experiments that prove the possibility of substances moving directly from cell to cell without passing out into the intercellular medium. It was first shown that electrical currents may flow from cell to cell, which involves essentially the transmission of sodium and potassium ions. Subsequently it was found that larger molecules, of molecular weight over 1000, but not larger than 10,000, can also pass from cell to cell by simple diffusion (Loewenstein, 1970). The passage of these molecules (as well as of simple ions) is effected through a special type of junction, which can be established between most kinds of cells where they come in contact with one another. The junctions have been called "tight junctions." We have seen that the regeneration-activating substances extracted from hydras have a molecular weight of about 900; they are thus well within the range of molecules that can pass through tight junctions between cells. The molecular weight of the substances repressing regeneration has not been accurately determined, but the possibility of their passing through tight junctions is quite likely.

What has been said here of the movement of substances from cell to cell, as a means of establishing and maintaining morphogenetic gradients, may well also apply to the early stages of embryonic development, such as to the gradient system in sea urchin eggs (dealt with in Section 6–8). It has been found that tight junctions, permeable to intracellular substances, are established in the morula-early blastula stage in amphibians, fishes, birds, and squids (Loewenstein, 1970).

There is one more point to be made. The existence in regeneration of substances produced by certain parts of the body, which repress the development of *the same* kind of structure, reminds us of the chalones, which are produced by differentiated tissues and which repress the growth of the same kind of tissue. Scattered observations have been made that in embryonic development as well extracts of organs may have a specific harmful influence on the development of the same kind of organ (Lenique, 1959; Rose, 1970). It is not known whether the inhibiting substances in the three cases are of the same nature.

19–8 RECONSTITUTION FROM ISOLATED CELLS

Related to regeneration is the reaggregation of isolated cells into a new whole animal. This remarkable phenomenon has been discovered in sponges (H. V. Wilson, 1907). A sponge may be rubbed through bolting silk, so that the entire organization of the sponge is broken up, and the tissues of the animal are reduced to a pulp consisting of isolated cells and cell debris. If the pulp is allowed to stand, the isolated cells begin crawling about and aggregating themselves into larger masses. These masses then become organized into new sponges. Among the isolated cells one can distinguish the archeocytes, the collar cells, and the dermal cells of the adult sponge. When the cells reaggregate, each type of cells sorts itself out and takes up in the aggregate a position which belongs to it because of its specific properties: The dermal cells cover the whole aggregate from without; the collar cells join together and rebuild the collar cell chambers in the inside. The archeocytes take up their normal position, and they also play an active part in the formation of the complex, owing to their greater ability for ameboid movement; their rate of movement is 0.6 to 3.5 μ per minute. Moving about, the archeocytes help the other cells to aggregate. About 2000 cells are necessary to produce a new individual (Galtsoff, 1925). The whole process, up to the opening of new oscula, takes about three weeks.

The phenomenon has been called **reconstitution.** Although reconstitution starts from individual cells, it has nothing in common with embryonic development, as the individual cells into which the sponge is broken up each retain their specific histologic character, and the whole process rests mainly on a rearrangement of the cells in space rather than on a progressive differentiation, although minor readjustments may possibly occur. The whole process can thus best be compared to regeneration by morphallaxis (Section 19–2).

Reconstruction of whole and complete animals from disaggregated cells has been also achieved with hydras. Hydras were disrupted into single cells, and the cells were allowed to form aggregates. These were shapeless clumps to begin with, but soon the ectoderm and the endoderm in the clumps sorted themselves out, forming double-walled bodies. Tentacles and hypostomes appeared on the surface (high gradient level structures having ability for self-differentiation!). Originally several "head" ends were formed, but later each "head" end shaped a body from the adjacent parts of the clump, and eventually each individual separated itself from the others as a hydra of normal structure (Gierer, 1974).

Reconstitution by reaggregation of isolated cells as observed in adult sponges and hydras bears an obvious relationship to the reaggregation of embryonic cells, which has been described in Sections 9–1 and 16–5, and shows that the mechanisms of differential cellular affinity are present and active throughout the life of a metazoon, even if we do not always observe them. It would probably not be too far-fetched to say that these mechanisms, after having participated in producing a multicellular organism, serve to maintain it and preserve its integrity.

Chapter 20

ASEXUAL REPRODUCTION

Asexual reproduction is the development of a new individual without the participation of any stages of the sexual cycle, that is, without maturation and copulation of sex cells with concomitant reduction of the number of chromosomes in meiosis. As applied to multicellular animals, asexual reproduction means development of new individuals at the expense of **somatic** cells — cells of the **soma,** or body, as opposed to generative or sex cells **(gametes).**

In this chapter we will not be concerned with the evolution and adaptive significance of asexual reproduction; neither will we deal with the factors regulating the alternation of sexual and asexual reproductive cycles. What will interest us is how a part of an already differentiated organism may again embark on an active process of morphogenesis. The part of the parental organism giving rise to a new individual in asexual reproduction may be called a **blastema,** and it always consists of a group of cells, whereas in sexual reproduction the new individual develops from one cell, the fertilized or parthenogenetically activated ovum. The development that starts from a blastema may be termed **blastogenesis,** as opposed to **embryogenesis,** or development from the ovum. The individuals resulting from asexual reproduction are often referred to as **blastozooids;** individuals developing from an egg are then termed **oozooids.**

The blastozooids may have the same general organization as oozoids, or they may even be indistinguishable from the latter. How this is achieved, in spite of the profoundly different initial stages, is the second major problem in relation to asexual reproduction.

20−1 OCCURRENCE AND FORMS OF ASEXUAL REPRODUCTION

Asexual reproduction may take a variety of forms, largely because of the amount of tissue set aside for the production of a new individual and, correlated with this, the degree of organization of this tissue. Taking the size of the fragment and the degree of its organization as a guiding principle, we may subdivide the infinite multiplicity of modifications occurring in different Metazoa into three main types:

1. Fission: The new individual is formed from a relatively large portion of the body of the maternal organism, and differentiated organs and tissues or their parts are passed on to the offspring.
2. Budding: The new individual develops from a small outgrowth on the surface of the parent. No organs of the parent are passed on as such to the offspring, but the bud is supported by the parental organism at least during the initial stages of its development.
3. Gemmule formation: The new individual develops from groups of cells which become completely cut off from the maternal individual and disseminated, so

that the development is, from the start, quite independent of the maternal body.

Fission.   The simplest form of fission is the separation of an adult individual into two parts of approximately equal size, similar to the binary fission in protozoans. In Metazoa it occurs in the most typical form in coelenterates and worms. In some corals (Anthozoa) fission occurs in the rather rare form of longitudinal fission, with the division plane in the long axis of the body. Fission also occurs in some brittle stars, such as in *Ophiactis savignyi,* in which parental individuals break across the disc, and each half then proceeds to regenerate the missing half of the disc and three arms. In the worms (Berrill, 1952), the planes of division are transverse to the body axis, so that the worm is divided into anterior and posterior halves. This type of reproduction is found in rhabdocoele turbellarians and in annelids, both polychaetes and oligochaetes. Both halves inherit from the parent the skin, a section of the alimentary canal, sections of the nerve cord, and in annelids, a number of mesodermal segments, with muscles and nephridia, and correspondingly, sections of parenchymatous mesoderm in rhabdocoeles. The posterior individual is at first devoid of a head and therefore initially lacks the supraesophageal ganglion, sense organs (eyes), tentacles, and mouth parts, where these occur. The head is produced either subsequent to division or, more often, in preparation for the division, so that the posterior individual is fully developed when the two blastozooids separate.

In rhabdocoeles and some oligochaetes, new fissions may be started by one or both filial individuals even before they are separated from each other (Fig. 460). The result is a chain of blastozooids, some of them in different stages of reconstitution (degree of head development). The division may also be very unequal, so that of the two or sever-

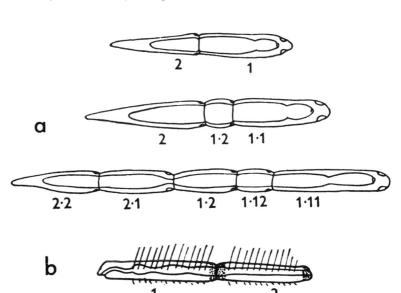

Figure 460. Reproduction by transverse fission in worms: *a,* in the rhabdocoele turbellarian *Stenostomum* (after Child); *b,* in the oligochaete *Nais* (after Stolte). The numbers show the succession in the formation of the individual blastozooids. (From Berrill, 1952.)

al zooids, one may be clearly distinguishable as the parent and the other, or others, as offspring.

Cases in which subsequent transverse divisions occur prior to the earlier divisions being completed lead to a special form of transverse fission, known as **strobilation,** in which numerous transverse divisions occur more or less simultaneously or in close succession, giving rise to a number of filial blastozooids. This is the classical method of propagation by which, in Scyphozoa, the asexual polypoid generation gives rise to the sexual medusoid generation. Similar multiple transverse division occurs in some polychaetes and also in ascidians. (See Berrill, 1951.) In the latter, the part undergoing transverse division is the abdominal section of the body, which is devoid of the branchial chamber but contains the intestinal loop.

Both in worms and in ascidians, the initial separation of the zooids is performed through the activity of the ectodermal epidermis. The epidermis forms a circular constriction which cuts inward, severs the internal organs, and eventually cuts the body of the animal in two (Fig. 461; Berrill, 1951, 1952).

Budding. The most typical examples of budding are found in coelenterates and tunicates. Superficially, the early bud appears as a small swelling or nodule on the lateral surface of the body of the parental animal. The nodule grows, takes shape, and develops a mouth and a whorl of perioral tentacles, in the case of the coelenterates, or in the tunicates, develops a branchial chamber and other associated structures including the atrial cavity and the oral and atrial siphons. The bud may become completely separated from the maternal body or may remain permanently in connection with the latter, thus leading to the formation of a colony. In the freshwater hydra, which normally exists in the form of single polyps, the daughter individuals may remain connected to the maternal body and occasionally even start forming secondary buds, but eventually this temporary colony splits into single individuals.

Not always are the buds directly formed on the body of the parental zooids. Often the buds develop on special outgrowths of the maternal animals; the outgrowths are then called **stolons.** These are typically present in many tunicates, but the branches of a hydrozoan colony on which the polyps develop have the same significance.

Gemmule Formation. This form of asexual reproduction is found in freshwater sponges and in bryozoans, and in both cases bodies are produced — called **gemmules** in sponges and **statoblasts** in bryozoans — that can survive after the maternal individual (or rather maternal colony) dies off during an unfavorable season (winter in temperate countries, periods of drought in warmer climates). The gemmules and statoblasts are formed in the interior of the parental body from a number of undifferentiated cells, which become enclosed in a shell with special spicules in sponges (Fig. 462) or in a chitinous enevelope in bryozoans. Upon destruction of the parental body, the gemmules or statoblasts are set free, and after conditions have become favorable, the cells contained inside burst out and develop a new individual. For the gemmules of sponges,

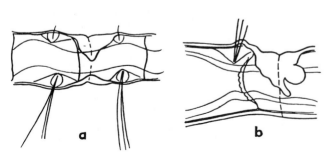

Figure 461. The initial stages of transverse division in annelids. *a, Pristina; b, Chaetogaster.* The epidermal thickenings are clearly indicated. (From Berrill, 1952.)

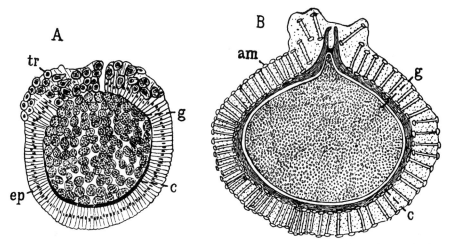

Figure 462. Gemmule of the sponge *Ephydatia*: *A*, in the process of formation; *B*, in the mature stage. am, Amphidiscs; c, cuticular membrane; ep, epithelium; g, archeocytes; tr, trophocytes. (After Evans, from Bounoure, 1940.)

such favorable conditions are created by the temperature rising above 16°C. (Brien, 1932).

20–2 SOURCES OF CELLULAR MATERIAL IN ASEXUAL REPRODUCTION

The major problem with which we are faced in studying asexual reproduction is, what resources does a differentiated organism have in order to start building new parts in a new burst of morphogenetic activity?

The least difficulty in interpreting the development of new parts is encountered in asexual reproduction by **fission.** It will be evident from what has been said previously that reproduction by fission is closely related to regeneration, or to put it another way, regeneration is an essential part of asexual reproduction by fission. In both cases, a comparatively large part of the body is isolated and reconstructed into a new complete whole by the regeneration of the missing parts. There is every reason to believe that the actual mechanism of development of the missing parts is the same in the two cases. We have seen that the organs which are produced anew during regeneration in annelids and in turbellarians owe their origin in large measure to undifferentiated "reserve" cells: neoblasts in annelids, parenchymal cells in turbellarians (pp. 575–576). The same cells are available for the development of missing parts in daughter zooids. Neoblasts are actually reported to concentrate in the fission zone of oligochaetes. As in regeneration, the epidermis gives rise to the nervous system and sense organs in the new head of the posterior zooid.

Where transverse fission occurs as a form of asexual reproduction in ascidians such as *Eudistoma* (Berrill, 1951), the daughter zooids consist of a section of the parent's body, including epidermis, mesenchyme, and a section of the intestinal tube, but do not include the branchial chamber and associated structures such as the atrial cavity, the nerve ganglia, and the siphons. The missing parts are replaced by regeneration.

In the process of regeneration, a most important role is played by the **epicardium**, a mesodermal tube lying ventral to the intestine (Fig. 463). The mesodermal lining of the epicardium has the properties of an undifferentiated reserve tissue, at the expense of which most of the internal organs in the anterior part of the new zooid develop.

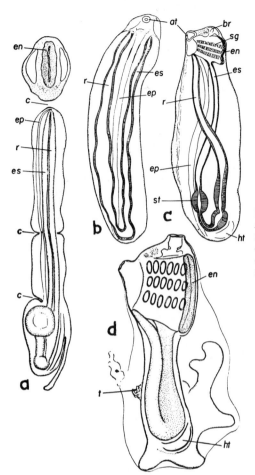

Figure 463. Reproduction by transverse fission in the tunicate *Eudistoma*. *a*, Individual in the process of fission. *b, c*, Blastozooid regenerating the anterior end. *d*, Newly metamorphosed oozooid. at, Atrial siphon; br, oral siphon; c, constrictions; en, endostyle; ep, epicardium; es, esophagus; ht, heart; r, rectum; sg, gill clefts; st, stomach; t, remnant of tail. (From Berrill, 1947.)

Remarkably enough, the branchial chamber does not regenerate at the expense of the remaining part of the endodermal alimentary canal but is also produced by the epicardium and secondarily becomes fused onto the endodermal esophagus.

In **budding,** very little of the organization of the parent is passed on to the daughter zooid. In a scyphozoan polyp, the first sign of impending budding is the formation of a pocket-like evagination of the endodermal epithelium which pushes through the mesoglea and reaches the ectodermal layer. The ectodermal layer then becomes activated, and both ectoderm and endoderm form a conical protrusion on the surface of the body. This protrusion gradually increases and elongates and becomes a cylindrical body with an internal cavity which is in open communication with the gastric cavity of the parent (Gilchrist, 1937). At a later stage, a mouth is formed at the tip of the bud, and a whorl of tentacles develops around it; the bud has become a new polyp.

The budding in hydrozoan polyps is essentially similar. From this one would tend to draw the conclusion that if not the pattern of organization, then at least the two main layers of the body, the ectoderm and the endoderm of the daughter zooid, are directly taken over from the parental organism. Some students of coelenterate development and reproduction believe, however, that the process is not nearly so simple. (See Hadzi, 1909; Schulze, 1918; Weiler-Stolt, 1960). We have seen that regeneration in the hydra is due in part to the activity of so-called interstitial cells, small undifferentiated cells located between the larger functioning cells both of the ectoderm and endoderm. (Compare p. 573.) These cells serve for the replacement of differentiated cells, especially of the cnidoblast cells, and in cases of injury the interstitial cells assemble in the regeneration blastema. It has been claimed that the same interstitial cells are the main source for the formation of the bud in asexual reproduction. The interstitial cells either accumulate locally and form the first thickening which is to become the bud, or possibly they infiltrate the bud in the process of its formation and continue flowing into the bud and increasing its size even after it has been formed. Eventually the whole or almost the whole of the daughter zooid is built from interstitial cells, the differentiated ectodermal and endodermal cells taking little, if any, part in its formation (Fig. 464; Weiler-Stolt, 1960).

In Hydromedusae (*Lizzia claparedi, Rathkea octopunctata* — see Bounoure, 1940), it was observed that endoderm does not contribute to the formation of the bud at all. The bud is at first represented by a nodule of cells in the ectodermal layer. Some of the cells lying in the interior of the nodule become arranged in the form of a vesicle, and it is from this vesicle that the endoderm of the daughter medusa develops (Fig. 465). This observation can be better understood if it is accepted that in the hydrozoans a new zooid is derived neither from the endoderm nor from the ectoderm of the parent but from the undifferentiated interstitial cells, an interpretation which has been supported by direct observation (Weiler-Stolt, 1960).

The budding in tunicates presents an almost infinite variety of forms, but in the great majority of these animals, in particular in the ascidians, there is a distinct common pattern which presents some remarkable features. The bud is essentially an outgrowth of the ectodermal epidermis, supported by another tissue which forms a hollow vesicle inside the outgrowth. In some cases, parts of the gonads are included as a third element. The epidermis of tunicates is a very specialized tissue which produces the mantle, the cellulose layer covering the surface of the skin in these animals. In budding, the epidermis produces only more epidermis and provides the external covering for the

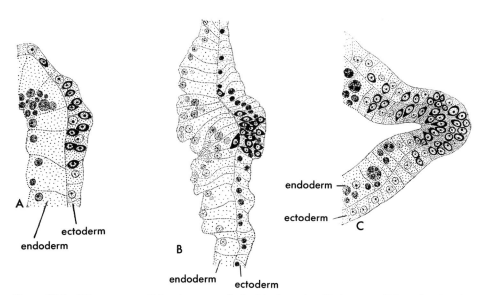

Figure 464.   Three stages of development of a bud in the polyp *Cladonema radiatum*, showing the aggregation of interstitial cells (black) in the bud. The endoderm is initially free of interstitial cells (*A*) but becomes invaded by these cells later (in *B* and *C*). (From Weiler-Stolt, 1960.)

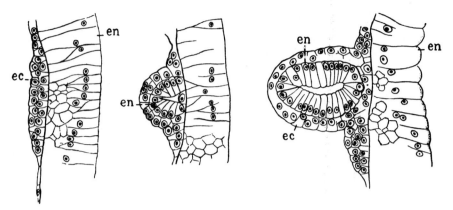

Figure 465. Development of a bud in the medusa *Rathkea*. ec, Ectoderm; en, endoderm. (After Chun, from Bounoure, 1940.)

blastozooid. The internal organs are developed from the previously mentioned vesicle formed inside the bud. Now, it is most remarkable that the inner vesicle may be derived from a number of different tissues, namely: (1) an outgrowth of the atrial cavity *(Botryllus);* (2) an outgrowth from the pharyngeal wall at the posterior end of the endostyle *(Salpa, Pyrosoma);* (3) an outgrowth of the epicardium, which has already been mentioned *(Distaplia);* or (4) a group of blood cells *(Botryllus;* Oka and Watanabe, 1957). All these types of development are found in cases in which the bud is developed on the body of the parent or on a short stolon connected to the body.

In some ascidians *(Clavellina, Perophora)*, the buds develop on long rhizoid-like stolons. These consist only of epidermis and a longitudinal mesodermal septum, splitting the cavity of the stolon into two canals which allow for the backward and forward circulation of blood. Buds developing on such a stolon consist of the epidermal covering epithelium and the inner vesicle, which in this case is produced from the cells of the mesodermal septum (Fig. 466; Berrill, 1951).

Regardless of the origin of the internal vesicle, it produces a variety of organs and tissues: the alimentary canal with the branchial chamber, the atrial cavities, the heart,

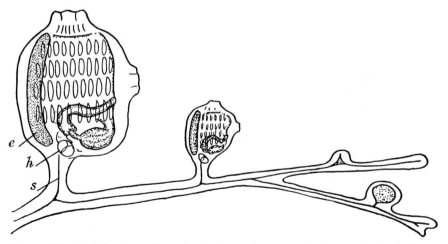

Figure 466. Budding from the stolon in the tunicate *Perophora*. e, Endostyle; h, heart; s, stolonial septum. (From Berrill, 1935.)

the nervous system (except in *Salpa*), and sometimes also the gonads. It is thus evident that several parts in the fully differentiated body of an ascidian retain very broad potentialities for development. These potentialities embrace practically the whole organization of the ascidian except the epidermis, and even of this limitation we cannot be sure, as the epidermis is always provided; so there would appear to be no incentive for the inner vesicle to produce the epidermis as well. It may well be that the potentiality for forming epidermis is not lacking, and then the cells of the inner vesicle would be truly **totipotent**—capable of producing every differentiation found in the animal species in question.

The existence of really totipotent cells is indisputable in the case of **gemmule formation** in the freshwater sponges. The first rudiment of a gemmule in the interior of the sponge appears in the form of an accumulation of archeocytes—undifferentiated ameboid cells which are dispersed in between the differentiated elements of the body of a sponge. Around the central core, consisting of archeocytes, other migrating cells are lodged and become arranged in the form of a columnar epithelium. The role of the epithelium is to produce the shell on the surface of the gemmule. This is strengthened by the action of special skeleton-building cells, **scleroblasts,** which deposit in the shell peculiar spicules, the amphidiscs (Fig. 462). Both the columnar epithelial cells and the scleroblasts are later withdrawn, or they degenerate. A small opening, the micropyle, remains temporarily open in the shell. It has been claimed that feeder cells, **trophocytes,** enter the interior of the developing gemmule and pass on food reserves to the archeocytes and in their turn disappear. The archeocytes, on the other hand, accumulate in their cytoplasm food reserves consisting of glycoproteins in the form of discoidal platelets (Fig. 469). When the development of the gemmule is completed, the archeocytes are the only living cells left in it, and all the tissues and cell types in the new sponge which develop eventually from the gemmule, including its sex cells, are derived from the archeocytes.

Summing up our review of the sources of materials for asexual reproduction, we see that the renewal of morphogenetic processes depends on the persistence, in the body of the adult animals, of cells which do not become as highly differentiated as the rest and are capable of developing into a variety of cell types. The role of these undifferentiated or "reserve" cells is the greater, the smaller the part of the parental body that is used in asexual propagation. In extreme cases, the undifferentiated cells are totipotent and are in this respect equivalent to the egg cell, a similarity which may be further enhanced by the presence of deutoplasmic (food reserve) inclusions, as in the case of the gemmules of sponges.

Although asexual reproduction is a special form of morphogenesis which has been acquired, probably independently, only in some groups of the animal kingdom, it is evident from the preceding that it falls back on a very general property of all developing organisms: the potential equivalence of the daughter cells resulting from the cleavage of the fertilized egg. We have seen (Section 6–5) that the cleavage cells retain the full complement of hereditary factors (genome) independently of the prospective significance of the individual cells and that the nuclei are unrestricted in their potencies even after determination of the various areas of the embryo has set in (p. 122). The cytoplasm of cleavage cells may show differences in early stages, and later the elaboration of special cytoplasmic mechanisms (histological differentiation) may go so far that the specific types of cells lose the ability to change the direction of their development. The loss of plasticity, however, is not inevitable, nor does it proceed at an equal tempo in all parts and in all cells of a differentiated organism. Cells which do not achieve a high degree of differentiation retain their plasticity and, being in possession of the complete genome, can be made use of in asexual reproduction. (See Bounoure, 1940.)

20-3　COMPARISON OF BLASTOGENESIS AND EMBRYOGENESIS

In the general survey of ontogenetic development (Section 1–2), we have found it useful to consider what tasks have to be performed by the embryo before the final condition (the development of the new adult individual) is achieved. If from this same viewpoint we compare embryogenesis (development of the egg) with blastogenesis (development from a blastema in asexual reproduction), we see at once that the task is very much simpler in the latter case.

The process of producing a new individual is simplest in reproduction by fission, when the blastozooid is derived from half the parental organism and in this way is provided with a large proportion of the organs and parts which are necessary for making the new individual self-sufficient. What has to be done is the regeneration of missing parts. The whole mechanism of regeneration, as considered in Chapter 19, is brought into play, including the factors determining the regenerating parts (pp. 581–590). The remnant of the old individual determines the nature, position, and orientation of the newly differentiated organs. The polarity and bilaterality of the parent organism prevail in the blastozooids.

The task of development is more complicated in the case of budding, since all organs and differentiated parts of the blastozooid have to be produced anew. Nevertheless, the initial system, the bud, always has a higher degree of complexity than a fertilized egg or even than a blastula as it occurs in embryogenesis. A typical bud, as we have already explained, always consists of two layers of epithelial cells. The young zooid is thus already in possession of the concentric stratification of body layers, a condition which in embryogenesis is achieved only after gastrulation. It is very noteworthy, however, that the layers formed in the blastozooid do not necessarily correspond to the germinal layers developing in embryogenesis.

In the case of budding in coelenterates, there is a closer correspondence between the outer and inner layers of the bud and the ectoderm and endoderm of the gastrula. The fate of the two layers is the same, but we have seen that the inner layer may be derived not from the endodermal epithelium of the parent but from a thickening in the ectodermal epithelium.

When we turn to the buds of tunicates, we find that the inner vesicle corresponds to the endoderm neither in its origin nor its fate. It has been shown (p. 598) that, although in some tunicates the inner vesicle may be derived from the endoderm of the parent (in the form of an outgrowth from the pharyngeal epithelium), it can also be derived from mesoderm (epicardium, mesodermal septum of the stolon, blood cells) or even ectoderm (lining of the atrial cavity). What is even more important is that the inner vesicle gives rise to parts derived in embryogenesis from any of the three germinal layers. In spite of the diversity of origin of the inner vesicle, its later differentiation shows a considerable degree of uniformity in different tunicates. After the bud has grown to a certain degree—a minimal size, varying of course in different species, is essential (Berrill, 1941)—it becomes constricted at least partially from the parent zooid. Then folds start subdividing the inner vesicle into sections. (The following description refers to the development of the blastozooid in *Botryllus*.)

First, two folds cut in from what will be the distal part of the new zooid, subdividing the inner vesicle into a median part, which will become the branchial chamber, and two lateral parts, which give rise to the atrial cavity (Fig. 467). (In embryogenesis, the atrial cavity develops as a pair of invaginations of the ectoderm which partially fuse and later have a common opening to the exterior.) In addition to these three main subdivisions, further smaller pocket-like evaginations of the vesicle appear. One evagination, near the anterior end of the central cavity, gives rise to the nerve center. (In embryo-

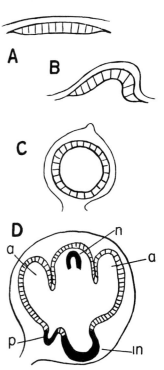

Figure 467. Development of the bud in *Botryllus*. a, Rudiment of atrial cavity; in, rudiment of intestine; n, rudiment of ganglion; p, rudiment of pericardium. (From Berrill, 1951.)

genesis, the nerve ganglion is a remnant of the neural tube formed by infolding of the ectodermal neural plate, as in *Amphioxus* and in vertebrates.) Two pockets at the posterior end give rise to the intestinal loop (including the esophagus and the stomach) and to the pericardium, respectively. (In embryogenesis the intestine is, of course, of endodermal origin, and the pericardium develops from mesodermal mesenchyme.) Because the initial state of the system in blastogenesis is not the same as that in embryogenesis, the course of development is different. The morphogenetic processes in budding appear to be simpler and more straightforward than in embryogenesis. The actively developing part, which is the inner vesicle (the epidermis of the bud is a differentiated tissue all the time, specialized in secreting the cellulose mantle), proceeds directly to the formation of organ rudiments, omitting the stage of germ layer development.

In one further respect the morphogenetic processes in budding are simpler than in embryogenesis: The new individual inherits its polarity directly from the parent zooid. The point of attachment of the bud to the maternal body or the stolon always becomes the proximal end of the blastozooid.

In the development of gemmules, the task of producing a new individual becomes most complicated and approaches that of the development of the egg. The special difficulties encountered are:

1. The germination of the gemmule occurs after the death and decomposition of the parent animal; consequently, the polarity of the new individual has to be worked out by itself, and the parent body is no longer there to influence the polarity of the offspring.
2. The complete homogeneity of the contents of the gemmule (in the case of the gemmules of sponges) deprives the new individual of any remnant of morphological organization. The structure of the new sponge has to be established by the interaction of practically independent cells.

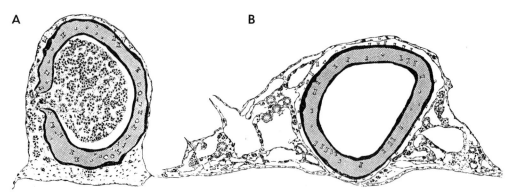

Figure 468. Germination of gemmules. *a*, Contents of the gemmule of *Spongilla*, leaving the shell. *b*, Transformation of the contents of the gemmule of *Ephydatia* into a new sponge. (From Brien, 1932.)

As has already been stated (p. 599), the interior of the gemmule consists of only one type of cell, the archeocytes, which are rather large cells containing platelets of glycoprotein. Even before the germination of the gemmule, some of the archeocytes become activated; they start dividing and in so doing give rise to smaller and smaller cells, very much like the blastomeres which diminish in size as cleavage progresses. The glycoprotein platelets gradually disappear; the nuclei become richer in chromatin, a usual characteristic in actively growing and metabolizing cells. These small and active cells have been referred to as **histioblasts**—cells producing tissues (Brien, 1932). When the gemmule germinates, the contents of the gemmule crawl out through the micropyle and form an irregular mass, surrounding the empty shell of the gemmule (Fig. 468*a*). Both the histioblasts and the remaining glycoprotein-containing archeocytes leave the shell. Outside the shell, the division of archeocytes and their conversion into histioblasts continue (Fig. 469). The histioblasts now become arranged into an irregular meshwork (Fig. 468 *b*), cavities appear, and some of the histioblasts surrounding these cavities differentiate into choanocytes. Other histioblasts become epidermal cells, scleroblasts, pore cells, or mesenchymal cells arranged in a typical way, so that the mass of cells soon becomes a small sponge, with a system of internal canals with ciliated chambers, ostia, etc.

It may be significant that the sponges are capable of reconstituting their structure after complete dissaggregation, as described in Section 19–8. The development of the gemmule proceeds along very much the same lines; the development is direct in the extreme, the individual cells differentiating and taking up their positions in the whole.

A review of different forms of asexual reproduction reveals a principle of general significance. It shows that the complexity of morphogenetic processes is largely deter-

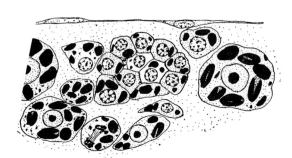

Figure 469. Transformation of archeocytes into histioblasts during development of a new sponge from a gemmule. Archeocytes still with deutoplasmic inclusions; one is in mitosis. (From Brien, 1932.)

mined by the degree of difference that exists between the initial stage of development and the final condition. The initial stage in embryogenesis is a single cell; therefore a period of cleavage is necessary, which brings the system into a multicellular condition. In asexual reproduction, the initial system is already multicellular, and cleavage falls away. (Something resembling cleavage occurs, as has been indicated, in the gemmules of sponges, where the archeocytes accumulate food reserves, similarly to oocytes in the ovary.) In asexual reproduction the initial system may have the cells arranged in more than one layer, and this makes gastrulation dispensable. What remains of the main periods of development outlined in Section 1–2 are organogenesis, differentiation, and growth.

A second very suggestive fact which emerges from a study of asexual reproduction is that, given a normal environment, the organization of the animal's body is entirely determined, in the last instance, by the hereditary constitution of the species-specific cells. The structure of the egg cell with its polarity and heterogeneous arrangement of cytoplasmic substances is a mechanism which provides for the orderly course of differentiation of the cells in a developing embryo. This mechanism is, however, dispensable; the same end product can be attained starting from a different initial constellation, provided that the cells have the same hereditary constitution. In principle, the orderly organization of an animal's body should be attainable by the interaction of an assortment of different types of cells produced on the basis of species-specific hereditary potentialities (compare the reaggregation of cell suspensions, Section 16–5). In practice, this is possible only in relatively very simple biological systems.

REFERENCES*

ABELOOS, M., 1956, Les métamorphoses. Armand Collin, Paris.

ABELSON, J., 1963, Transfer of genetic information. Science, N.Y. *139*, 774–776.

ABERCROMBIE, M., and HEAYSMAN, J. E., 1953. Observations on the social behaviour of cells in tissue culture. I. Speed of movement of chick heart fibroblasts in relation to their mutual contacts. Exp. Cell Res. *5*, 111–131.

ABRAMS, R., 1951. Synthesis of nucleic acid purines in the sea urchin embryo. Exp. Cell Res. *2*, 235–242.

ADELMANN, H. B., 1932. The development of the prechordal plate and mesoderm of *Amblystoma punctatum*. J. Morph. *54*, 1–67.

———, 1936. The problem of cyclopia. I. II. Q. Rev. Biol. *11*, 161–182, 284–304.

———, 1937. Experimental studies on the development of the eye. IV. The effect of the partial and complete excision of the prechordal substrate on the development of the eyes of *Amblystoma punctatum*. J. exp. Zool. *75*, 199–237.

ADLER, L., 1914. Metamorphosestudien an Batrachierlarven. Roux Arch. *39*, 21–45.

AFZELIUS, B. A., 1956. The ultrastructure of the cortical granules and their products in the sea urchin egg as studied with the electron microscope. Exp. Cell Res. *10*, 257–285.

———, 1959. Electron microscopy of the sperm tail. Results obtained with a new fixative. J. biophys. biochem. Cytol. *5*, 269–278.

ALBAUM, H. G., and NESTLER, H. A., 1937. Xenoplastic ear induction between *Rana pipiens* and *Amblystoma punctatum*. J. exp. Zool. *75*, 1–9.

ALDERMAN, A. L., 1935. The determination of the eye in the anuran *Hyla regilla*. J. exp. Zool. *70*, 205–232.

———, 1938. A factor influencing the bilaterality of the eye rudiment in *Hyla regilla*. Anat. Rec. *72*, 297–302.

ALFERT, M., and SWIFT, H., 1953. Nuclear DNA constancy: A critical evaluation of some exceptions reported by Lison and Pasteels. Exp. Cell Res. *5*, 455–460.

ALLEN, B. M., 1918. The results of thyroid removal in larvae of *Rana pipiens*. J. exp. Zool. *24*, 499–519.

———, 1929. The influence of the thyroid and hypophysis upon growth and development of amphibian larvae. Q. Rev. Biol. *4*, 325–352.

———, 1938. The endocrine control of amphibian metamorphosis. Biol. Rev. *13*, 1–19.

ALLFREY, V. G., and MIRSKY, A. E., 1961. How cells make molecules. Scient. Am. *205*(3), 74–82.

———, LITTAU, V. C., and MIRSKY, A. E., 1963. On the role of histones in regulating ribonucleic acid synthesis in the cell nucleus. Proc. natn. Acad. Sci., U.S.A. *49*, 414–421.

ANCEL, P., and VINTEMBERGER, P., 1948. Recherches sur le déterminisme de la symétrie bilaterale dans l'oeuf des amphibiens. Bull. biol. Suppl. *31*, 1–182.

ANDERSON, E., 1968. Oocyte differentiation in the sea urchin, *Arbacia punctulata*, with particular reference to the origin of cortical granules and their participation in the cortical reaction. J. Cell Biol. *37*, 514–539.

———, and BEAMS, H. W., 1960. Cytological observations on the fine structure of the guinea pig ovary with special reference to the oogonium, primary oocyte and associated follicle cells. J. Ultrastruct. Res. *3*, 432–446.

---

*For brevity the title of the journal "Wilhelm Roux' Archiv für Entwicklungsmechanik der Organismen," previously "Archiv für Entwicklungsmechanik der Organismen," is cited as "Roux Arch." Other abbreviations are according to P. Brown and G. B. Stratton, 1963–1965. World List of Scientific Periodicals, 4th edition, Butterworths, London.

ANDRÉ, J., and ROUILLER, C., 1957. The ultrastructure of the vitelline body in the oocyte of the spider *Tegenaria parietina*. J. biophys. biochem. Cytol. *3*, 977–984.

ANDRES, G., 1953. Experiments on the fate of dissociated embryonic cells (chick) disseminated by the vascular route. Part II. Teratomas. J. exp. Zool. *122*, 507–540.

ANDREW, A., 1963. A study of the developmental relationship between enterochromaffin cells and the neural crest. J. Embryol. exp. Morph. *11*, 307–324.

ANTON, H. J., 1961. Zur Frage der Aktivierung der Gewebe im Extremitätenstumpf bei Urodelen vor der Blastembildung. Roux Arch. *153*, 363–369.

AREY, L. B., 1947. Developmental anatomy. 5th ed. Saunders, Philadelphia.

———, 1954. Developmental anatomy. 6th ed. Saunders, Philadelphia.

ARMSTRONG, P. B., and PARENTI, D., 1973. Scanning electron microscopy of the chick embryo. Devl. Biol. *33*, 457–462.

ARONSON, A. I., and WILT, F. H., 1969. Properties of nuclear RNA in sea urchin embryos. Proc. natn. Acad. Sci., U.S.A. *62*, 186–193.

AUSTIN, C. R., 1961. The mammalian egg. Blackwell Scientific Publications, Oxford.

———, 1965. Fertilization. Prentice-Hall, Inc., Englewood Cliffs, N.J.

———, 1968. Ultrastructure of fertilization. Holt, Rinehart and Winston International, New York.

——— (Ed.), 1973. The mammalian fetus *in vitro*. Chapman & Hall, London.

BACA, M., and ZAMBONI, L., 1967. The fine structure of human follicular oocytes. J. Ultrastruct. Res. *19*, 354–381.

BACHVAROVA, R., DAVIDSON, E. H., ALLFREY, V. G., and MIRSKY, A. E., 1966. Activation of RNA synthesis associated with gastrulation. Proc. natn. Acad. Sci., U.S.A. *55*, 358–365.

BACON, R. L., 1945. Self-differentiation and induction in the heart of *Amblystoma*. J. exp. Zool. *98*, 87–121.

BAER, K. E. VON, 1828. Ueber Entwicklungsgeschichte der Tiere, Beobachtung und Reflexion. Königsberg.

BAKER, P. C., and SCHROEDER, T. E., 1967. Cytoplasmic filaments and morphogenetic movement in the amphibian neural tube. Devl. Biol. *15*, 432–450.

BALFOUR, F. M., 1880. A treatise on comparative embryology. Macmillan Co., London.

BALINSKY, B. I., 1925. Transplantation des Ohrbläschens bei *Triton*. Roux Arch. *105*, 718–731.

———, 1931. Zur Dynamik der Extremitätenknospenbildung. Roux Arch. *123*, 565–648.

———, 1933. Das Extremitätenseitenfeld, seine Ausdehnung und Beschaffenheit. Roux Arch. *130*, 704–747.

———, 1939. Experiments on total extirpation of the whole entoderm in *Triton* embryos. C.R. Acad. Sci. URSS *23*, 196–198.

———, 1947. Kinematik des entodermalen Materials bei der Gestaltung der wichtigsten Teile des Darmkanals bei den Amphibien. Roux Arch. *143*, 126–166.

———, 1948. Korrelationen in der Entwicklung der Mund- und Kiemenregion und des Darmkanals bei Amphibien. Roux Arch. *143*, 365–395.

———, 1950. On the developmental processes in mammary glands and other epidermal structures. Trans. R. Soc. Edinb. *62*, 1–31.

———, 1951. On the eye cup–lens correlation in some South African amphibians. Experientia *7*, 180.

———, 1956. Discussion of Dr. Zwilling's paper. Cold Spring Harb. Symp. quant. Biol., *21*, 354.

———, 1958. On the factors controlling the size of the brain and eyes in anuran embryos. J. exp. Zool. *139*, 403–442.

———, 1961a. The role of cortical granules in the formation of the fertilization membrane and the surface membrane of fertilized sea urchin eggs. Symp. on Germ Cells and Development, Inst. Internat. d'Embryologie and Fondazione A. Baselli, Pavia, 205–219.

———, 1961b. Ultrastructural mechanisms of gastrulation and neurulation. Symp. on Germ Cells and Development, Inst. Internat. d'Embryologie and Fondazione A. Baselli, Pavia, 550–563.

———, 1974. Supernumerary limb induction in the Anura. J. exp. Zool. *188*, 195–202.

———, and DEVIS R. J., 1963. Origin and differentiation of cytoplasmic structures in the oocytes of *Xenopus laevis*. Acta Embryol. Morph. exp. *6*, 55–108.

———, and WALTHER, H., 1961. The immigration of presumptive mesoblast from the primitive streak in the chick as studied with the electron microscope. Acta Embryol. Morph. exper. *4*, 261–283.

BALTZER, F., 1940. Ueber erbliche letale Entwicklung und Austauschbarkeit artverschiedener Kerne bei Bastarden. Naturwissenschaften *28*, 177–187, 196–206.

BALTZER, F., 1941. Ueber die Pigmentierung merogonisch-haploider Bastarde zwischen der schwarzen und weissen Axolotlrasse. Verh. schweiz. naturf. Ges. 121 Jahresvers, Basel, 169–179.

BANDIER, J., 1937. Histologische Untersuchungen über die Regeneration von Landplanarien. Roux Arch. *135*, 316–348.

BANKI, O., 1929. Die Entstehung der äusseren Zeichen der bilateralen Symmetrie am Axolotlei nach Versuchen mit örtlicher Vitalfärbung. Verh. X. int. Zool. Kongr. Budapest.

BARTH, L. G., 1938. Quantitative studies of the factors governing the rate of regeneration. Biol. Bull. mar. biol. Lab., Woods Hole *74*, 155–177.

———, and BARTH, L. J., 1954. The energetics of development. Columbia University Press, New York.

BATAILLON, E., 1910. L'embryogenèse complète provoquée chez les Amphibiens par picûre de l'oeuf vierge, larves parthénogénétiques de *Rana fusca*. C. R. Acad. Sci. Paris *150*, 996.

BATTEN, E. H., 1958. The origin of the acoustic ganglion in the sheep. J. Embryol. exp. Morph. *6*, 597–615.

BAUTZMANN, H., 1926. Experimentelle Untersuchungen zur Abgrenzung des Organisationszentrums bei *Triton taeniatus*, mit einem Anhang: Ueber Induktion durch Blastulamaterial. Roux Arch. *108*, 283–321.

———, HOLTFRETER, J., SPEMANN, H., and MANGOLD, O., 1932. Versuche zur Analyse der Induktionsmittel in der Embryonalentwicklung. Naturwissenschaften 971–974.

BEATTY, R. A., 1957. Parthenogenesis and polyploidy in mammalian development. University Press, Cambridge.

BECKER, H. J., 1962. Die Puffs der Speicheldrüsenchromosomen von *Drosophila melanogaster*. II Mitteilung. Die Auslösung der Puffbildung, ihre Spezifität und ihre Beziehung zur Funktion der Ringdrüse. Chromosoma, *13*, 341–384.

BEDFORD, J. M., 1967. The influence of the uterine environment on spermatozoa and eggs. In Lamming, G. E., and Ambroso, E. C. (Eds.): Reproduction in the female mammal. Butterworths, London, 478–499.

BEER, G. R. DE, 1947. The differentiation of neural crest cells into visceral cartilages and odontoblasts in *Amblystoma*, and a re-examination of the germ layer theory. Proc. R. Soc. B. *134*, 377–398.

BEERMANN, W., and CLEVER, U., 1964. Chromosome puffs. Scient. Am. *210*(4), 50–58.

BELLAIRS, R., 1953. Studies on the development of the foregut in the chick blastoderm. 2. The morphogenetic movements. J. Embryol. exp. Morph. *1*, 369–385.

———, 1961. The structure of the yolk of the hen's egg as studied by electron microscopy. I. The yolk of the unincubated egg. J. biophys. biochem. Cytol. *11*, 207–225.

———, 1963. Personal communication.

———, 1964. Biological aspects of the yolk in the hen's egg. In Abercrombie, M., and Brachet, J. (Eds.): Advances in morphogenesis, *4*. Academic Press, New York, 217–272.

———, 1967. Aspects of the development of yolk spheres in the hen's oöcyte, studied by electron microscopy. J. Embryol. exp. Morph. *17*, 267–281.

———, 1971. Developmental processes in higher vertebrates. Logos Press, Plainfield, N.J.

———, HARKNESS, M., and HARKNESS, R. D., 1963. The vitelline membrane of the hen's egg: A chemical and electron microscopic study. J. Ultrastruct. Res. *8*, 339–359.

BERG, W. E., and HUMPHREYS, W. J., 1960. Electron microscopy of four-cell stages of the ascidians *Ciona* and *Styela*. Devl. Biol. *2*, 42–60.

BERN, H. A., NICOL, C. S., and STROHMAN, R. C., 1967. Prolactin and tadpole growth. Proc. Soc. exp. Biol. Med. 126, 518–520.

BERRILL, N. J., 1935. Studies on tunicate development. III. Differential retardation and acceleration. Phil. Trans. R. Soc. B *225*, 255–379.

———, 1941. Size and morphogenesis in the bud of *Botryllus*. Biol. Bull. mar. biol. Lab., Woods Hole *80*, 185–193.

———, 1947. The structure, development and budding of the ascidian *Eudistoma*. J. Morph. *81*, 269–281.

———, 1951. Regeneration and budding in tunicates. Biol. Rev. *26*, 456–475.

———, 1952. Regeneration and budding in worms. Biol. Rev. *27*, 407–438.

BERTALANFFY, L. VON, 1948. Das organische Wachstum und seine Gesetzmässigkeiten. Experientia *4*, 255.

———, 1949. Problems of organic growth. Nature, Lond. *163*, 156–158.

BIER, K., 1963. Autoradiographische Untersuchungen über die Leistungen des Follikelepithels und der Nährzellen bei der Dotterbildung und Eiweisssynthese im Fliegenovar. Roux Arch. *154*, 552–575.

BIJTEL, J. H., 1931. Ueber die Entwicklung des Schwanzes bei Amphibien. Roux Arch. *125*, 448–486.

———, 1936. Die Mesodermbildungspotenzen der hinteren Medullarplattenbezirke bei *Amblystoma mexicanum* in bezug auf die Schwanzbildung. Roux Arch. *134*, 262–282.

BIRBECK, M. S. C., and MERCER, E. H., 1957. Electron microscopic, x-ray and birefringence studies on the proteins of the hair follicle. In SJÖSTRAND, F. S., and RHODIN, J.: Electron microscopy. Almqvist and Wiksell. Stockholm, 158–160.

BLACKLER, A. W., 1958. Contribution to the study of germ-cells in the anura. J. Embryol. exp. Morph. *6*, 491–503.

——, 1962. Transfer of primordial germ-cells between two subspecies of *Xenopus laevis*. J. Embryol. exp. Morph. *10*, 641–651.

——, and FISCHBERG, M., 1961. Transfer of primordial germ-cells in *Xenopus laevis*. J. Embryol. exp. Morph. *9*, 634–641.

BLANCHETTE, E. J., 1961. A study of the fine structure of rabbit primary oocyte. J. Ultrastruct. Res. *5*, 349–363.

BLOCH, D. P., 1962. On the derivation of histone specificity. Proc. natn. Acad. Sci., U.S.A. *48*, 324–326.

BOELL, E. J., 1945. Functional differentiation in embryonic development. II. Respiration and cytochrome oxidase activity in *Amblystoma punctatum*. J. exp. Zool. *100*, 331–352.

——, 1948. Biochemical differentiation during amphibian development. Ann. N.Y. Acad. Sci. *49*, 773–800.

——, 1955. Energy exchange and enzyme development during embryogenesis. In WILLIER, B. H., WEISS, P. A., and HAMBURGER, V. (Eds.): Analysis of development. Saunders, Philadelphia, 520–555.

BONNER, J., 1971. In interview, quoted in Editorial. New Scientist and Science J., June, 700–702.

BONNET, C., 1745. Traité d'Insectologie, Paris. Quoted after J. Needham, 1959.

BORGHESE, E., 1950. The development *in vitro* of the submandibular and sublingual glands of *Mus musculus*. J. Anat. *84*, 287–302.

BOUNHIOL, J., 1937. La métamorphose des insectes serait inhibée dans leur jeune age par les corpora allata? C.R. Soc. Biol. Paris *126*, 1189–1191.

BOUNOURE, L., 1934. Recherches sur la lignée germinale chez la grenouille rousse aux premier stades au développement. Ann. Sci. natur. zool. 10 ser. *17*, 67–248.

——, 1939. L'origine des cellules reproductrices et le problème de la lignée germinale. Gauthier-Villars, Paris.

——, 1940. Continuité germinale et reproduction agame. Gauthier-Villars, Paris.

BOYCOTT, A. E., DIVER, C., GARSTANG, S. L., and TURNER, F. M., 1930. The inheritance of sinistrality in *Limnaea peregra* (Mollusca, Pulmonata). Phil. Trans. R. Soc. B *219*, 51–131.

BRACHET, A., 1935. Traité d'embryologie des vertébrés. 2nd ed., revised by A. DALCQ and P. GERARD, Masson, Paris.

BRACHET, J., 1941. La localisation des acides pentosenucléiques dans les tissus animaux et les oeufs d'Amphibiens en voie de développement. Archs. Biol. Paris *53*, 207–257.

——, 1947. The metabolism of nucleic acids during embryonic development. Cold Spring Harb. Symp. quant. Biol. *12*, 18–27.

——, 1950a. Les caractéristiques biochémique de la compétence et de l'induction. Rev. Suisse Zool. *57*, Fascicule suppl. 1, 57–75.

——, 1950b. Chemical embryology. Interscience Publishers, New York.

——, 1969. Acides nucleiques et differenciation embryonnaire. Annales d'Embryologie et de Morphogenese, Suppl. 1, 21–37.

——, DECROLY, M., FICQ, A., and QUERTIER, J., 1963. Ribonucleic acid metabolism in unfertilized and fertilized sea urchin eggs. Biochim. biophys. Acta *72*, 660–662.

BRAMBELL, W. F. R., HEMMINGS, W. A., and HENDERSON, M., 1951. Antibodies and embryos. Athlone Press, London.

BRAUER, A., 1902. Beiträge zur Kenntniss der Entwicklung und Anatomie der Gymnophionen. III. Die Entwicklung der Excretionsorgane. Zool. Jb. Abt. Anat. u. Ont. *16*, 1–176.

BRENNER, S., JACOB, F., and MESELSON, M., 1961. An unstable intermediate carrying information from genes to ribosomes for protein synthesis. Nature, Lond. *190*, 576–581.

BRETSCHNEIDER, L. H., and RAVEN, C. P., 1951. Structural and topochemical changes in the egg cells of *Limnaea stagnalis* L. during oogenesis. Arch. néerl. Zool. *10*, 1–31.

BRIEN, P., 1932. Contribution a l'étude de la régénération naturelle chez les Spongilldae. Archs. Zool. exp. gén. *74*, 461–506.

BRIGGS, R., and CASSENS, J., 1966. Accumulation in the oocyte nucleus of a gene product essential for embryonic development beyond gastrulation. Proc. natn. Acad. Sci., U.S.A. *55*, 1103–1109.

——, and JUSTUS, J. T., 1967. Partial characterization of the component from normal eggs which corrects the maternal effect of gene o in the mexican axolotl (*Amblystoma mexicanum*). J. exp. Zool. *167*, 105–116.

——, and KING, T. J., 1952. Transplantation of living nuclei from blastula cells into enucleated frogs' eggs. Proc. natn. Acad. Sci., U.S.A. *38*, 455–463.

——, and ——, 1953. Factors affecting the transplantability of nuclei of frog embryonic cells. J. exp. Zool. *122*, 485–506.

BRIGGS, R., and KING, T. J., 1957. Changes in the nuclei of differentiating endoderm cells as revealed by nuclear transplantation. J. Morph. *100*, 269–312.

BRIMACOMBE, R., TRUPIN, J., NIRENBERG, M., LEDER, P., BERNFIELD, M., and JAOUNI, T., 1965. RNA codewords and protein synthesis. VIII. Nucleotide sequences of synonym codons for arginine, valine, cysteine and alanine. Proc. natn. Acad. Sci., U.S.A. *54*, 954–960.

BRITTEN, R. J., and DAVIDSON, E. H., 1969. Gene regulation for higher cells: A theory. Science, N.Y. *165*, 349–357.

———, and KOHNE, D. A., 1970. Repeated segments of DNA. Scient. Am. *222*(4), 24–31.

BROMLEY, N. W., and ORECHOWITSCH, W. N., 1934. Ueber die Proteolyse in den regenerierenden Geweben. II. Die Aktivität der Gewebeprotease in verschiedenen Gebieten des Regenerats. Biochem. Z. *272*, 324–331.

BROWN. D. D., 1966. The nucleolus and synthesis of ribosomal RNA during oogenesis and embryogenesis of *Xenopus laevis*. Natn. Cancer Inst. Monogr. *23*, 297–309.

———, 1973. The isolation of genes. Scient. Am. *229* (2), 20–29.

———, and Dawid, I. B., 1968. Specific gene amplification in oocytes. Science, N.Y. *160*, 272–280.

———, and GURDON, J. B., 1964. Absence of ribosomal RNA synthesis in the anucleolate mutant of *Xenopus laevis*. Proc. natn. Acad. Sci., U.S.A. *51*, 139–146.

———, and LITTNA, E., 1966. Synthesis and accumulation of DNA-like RNA during embryogenesis of *Xenopus laevis*. J. molec. Biol. *20*, 81–94.

BRUNST, V. V., 1950. Influence of x-rays on limb regeneration in urodele amphibians. Q. Rev. Biol. *25*, 1–29.

———, and CHÉRÉMÉTIEVA, E. A., 1936. Sur la perte locale du pouvoir régénérateur chez le triton et l'axolotl causée par l'irradiation avec les rayons x. Archs. Zool. exp. gén. *78*, 57–67.

———, and SCHEREMETJEWA, E. A., 1933. Untersuchung des Einflusses von Röntgenstrahlen auf die Regeneration der Extremitäten beim Triton. Roux Arch. *128*, 181.

BRUYN, P. P. H. DE, 1945. The motion of migrating cells in tissue cultures of lymph nodes. Anat. Rec. *93*, 295–315.

BUEKER, E. D., 1947. Limb ablation experiments. on the embryonic chick and its effect as observed on the mature nervous system. Anat. Rec. *97*, 157–174.

BULLOUGH, W. S., 1967. The evolution of differentiation. Academic Press, London.

———, 1973. The chalones: A review. Natn. Cancer Inst. Monogr. *38*, 5–15.

———, and LAURENCE, E. B., 1960. The control of epidermal mitotic activity in the mouse. Proc. R. Soc. Lond. (Biol.) *151*, 517–536.

BURGOS, M. H., and FAWCETT, D. W., 1955. Studies on the fine structure of the mammalian testis. I. Differentiation of the spermatids in the cat *(Felis domestica)*. J. biophys. biochem. Cytol. *1*, 287–300.

———, and ———, 1956. An electron microscope study of spermatid differentiation in the toad, *Bufo arenarum* Hensel. J. biophys. biochem. Cytol. *2*, 223–240.

BURKE, V., SULLIVAN, N. P., PETERSEN, H., and WEED, R., 1944. Ontogenetic change in antigenic specificity of the organs of the chick. J. infect. Dis. *74*, 225–233.

BURNET, M., 1962. The thymus gland. Scient. Am. *207*(5), 50–57.

BURNETT, A. L., 1962. The maintenance of form in Hydra. Regeneration. 20th Symp. of the Soc. for the Study of Develop. and Growth. D. Rudnick (Ed.). Ronald Press Co., New York, 27–52.

BURNS, R. K., 1961. Role of hormones in the differentiation of sex. In Young, W. C.: Sex and internal secretions. Williams & Wilkins Co., Baltimore, 76–160.

BURR, H. S., 1916. The effect of removal of the nasal pits in *Amblystoma* embryos. J. exp. Zool. *20*, 27–57.

———, 1930. Hyperplasia in the brain of *Amblystoma*. J. exp. Zool. *55*, 171–191.

BUTENANDT, A., and KARLSON, P., 1954. Ueber die Isolierung eines Metamorphose-Hormons der Insecten in kristallisierter Form. Z. Naturf. *9b*, 389–391.

BUTLER, E. G., 1933. The effects of x-radiation on the regeneration of the forelimb of *Amblystoma* larvae. J. exp. Zool. *65*, 271–315.

———, 1935. Studies on limb regeneration in x-rayed *Amblystoma* larvae. Anat. Rec. *62*, 295–307.

BYTINSKI-SALZ, H., 1936. Lo sviluppo della coda negli anfibi. II. Alterazioni della correlazioni fra i territori costituenti l'abbozzo codale e comportamento dell' ectoderma della pinna codale. R. C. Accad. Lincei, *24*, 82–88.

———, 1937. Trapianti di "organizzatore" nelle ouva di Lampreda. Archo. ital. Anat. Embriol. *39*, 177,–228.

CAMPBELL, R. D., 1965. Cell proliferation in Hydra: An autoradiographic approach. Science, N.Y. *148*, 1231–1232.

CAMPBELL, R. D., 1967. Tissue dynamics of steady state growth in *Hydra littoralis*. I. Patterns of cell division. Devl. Biol. *15*, 487–502.

CAMPENHOUT, E. VAN, 1935. Experimental researches on the origin of the acoustic ganglion in amphibian embryos. J. exp. Zool. *72*, 175–193.

CARASSO, N., and FAVARD, P., 1960. Vitellogenèse de la planorbe. Ultrastructure des plaquettes vitellines. 4th Internat. Conference on Electron Microscopy. Berlin, Vol. II. Berlin, Göttingen, Heidelberg, 431–435.

CARTER, T. C., 1954. The genetics of luxate mice. IV. Embryology. J. Genet. *52*, 1–35.

CHALKLEY, D. T., 1954. A quantitative histological analysis of forelimb regeneration in *Triturus viridescens*. J. Morph. *94*, 21–70.

CHANG, M. C., 1954. Development of parthenogenetic rabbit blastocysts induced by low temperature storage of unfertilized ova. J. exp. Zool. *125*, 127–149.

CHICA SCHALLER, H., 1973. Isolation and characterization of a low-molecular-weight substance activating head and bud formation in hydra. J. Embryol. exp. Morph. *29*, 27–38.

CHILD, C. M., 1929. Physiological dominance and physiological isolation in development and reconstitution. Roux Arch. *117*, 21–66.

———, 1936. Differential reduction of vital dyes in the early development of echinoderms. Roux Arch. *135*, 426–456.

———, 1941. Patterns and problems of development. University of Chicago Press, Chicago.

———, 1948. Exogastrulation by sodium azide and other inhibiting conditions in *Strongylocentrotus purpuratus*. J. exp. Zool. *107*, 1–38.

———, and WATANABE, Y., 1935. The head frequency gradient in *Euplanaria dorotocephala*. Physiol. Zool. *8*, 1–40.

CLAYTON, R. M., 1951. Antigens in the developing newt embryo. Nature, Lond., *168*, 120–121.

CLERMONT, Y., and LEBLOND, C. P., 1955. Spermiogenesis of man, monkey, ram and other mammals as shown by the "periodic acid-Schiff" technique. Am. J. Anat. *96*, 229–253.

CLEVER, U., 1961. Genaktivitäten in den Riesenchromosomen von *Chironomus tentans* und ihre Beziehung zur Entwicklung. I. Genaktivierung durch Ecdyson. Chromosoma *12*, 607–675.

———, 1965. The effect of ecdysone on gene activity patterns in giant chromosomes. In Karlson, P. (Ed.): Mechanisms of hormone action. Academic Press, New York, 142–148.

———, and KARLSON, P., 1960. Induktion von Puff-Veränderungen in den Speicheldrüsenchromosomen von *Chironomus tentans* durch Ecdyson. Expl. Cell Res. 20, 623–626.

COLWIN, A. L., and COLWIN, L. H., 1957. Morphology of fertilization: Acrosome filament formation and sperm entry. In TYLER, A., BORSTEL, R. C. VON, and METZ, C. B.: The beginnings of embryonic development. American Association for the Advancement of Science, Washington.

———, and ———, 1961. Fine structure of the spermatozoon of *Hydroides hexagonus* (Annelida) with special reference to the acrosomal region. J. biophys. biochem. Cytol. *10*, 211–230.

———, and ———, 1963. Role of the gamete membranes in fertilization in *Saccoglossus kowalevskii* (Enteropneusta). I. The acrosomal region and its changes in early stages of fertilization. J. Cell Biol. *19*, 477–500.

COLWIN, L. H., and COLWIN, A. L., 1961. Changes in the spermatozoon during fertilization in *Hydroides hexagonus* (Annelida). I. Passage of the acrosomal region through the vitelline membrane. J. biophys. biochem. Cytol. *10*, 231–254.

———, and ———, 1963. Role of the gamete membranes in fertilization in *Saccoglossus kowalevskii* (Enteropneusta). II. Zygote formation by gamete membrane fusion. J. Cell Biol. *19*, 501–518.

CONKLIN, E. G., 1905. The orientation and cell-lineage of the ascidian egg. J. Acad. nat. Sci. Philad. Ser. 2, *13*.

———, 1931. The development of centrifuged eggs of ascidians. J. exp. Zool. *60*, 1–119.

———, 1932. The embryology of *Amphioxus*. J. Morph. *54*, 69–118.

COOPER, R. S., 1948. Antigens in development. J. exp. Zool. *107*, 397–433.

COPENHAVER, W. M., 1926. Experiments on the development of the heart of *Amblystoma punctatum*. J. exp. Zool. *43*, 321–371.

———, 1930. Results of heteroplastic transplantation of anterior and posterior parts of the heart rudiment in *Amblystoma* embryos. J. exp. Zool. *55*, 293–318.

———, 1933. Transplantation of heart and limb rudiments between *Amblystoma* and *Triton* embryos. J. exp. Zool. *65*, 131–157.

CURTIS, A. S. G., 1960. Cortical grafting in *Xenopus laevis*. Embryol. exp. Morph. *8*, 163–173.

———, 1962. Cell contact and adhesion. Biol. Rev. *37*, 82–129.

———, 1963. The cell cortex. Endeavour *22* (No. 87), 134–137.

CUSIMANO, T. FAGONE, A. and REVERBERI, G., 1962. On the origin of the larval mouth in the anurans. Acta Embryol. Morph. exp. *5*, 82–103.

CZIHAK, G., 1965. Entwicklungsphysiologische Untersuchungen an echiniden Ribonucleinsäure-synthese in den Micromeren und Entodermdifferenzierung. Ein Beitrag zum Problem der Induktion. Roux Arch. *156*, 504–524.

———, and HÖRSTADIUS, S., 1970. Transplantation of RNA-labeled micromeres into animal halves of sea urchin embryos. A contribution to the problem of embryonic induction. Devl. Biol. *22*, 15–30.

———, WITTMANN, H. G., and HIDENNACH, I., 1967. Uridineinbau in die Nucleinsäuren von Furchungsstadien der Eier des Seeigels *Paracentrotus lividus*. Zeitschr. Naturforsch. *22*, 1176–1182.

DALCQ, A. M., 1954. Nouvelles données structurales et cytochimiques sur l'oeuf des mammifères, Revue gèn. Sci. pur. appl. *61*, 19–41.

———, 1960. Réflexions sur la morphologie du cou. Soc. oto-rhino-laryngologica latina, 30 conventus, Brussels, 15–22.

———, and PASTEELS, J., 1937. Une conception nouvelle des bases physiologique de la morphogénèse. Arch. Biol. Liège, *48*, 121–147.

DAN, J. C., and WADA, S. K., 1955. Studies on the acrosome. IV. The acrosome reaction in some bivalve spermatozoa. Biol. Bull. mar. biol. Lab., Woods Hole *109*, 40–55.

DANFORTH, C. H., 1930. Developmental anomalies in a special strain of mice. Am. J. Anat. *45*, 275–288.

DANIELLI, J. F., 1953. Cytochemistry—a critical approach. Chapman & Hall, London.

DANTSCHAKOFF, V., 1941. Der Aufbau des Geschlechts beim höheren Wirbeltier. Gustav Fischer, Jena.

DAREVSKII, I. S., and KULIKOWA, W. N., 1961. Natürliche Parthenogenese in der polymorphen Gruppe der Kaukasischen Feldeidechse (*Lacerta saxicola* Eversmann). Zool. Jb. (Syst.) *89*, 119–176.

DARNELL, J. E., JELINEK, W. R., and MOLLOY, G. R., 1973. Biogenesis of mRNA: Genetic regulation in mammalian cells. Science, N.Y. *181*, 1215–1221.

DAVENPORT, C. B., 1895. Studies in morphogenesis. IV. A preliminary catalogue of the processes concerned in ontogeny. Bull. Mus. comp. Zool. Harv. 27, 173–199.

DAVIDSON, E. H., and HOUGH, B. R., 1971. Genetic information in oocyte RNA. J. molec. Biol. *56*, 491–506.

DAVIDSON, J. N., 1947. Some factors influencing the nucleic acid content of cells and tissues. Cold Spring Harb. Symp. quant. Biol. *12*, 50–59.

———, and LESLIE, I., 1950. A new approach in the biochemistry of growth and development. Nature, Lond. *165*, 49–53.

DAVIS, C. L., 1923. Description of a human embryo having twenty paired somites. Contr. Embryol. Carneg. Instn. *15*, 3–51.

DAWID, I. B., 1970. Cytoplasmic DNA. In Biggers, J. D., and Schuetz, A. W. (Eds.): Oogenesis. University Park Press, Baltimore.

DELAGE, Y., 1884. Evolution de la Sacculine. Archs. Zool. exp. gén. 2, 417–736.

DE PETROCELIS, B., and MONROY, A., 1974. Regulatory processes of DNA synthesis in the embryo. Endeavour *33* (No. 119), 92–98.

DE REUCK, A. V. S., and KNIGHT, J. (Eds.), 1966. Ciba Found. Study Group No 24. Histones, their role in the transfer of genetic information. Churchhill Ltd., London.

DETWILER, S. R., 1918. Experiments on the development of the shoulder girdle and the anterior limb of *Amblystoma punctatum*. J. exp. Zool. *25*, 499–538.

———, 1920. Experiments on the transplantation of limbs in *Amblystoma*. The formation of nerve plexuses and the function of the limbs. J. exp. Zool. *31*, 117–169.

———, 1926a. The effect of reduction of skin and muscle on the development of spinal ganglia. J. exp. Zool. *45*, 399–414.

———, 1926b. Experimental studies on morphogenesis in the nervous system. Q. Rev. Biol. *1*, 61–86.

———, 1930. Observations upon the growth, function and nerve supply of limbs when grafted to the head of salamander embryos. J. exp. Zool. *55*, 319–379.

———, 1934. An experimental study of spinal nerve segmentation in *Amblystoma* with reference to the pluri-segmental contribution to the brachial plexus. J. exp. Zool. *67*, 395–441.

———, 1936. Neuroembryology: An experimental study. Macmillan Co., New York.

———, 1949. The swimming capacity of *Amblystoma* larvae following reversal of the embryonic hindbrain. J. exp. Zool. *111*, 79–94.

———, and VAN DYKE, R. H., 1934. Further observations upon abnormal growth responses of spinal nerves in *Amblystoma* embryos. J. exp. Zool. *69*, 137–164.

Di Berardino, M. A., and King, T. J., 1967. Development and cellular differentiation of neural nuclear-transplants of known karyotype. Devl. Biol. *15*, 102–128.

———, ———, and McKinnel, R. G., 1963. Chromosome studies of a frog renal adenocarcinoma line carried by serial intraocular transplantation. J. natn. Cancer. Inst. *31*, 769–789.

Doljanski, L., and Roulet, F., 1934. Zur Frage der Entstehung der bindegewebigen Strukturen. Roux Arch. *131*, 512–531.

Dollander, A., 1953. Observations relatives à certaines propriétés du cortex de l'oeuf d'amphibien. Archs. Anat. microsc. Morph. exp. *42*, 185–193.

Dorris, F., 1935. The development of structure and function in the digestive tract of *Amblystoma punctatum*. J. exp. Zool. *70*, 491–527.

Dragomirow, N., 1929. Ueber die Faktoren der embryonalen Entwicklung der Linse bei Amphibien. Roux Arch. *116*, 633–668.

———, 1933. Ueber Koordination der Teilprocesse in der embryonalen Morphogenese des Augenbechers. Roux Arch. *129*, 522–560.

———, 1936. Ueber Induktion sekundärer Retina im transplantierten Augenbecher bei *Triton* und *Pelobates*. Roux Arch. *134*, 716–737.

Drew, A. H., 1923. Growth and differentiation in tissue cultures. Br. J. exp. Path. *4*, 46–52.

Driesch, H., 1891. Entwicklungsmechanische Studien. I–II. Z. wiss. Zool. *53*, 160–182.

Dubois, F., and Wolff, E., 1947. Sur une méthode d'irradiation localisée permettant de mettre en évidence la migration des cellules de régénération chez les Planaires. C. r. Séanc. Soc. Biol. *141*, 903–906.

Dürken, B., 1913. Ueber einseitige Augenexstirpation bei jungen Froschlarven. Z. wiss. Zool. *105*, 192–242.

Du Shane, G. P., 1935. An experimental study of the origin of pigment cells in amphibia. J. exp. Zool. *72*, 1–31.

Eakin, R. M., 1961. Photoreceptors in the amphibian frontal organ. Proc. natn. Acad. Sci., U.S.A. *47*, 1084–1088.

———, 1963. Ultrastructural differentiation of the oral sucker in the treefrog *Hyla regilla*. Devl. Biol. *7*, 169–179.

Edström, J. E., and Beermann, W., 1962. The base composition of nucleic acids in chromosomes, puffs, nucleoli and cytoplasm of *Chironomus* salivary gland cells. J. Cell Biol. *14*, 371–379.

Edwards, R. G., and Fowler, R. S., 1970. Human embryos in the laboratory. Scient. Am. *223* (6), 44–54.

Ekman, G., 1925. Experimentelle Beiträge zur Herzentwicklung der Amphibien. Roux Arch. *106*, 320–352.

Endo, Y., 1952. The role of the cortical granules in the formation of the fertilization membrane in the eggs from Japanese sea urchins. Expl. Cell Res. *3*, 406–418.

Endres, H., 1895. Ueber Anstich- und Schnürversuche an Eiern von *Triton taeniatus*. Jber. Schles. Ges. vaterländ. Kultur. *73*.

Evans, H. M., Simpson, M. E., Meyer, R. K., and Reichert, 1933. The growth and gonadstimulating hormones of the anterior hypophysis. Mem. Univ. Calif. *11*, 1–446.

Eyal-Giladi, H., and Wolk, M., 1970. The inducing capacities of the primary hypoblast as revealed by transfilter induction studies. Roux Arch. *165*, 226–241.

Fales, D. E., 1935. Experiments on the development of the pronephros of *Amblystoma punctatum*. J. exp. Zool. *72*, 147–173.

Faure-Fremeit, E., 1925. La cinétique du developpement. Presses Universit. de France, Paris.

Fawcett, D. W., and Porter, K. R., 1954. A study of the fine structure of ciliated epithelia. J. Morph. *94*, 221–281.

Ficq, A., 1954. Analyse de l'induction neurale chez les Amphibiens au moyen d'organisateurs marqués. J. Embryol. exp. Morph. *2*, 194–203.

Filatoff, D., 1916. The removal and transplantation of the auditory vesicle of the embryo of *Bufo* (the correlations at the formation of the cartilaginous skeleton). Russk. zool. Zh. *1*, 48–54.

Fischel, A., 1919. Ueber den Einfluss des Auges auf die Entwicklung und Erhaltung der Hornhaut. Klin. Mbl. Augenheilk *62*, 1–5.

Flickinger, R. A., 1959. A gradient of protein synthesis in *Planaria* and reversal of axial polarity of regenerates. Growth *23*, 251–271.

FLICKINGER, R. A., 1961. Formation, biochemical composition and utilization of amphibian egg yolk. Symp. on Germ Cells and Development. Inst. Internat. d'Embryologie and Fondazione A. Baselli, Pavia, 29–48.

——, and COWARD, S. J., 1962. The induction of cephalic differentiation in regenerating *Dugesia dorotocephala* in the presence of the normal head and in unwounded tails. Devl. Biol. *5*, 179–204.

——, and ROUNDS, D. E., 1956. The maternal synthesis of egg yolk proteins as demonstrated by isotopic and serological means. Biochim. biophys. Acta *22*, 38–72.

FLYNN, T. T., and HILL, J. P., 1939. The development of the Monotremata. IV. Growth of the ovarian ovum, maturation, fertilization, and early cleavage. Trans. zool. Soc. Lond. *24*, 445–582.

FRANCHI, L. L., 1960. Electron microscopy of oocyte-follicle cell relationships in the rat ovary. J. biophys. biochem, cytol. *7*, 397–399.

FRASER, E. A., 1950. The development of the vertebrate excretory system. Biol. Rev. *25*, 159–187.

FRASER, R. C., 1954. Studies on the hypoblast of the young chick embryo. J. exp. Zool. *126*, 349–399.

FRENSTER, J. H., 1965. A model of specific de-repression within interphase chromatin. Nature, Lond. *206*, 1269–1270.

FRIEDEN, E., 1963. The chemistry of amphibian metamorphosis. Scient. Am *209*(5), 110–118.

FUJITA, S., 1963. The matrix cell and cytogenesis in the developing nervous system. J. comp. Neurol. *120*, 37–42.

FULLER, W., and HODGSON, A., 1967. Conformation of the anticodon loop in RNA. Nature, Lond. *215*, 817–821.

GABRIEL, M. L., and FOGEL, S., 1955. Great experiments in biology. Prentice-Hall, Englewood Cliffs, N.J.

GAILLARD, P. J., 1942. Hormones regulating growth and differentiation in embryonic explants. Hermann and Co., Paris.

GALL, J. G., 1963. Chromosomes and cytodifferentiation. In LOCKE, M. (Ed.): Cytodifferentiation and macromolecular synthesis. Academic Press, New York, 119–143.

GALLERA, J., NICOLET, G., and BAUMANN, M., 1968. Induction neurale chez les oiseaux à travers un filtre millipore: Etude au microscope optique et electronique. J. Embryol. exp. Morph. *19*, 439–450.

GALTSOFF, P. S., 1925. Regeneration after dissociation (an experimental study on sponges). I. Behavior of dissociated cells of *Microciona prolifera* under normal and altered conditions. J. exp. Zool. *42*, 183–255.

——, and PHILPOTT, D. E., 1960. Ultrastructure of the spermatozoon of the oyster *Crassostrea virginica*. J. Ultrastruct. Res. *3*, 241–253.

GARDNER, R. L., 1972. An investigation of inner cell mass and trophoblast tissues following their isolation from the mouse blastocyst. J. Embryol. exp. Morph. *28*, 279–312.

GEBHARDT, H., 1926. Untersuchungen über die Determination bei Planarienregeneraten. Roux Arch. *107*, 684–726.

GERALD, P. S., 1961. The abnormal haemoglobins. In PENROSE, L. A. (Ed.): Recent advances in human genetics. J. & A. Churchill, Ltd., London.

GIBBINS, J. R., TILNEY, L. G., and PORTER, K. R., 1969. Microtubules in the formation and development of the primary mesenchyme in *Arbacia punctulata*. I. The distribution of microtubules. J. Cell Biol. *41*, 201–226.

GIERER, A., 1974. Hydra as a model for the development of biological form. Scient. Am. *231*(6), 44–54.

GILCHRIST, F. G., 1937. Budding and locomotion in the Scyphostomas of *Aurelia*. Biol. Bull. mar. biol. Lab., Woods Hole *72*, 99–124.

GINSBURG, A., 1953. The origin of bilateral symmetry in the eggs of acipenserid fishes. C.R. Acad. Sci. URSS *90*, 477–480.

——, and DETTLAFF, T., 1944. Experiments on transplantation and removal of organ rudiments in embryos of *Acipenser stellatus* in early developmental stages. C. R. Acad. Sci. URSS *44*, 209–212.

GLATTHAAR, E., and TONDURY, G., 1950. Untersuchungen an abortierten Früchten nach Rubeolaerkrankung der Mutter in der Frühschwangerschaft. Gynaecologia *129*, 315–320.

GLUECKSOHN-SCHOENHEIMER, S., 1943. The morphological manifestation of a dominant mutation in mice affecting tail and urogenital system. Genetics *28*, 341–348.

——, 1945. The embryonic development of mutants of the Sd-strain in mice. Genetics *30*, 29–38.

GLUECKSOHN-SCHOENHEIMER, S., 1949. The effects of a lethal mutation responsible for duplications and twinning in mouse embryos. J. exp. Zool. *110*, 47–76.

GOERTTLER, K., 1928. Die Bedeutung der ventrolateralen Mesodermbezirke für die Herzanlage der Amphibienkeime. Anat. Anz. Erg. Heft *66*, 132–139.

GOETZ, R. H., 1938. On the early development of the Tenrecoidea *(Hemicentetes semispinosus)*. Biomorphosis *1*, 67–79.

GOODRICH, E. S., 1958. Studies on the structure and development of vertebrates. Dover Publications, New York.

GORBMAN, A., and BERN, H. A., 1962. A textbook of comparative endocrinology. John Wiley & Sons, New York.

GORBUNOVA, G. P., 1939. On the inducing properties of the medulla oblongata in amphibian embryos. C.R. Acad. Sci. URSS *23*, 298–301.

GRANHOLM, N. H., and BAKER, J. R., 1970. Cytoplasmic microtubules and the mechanism of avian gastrulation. Devl. Biol. *23*, 563–584.

GRANT, P., 1953. Phosphate metabolism during oogenesis in *Rana temporaria*. J. exp. Zool. *124*, 513–543.

———, 1958. The incorporation of P^{32} and glycine-2-C^{14} into nucleic acids during early embryonic development of *Rana pipiens*. J. cell. comp. Physiol. *52*, 249–268.

GROBSTEIN, C., 1953a. Analysis *in vitro* of the early organization of the rudiment of the mouse sub-mandibular gland. J. Morph. *93*, 19–44.

———, 1953b. Epithelio-mesenchymal specificity in the morphogenesis of mouse sub-mandibular rudiments *in vitro*. J. exp. Zool. *124*, 383–413.

———, 1953c. Morphogenetic interaction between embryonic mouse tissues separated by a membrane filter. Nature, Lond. *172*, 869.

———, 1955. Inductive interaction in the development of the mouse metanephros. J. exp. Zool. *130*, 319–339.

———, 1957. Some transmission characteristics of the tubule-inducing influence on mouse metanephrogenic mesenchyme. Expl. Cell Res. *13*, 575–587.

———, and DALTON, A. J., 1957. Kidney tubule induction in mouse metanephrogenic mesenchyme without cytoplasmic contact. J. exp. Zool. *135*, 57–73.

GROSSER, O., 1945. Grundriss der Entwicklungsgeschichte des Menschen. Springer, Berlin.

GRUENWALD, P., 1952. Development of the excretory system. Ann. N.Y. Acad. Sci. *55*, 142–146.

GUARESCHI, C., 1935. Studi sulla determinazione dell' orechio interno degli anfibi anuri. Archo. ital. Anat. Embriol. *35*, 97–129.

GUDERNATSCH, F., 1912. Feeding experiments on tadpoles. Roux Arch. *35*, 457–483.

GUDERNATSCH, J. F., 1913. Concerning the mechanism and direction of the embryonic foldings. Anat. Rec. *7*, 411–431.

GURDON, J. B., 1960. The developmental capacity of nuclei taken from differentiating endoderm cells of *Xenopus laevis*. J. Embryol. exp. Morph. *8*, 505–526.

———, 1968a. Nucleic acid synthesis in embryos and its bearing on cell differentiation. Essays in Biochemistry *4*, 25–68.

———, 1968b. Transplanted nuclei and cell differentiation. Scient. Am. *219*(6), 24–35.

———, 1968c. Changes in somatic cell nuclei inserted into growing and maturing amphibian oocytes. J. Embryol. exp. Morph. *20*, 401–414.

———, 1969. Intracellular communication in early animal development. Devl. Biol., Suppl. *3*, 59–82.

———, and BROWN, D. D., 1965. Cytoplasmic regulation of RNA synthesis and nucleolus formation in developing embryos of *Xenopus laevis*. J. molec. Biol. *12*, 27–35.

———, and GRAHAM, C. F., 1967. Nuclear changes during cell differentiation. Sci. Prog. Oxf. *55*, 259–277.

———, LANE, C. D., WOODLAND, H. R., and MARBAIX, G., 1971. Use of frog eggs and oocytes for the study of messenger RNA and its translation in living cells. Nature, Lond. *233*, 177–182.

———, and LASKEY, R. A., 1970. The transplantation of nuclei from single cultured cells into enucleated frogs' eggs. J. Embryol. exp. Morph. *24*, 277–248.

———, and UEHLINGER, V., 1966. "Fertile" intestinal nuclei. Nature, Lond. *210*, 1240–1241.

GUSTAFSON, T., 1950. Survey of the morphological action of the lithium ion and the chemical basis of its action. Revue suisse Zool. *57*, Suppl. 1, 77–92.

———, and HJELTE, M., 1951. The amino-acid metabolism of the developing sea urchin egg. Exp. Cell Res. *2*, 474–490.

———, and KINNANDER, H., 1956. Microaquaria for time lapse cinematographic studies of morphogenesis in swimming larvae and observations on sea urchin gastrulation. Expl. Cell Res. *11*, 36–51.

———, and WOLPERT, L., 1961. The forces that shape the embryo. Discovery, Lond. *22*, 470–477.

GUYÉNOT, E., and PONSE, K., 1930. Territoires de régénération et transplantations. Bull. biol. Fr. Belg. *64*, 251–287.

HADEK, R., 1963a. Submicroscopic study on the cortical granules in the rabbit ovum. J. Ultrastruct. Res. *8*, 170–175.

———, 1963b. Submicroscopic study on the sperm-induced cortical reaction in the rabbit ovum. J. Ultrastruct. Res. *9*, 99–109.

HADORN, E., 1932. Ueber Organentwicklung und histologische Differenzierung in transplantierten merogonischen Bastardgeweben *(Triton palmatus (♀) × Triton cristatus (♂))*. Roux Arch. *125*, 495–565.

HADZI, J., 1909. Die Entstehung der Knospe bei *Hydra*. Arb. zool. Inst. Wien, *18*.

HAECKEL, E., 1868. Natürliche Schöpfungsgeschichte, Berlin.

HALL, T. S., 1951. A source book in animal biology. McGraw-Hill, New York.

HAMBURGER, V., 1929. Experimentelle Beiträge zur Entwicklungsphysiologie der Nervenbahnen in der Froschextremität. Roux Arch. *119*, 47–99.

———, 1934. The effect of wing bud extirpation on the development of the central nervous system in chick embryos. J. exp. Zool. *68*, 449–494.

———, 1947. A manual of experimental embryology. (2nd imp.) University Press, Chicago.

———, 1956. Developmental correlations in neurogenesis. In RUDNICK, D. (Ed.): Cellular mechanisms in differentiation and growth. Princeton University Press, 191–212.

———, 1961. Experimental analysis of the dual origin of the trigeminal ganglion in the chick embryo. J. exp. Zool. *148* (2), 91–124.

———, 1962. Specificity in neurogenesis. J. cell. comp. Physiol. *60*, No. 2, Suppl. 1, 81–92.

———, and HAMILTON, H. L., 1951. A series of normal stages in the development of the chick embryo. J. Morph. *88*, 49–92.

———, and LEVI-MONTALCINI, R., 1949. Proliferation, differentiation and degeneration in the spinal ganglia of the chick embryo under normal and experimental conditions. J. exp. Zool. *111*, 457–501.

HAMILTON, W. J., BOYD, J. D., and MOSSMAN, H. W., 1947. Human embryology. Heffer & Sons, Cambridge.

HAMMOND, W. S., and YNTEMA, C. L., 1947. Depletions in the thoracolumbar sympathetic system following removal of neural crest in the chick. J. comp. Neurol. *86*, 237–265.

———, and ———, 1958. Origin of ciliary ganglia in the chick. J. comp. Neurol. *110*, 367–390.

HARDING, C. V., HARDING, D., and PERELMAN, P., 1954. Antigens in sea urchin embryos. Expl. Cell Res. *6*, 202–210.

HARDY, M. H., 1952. The histochemistry of hair follicles in the mouse. J. Anat. *90*, 285–337.

———, 1953. Vaginal cornification of the mouse produced by oestrogens *in vitro*. Nature, Lond. *172*, 1196.

HARRISON, R. G., 1903. Experimentelle Untersuchungen über die Entwicklung der Sinnesorgane der Seitenlinie bei den Amphibien. Arch. mikr. Anat. *63*, 35–149.

———, 1908. Embryonic transplantation and development of the nervous system. Anat. Rec. *2*, 385.

———, 1918. Experiments on the development of the forelimb of *Amblystoma*, a self-differentiating equipotential system. J. exp. Zool. *25*, 413–461.

———, 1921a. Experiments on the development of the gills in the amphibian embryos. Biol. Bull. mar. biol. Lab., Woods Hole *41*, 156–168.

———, 1921b. On relations of symmetry in transplanted limbs. J. exp. Zool. *32*, 1–136.

———, 1925a. The development of the balancer in *Amblystoma*, studied by the method of transplantation and in relation to the connective tissue problem. J. exp. Zool. *41*, 349–427.

———, 1925b. The effect of reversing the medio-lateral or transverse axis of the forelimb bud in the salamander embryo *(Amblystoma punctatum)*. Roux Arch. *106*, 469–502.

———, 1929. Correlation in the development and growth of the eye studied by means of heteroplastic transplantation. Roux Arch. *120*, 1–55.

———, 1935. Factors concerned in the development of the ear in *Amblystoma punctatum*. Anat. Rec. *64*, 38–39.

———, 1969. Organization and development of the embryo. Yale University Press, New Haven.

HARVEY, E. B., 1936. Parthenogenetic merogony or cleavage without nuclei in *Arbacia punctulata*. Biol. Bull. mar. biol. Lab., Woods Hole *17*, 101–121.

———, 1946. Structure and development of the clear quarter of the *Arbacia punctulata* egg. J. exp. Zool. *102*, 253–275.

———, 1956. The American *Arbacia* and other sea urchins. Princeton University Press, Princeton.

HAUENSCHILD, C., 1964. Postembryonale Entwicklungssteuerung durch ein Gehirn-Hormon bei *Platynereis dumerilii*. Zool. Anz. *27*, Suppl., 111–120.

HAY, E. D., 1962. Cytological studies of dedifferentiation and differentiation in regenerating amphibian limbs. In RUDNICK, D. (Ed.): Regeneration. Symp. Dev. Growth. Ronald Press Co., New York, 177–210.

HAY, E. D., and FISCHMAN, D. A., 1961. Origin of the blastema in regenerating limbs of the newt *Triturus viridescens*. An autoradiographic study using tritiated thymidine to follow cell proliferation and migration. Devl. Biol. *3*, 26–59.

HAYASHI, Y., 1955. Inductive effect of some fractions of tissue extracts after removal of pentose nucleic acid, tested on the isolated ectoderm of *Triturus* gastrula. Embryologia *2*, 145–162.

———, 1956. Morphogenetic effects of pentose nucleoprotein from the liver upon the isolated ectoderm. Embryologia *3*, 57–67.

———, 1958. The effects of pepsin and trypsin on the inductive ability of pentose nucleoprotein from guinea pig liver. Embryologia *4*, 33–53.

HAYNES, J., and BURNETT, A. L., 1963. Dedifferentiation and redifferentiation of cells in *Hydra viridis*. Science, N.Y. *142*, 1481–1483.

HELFF, O. M., 1928. Studies on amphibian metamorphosis. III. Physiol. Zoöl. *1*, 463–495.

HERBST, C., 1893. Experimentelle Untersuchungen über den Einfluss der veränderten chemischen Zusammensetzung des umgebenden Mediums auf die Entwicklung der Tiere. II. Mitt. zool. Stn. Neapel. *11*, 136–220.

HERRICK, C. J., 1934. An introduction to neurology. Saunders, Philadelphia.

HERRMANN, H., and NICHOLAS, J. S., 1948. Quantitative changes in muscle protein fractions during development. J. exp. Zool. *107*, 165–176.

———, and ———, 1949. Nucleic acid content of whole homogenates and of fractions of developing rat muscle. J. exp. Zool. *112*, 341–360.

HERTIG, A. T., and ROCK, J., 1941. Two human ova of the pre-villous stage, having an ovulation age of about eleven and twelve days respectively. Contr. Embryol. Carneg. Instn. *29*, 127–156.

———, and ———, 1945. Two human ova in the pre-villous stage, having a developmental age of about seven and nine days respectively. Contr. Embryol. Carneg. Instn. *31*, 67–84.

HERTWIG, O., 1906. Handbuch der vergleichenden und experimentellen Entwicklungslehre der Wirbeltiere. G. Fischer, Jena.

HERTWIG, R., 1896. Ueber die Entwicklung des unbefruchteten Seeigeleies. Festschr. f. Gegenbaur. Leipzig.

HEUSER, C. H., 1923. The branchial vessels and their derivatives in the pig. Contr. Embryol. Carneg. Instn. *15*, 123–139.

———, and CORNER, G. W., 1957. Developmental horizons in human embryos. Description of age group X, 4 to 12 somites. Contr. Embryol. Carneg. Instn. *36*, 31–39.

———, and STREETER, G. L., 1928. Early stages of the development of the pig from the period of the initial cell cleavage to the time of appearance of limb buds. Contrib. Embryol. Carneg. Inst. *20*, 1–30.

———, and ———, 1941. Development of the macaque embryo. Contr. Embryol. Carneg. Instn. *29*, 17–56.

HILL, J. P., 1918. Some observations on the early development of *Didelphis aurita*. Q. Jl. microsc. Sci. *63*, 91–140.

HIRAMOTO, Y., 1957. The thickness of the cortex and the refractive index of the protoplasm in sea urchin eggs. Embryologia *3*, 361–374.

HOAGLAND, M. B., 1959. Nucleic acids and proteins. Scient. Am. *201*(6), 55–61.

HOLTFRETER, J., 1934a. Der Einfluss thermischer, mechanischer und chemischer Eingriffe auf die Induzierfähigkeit von *Triton*-Keimteilen. Roux Arch. *132*, 225–306.

———, 1934b. Ueber die Verbreitung induzierender Substanzen und ihre Leistungen im *Triton*-Keim. Roux Arch. *132*, 307–383.

———, 1938a. Differenzierungspotenzen isolierter Teile der Urodelengastrula. Roux Arch. *138*, 522–656.

———, 1938b. Differenzierungspotenzen isolierter Teile der Anurengastrula. Roux Arch. *138*, 657–738.

———, 1939a. Gewebeaffinität, ein Mittel der embryonalen Formbildung. Arch. exp. Zellforsch *23*, 169–209.

———, 1939b. Studien zur Ermittlung der Gestaltungsfaktoren in der Organentwicklung der Amphibien. I and II. Roux Arch. *139*, 110–190, 227–273.

———, 1943a. Properties and functions of the surface coat in amphibian embryos. J. exp. Zool. *93*, 251–323.

———, 1943b. A study of the mechanics of gastrulation. I. J. exp. Zool. *94*, 261–318.

———, 1943c. Experimental studies on the development of the pronephros. Revue can. Biol. *3*, 220–250.

———, 1946. Experiments on the formed inclusions of the amphibian egg. I. The effect of pH and electrolytes on yolk and lipochondria. J. exp. Zool. *101*, 355–405.

———, 1947a. Changes of structure and the kinetics of differentiating embryonic cells. J. Morph. *80*, 57–62.

———, 1947b. Neural induction in explants which have passed through a sublethal cytolysis. J. exp. Zool. *106*, 197–222.

HOLTFRETER, J., 1968. Mesenchyme and epithelia in inductive and morphogenetic processes. In Fleischmajer, R. and Billingham (Eds.): Epithelial Mesenchymal Interactions. William and Wilkins Co., Baltimore. 21.

HOLTZER, H., 1961. Aspects of chondrogenesis and myogenesis. In RUDNICK, D. (Ed.): Synthesis of molecular and cellular structure. Symp. of the Soc. for the Study of Develop. and Growth. Ronald Press Co., New York.

———, 1963. Mitosis and cell transformations. In MAZIA, D., and TYLER, A. (Eds.): General Physiology of Cell specialization. McGraw-Hill, New York, 80–90.

———, and DETWILER, S. R., 1953. An experimental analysis of the development of the spinal column. III. Induction of skeletogenous cells. J. exp. Zool. *123*, 335–368.

———, WEINTRAUB, H., MAYNE, R., and MACHAN, B., 1972. The cell cycle, cell lineages, and cell differentiation. Current topics in developmental Biol. *7*, 229–256.

HORST, C. J. VAN DER, 1942. Early stages in the development of *Elephantulus*. S. Afr. J. med. Sci. *7*, Biol. Suppl., 55–65.

HÖRSTADIUS, S., 1928. Ueber die Determination des Keimes der Echinodermen. Acta zool. Stockh. *9*, 1–192.

———, 1935. Ueber die Determination im Verlaufe der Eiachse bei Seeigeln. Pubbl. Staz. zool. Napoli. *14*, 251–479.

———, 1944. Ueber die Folgen von Chordaexstirpation an spaeten Gastrulae und Neurulae von *Amblystoma punctatum*. Acta zool. Stockh. *25*, 75–87.

———, 1950. The neural crest. Oxford University Press, London.

———, 1952. Induction and inhibition of reduction gradients by the micromeres in the sea urchin egg. J. exp. Zool. *120*, 421–436.

———, 1953a. Influence of implanted micromeres on reduction gradients and mitochondrial distribution in developing sea urchin eggs. J. Embryol. exp. Morph. *1*, 257–259.

———, 1953b. Vegetalization of the sea-urchin egg by dinitrophenol and animalization by trypsin and ficin. J. Embryol. exp. Morph. *1*, 327–348.

———, 1955. Reduction gradients in animalized and vegetalized sea urchin eggs. J. exp. Zool. *129*, 249–256.

———, and JOSEFSSON, L., 1972. Morphogenetic substances from sea urchin eggs. Isolation of animalizing substances from developing eggs of *Paracentrotus lividus*. Acta Embryol. exp., 7–23.

———, and SELLMAN, S., 1946. Experimentelle Untersuchungen über die Determination des knorpeligen Kopfskelettes bei Urodelen. Nova Acta. R. Soc. Scient. Upsal. *13*.

———, and WOLSKY, A., 1936. Studien über die Determination der Bilateralsymmetrie des jungen Seeigelkeimes. Roux Arch. *135*, 69–113.

HSU, Y. C., 1973. Differentiation *in vitro* of mouse embryos to the stage of early somite. Devl. Biol. *33*, 403–411.

HUANG, R. C., and BONNER, J., 1962. Histone, a suppressor of chromosome RNA synthesis. Proc. natn. Acad. Sci., U.S.A. *48*, 1216–1222.

HUMPHREYS, W. J., 1962. Electron microscope studies on eggs of *Mytilus edulis*. J. Ultrastruct. Res. *7*, 467–487.

HUNT, T. E., 1937. The origin of entodermal cells from the primitive streak of the chick embryo. Anat. Rec. *68*, 449–460.

HURWITZ, J., and FURTH, J. J., 1962. Messenger RNA. Scient. Am. *206*(2), 41–49.

HUXLEY, J. S., 1932. Problems of relative growth. Methuen and Co., London.

———, and DE BEER, G. R., 1934. The elements of experimental embryology. University Press, Cambridge.

HYMAN, L. H., 1951. The invertebrates, Vol. 3. McGraw-Hill, New York.

ILLMENSEE, K., 1968. Transplantation of embryonic nuclei into unfertilized eggs of *Drosophila melanogaster*. Nature, Lond. *219*, 1268–1269.

———, 1972. Developmental potencies of nuclei from cleavage, preblastoderm, and syncytial blastoderm transplanted into unfertilized eggs of *Drosophila melanogaster*. Roux Arch. *170*, 267–298.

———, 1973. The potentialities of transplanted early gastrula nuclei of *Drosophila melanogaster*. Production of their imago descendants by germ-line transplantation. Roux Arch. *171*, 331–343.

JACOB, F., and MONOD, J., 1961. On the regulation of gene activity. Cold Spring Harb. Symp. quant. Biol. *26*, 193–211.

JACOBSON, A. G., 1955. The roles of the optic vesicle and other head tissues in lens induction. Proc. natn. Acad. Sci., U.S.A. *41*, 522–525.

JACOBSON, A. G., 1960. Influences of ectoderm and endoderm on heart differentiation in the newt. Devl. Biol. *2*, 138–154.

———, and DUNCAN, J. T., 1968. Heart induction in salamanders. J. exp. Zool. *167*, 79–103.

JOHANNSEN, O. A., and BUTT, F. H., 1941. Embryology of insects and myriapods. McGraw-Hill, New York.

JOHNEN, A. G., 1956. Experimental studies about the temporal relationships in the induction process. Experiments on *Amblystoma mexicanum*. Proc. Koninkl. Nederl. Akad. Wetensch., Series C, *59*(4), 554–660.

———, 1964. Experimentelle Untersuchungen über die Bedeutung des Zeitfaktors beim Vorgang der neuralen Induktion. II. Roux Arch. *155*, 302–313.

———, 1970. Der Einfluss von Li- und SCN-Ionen auf die Differenzierungsleistungen des *Ambystoma* Ektoderms und ihre Veränderung bei kombinierter Einwirkung beider Ionen. Roux Arch. *165*, 150–162.

JONES, H. O., and BREWER, J. I., 1941. A human embryo in the primitive-streak stage. Contr. Embryol. Carneg. Instn. *29*, 157–165.

JONES-SEATON, A., 1950. Étude de l'organisation cytoplasmique de l'oeuf des rongeurs, principalement quant à la basophilie ribonucléique. Archs. Biol. Paris. *61*, 291–444.

JOST, A., 1953. Problems of fetal endocrinology: The gonadal and hypophyseal hormones. Recent Prog. Horm. Res. *8*, 379–418.

KAMER, J. C. VAN DE, 1949. Over de ontwikkeling de determinatie en de betekenis van de epiphyse en de paraphyse van de amphibiën. Dissertation. Van der Weil, Arnheim.

KANAJEW, J., 1926. Ueber die histologischen Vorgänge bei der Regeneration von *Pelmatohydra oligactis* Pall. Zool. Anz. *65*, 217–226.

KARASAKI, S., 1957. On the mechanism of the dorsalization in the ectoderm of *Triturus* gastrulae caused by precytolytic treatments. I. Embryologia *3*, 317–334.

———, 1963. Studies on amphibian yolk. I. The ultrastructure of the yolk platelet. J. Cell Biol. *18*, 135–151.

———, 1967. An electron microscope study on the crystalline structure of the yolk platelets of the lamprey eggs. J. Ultrastruct. Res. *18*, 377–390.

KARFUNKEL, P. R., 1970. The role of microtubules and microfilaments in neurulation. Anat. Rec. *166*, 328.

KARLSON, P., and HOFFMEISTER, H., 1963. Zur Chemie des Ecdysons. Just. Liebig's Annln. Chem. *662*, 1–20.

KAYE, J. S., 1962. Acrosome formation in the house cricket. J. Cell. Biol. *12*, 411–432.

KEDES, L. H., and GROSS, P. R., 1969. Identification in cleaving embryos of three RNA species serving as templates for the synthesis of nuclear proteins. Nature, Lond. *223*, 1335–1339.

KELLEY, R. O., 1969. An electron microscopic study of chordomesoderm-neuroectoderm association in gastrulae of a toad, *Xenopus laevis*. J. exp. Zool. *172*, 153–179.

KEMP, N. E., 1953. Synthesis of yolk in oocytes of *Rana pipiens* after induced ovulation. J. Morph. *92*, 487–511.

———, 1956. Electron microscopy of growing oocytes of *Rana pipiens*. J. biophys. biochem. Cytol. *2*, 281–292.

KERR, J. G., 1919. Textbook of embryology. Vol. II. Vertebrata with the exception of mammalia. Macmillan, London.

KESSEL, R. G., 1963. Electron microscope studies on the origin of annulate lamellae in oocytes of *Necturus*. J. Cell Biol. *19*, 391–414.

———, 1968a. Mechanism of protein yolk synthesis and deposition in crustacean oocytes. Z. Zellforsch. *89*, 17–38.

———, 1968b. Electron microscope studies on developing oocytes of a Coelenterate medusa with special reference to vitellogenesis. J. Morph. *126*, 211–248.

———, 1968c. Annulate lamellae. J. Ultrastr. Res., Suppl. 10, 1–82.

KIJIMA, S., and WILT, F. H., 1969. Rate of nuclear ribonucleic acid turnover in sea urchin embryos. J. molec. Biol. *40*, 235–246.

KING, R. C., 1960. Oogenesis in adult *Drosophila melanogaster*. IX. Studies on the cytochemistry and ultrastructure of developing oocytes. Growth *24*, 265–323.

———, and DEVINE, R. L., 1958. Oogenesis in adult *Drosophila melanogaster*. VII. The submicroscopic morphology of the ovary. Growth *22*, 299–326.

———, and KOCH, E. A., 1963. Studies on the ovarian follicle cells of *Drosophila*. Q. Jl. microsc. Sci. *104*, 297–320.

———, and MILLS, R. P., 1962. Oogenesis in adult *Drosophila*. XI. Studies of some organelles of the nutrient stream in egg chambers of *D. melanogaster* and *D. willistoni*. Growth *26*, 235–253.

KING, T. J., and BRIGGS, R., 1954. Transplantation of living nuclei of late gastrulae into enucleated eggs of *Rana pipiens*. J. Embryol. exp. Morph. *2*, 73–80.

———, and ———, 1956. Serial transplantation of embryonic nuclei. Cold Spring Harb. Symp. quant. Biol. *21*, 271–290.

KING, T. J., and DI BERARDINO, M. A., 1965. Transplantation of nuclei from the frog renal adenocarcinoma. I. Development of tumor nuclear-transplant embryos. Ann. N.Y. Acad. Sci. *126*, 115–126.

KINNANDER, H., and GUSTAFSON, T., 1960. Further studies on the cellular basis of gastrulation in the sea urchin larva. Expl. Cell Res. *19*, 278–290.

KITCHIN, I. C., 1949. The effect of notochordectomy in *Amblystoma mexicanum*. J. exp. Zool. *112*, 393–416.

KOGAN, R. E., 1939. Inducing action of the medulla oblongata on the trunk epithelium in amphibia. C.R. Acad. Sci. URSS *23*, 307–310.

KORSCHELT, E., 1936. Vergleichende Entwicklungsgeschichte der Tiere. G. Fischer, Jena.

KOWALEVSKY, A., 1866. Entwicklungsgeschichte der einfachen Ascidien. Mem. Acad. Sci. St. Petersb. *10*.

KROEGER, H., 1963. Chemical nature of the system controlling gene activities in insect cells. Nature, Lond. *200*, 1234–1235.

———, 1964. Zellphysiologische Mechanismen bei der regulation von genaktivitäten in den Riesenchromosomen von *Chironomus thummi*. Chromosoma *15*, 36–70.

KUHL, W., 1941. Untersuchungen über die Cytodynamik der Furchung und Frühentwicklung des Eies der weissen Maus. Abh. senckenb. naturforsch. Ges. *456*, 1–17.

KÜHN, A., 1939. Zur Entwicklungsphysiologie der Schmetterlingsmetamorphose. Verh. 7 internat. Kongr. Ent. Berlin, 780–796.

———, 1955. Vorlesungen über Entwicklungsphysiologie. Springer, Berlin.

———, and PIEPHO, H., 1936. Ueber hormonale Wirkungen bei der Verpuppung der Schmetterlinge. Nachr. Ges. Wiss. Göttingen *2*, 141–154.

KÜNTZEL, H., 1969. Proteins of mitochondrial and cytoplasmic ribosomes from *Neurospora crassa*. Nature, Lond. 222, 142–146.

KUYPER, C. M. A., SMETS, L. A., and PIECK, A. C. M., 1962. The life cycle of a strain of liver cells cultivated in vitro. Expl. Cell Res. *26*, 217–219.

LALLIER, R., 1956. Les ions de métaux lourds et le problème de la dètermination embryonnaire chez les Echinodermes. J. Embryol. exp. Morph. *4*, 265–278.

———, 1957. Recherches sur l'animalisation de l'oeuf de l'oursin *Paracentrotus lividus* par les dérivés polysulfoniques. Pubbl. Staz. zool. Napoli *30/2*, 185–209.

LANDACRE, F. L., 1910. The origin of the cranial ganglia in *Ameiurus*. J. Comp. Neurol. 20, 309–411.

LANE, C. D., MARBAIX, G., and GURDON, J. B., 1971. Rabbit haemoglobin synthesis in frog cells: The translation of reticulocyte 9s RNA in frog oocytes. J. molec. Biol. *61*, 73–91.

LANZAVECCHIA, G., 1960. The formation of the yolk in frog oocytes. The proceedings of the European Regional Conference on Electron Microscopy, Delft, *2*, 746.

LASKEY, R. A., and GURDON, J. B., 1970. Genetic content of adult somatic cells tested by nuclear transplantation from cultured cells. Nature, Lond. *228*, 1332–1334.

LAURENCE, E. B., 1973. Experimental approach to the epidermal chalone. Natl. Cancer Inst. Monogr. *38*, 37–45.

LEHMANN, F. E., 1937. Mesodermisierung des präsumptiven Chordamaterials durch Einwirkung von Lithiumchlorid auf die Gastrula von *Titon alpestris*. Roux Arch. *136*, 112–146.

———, 1945. Einführung in die physiologische Embryologie. Birkhäuser, Basel.

LENIQUE, P., 1959. Studies on homologous inhibition in the chick embryo. Acta Zool. *40*, 141–202.

LENTZ, T. L., and TRINKAUS, J. P., 1967. A fine structural study of cytodifferentiation during cleavage, blastula and gastrula stages of *Fundulus heteroclitus*. J. Cell Biol. *32*, 121–138.

LEVI-MONTALCINI, R., and AMPRINO, R., 1947. Recherches experimentales sur l'origine du ganglion ciliaire dans l'embryon de poulet. Arch. Biol., Paris *58*, 265–288.

LEVITT, M. M., 1932. On the post-embryonic growth of larvae of some Lepidoptera. Trav. Inst. Biol. Kiev. *5*, 451–468.

LEWIS, W. H., 1904. Experimental studies on the development of the eye in Amphibia. I. On the origin of the lens in *Rana palustris*. Am. J. Anat. *3*, 505–536.

———, 1907. On the origin and differentiation of the otic vesicle in amphibian embryos. Anat. Rec. *1*, 141–145.

———, and HARTMANN, C. G., 1933. Early cleavage stages of the egg of the monkey *(Macacus rhesus)*. Contr. Embryol. Carneg. Instn. *24*, 189–202.

LIEDKE, K. B., 1951. Lens competence in *Amblystoma punctatum*. J. exp. Zool. *17*, 573–591.

———, 1955. Studies on lens induction in *Amblystoma punctatum*. J. exp. Zool. *130*, 353–379.

LILLIE, F. R., 1911. Studies on fertilization in *Nereis*. I. The cortical changes in the egg. II. Partial fertilization. J. Morph. *22*, 361–394.

LILLIE, F. R., 1919a. Problems of fertilization. University of Chicago Press, Chicago.
_____, 1919b. The development of the chick. 2nd ed. Henry Holt, New York.
LINDAHL, P. E., 1933. Ueber "animalisierte" und "vegetativisierte" Seeigellarven. Roux Arch. *128*, 661–664.
_____, 1936. Zur Kenntnis der physiologischen Grundlage der Determination im Seeigel-keim. Acta Zool. Stockh. *17*, 179–395.
LIOSNER, L. D., and WORONZOWA, M. A., 1936. Regeneration des Organs mit transplan-tierten ortsfremden Muskeln. 2. Mitteilung. Zool. Anz. *115*, 55–58.
LITTLE, C. C., and BAGG, H. J., 1924. The occurrence of four inheritable morphological variations in mice and their possible relation to treatment with x-rays. J. exp. Zool. *41*, 45–92.
LOCATELLI, P., 1924. Sulla formazione di arti sopranumerari. Boll. Soc. med-chir. Pavia *36*.
LOEB, J., 1913. Artificial parthenogenesis and fertilization. University of Chicago Press, Chi-cago.
LOEWENSTEIN, W. R., 1970. Intercellular communication. Scient. Am. *222*(5), 78–86.
LOGACHEV, E. D., 1956. On the mutual relations between the nucleus and the cytoplasm in growing egg-cells of Platyhelminthes. C.R. Acad. Sci. URSS *111*, 507–509.
LONGO, F. J., and ANDERSON, E., 1968. The fine structure of pronuclear development and fusion in the sea urchin *Arbacia punctulata*. J. Cell Biol. *39*, 339–368.
LOPASHOV, G. V., 1950. Experimental investigations of the sources of cellular material and condition of formation of the pectoral fins in teleost fishes. C.R. Acad. Sci. URSS *70*, 137–140.
_____, 1956. Mechanisms of formation and origin of the choroid coat in the amphibian eye. C.R. Acad. Sci. URSS *109*, 653–656.
LUDUENA, M. A., and WESSELLS, N. K., 1973. Cell locomotion, nerve elongation and micro-filaments. Devl. Biol. *30*, 427–440.
LUTHER, W., 1935. Entwicklungsphysiologische Untersuchungen am Forellenkeim: Die Rolle des Organizationszentrums bei der Entstehung der Embryonalanlage. Biol. Zbl. *55*, 114–137.
LYNCH, W. F., 1961. Extrinsic factors influencing metamorphosis in bryozoan and ascidian larvae. Am. Zoologist *1*, 59–66.

MACHEMER, H., 1929. Differenzierungsfähigkeit der Urnierenanlage von *Triton alpestris*. Roux Arch. *118*, 200–251.
_____, 1932. Experimentelle Untersuchung über die Induktionsleistungen der oberen Urmundlippe in älteren Urodelenkeimen. Roux Arch. *126*, 391–456.
MAGGIO, R., and CATALANO, C., 1963. Activation of amino acids during sea urchin develop-ment. Arch. Biochem. Biophys. *103*, 164–167.
MAKINO, F., and TSUZUKI, J., 1971. Absence of histone in the blue-green alga *Anabaena cylin-drica*. Nature, Lond. *231*, 446–447.
MANCHOT, E., 1929. Abgrenzung des Augenmaterials und anderer Teilbezirke der Medul-larplatte; die Teilbewegungen wärend der Auffaltung (Farbmarkierungsversuche an Keimen von Urodelen). Roux Arch. *116*, 689–708.
MANGOLD, O., 1923. Transplantationsversuche zur Frage der Spezifität und der Bildung der Keimblätter bei *Triton*. Arch. micr. Anat. u. Entw. mech. *100*, 198–301.
_____, 1928. Das Determinationsproblem. I. Das Nervensystem und die Sinnesorgane der Seitenlinie unter spezieller Berücksichtigung der Amphibien. Ergebn. Biol. *3*, 152–227.
_____, 1929. Das Determinationsproblem. II. Die paarigen Extremitäten der Wirbeltiere in der Entwicklung. Ergebn. Biol. *5*, 290–404.
_____, 1931a. Versuche zur Analyse der Entwicklung des Haftfadens bei Urodelen: Ein beispiel für die Induktion artfremder Organe. Naturwissenschaften *19*, 905–911.
_____, 1931b. Das Determinationsproblem. III. Das Wirbeltierauge in der Entwicklung und Regeneration. Ergebn. Biol. *7*, 193–403.
_____, 1932. Autonome und komplementäre Induktionen bei Amphibien. Naturwissen-schaften *20*, 371–375.
_____, 1936. Experimente zur Analyse der Zusammenarbeit der Keimblätter. Naturwissen-schaften *24*, 753–760.
_____, 1949. Totale Keimblattchimären bei *Triton*. Naturwissenschaften *36*, 112–120.
_____, and SPEMANN, H., 1927. Ueber Induktion von Medullarplatte durch Medullarplatte im jüngeren Keim, ein Beispiel homeogenetischer oder assimilatorischer Induktion. Roux Arch. *111*, 341–422.
MARKERT, C. L., 1961. Isozymes in kidney development. In METCOFF, J. (Ed.): Hereditary, developmental and immunological aspects of kidney diseases. Northwestern Univ. Press, Evanston.
_____, 1963. Lactate dehydrogenase isozymes: Dissociation and recombination of subunits. Science, N.Y. *140*, 1329–1330.

MARKS, R., 1972. Comment: The role of chalones in epidermal homeostasis. Br. J. Derm. *86*, 543–548.

MARTINI, E., 1912. Studien über die Konstanz histologischer Elemente. 3. *Hydatina senta.* Z. wiss. Zool. *102*, 425–645.

MARX, J. L., 1973. Embryology: Out of the womb—into test tube. Science, N.Y. *182*, 811–814.

MARX, L., 1935. Bedingungen für die Metamorphose des Axolotls. Ergebn. Biol. *11*, 244–334.

MAUGH, T. H., 1973. Molecular biology: A better artificial gene. Science, N.Y., 1235.

MAZIA, D., 1974. The cell cycle. Scient. Am. *230* (1), 55–64.

McCONNELL, C. H., 1936. Mitosis in *Hydra.* Mitosis in the indifferent interstitial cells of *Hydra.* Roux Arch. *135*, 202–210.

McKEEHAN, M. S., 1951. Cytological aspects of embryonic lens induction in the chick. J. exp. Zool. *117*, 31–64.

_____, 1956. The relative ribonucleic acid content of lens and retina during lens induction in the chick. Am. J. Anat. *99*, 131–156. ·

_____, 1958. Induction of portions of the chick lens without contact with the optic cup. Anat. Rec. *132*, 297–306.

McMASTER, R. D., 1955. Desoxyribose nucleic acid in cleavage and larval stages of the sea urchin. J. exp. Zool. *130*, 1–27.

MERCER, E. H., and WOLPERT, L., 1962. An electron microscope study of the cortex of the sea urchin *(Psammechinus miliaris)* egg. Expl. Cell Res. *27*, 1–13.

MERRIAM, R. W., 1959. The origin and fate of annulate lamellae in maturing sand dollar eggs. J. biophys. biochem. Cytol. *5*, 117–122.

MESTSCHERSKAIA, K. A., 1935. (cited after NEEDHAM, J., 1942).

METZ, C. B., 1957. Specific egg and sperm substances and activation of the egg. In TYLER, A., BORSTEL, R. C. VON, and METZ, C. B.: The beginnings of embryonic development Amer. Ass. Adv. Sci., Washington.

_____, and MONROY, A., 1967. Fertilization. Vol. 1. Academic Press, New York and London.

MILAIRE, J., 1956. Contribution à l'étude morphologique et cytochimique des bourgeons de membres chez le rat. Archs. Biol., Paris *67*, 297–391.

MILLARD, N., 1945. The development of the arterial system of *Xenopus laevis,* including experiments on the destruction of the larval aortic arches. Trans. R. Soc. S. Afr. *30*, 217–234.

MILLER, J. A., 1937. Some effect of oxygen on polarity in *Tubularia crocea.* Biol. Bull. mar. biol. Lab., Woods Hole *73*, 369.

_____, 1959. Nutritive substances and reconstruction in *Tubularia.* Proc. Soc. exp. Biol. Med. *100*, 186–189.

MILLER, O. L., 1973. The visualization of genes in action. Scient. Am. *228* (3), 34–42.

MILLER, R. L., 1966. Chemotaxis during fertilization in the hydroid *Campanularia.* J. exp. Zool. *162*, 23–44.

MINOT, C. S., 1891. Senescence and rejuvenation. J. Physiol. *12*, 97–153.

_____, 1908. The problem of age, growth and death. London.

MINTZ, B., 1960. Embryological phases of mammalian gametogenesis. J. cell. comp. Physiol. *56*, Suppl. 1, 31–47.

_____, 1962. Formation of genotypically mosaic mouse embryos. Amer. Zool. *2*, 432.

_____, 1964. Formation of genetically mosaic mouse embryos, and early development of 'lethal (t^{12}/t^{12}) normal' mosaics. J. exp. Zool. *157*, 273–291.

_____, 1967. Gene control of mammalian pigmentary differentiation. I. Clonal origin of melanocytes. Proc. natn. Acad. Sci., U.S.A. *58*, 344–351.

_____, 1968. Hermaphroditism, sex chromosomal mosaicism and germ cell selection in allophenic mice. J. Animal Science *27* (Suppl. I), 51–60.

_____, and SILVERS, W. K., 1967. "Intrinsic" immunological tolerance in allophenic mice. Science *158*, 1484–1487.

MIRSKY, A. E., and RIS, H., 1949. Variable and constant components of chromosomes. Nature, Lond. *163*, 666–667.

MITCHISON, J. M., 1956. The thickness of the cortex of the sea urchin egg and the problem of the vitelline membrane. Q. Jl. microsc. Sci. *97*, 109–121.

MIZELL, M., 1968. Limb regeneration: Induction in the newborn opossum. Science, N.Y. *161*, 283–286.

MOAR, V. A., GURDON, J. B., LANE, C. D., and MARBAIX, G., 1971. Translational capacity of living frog eggs and oocytes, as judged by messenger RNA injection. J. molec. Biol. *61*, 93–104.

MOK, C. C., MARTIN, W. J., and COMMON, R. H., 1961. A comparison of phosvitins prepared from hen's serum and from hen's egg yolk. Can. J. Biochem. Physiol. *39*, 109–117.

MONROY, A., 1939. Sulla localizzazione delle celluli genitali primordiali in fasi precoci di sviluppo. Richerche sperimentali in Anfibi Anuri. Archo. ital. Anat. Embriol. *41*, 368–389.

MONROY, A., MAGGIO, R., and RINALDI, A. M., 1965. Experimentally induced activation of the ribosomes of the unfertilized sea urchin egg. Proc. natn. Acad. Sci., U.S.A. *54*, 107–111.

———, and TYLER, A., 1963. Formation of active ribosomal aggregates (polysomes) upon fertilization and development of sea urchin eggs. Archs. Biochem. Biophys. *103*, 431–435.

———, and ———, 1967. The activation of the egg. In METZ, C. B., and MONROY, A. (Eds.): Fertilization. Academic Press, New York.

———, and VITORELLI, M. L., 1962. Utilization of C¹⁴-glucose for amino acids and protein synthesis by the sea urchin embryo. J. cell. comp. Physiol. *60*, 285–287.

MOOG, F., 1946. The physiological significance of the phosphomono-esterases. Biol. Rev. *21*, 41–59.

MOORE, A. B. C., 1950. The development of reciprocal androgenetic frog hybrids. Biol. Bull. mar. biol. Lab., Woods Hole *99*, 88–111.

MOORE, A. R., 1933. Is cleavage rate a function of the cytoplasm or of the nucleus? J. exp. Biol. *10*, 230–236.

MOORE, J. A., 1946. Studies in the development of frog hybrids. I. Embryonic development in the cross *Rana pipiens* ♀ × *Rana sylvatica* ♂ J. exp. Zool *101*, 173–213.

MORGAN, T. H., 1901. Regeneration. Macmillan Co., New York.

———, 1919. The physical basis of heredity. J. B. Lippincott Co., Philadelphia.

———, 1927. Experimental embryology. Columbia University Press, New York.

———, 1933. The formation of the antipolar lobe in *Ilyanassa*. J. exp. Zool. *64*, 433–467.

MOSCONA, A., 1952. Cell suspensions from organ rudiments of chick embryos. Expl. Cell Res. *3*, 535.

———, 1956. Development of heterotypic combinations of dissociated embryonic chick cells. Proc. Soc. exp. Biol. Med. *92*, 410–416.

MOSES, M. J., 1961. Spermiogenesis in the crayfish *(Procambarus clarkii)*. I. Structural characterization of the mature sperm. J. biophys. biochem. Cytol. *9*, 222–228.

MOSSMAN, H. W., 1937. Comparative morphogenesis of the foetal membranes and accessory uterine structures. Contr. Embryol. Carneg. Instn. *26*, 133–246.

MUCHMORE, W. B., 1951. Differentiation of the trunk mesoderm in *Amblystoma maculatum*. J. exp. Zool. *118*, 137–185.

MÜLLER, F., 1864. Für Darwin. Leipzig.

MUNRO, A. F., 1939. Nitrogen excretion and arginase activity during amphibian development. Biochem. J. *33*, 1957–1965.

NAGANO, T., 1962. Observations on the fine structure of the developing spermatid in the domestic chicken. J. Cell Biol. *14*, 193–205.

NAKAMURA, O., and TAHARA, Y., 1953. Formation of the stomach in Anura. Mem. Osaka Univ. Lib. Arts & Ed. *2*, 1–8.

NASSONOV, N. V., 1930. Die Regeneration der Axolotl Extremitäten nach Ligaturanlegung. Roux Arch. *121*, 639–657.

———, 1936. Influence of various factors on morphogenesis following homotopical subcutaneous insertions of cartilage in the axolotl. C.R. Acad. Sci. URSS *4*, 97–100.

NAWAR, G., 1956. Experimental analysis of the origin of the autonomic ganglia in the chick embryo. Am. J. Anat. *99*, 473–506.

NEEDHAM, A. E., 1941. Some experimental biological uses of the element beryllium (glucinum). Proc. zool. Soc. Lond. A. *111*, 59–85.

———, 1952. Regeneration and wound-healing. Methuen, London.

NEEDHAM, J., 1931. Chemical embryology. University Press, Cambridge.

———, 1942. Biochemistry and morphogenesis, University Press, Cambridge.

———, 1959. A history of embryology. 2nd ed. University Press, Cambridge.

NELSEN, O. E., 1953. Comparative embryology of the vertebrates. Blakiston, New York.

NICHOLAS, J. S., 1926. Extirpation experiments upon the embryonic forelimb of the rat. Proc. Soc. exp. Biol. Med. *23*, 436–439.

———, 1950. Development of contractility. Proc. Am. phil. Soc. *94*, 175–183.

———, and RUDNICK, D., 1933. The development of embryonic rat tissues upon the chick chorioallantois. J. exp. Zool. *66*, 193–261.

NICOLET, G., 1970a. Analyse autoradiographique de la localisation des differentes ébauches presomptives dans la ligne primitive de l'embryon de poulet. J. Embryol. exp. Morph. *23*, 79–108.

———1970b. Determination et controle de la differentiation des somites. Médicine et Hygiene, No. 932, 1–8.

NIEUWKOOP, P. D., 1946. Experimental investigations on the origin and determination of the germ cells, and on the development of the lateral plates and germ ridges in urodeles. Archs. néerl. Zool. *8*, 1–205.

Nieuwkoop, P. D. and Faber, J., 1956. Normal table of *Xenopus laevis* (Daudin). North Holland Publishing Co., Amsterdam.

———, et al., 1952. Activation and organization of the central nervous system in amphibians. J. exp. Zool. *120*, 1–108.

Niu, M. C., 1947. The axial organization of the neural crest, studied with particular reference to the pigmentary component. J. exp. Zool. *105*, 79–114.

———, 1956. New approaches to the problem of embryonic induction. In Rudnick, D. (Ed.): Cellular mechanisms in differentiation and growth. University Press, Princeton, 155–171.

———, and Twitty, V. C., 1953. The differentiation of gastrula ectoderm in medium conditioned by axial mesoderm. Proc. natn. Acad. Sci., U.S.A. *39*, 985–989.

Nordenskiöld, E., 1929. The history of biology. Kegan Paul, Trench and Trubner, London.

Noronha, J. M., Sheys, G. H., and Buchanan, J. M., 1972, Induction of a reductive pathway for deoxyribonucleotide synthesis during early embryogenesis of the sea urchin. Proc. natn. Acad. Sci. U.S.A, *69*, 2006–2010.

O'Connor, R. J., 1939. Experiments on the development of the amphibian mesonephros. J. Anat. *74*, 34–44.

Odor, D. L., 1960. Electron microscopic studies on ovarian oocytes and unfertilized tubal ova in the rat. J. biophys. biochem. Cytol. *7*, 567–577.

Oka, H., 1943. Metamorphosis of *Polycitor mutabilis* (Ascidiae compositae). Annot. zool. jap. *22*, 54–58.

———, 1958. Eksperimentaj studoj pri metamorfozo de ascidioj. Sciencaj Studoj Kopenhago, 217–220.

———, and Watanabe, H., 1957. Vascular budding. A new type of budding in *Botryllus*. Biol. Bull. mar. biol. Lab., Woods Hole *112*, 225–240.

Okada, E. W., 1955. Isolationsversuche zur Analyse der Knorpelbildung aus Neuralleistenzellen bei Urodelenkeim. Mem. Coll. Sci., Kyoto Univ. *22*, 23–28.

———, and Waddington, C. H., 1959. The submicroscopic structure of the *Drosophila* egg. J. Embryol. exp. Morph. *7*, 583–597.

Okada, T. S., 1955a. Experimental studies on the differentiation of the endodermal organs in amphibia. III. The relation between the differentiation of pharynx and head-mesenchyme. Mem. Coll. Sci., Kyoto Univ. *22*, 17–22.

———, 1955b. Experimental studies on the differentiation of the endodermal organs in amphibia. IV. The differentiation of the intestine from the foregut. Annot. zool. jap. *28*, 210–214.

Okuneff, N., 1928. Ueber einige physiko-chemische Erscheinungen während der Regeneration. I. Messung der Wasserstoffionenkonzentration in regenerierenden Extremitäten des Axolotl. Biochem. Z. *195*, 421–427.

———, 1933. Ueber einige physiko-chemische Erscheinungen während der Regeneration. V. Ueber den Milchsäuregehalt regenerierender Axolotlextremitäten. Biochem. Z., *257*, 242–244.

Olsen, M. W., 1960a. Nine-year summary of parthenogenesis in turkeys. Proc. Soc. exp. Biol. Med., 279–281.

———, 1960b. Performance record of a parthenogenetic turkey male. Science, N.Y. *132*, 1661.

O'Malley, B. M., and Means, A. R., 1974. Female steroid hormones and target cell nuclei. Science, N.Y. *183*, 610–620.

Oppenheimer, J. M., 1936. Transplantation experiments on developing teleosts (*Fundulus* and *Perca*). J. exp. Zool. *72*, 409–437.

———, 1947. Organization of the teleost blastoderm. Q. Rev. Biol. *22*, 105–118.

———, 1967. Essays on the history of embryology and biology. The M.I.T. Press, Cambridge, Mass.

Orechowitsch, W. N., and Bromley, N. W., 1934. Die histolysierenden Eigenschaften des Regenerationsblastems. Biol. Zbl. *54*, 524–535.

———, ———, and Kosmina, N. A., 1935. Ueber die Proteolyse in den regenerierenden Geweben. III. Die Veränderung der Aktivität der Gewebeprotease während des Regenerationsprozesses der Organe von Amphibien. Biochem. Z. *277*, 186.

Pasteels, J. J., 1937. Études sur la gastrulation des vertébrés méroblastiques. II. Reptiles. Archs. Biol. Paris *48*, 105–184.

———, 1940. Un aperçu comparatif de la gastrulation chez les chordés. Biol. Rev. *15*, 59–106.

———, 1945. On the formation of the primary entoderm of the duck (*Anas domestica*) and on the significance of the bilaminar embryo in birds. Anat. Rec. *93*, 5–21.

PASTEELS, J. J., 1957. La formation de l'endophylle et de l'endoblaste vitelline chez les reptiles, chéloniens et lacertiliens. Acta anat. *30*, 601–612.

———, 1961. La réaction corticale de fécondation ou d'activation. Bull. Soc. zool. Fr. *86*, 600–629.

———, 1962. Gastrulation du *Protopterus dolloi* Blgr. Annales du Musée Royal de l'Afrique Centrale, Tervuren (Belgique) Ser. 8, Sciences Zoologique No. 108, 175–183.

———, and DE HARVEN, E., 1962. Étude au microscope électronique du cortex de l'oeuf de *Barnea candida* (mollusque bivalve), et son évolution au moment de la fécondation, de la maturation, et de la segmentation. Archs. Biol. Paris *73*, 465–490.

PATTEN, B. M., 1944. The embryology of the pig. 2nd ed. Blakiston, Philadelphia.

———, 1957. Early embryology of the chick. 4th ed. McGraw-Hill, New York.

———, 1958. Foundations of embryology. McGraw-Hill, New York.

PATTERSON, J. T., 1910. Studies on the early development of the hen's egg. I. History of the early cleavage and the accessory cleavage. J. Morph. *21*, 101–134.

PAUL, J. and HUNTER, J. A., 1968. DNA synthesis is essential for increased haemoglobin synthesis in response to erythropoietin. Nature, Lond. *219*, 1362–1363.

———, and ———, 1969. Synthesis of macromolecules during induction of haemoglobin synthesis by erythropoietin. J. molec. Biol. *42*, 31–41.

PAULING, L., ITANO, H. A., SINGER, S. J., and WELLS, I. C., 1949. Sickle cell anemia, a molecular disease. Science, N.Y. *110*, 543–548.

PELTRERA, A., 1940. La capacita regolative dell' uovo di *Aplysia limacina* L., studiate con la centrifugazione e con le reazioni vitali. Pubbl. Staz. zool. Napoli *18*, 20.

PERLMAN, P., and GUSTAFSON, T., 1948. Antigens in the egg and early development stages of the sea urchin. Experientia *4*, 481–483.

PERRI, T., 1951. Richerche sperimentali sull' induzione di arti sopranumerari negli Anfibi Anuri (*Bufo vulgaris*). Atti dell' Acad. Naz. dei Lincei. *3*, 41–113.

PERRY, M. M., and WADDINGTON, C. H., 1966. Ultrastructure of the blastoporal cells in the newt. J. Embryol. exp. Morph. *15*, 317–330.

PERUTZ, M. F., 1961. Proteins and nucleic acids. Structure and function. Elsevier Publ. Co., Amsterdam, London, New York.

PIATT, J., 1948. Form and causality in neurogenesis. Biol. Rev. *23*, 1–45.

———, 1951. Transplantation experiments between pigmentless and pigmented eggs of *Ambystoma punctatum*. J. exp. Zool. *118*, 101–135.

PIEPHO, H., and MEYER, H., 1951. Reactionen der Schmetterlingshaut auf Häutungshormone. Biol. Zbl. *70*, 252–260.

PINCUS, G., 1936. The eggs of mammals. Macmillan, New York.

———, 1939. The comparative behavior of mammalian eggs *in vivo* and *in vitro*. IV. The development of fertilized and artificially activated rabbit eggs. J. exp. Zool. *82*, 85–129.

———, and SHAPIRO, H., 1940. The comparative behavior of mammalian eggs *in vivo* and *in vitro*. VII. Further studies on the activation of rabbit eggs. Proc. Am. phil. Soc. *83*, 631–647.

PITT-RIVERS, R., and TATA, J. R., 1959. The thyroid hormones. Pergamon Press, London, New York, Paris.

POLEZHAYEV, L. W., 1946. The loss and restoration of regenerative capacity in the limbs of tailless amphibia. Biol. Rev. *21*, 141–147.

POLLISTER, A. W., and MOORE, J. A., 1937. Tables for the normal development of *Rana sylvatica*. Anat. Rec. *68*, 486–496.

PRENTISS, C. W., and AREY, L. B., 1917. A laboratory manual and text-book of embryology. 2nd ed. Saunders, Philadelphia.

PRESCOTT, D. M., 1957. Relations between cell growth and cell division. In RUDNICK, D. (Ed.): Rhythmic and synthetic processes in growth. University Press, Princeton, 59–74.

———, and BENDER, M. A., 1962. Synthesis of RNA and protein during mitosis in mammalian tissue culture cells. Expl. Cell Res. *26*, 260–268.

———, and ———, 1963. Autoradiographic study of chromatid distribution of labeled DNA in two types of mammalian cells *in vitro*. Expl. Cell Res. *29*, 430–442.

———, and KIMBALL, R. F., 1961. Relation between RNA, DNA and protein synthesis in the replicating nucleus of *Euplotes*. Proc. natn. Acad. Sci., U.S.A. *47*, 686–693.

PRESS, N., 1959. An electron microscopic study of a mechanism for the delivery of follicular cytoplasm to an avian egg. Expl. Cell Res. *18*, 194–196.

PTASHNE, M., and GILBERT, W., 1970. Genetic repressors, Scient. Am. *222* (6), 36–44.

PULLINGER, B. D., 1949. Squamous differentiation in mouse mammae: Spontaneous and induced. Br. J. Cancer *3*, 494–501.

RAFF, R. A., COLST, H. V., SELVIG, S. E., and GROSS, P. R., 1972. Oogenetic origin of messenger RNA for embryonic synthesis of microtubule protein. Nature, Lond. *235*, 211–214.

RANZI, S., 1957. Early determination of development under normal and experimental conditions. In TYLER, A., BORSTEL, R. C. VON, and METZ, C. B.: The beginnings of embryonic development. Amer. Ass. Adv. Sci. Washington, 291–318.

———, 1968. Considerations on transcription and translation in embryonic development. Accademia Nazionale dei Lincei, Quaderno *104*, 227–232.

———, and TAMINI, E., 1939. Die Wirkung von NaSCN auf die Entwicklung von Froschembryonen. Naturwissenschaften *27*, 566–567.

RAVEN, C. P., 1931. Zur Entwicklung der Ganglienleiste. I. Die Kinematik der Ganglienleistenentwicklung bei den Urodelen. Roux Arch. *125*, 210–292.

———, 1936. Zur Entwicklung der Ganglienleiste. V. Ueber die Differenzierung des Rumpfganglienleistenmaterials. Roux Arch. *134*, 122–146.

———, 1958. Morphogenesis: The analysis of molluscan development. Pergamon Press, London, New York, Paris.

———, 1961. Oogenesis: The storage of developmental information. Pergamon Press, London, New York, Paris.

RAWLES, M. E., 1947. Origin of pigment cells from the neural crest in the mouse embryo. Physiol. Zool. *20*, 248–265.

———, 1948.. Origin of melanophores and their role in development of the color patterns in vertebrates. Physiol. Rev. *28*, 383–408.

REBHUN, L. I., 1961. Some electron microscope observations on membranous basophilic elements of invertebrate eggs. J. Ultrastruct. Res. *5*, 208–225.

———, 1962. Electron microscope studies on the vitelline membrane of the surf clam *Spisula solidissima.* J. Ultrastruct. Res. *6*, 107–122.

RECOURT, A., 1961. Elektronenmicroscopisch onderzoek naar de oogenese bij *Limnaea stagnalis* L. Thesis, Utrecht.

REVERBERI, G., and MANCUSO, V. 1961. The constituents of the egg of *Mytilus* as seen at the electron microscope. Acta Embryol. Morph. exper. *4*, 102–121.

———, ORTOLANI, G., and FARINELLA-FERRUZZA, N., 1960. The causal formation of the brain in the ascidian larva. Acta. Embryol. Morph. exper. *3*, 296–336.

REYER, R. W., 1962. Regeneration in the amphibian eye. Symp. Soc. Devel. Growth *20*, 211–265.

ROBERTSON, T. B., 1908. On the normal rate of growth of an individual and its biochemical significance. Roux Arch. *25*, 571–614.

ROCHE, V. DE, 1937. Differenzierungen von Geweben und ganzen Organen in Transplantaten der bastardmerogonischen Kombination *Triton alpestris* (♀) × *Triton palmatus* ♂. Roux Arch. *135*, 620–663.

ROLLINS, J. W., and FLICKINGER, R. A., 1972. Collagen synthesis in *Xenopus* oocytes after injection of nuclear RNA of frog embryos. Science, N.Y. *178*, 1204–1205.

ROMANOFF, A. L., 1960. The avian embryo, structural and functional development. Macmillan Co., New York.

———, and ROMANOFF, A. J., 1949. The avian egg. John Wiley & Sons, New York.

ROSE, S. M., 1942. A method of inducing limb regeneration in adult Anura. Proc. Soc. exp. Biol. Med., N.Y. *49*, 408–410.

———, 1970. Regeneration, a key to understanding normal and abnormal growth and development. Appleton-Century-Crofts, New York.

ROSENQUIST, G. C., 1966. A radioautographic study of labeled grafts in the chick blastoderm. Development from primitive-streak stages to stage 12. Contr. Embryol. Carneg. Instn. *38*, 71–110.

ROTH, J. S., 1964. Biological information in a single strand of deoxyribonucleic acid. Nature, Lond. *202*, 182–183.

ROTH, P. C. J., 1955. Les métamorphoses des batraciens. Dunod, Paris.

ROTHSCHILD, L., 1951. Sea urchin spermatozoa. Biol. Rev. *26*, 1–27.

———, 1956. Fertilization. Methuen, London.

ROTMANN, E., 1935. Der Anteil von Induktor und reagierenden Gewebe an der Entwicklung des Haftfadens. Roux Arch. *133*, 193–224.

ROUNDS, D. E., and FLICKINGER, R. A., 1958. Distribution of ribonucleoprotein during neural induction in the frog embryo. J. exp. Zool. *137*, 479–500.

ROUX, W., 1885. Ueber die Bestimmung der Hauptrichtungen des Froschembryo im Ei und über die erste Theilung des Froscheies. Breslauer ärztl. Zeitschr., 1–54.

———, 1888. Beiträge zur Entwicklungsmechanik des Embryo. 5. Ueber die küntsliche Hervorbringung halber Embryonen durch Zerstörung einer der beiden ersten Furchungskugeln, sowie über die Nachentwicklung (Post-generation) der fehlenden Köperhälfte. Virchows Arch. path. Anat. Physiol. *64*, 113–154, 246–291.

———, 1905. Die Entwicklungsmechanick, ein neuer Zweig der biologischen Wissenschaft, Leipzig.

RUDDLE, F. H., and KUCHERLAPATI, R. S., 1974. Hybrid cells and human genes. Scient. Am. *231*(1), 36–44.

RUDNICK, D., 1948. Prospective areas and differentiation potencies in the chick blastoderm. Ann. N.Y. Acad. Sci. *49*, 761–772.

RUDNICK, D., 1952. Development of the digestive tube and its derivatives. Ann. N.Y. Acad. Sci. *55*, 109–116.

——, and RAWLES, M. E., 1937. Differentiation of the gut in chorio-allantoic grafts from chick blastoderms. Physiol. Zoöl. *10*, 381–395.

RUGH, R., 1948. Experimental embryology. Burgess, Minneapolis.

RUNNSTROM, J., 1928. Plasmabau und Determination bei dem Ei von *Paracentrotus lividus* Lk. Roux Arch. *113*, 556–581.

——, 1952. The cell surface in relation to fertilization. Symp. Soc. exp. Biol. *6*, 39–88.

——, HAGSTRÖM, B. E., and PERLMAN, P., 1959. Fertilization. In BRACHET, J., and MIRSKY, A. E.: The cell, Vol. 1. Academic Press, London and New York, 327–397.

RUTH, R. F., ALLEN, C. P., and WOLFE, H. R., 1964. The effect of thymus on lymphoid tissue. In GOOD R. A., and GABRIELSEN, A. E. (Eds.): The thymus in immunobiology. Harper & Row, New York.

RUUD, G., 1929. Heteronom-orthotopische Transplantationen von Extremitätenanlagen bei Axolotlembryonen. Roux Arch. *118*, 308–351.

SADOV, I. A., 1956. Micropyle formation in oocytes of Acipenseridae. C.R. Acad. Sci. URSS *111*, 1400–1402.

SANTOS, F., 1931. Studies on transplantation in Planaria. Physiol. Zoöl. *4*, 111–164.

SARVELLA, P., 1973. Adult parthenogenetic chickens. Nature, Lond. *243*, 171.

SAUER, M. E., and WALKER, B. E., 1959. Radioautographic study of interkinetic nuclear migration in the neural tube. Proc. Soc. exp. Biol. Med. *101*, 557–560.

SAUNDERS, J. W., 1948. The proximo-distal sequence of origin of the parts of the chick wing and the role of the ectoderm. J. exp. Zool. *108*, 363–403.

——, 1966. Death in embryonic systems. Science, N.Y. *154*, 604–612.

——, and FALLON, J. F., 1967. Cell death in morphogenesis. In Locke, M. (Ed.): Major problems in developmental biology. Academic Press, New York, 289–314.

SAXÉN, L., 1961. Transfilter neural induction of amphibian ectoderm. Devl. Biol. *3*, 140–152.

——, and TOIVONEN, S., 1962. Primary embryonic induction. Logos Press, London.

SCHALLER, H., and GIERER, A., 1973. Distribution of the head-activating substance in hydra and its localization in membranous particles in nerve cells. J. Embryol. exp. Morph. *29*, 39–52.

SCHEREMETJEWA, E. A., and BRUNST, V. V., 1938. Preservation of the regeneration capacity in the middle part of the limb of newt and its simultaneous loss in the distal and proximal parts of the same limb. Bull. Biol. Méd. exp. URSS *6*, 723–724.

SCHMALHAUSEN, I. I., 1927. Beiträge zur quantitativen Analyse der Formbildung. I. Ueber Gesetzmässigkeiten des embryonalen Wachstums. Roux Arch. *109*, 455–512.

——, 1930a. Ueber Wachstumsformeln und Wachstumstheorien. Biol. Zbl. *50*, 292–307.

——, 1930b. Das Wachstumsgesetz als Gesetz der progressiven Differenzierung. Roux Arch. *123*, 153–178.

——, and BORDZILOWSKAJA, N., 1930. Das Wachstum niederer Organismen. I. Das Individuelle Wachstum der Bakterien und Hefe. Roux Arch. *121*, 726.

——, and SYNGAJEWSKAJA, E., 1925. Studien über Wachstum und Differenzierung. I. Die individuelle Wachstumskurve von *Paramaecium caudatum.* Roux Arch. *105*, 711–717.

SCHMALHAUSEN, O. I., 1939. The role of the olfactory sac in the development of the cartilaginous capsule of the olfactory organ in urodeles. C.R. Acad. Sci. URSS *23*, 395–397.

——, 1950. A comparative experimental investigation of the early stages of development of the olfactory rudiments in amphibians. C.R. Acad. Sci. URSS *74*, 863–865.

SCHMIDT, G. A., 1933. Schnürungs- und Durchschneidungsversuche am Amphibienkeim. Roux Arch. *129*, 1–44.

SCHNEIDERMAN, H. A., and GILBERT, L. I., 1964. Control of growth and development in insects. Science, N.Y. *143*, 325–333.

SCHOTTÉ, O. E., 1923. Influence de la section tardive des nerfs sur les pattes de Tritons en régénération. C.R. Soc. Phys. Hist. nat. Genéve *40*, 86–88.

——, 1926. Systéme nerveux et régénération chez le *Triton.* Revue suisse Zool. *33*, 1–211.

——, 1930. Der Determinationszustand der Anurengastrula im Transplantationsexperiment. Roux Arch. *122*, 633–664.

——, and BUTLER, E. G., 1941. Morphological effects of denervation and amputation of limbs in urodele larvae. J. exp. Zool. *87*, 279–322.

——, and HARLAND, M., 1943. Effects of denervation and amputation of hindlimbs in Anuran tadpoles. J. exp. Zool. *93*, 453–493.

SCHROEDER, T. E., 1972. The contractile ring. II. Determining its brief existence, volumetric changes, and vital role in cleaving *Arbacia* eggs. J. Cell Biol. *53*, 419–434.

SCHULZE, P., 1918. Die Bedeutung der interstitiellen Zellen, etc. Sitz. Ber. Ges. Nat. Freunde, Berlin.

SCHWIDEFSKY, G., 1934. Entwicklung und determination der Extremitätanregenerate bei den Molchen. Roux Arch. *132*, 57–114.

SCHWIND, J., 1933. Tissue specificity at the time of metamorphosis in frog larvae. J. exp. Zool. *66*, 1–14.

SEIDEL, F., 1932. Die Potenzen der Furchungskerne im Libellenei und ihre Rolle bei der Aktivierund des Bildungszentrums. Roux Arch. *126*, 213–276.

———, 1952. Die Entwicklungspotenzen einer isolierten Blastomere des Zweizellenstadiums im Säugetierei. Naturwissenschaften *39*, 355–356.

———, 1960. Die Entwicklungsfähigkeiten isolierter Furchungszellen aus dem Ei des Kaninchens *Oryctolagus cuniculus*. Roux Arch. *152*, 43–130.

SEVERINGHAUS, A. E., 1930. Gill development in *Amblystoma punctatum*. J. exp. Zool. *56*, 1–30.

SEWERTZOFF, A. N., 1931. Morphologische Gesetzmässigkeiten der Evolution. G. Fischer, Jena.

SHAVER, J. R., 1953. Studies on the initiation of cleavage in the frog egg. J. exp. Zool. *122*, 169–192.

SHEARER, R. W., and McCARTHY, B. J., 1967. Evidence for ribonucleic acid molecules restricted to the cell nucleus. Biochemistry *6*, 283–289.

SHUMWAY, W., 1940. Stages in the normal development of *Rana pipiens*. I. External form. Anat. Rec. *78*, 139–148.

SINGER, C., 1931. A short history of biology. Clarendon Press, Oxford.

SINGER, M., 1950. Induction of regeneration of the limb of the adult frog by augmentation of the nerve supply. Anat. Rec. *108*, 518–519.

SIRLIN, J. L., and BRAHMA, S. K., 1959. Studies on embryonic induction using radioactive tracers. II. The mobilization of protein components during induction of the lens. Devl. Biol. *1*, 234–246.

———, ———, and WADDINGTON, C. H., 1956. Studies on embryonic induction using radioactive tracers. J. Embryol. exp. Morph. *4*, 248–253.

SMITH, A. E. S., 1964. The localization of acid phosphatase in the eggs of several species of invertebrates. Biol. Bull. mar. biol. Lab., Woods Hole *127*, 389–390.

SMITH, P. E., and MacDOWELL, E. C., 1930. An hereditary anterior-pituitary deficiency in the mouse. Anat. Rec. *46*, 249–257.

SNELL, J. D. (Ed.), 1941. Biology of the laboratory mouse. Dover, New York.

SOBELL, H. M., 1974. How actinomycin binds to DNA. Scient. Am. *231*(2), 82–91.

SOLL, D., OHTSUKA, E., JONES, D. S., LOHRMANN, H., NAYATSU, H., NISHIMURA, S., and KHORANA, H. G., 1965. Studies on polynucleotides. XLIX. Stimulation of the binding of aminoacyl-sRNA's to ribosomes by ribotrinucleotides and a survey of codon assignments for 20 amino acids. Proc. natn. Acad. Sci., U.S.A. *54*, 1378–1385.

SOTELO, J. R., and PORTER K. R., 1959. An electron microscope study of the rat ovum. J. biophys. biochem. Cytol. *5*, 327–342.

———, and TRUJILLO-CENOZ, O., 1957. Electron microscope study of the vitelline body of some spider oocytes. J. biophys. biochem. Cytol. *3*, 301–310.

SPEIR, R. S., 1964. How cells attack antigens. Scient. Am. *210*, 58–64.

SPEMANN, H., 1901. Ueber Korrelationen in der Entwicklung des Auges. Verh. anat. Ges. Jena Verslg. Bonn *15*, 61–79.

———, 1901/1903. Entwicklungsphysiologische Studien am Tritonei. I, II, III. Roux Arch. *12*, 224–264; *15*, 448–534; *16*, 551–631.

———, 1912a. Zur Entwicklung des Wirbeltierauges. Zool. Jb., Abt. allg. Zool. *32*, 1–98.

———, 1912b. Ueber die Entwicklung umgedrehter Hirnteile bei Amphibienembryonen. Zool. Jb. J. Suppl. *15*, 1–48.

———, 1919. Experimentelle Forschungen zum Determinations- und Individualitätsproblem. Naturwissenschaften *32*, 1–33.

———, 1921. Die Erzeugung tierischer Chimären durch heteroplastische embryonale Transplantation zwischen *Triton cristatus* und *taeniatus*. Roux Arch. *48*, 533–570.

———, 1928. Die Entwicklung seitlicher und dorso-ventraler Keimhälften bei verzögerter Kernversorgung. Z. wiss. Zool. *132*, 105–134.

———, 1931. Ueber den Anteil von Implantat und Wirtskeim an der Orientierung und Beschaffenheit der induzierten Embryonalanlage. Roux Arch. *123*, 389–517.

———, 1936. Experimentelle Beiträge zu einer Theorie der Entwicklung. Springer, Berlin.

———, 1938. Embryonic development and induction. Yale University Press, New Haven.

———, and MANGOLD, H., 1924. Ueber Induktion von Embryonalanlagen durch Implantation artfremder Organisatoren. Arch. mikrosk. Anat. Entwmech. *100*, 599–638.

———, and SCHOTTÉ, O., 1932. Ueber xenoplastische Transplantation als Mittel zur Analyse der embryonalen Induktion. Naturwissenschaften *20*, 463–467.

SPIEGELMAN, S., 1948. Differentiation as the controlled production of unique enzymatic patterns. Symp. Soc. exp. Biol. *2*, 286–325.

SPOONER, B. S., YAMADA, K. M., and WESSELLS, N. K., 1971. Microfilaments and cell locomotion. J. Cell Biol. *49*, 595–613.

Spratt, N. T., Jr., 1946. Formation of the primitive streak in the explanted chick blastoderm marked with carbon particles. J. exp. Zool. *103*, 259–304.

———, 1947. Regression and shortening of the primitive streak in the explanted chick blastoderm. J. exp. Zool. *104*, 69–100.

———, and Haas, H., 1960. Morphogenetic movements in the lower surface of the unincubated and early chick blastoderm. J. exp. Zool. *144*, 139–158.

Stebbins, R. C., and Eakin, R. M., 1958. The role of the "third eye" in reptilian behavior. American Museum Novitates No. 1870, 1–40.

Stedman, E., and Stedman, E., 1950. Cell specificity of histones. Nature, Lond. *166*, 780–781.

Steinberg, M. S., 1963. Reconstruction of tissues by dissociated cells. Science, N.Y. *141*, 401–408.

———, 1964. The problem of adhesive selectivity in cellular interactions. In Locke, M. (Ed.): Cellular membranes in development. Academic Press, New York, 321–366.

———, 1970. Does differential adhesion govern self-assembly processes in histogenesis? Equilibrium configurations and the emergence of a hierarchy among populations of embryonic cells. J. exp. Zool. *173*, 395–434.

———, and Wiseman, L. L., 1972. Do morphogenetic tissue rearrangements require active cell movements? J. Cell Biol. *55*, 606–615.

Stéphan, F., 1949. Les suppléances obtenues expérimentalement dans le systéme des arcs aortiques de l'embryon d'oiseau. C.R. Ass. Anat. *36*, 647.

Stevens, L. C., 1954. The origin and development of chromatophores of *Xenopus laevis* and other anurans. J. exp. Zool. *125*, 221–246.

Stöhr, P., Jr., 1924. Experimentelle Studien an embryonalen Amphibienherzen. I. Ueber Explantation embryonaler Amphibienherzen. Arch. mikrosk. Anat. u. Entwmech. *102*, 426–451.

Stone, L. S., 1922. Experiments on the development of the cranial ganglia and the lateral line sense organs in *Amblystoma punctatum*. J. exp. Zool. *35*, 421–496.

———, 1926. Further experiments on the extirpation and transplantation of mesectoderm in *Amblystoma punctatum*. J. exp. Zool. *44*, 95–131.

Streeter, G. L., 1906. On the development of the membranous labyrinth and the acoustic and facial nerves in the human embryos. Amer. J. Anat. *6*, 139–165.

———, 1942, 1945, 1948, 1949, 1951. Developmental horizons in human embryos. Contr. Embryol. Carneg. Instn. *30*, 211–245; *31*, 29–64; *32*, 133–203; *33*, 149–167; *34*, 165–196.

Ströer, W. F. H., 1933. Experimentelle Untersuchungen über die Mundentwicklung bei den Urodelen. Roux Arch. *130*, 131–186.

Suzuki, A., 1968a. Studies on primary induction of *Triturus* embryo. I. Changes of inducing potency and metabolism of dorsal mesoderm of early *Triturus* gastrula. Kumamoto J. Sc., Ser. B. Sec. 2, *9*, 1–8.

———, 1968b. Studies on primary induction of *Triturus* embryo. II. The definite transmission of inductive information and the transference of the labeled organizer material into the reacting ectoderm. Kumamoto J. Sc., Ser. B, Sec. 2, *9*, 9–16.

Suzuki, Y., and Brown, D. D., 1972. Isolation and identification of the messenger RNA for silk fibroin from *Bombyx mori*. J. molec. Biol. *63*, 409–429.

———, Gage, L. P., and Brown, D. D., 1972. The genes for silk fibroin in *Bombyx mori*. J. molec. Biol. *70*, 637–649.

Swanson Beck, J., and Walker, P. J., 1964. Antigenicity of Trypanosome nuclei: Evidence that DNA is not coupled to histone in the Protozoa. Nature, Lond. *204*, 194–195.

Swett, F. H., 1926. On the production of double limbs in amphibians. J. exp. Zool. *44*, 419–473.

———, 1927. Differentiation of the amphibian limb. J. exp. Zool. *47*, 385–432.

———, 1937. Determination of limb-axes. Q. Rev. Biol. *12*, 322–339.

Swift, C. H., 1914. Origin and early history of the primordial germ-cells in the chick. Am. J. Anat. *15*, 483–516.

———, 1916. Origin of the sex-cords and definitive spermatogonia in the male chick. Am. J. Anat. *20*, 375–410.

Swift, H. H., 1950. The desoxyribose nucleic acid content of animal nuclei. Physiol. Zoöl. *23*, 169–198.

Syngajewskaja, E., 1935. The individual growth of Protozoa: *Blepharisma lateritia* and *Actinophrys* sp. Trav. de l'Inst. Zool. Biol. Acad. Sci. Ukr. *8*, 151–157.

Syngajewskaja, K., 1936. Die Wachstumsgeschwindigkeit bei der Regeneration der Extremitäten bei *Siredon pisciformis*. Zool. Jb. Abt. allg. Zool. *56*, 487–500.

———, 1937. Die reaktiven Eigenschaften des axialen Mesoderms. Trav. de l'Inst. Zool. Biol. Acad. Sci. Ukr. *17*, 41–59.

Sze, L. C., 1953. Changes in the amount of desoxyribonucleic acid in the development of *Rana pipiens*. J. exp. Zool. *122*, 577–601.

TAHARA, Y., 1962. Formation of the independent lens in Japanese amphibians. Embryologia 7, 127–149.

———, and NAKAMURA, O., 1961. Topography of the presumptive rudiments in the endoderm of the anuran neurula. J. Embryol. exp. Morph. 9, 138–158.

TARDENT, P. E., 1960. Principles governing the process of regeneration in Hydroids. Symp. Soc. Devel. Growth 18, 21–43.

———, 1964. Der Sauerstoff-Verbrauch normaler und regenerierender hydrocauli von Tubularia. Rev. Suisse Zool. 71, 167–181.

———, and TARDENT, T., 1956. Wiederholte Regeneration bei Tubularia. Pubbl. Staz. Zool. Napoli 28, 367–396.

TARKOWSKI, A. K., 1959. Experiments on the development of isolated blastomeres of mouse eggs. Nature, Lond. 184, 1286–1287.

———, 1961. Mouse chimaeras developed from fused eggs. Nature, Lond. 190, 857–860.

———, and WROBLEWSKA, J., 1967. Development of blastomeres of mouse eggs isolated at the 4- and 8-celled stage. J. Embr. exp. Morph. 18, 155–180.

TAYLOR, J. H., 1960a. Asynchronous duplication of chromosomes in cultured cells of Chinese hamster. J. biophys. biochem. Cytol. 7, 455–464.

———, 1960b. Nucleic acid synthesis in relation to the cell division cycle. Ann. N.Y. Acad. Sci. 90, 409–421.

———, WOODS, P. S., and HUGHES, W. L., 1957. The organization and duplication of chromosomes as revealed by autoradiographic studies using tritium-labeled thymidine. Proc. natn. Acad. Sci., U.S.A. 43, 122–128.

TELFER, W. H., 1954. Immunological studies of insect metamorphosis. II. The role of a sex limited blood protein in egg formation of the cecropia silkworm. J. gen. Physiol. 37, 539–558.

TELKKA, A., FAWCETT, D. W., and CHRISTENSEN, A. K., 1961. Further observations on the structure of the mammalian sperm tail. Anat. Rec. 141, 231–245.

TEN CATE, G., 1953. The intrinsic development of amphibian embryos. Dissertation. North Holland Publishing Co., Amsterdam.

———, and DOORENMAALEN, W. J. VAN, 1950. Analysis of the development of the eye-lens in chicken and frog embryos by means of the precipitin reaction. Proc. K. ned. Akad. Wet. 53, 894–909.

TENNENT, D. H., 1914. The early influence of the spermatozoon upon the characters of Echinoid larvae. Pap. Tortugas Lab. 182.

TERENTIEV, I. B., 1941. On the role played by the neural crest in the development of the dorsal fin in Urodela. C.R. Acad. Sci. URSS 31, 91–94.

TERNI, T., 1934. Studio sperimentale della capacita pinnoformativa dei cercini midollari. Archo. ital. Anat. Embriol. 33, 667–692.

THORELL, B., 1947. The relation of nucleic acids to the formation and differentiation of cellular proteins. Cold Spr. Harb. Symp. quant. Biol. 12, 247–255.

TIEDEMANN, H., 1961. Ueber die chemische Natur der organdeterminierenden Stoffe beim Organizator-Effekt Spemanns. Verh. d. Deutschen Zool. Ges., 251–258.

———, 1962. Biochemische Untersuchungen ueber die Induktionsstoffe und die Determination der ersten Organanlagen bei Amphibien. 13 Colloquium d. Ges. f. physiolog. Chemie, Mosbach, 177–204.

———, 1968. Factors determining embryonic differentiation. J. cellular Physiol., 72, Suppl. 1, 129–144.

———, BECKER, U., and TIEDEMANN, H., 1961. Ueber die primären Schritte bei der embryonalen Induktion. Embryologia 6, 204–218.

———, ———, and ———, 1963. Chromatographic separation of a hind-brain inducing substance into mesodermal- and neural-inducing subfractions. Biochem. Biophys. Acta 74, 557–560.

———, BORN, J., KOCHER-BECKER, U., and TIEDEMANN, H., 1965. Anreicherung des mesodermalen Induktionsfactors durch Electrophorese in Dextrangelen. Z. Naturf. 20b, 608–609.

TING, H. P., 1951. Diploid androgenetic and gynogenetic haploid development in anuran hybridization. J. exp. Zool. 116, 21–57.

TOIVONEN, S., 1940. Ueber die Leistungspezifität der abnormen Induktoren im Implantatversuch bei Triton. Ann. Acad. Sci. Fenn. A. 55, 1–150.

———, 1945. Zur Frage der Induktion selbständiger Linsen durch abnorme Induktoren im Implantationsversuch bei Triton. Ann. Soc. zool.-bot. Fenn. Vanamo, 11, 1–28.

———, 1949. Zur Frage der Leistungspezifität abnormer Induktoren. Experientia 5, 323.

———, 1950. Stoffliche Induktoren. Rev. suisse Zool. 57, Suppl. 1, 41–56.

———, 1953. Bone-marrow of the guinea-pig as a mesodermal inductor in implantation experiments with embryos of Triturus. J. Embryol. exp. Morph. 1, 97–104.

———, 1958. The dependence of the cellular transformation of the competent ectoderm on temporal relationships in the induction process. J. Embryol. exp. Morph. 6, 479–485.

TOIVONEN, S., 1961. An experimentally produced change in the sequence of neutralizing and mesodermalizing inductive actions. Experientia *17*, 87.

———, and SAXÉN, L., 1955. Ueber die Induktion des Neuralrohrs bei Trituruskeimen als simultane Leistung des Leber- und Knochenmarkgewebes vom Meerschweinchen. Ann. Acad. Sci. Fenn. A. *30*, 1–29.

———, VAINIO, T., and SAXEN, L., 1964. The effect of actinomycin D on primary embryonic induction. Rev. Suisse Zool. *71*, 139–145.

TOKIN, B., and GORBUNOWA, G., 1934. Untersuchungen über die Ontogenie der Zellen. Biol. Zh. *3*, 294–306.

TOWNES, P. L., and HOLTFRETER, J., 1955. Directed movements and selective adhesion of embryonic amphibian cells. J. exp. Zool. *128*, 53–120.

TRINKAUS, J. P., 1951. A study of the mechanism of epiboly in the egg of *Fundulus heteroclitus.* J. exp. Zool. *118*, 269–320.

———, 1967. Morphogenetic cell movements. In LOCKE, M. (Ed.): Major problems in developmental biology. Academic Press, New York, 125–176.

———, 1969. Cells into organs. The forces that shape the embryo. Prentice-Hall, Englewood Cliffs, N.J.

———, 1973. Surface activity and locomotion of *Fundulus* deep cells during blastula and gastrula stages. Devl. Biol. *30*, 68–103.

TRUJILLO-CENÓZ, O., and SOTELO, J. R., 1959. Relationships of the ovular surface with follicle cells and origin of the zona pellucida in rabbit oocytes. J. biophys. biochem. Cytol. *5*, 347–350.

TUFFREY, M., BISHUN, N. P., and BARNES, R. D., 1969. Porosity of the mouse placenta to maternal cells. Nature, Lond. *221*, 1029–1030.

TUNG, T. C., WU, S. C., and TUNG, Y. Y. F., 1962a. The presumptive areas of the egg of *Amphioxus.* Scientia Sinica *11*, 629–644.

———, ———, and ———, 1962b. Experimental studies on the neural induction in *Amphioxus.* Scientia Sinica *11*, 805–820.

TWITTY, V. C., 1949. Developmental analysis of amphibian pigmentation. Symp. Dev. Growth. *9*, 133–161.

——— and NIU, M. C., 1948. Causal analysis of chromatophore migration. J. exp. Zool. *108*, 405–437.

TYLER, A., 1948. Fertilization and immunity. Physiol. Rev. *28*, 180–219.

———, 1967. Masked messenger RNA and cytoplasmic DNA in relation to protein synthesis and processes of fertilization and determination in embryonic development. Devl. Biol. Suppl. 1, 170–226.

———, BORSTEL, R. C. VON, and METZ, C. B. (Eds.), 1957. The beginnings of embryonic development. Amer. Ass. Adv. Sc., Washington.

UMANSKI, E., 1935. Ueber die gegenseitige Vertretbarkeit präsumptiver Anlagen der Rückenmark- und Gehirnteile bei den Amphibien. Zool. Anz. *110*, 25–30.

VAHS, W., 1962. Quantitative cytochemische Untersuchungen über die Veränderungen des Ribonucleoproteid-Status im heterogenen Induktor und im Reaktionssystem des *Triturus* Embryos während der Induktions- und frühen Differenzierungsphase. Roux Arch. *153*, 403–550.

VAINIO, T., SAXÉN, S., TOIVONEN, S., and RAPOLA, J., 1962. The transmission problem in primary embryonic induction. Expl. Cell Res. *27*, 527–538.

VAKAET, L., 1962. Some new data concerning the formation of the definitive endoblast in the chick embryo. J. Embryol. exp. Morph. *10*, 38–57.

VENDRELY, R., and VENDRELY, C., 1949. La teneur du noyau cellulaire en acide désoxyribonucléique á travers les organes, les individus et les espéces animales. Experientia *5*, 327–329.

VINCENTIIS, M. DE, 1954. Ulteriori indagini sull' organogenesi del cristallino. Riv. Biol. *36*.

VOGT, W., 1925. Gestaltungsanalyse am Amphibienkeim mit örtlicher Vitalfärbung. Vorwort über Wege und Ziele. I. Methodik und Wirkungsweise der ortlichen Vitalfärbung mit Agar als Farbträger. Roux Arch. *106*, 542–610.

———, 1929. Gestaltungsanalyse am Amphibienkeim mit örtlicher Vitalfärbung. II. Gastrulation und Mesodermbildung bei Urodelen und Anuren. Roux Arch. *120*, 385–706.

WADDINGTON, C. H., 1933. Induction by the endoderm in birds. Roux Arch. *128*, 502–521.

———, 1934. Experiments on embryonic induction. III. A note on inductions by chick primitive streak transplanted to the rabbit embryo. J. exp. Biol. *11*, 224–226.

WADDINGTON, C. H., 1936. Organizers in mammalian development. Nature, Lond. *138*, 125.

———, 1938. The morphogenetic function of a vestigial organ in the chick. J. exp. Biol. *15*, 371–376.

———, 1952. The epigenetics of birds. University Press, Cambridge.

———, and MULHERKAR, L., 1957. The diffusion of substances during embryonic induction in the chick. Proc. Zool. Soc., Calcutta, Mookerjee Memor. Vol., 141–147.

———, NEEDHAM, J., and BRACHET, J., 1936. Studies on the nature of the amphibian organization centre. III. The activation of the evocator. Proc. R. Soc., B *120*, 173–198.

———, ———, NOWINSKY, W. W., NEEDHAM, D. M., and LEMBERG, R., 1934. Active principle of the amphibian organization centre. Nature, Lond. *134*, 103.

———, and PERRY, M. M., 1966. A note on the mechanism of cell deformation in the neural folds of the amphibian. Exp. Cell Res. *41*, 691–693.

———, and SCHMIDT, G. A., 1933. Induction by heteroplastic grafts of the primitive streak in birds. Roux Arch. *128*, 522–563.

WALLACE, R. A., 1963. Studies on amphibian yolk. IV. An analysis of the main-body component of yolk platelets. Biochim. biophys. Acta *74*, 505–518.

———, 1964. Studies on amphibian yolk. VI. A protein kinase from the ovary of *Rana pipiens*. Biochim. biophys. Acta *86*, 286–294.

WARBURG, O., 1908. Beobachtungen über die Oxydationsprozesse im Seeigelei. Hoppe-Seyler's Z. physiol. Chem. *57*, 1–16.

WARD, R. T., 1962. The origin of protein and fatty yolk in *Rana pipiens*. II. Electron microscopical and chemical observations of young and mature oocytes. J. Cell Biol. *14*, 309–341.

WARTENBERG, H., 1962. Elektronenmikroskopische und histochemische Studien über die Oogenese der Amphibieneizelle. Z. Zellforsch. *58*, 427–482.

———, and SCHMIDT, W., 1961. Elektronenmikroskopische Untersuchungen der strukturellen Veränderungen im Rindenbereich des Amphibieneies im Ovar und nach der Befruchtung. Z. Zellforsch. *54*, 118–146.

WATERMAN, A. J., 1936. Developmental capacities of transplanted hepatic, pancreatic and lung tissues of the rabbit embryo. Am. J. Anat. *58*, 2–57.

WATSON, J. D., and CRICK, F. H. C., 1953. Molecular structure of nucleic acids. A structure for deoxyribose nucleic acid. Nature, Lond. *171*, 737–738.

WATTERSON, R. L., 1965. Structure and mitotic behaviour of the early neural tube. In DE HAAN, R. L., and URSPRUNG, H. (Eds.): Organogenesis. Holt, Rinehart and Winston, New York, 129–159.

WEBER, R., and BOELL, E. J., 1955. Ueber die Cytochromoxydaseaktivität der Mitochondrien von frühen Entwicklungsstadien des Krallenfrosches (*Xenopus laevis* Daud.). Rev. suisse Zool. *62*, 260–268.

WEHMEIER, E., 1934. Versuche zur Analyse der Induktionsmittel bei der Medullarplatteninduktion von Urodelen. Roux Arch. *132*, 384–423.

WEILER-STOLT, B., 1960. Ueber die Bedeutung der interstitiellen Zellen für die Entwicklung und Fortpflanzung mariner Hydroiden. Roux Arch. *152*, 398–454.

WEISS, P., 1925. Unabhängigkeit der Extremitätenregeneration vom Skelett (bei *Triton cristatus*). Roux Arch. *104*, 359–394.

———, 1926a. Morphodynamik. Bornträger, Berlin.

———, 1926b. Ganzregenerate aus halbem Extremitätenquerschnitt. Roux Arch. *107*, 1–53.

———, 1927. Potenzprüfung am Regenerationsblastem. I. Extremitätenbildung aus Schwanzblastem im Extremitätenfeld bei *Triton*. Roux Arch. *111*, 317.

———, 1928. Experimentelle Untersuchungen zur Metamorphose der Ascidien; Beschleunigung des Metamorphoseeintrittes durch Thyreoideabehandlung der Larve. Biol. Zbl. *48*, 69–79.

———, 1929. Erzwingung elementarer Strukturverschiedenheiten am in vitro wachsenden Gewebe (Die Wirkung mechanischer Spannug auf Richtung und Intensität des Gewebewachstums und ihre Analyse). Roux Arch. *116*, 438–554.

———, 1934. *In vitro* experiments on the factors determining the course of the outgrowing nerve fiber. J. exp. Zool. *68*, 393–448.

———, 1939. Principles of development. Henry Holt, New York.

———, 1947. The problem of specificity in growth and development. Yale J. Biol. Med. *19*, 235–278.

———, 1950. The deplantation of fragments of nervous system in amphibians. I. Central reorganization and the formation of nerves. J. exp. Zool. *113*, 397–461.

———, 1955. Nervous system. In WILLIER, B. H., WEISS, P. A., and HAMBURGER, V.: Analysis of development. Saunders, Philadelphia, 346–401.

———, and ANDRES, G., 1952. Experiments on the fate of embryonic cells (chick) disseminated by the vascular route. J. exp. Zool. *121*, 449–488.

———, and JAMES, R., 1955. Skin metaplasia in vitro induced by brief exposure to vitamin A. Expl. Cell Res., Suppl. *3*, 381–394.

WEISSMANN, A., 1904. Vorträge über Descendenztheorie. Jena. 2 vols.

WENRICH, D. H., 1916. The spermatogenesis of *Phrynotettix magnus* with special reference to synapsis and the individuality of the chromosomes. Bull. Mus. comp. Zool. Harv. *60*, No. 3, 57–133.

WESSELLS, N. K., 1968. Problems in the analysis of determination, mitosis and differentiation. In FLEISCHMAJER (Ed.): Epithelial-mesenchymal interactions. Williams and Wilkins Co., Baltimore. 132–152.

———, 1971. How living cells change shape. Scient. Am. *225* (4), 77–82.

———, 1973. Tissue interactions in development. An Addison-Wesley Module in Biology, No. 9, 1–43.

———, and ROESSNER, K. D., 1965. Nonproliferation in dermal condensations of mouse vibrissae and pelage hairs. Devl. Biol. *12*, 419–433.

———, and RUTTER, W. J., 1969. Phases in cell differentiation. Scient. Am. *220* (3), 36–44.

———, SPOONER, B. S., ASH, J. F., BRADLEY, M. O., LUDUEÑA, M. A., TAYLOR, E. L., WRENN, J. T., and YAMADA, K. M., 1971. Microfilaments in cellular and developmental processes. Science, N.Y. *171*, 135–143.

WESTON, J. A., 1963. A radioautographic analysis of the migration and localization of trunk neural crest cells in the chick. Devl. Biol. *6*, 279–310.

WETZEL, R., 1929. Untersuchungen am Hühnchen. Die Entwicklung des Keims während der ersten beiden Bruttage. Roux Arch. *119*, 118–321.

WHITE, E. L., 1948. An experimental study of the relationship between the size of the eye and the size of the optic tectum in the brain of the developing teleost *Fundulus heteroclitus.* J. exp. Zool. *108*, 439–469.

WHITELEY, A. H., McCARTHY, B. J., and WHITELEY, H. R., 1966. Changing populations of messenger RNA during sea urchin development. Proc. natn. Acad. Sci., U.S.A. *55*, 519–525.

WHITTINGHAM, D. G., LEIBO, S. P., and MAZUR, P., 1972. Survival of mouse embryos frozen to −196° and −269°C. Science, N.Y. *178*, 411–414.

WIEMAN, H. L., 1949. An introduction to vertebrate embryology. McGraw-Hill, New York.

WIGGLESWORTH, V. B., 1939. The principles of insect physiology. Methuen, London.

———, 1954. The physiology of insect metamorphosis. University Press, Cambridge.

WILENS, S., 1955. The migration of heart mesoderm and associated areas in *Amblystoma punctatum.* J. exp. Zool. *129*, 579–606.

WILKINS, L., 1960. The thyroid gland. Scient. Am. *202* (3), 119–129.

WILLIAMS, C. M., 1946. Physiology of insect diapause; the role of the brain in the production and termination of pupal dormancy in the giant silkworm *Platysamia cecropia.* Biol. Bull. mar. biol. Lab., Woods Hole *90*, 234–243.

———, 1947. Physiology of insect diapause. II. Interaction between the pupal brain and prothoracic glands in the metamorphosis of the giant silkworm *Platysamia cecropia.* Biol. Bull. mar. biol. Lab., Woods Hole *93*, 89–98.

———, MOORHEAD, L. V., and PULIS, J., 1959. Juvenile hormone in thymus, human placenta and other mammalian organs. Nature, Lond. *183*, 405.

WILLIER, B. H., 1952. Cells, feathers and colors. Bios *23*, 109–125.

———, 1955. Ontogeny of endocrine correlation. In WILLIER, B. H., WEISS, P. A., and HAMBURGER, V. (Eds.): Analysis of development. Saunders, Philadelphia, 574–619.

———, and RAWLES, M. E., 1931. Developmental relations of heart and liver in chorio-allantoic grafts of whole chick blastoderms. Anat. Rec. *48*, 277–301.

———, WEISS, P. A., and HAMBURGER, V. (Eds.), 1955. Analysis of development. Saunders, Philadelphia.

WILSON, E. B., 1904. Experimental studies on germinal localization. I. The germ regions in the egg of *Dentalium.* II. Experiments on the cleavage-mosaic in *Patella* and *Dentalium.* J. exp. Zool. *1*, 1–72.

———, 1925. The cell in development and heredity. 3rd ed. Macmillan, New York.

WILSON, H. V., 1907. On some phenomena of coalescence and regeneration in sponges. J. exp. Zool. *5*, 245–258.

WISCHNITZER, S., 1960. Observations on the annulate lamellae of immature amphibian oocytes. J. biophys. biochem. Cytol. *8*, 558–563.

WITSCHI, E., 1929. Studies on sex differentiation and sex determination in amphibians. I. Development and sexual differentiation of the gonads of *Rana sylvatica.* J. exp. Zool. *52*, 235–263.

———, 1948. Migration of the germ cells of human embryos from the yolk sac to the primitive gonadal folds. Contr. Embryol. Carneg. Instn. *32*, 67–80.

———, 1956. Development of vertebrates. Saunders, Philadelphia.

WITTEK, M., 1952. La vitellogénèse chez les Amphibiens. Archs. Biol. Paris. *63*, 133–198.

WOERDEMAN, M. W., 1933. Ueber den Glykogenstoffwechsel des Organisationszentrums in der Amphibiengastrulation. Proc. K. ned. Akad. Wet *36*, 189–193.

WOLFF, C. F., 1759. Theoria generationis (cited after NEEDHAM, J., 1931).

WOLPERT, L., and MERCER, E. H., 1961. An electron microscope study of fertilization of the sea urchin egg *Psammechinus miliaris.* Expl. Cell Res. *22*, 45–55.

WRENN, J. T., 1971. An analysis of tubular gland morphogenesis in chick oviduct. Devl. Biol. *26*, 400–415.

WRIGHT, S., and WAGNER, K., 1934. Types of subnormal development of the head from inbred strains of guinea pigs and their bearing on the classification of vertebrate monsters. Am. J. Anat. *54*, 383–447.

YAMADA, T., 1937. Der Determinationszustand des Rumpfmesoderms im Molchkeim nach der Gastrulation. Roux Arch. *137*, 151–270.

———, 1938. Induktion der sekundären Embryonalanlage im Neunaugenkeim. Okajimas Folia anat. jap. *17*, 369–388.

———, 1950. Dorsalization of the ventral marginal zone of the *Triturus* gastrula. I. Ammonia treatment of the medio-ventral marginal zone. Biol. Bull. mar. biol. Lab., Woods Hole *98*, 98–121.

———, 1958. Induction of specific differentiation by samples of proteins and nucleoproteins in the isolated ectoderm of *Triturus* gastrulae. Experientia *14*, 81–87.

———, 1962. The inductive phenomenon as a tool for understanding the basic mechanism of differentiation. J. cell. comp. Physiol. *60*, Suppl. 1, 49–64.

———, and TAKATA, K., 1961. A technique for testing macromolecular samples in solution for morphogenetic effects on the isolated ectoderm of the amphibian gastrula. Devl. Biol. *3*, 411–423.

YAMAMOTO, K., and OOTA, I., 1967. An electron microscope study of the formation of the yolk globule in the oocyte of Zebrafish, *Brachydanio rerio.* Bull. Fac. Fish. Hokkaido Univ. *17*, 165–174.

YAMAMOTO, T., 1954. Physiological studies on fertilization and activation of fish eggs. V. The role of calcium in activation of *Oryzias latipes.* Expl. Cell Res. *6*, 56–68.

———, 1961. Physiology of fertilization of fish eggs. Int. Rev. Cytol. *12*, 361–405.

YASUZUMI, G., 1956. Spermatogenesis in animals as revealed by electron microscopy. I. Formation and submicroscopic structure of the middle-piece of the albino rat. J. biophys. biochem. Cytol. *2*, 445–450.

———, TANAKA, H., and TEZUKA, O., 1960. Spermatogenesis in animals as revealed by electron microscopy. VIII. Relation between nutritive cells and the developing spermatids in a pond snail *Cipangopaludina malleata* Reeve. J. biophys. biochem. Cytol *7*, 499–504.

YNTEMA, C. L., 1950. An analysis of induction of the ear from foreign ectoderm in the salamander embryo. J. exp. Zool. *113*, 211–244.

———, and HAMMOND, W. S., 1947. The development of the autonomic nervous system. Biol. Rev. *22*, 344–359.

ZHINKIN, L., 1934. Ueber die Wirkung der Röntgenstrahlen auf die Regeneration bei *Lumbriculus variegatus* Gr.. Trav. Lab. Zool. exp. Morph. Anim. *3*, 71–97.

ZWILLING, E., 1940. An experimental analysis of the development of the anuran olfactory organ. J. exp. Zool. *84*, 291–323.

———, 1955. Ectoderm-mesoderm relationship in the development of the chick embryo limb bud. J. exp. Zool. *18*, 423–441.

———, 1956. Interaction between limb bud ectoderm and mesoderm in the chick embryo. IV. Experiments with a wingless mutant. J. exp. Zool. *132*, 241–253.

———, 1963. Formation of endoderm from ectoderm in *Cordylophora.* Biol. Bull. mar. biol. Lab., Woods Hole *124*, 368–378.

INDEX

Accessory ganglion, 351
Acid, amino. See *Amino acid.*
  deoxyribonucleic. See *Deoxyribonucleic acid.*
  ribonucleic. See *Ribonucleic acid.*
Acipenser, cleavage in, 111
Acoustic ganglion, 350
Acoustic nerve, 351
Acrosomal cone, 28
Acrosomal filament, 28, 79
Acrosomal granule, 27
Acrosome, 25
  and production of sperm lysins, 79
  changes caused by egg, 79
Actinomycin D, hemoglobin synthesis and, 492
"Activating agent," 220
Activation, of egg. See *Egg, activation.*
Adenine, in deoxyribonucleic acid, 14(t)
  in ribonucleic acid, 14(t)
Adhesive organ, development, 373
Agglutination, 77
Aging, 526
Alimentary canal, accessory organs, development, 457–461
  archenteron and, 440–448
  primitive streak and, 446
Alkaline phosphatase, appearance in tissues, 456
Allantoic diverticulum, in man, 446
Allantoic stalk, in human and monkey embryos, 279
Allantoic vesicle, in man, 277, 279
Allantois, formation, in mammalian embryo, 277
  in birds and reptiles, 258
Ambystoma mexicanum, maternal genes and morphogenetic processes, 145
Ambystoma punctatum embryo, development of gastric glands, 497
  limb differentiation experiments, 390
Amia, cleavage in, 111
Amino acids, activation of, 14
  and protein synthesis, 476
  in egg, activation by fertilization, 89
  nonessential, synthesis, 476
Amnion, in birds and reptiles, 256
Amniote, alimentary canal, development, 445
  aortic arches, 418
  development of paired limbs, 384
  liver development, 460

Amniote *(Continued)*
  midgut, 445
  shape during development, 298
Amniotic cavity, formation, in mammalian embryo, 275
  in birds and reptiles, 256
Amniotic folds, in birds and reptiles, 256
Amphibian, development of ovary, 434
  gill clefts, 8
  metamorphosis, causation, 543–544
    changes of organization, 536–540
    comparison with insect metamorphosis, 561
    hypophysis, 542
    induction, 544–545
    physiological changes, 538
    tissue reactivity, 543, 544, 618
  nerves and regeneration, 567
  pharyngeal pouches, 8
  shape during development, 297
Amphibian egg, contents, 45
  cytoplasmic changes after fertilization, 98
  membrane of, 64
  pigment in, 54
  vitellogenesis in, 50
Amphibian embryo, fate map, 152
  gastrulation, 161–170
  heart development, 402
  hindgut, 442
  lateral line organ cells, 365
  limbs, development, 382
    differentiation, 389
    supernumerary, 386
  liver development, 459
  mesocardium, 405
  midgut, 442
  myocardium, 406
  organogenesis, primary, 170–175
  pericardium, 406
  primordial germ cells, 425
  selective affinities and cellular rearrangements, 230
  septum transversum, 405
Amphibian hybrid, cleavage in, 196
  gastrulation in, 196
Amphimixis, 75
Amphioxus, archenteron and alimentary canal, 440
  gastrulation, 153–158
  organogenesis, primary, 158–161

635

Androgenesis, 195
Andromerogone, 195
Angioblasts, 409
Animal gradient, 138
Animal polar plasm, 56
Animal pole, 54
Animal region, fate of, 152
Animal-vegetal gradient concept, 138
Animalization, chemical agents and, 142–143
Annelids, blastopore, 6
    regeneration, 564
        nerves and, 569
Antibodies, 500
    thymus and, 455
Antifertilizins, 77
Antigens, 77, 196, 500
Anurans, external gills, 456
    metamorphosis, 536
Aorta, dorsal, 410
    ventral, 405, 410, 418
Aortic arches, 410
Aplysia limacina egg, centrifugation during
    cleavage, 133
Apoplasmic substances, synthesis, 467
Apterygota, molting, 546
Aqueduct of Sylvius, 344
Arbacia eggs, centrifugation during cleavage,
    132
Archencephalic inductor, 203
Archenteron, 5, 156
    and alimentary canal, 440–448
    and gastrulation, 5
Archenteron roof, and neural plate, 332
Archeocytes, 599
Arcualia, 377
Area opaca, 117
Area pellucida, 116–117
Area vasculosa, in chick embryo, 254
Area vitellina, in chick embryo, 254
Artery, carotid, internal, 416
    innominate, 418
Arthropod, blastopore, 6
    regenerative ability, 565
Ascaris, cleavage in, 109, 110
Ascidian, budding, 598
Ascidian egg, cytoplasmic changes after
    fertilization, 97
Ascidian tadpole, metamorphosis, 561
Asexual reproduction, 4, 7, 592–603
    forms, 592–603
    occurrence, 592–595
    sources of cellular material, 595–600
Atrium, in mammalian embryo, 409
Autonomic nervous system, development, 352
Autonomy, 565
Autopodium, 389
Axial body, 28
Axial filament, 28
Axial skeleton, development, 377–382

Baer's law, 7
Balancers development, 373
    in salamander larva, 512
Balbiani, yolk nucleus of, 49
Barnacles, larva, 9

Basal plate, of skull, 382
Beryllium nitrate, and regeneration, 581
Bidder's organ, 434
Bile duct, in amphibian embryo, 460
Biogenetic law, 8
Bird, development of alimentary canal, 443
    development of ovary, 434
    egg membranes of, 65
    extraembryonic structures, 254–259
    gastrulation and primary organogenesis,
        178–188
    induction, 205
    pharyngeal pouches, 8
    primary organizer, 205
    regenerative ability, 566
    yolk sac, 254
Bivalent, 23
Blastema, 592
    regeneration. See *Regeneration blastema.*
Blastocoele, 5, 115
Blastocyst, in mammalian embryo, 266
    in man, 276, 277
Blastoderm, 5, 154, 216
Blastodisc, 116
Blastogenesis, 4, 592
    comparison with embryogenesis, 600–603
Blastomere, 5, 101
    division of, 101
    nucleus, 103, 119–125
        transplantation, 124
    separation of, *12, 13*
Blastopore, 6, 156, 163, 172, 186
    closure of, 163
    rim of, 163
Blastopore lip, 156, 219
Blastozooids, 592
Blastula, 5, 114–119
    polarity, 115
Blood islands, in chick embryo, 254
Blood vessels, development, 409–423
Body folds, in chick embryo, 256
Bombyx mori, messenger RNA and, 482
Botryllus, development of blastozooid, 600
    development of esophagus, 600
Bouquet stage, 21
Bowman's capsule, 394, 399
Brachial plexus, 337
Brachiolaria, larva, 9
Brain, development, 341
    in human embryo, 351
    insect, and molting metamorphosis, 551–
        553
Branchial aperture, 456
Branchial arch, formation, 456
Branchial grooves, epidermal, and pharyn-
    geal pouches, 456
Branchial membrane, 454
Branchial pouches. See *Pharyngeal pouches.*
Branchial region, development, 453–457
Budding, development in, 600
    in hydromedusa, 597
    in hydrozoan polyp, 596
    in scyphozoan polyp, 596
    in tunicates, 597
    occurrence, 594
    sources of cellular material, 596
Bursa fabricii, development of, 461

Campanularia, fertilization of, 76
Carotids, common, 416
    external, 416
"Caudalizing agent," 220
Cavitation, 276
Cell(s), cleavage. See *Blastomere.*
    division, in cleavage, 101–104
    epidermal, degeneration, 546
    formative, in mammalian embryo, 270
    generative. See *Gamete.*
    germ. See *Primordial germ cells.*
    interstitial, 573
    neural crest, 361–364
    neurosecretory, 551
    nurse, 40
    reproduction, differentiation and, 476–477
        mechanisms of, 469–476
    selective affinities of, 230–234
    Sertoli, 20
    shape of, change mechanism, 240–251
    somatic, 4, 592
    totipotent, 599
Cellular rearrangements, selective affinities
    in, 230–234
Central nervous system, development, 330–
    353
Centriole, distal, 28
    proximal, 28
    ring, 30
Centrolecithal egg, 48
    cleavage in, 112
Cerebellum, 330
    development, 345
Cerebral hemispheres, cortex, 343
Cerebral peduncles, 345
Chaetopterus embryo, deoxyribonucleic acid
    in nuclei, 103
Chiasma, 23
Chick embryo, dedifferentiation experi-
        ments, 502
    determination of endodermal organs, 463
    developmental anatomy of, 300, 302–328
    extraembryonic structure, 254
    head fold, 256
    primordial germ cells, 424
    umbilical cord, 256
    vitamin A and differentiation, 506
Child's theory, 585
Chonanae, 367
Chordates, blastopore, 6
Chordo-mesodermal mantle, 166
Chorion, 62
    in birds and reptiles, 256
    in insects, 62
Choroid coat, 360
Choroid fissure, 354
Choroid plexus, 344
    posterior, 345
Chromosome(s), 13
    daughter, 23
    duplication, 470
    giant, 477, 633
    haploid set, 24
    histones in, 486
    homologous, 20
    in transplanted nuclei, 124
    lamp brush, 33

Chromosome number, reduction of. See
    Meiosis.
Chymotrypsin, and induction, 216
Ciona intestinalis, 9
Cirripedia, larva, 9
Cleavage, 5, 101–147
    bilateral, 108
    chemical changes, 104–106
    determinate, 109
    dextral, 108
    incomplete, 111
    mammalian embryo, 261
    monkey embryo, 265
    patterns, 106–114
    rabbit embryo, 266
    radial, 107, *107*
    sea urchin hybrids, 195
    sinistral, 108
    spiral, *107,* 108, *108*
    superficial, 114
    synthesis of ribonucleic acids during, 105
    yolk and, 109
Cleavage cell. See *Blastomere.*
Cleithrum, 393
Cloaca, external, 448
Cloacal membrane, 448
Cnidoblasts, 573
Coding, analytical embryology and, 13
Coelenterates, blastopore, 6
    regenerative ability, 564
Coelom, 5, 159, 170, 188, 394
    extraembryonic, in rodent embryo, 282
Coelomic cavity, 170, 235, 322
    extraembryonic, 258
Coelomic spaces, 463
Coloboma, of iris, 355
Commissura mollis, 344
Competence, 202
    and metamorphosis, 543
Connecting stalk, in human and monkey
    embryo, 279
Cornea, 360
Corona radiata, 63
Corpora allata, 551
    and insect metamorphosis, 551
Corpora cardiaca, 551
Corpora quadrigemina, 344
Corpus luteum, 292
"Corrective factor," maternal genes and, 146
Cortical granules, 57
    changes at activation, 83
Cortical reaction, 83
    in lamellibranch mollusc egg, 87
    in mammalian egg, 86
Cotyledon, 289
Covering layer, 365
Cranial ganglion, development, 349
Cranial nerves, development, 346
    sensory, 346
Cranium, 380
Crossing over, 23
Crustaceans, regenerative ability, 565
Curtis, experiments with gray crescent of
    Xenopus laevis, 203
Cuticulin, 546
Cyclopia, 225
    in guinea pigs, 225

Cyclopia *(Continued)*
  in man, 227
Cyclostomes, pharyngeal pouches, 8
Cynthia partita egg, 54. See also *Styela partita.*
Cytoplasm, and differentiation, 468
  egg. See *Egg cytoplasm.*
Cytoplasmic substances, distribution during
  cleavage, 125–130
Cytosine, in deoxyribonucleic acid, 14, 491
  in ribonucleic acid, 14(t)
Cytotrophoblast, 285

Daughter chromosome, 23
Dedifferentiation, 502
  during regeneration, 572
  in chick embryo, 502
Degeneration, epidermal cells, 546
  muscle fibers, 220
Degrowth, 468
  during amphibian metamorphosis, 539
Dentalium egg, cleavage, 126, *127*
Dentalium oocyte, cytoplasm, 55
Dentine, 450
Deoxyribonucleic acid (DNA), 13–15
  analytical embryology and, 13
  and growth, 469
  and mitosis, 469, 490
  chromosomal, messenger RNA and, 484
  histones and, 486
  in *Chaeopterus* embryo nuclei, 103(t)
  in oocyte cytoplasm, 36
  in oocyte nucleus, 34
  in sea urchin nuclei, 103(t)
  increase during cleavage, 104
  messenger RNA and, 14, 489
  replication, 470
  sequence, 491
  synthesis, 491
  Watson and Crick model, 13
Deoxyribonucleic acid molecule, duplication, 23
Deoxyribose sugar, in deoxyribonucleic acid,
  13
Dermatone, 375
Dermis, 365
Desmosomes, 37
Determinants, 119
Determinate cleavage, 109
Determination, 200
  experiments in newt embryo, 200
  of endodermal organs, 461–463
  of organ rudiments, 197–202
Deuterencephalic inductor, 203
Deuterostomia, blastopore, 6
Development, control by gradients, 224
  definition, 3
  functional stage, 467
  in budding, 600
  in fission, 600
  in gemmule formation, 601
  of mammals, embryonic adaptation, 253–
    295
  ontogenetic, 3, 4–7
  phylogenetic, 3
  prefunctional stage, 467
  sequence of gene action, 515–519
  stages of, 299

Diakinesis, *23,* 24
Diapause, 551
Diencephalon, 330, 344
Differentiation, 468
  and growth, general considerations, 467–
    477
  and growth rate, 526, 531
  cell, inducing substances and, 218
    neural tube, 336
  cell reproduction and, 476–477
  chemical basis, 478–482
  control of, by chemical substances, 505
    by intraorganismic environment, 502–
      512
  definition, 468
  digits, 388
  gene quantity and, 493
  histological, 6, 468
  hormones and, 505
  limbs, 388
    in salamander embryo, 390–391
  mitosis and, 492
  neural tube, 331
  of spermatozoon, 24–31
  proteins and, 478
  reversibility, 502
  time factor, 494
  vitamin A and, 507
Digits, differentiation, 388
Diploid number, 20
Diploid state, 4
Diplonema stage, 23
DNA. See *Deoxyribonucleic acid.*
Donor, 198
"Dorsalizing agent," 220
Dragonfly, blastomere nucleus, *122*
Drosophila, nurse cells of, 39
  transplantation of nuclei in, 124
Duct of Botalli, 417
Duct of Cuvier, 416
Ductus venosus, in reptiles and birds, 423
  mammalian, 423
Duodenum, 443
Dwarfism, in man, 528

Ear, capsule, 381
  development of, 367, 368, 416
  middle, development, 370
  vesicle, 367
Earthworm, heteromorphosis, 582
  super-regeneration, 579
Ecdysone, 551
  and puffs, 560
  chemical formula, 557
Echinoderms, blastopore, 6
  regenerative ability, 565
Ectoderm, 5, 205
  and limb development, 384
  expansion, 163
  gastrulation and, 5
  in mouth development, 449
Ectodermal organs, development in vertebrates, 353–374
Ectoplacental cone, in rodent embryo, 282
Egg. See also *Oocyte.*
  activation, 75
    cortical reaction, 83

Egg *(Continued)*
　physiology, 88–90
　centrifugation during cleavage, 131
　centrolecithal, 48
　　cleavage in, 112
　cleidoic, 69
　cortex and cleavage, 131–134
　　organizer properties, 203
　cytoplasm, and differentiation, 488
　　changes after fertilization, 96–100
　DNA in, 36
　　in sea urchin, 135
　　morphogenetic gradients, 134–143
　　organization of, 54–60
　holoblastic, 111
　membrane(s), 38, 62–66
　meroblastic, 111
　messenger RNA in, after fertilization, 89
　oligolecithal, 45
　pigment granules, 54
　polarity, 60
　shape of, 59
　spermatozoon in, 92–96
　　approach of, 75–81
　　　reaction, 81–88
　　　penetration, 79
　teleolecithal, 46
　water, 77
Embryo, animalized, 138
　definition, 3
　meroblastic, heart development, 408
　nutrition, 67–71
　parts of, diversification, 189–227
　vegetalized, 138
　water absorption by, 68
Embryogenesis, 4
　comparison with blastogenesis, 600–603
Embryology, analytical, 13
　comparative, 7–11
　definition, 3–4
　descriptive, 7–11
　experimental, 11–13
Embryonic adaptations, 253–293
Embryonic membranes, in birds and reptiles, 256
Embryonic shield, 176
Enamel organ, 450
Endocardium, in amphibian embryo, 405
Endocuticle, 546
Endoderm, 5, 209
　extraembryonic, and primordial germ cells, 425
　gastrulation and, 5
　heart development and, 404
Endodermal organs, development in vertebrates, 440–463
Endopterygota, wing development, 548
Enveloping layer, in mammalian embryo, 270
Enzymes, and differentiation, 478
　digestive, development, 497
　proteolytic, and induction, 216
Ependyma, 334
Ephestia kühniella, molting experiments, 555
Epiblast, 117
Epiboly, 177
　epithelia and, 234
Epicardium, 595
Epicuticle, 546
Epidermis, accretionary growth, 523

Epidermis *(Continued)*
　and limb development, 385
　insect, in molting, 546
　structures derived from, 364–374
Epididymis, 435
Epigenesis, theory, 10–11
Epimorphosis, 564
Epithelial layers, folding of, 236
　mesenchyme and, 239–240
　separation of, 235–236
　thickening of, 234–235
　thinning of, 234
　tube formation and, 238–239
Epithelium, 234
　differentiation, dependence on mesenchyme, 508
　morphogenetic movements of, 234–240
Erythropoiesis, 489
Erythropoietin, 489
Esophagus, 441, 446
　Botryllus, 600
Estrogens, parturition and, 293
Eudistoma, transverse fission, 595
Eukaryotes, messenger RNA and, 484
Eustachian tube, origin, 456
Evolution, 3
Exocoelom, in rodent embryo, 282
Exocuticle, 546
Exogastrulation, 136–137
Exopterygota, wing development, 547
Extraembryonic coelom, in rodent embryo, 282
Extraembryonic endoderm, and primordial cells, 425
Extraembryonic structure, in mammalian embryo, 275
　in reptiles and birds, 254–259
Eye, accessory structures, 360. See also *Optic.*
　capsule, 381
　choroid coat, 360
　cornea, 360
　development, 353–361
　iris, 354
　lens, 357
　　induction, 358
　sclera, 360, 381
　sensory retina, 353

Facial nerve, 350
Fate maps, 151–153
Fertilization, 4, 75–100
　metabolic changes, 89
　monospermic, 89
　polyspermic, 88
Fertilization cone, 82
Fertilization membrane, 83
Fertilizin(s), 77
Fertilizin-antifertilizin reaction, and activation of egg, 88
Fiddler crab, allometric growth, 531
Fin fold, development, 370
　in salamander larva, 513
Fish(es), ganoid, cleavage in, 111
　muscle buds, 393
　parabolic growth, 524
　pharyngeal pouches, 8

Fish(es) *(Continued)*
  regenerative ability, 566
  shape during development, 299
  swim bladder, 459
  ventral aorta, 418
  yolk sac, 253, 254
Fish embryo, gastrulation and primary
    organogenesis, 175–178
  yolk during gastrulation, 253
Fission, development in, 600
  occurrence, 593
  regeneration in, 595
Flagellum, of spermatozoon, 25
Flexure, 351
  cephalic, 345, 351
  cervical, 351
  pontine, 351
Follicle cell, 37
Follicle stimulating hormone, ovulation
    and, 260, 291
Food, and regeneration, 567
Foramen, of Monro, 342, 344
Foregut, 170, 440
  in amniotes, 445
Form, 229
  development of, 297–299
  gastrulation and, 229–252
Frog, artificial parthenogenesis, 91
  nuclear transplantation in, 124
  oocyte, 32
  regeneration, 565
Frog egg, cleavage in, 110, *110*
Frog embryo, adhesive organ, development,
    373
  blastula, 115
  developmental anatomy of, 302–328
  external gills, induction, 513
  mouth induction, 513
  neurulation stage, 304
  oral ectoderm and mouth development,
    449
  oxygen consumption during gastrulation,
    190, 191
  pancreas development, 461
  pharynx development, 441
  subchorda, 304
  synthesis of nucleic acids in, 105
  tailbud stage, 303
  tympanic membrane development, 544
Frontal process, medial, 452
FSH. See *Follicle stimulating hormone.*

Gallbladder, development, 460
Gamete, 3
  maturation of. See *Meiosis.*
Gametogenesis, 4, 17–71
Ganglion, accessory, 351
  acoustic, 350
  cranial, development, 346
  geniculate, 350
  petrosal, 351
  "root," 351
  spinal, 337
Ganglion jugulare, 351
Ganglion nodosum, 351
Ganglion semilunare, 350

Gastrula, 5, 156
Gastrulation, 5
  definition, 149
  embryonic diversification and, 189–227
  form and, 229–252
  gene activity, 192–197
  glycogen consumption during, 189
  in fish embryo, 175–178
  in mammalian embryo, 271
  metabolism during, 189–192
  morphology, 151–189
  oxygen consumption during, 190, 191
  protein synthesis during, 191
  yolk and, in fish embryo, 253
Gemmule, 594
Gemmule formation, development in, 601
  occurrence, 594
  sources of cellular material, 599
Generative cells. See *Gamete.*
Generative layer, 365
Genes, 13
  activity during gastrulation, 192–194
  and growth, 517
  and organ rudiments, 517
  and protein structure in hemoglobin, 480
  in organogenesis, 516
  maternal, in early stages of development,
    143–147
  number, differentiation and, 493
  sequence of action, in development, 515–
    519
Genetic code, 14(t)
Genetic information, transmission of, 13–15
Geniculate ganglion, 350
Genital ducts, development, 435
Genotype, and induction, 512
  and reactive ability of tissues, 512–515
  maternal, and morphogenetic processes,
    146
    influence of cleavage, 144
Germ cell. See *Primordial germ cells.*
Germ plasm theory, of Weismann, 119
Germinal layer, 5
  formation, in mammalian embryo, 271
Germinal plasm, 428
Germinal ridge, 423
Germinal vesicles, 33
Gill cleft, 454
Gills, external, 456
  development, 371
  induction in frogs, 513
  internal, 454
Gland(s), gastric, development in salamander
    embryo, 497
  parathyroid, 456
  pituitary, 344
  prothoracic, 551
    insect, and molting and
      metamorphosis, 551
  shell, 64
  thymus, 456
  thyroid. See *Thyroid gland.*
Glomerulus, 395
  external, 394
Glomus, 395
Glossopharyngeal nerve, 351
Glycogen, consumption during gastrula-
    tion, 189

Glycogen *(Continued)*
  in amphibian egg, 47
Golgi bodies, 27
  and production of sperm lysins, 79
  and vitellogenesis, 50
  in oocyte, 43
Golgi rest, 27
Gonad, 432
Graafian follicle, 37
Gradient(s), and control of development, 224
  and determination of primary organ rudiments, 220–228
  animal, 138
  anteroposterior, 220
  dorsoventral, 220
  morphogenetic, in egg cytoplasm, 134–143
  vegetal, 138
Gray crescent, 98, 125
  in *Xenopus laevis,* 203
  organizer properties, 203
Gray matter, of spinal cord, 337
Growth, 6, 467, 519–532
  absolute increase, 519
  accretionary, 523
  allometric, 529
    in mammals, 531
  auxetic, 522
  cellular, 521–523
    and organismic, 521–523
  decline of, 525
  definition, 467
  deoxyribonucleic acid and, 469
  exponential, 522, 524
    and mitosis, 521
  formula, 520
  genes and, 517
  hormone regulation, 528
  increments, 519
  isometric, 529
  limit, 525
  logistic equation, 525
  multiplicative, 522
  organismic, 521–523
  parabolic, 524
  ribonucleic acid and, 474
  ribosomal RNA and, 475
  sigmoid curve, 519, 524
Growth curve, 519
  interpretation, 523–529
Growth rate, differentiation and, 526, 531
  potential, 524
Growth ratio, 529
Guanine, in deoxyribonucleic acid, 14(t)
  in ribonucleic acid, 14(t)
Guanosine, DNA sequence, 491
Guinea pig, cyclopia, 517
Gut, primary, 156

Haploid number, 20
Haploid set, of chromosomes, 24
Haploid state, 4
Head fold, in chick embryo, 256
Head inductor, 203
Head process, 183
Heart development, 402–409
Hemal arch, 377

Hemoglobin, differentiation and, 489
  fetal, 495
  genes and protein structure, 480
Hemoglobin synthesis, actinomycin D and, 492
Hensen's node, 181
  in mammalian embryo, 272, 273
Hepatic cavity, primary, in amphibian embryo, 459
Hepatic diverticulum, 440
Hepatic portal vein, 420
Heterochrony, 302
Heteromorphosis, 582
Hindgut, 440
  in amniotes, 445, 446
  in amphibian embryo, 442
Histioblasts, 602
Histological differentiation, 6, 468
Histones, and DNA, 486
  and ribonucleic acid synthesis, 486
  regeneration and, 589
Holoblastic egg, 111
Hormone(s), and differentiation, 505
  and growth regulation, 528
  follicle stimulating, ovulation and, 260, 291
  insect, mechanism of action, 557–558
  molting and metamorphosis, 551, 555–557
  juvenile, 554, 557
  luteinizing, ovulation and, 260, 291
  thyroid. See *Thyroid hormone.*
  thyrotropic, 542
Host, 198
Human embryo, brain development, 351
  development of, 301
  early development, 277
Hyaline layer, 85
Hyaluronidase, 79
Hybrid, true, 195
Hybridization, 195
Hydra, regenerative ability, 564
Hydranth, 586
Hydrocaulus, 586
Hydroid, oxygen and regeneration, 581
Hydroid polyp, regeneration, 572
Hydromedusa, budding, 597
Hydrozoan polyp, budding, 597
Hypoblast, 117
  in mammalian embryo, 269
Hypoglossal nerve, 351
Hypophysis, 344, 450
  and amphibian metamorphosis, 542
Hypothalamus, 344

Ilyanassa egg, centrifugation during cleavage, 132
Imaginal discs, 548
Imago, 547
Immigration, 181
Immunity, thymus and, 455
Implantation, 275
  interstitial, 285
Inducing substances, 512
  chemical analysis, 215
  chemical nature, 215
  emission by inductors, 213
  mechanism of action, 218–220

Induction, 202, 208–218
  and genotype, 513
  by substances of known chemical composition, 208
  definition, 202
  during amphibian metamorphosis, 544–545
  of lens, 358
  of mouth in frog, 513
  pepsin and, 216
  ribonucleic acid and, 216, 219
  trypsin and, 216
Inductor, 202
  abnormal, 208
  archencephalic, 203
  deuterencephalic, 203
  emission of inducing substance, 213
  head, 203
  spinocaudal, 203
  trunk, 203
Infundibulum, 344
Inhibitors, regeneration and, 589
Inner cell mass, in mammalian embryo, 265
Insects, metamorphosis, 545–550
    causation, 550–555
    comparison with amphibian, 561
    corpora allata and, 551
    nature of hormones, 555–557
Intestinal portal, 446
Intestine, 443
Invagination, 5, 156
Iodine, and amphibian metamorphosis, 541
Iris, 354
Isozymes, lactate dehydrogenase, 496

Jacobson's organ, 367
Juvenile hormone, 554, 557

Karyomeres, 95
Keratin, 65
Kidney, development, 394

Labyrinth, 368
Lactate dehydrogenase, 496
Lacunae, trophoblastic, 285
Laerta saxicola armeniaca, parthenogenesis of, 92
Lamina terminalis, 342
Larva, 6, 67
  in comparative embryology, 9
  occurrence, 535
Larval molt, 546, 550
Lateral line sense organs, 365
Lateral plates, 170
Lens, development, 357
  induction, 358
Lepidosteus, cleavage in, 101
Leptonema stage, 21
LH. See *Luteinizing hormone.*
Ligamentum venosum, 423
Limb-bud, 383
Limb girdle, 392

Limbs, development, 382–394
  differentiation, 388
Limiting groove, in development of spinal cord, 334
Lipochondria, 46
Lipoid, in amphibian egg, 46
Lipovitellin, 46
  in amphibian egg yolk platelets, 46
Lithium chloride, and gradients, 227
Liver. See also *Hepatic.*
  and vitellogenesis, 40
  development, 459
  diverticulum, 440
Liver, diverticulum, 440
Lizards, regenerative ability, 566
Lumbar plexus, 337
Lungs, development, 456
Luteinizing hormone, ovulation and, 260, 291
Luteotropin, 292

Macromeres, 135
Malphigian layer, 365
Mammal(s), allometric growth, 531
  artificial parthenogenesis, 92
  development of, embryonic adaptation, 253–295
  oocyte cytoplasm, 56
  pharyngeal pouches, 8
  regenerative ability, 566
Mammalian egg, cortical reaction, 86
  development, 259–263
    morula, 289
  spermatozoon penetration, 79
  yolk, 260, 263
Mammalian embryo, cleavage, 261
  determination of endodermal organs, 463
  enveloping layer, 270
  formative cells, 270
  germinal layer, 271
  Hensen's node, 272, 273
  hypoblast, 269
  inner cell mass, 265
  organizer, 206
  primitive streak, 271, 272, 273, 297
  Rauber's layer, 270
  tooth development, 450
  trophoblast, 265
Man, allantoic diverticulum, 446
  allantoic vesicle, 277
  blastocyst, 276, 277
  cyclopia, 227
  dwarfism in, 528
  embryo. See *Human embryo.*
  notochordal rudiment in, 273
  placenta in, 289
Manchette, 29
Mandibular arch, development, 451
Mandibular process, 452
Mangold, induction experiment, 200
Mantle, of spinal cord, 336
Marginal zone, fate of, 152
Marsupial, placenta in, 284, 290
Marsupial eggs, 260
Maturation, 4, 93
Maxillary process, 452
Meckel's cartilage, 363
Median fissure, ventral, 334

Median septum, dorsal, 334
Median sulcus, 345
Medulla, of gonad, 432
Medulla oblongata, 330, 345
Meiosis, 3
    in spermatozoon, 20–24
    of oocyte, 60
Meiotic prophase, stages of, 21–22
Membrane, branchial, 454
    cloaca, 448
    egg, 38, 62–66, 94–98
    embryonic, in birds and reptiles, 256
    fertilization, 83
    oropharyngeal, 448
    shell, 65
    tympanic, 544
    vitelline, 62
Meroblastic egg, 111
Meroblastic embryo, heart development, 408
Merogony, 195
Mesencephalon, 330
Mesenchyme, and epithelium differentiation,
    508
    in mouth development, 453
Mesocardium, in amphibian embryo, 405
Mesoderm, 165
    frog embryo and, 312
    gastrulation and, 5
Mesodermal crescent, 97
Mesodermal organs, development in
    vertebrates, 375–439
Mesodermal segments, 158
"Mesodermalizing agent," 220
Mesomeres, 135
Mesonephric duct, 399, 435
    ridges, 383
Mesonephric tubules, 399
Mesonephros, 398
Mesorchium, 423
Mesovarium, 423
Messenger RNA, 14
    and deoxyribonucleic acid, 14, 489
    and protein synthesis, 475, 490
        during gastrulation, 192
    in egg, after fertilization, 89
    metazoa and, 482–485
    synthesis of, during cleavage, 106
Metabolic changes, after fertilization, 88
Metabolism, during gastrulation, 189–192
Metamorphosis, 7, 535–561
Metanephros, 400
Metazoa, messenger RNA and, 482–485
Metencephalon, 330, 345
Microfilaments, 241–251
Micromeres, 135
Microtubules, 241–251
Microvilli, 37
Midbrain, 344
Midgut, 170
    in amniote embryo, 445
    in amphibian embryo, 442
Mitochondria, and vitellogenesis, 50
    in oocyte, 43
    nucleolus, 36
    of spermatozoon, 25
    organization, 43
    protein synthesis in, 36
    secondary, 62

Mitosis, 20–24
    and deoxyribonucleic acid, 469
    and exponential growth, 521
    differentiation and, 476, 492
    in neural tube, 336
    in regeneration, 578
Mollusc, blastopore, 6
    elephant tusk, oocyte cytoplasm, 55
    gastropod, maternal genes and cleavage,
        144
        vitellogenesis in, 50
    lamellibranch, cortical reaction in egg, 87
    regenerative ability, 565
Molting, 546, 557
    causation, 550–555
    nature of hormones, 555–557
Monkey embryo, cleavage, 265
    early development, 277
Morphallaxis, 564
Morphogenesis, in epithelia, 234–240
    maternal genes and, 145
Morphogenetic gradients, in egg cytoplasm,
    134–143
Morphogenetic movements, 229–230
Morphological plan, 467
Morula, 114–119
    in mammalian embryo, 265
Mouse embryo, alkaline phosphates in hair
        papillae, 498
    genes and duplication of body, 516
        excessive development, 517
        growth, 517
        organogenesis, 517
        primordial germ cells, 431
Mouth, development, 448–453
    induction, in frog, 512
        in salamander larva, 512
Müller-Haeckel's biogenetic law, 8
Müllerian ducts, 436
Muscle, and determination of regenerating
        organ, 584
    buds, 393
    somatic, origin, 375–377
Myelencephalon, 330
Myocardium, in amphibian embryo, 406
    plate mesoderm and, 312
Myocoele, of somite, 375
Myosin, amount in muscle tissue during
    development, 496
Myotome, 375

Nasal process, 452. See also *Olfactory*.
Nasolacrimal duct, 452
Nematodes, cleavage in, *109*
    regenerative ability, 565
Nemerteans, regenerative ability, 564
Neoblasts, 576
Nephrocoele, 394
Nephrogenic cord, 399
Nephrostomes, 394
Nephrotomes, 394
Nerve(s), acoustic, 350
    cranial, 346
    facial, 350
    glossopharyngeal, 351

Nerve(s) *(Continued)*
  hypoglossal, 351
  oculomotor, 397
  olfactory, 346
  optic, 346
  peripheral organs and, 337
  spinal, 337
  spinal accessory, 351
  trigeminal, 350
  trochlear, *347*
  vagus, 351
Nerve fibers, in spinal cord, 337
Nervous system, and regeneration, 567
  autonomic, development, 352
  central, development, 330–353
  peripheral, neural crest cells and, 362
Neural arch, 377
  segmentation, 377
Neural crest, 173
Neural crest cells, 361–364
Neural fold, 172
  transverse, and mouth development, 450
Neural plate, 160, 172
  and optic vesicle, 331
Neural tube, 161
  and central nervous system, 330
  cell differentiation, 336
  differentiation, 331, 336
  migration of cells, 336
  proliferation of cells, 336
"Neuralizing agent," 220
Neurenteric canal, 161
Neuromeres, 330
Neurons, preganglionic, 353
Neuropore, 161
Neurosecretory cells, 551
Neurula, 172
Neurulation, 172
  frog embryo and, 304
Newt, regeneration blastema, 576
  regeneration experiments, 584
  regenerative ability, 565
Newt cleavage, 126, *126*
Newt embryo, blastomere nucleus, 120
  determination experiments, 198–202
  mouth development, 450
  protein synthesis during gastrulation, 191
Nose, capsule, 381. See also *Olfactory.*
Notochord, 158, 164
  and vertebral cartilages, 377
Notochordal crescent, 97
Notochordal process, 183
Notochordal rudiment, in man, 273
Nuclear membrane, rupture of, 60
Nuclear sap, 57, 60
Nuclei, transplantation of, 124
Nucleic acids, synthesis, 476
  in frog embryo, 105
Nucleic acid precursor, entry into oocyte, 42
Nucleolus oocyte, 34
Nurse cells, 40
  of *Drosophila,* 40
Nutrition of embryo, 67–71
Nymphymal stages, 547

Oculomotor nerve, 347
Olfactory nerve, *347*
Olfactory organ, development, 367

Olfactory organ *(Continued)*
  telencephalon and, 342
Olfactory placodes, 367
Oligolecithal egg, 45
Ontogenetic development, 4–7
Ontogeny, 8
Oocyte, 4. See also *Egg.*
  cortex, 57
  entry of protein precursors, 42
  frog, 32
    messenger RNA and, 482
  Golgi bodies, 43
  growth of, 32–42
  meiosis, 60
  mitochondria, 43
Oocyte cytoplasm, DNA in, 36
  food supply, 42–54
  in elephant tusk mollusc, 55
  in mammal, 56
Oocyte nucleus, deoxyribonucleic acid in, 33
  growth, 33
  ribonucleic acid synthesis, 33
Oogenesis, 4, 34–66
  comparison to spermatogenesis, 62, *63*
Oogonia, 4, 59
Oozooids, 592
Opercular fold, 456
Opistonephros, 400
Optic chiasma, 357. See also *Eye.*
Optic cup, 353
  and lens, 357
Optic nerve, 346
Optic stalk, 330
Optic thalamus, 344
Optic vesicle, 330, 355
Oral sucker, 373
Organ(s), adhesive, development, 373
  ectodermal, development in vertebrates,
    330–374
  enamel, 450
  endodermal, development in vertebrates,
    440–463
  formation. See *Organogenesis.*
  mesodermal, development in vertebrates,
    375–439
  olfactory, and telencephalon, 342
    development, 367
  peripheral, and development of nerves, 337
  proportional and disproportional growth,
    529–539
  rudiment, 6
    determination of, 197–202
    genes and, 517
  sense. See *Sense organs.*
Organism, growth, 521–523
Organizer, 202–207
Organogenesis, 6, 151–189, 295–329
  genes and, 516
Oropharyngeal membrane, 310, 448
Ostium tubae, 436
Oviducts, 436
Oviparity, 70
Ovoviviparity, 70
Ovulation, and pregnancy, hormonal control
    of, 291–293
  hormones and, 260
  in mammal eggs, 260
Ovum, 4
Oxygen, consumption, during gastrulation,
  190, 191

Pachynema stage, 22
Palinurus, heteromorphosis, 582
Pancreas, development, 460
Papilla, 450
Paracentrotus lividus egg, cytoplasmic changes
    after fertilization, 98
  pigment, 54
Parachordals, 381
Paramesonephric ducts, 436
Parathyroid gland, 455
Parietal layer, 170
Parthenogenesis, 90–92
Parturition, hormones and, 292
Penetration path, 95
Pepsin, and induction, 216
  in salamander embryo, 498
Periblast, 112, 175
Pericardium, in amphibian embryo, 406
  plate mesoderm and, 312
Periderm, 365
Perivitelline space, 63
Petrosal ganglion, 351
Pharyngeal pouches, 8, 454
Pharynx, development in frog embryo, 441
Phosphoribonucleotides, synthesis, 476, 543
Phosvitin, 53
  in amphibian yolk platelets, 46
Phylogenetic development, 3
Phylogeny, 8
Phylum, 7
Pigment, in egg, 54
Pigment coat, 353
Pigmentation, neural crest cells and, 362
Pinocytosis, 38
Pituitary gland, 344
Placenta, 71, 284
  bidiscoidal, 289
  chorioallantoic, 284
  choriovitelline, 284
  cotyledon, 289
  deciduous, 285
  diffuse, 289
  discoidal, 289
  endotheliochorial, 290
  epitheliochorial, 290
  hemochorial, 290
  in man, 289
  in marsupial, 284, 290
  nondeciduous, 290
  physiology of, 290–291
  zonary, 289
Placental barrier, 290
Placentation, 284–289
Placode, 349, 365
  auditory, 365, 366
  olfactory, 367
Planaria lugubris, food and regeneration, 567
Planaria torva, temperature and regeneration,
  567
Planarians, heteromorphosis, 582
  regeneration, 564, 575
  super-regeneration, 580
Plasmalemma, 62
Platycnemis pennipes, blastomere nucleus, 120
Platysamia cecropia, causation of molting, 551
Pluteus, 67, 136
Polar lobe, 127
Polarity, 586
  regeneration and, 585–590

Pons varolli, 345
Pontine flexure, 351
Potency, prospective, 199
Prechordal plate, 165
Preformation theory, 10–11
Pregnancy, ovulation and, hormonal
    control of, 291–293
Primary gut, 156
Primary organ rudiments, 295
Primary organizer, 202–207
Primitive groove, 181
Primitive knot, 181
Primitive streak, 181
  formation, 206
  in development of alimentary canal,
    446
  in mammalian embryo, 271, 272, 273
Primordial germ cells, and extraembryonic
    endoderm, 425
  in chick embryo, 424
  in mouse embryo, 431
  in salamander embryo, 429
  origin, 424
    in human embryo, 431
Proacrosomal granule, 27
Proctodeum, 448
Progesterone, action of, 292
Pronephric duct, 394, 398
  and development of mesonephros, 394
Pronephric tubules, 394
Pronephros, 395
Pronucleus, 93
  fusion, 93
Prosencephalon, 330
Proteins, and differentiation, 478
  in yolk, 54
  synthesis, after fertilization, 89
    amino acids and, 476
    changes during development, 495–502
    during cleavage, 106
    during gastrulation, 191
    in oocyte, 33
    messenger RNA and, 476
      during gastrulation, 192
    protein yolk and, 191
    regeneration gradients and, 586
    transfer RNA and, 476
Protein precursors, entry to oocyte, 42
Protein structure, and genes, in hemoglobin,
    480
Protein yolk, 54
  and protein synthesis, 191
Prothoracic gland, 551
  insect, and molting and
    metamorphosis, 551
Protocerebrum, 551
Protoplasm, synthesis, 467
Protostomia, blastopore, 6
Pseudopregnancy, 262
Pterygota, metamorphosis, 547
Puffs, 560
  ecdysone and, 560
Pulmonary arch, 417
Pupa, 549

Quadrate, 363

Rabbit embryo, cleavage experiments, 266
Ramus communicans, 337
Rana palustris embryo, lateral line sense organs, 365
Rana pipiens, oocyte, 32
Rana sylvatica embryo, lateral line sense organs, 365
Rana temporaria egg, cytoplasmic changes after fertilization, 98
Rathke's pocket, 450
Rauber's layer, in mammalian embryo, 270
Reconstitution, from isolated cells, 590–591
Reduplication, 392
Regeneration, 7, 562–591
    control of, 567–571
    epimorphic, 564
    histological processes, 571–579
    in fission, 595
    morphallactic, 564
    new part and organism, relation, 581–585
    polarity and gradients in, 585–590
    release of, 579–581
Regeneration blastema, 562
    capacities of cells, 584
    formation, 672
    in newt, 576
    in salamander, 576
Regeneration bud. See *Regeneration blastema.*
Regeneration territory, 582
Regenerative ability, 564–567
Renal tubules, development, 510
Reproduction, asexual. See *Asexual reproduction.*
    sexual, 4
    virginal, 90
Reproductive organs, development, 423–439
Reptiles, extraembryonic structures, 254–259
    parabolic growth, 524
    pharyngeal pouches, 8
    yolk sac, 254
Rete testis, 433
Reticulocyte, messenger RNA and, 482
Rhombencephalon, 330, 345
Ribonucleic acid (RNA), 13–14
    analytical embryology and, 13
    and development of lens, 358
    and growth, 474
    and induction, 216, 219
    insoluble. See *Ribosomal RNA.*
    messenger. See *Messenger RNA.*
    ribosomal. See *Ribosomal RNA.*
    soluble. See *Transfer RNA.*
    synthesis, 476, 486
        during cleavage, 105, 142
        histones and, 486
        in oocyte nucleus, 33
        on puffs, 560
    transfer. See *Transfer RNA.*
Ribosomal RNA, 14
    during growth, 475
    synthesis of, cleavage and, 105
Ribosomes, 14
    in egg, after fertilization, 89
RNA. See *Ribonucleic acid.*
Rodent embryo, early development, 279
    ectoplacental cone, 279
    extraembryonic coelom, 282

"Root" ganglion, 351
Roux, Wilhelm, 12
Rudiments, organ, 295

Sacculina, larva, 9
Salamander, limb regeneration, 562–564, 565
    regenerative blastema, 578
    super-regeneration, 580
Salamander embryo, development of gastric glands, 497
    limb differentiation experiments, 390–391
    pepsin in, 498
    primordial germ cells, 429
    trypsin in, 498
Salamander larva, induction of balancers, 512
    induction of fin fold, 513
    mouth induction, 512
Sclera, 360, 381
Scleroblasts, 599
Sclerotome, 375
    and axial skeleton, 375
Scyphozoan polyp, budding, 596
Sea urchin egg, animal-vegetal gradient system, 139
    artificial parthenogenesis, 90
    cortical reaction, 83
    cytoplasm, 136, *136*
    cytoplasmic changes after fertilization, 98
    fertilization of, 75, *78*
    larva of, 67
Sea urchin embryo, deoxyribonucleic acid in nuclei, 103(t)
Sea urchin hybrids, cleavage in, 195
Secondary organ rudiments, 295
Self-regulation, 356
Seminiferous tubules, development, 433
    spermatozoa in, 19–20
Sense organs, capsules, 381
    lateral line, 365
Sensory layer, 365
Sensory retina, 353
Septum transversum, in amphibian embryo, 405
Serosa, in birds and reptiles, 258
Sertoli cells, 20
    development, 433
Sex cords, primitive, 432
Shape, body, during development, 297
Shell glands, 64
    membranes, 65
    of bird egg, 64
Significance, prospective, 199
Skeleton, and determination of regenerating organs, 583
    axial, development, 377–382
    visceral, neural crest cells and, 363
Skull, 380
Sodium thiocyanate, and gradients, 227
Somatic cells, 4, 592
Somatic muscles, origin, 375–377
Somites, 170
    fate of, 375–377
    myocoele, 375
    stalk, 170
Spemann, cleavage experiments, 120, *121*

Spemann (*Continued*)
  primary organizer, 202–207
Sperm lysins, 79
Spermatid, 24
  centrosome of, 28
  nucleus of, *25*
Spermatocyte, 4, 20
  primary, 20
  secondary, 24
Spermatogenesis, 4, 19–31
  comparison to oogenesis, 62, *63*
Spermatogonia, 4, *19*, 20
Spermatozoon, 4
  differentiation, 24–31
  egg, approach to, 75–100
    inside, 92–96
    penetration of, 79
    reaction of, 81–88
  function of, 24
  in seminiferous tubules, 19–20
  mitochondria, 25
  nonflagellate, 31
Spinal accessory nerve, 351
Spinal cord, development, 334
  mantle, 336
  marginal layer, 337
  nerve fibers, 337
  ventral columns, 337
  white matter, 337
Spinal ganglion, 337
Spinal nerve, 337
Spinocaudal inductor, 203
Sponges, reconstitution, 590
Stages, anatomy of, 302–328
  of development, 299–302
Starfish, larva, 9
Statoblast, 594
Stolon, 594
Stomodeum, 448
Strobilation, 594
Styela partita egg, cleavage, 128
  cytoplasmic changes after fertilization, 97
  pigment, 54
Stylopodium, 389
Subchorda, 304
Subgerminal cavity, 117
Subimago, 547
Super-regeneration, 580
Swim bladder, in fishes, 459
Synapsis, 22
Syncytiotrophoblast, 285

Tadpoles, development of, from transplanted
  nuclei, 122, *123*
Tailbud, 295
Tapetum nigrum, 353
Tectum, 344
Telencephalon, 330, 342
Teleolecithal egg, 46
Temperature, and molting in insects, 551
  and regeneration, 567
Teratoma, 504
Testes, 19
Thymus, 455
Thyroglobulin, 541

Thyroid gland, 455
  and amphibian metamorphosis, 540
Thyroid hormone, and amphibian meta-
  morphosis, 540
Thyrotropic hormone, 542
Thyroxine, 514
Tiedmann, experiments with inducing
  substance, 215
Tooth, development, 450
  visceral arches and, 456
Trabeculae, 380
Transfer RNA, and protein synthesis, 475
  synthesis of, during cleavage, 105
"Transforming agent," 220
Trigeminal nerve, 350
Tri-iodothyronine, 541
Triturus embryo, blastomere nucleus, 120
  protein synthesis during gastrulation, 191
Trochlear nerve, *347*
Trophoblast, in mammalian embryo, 265, 285
Trophocytes, 599
Trunk inductor, 203
Trypsin, and induction, 216
  in salamander embryo, 498
Tunicates, budding, 597
Tympanic cartilage, 544
Tympanic membrane, 544

Uca pugnax, allometric growth, 530
Umbilical cord, in chick embryo, 256
Uracil, in ribonucleic acid, 14(t)
Ureteric bud, 400
Urinary system, development of, 394–402
Urodeles, balancers, development, 373
  metamorphosis, 538
  nervous system and regeneration, 567
  regenerative ability, 565

Vagus nerve, 351
Vas deferens, 436
Vegetal gradient, 138
Vegetal polar plasm, 56
Vegetal pole, 54
Vegetal region, fate of, 153
Vegetalization, chemical agents and, 142–143
Vein, cardinal, 415
  hepatic portal, 420
  mesenteric, 420
  postcardinal, 422
  subcardinal, 420
Vena cava, posterior, 422
Ventricle, brain, 342
  fourth, 345
  second, 342
  third, 344
Vermis, of cerebellum, 345
Vertebral column, development, 377
Vertebrates, ectodermal organs, develop-
  ment, 330–374
  endodermal organs, development, 440–463
  holoblastic, archenteron and alimentary
    canal, 440

Vertebrates *(Continued)*
  meroblastic, archenteron and alimentary
    canal, 443
  mesodermal, development, 375–439
Villi, 288
Visceral arches, and tooth development, 456
Visceral layer, 170
Visceral skeleton, neural crest cells and, 363
Vitamin A, and differentiation, 506
Vitelline membrane, 62
Vitellogenesis, 44, 50
  Golgi bodies and, 50
  in gastropod mollusc, 52
  liver and, 40
  mitochondria and, 50
Viviparity, 71

Water, absorption by embryo, 68
  egg, 77
Watson and Crick model, deoxyribonucleic
    acid, 13
Weismann, germ plasm theory, 119
White matter, of spinal cord, 337
Wolff, Casper Friedrich, 11
Wolffian duct. See *Mesonephric duct.*

Xenopus embryo, lung development, 458
Xenopus laevis embryo, gray crescent, 203

X-rays, and regeneration, 571
  annelid, 576
  planarian, 575
  salamander, 576

Yamanda, experiments with inducing sub-
    stance, 215
Yolk, 44
  and cleavage, 109
  fatty, 46
  in mammalian eggs, 260, 263
  platelets, 46, 54
    solubilization, 192
  plug, 163
  production. See *Vitellogenesis.*
  protein, 46, 54
    and protein synthesis, 191
  sac, 186
    in birds and reptiles, 254
    in fishes, 253, 254
  stalk, 446
  utilization during cleavage, 105
Yolk nucleus of Balbiani, 49

Zeugopodium, 389
Zona pellucida, 62
Zona radiata, 38
Zygonema stage, 22
Zygote, 4